Optimal and Robust State Estimation

IEEE Press
445 Hoes Lane
Piscataway, NJ 08854

Optimal and Robust State Estimation

Finite Impulse Response (FIR) and Kalman Approaches

Yuriy S. Shmaliy
Universidad de Guanajuato, Mexico

Shunyi Zhao
Jiangnan University, China

IEEE PRESS
WILEY

Published by John Wiley & Sons, Inc., Hoboken, New Jersey.
Published simultaneously in Canada.

Library of Congress Cataloging-in-Publication Data

Names: Shmaliy, Yuriy, author. | Zhao, Shunyi, author.
Title: Optimal and robust state estimation : finite impulse response (FIR) and Kalman approaches / Yuriy S. Shmaliy, Shunyi Zhao.
Description: Hoboken, NJ : Wiley-IEEE Press, 2022. | Includes bibliographical references and index.
Identifiers: LCCN 2022016217 (print) | LCCN 2022016218 (ebook) | ISBN 9781119863076 (cloth) | ISBN 9781119863083 (adobe pdf) | ISBN 9781119863090 (epub)
Subjects: LCSH: Observers (Control theory) | Systems engineering.
Classification: LCC QA402.3 .S53 2022 (print) | LCC QA402.3 (ebook) | DDC 629.8/312–dc23/eng20220628
LC record available at https://lccn.loc.gov/2022016217
LC ebook record available at https://lccn.loc.gov/2022016218

Cover Design: Wiley
Cover Image: © Science Photo Library/Getty Images

Set in 9.5/12.5pt STIXTwoText by Straive, Chennai, India

To our families

Contents

Preface

The *state estimation* approach arose from the need to know the internal state of a real system, given that the input and output measurements are known. The corresponding structure is called a *state estimator*, and in control theory it is also called a *state observer*. In signal processing, the problem is related to the *process state* and its transition from one point to another. In contrast to parameter estimation theory, which deals with the estimation of the parameters of the fitting function, the state estimation approach is more suitable for engineering applications and the development of end-to-end algorithms.

Knowing the system state helps to solve many engineering problems. In systems, the state usually cannot be observed directly, but its indirect observation can be provided by way of the system outputs. In control, it is used to stabilize a system via state feedback. In signal processing, the direct, inverse, and identification problems are solved by applying state estimators (filters, smoothers, and predictors) to linear and nonlinear processes. In biomedical applications, state estimators facilitate extracting required process features.

The most general state estimator is a *batch estimator*, which requires data and input over a time horizon and has either infinite impulse response (IIR) or finite impulse response (FIR). Starting with the seminal works of Kalman, recursive state estimates have found an enormous number of applications. However, since recursions are mostly available for white noise, they are less accurate when noise is not white. The advantage is that recursions are computationally easy. But, unlike in Kalman's days, the computational complexity is no longer an issue for modern computers and microprocessors, and interest in batch optimal and robust estimators is growing.

To immediately acquaint the reader with the FIR approach, suppose that discrete measurements on a finite horizon are collected in the vector Y and the gain matrix $\mathcal{H}$, which contains the impulse response values, is defined in some sense (optimal or robust). Then the discrete convolution-based batch FIR estimate $\hat{x} = \mathcal{H}Y$, which can be easily computed recursively, will have three advantages over Kalman recursions:

- Bounded input bounded output stability, which means there is no feedback and additional constraints to ensure stability and avoid divergence.
- Better accuracy in colored noise due to the ability to work with full block error matrices; recursive forms require such matrices to be diagonal.
- Higher robustness as uncertainties beyond the averaging horizon are not projected onto the current estimate.

This book is the first systematic investigation and analysis of batch state estimators and recursive forms. To elucidate the theory of optimal and robust FIR state estimators in continuous and discrete time, the book is organized as follows. Chapter 1 introduces the reader to the state estimation approach, discusses the properties of FIR state estimators, provides a brief historical overview, and observes the most noticeable works on the topic. Chapter 2 gives the basics of probability and stochastic processes. Chapter 3 discusses the available linear and nonlinear state estimators. Chapter 4 deals with optimal FIR filtering and considers *a posteriori* and *a priori* optimal, optimal unbiased, ML, and limited memory batch and recursive algorithms. Chapter 5 solves the q-lag FIR smoothing problem. Chapter 6 presents an unbiased FIR state estimator. Chapter 7 introduces the receding horizon (RH) FIR state estimation approach. Chapter 8 develops the theory of FIR state estimation under disturbances and Chapter 9 for uncertain systems. Chapter 10 lists several additional topics in FIR state estimation. Chapter 11 provides several applications, where the FIR state estimators are used effectively. The remainder of the book is built with Appendix A, which presents matrix forms and relationships; Appendix B, which introduces the norms; and Appendix C, which contains Matlab-based codes of FIR state estimators.

The authors appreciate the collaboration with Prof. Choon Ki Ahn of Korea University, South Korea, with whom several results were co-authored. Yuriy Shmaliy appreciates the collaboration with Prof. Dan Simon of Cleveland University, Prof. Wojciech Pieczynski of Institut Polytechnique de Paris (Telecom SudParis), and Dr. Yuan Xu of University of Jinan, China, as well as the support of Prof. Oscar Ibarra-Manzano and Prof. José Andrade-Lucio of Universidad de Guanajuato, and contributions of his former and present Ph.D. and M.D. students Dr. Jorge Muñoz-Minjares, Dr. Miguel Vázquez-Olguín, Dr. Carlos Lastre-Dominguez, Sandra Márquez-Figueroa, Karen Uribe-Murcia, Jorge Ortega-Contreras, Eli Pale-Ramon, and Juan José López Solórzano to the development of FIR state estimators for various signal processing and control areas. Shunyi Zhao appreciates the collaboration and support of Prof. Fei Liu of the Jiangnan University, China, and Prof. Biao Huang of the University of Alberta, Canada.

Yuriy S. Shmaliy
Shunyi Zhao

Foreword

I had the privilege and pleasure of meeting Yuriy Shmaliy several years ago when he visited me at Cleveland State University. We spent the day together talking about state estimation, and he gave a well-attended and engaging seminar about finite impulse response (FIR) filtering to enthusiastic CSU graduate students and faculty. I was fascinated by his approach to state estimation. At the time, I was thoroughly immersed in the Kalman filtering paradigm, and his FIR methods were new to me. I could immediately see how they could address the problems that the typical Kalman filter has with stability and robustness. I already knew all about the approaches for addressing these Kalman filter problems—in fact, I had studied and published several such approaches myself. But I was always left with the nagging thought that no matter how much the Kalman filter is modified for enhanced stability and robustness, it is still the Kalman filter, which was not designed with stability and robustness in mind, so any attempts to enhance its stability and robustness will always be ad hoc. The FIR filter, in contrast, is designed from the outset for stability and robustness. Does this mean the FIR filter is better than the Kalman filter? That question is too simplistic to even be coherent. As we know, all optimization is multi-objective, so claiming that one filter is better than the other is ill-advised. But we can definitely say that some filters are "better" than other filters from certain perspectives, and the FIR filter has clearly established itself as an approach that is better than other filters (including the Kalman filter) from certain perspectives.

This textbook deals with state estimation using FIR filters. Anyone who's seriously interested in state estimation theory, research, or application should study this book and make its algorithms a part of his or her toolbox. There are many ways to estimate the state of a system, with the Kalman filter being the most tried-and-true method. The Kalman filter became the standard in state estimation after its invention around 1960. Its advantages, including its theoretical rigor and relative ease of implementation, have overcome its well-known disadvantages, which include a notorious lack of robustness and frequent problems with stability.

FIR filters have arisen as a viable alternative to Kalman filters in a targeted attempt to address the disadvantages of Kalman filtering. One of the advantages of Kalman filtering is its recursive nature, which makes it an infinite impulse response (IIR) filter, but this feature creates an inherent disadvantage, which is a tendency toward instability. The FIR filter is specifically formulated without feedback, which provides it with inherent stability and improved robustness.

Based on their combined 30 years of research in this field, the authors have compiled a thorough and systematic investigation and analysis of FIR state estimators. Chapter 1 introduces the concept and the basic approaches of state estimation, including a review of properties such as optimality, unbiasedness, noise distributions, performance measures, stability, robustness, and computational complexity. Chapter 1 also presents a brief but interesting historical review that traces FIR filtering all the way back to Johannes Kepler in 1601. Chapter 2 reviews the basics of probability and stochastic processes, and culminates with an overview of stochastic differential equations. Chapter 3 reviews state space modeling theory and summarizes some of the popular approaches to state estimation, including Bayesian estimation, maximum likelihood estimation, least squares estimation, Kalman filtering and smoothing, extended Kalman filtering, unscented Kalman filtering, particle filtering, and H-infinity filtering. The overview of Kalman filtering in this chapter is quite good and delves into many theoretical considerations, such as optimality, unbiasedness, the effects of initial condition errors and noise covariance errors, noise correlations, and colored noise.

As good as the first three chapters are, the meat of the book really begins in Chapter 4, which derives the FIR filter by combining the forward discrete-time system model with the backward model into a single matrix equation that can be handled with a single batch of measurements. The FIR filter is derived in both *a priori* and *a posteriori* forms. Although the FIR filter is not recursive, the batch arrangement of the filter can be rewritten in a recursive form for computational savings. The authors show how unbiasedness and maximum likelihood can be incorporated into the FIR filter. The end of the chapter extends FIR filter theory to continuous time systems.

Chapter 5 derives several different FIR smoother formulations. Chapter 6 discusses a specific FIR filter, which is the unbiased FIR (UFIR) filter. The UFIR filter uses an optimal horizon length to minimize mean square estimation error. The authors also extend the UFIR filter to smoothing and to nonlinear systems. Chapter 7 discusses prediction using the FIR approach and the special case of one-step prediction, which is called receding-horizon FIR prediction. Chapter 8 derives the FIR filter that is maximally robust to noise statistics and system modeling errors while constraining the bias. This chapter discusses robustness from the H_2, H_∞, hybrid H_2/H_∞, L_1, and L_∞ perspectives. Chapter 9 rederives many of the previous results while considering uncertainty in the system model. Chapter 10 is titled "Advanced Topics" and considers problems such as distributed FIR filtering, correlated noise, hybrid Kalman/FIR filtering, and delayed and missing measurements. Chapter 11 presents several case studies of FIR filtering to illustrate the design decisions, trade-offs, and implementation issues that need to be considered during application. The examples include the estimation of clock states (with a special consideration for GPS receiver clocks), clock synchronization, localization of wireless sensor networks, localization over RFID tag grids, quadrotor localization, ECG signal noise reduction, and EMG waveform estimation.

The book includes 33 examples scattered throughout, including careful comparisons between Kalman and FIR results. These examples are in addition to the more comprehensive case studies in the final chapter. The book also includes 26 pseudocode listings to assist the student and researcher in their implementation of FIR algorithms and about 200 end-of-chapter problems for self-study or coursework. This is a book that I would have loved to have read as a student or early-career researcher. I think that any researcher who studies it will be well-rewarded for their effort.

Cleveland State University *Daniel J. Simon*

Acronyms

AWGN	Additive white Gaussian noise
BE	Backward Euler
BFBM	Backward filter backward model
BFFM	Backward filter forward model
BIBO	Bounded input bounded output
CMN	Colored measurement noise
CPN	Colored process noise
DARE	Discrete algebraic Riccati equation
DARI	Discrete algebraic Riccati inequality
DDRE	Discrete dynamic (difference) Riccati equation
DFT	Discrete Fourier transform
DSM	Discrete Shmaliy moments
EKF	Extended Kalman filter
EOFIR	Extended optimal finite impulse response
FE	Forward Euler method
FF	Fusion filter
FFBM	Forward filter backward model
FFFM	Forward filter forward model
FH	Finite horizon
FIR	Finite impulse response
FPK	Fokker-Plank-Kolmogorov
GKF	General Kalman filter
GNPG	Generalized noise power gain
GPS	Global Positioning System
IDFT	Inverse discrete Fourier transform
IIR	Infinite impulse response
KBF	Kalman-Bucy filter
KF	Kalman filter
KP	Kalman predictor
LMF	Limited memory filter
LMKF	Limited memory Kalman filter
LMI	Linear matrix inequality
LMP	Limited memory predictor
LUMV	Linear unbiased minimum variance
LS	Least squares

LTI	Linear time invariant
LTV	Linear time varying
MBF	Modified Bryson-Frazier
MC	Monte Carlo
MIMO	Multiple input multiple output
ML	Maximum likelihood
MPC	Model predictive control
MSE	Mean square error
MVF	Minimum variance FIR
MVU	Minimum variance unbiased
NARE	Nonsymmetric algebraic Riccati equation
NPG	Noise power gain
ODE	Ordinary differential equation
OFIR	Optimal finite impulse response
OUFIR	Optimal unbiased finite impulse response
PF	Particle filter
PMF	Point mass filter
PSD	power spectral density
RDE	Riccati differential equation
RH	Receding horizon
RKF	Robust Kalman filter
RMSE	Root mean square error
ROC	Region of convergence
RTS	Rauch-Tung-Striebel
SDE	Stochastic differential equation
SIS	Sequential importance sampling
SPDE	Stochastic partial differential equation
UFIR	Unbiased finite impulse response
UKF	Unscented Kalman filter
UT	Unscented transformation
WGN	White Gaussian noise
WLS	Weighted least squares
WSN	Wireless sensor network
cdf	cumulative distribution function
cf	characteristic function
cKF	centralized Kalman filter
cUFIR	centralized unbiased finite impulse response
dKF	distributed Kalman filter
dUFIR	distributed unbiased finite impulse response
μKF	micro Kalman filter
μUFIR	micro unbiased finite impulse response
pdf	probability density function
pmf	probability mass function

1 Introduction

The limited memory filter appears to be the only device for preventing divergence in the presence of unbounded perturbation.

Andrew H. Jazwinski [79], p. 255

The term *state estimation* implies that we want to estimate the state of some process, system, or object using its measurements. Since measurements are usually carried out in the presence of noise, we want an accurate and precise estimator, preferably optimal and unbiased. If the environment or data is uncertain (or both) and the system is being attacked by disturbances, we also want the estimator to be robust. Since the estimator usually extracts state from a noisy observation, it is also called a *filter*, *smoother*, or *predictor*. Thus, a state estimator can be represented by a certain block (hardware or software), the operator of which allows transforming (in some sense) input data into an output estimate. Accordingly, the linear state estimator can be designed to have either infinite impulse response (IIR) or finite impulse response (FIR). Since IIR is a feedback effect and FIR is inherent to transversal structures, the properties of such estimators are very different, although both can be represented in batch forms and by iterative algorithms using recursions. Note that effective recursions are available only for delta-correlated (white) noise and errors.

In this chapter, we introduce the reader to FIR and IIR state estimates, discuss cost functions and the most critical properties, and provide a brief historical overview of the most notable works in the area. Since IIR-related recursive Kalman filtering, described in a huge number of outstanding works, serves in a special case of Gaussian noise and diagonal block covariance matrices, our main emphasis will be on the more general FIR approach.

1.1 What Is System State?

When we deal with some stochastic dynamic system or process and want to predict its further behavior, we need to know the system characteristics at the present moment. Thus, we can use the fundamental concept of *state variables*, a set of which mathematically describes the state of a system. The practical need for this was formulated by Jazwinski in [79] as "…the engineer must know what the system is "doing" at any instant of time" and "…the engineer must know the state of his system."

Obviously, the set of state variables should be sufficient to predict the future system behavior, which means that the number of state variables should not be less than practically required. But the

Optimal and Robust State Estimation: Finite Impulse Response (FIR) and Kalman Approaches, First Edition.
Yuriy S. Shmaliy and Shunyi Zhao.

number of state variables should also not exceed a reasonable set, because redundancy, ironically, reduces the estimation accuracy due to random and numerical errors. Consequently, the number of useful state variables is usually small, as will be seen next.

When tracking and localizing mechanical systems, the coordinates of location and velocities in each of the Cartesian coordinates are typical state variables. In precise satellite navigation systems, the coordinates, velocities, and accelerations in each of the Cartesian coordinates are a set of nine state variables. In electrical and electronic systems, the number of state variables is determined by the order of the differential equation or the number of storage elements, which are inductors and capacitors.

In periodic systems, the amplitude, frequency, and phase of the spectral components are necessary state variables. But in clocks that are driven by oscillators (periodic systems), the standard state variables are the time error, fractional frequency offset, and linear frequency drift rate.

In thermodynamics, a set of state variables consists of independent variables of a state function such as internal energy, enthalpy, and entropy. In ecosystem models, typical state variables are the population sizes of plants, animals, and resources. In complex computer systems, various states can be assigned to represent processes.

In industrial control systems, the number of required state variables depends on the plant program and the installation complexity. Here, a state observer provides an estimate of the set of internal plant states based on measurements of its input and output, and a set of state variables is assigned depending on practical applications.

1.1.1 Why and How Do We Estimate State?

The need to know the system state is dictated by many practical problems. An example of signal processing is system identification over noisy input and output. Control systems are stabilized using state feedback. When such problems arise, we need some kind of model and an estimator.

Any stochastic dynamic system can be represented by the first-order linear or nonlinear vector differential equation (in continuous time) or difference equation (in discrete time) with respect to a set of its states. Such equations are called *state equations*, where state variables are usually affected by internal noise and external disturbances, and the model can be uncertain.

Estimating the state of a system with random components represented by the state equation means evaluating the state approximately using measurements over a finite time interval or all available data. In many cases, the complete set of system states cannot be determined by direct measurements in view of the practical inability of doing so. But even if it is possible, measurements are commonly accompanied by various kinds of noise and errors. Typically, the full set of state variables is observed indirectly by way of the system output, and the observed state is represented with an *observation equation*, where the measurements are usually affected by internal noise and external disturbances. The important thing is that if the system is observable, then it is possible to completely reconstruct the state of the system from its output measurements using a state observer. Otherwise, when the inner state cannot be observed, many practical problems cannot be solved.

1.1.2 What Model to Estimate State?

Systems and processes can be both nonlinear or linear. Accordingly, we recognize nonlinear and linear state-space models. Linear models are represented by linear equations and Gaussian noise. A model is said to be nonlinear if it is represented by nonlinear equations or linear equations with non-Gaussian random components.

Nonlinear Systems

A physical nonlinear system with random components can be represented in continuous time t by the following time-varying *state space model*,

$$\dot{x}(t) = f[x(t), u(t), w(t), t] , \tag{1.1}$$

$$y(t) = h[x(t), w(t), v(t), t] , \tag{1.2}$$

where the nonlinear differential equation (1.1) is called the *state equation* and an algebraic equation (1.2) the *observation equation*. Here, $x(t) \in \mathbb{R}^K$ is the system state vector; $\dot{x}(t) \triangleq \frac{d}{dt}x(t)$, $u(t) \in \mathbb{R}^L$ is the input (control) vector; $y(t) \in \mathbb{R}^P$ is the state observation vector, $w(t) \in \mathbb{R}^M$ is some system error, noise, or disturbance; $v(t) \in \mathbb{R}^H$ is an observation error or measurement noise; $f(t)$ is a nonlinear system function; and $h(t)$ is a nonlinear observation function. Vectors $w(t)$ and $v(t)$ can be Gaussian or non-Gaussian, correlated or noncorrelated, additive or multiplicative. For time-invariant systems, both nonlinear functions become constant.

In discrete time t_k, a nonlinear system can be represented in state space with a time step $\tau_k = t_k - t_{k-1}$ using either the *forward Euler* (FE) *method* or the *backward Euler* (BE) *method*. By the FE method, the discrete-time state equation turns out to be predictive and we have

$$x_{k+1} = f_k(x_k, u_k, w_k) , \tag{1.3}$$

$$y_k = h_k(x_k, w_k, v_k) , \tag{1.4}$$

where $x_k \in \mathbb{R}^K$ is the state, $u_k \in \mathbb{R}^L$ is the input, $y_k \in \mathbb{R}^P$ is the observation, $w_k \in \mathbb{R}^M$ is the system error or disturbance, and $v_k \in \mathbb{R}^H$ is the observation error. The model in (1.3) and (1.4) is basic for digital control systems, because it matches the predicted estimate required for feedback and model predictive control.

By the BE method, the discrete-time nonlinear state-space model becomes

$$x_k = f_k(x_{k-1}, u_k, w_k) , \tag{1.5}$$

$$y_k = h_k(x_k, w_k, v_k) \tag{1.6}$$

to suit the many signal processing problem when prediction is not required. Since the model in (1.5) and (1.6) is not predictive, it usually approximate a nonlinear process more accurately.

Linear Systems

A *linear time-varying* (LTV) physical system with random components can be represented in continuous time using the following *state space model*

$$\dot{x}(t) = A(t)x(t) + E(t)u(t) + B(t)w(t) , \tag{1.7}$$

$$y(t) = C(t)x(t) + D(t)w(t) + v(t) , \tag{1.8}$$

where the noise vectors $w(t)$ and $v(t)$ can be either Gaussian or not, correlated or not. If $w(t) \sim \mathcal{N}(0, \mathcal{R}_w)$ and $v(t) \sim \mathcal{N}(0, \mathcal{R}_v)$ are both zero mean, uncorrelated, and white Gaussian with the covariances $\mathcal{R}_w = \mathcal{E}\{w(\theta_1)w^T(\theta_2)\} = \mathcal{S}_w\delta(\theta_1 - \theta_2)$ and $\mathcal{R}_v = \mathcal{E}\{v(\theta_1)v^T(\theta_2)\} = \mathcal{S}_v\delta(\theta_1 - \theta_2)$, where $\mathcal{S}_w$ and $\mathcal{S}_v$ are the relevant power spectral densities, then the model in (1.7) and (1.8) is said to be linear. Otherwise, it is nonlinear. Note that all matrices in (1.7) and (1.8) become constant as A, B, C, D, E when a system is *linear time-invariant* (LTI). If the order of the disturbance $w(t)$ is less than the order of the system, then $D(t) = 0$, and the model in (1.7) and (1.8) becomes standard for problems considering vectors $w(t)$ and $v(t)$ as the system and measurement noise, respectively.

By the FE method, the linear discrete-time state equation also turns out to be predictive, and the state-space model becomes

$$x_{k+1} = F_k x_k + E_k u_k + B_k w_k \,, \tag{1.9}$$

$$y_k = H_k x_k + D_k w_k + v_k \,, \tag{1.10}$$

where $F_k \in \mathbb{R}^{K\times K}$, $E_k \in \mathbb{R}^{K\times L}$, $B_k \in \mathbb{R}^{K\times M}$, $H_k \in \mathbb{R}^{P\times K}$, and $D_k \in \mathbb{R}^{P\times M}$ are time-varying matrices. If the discrete noise vectors $w_k \sim \mathcal{N}(0, Q_k)$ and $v_k \sim \mathcal{N}(0, R_k)$ are zero mean and white Gaussian with the covariances $Q_k = \mathcal{E}\{w_k w_k^T\}$ and $R_k = \mathcal{E}\{v_k v_k^T\}$, then this model is called linear.

Using the BE method, the corresponding state-space model takes the form

$$x_k = F_k x_{k-1} + E_k u_k + B_k w_k \,, \tag{1.11}$$

$$y_k = H_k x_k + D_k w_k + v_k, \tag{1.12}$$

and we notice again that for LTI systems all matrices in (1.9)–(1.12) become constant.

Both the FE- and BE-based discrete-time state-space models are employed to design state estimators with the following specifics. The term with matrix D_k is neglected if the order of the disturbance w_k is less than the order of the system, which is required for stability. If noise in (1.9)–(1.12) with $D_k = 0$ is Gaussian and the model is thus linear, then the optimal state estimation is provided using the batch optimal FIR filtering and recursive optimal Kalman filtering. When w_k and/or v_k are non-Gaussian, then the model becomes nonlinear and other estimators can be more accurate. In some cases, the nonlinear model can be converted to linear, as in the case of colored Gauss-Markov noise. If w_k and v_k are unknown and bounded only by the norm, then the model in (1.9–1.12) can be used to derive different kinds of estimators called *robust*.

1.1.3 What Are Basic State Estimates in Discrete Time?

Before discussing the properties of state estimators fitting various cost functions, it is necessary to introduce baseline estimates and errors, assuming that the observation is available from the past (not necessarily zero) to the time index n inclusive. The following *filtering* estimates are commonly used:

- $\hat{x}_k \triangleq \hat{x}_{k|k}$ is the *a posteriori* estimate.
- $\hat{x}_k^- \triangleq \hat{x}_{k|k-1}$ is the *a priori* or predicted estimate.
- $P_k \triangleq P_{k|k} = \mathcal{E}\{\varepsilon_k \varepsilon_k^T\}$ is the *a posteriori* error covariance.
- $P_k^- \triangleq P_{k|k-1} = \mathcal{E}\{\varepsilon_k^- \varepsilon_k^{-T}\}$ is the *a priori* or predicted error covariance,

where $\hat{x}_{k|n}$ means an estimate at k over data available from the past to and including at time index n, $\varepsilon_k = x_k - \hat{x}_k$ is the *a posteriori* estimation error, and $\varepsilon_k^- = x_k - \hat{x}_k^-$ is the *a priori* estimation error. Here and in the following, $\mathcal{E}\{\cdot\}$ is an operator of averaging.

Since the state estimates can be derived in various senses using different performance criteria and cost functions, different state estimators can be designed using FE and BE methods to have many useful properties. In considering the properties of state estimators, we will present two other important estimation problems: *smoothing* and *prediction*.

If the model is linear, then the optimal estimate is obtained by the batch optimal FIR (OFIR) filter and the recursive Kalman filter (KF) algorithm. The KF algorithm is elegant, fast, and optimal for the white Gaussian approximation. Approximation! Does this mean it has nothing to do with the real world, because white noise does not exist in nature? No! Engineering is the science of approximation, and KF perfectly matches engineering tasks. Therefore, it found a huge number of applications, far more than any other state estimator available. But is it true that KF should always be used when we need an approximate estimate? Practice shows no! When the environment is strictly non-Gaussian and the process is disturbed, then batch estimators operating with full block covariance and error matrices perform better and with higher accuracy and robustness. This is why, based on practical experience, F. Daum summarized in [40] that "Gauss's batch least squares ...often gives accuracy that is superior to the best available extended KF."

1.2 Properties of State Estimators

The state estimator performance depends on a number of factors, including cost function, accurate modeling, process suitability, environmental influences, noise distribution and covariance, etc. The linear optimal filtering theory [9] assumes that the best estimate is achieved if the model adequately represents a system, an estimator is of the same order as the model, and both noise and initial values are known. Since such assumptions may not always be met in practice, especially under severe operation conditions, an estimator must be stable and sufficiently robust. In what follows, we will look at the most critical properties of batch state estimators that meet various performance criteria. We will view the real-time state estimator as a filter that has an observation y_k and control signal u_k in the input and produces an estimate in the output. We will also consider smoothing and predictive state estimation structures. Although we will refer to all the linear and nonlinear state-space models discussed earlier, the focus will be on discrete-time systems and estimates.

1.2.1 Structures and Types

In the time domain, the general operator of a linear system is *convolution*, and a convolution-based linear state estimator (filter) can be designed to have either IIR or FIR. In *continuous time*, linear and nonlinear state estimators are electronic systems that implement differential equations and produce output electrical signals proportional to the system state. In this book, we will pay less attention to such estimators.

In *discrete time*, a *discrete convolution*-based state estimator can be designed to perform the following operations:

- *Filtering*, to produce an estimate $\hat{x}_{k|k}$ at k
- *Smoothing*, to produce an estimate $\tilde{x}_{k-q|k}$ at $k-q$ with a delay lag $q>0$
- *Prediction*, to produce an estimate $\tilde{x}_{k+p|k}$ at $k+p$ with a step $p>0$
- *Smoothing filtering*, to produce an estimate $\tilde{x}_{k|k+q}$ at k taking values from q future points
- *Predictive filtering*, to produce an estimate $\tilde{x}_{k|k-p}$ at k over data delayed on p points

These operations are performed on the horizon of N data points, and there are three procedures most often implemented in digital systems:

- *Filtering* at k over a data horizon $[m,k]$, where $m=k-N+1$, to determine the current system state
- *One-step prediction* at $k+1$ over $[m,k]$ to predict future system state
- *Predictive filtering* at k over $[m-1,k-1]$ to organize the *receding horizon* (RH) state feedback control or *model predictive control* (MPC)

It is worth noting that if *discrete convolution* is long, then the computational problem may arise and batch estimation becomes impractical for real-time applications.

Nonlinear Structures

To design a batch estimator, observations and control signals collected on a horizon $[m, k]$, from $m = k - N + 1$ to k, can be united in extended vectors $Y_{m,k} = [\, y_m^T \; y_{m+1}^T \; \cdots \; y_k^T \,]^T$ and $U_{m,k} = [\, u_m^T \; u_{m+1}^T \; \cdots \; u_k^T \,]^T$. Then the nonlinear state estimator can be represented by the time-varying operator $\mathcal{O}_k(Y_{m,k}, U_{m,k})$ and, as shown in Fig. 1.1, three basic p-shift state estimators recognized to produce the filtering estimate, if $p = 0$, q-lag smoothing estimate, if $q = -p > 0$, and p-step prediction, if $p > 0$:

- FIR state estimator (Fig. 1.1a), in which the initial state estimate $\hat{x}_m$ and error matrix P_m are variables of $\mathcal{O}_k$
- IIR limited memory state estimator (Fig. 1.1b), in which the initial state x_{m-1} is taken beyond the horizon $[m, k]$ and becomes the input
- RH FIR state estimator (Fig. 1.1c) that processes one-step delayed inputs and in which $\hat{x}_{m-1}$ and P_{m-1} are variables of $\mathcal{O}_k$

Due to different cost functions, the nonlinear operator $\mathcal{O}_k$ may or may not require information about the noise statistics, and the initial values may or may not be its variables. For time-invariant models, the operator $\mathcal{O}$ is also time-invariant. Regardless of the properties of $\mathcal{O}_k$, the p-dependent structures (Fig. 1.1) can give either a filtering estimate, a q-lag smoothing estimate, or a p-step prediction.

In the FIR state estimator (Fig. 1.1a) the initial $\hat{x}_m$ and P_m represent the supposedly known state x_m at the initial point m of $[m, k]$. Therefore, $\hat{x}_m$ and P_m are variables of the operator $\mathcal{O}_k$. This estimator has no feedback, and all its transients are limited by the horizon length of N points.

In the limited memory state estimator (Fig. 1.1b), the initial state x_{m-1} is taken beyond the horizon $[m, k]$. Therefore, x_{m-1} goes to the input and is provided through estimator state feedback, thanks to which this estimator has an IIR and long-lasting transients.

The RH FIR state estimator (Fig. 1.1c) works similarly to the FIR estimator (Fig. 1.1a) but processes the one-step delayed inputs. Since the predicted estimate $\hat{x}_{k|k-1}$ by $p = 0$ appears at the output of this estimator before the next data arrive, it is used in state feedback control. This property of RH FIR filters is highly regarded in the MPC theory [106].

Linear Structures

Due to the properties of *homogeneity* and *additivity* [167], data and control signal in linear state estimators can be processed separately by introducing the *homogeneous gain* $\mathcal{H}^{\mathrm{h}}_{m,k}$ and *forced gain* $\mathcal{H}^{\mathrm{f}}_{m,k}$ for LTV systems and constant gains $\mathcal{H}^{\mathrm{h}}_N$ and $\mathcal{H}^{\mathrm{f}}_N$ for LTI systems. The generalized structures

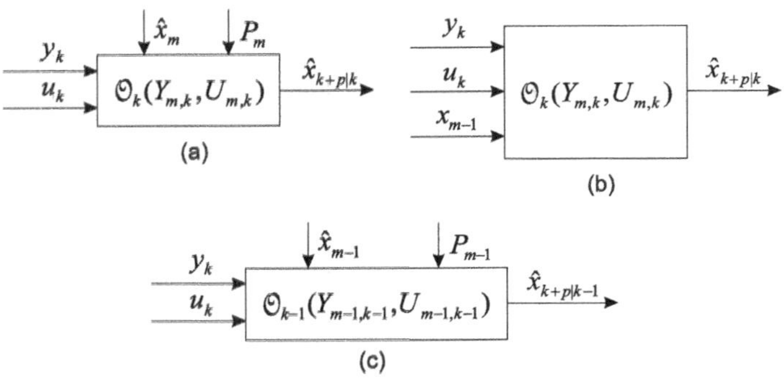

Figure 1.1 Generalized structures of nonlinear state estimators: (a) FIR, (b) IIR limited memory, and (c) RH FIR; filter by $p = 0$, q-lag smoother by $p = -q$, and p-step predictor by $p > 0$.

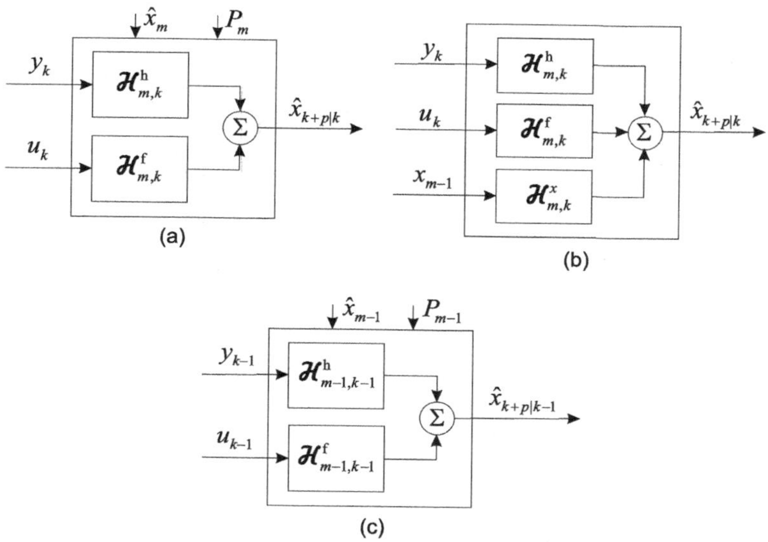

Figure 1.2 Generalized structures of linear state estimators: (a) FIR, (b) limited memory IIR, and (c) RH FIR; filter by $p = 0$, q-lag smoother by $p = -q$, and p-step predictor by $p > 0$. Based on [174].

of state estimators that serve for LTV systems are shown in Fig. 1.2 and can be easily modified for LTI systems using $\mathcal{H}^{\text{h}}_N$ and $\mathcal{H}^{\text{f}}_N$.

The p-shift linear FIR filtering estimate corresponding to the structure shown in Fig. 1.2a can be written as [173]

$$\hat{x}_{k+p|k} = \mathcal{H}^{\text{h}}_{m,k}(p)Y_{m,k} + \mathcal{H}^{\text{f}}_{m,k}(p)U_{m,k} , \tag{1.13}$$

where the p-dependent gain $\mathcal{H}^{\text{h}}_{m,k}(p)$ is defined for zero input, $U_k = 0$, and $\mathcal{H}^{\text{f}}_{m,k}(p)$ for zero initial conditions. For Gaussian models, the OFIR estimator requires all available information about system and noise, and thus the noise covariances, initial state $\hat{x}_m$, and estimation error P_m become variables of its gains $\mathcal{H}^{\text{h}}_{m,k}(p)$ and $\mathcal{H}^{\text{f}}_{m,k}(p)$. It has been shown in [229] that iterative computation of the batch OFIR filtering estimate with $p = 0$ is provided by Kalman recursions. If such an estimate is subjected to the unbiasedness constraint, then the initial values are removed from the variables. In another extreme, when an estimator is derived to satisfy only the unbiasedness condition, the gains $\mathcal{H}^{\text{h}}_{m,k}(p)$ and $\mathcal{H}^{\text{f}}_{m,k}(p)$ depend neither on the zero mean noise statistics nor on the initial values. It is also worth noting that if the control signal u_k is tracked exactly, then the forced gain can be expressed via the homogeneous gain, and the latter becomes the *fundamental gain* $\mathcal{H}_{m,k}(p) \triangleq \mathcal{H}^{\text{h}}_{m,k}(p)$ of the FIR state estimator.

The batch linear limited memory IIR state estimator appears from Fig. 1.2b by combining the subestimates as

$$\hat{x}_{k+p|k} = \mathcal{H}^{\text{h}}_{m,k}(p)Y_{m,k} + \mathcal{H}^{\text{f}}_{m,k}(p)U_{m,k} + \mathcal{H}^{x}_{m,k}(p)x_{m-1} , \tag{1.14}$$

where the initial state x_{m-1} taken beyond the horizon $[m, k]$ is processed with the gain $\mathcal{H}^{x}_{m,k}(p)$. As will become clear in the sequel, the *limited memory filter* (LMF) specified by (1.14) with $p = 0$ is the *batch KF*.

The RH FIR state estimator (Fig. 1.2c) is the FIR estimator (Fig. 1.2a) that produces a p-shift state estimate over one-step delayed data and control signal as

$$\hat{x}_{k+p|k-1} = \mathcal{H}^{\text{h}}_{m-1,k-1}(p)Y_{m-1,k-1} + \mathcal{H}^{\text{f}}_{m-1,k-1}(p)U_{m-1,k-1} . \tag{1.15}$$

By $p = 0$, this estimator becomes the RH FIR filter used in state feedback control and MPC. The theory of this filter has been developed in great detail by W. H. Kwon and his followers [91].

It has to be remarked now that a great deal of nonlinear problems can be solved using linear estimators if we approximate the nonlinear functions between two neighboring discrete points using the Taylor series expansion. State estimators designed in such ways are called *extended*. Note that other approaches employing the Volterra series and describing functions [167] have received much less attention in state space.

1.2.2 Optimality

The term *optimal* is commonly applied to estimators of linear stochastic processes, in which case the trace of the error covariance, which is the mean square error (MSE), is convex and the optimal gain $\mathcal{H}_{m,k}(p) \triangleq \mathcal{H}^{\mathrm{h}}_{m,k}(p)$ is required to keep it to a minimum. It is also used when the problem is not convex and the estimation error is minimized in some other sense.

The estimator *optimality* is highly dependent on noise distribution and covariance. That is, an estimator must match not only the system model but also the noise structure. Otherwise, it can be improved and thus *each type of noise requires its own optimal filter*.

Gaussian Noise

If a nonlinear system is represented with a nonlinear *stochastic differential equation* (SDE) (1.1), where $w(t)$ is white Gaussian, then the optimal filtering problem can be solved using the approach originally proposed by Stratonovich [193] and further developed by many other authors. For linear systems represented by SDE (1.7), an optimal filter was derived by Kalman and Bucy in [85], and this is a special case of Stratonovich's solution.

If a discrete-time system is represented by a stochastic difference equation, then an optimal filter (Fig. 1.1) can be obtained by minimizing the MSE, which is a trace of the error covariance P_k. The optimal filter gain $\mathcal{H}_{m,k}(p)$ can thus be determined by solving the minimization problem

$$\mathcal{H}_{m,k}(p) = \underset{\mathcal{H}_{m,k}}{\arg\min}\, \mathrm{tr}\, P_k(p) \tag{1.16}$$

to guarantee, at $k + p$, an optimal balance between random errors and bias errors, and as a matter of notation we notice that the *optimal estimate is biased*. A solution to (1.16) results in the batch p-shift OFIR filter [176]. Given $p = 0$, the OFIR filtering estimate can be computed iteratively using Kalman recursions [229]. Because the state estimator derived in this way matches the model and noise, then it follows that there is no other estimator for Gaussian processes that performs better than the OFIR filter and the KF algorithm.

In the transform domain, the FIR filter optimality can be achieved for LTI systems using the H_2 approach by minimizing the squared Frobenius norm $\|\overline{\mathcal{T}}(z)\|_F^2$ of the noise-to-error weighted transfer function $\overline{\mathcal{T}}(z)$ averaged over all frequencies [141]. Accordingly, the gain $\mathcal{H}_N(p)$ of the OFIR state estimator can be determined by solving the minimization problem

$$\mathcal{H}_N(p) = \underset{\mathcal{H}_N}{\arg\min}\, \|\overline{\mathcal{T}}(z,p)\|_F^2 , \tag{1.17}$$

where the weights for $\overline{\mathcal{T}}(z)$ are taken from the error covariance, which will be discussed in detail in Chapter 8. Note that if we solve problem (1.17) for unweighed $\mathcal{T}(z)$, as in the early work [109], then gain $\mathcal{H}_N(p)$ will be valid for unit-intensity noise. It is also worth noting that, by Parseval's theorem and under the same conditions, the gains obtained from (1.16) and (1.17) become equivalent. The disadvantage of (1.17) is stationarity. However, (1.17) does not impose any restrictions on the bounded noise, which thus can have any distribution and covariance, which is a distinct advantage.

It follows that the optimality in state estimates of Gaussian processes can be achieved if accurate information on noise covariance and initial values is available. To avoid the requirement of the initial state, an estimator is often derived to be optimal unbiased or maximum likelihood. The same approach is commonly used when the noise is not Gaussian or even unknown, so the estimator acquires the property of *unbiased optimality*.

1.2.3 Unbiased Optimality (Maximum Likelihood)

We will use the term *unbiased optimality* to emphasize that the optimal estimate subject to the unbiasedness constraint becomes *optimal unbiased* or an estimator that involves information about noise is designed to track the most probable process value under the assumed statistical model. In statistics, such an estimator is known as the *maximum likelihood* (ML) estimator, and we will say that it has the property of *unbiased optimality*. For example, the *ordinary least squares* (OLS) estimator maximizes the likelihood of a linear regression model and thus has the property of unbiased optimality. From the standpoint of Bayesian inference, the ML estimator is a special case of the maximum *a posteriori* probability estimator under the uniform *a priori* noise or error distribution, in which case the ML estimate coincides with the most probable Bayesian estimate. In frequentist inference, the ML estimator is considered as a special case of the extremum estimator. The ML estimation approach is also implemented in many artificial intelligence algorithms such as machine learning, supervised learning, and artificial neural networks.

For further study, it is important that the various types of state estimators developed using the ML approach do not always have closed-form engineering solutions in state space (or at least reasonably simple closed-form solutions) due to challenging nonlinear and nonconvex problems.

Gaussian Noise

As was already mentioned, the property of unbiased optimality can be "inoculated" to an optimal state estimate of Gaussian processes if the unbiasedness condition $\mathcal{E}\{\hat{x}_k\} = \mathcal{E}\{x_k\}$ is obeyed in the derivation. The FIR filter derived in this way is called the *a posteriori* optimal unbiased FIR (OUFIR) filter [222]. The p-dependent gain $\mathcal{H}_{m,k}(p)$ can be determined for the OUFIR filter by solving the minimization problem

$$\mathcal{H}_{m,k}(p) = \underset{\substack{\mathcal{H}_{m,k} \\ \text{subject to } \mathcal{E}\{\hat{x}_k\}=\mathcal{E}\{x_k\}}}{\arg\min} \, \mathrm{tr} P_k(p) \ , \tag{1.18}$$

and we notice that (1.18) does not guarantee optimality in the MSE sense, which means that the unbiased OUFIR estimate is less accurate than the biased OFIR estimate. A distinct advantage is that a solution to (1.18) ignores the initial state and error covariance. It was also shown in [223] that the ML FIR estimate is equivalent to the OUFIR estimate and the minimum variance unbiased (MVU) FIR estimate [221], and thus the property of unbiased optimality is achieved in the following canonical ML form

$$x_{k+p|k} = [C^T_{m,k}(p)\Sigma^{-1}_{m,k}C_{m,k}(p)]^{-1}C^T_{m,k}(p)\Sigma^{-1}_{m,k}Y_{m,k} \ , \tag{1.19}$$

where matrix $C_{m,k}$ is combined with matrices F_i and H_i, $i \in [m,k]$, and weight $\Sigma_{m,k}$ is a function of the noise covariances Q_i and R_i, $i \in [m,k]$.

It turns out that the recursive algorithm for the batch *a posteriori* OUFIR filter [222] is not the KF algorithm that serves the batch *a posteriori* OFIR filter. The latter means that the *KF is optimal and not optimal unbiased.*

In the transform domain, the property of unbiased optimality can be achieved by applying the H_2 approach to LTI systems if we subject the minimization of the squared Frobenius norm $\|\overline{\mathcal{T}}(z)\|_F^2$ of the noise-to-error weighted transfer function $\overline{\mathcal{T}}(z)$ averaged over all frequencies to the unbiasedness condition $\mathcal{E}\{\hat{x}_k\} = \mathcal{E}\{x_k\}$ as

$$\mathcal{H}_N(p) = \underset{\substack{\mathcal{H}_N \\ \text{subject to } \mathcal{E}\{\hat{x}_k\}=\mathcal{E}\{x_k\}}}{\arg\min} \|\overline{\mathcal{T}}(z,p)\|_F^2 \,, \tag{1.20}$$

and we notice that, by Parseval's theorem and under the same conditions, gains produced by (1.18) and (1.20) become equivalent.

Laplace Noise

The heavy-tailed Laplace distribution may better reflect measurement noise associated with harsh environments such as industrial ones. The Laplace noise is observed in underlying signals in radar clutter, ocean acoustic noise, and multiple access interference in wireless system communications [13, 86, 134]. To deal with heavy-tailed noise, a special class of ML estimators called *M-estimators* was developed in the theory of robust statistics, and it was shown that the nonlinear *median filter* is an ML estimator of location for Laplace noise distribution [90]. For multivariate Laplace measurement noise, the median approach [13] can be applied in state space if we consider the sum of absolute errors $|\varepsilon_k|$ on $[m, k]$ and determine the p-dependent gain of the median ML FIR filter by solving the minimization problem in the infinum as

$$\mathcal{H}_{m,k}(p) = \inf_{\mathcal{H}_{m,k}} \sum_{i=m}^{k} |\varepsilon_i(p)| \,. \tag{1.21}$$

It is worth noting that the nonlinear minimization problem (1.21) cannot be solved by applying the derivative with respect to $\mathcal{H}_{m,k}$, and the exact analytic form of the median FIR filter gain is thus unavailable. However, a numerical solution can be found if we solve the minimization problem (1.21) by approximating $|\varepsilon_k(p)|$ with a differentiable function.

When the heavy-tailed measurement noise is represented by a ratio of two independent zero mean Laplacian random variables, it acquires the *meridian* distribution [13]. Following the meridian strategy, the p-shit meridian ML FIR filter gain can be determined by solving the minimization problem

$$\mathcal{H}_{m,k}(p) = \inf_{\mathcal{H}_{m,k}} \sum_{i=m}^{k} \log\{\Delta + |\varepsilon_i(p)|\} \,, \tag{1.22}$$

where Δ is referred to as the *medianity* parameter [13], and we notice the same advantages and drawbacks as in the case of the median ML FIR filter.

Cauchy Noise

Another type of heavy-tailed noise has the Cauchy distribution [128], and the corresponding nonlinear filtering approach based on the ML estimate of location under Cauchy statistics is called *myriad* filtering [60]. Under the Cauchy distributed measurement noise, the p-dependent myriad ML FIR filter gain can be determined by solving the nonlinear minimization problem

$$\mathcal{H}_{m,k}(p) = \inf_{\mathcal{H}_{m,k}} \sum_{i=m}^{k} \log\{\Upsilon + \varepsilon_i(p)\varepsilon_i^T(p)\} \,, \tag{1.23}$$

where Υ is called the *linearity* parameter [60]. This nonlinear problem also cannot be solved analytically with respect to the filter gain, but approximate numerical solutions can be feasible for implementation.

1.2.4 Suboptimality

The property of *suboptimality* is inherent to minimax state estimators, where gains do not exist in closed analytic forms but can be determined numerically by solving the *discrete algebraic Riccati inequality* (DARI) or *linear matrix inequality* (LMI). The most elaborated estimators of this type minimize the disturbance-to-error (ς-to-ϵ) transfer function $\mathcal{T}$ for maximized norm-bounded random components and are called *robust*. Using the LMI approach, the FIR filter gain is computed for the smallest possible error-to-disturbance ratio γ. It is important to note that although the LMI problem is convex, the numerical solution does not guarantee the exact optimal value of γ_{opt}. Moreover, estimators of this type may fail the robustness test if they are too sensitive to tuning factors [180].

H_2 Performance

We have already mentioned that a solution to the H_2 problem can be found analytically and, for LTI systems and Gaussian noise, the H_2 FIR filter is equivalent to the OFIR filter. For arbitrary positive definite symmetric error matrices, the time-invariant gain $\mathcal{H}_N$ of the suboptimal H_2 FIR filter can be found numerically to match the hybrid H_2/H_∞ FIR structure [109].

The suboptimal H_2 filter performance can be obtained by considering the error matrix P_k and introducing an auxiliary matrix $\mathcal{Z}$ such that $\mathcal{Z} > P_k$. Then a nonlinear inequality $\mathcal{Z} - P_k > 0$ can be represented as a function of $\mathcal{H}_N$ in the LMI form, where structure and complexity depend on disturbances and uncertainties. The H_2 FIR filter gain $\mathcal{H}_N(p)$ can finally be computed numerically by solving the minimization problem

$$\mathcal{H}_N(p) = \underset{\text{subject to LMI by } \mathcal{Z}-P_k(p)>0}{\min_{\mathcal{H}_N,\mathcal{Z}} \operatorname{tr} \mathcal{Z}} . \tag{1.24}$$

It is worth noting that the LMI approach can be used to determine suboptimal gains for all kinds of H_2 FIR state estimators applied to LTI systems affected by bounded errors, model uncertainties, and external disturbances.

H_∞ Performance

The H_∞ estimation performance is reached in state space by minimizing the induced norm of $\mathcal{T}$ for the maximized disturbance ς in what is known as the $\mathcal{L}_2$-to-$\mathcal{L}_2$ or energy-to-energy filter. The approach is developed to minimize the H_∞ norm $\|\mathcal{T}\|_\infty = \sup \sigma_{\max}[\mathcal{T}(z)]$, where $\sigma_{\max}$ is the maximum singular value of $\mathcal{T}(z)$. In the Bode plot, the H_∞ norm minimizes the highest peak value of $\mathcal{T}(z)$. In the designs of H_∞ estimators, the induced H_∞ norm $\|\mathcal{T}\|_\infty = \sup_{\varsigma \neq 0} \frac{\|\mathcal{T}\varsigma\|_2}{\|\varsigma\|_2} = \sup_{\varsigma\neq 0} \frac{\|\epsilon\|_2}{\|\varsigma\|_2}$ of $\mathcal{T}$ is commonly minimized using Parseval's theorem [70], where the ratios of the squared norms have meanings the ratios of the energies. Since the H_∞ norm reflects the worst estimator case, the H_∞ estimator is called *robust*.

The optimal H_∞ FIR filtering problem implies that the H_∞ FIR filter gain $\mathcal{H}_N(p)$ can be found by solving on $[m, k]$ the following optimization problem,

$$\mathcal{H}_N(p) = \inf_{\mathcal{H}_N} \sup_{\varsigma \neq 0} \frac{\sum_{i=m}^{k} \varepsilon_i^T P_\varepsilon(p) \varepsilon_i}{\sum_{i=m}^{k} \varsigma_i^T P_\varsigma \varsigma_i} , \tag{1.25}$$

where P_ε and P_ς are some proper weights. Unfortunately, closed-form optimal solutions to (1.25) are available only in some special cases. Therefore, the following suboptimal algorithm can be used to determine the gain $\mathcal{H}_N(p)$ numerically,

$$\mathcal{H}_N(p) \Leftarrow \sup_{\varsigma \neq 0} \frac{\sum_{i=m}^{k} \varepsilon_i^T P_\varepsilon(p)\varepsilon_i}{\sum_{i=m}^{k} \varsigma_i^T P_\varsigma \varsigma_i} < \gamma^2 , \tag{1.26}$$

where a small enough factor γ^2 indicates a part of the disturbance energy that goes to the estimator error. The solution to (1.26) is commonly found using LMI and the bounded real lemma.

Hybrid H_2/H_∞ Performance

Hybrid suboptimal H_2/H_∞ FIR state estimation structures are developed to improve the robustness by minimizing simultaneously the trace of the average weighted transfer function $\mathcal{T}$ using the H_2 approach and the peak value of $\mathcal{T}$ using the H_∞ approach. An example is the H_2/H_∞ FIR filter, where gain obeys both the H_2 and H_∞ constraints and can be determined by solving the following minimization problem

$$\mathcal{H}_N(p) = \min_{\substack{\mathcal{H}_N, \mathcal{Z} \\ \text{subject to LMI by } \mathcal{Z} - P_k(p) > 0 \text{ and (1.26)}}} \operatorname{tr} \mathcal{Z} \; . \tag{1.27}$$

Similarly, a hybrid H_∞/H_2 FIR state estimator can be developed.

Generalized H_2 Performance

In *generalized H_2 filtering*, the energy-to-peak transfer function $\mathcal{T}$ is minimized and the generalized H_2 performance is achieved by minimizing the peak error for the maximized disturbance energy [213] in what is called the *energy-to-peak* or $\mathcal{L}_2$-to-$\mathcal{L}_\infty$ estimation algorithm. Because an optimal solution to the generalized H_2 filtering problem does not exist in closed form, suboptimal algorithms were elaborated in [188] using LMI and the energy-to-peak lemma and then developed by many authors [5, 31, 144, 159].

The energy-to-peak filtering approach [213] implies minimizing the infinity error norm[1] $\|\varepsilon\|_\infty \triangleq \sup_{i\in[m,k]} \|\varepsilon_i\|_2 = \sqrt{\sup_{i\in[m,k]} (\varepsilon_i^T \varepsilon_i)}$ [188] over the maximized disturbance norm $\|w\|_2 = \sqrt{\sum_{i=m}^{k} w_i^T w_i}$. Accordingly, the suboptimal FIR filter gain $\mathcal{H}_N(p)$ can be determined by solving the optimization problem

$$\mathcal{H}_N(p) = \inf_{\mathcal{H}_N} \sup_{\|w\|_2 < \infty} \frac{\|\varepsilon(p)\|_\infty}{\|w\|_2} \tag{1.28}$$

using the energy-to-peak lemma [188]. In Chapter 8, we will show that this lemma is universally applicable to BE- and FE-based state-space models.

$\mathcal{L}_1$ Performance

In some applications, it is required to minimize the peak error with respect to the peak disturbance using the *peak-to-peak*, $\mathcal{L}_\infty$-to-$\mathcal{L}_\infty$, or most generally $\mathcal{L}_1$ state estimator. The need to exploit such estimators arises when a system operates under impulsive external attacks and this is not enough to minimize the peak estimation error with respect to the disturbance energy. The approach was originally proposed in [208] and then developed by many authors [1, 133, 144, 209]. It suggests to solve the *peak-to-peak* filtering problem by minimizing the $\mathcal{L}_1$ norm of the peak disturbance ς to peak error ε transfer function $\mathcal{T}$ on $[m, k]$ as

1 Note that the infinity error norm has another meaning than the standard vector infinity norm.

$\|\mathcal{T}\|_{\mathcal{L}_1} = \sup_{\|\varsigma\|<\infty} \frac{\|\varepsilon\|_\infty}{\|\varsigma\|_\infty} = \sup_{0<\|\varsigma\|<\infty} \left(\sqrt{\sup_{i\in[m,k]} \|\varepsilon_i\|_2} \Big/ \sqrt{\sup_{i\in[m,k]} \|\varsigma_i\|_2} \right)$. Accordingly, the FIR filter gain $\mathcal{H}_N(p)$ can be computed numerically by solving the optimization problem

$$\mathcal{H}_N(p) = \inf_{\mathcal{H}_N} \sup_{\|\varsigma\|<\infty} \frac{\sup_{i\in[m,k]} \|\varepsilon_i(p)\|_2^2}{\sup_{i\in[m,k]} \|\varsigma_i\|_2^2} . \tag{1.29}$$

It can be shown that an optimal solution to (1.29) is mathematically untractable, but a suboptimal $\mathcal{H}_N(p)$ can be found under the following assumptions. Since $\|\varepsilon_k\|$ at k does not exceed the upper bound $\sup_{i\in[m,k]} \|\varepsilon_i\|_2$, the following substitution can be made: $\sup_{i\in[m,k]} \|\varepsilon_i\|_2^2 = \alpha_\varepsilon \|\varepsilon_k\|_2^2$, where α_ε is a scaling scalar. Likewise, substitute $\sup_{i\in[m,k]} \|\varsigma_i\|_2^2 = \alpha_\varsigma \|\varsigma_k\|_2^2$, where α_ς is a scaling scalar. A suboptimal $\mathcal{H}_N(p)$ can then be determined by solving the optimization problem

$$\mathcal{H}_N(p) \Leftarrow \sup_{0<\|\varsigma\|<\infty} \frac{\|\varepsilon_k(p)\|_2^2}{\mu\|\varsigma_k\|_2^2} < \gamma^2 , \tag{1.30}$$

where $\mu = \alpha_\varsigma / \alpha_\varepsilon$ is a scalar and $\gamma > 0$ is small enough. In Chapter 8, we will show that the problem (1.30) can be solved numerically using the peak-to-peak lemma and LMI.

1.2.5 Unbiasedness

The property of *unbiasedness* is inherent to state estimators, which satisfy the *unbiasedness condition*

$$\mathcal{E}\{\hat{x}_{k+p|k}\} = \mathcal{E}\{x_{k+p}\} . \tag{1.31}$$

Because the operator of averaging removes zero mean noise from the model and zero mean random errors from the estimate, then it follows that any estimator will be *unbiased* if the test by (1.31) is positive. Estimators, which belong to the family of ML state estimators, produce the optimal unbiased estimate and thus obey (1.31). A purely unbiased estimator can be obtained by ignoring the zero mean noise in the system and the observation. That leads to the p-shift unbiased FIR (UFIR) filtering estimate [179],

$$x_{k+p|k} = [C_{m,k}^T(p)C_{m,k}(p)]^{-1}C_{m,k}^T(p)Y_{m,k} , \tag{1.32}$$

which belongs to the class of OLS and also appears from the canonic ML estimate (1.19) by setting $\Sigma_{m,k} = I$. Note that (1.32) does not pretend to minimize error covariance. Even so, this estimator is also able to minimize the MSE on the optimal horizon of N_{opt} points.

1.2.6 Deadbeat

If there is no noise in the state space model, then the state x_{k+p} and its estimate $\hat{x}_{k+p|k}$ become deterministic and the unbiasedness condition (1.31) degenerates to

$$\hat{x}_{k+p|k} = x_{k+p} , \tag{1.33}$$

which is known as the *deadbeat property*. Any estimator can be tested for (1.33) if we ignore noise in the model. Inherently, the OFIR estimator [176], ML estimator (1.19), and UFIR estimator (1.32) are *deadbeat* estimators. However, it cannot be said that KF has this property, because the noise covariances cannot be ignored in its recursions.

1.2.7 Denoising (Noise Power Gain)

As a measure of noise reduction in FIR filters, the *noise power gain* (NPG) was introduced by W. F. Trench in [198] as the ratio of the output noise variance σ_{out}^2 to the input noise variance σ_{in}^2,

$$\text{NPG} = \frac{\sigma_{\text{out}}^2}{\sigma_{\text{in}}^2} \,. \tag{1.34}$$

The NPG is thus akin to the noise figure in wireless communications. For white Gaussian noise, the NPG is equal to the sum of the squared coefficients of the impulse response h_k of the FIR filter or the squared norm of h_k. Thus, for white Gaussian noise it can be computed as $\text{NPG} = \sum_{k=0}^{N-1} h_k^2$, which does not apply for other types of noise.

In state space, the homogeneous gain $\mathcal{H}_{m,k}$ represents the impulse response coefficients of the FIR filter for each state. Therefore, in [186] a symmetric square matrix $\mathcal{G}_k$ of dimensions $K \times K$ was introduced and called the *generalized NPG (GNPG)*,

$$\text{GNPG} = \mathcal{G}_k = \mathcal{H}_{m,k}^{\text{h}} \mathcal{H}_{m,k}^{\text{h}^T} \,. \tag{1.35}$$

The value of this measure resides in the fact that it shows which of the estimators is prone to best noise reduction. Indeed, the main diagonal components of $\mathcal{G}_k$ are proportional to the NPGs in the system states, and the remaining components to the cross NPGs. Therefore, the estimator whose GNPG has the smallest components generates the least random noise at the output. For example, the UFIR filter has a larger GNPG than the OFIR filter and thus provides poorer noise reduction. It then follows that GNPG (1.35) can be used to compare different types of FIR state estimator in terms of noise reduction, provided that gains $\mathcal{H}_{m,k}^{\text{h}}$ of these estimators are known.

1.2.8 Stability

All FIR structures are *bounded input bounded output* (BIBO) stable by design. In contrast, LMF (Fig. 1.2b) and KF having IIR are not BIBO stable estimators due to the feedback. The stability of these estimators must be assured in each specific application.

1.2.9 Robustness

George E. P. Box was the first to introduce the term *robustness* in statistics [21]. Years later, P. J. Huber summarized that *robustness signifies insensitivity to small deviations from the assumptions* [74], and many other definitions of robustness have been adopted over time for specific applications. In this book, we will view *robustness* as *the ability of an estimator to be unresponsive to model errors, uncertainties, and disturbances*. This, however, has two different meanings. Solving the identification problem requires an estimator to be robust to all kinds of internal and external errors. In contrast, estimating the disturbed state requires an estimator to be robust only to observation errors. In both cases, the estimator must be robust to temporary uncertainties and impulsive attacks.

Obviously, a robust estimator is not one that is called robust, but which passes the test for robustness. The estimator robustness is highly dependent on the tuning factors and variables such as initial state, noise covariances, error matrices, scaling of norm-bounded estimators, etc. Since each tuning factor is introduced to improve performance, it follows that the performance can degrade dramatically if tuning is not done properly. Thus, the rule of thumb for robustness to tuning errors

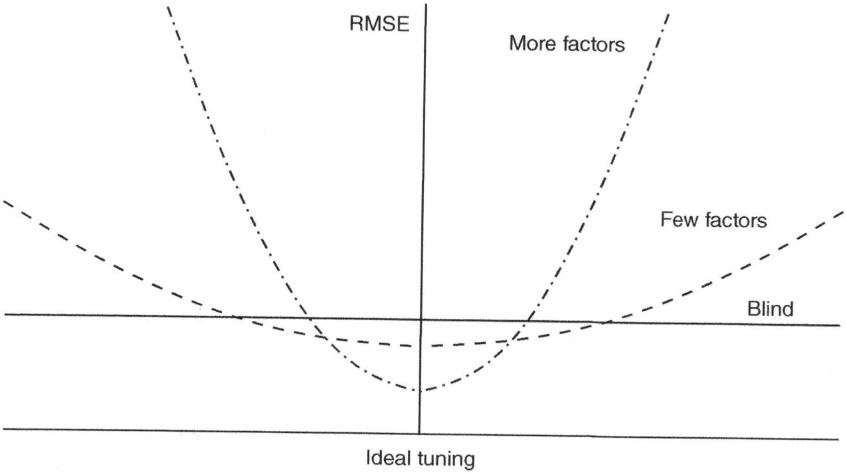

Figure 1.3 Worst-case effect of tuning errors on the estimator accuracy.

can be formulated as follows. An estimator with a small number of tuning factors is more robust than an estimator with a large number of tuning factors.

The effect of tuning errors on the estimation accuracy is sketched in Fig. 1.3 for the worst case, when these errors do not compensate for each other and can occur in opposite directions.

At the tuning phase, the blind estimator, which has no tuning factor, gives the largest errors. But this estimator is truly robust because its output is not affected by tuning. Several tuning factors can dramatically improve accuracy. However, tuning errors will degrade its performance. More tuning factors can further improve accuracy. But this can be accompanied by significant performance degradation due to tuning errors. The reader is referred to [180], where the trade-off in robustness between the UFIR, Kalman, and game theory H_∞ filters was investigated.

How can we tune an estimator to make it the most robust? The theory suggests that an estimator becomes robust if it is optimally tuned for maximized disturbance. If this is the case, then any deviation from the tuning point will lead to an increase in errors and a decrease in disturbance, which compensate for each other, and hence robustness.

1.2.10 Computational Complexity

The *computational complexity* of digital estimators is commonly associated with the *computation time*. In this regard, the most accurate batch FIR estimators, operating with data on the horizon $[m, k]$ of N points, have the highest computational complexity $\mathcal{O}(N^2)$ and largest computation time. An iterative algorithm for computing OFIR estimates using KF recursions has complexity $\mathcal{O}(N)$ and medium computation time. The KF algorithm has the lowest computational complexity $\mathcal{O}(1)$ and is the fastest state estimator. But the computation time consumed by iterative FIR algorithms can be significantly reduced using parallel computing.

Thus, the obvious reason for getting recursive forms is to reduce complexity when it is critical for real-time applications. But in trying to find suitable recursions, one realizes that they are available only for Gaussian processes and diagonal block error matrices. To illustrate this issue, let us consider the block measurement noise covariance matrix $\mathcal{R}_{m,k}$. For delta-correlated noise, this matrix is diagonal and can readily be represented recursively as

$$\mathcal{R}_{m,k} = \begin{bmatrix} R_m & 0 & \dots & 0 \\ 0 & R_{m+1} & \dots & 0 \\ \vdots & \vdots & \ddots & \vdots \\ 0 & 0 & \dots & R_k \end{bmatrix} = \begin{bmatrix} \mathcal{R}_{m,k-1} & 0 \\ 0 & R_k \end{bmatrix} \tag{1.36}$$

that, for OFIR filter with diagonal $\mathcal{Q}_{m,k}$, leads to Kalman recursions [229]. It can be shown that recursions for nondiagonal $\mathcal{R}_{m,k}$ and $\mathcal{Q}_{m,k}$ are complex, which can increase the computational complexity so that the original batch estimate can become more efficient. This observation explains the fact that the optimal recursions derived by Kalman for white Gaussian noise and serving the OFIR filter with diagonal $\mathcal{R}_{m,k}$ and $\mathcal{Q}_{m,k}$ and the unbiased recursions derived by Shmaliy for the UFIR filter by ignoring zero mean noise appear to be the only low-complexity engineering schemes used in practice.

1.2.11 Memory Use

The memory consumed by digital estimators typically depends on computational complexity. Batch state estimators, which simultaneously process N measurements, require about N^2 times more memory than KF. Iterative algorithms that compute batch estimates using recursions utilize about N times more memory than KF. Iterative FIR algorithms implemented using parallel computing require much more memory. However, memory is no longer an issue for modern computers.

1.3 More About FIR State Estimators

Now, it might be worthwhile to outline the most general features of FIR state estimators to make the rest of this book easier to read.

First, recall that the FIR state estimator is a batch estimator (Fig. 1.1 and Fig. 1.2) and, as such, is generally more accurate than a recursive computation algorithm. This is because the batch structure operates with full block noise covariance and error matrices, while suitable recursive forms are available when such matrices are diagonal. An exception is white Gaussian processes, where the block noise covariance matrices are diagonal and the Kalman recursions exactly compute the batch OFIR estimates.

Next, the folk belief that the batch is computationally less efficient than the recursive form is true only for Gaussian processes. For non-Gaussian processes, low-complexity recursive forms are commonly not available. Furthermore, in the modern world of fast computers and large memory, computational complexity no longer matters much, unlike in Kalman's days. Consequently, there is growing interest in batch FIR state estimators that produce the highest accuracy with the highest robustness.

Jazwinski was the first to draw attention in [79] to another fact that *finite horizon* (FH) control (read: RH FIR control) is more robust to temporary uncertainties than infinite horizon control (read: KF-based control). His argument was that the FH estimator does not project old errors beyond the horizon to the current point, and therefore the LMF is preferable. The general idea behind the FH approach was then formulated by Schweppe [161] as an old estimate updated not over all data on $[0, k]$ but over an FH $[m, k]$ of N most recent observations. It was later rephrased by Maybeck [126] that it is preferable to rerun the growing memory filter over the data FH for each k. In the sequel, FH estimators were developed using KF as LMFs for several control theories. In the general formulation of the FH estimation problem, it is required to minimize the cost function on $[m, k]$ at $k + p$, $p \lesseqgtr 0$, taking into account the input signal u_k. In the RH FIR control or MPC [106], this goal is achieved by predicting the state at k over most recent past data taken from $[m - 1, k - 1]$. The most elaborated RH *minimum variance FIR* (MVF) filter was designed in [105] to be bias-constrained or ML (not optimal).

Finally, it should be noted that the FIR approach recognizes the difference between the linear FE-based state model $x_k = F_{k-1}x_{k-1} + u_{k-1} + w_{n-1}$ used in control and the BE-based $x_k = F_k x_{k-1} + u_k + w_k$ used when prediction is not required. For example, the *a posteriori* p-shift OFIR filter [176] uses the BE-based state model, and the MVF filter [105] is based on the FE-based state model. This difference is essential when developing batch estimators but is poorly recognized in recursive schemes.

1.4 Historical Overview and Most Noticeable Works

The history of the development of methods of batch FIR state estimation related to filtering, smoothing, and prediction is rooted in a wide area of estimation theory. Probably the first and rather awkward batch method was used by J. Kepler[2] in 1601 [1.1] to estimate the orbit of Mars from 12 observations. The next noticeable step was made in 1795 by C. F. Gauss[3] [1.2], who contributed significant theoretical advances used to predict the planetary orbits with reasonable accuracy for a more elegant batch estimation approach, now referred to as the least squires.

In the 1930s, A. N. Kolmogorov[4] developed the least squires approach for discrete stationary random processes from purely mathematical consideration [1.3]. Soon after, in the early 1940s, N. Wiener[5] reconsidered the problem from the engineering perspective, extended the least squires approach to continuous-time stationary random processes, and derived an optimal filter, now known as the Wiener filter [1.4]. The optimal Wiener filter was then applied to a wide range of applications, including weather forecasting, economics, and communications. Further, Wiener investigated and solved several smoothing problems including working in the frequency domain with signals characterized by their power spectral densities and estimators derived in terms of their transfer functions.

Since the class of random stationary processes is relatively narrow, further progress has been achieved for dynamic systems governed by white Gaussian noise. This made it possible to conveniently simulate nonstationary processes in state space, both in continuous and discrete time, and subsequently to elegantly solve the optimal state estimation problem. The most significant works on the optimal state estimation of linear nonstationary stochastic processes with white Gaussian noise were published in 1960–1961 by R. E. Kalman.[6] The approach is based on the use of Bayes[7] formula, *a priori* and *a posteriori* probability distributions, and likelihood function. Despite the fact that Bayes' rule cannot be applied directly to the filtering problem, the optimality conditions imposed on it lead to an optimal filtering algorithm. The Kalman approach in discrete time

2 Johannes Kepler (December 27, 1571 to November 15, 1630), German astronomer, mathematician, and astrologer.
3 Johann Carl Friedrich Gauss (April 30, 1777 to Feb 23, 1855), German mathematician and physicist.
4 Andrey Nikolaevich Kolmogorov (April 25, 1903 to October 20, 1987), Soviet mathematician.
5 Norbert Wiener (November 26, 1894 to March 18, 1964), American mathematician and philosopher.
6 Rudolf Emil Kálmán (May 19, 1930 to July 2, 2016), Hungarian-American electrical engineer, mathematician, and inventor.
7 Thomas Bayes (1701 to April 7, 1761), English statistician, philosopher, and Presbyterian minister.

[1.5] and the Kalman-Bucy approach in continuous time [1.6] are known now as Kalman filtering, and the corresponding algorithms are referred to as the Kalman and Kalman-Bucy filters, respectively. Moreover, since this approach covers both stationary and nonstationary stochastic processes, the linear optimal filtering theory is called simply linear optimal Kalman filtering. About the time the Kalman filtering theory appeared, fundamental results on filtering of non-Gaussian nonlinear random processes were presented in 1955–1965 by R. L. Stratonovich[8] [1.7], whose nonlinear approach generalizes Kalman filtering in the linear case.

Soon after the optimal KF became a powerful tool for state estimation in linear systems, many authors also showed that its operation conditions are not always met in practice. Accordingly, the KF algorithm has been modified for colored Gauss-Markov noise, extended to nonlinear cases using Taylor series and unscented transform, reorganized to solve various smoothing problems, and robustified. These and other developments of the KF algorithm can be found in D. Simon's book [1.8] and many other outstanding works.

In 1968, A. H. Jazwinski has made a fundamental observation [1.9], which, decades later, arouses great interest in state estimation using FIR structures. He stated [1.10] that the *LMF appears to be the only device for preventing divergence in the presence of unbounded perturbation* in the system and concluded that FH control is more robust than infinite horizon control. It then turned out that the initial conditions for optimal LMF operation could not always be set properly, and the approach was not widely adopted, although the basic idea was latter used in FIR filtering.

The batch FIR estimator can be viewed as a Gauss least squares estimator, although its roots can rather be found in Wiener's filtering theory. In [1.11], N. Levinson used Wiener's MSE criterion to design a discrete-time filter and predictor in recursive form on the FH of past data. The FIR modification of the Wiener filter was made by L. A. Zadeh[9] and J. R. Ragazzini in [1.12]. In [1.13], K. R. Johnson extended the Zadeh-Ragazzini results to discrete time. The first RH UFIR filter based on discrete convolution was obtained by P. Heinonen and Y. Neuvo in [1.14], and the UFIR filter was derived by Y. S. Shmaliy in [1.16]. A general approach to FIR estimation using convolution and Hankel operator norms was developed by D. Wilson in [1.17].

In state space, the theory of various RH FIR structures and recursive forms was first developed in the works of W. H. Kwon *et al.* [1.18,1.19]. The theory of p-shift OFIR filtering was developed by Y. S. Shmaliy [1.20,1.21]. In the batch form, the first RH UFIR filter was shown by P. S. Kim and M. E. Lee in [1.22], and the p-shift UFIR filter was derived by Y. S. Shmaliy in [1.20]. The UFIR filter was then represented with an iterative algorithm using recursions by Y. S. Shmaliy in [1.23] to become a robust alternative to KF [1.24]. Several H_2, H_∞, and hybrid H_2/H_∞ FIR structures were first proposed by W. H. Kwon *et al.* in [1.19, 1.25].

In most recent years, significant progress in FIR state estimation has been achieved in works of Y. S. Shmaliy, S. Zhao, and C. K. Ahn. The first- and second-order extended UFIR filters were proposed for nonlinear models in [1.26]. The optimal horizon N_{opt} for the UFIR filter is justified in [1.27]. In [1.28], it was shown that recursions for OUFIR and ML FIR filters are not Kalman recursions, and therefore KF is optimal, not optimal unbiased, as previously thought. It was shown in [1.29] that all types of bias-constrained OFIR filters belong to the class of ML state estimators. In [1.30], it was proved that the error covariance of the UFIR filter can be computed iteratively. In [1.31], an improved frequency efficient RH H_∞ FIR filtering algorithm was proposed. The UFIR filtering approach has been extended in [1.32] to time-stamped delayed and missing data. In [1.33], a new H_2 FIR state estimation approach was developed under disturbances, data errors, and uncertain

8 Ruslan Leont'evich Stratonovich (May 31, 1930 to January 13, 1997), Russian physicist, engineer, and probabilist.
9 Lotfi Aliasker Zadeh (February 4, 1921 to September 6, 2017), Iranian-American mathematician, computer scientist, electrical engineer.

initial conditions. It was shown in [1.34] that the batch OFIR filtering estimate can be computed using Kalman recursions, and therefore the OFIR and Kalman approaches are not two different methods as previously thought. Chapter 8 and Chapter 9 of this book contribute to the current view of FIR state estimation under disturbances and for uncertain systems, and other important publications can be found in the references.

··········◁ ○ ▷··········

[1.1] Kepler J (1602) De Fundamentis Astrologiae Certioribus. Typis Schumanianis, Pragae Bohemorum

[1.2] Plackett RL (1972) Studies in the History of Probability and Statistics. XXIX: The discovery of the Method of least squares. Biometrika 59(2):239–251

[1.3] Publications of A. N. Kolmogorov (1989) Annals of Probability 17(3):945–964

[1.4] Wiener N (1942) Extrapolation, Interpolation, and Smoothing of Stationary Time Series with Engineering Applications. MIT Press, Boston MA

[1.5] Kalman RE (1960) A new approach to linear filtering and prediction problems. Trans. ASME (J. Basic Eng.) D82(1):35–45

[1.6] Kalman RE, Bucy RS (1961) New results in linear filtering and prediction theory. Trans. ASME (J. Basic Eng.) D83(1):95–108

[1.7] Stratonovich RL Topics in the Theory of Random Noise, Vol. 1 (1963), Vol. 2 (1967). Gordon and Breach, NY

[1.8] Simon D (2006) Optimal State Estimaiton: Kalman, H_∞, and Nonlinear Approaches. John Wiley & Sons, Hoboken, NJ

[1.9] Jazwinski AH (1968) Limited memory optimal filtering. IEEE Trans. Automat. Contr. 13(5):558–563

[1.10] Jazwinski AH (1970) Stochastic Processes and Filtering Theory. Academic Press

[1.11] Levinson N. (1947) The Wiener RMS error criterion in filter design and prediction. J. Math. Phys. 25:261–278.

[1.12] Zadeh LA, Ragazzini JR (1950) An Extension of Wiener's Theory of Prediction. J. Appl. Phys. 21:645–655

[1.13] Johnson KR (1956) Optimum, linear, discrete filtering of signals containing a nonrandom component. IRE Trans. Inform. Theory 2(2):49–55

[1.14] Friedlander B, Morf M, Kailath T, Ljung L (1978) Extended Levinson- and Chandra-Sekhar-type equations for a general discrete time linear estimation problem. IEEE Trans. Automat. Contr. 23(4):653–659

[1.15] Heinonen P, Neuvo Y (1988) FIR-median hybrid filter with predictive FIR substructures, IEEE Trans. Acoust. Speech Signal Process. 36(6):892–899

[1.16] Shmaliy YS (2006) An unbiased FIR filter for TIE model of a local clock in applications to GPS-based timekeeping, IEEE Trans. on Ultrason. Ferroelec. Freq. Contr. 53(5):862–8706

[1.17] Wilson D (1989) Convolution and Hankel operator norms for linear systems, IEEE Trans. Autom. Contr. 34(1):94–97

[1.18] Kwon OK, Kwon WH, Lee KS (1989) FIR filters and recursive forms for discretetime state-space models. Automatica 25(5):715–728

[1.19] Kwon WH, Han S (2005) Receding Horizon Control: Model Predictive Control for State Models. Springer, London

[1.20] Shmaliy YS (2010) Linear optimal FIR estimation of discrete time-invariant state space models, IEEE Trans. Signal Process. 58(6):3086–3096

[1.21] Shmaliy YS, Ibarra-Manzano OG, Time-variant linear optimal finite impulse response estimator for discrete state-space models, Int. J. Adapt. Contr. Signal Process. 26(2):95–104

[1.22] Kim PS, and Lee ME (2007) A new FIR filter for state estimation and its applications, J. Comput. Sci. Technol. 22(5):779–784

[1.23] Shmaliy YS (2011) An iterative Kalman-like algorithm ignoring noise and initial conditions, IEEE Trans. Signal Process. 59(6):2465–2473

[1.24] Shmaliy YS, Zhao S, Ahn CK (2017) Unbiased FIR filtering: an iterative alternative to Kalman filtering ignoring noise and initial conditions, IEEE Contr. Syst. Mag. 37(5):70–89

[1.25] Lee YS, Han SH, Kwon WH (2006) H_2/H_∞ FIR filters for discrete-time state space models. Int. J. Control, Autom. Syst. 4(5):645–652

[1.26] Shmaliy YS (2012) Suboptimal FIR filtering of nonlinear models in additive white Gaussian noise, IEEE Trans. Signal Process. 60(10):5519–5527

[1.27] Ramirez-Echeverria Sarr FA, Shmaliy YS (2014) Optimal memory for discrete-time FIR filters in state space, IEEE Trans. Signal Process 62(3):557–561

[1.28] Zhao S, Shmaliy YS, Liu (2016) Fast Kalman-like optimal unbiased FIR filtering with applications, IEEE Trans. Signal Process. 64(9):2284–2297

[1.29] Zhao S, Shmaliy YS, Unified maximum likelihood form for bias constrained FIR filters, IEEE Signal Process. Lett. 23(12):1848–1852

[1.30] Zhao S, Shmaliy YS, Liu F (2017) On the iterative computation of error matrix in unbiased FIR filtering, IEEE Signal Process. Lett. 24(5):555–558

[1.31] Ahn CK, Zhao S, Shmaliy YS (2017) Frequency efficient receding horizon H_∞ FIR filtering in discrete-time state-space, IEEE Trans. Circ. Syst. I: Regular Papers 64(11):2945–2953

[1.32] Uribe-Murcia K, Shmaliy YS, Ahn CK, Zhao S (2020) Unbiased FIR filtering for time-stamped discretely delayed and missing data. IEEE Trans. Automat. Contr. 65(5):2155–2162

[1.33] Ortega-Contreras J, Eli Pale-Ramon E, Shmaliy YS, Xu Y (2021) A novel approach to H_2 FIR prediction under disturbances and measurement errors, IEEE Signal Process. Lett. 28:150–154

[1.34] Zhao S, Shmaliy YS, Liu F (2021) Optimal FIR filter for discrete-time LTV systems and fast iterative algorithm, IEEE Trans. Circ. Syst. II, Express Briefs

1.5 Summary

Unlike the classical estimation theory, which was developed to estimate parameters of the approximating function based on measurements with random components, the state estimation theory deals with the internal state of a system. Hence, the state estimator can be thought of as an input-to-output structure (hardware or software) that has an IIR or FIR. The set of state variables should be sufficient to predict the future behavior of a system and should not be redundant to avoid excess errors. Physical systems are represented in state space via nonlinear or linear differential equations with random terms. In discrete-time state-space, noisy systems are represented via difference equations, which appear from the corresponding differential equations by applying the FE method or BE method.

The theory of linear systems with Gaussian random components is most developed. If a model is nonlinear, then the most common way is to linearize it using the Taylor series expansion and apply methods of linear state estimation. Regardless of structure, the state estimator can be designed to be a filter, q-lag smoother, p-step predictor, smoothing filter, or predictive filter.

The term *optimal* most often refers to estimators of linear stochastic processes, in which case an optimal balance in the MSE sense is obtained between regular bias and random errors. It also applies to systems with random components when the estimation error is minimized in some other sense. In general, each noise requires its own optimal filter: OFIR filter and KF for Gaussian noise, median filter for Laplace noise, myriad filter for Cauchy noise, etc. If the derivation of the optimal estimator obeys the unbiasedness constraint, then the estimate has the property of unbiased optimality. All linear bias-constrained FIR state estimators belong to the class of ML estimators and have the canonic ML FIR form (1.19).

The property of suboptimality is inherent to norm-bounded filters, the gains of which cannot be obtained analytically and are determined numerically using LMI. The following suboptimal FIR state estimators having a disturbance-to-error transfer function $\mathcal{T}$ can be designed using LMI: H_2 by minimizing the squared Frobenius norm $\|\mathcal{T}\|_F^2$, H_∞ by minimizing the induced norm of $\mathcal{T}$ in the $\mathcal{L}_2$-to-$\mathcal{L}_2$ or energy-to-energy filter, H_2/H_∞ by combining the H_2 and H_∞ filtering algorithms, generalized H_2 by minimizing the energy-to-peak $\mathcal{T}$ in the $\mathcal{L}_2$-to-$\mathcal{L}_\infty$ structure, and $\mathcal{L}_1$ by minimizing the $\mathcal{L}_\infty$-to-$\mathcal{L}_\infty$ or peak-to-peak $\mathcal{T}$.

Unbiasedness is achieved when the state estimator satisfies the unbiasedness condition (1.31). This condition results in a batch UFIR filter whose canonic form (1.32) can be computed using recursions that are not Kalman recursions. An estimator is a deadbeat estimator if it satisfies the condition (1.33) by neglecting noise in the model. All FIR state estimators are BIBO stable, which is not the case for IIR state estimators such as LMF and KF. The state estimator with the lowest GNPG is the most noise canceling. The state estimator with the least number of tuning factors and the sensitivity to these factors is the most robust. In this sense, the UFIR filter, which requires only one tuning factor, the horizon length, is the most robust of all other linear estimators.

The KF algorithm has low computational complexity $\mathcal{O}(1)$. Iterative FIR algorithms operating on a horizon of N points have complexity $\mathcal{O}(N)$ and batch FIR state estimators $\mathcal{O}(N^2)$. A computationally complex estimator usually requires a lot of memory to obtain an estimate.

1.6 Problems

1 Why can a physical system but not an image be represented by a differential equation?

2 The power spectral density of a working motor has a major peak at the motor frequency and a lateral peak associated with the motor failures. Intuitively assign state variables to estimate motor normal operation and failures.

3 A maneuvering robot is represented in state space with the following nonlinear equations

$$
\begin{aligned}
x_{k+1} &= f(x_k) + w_k \,, \\
y_k &= h(x_k) + v_k \,.
\end{aligned}
$$

Apply the first-order Taylor series expansion to the nonlinear functions $f(x_k)$ and $h(x_k)$ and convert these equations to linear equations.

4 The continuous-time state-space model in (1.7) and (1.8) can be represented in discrete time using the FE method as (1.9) and (1.10) or BE method as (1.11) and (1.12). Analyze the difference between (1.9) and (1.11) and argue which model is 1) more accurate and 2) more suitable for feedback state control. Formulate the conditions under which these models can be transformed into each other.

5 Batch state estimators, such as those shown in Fig. 1.2, process data collected on $[m, k]$, which can cause a computational problem. Explain intuitively why recursive forms for batch estimators are mostly available when the block noise covariance matrices are diagonal.

6 Optimal solutions to the linear stochastic problem are the batch OFIR filter and the recursive KF algorithm. Why are these filters not optimal for non-white Gaussian noise? Why does each noise distribution require its own optimal filter? Find logical explanations.

7 There are three basic batch FIR state estimators for linear stochastic systems: OFIR, ML FIR, and UFIR. Test these estimators intuitively on the properties of state estimators and classify them in terms of 1) optimality, 2) noise reduction, and 3) robustness.

8 Generate a discrete-time two-state process $\dot{x}(t) = Ax(t) + w(t)$, where $A = \begin{bmatrix} 0 & 1 \\ 0 & 0 \end{bmatrix}$, $w(t) = [\ 0\ \xi(t)\ ]^T$, and $\xi(t) \sim \mathcal{N}(0, \sigma_\xi^2)$ is zero mean white Gaussian noise with the variance σ_ξ^2. Plot the first and second states of the system as functions of time for $\sigma_\xi = 0.1$, $\sigma_\xi = 1$, and $\sigma_\xi = 10$. Analyze the difference between functions.

9 A system is represented with the state-space model in (1.11) and (1.12), where $D_k = 0$ and $E_k = 0$. Assume that noise vectors w_k and v_k are correlated,

$$\mathcal{E}\begin{bmatrix} w_k \\ v_k \end{bmatrix}\begin{bmatrix} w_k^T & v_k^T \end{bmatrix} = \begin{bmatrix} Q_k & L_k \\ L_k^T & R_k \end{bmatrix} ,$$

where $Q_k = \mathcal{E}\{w_k w_k^T\}$, $R_k = \mathcal{E}\{v_k v_k^T\}$, and $L_k = \mathcal{E}\{w_k v_k^T\}$. Analyze intuitively how the correlation will affect the state estimate if we neglect it by setting $L_k = 0$ in the algorithm.

10 The general structures of FIR and limited memory IIR estimators are shown in Fig. 1.1a and Fig. 1.1b, respectively. Explain logically why the initial $\hat{x}_m$ and P_m serve as variables of the FIR estimator gain, while x_{m-1} is an input of the limited memory IIR estimator.

11 In a satellite navigation system, the geometric distance d_i between the receiver and the ith satellite is measured as

$$d_i = \sqrt{(x - x_i)^2 + (y - y_i)^2 + (z - z_i)^2} ,$$

where (x, y, z) are 3D ground coordinates and (x_i, y_i, z_i) are 3D coordinates of the ith satellite. High navigation accuracy is guaranteed by the estimation of distance, velocity, and acceleration. Determine the dimensions of the navigation state vector.

12 The dynamics of a DC motor are described by the following equations,

$$\frac{d^2}{dt^2}\theta = -\frac{b}{J}\frac{d}{dt}\theta + \frac{K}{J}i ,$$

$$\frac{d}{dt}i = -\frac{K}{J}\frac{d}{dt}\theta - \frac{R}{L}i + \frac{1}{L}V ,$$

where θ is the angle of rotation, i is the armature current, R is the electric resistance, L is the electric inductance, b is the motor viscous friction constant, J is the moment of inertia of the rotor, V is the driving voltage, and K represents the motor torque and back electromotive force constants. The θ measurements are performed in the presence of noise $v(t)$. Introduce the state vector $x = [\ \theta\ \dot{\theta}\ i\ ]^T$ and represent the motor equations and measurement in continuous time state space.

13 A dynamic physical system is represented in continuous time state space with the state space equations $\dot{x}(t) = Ax(t) + Bu(t)$ and $y(t) = Cx(t) + Du(t)$, where matrices are of the form

$$A = \begin{bmatrix} 0 & 1 & 0 \\ 0 & 0 & 1 \\ -3 & 1 & -2 \end{bmatrix}, \quad B = \begin{bmatrix} 0 \\ 0 \\ 1 \end{bmatrix}, \quad C = \begin{bmatrix} -\frac{3}{2} & \frac{3}{2} & -\frac{5}{2} \end{bmatrix}, \quad D = \begin{bmatrix} \frac{3}{2} \end{bmatrix}.$$

Determine the dimensions of vectors $x(t)$ and $u(t)$ and write the differential equation for this system.

14 Find the conditions under which the RH FIR estimator shown in Fig. 1.2c becomes a special case of the FIR estimator shown in Fig. 1.2a.

15 The autonomous system consists of two subsystems [211],

$$x_{k+1} = F_1 x_k + F_{d1} x_{k-d} + w_k\,,$$
$$x_{k+1} = F_2 x_k + F_{d2} x_{k-d} + w_k\,,$$

where $w_k \sim \mathcal{N}(0, 0.1)$ and matrices are specified as

$$F_1 = \begin{bmatrix} 0.4 & 0.05 \\ 0 & 0.35 \end{bmatrix}, \quad F_{d1} = \begin{bmatrix} 0.025 & 0 \\ -0.1 & -0.35 \end{bmatrix},$$
$$F_2 = \begin{bmatrix} -0.2 & 0 \\ 0.1 & 0.1 \end{bmatrix}, \quad F_{d2} = \begin{bmatrix} 0.05 & -0.1 \\ 0 & 0.15 \end{bmatrix}.$$

Investigate numerically the impulse responses of these systems.

16 Local clocks are characterized by three variables that represent their states: 1) x_1 is the time error, 2) x_2 is the fractional frequency offset, and 3) x_3 is the linear frequency drift rate. Variables are disturbed by independent noise sources. Write down the clock state model in discrete time.

17 An electronic oscillator is represented with the SDE

$$\ddot{V}(t) + 2\delta(1-K)\dot{V}(t) + \omega_0^2 V(t) = w(t)\,,$$

where $V(t)$ is the generated voltage, 2δ is the angular bandwidth, ω_0 is the oscillator angular frequency, and noise $w(t) \sim \mathcal{N}(0, \sigma^2)$ has the variance σ^2. Represent this equation in state space.

18 The block diagram of a stochastic LTV system, which is represented in continuous-time state space, is shown in Fig. 1.4. Write the state space equations of this system.

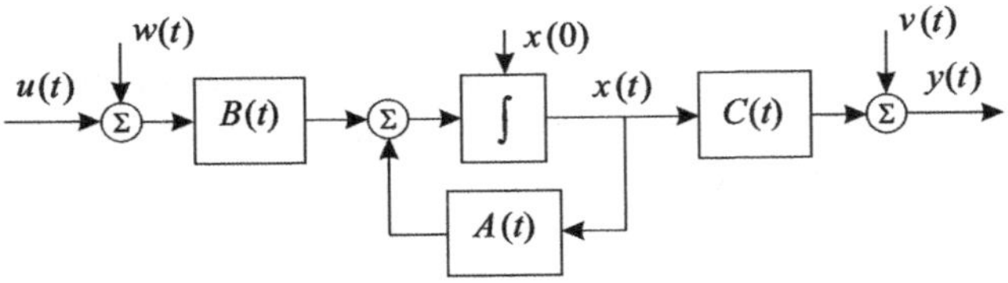

Figure 1.4 Block diagram of a stochastic LTV system observed in continuous time state space.

19 The system state x_m is observed over a horizon $[m, k]$ in the presence of zero mean measurement noise v_k as $Y_{m,k} = H_N x_m + V_{m,k}$, where

$$Y_{m,k} = \begin{bmatrix} y_m \\ y_{m+1} \\ \vdots \\ y_k \end{bmatrix}, H_N = \begin{bmatrix} H \\ H \\ \vdots \\ H \end{bmatrix}, V_{m,k} = \begin{bmatrix} v_m \\ v_{m+1} \\ \vdots \\ v_k \end{bmatrix}.$$

Show that the gain $\mathcal{H}_N$ of an UFIR filter $\hat{x}_k = \mathcal{H}_N Y_{m,k}$ that satisfies the unbiasedness condition (1.31) is given by

$$\mathcal{H}_N = (H_N^T H_N)^{-1} H_N^T .$$

Compare this gain to the canonic gain of the UFIR state estimator defined by (1.32) and analyze the differences.

20 The time-invariant gain $\mathcal{H}_N$ of the median filter is given by (1.21) with $p = 0$. Approximate the absolute error $|\varepsilon_k|$ with a differentiable function $f(\varepsilon_k)$ such that at the limit it is equal to $|\varepsilon_k|$. Minimize the sum $\sum_{i=m}^{k} f(\varepsilon_i)$ by $\mathcal{H}_N$ using the derivative and obtain a closed form for $\mathcal{H}_N$ in the limit.

2

Probability and Stochastic Processes

In stochastic processes the future is not uniquely determined, but we have at least probability relations enabling us to make predictions.

William Feller [47], p. 420.

Signals and messages containing information about electrical, mechanical, chemical, biological, and other processes are usually affected by various types of noise and disturbances, the values of which often cannot be ignored. In such cases, the deterministic approximation becomes too rough, and probabilistic methods are used to achieve the best results. Under the influence of noise, any process becomes random, and accurate information extraction about its features requires mathematical methods describing random variables, stochastic processes, and SDEs. This chapter provides a brief introduction to the concepts and foundations of the theory of probability and stochastic processes, which will be used later in the discussion of methods of state estimation.

2.1 Random Variables

In engineering practice we often deal with some kind of experiment and elements ξ of its random outcomes that cannot be used directly. For example, tracking distance d can be measured via time of arrival. Thus, to each d we can assign a real number $X(\xi)$, call it *random variable* [191], and describe $X(\xi)$ or simply X in terms of the probability theory. A collection of random variables sets up some *random process.*

Scalar random variable: Since X is random, it cannot be described in deterministic terms. But we may wonder how frequently the variable X occurs above some constant value of x, which leads to the concept of *probability*. Let us think that X occurs many times and assign an event A, which means that X is happening below x. The probability that $X \leqslant x$ is calculated as

$$P\{X \leqslant x\} = \frac{\text{Possible outcomes favoring event A}}{\text{Total possible outcomes}} \tag{2.1}$$

and X is represented by the *cumulative distribution function* (cdf),

$$F_X(x) = P\{X \leqslant x\} \ , \tag{2.2}$$

Optimal and Robust State Estimation: Finite Impulse Response (FIR) and Kalman Approaches, First Edition.
Yuriy S. Shmaliy and Shunyi Zhao.

which is equal to the probability $P\{X \leqslant x\}$ for X to take values below x with the following properties:

$$0 \leqslant F_X(x) \leqslant 1$$
$$F_X(\alpha) \leqslant F_X(\beta) \quad \text{if} \quad \alpha \leqslant \beta ,$$
$$P\{\alpha < X \leqslant \beta\} = F_X(\beta) - F_X(\alpha) ,$$
$$F_X(-\infty) = 0 ,$$
$$F_X(\infty) = 1 .$$

One might also be interested in a function that represents the concentration of X values around x. The corresponding function $p_X(x)$ is called the *probability density function* (pdf). But if the random variable X is discrete, it is represented with the *probability mass function* (pmf). An example is the Bernoulli distribution of two discrete random values 0 and 1.

Both $F_X(x)$ and $p_X(x)$ are positive-valued with the following fundamental properties:

$$\int_{-\infty}^{\infty} p_X(x)dx = 1 , \tag{2.3}$$

$$F_X(x) = \int_{-\infty}^{x} p_X(z)dz \quad \Leftrightarrow \quad p_X(x) = \frac{\mathrm{d}}{\mathrm{d}x} F_X(x) . \tag{2.4}$$

Another way to describe the properties of a random variable X is to define the expectation of its exponential measure as

$$\Phi_X(j\theta) = \mathcal{E}\{e^{j\theta X}\} = \int_{-\infty}^{\infty} p_X(x) e^{j\theta x} \,\mathrm{d}x , \tag{2.5}$$

which is the direct Fourier transform of pdf $p_X(x)$ with the variable θ playing the role of negative angular frequency. The $\Phi_X(j\theta)$ function, which is usually complex, is called the *characteristic function* (cf) and represents X in the transform domain. Provided $\Phi_X(j\theta)$, the pdf of X can be defined by inverse Fourier transform as

$$p_X(x) = E\{e^{j\theta X}\} = \frac{1}{2\pi} \int_{-\infty}^{\infty} \Phi_X(j\theta) e^{-j\theta x} \,\mathrm{d}\theta \geqslant 0 . \tag{2.6}$$

In some cases, one can also use the logarithm of the characteristic function or *log-characteristic function*

$$\Psi_X(j\theta) = \ln \Phi_X(j\theta) ,$$

which can be more convenient for some forms of $\Phi_X(j\theta)$.

Multiple random variables: Let us now consider two random variables X and Y such that, by (2.2), $F_X(x) = P\{X \leqslant x\}$ and $F_Y(x) = P\{Y \leqslant y\}$. The joint cdf of X and Y is defined by

$$F_{X,Y}(x, y) = P(X \leqslant x, Y \leqslant y) \tag{2.7}$$

to exist with the following main properties:

$$0 \leqslant F_{X,Y}(x, y) \leqslant 1 ,$$
$$F_{X,Y}(x, \infty) = F_X(x) ,$$
$$F_{X,Y}(\infty, y) = F_Y(y) ,$$
$$F_{X,Y}(\infty, \infty) = 1 ,$$

$$F_{X,Y}(x,-\infty) = F_{X,Y}(-\infty,y) = 0\ ,$$
$$F_{X,Y}(a,c) \leqslant F_{X,Y}(b,d) \quad \text{if} \quad a \leqslant b \text{ and } c \leqslant d\ .$$

The joint pdf and cdf of random X and Y relate to each other as

$$p_{X,Y}(x,y) = \frac{\partial^2 F_{X,Y}(x,y)}{\partial x \partial y}\ , \tag{2.8}$$

$$F_{X,Y}(x,y) = \int_{-\infty}^{x}\int_{-\infty}^{y} p_{X,Y}(v,u)\,\mathrm{d}v\,\mathrm{d}u \tag{2.9}$$

and $p_{X,Y}(x,y)$ has the following properties:

$$p_{X,Y}(x,y) \geqslant 0\ ,$$
$$\int_{-\infty}^{\infty}\int_{-\infty}^{\infty} p_{X,Y}(x,y)\,\mathrm{d}x\mathrm{d}y = 1\ ,$$
$$p_X(x) = \int_{-\infty}^{\infty} p_{X,Y}(x,y)\,\mathrm{d}y\ ,$$
$$p_Y(y) = \int_{-\infty}^{\infty} p_{X,Y}(x,y)\,\mathrm{d}x\ .$$

Given the distribution of a random variable X, the latter can be represented in general by an infinite number of quantitative measures, called *moments* associated with $p_X(x)$ and *cumulants* associated with $\Phi_X(j\theta)$. For example, the zero moment of X is its total probability (i.e., one), the first moment is the *mean* or *average*, the second central moment is the *variance*, the third standardized moment is the *skewness*, and the fourth standardized moment is the *kurtosis*.

2.1.1 Moments and Cumulants

Two kinds of special characteristics called *moments* have found applications in the representation of random variables: *raw (ordinary) moments* and *central moments*.

Raw Moments

The raw moment of the v-order of a random variable X is defined by the average of X^v as

$$\mathcal{E}\{X^v\} = m_v = \int_{-\infty}^{\infty} x^v p_X(x)\,\mathrm{d}x \tag{2.10}$$

and the infinite set of raw moments completely represents the variable X.

The v-order raw moment of a discrete population of X_i, $i \in [-N, N]$, can be represented with the raw moments calculated by

$$m_v \cong \frac{1}{2N+1}\sum_{i=-N}^{N} X_i^v \tag{2.11}$$

and it represents X exactly when $N \to \infty$.

For the population of X observed on a continuous time interval $[-T, T]$, the raw moment can also be calculated by

$$m_v \cong \frac{1}{2T}\int_{-T}^{T} X^v(t)\,\mathrm{d}t \tag{2.12}$$

and it represents a variable X exactly when $T \to \infty$.

The most commonly used raw moment is of the 1-order, $v = 1$, and is known as the *mean* or *average* of X,

$$m_1 = \bar{X} = \mathcal{E}\{X\} \; . \tag{2.13}$$

There are also two other special characteristics associated with the mean of X called *median* and *mode*.

Median: The median "med X" is the value separating the higher half of the population of X values from the lower half. Therefore, the median may be thought as the "middle" value. For example, given a set $X = \{1, 2, 3, 4, \mathbf{5}, 6, 7, 8, 9\}$, the median is 5, and for $X = \{1, 2, 3, \mathbf{4}, \mathbf{5}, 6, 7, 8\}$ it is 4.5. But for $X = \{1, 2, 3, 4, \mathbf{5}, 6, 7, 8, 90\}$ the median is also 5. Therefore, this statistic is robust against outliers.

Mode: The "mode X" of a population of X is the value that appears most often. The mode is thus the value of x at which the pmf $p_X(x)$ takes its maximum value. The mode is not unique if $p_X(x)$ takes the same maximum value at several points. The most extreme case is the uniform distributions, where all values occur equally.

> ***Mean, median, and mode:*** *The mean, median, and mode are the same in unimodal distributions, such as the Gaussian distribution, and they are different in skewed distributions.*

Central Moments

It is also desirable to have moments relative to the mean (2.13) of a random variable X. The corresponding v-order moment μ_v is called the *central moment* and is defined by the mean value of $(X - m_1)^v$ as

$$\mu_v = \mathcal{E}\{(X - \bar{X})^v\} = \int_{-\infty}^{\infty} (x - \bar{X})^v p(x) \, \mathrm{d}x \; . \tag{2.14}$$

For discrete and continuous populations of X, the v-order central moment can be determined by, respectively,

$$\mu_v = \lim_{N\to\infty} \frac{1}{2N+1} \sum_{i=-N}^{N} (X_i - \bar{X})^v \; , \tag{2.15}$$

$$\mu_v = \lim_{N\to\infty} \frac{1}{2T} \int_{-T}^{T} [X(t) - \bar{X}]^v \, \mathrm{d}t \; . \tag{2.16}$$

The most common central moment has the 2-order, $v = 2$, and is called the *variance* $\sigma_X^2 = \mu_2$, which can also be calculated by

$$\sigma_X^2 = m_2 - m_1^2 \; . \tag{2.17}$$

The root square of the variance is called the *standard deviation*

$$\sigma_X = \sqrt{\sigma_X^2} \geqslant 0 \; , \tag{2.18}$$

which always has positive values and is the degree of dissipation of a variable X with respect to its mean value $E\{X\}$.

Cumulants

The moments of a random variable X can also be determined using the expansion of cf $\Phi(j\theta)$ in a Maclaurin series. The corresponding coefficients of the Maclaurin series are called *semi-invariants* or *cumulants* κ_v of order v and are calculated as

$$\kappa_v = j^{-v} \left[\frac{\mathrm{d}^v \ln \Phi(j\theta)}{\mathrm{d}\theta^v} \right] \Bigg|_{\theta=0} .$$

For given cumulants κ_v, the characteristic function $\Phi(j\theta)$ is defined by

$$\Phi(j\theta) = \exp \left[\sum_{v=1}^{\infty} \frac{\kappa_v}{v!} (j\theta)^v \right] .$$

It follows from the definition of $\Phi(j\theta)$ that the raw moments of X can be specified via $\Phi(j\theta)$ as

$$m_v = j^{-v} \frac{\mathrm{d}^v}{\mathrm{d}\theta^v} \Phi(j\theta) \Bigg|_{\theta=0}$$

and $\Phi(j\theta)$ restored through raw moments as

$$\Phi(j\theta) = \sum_{v=0}^{\infty} \frac{m_v}{v!} (j\theta)^v .$$

What also comes is that cumulants can be represented through moments and vise versa as $\kappa_1 = m_1$, $\kappa_2 = m_2 - m_1^2 = \mu_2$, $\kappa_3 = m_3 - 3m_1 m_2 + 2m_1^3 = \mu_3$, … and $m_1 = \kappa_1$, $m_2 = \kappa_2 + \kappa_1^2$, $m_3 = \kappa_3 + 3\kappa_1\kappa_2 + \kappa_1^3$, …, due to exact transitions from one type of characteristics to another.

Skewness and Kurtosis

Symmetrically distributed X is not always the case in physics and engineering, since many quantities are positive-valued, such as distance, range, and magnitude. On the other hand, many symmetric distributions differ from the normal law by a greater rectangularity or longer tails. In such cases, two other statistics are used: *skewness* and *kurtosis*.

Skewness: If a real-valued random variable X is distributed asymmetrically about the mean, *skewness* is often used as a measure of asymmetry. The skewness value is calculated by

$$\gamma_1 = \frac{\mu_3}{\sqrt{\mu_2^3}} = \frac{\mu_3}{\sigma_X^3} = \frac{\kappa_3}{\sqrt{\kappa_2^3}} \tag{2.19}$$

and it can be positive, negative, undefined, or zero for symmetric distributions.

Figure 2.1 illustrates the effect of skewness on a Gaussian distribution, which is unimodal and symmetric. It is seen that the mean, median, and mode are equal in the Gaussian distribution

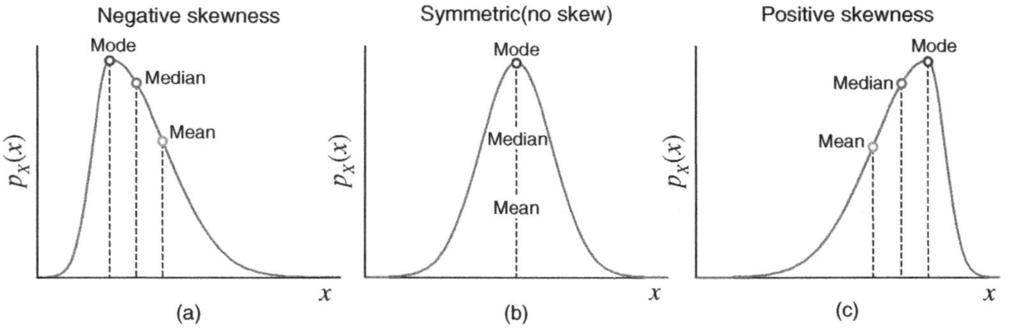

Figure 2.1 Effects of skewness on unimodal distributions: (a) negatively skewed, (b) normal (no skew), and (c) positively skewed.

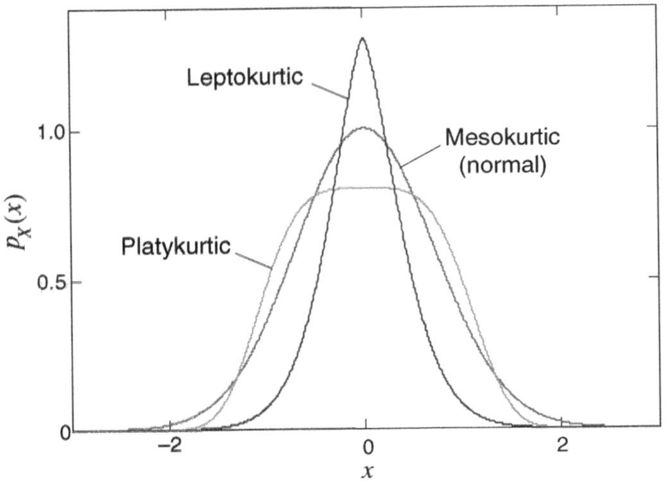

Figure 2.2 Common forms of kurtosis: mesokurtic (normal), platykurtic (higher rectangularity), and leptokurtic (longer tails).

(Fig. 2.1b). However, negative skewness (Fig. 2.1a) and positive skewness (Fig. 2.1c) matter, so mode > med > mean.

Kurtosis: In some cases, a real-valued random variable X is distributed with multiple outliers, and *kurtosis* is used as a measure of the "tailedness" or "peakedness" of its pdf. In this sense, kurtosis is called a descriptor of the pdf shape. There are different ways of quantifying kurtosis and estimating it from the population of X. The most widely used measure of kurtosis is the fourth standardized moment

$$\gamma_2 = \frac{\mu_4}{\sigma_X^4} = \frac{\kappa_4}{\kappa_2^2} \,. \tag{2.20}$$

Figure 2.2 illustrates three commonly recognized forms of kurtosis: *leptocurtic*, *mesokurtic* (normal), and *platycurtic*.

The difference is that the platykurtic has a higher squareness and the leptokurtic has longer tails compared to the mesokurtic, which is normal.

Example 2.1 ***Gaussian random variable.*** A random variable X, distributed according to the normal law

$$p_X(x) = \frac{1}{\sqrt{2\pi\sigma_X^2}} \exp\left[-\frac{(x-\bar{X})^2}{2\sigma_X^2}\right] , \tag{2.21}$$

is called a Gaussian random variable. It is the only random variable that is represented by two statistics: the mean $\bar{X}$ and the variance σ_X^2. Therefore, it is usually denoted as $X \sim \mathcal{N}(\bar{X}, \sigma_X^2)$. All other Gaussian statistics are zero, including skewness and kurtosis.

The cdf of a Gaussian variable X is given by

$$\begin{aligned} F_X(x) &= \int_{-\infty}^{x} p_X(z)\, dz \\ &= \frac{1}{\sqrt{2\pi}} \int_{-\infty}^{\frac{x-\bar{X}}{\sigma_X}} e^{-\frac{z^2}{2}}\, dz = \Phi\left(\frac{x-\bar{X}}{\sigma_X}\right) , \end{aligned} \tag{2.22}$$

where

$$\Phi(x) = \frac{1}{\sqrt{2\pi}} \int_{-\infty}^{x} e^{-\frac{z^2}{2}} dz = \frac{1}{2}\left[1 + \operatorname{erf}\left(\frac{x}{\sqrt{2}}\right)\right] \tag{2.23}$$

is the *probability integral* and erf(x) is the *error function* .

A Gaussian random variable exists from $-\infty$ to ∞, but the Gaussian pdf rapidly approaches zero when X deviates from its mean $\bar{X}$. Therefore, it is often desirable to specify the probability P_X that X exists in some interval $[a, b]$, and thereby estimate the error probability $P_E = 1 - P_X$ for X to exist outside this interval.

The probability that a Gaussian random variable exists in the interval $[a, b]$ is defined as

$$\begin{aligned} P_X\{a \leqslant x \leqslant b\} &= F_X(b) - F_X(a) \\ &= \Phi\left(\frac{b - \bar{X}}{\sigma_X}\right) - \Phi\left(\frac{a - \bar{X}}{\sigma_X}\right) \end{aligned} \tag{2.24}$$

via the probability integral (2.23). □

Example 2.2 ***Uniformly distributed random variable.*** Consider a random variable X that is uniformly distributed on $[a, b]$ with the pdf

$$p_X(x) = \begin{cases} \frac{1}{b-a} & a \leqslant x \leqslant b \\ 0 & \text{otherwise} \end{cases} . \tag{2.25}$$

By (2.7), the υth moment of X can be found to be

$$m_\upsilon = \frac{1}{\upsilon + 1} \sum_{i=0}^{\upsilon} a^i b^{\upsilon - i}$$

and the particular moments are $m_1 = \frac{a+b}{2}$, $m_2 = \frac{a^2+ab+b^2}{3}$, ... By (2.11), the variance becomes

$$\sigma_X^2 = \frac{1}{12}(b - a)^2 .$$

Because (2.25) is symmetric about m_1, the skewness is zero, $\gamma_1 = 0$, and kurtosis calculates $\gamma_2 = -\frac{6}{5}$. In a special case when X is distributed symmetrically about zero, $b = -a$, the υth moment is given by

$$m_\upsilon = \frac{1}{2a} \int_{-a}^{a} x^\upsilon dx = \frac{1}{2a(\upsilon + 1)} [a^{\upsilon+1} - (-a)^{\upsilon+1}] ,$$

and the particular moments are $m_1 = 0$, $m_2 = \frac{a^2}{3}$, $m_3 = 0$, $m_4 = \frac{a^4}{5}$, ... with all odd moments zeros. □

In Fig. 2.3 we summarize the previous analysis with connections between cdf $F_X(x)$, pdf $p_X(x)$, cf $\Phi_X(j\theta)$, raw moments m_υ, central moments μ_υ, and cumulants κ_υ of a random variable X.

2.1.2 Product Moments

Let a random variable $X(\xi)$ or simply X represent a random measurement ξ and $Y(\eta)$ or Y represent a random measurement η. Thus, we can think that the two variables are distributed with a joint pdf $p_{X,Y}(x, y)$.

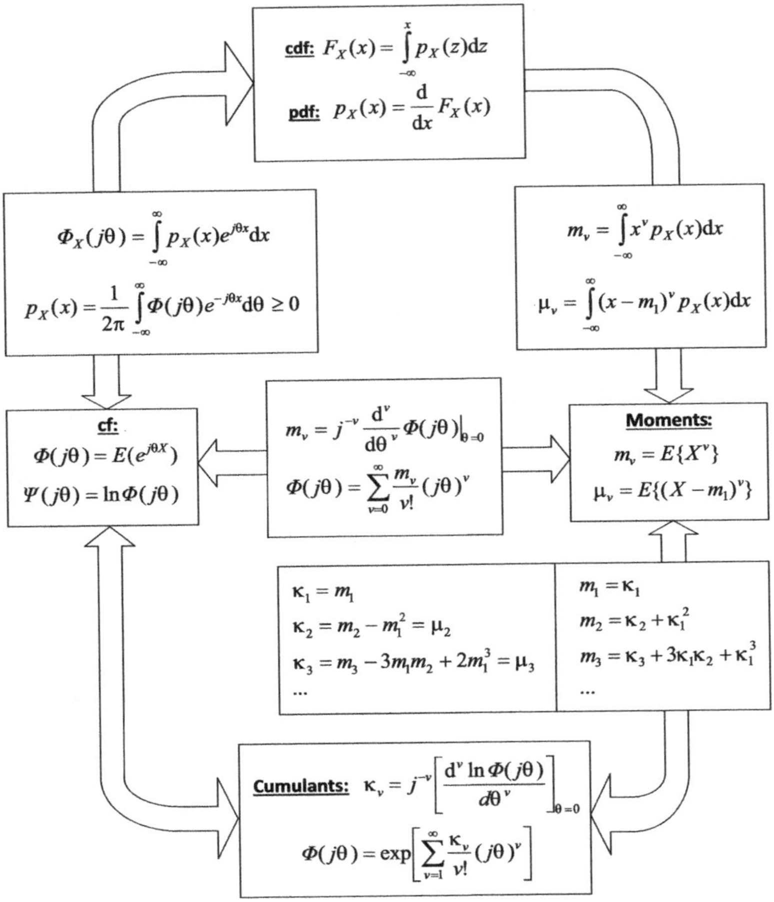

Figure 2.3 Relationships and connections between cdf $F_X(x)$, pdf $p_X(x)$, cf $\Phi_X(j\theta)$, raw moments m_υ, central moments μ_υ, and cumulants κ_υ of a random variable X.

Similarly to a singe variable, the product moment m_{11} of two variables X and Y is defined as follows:

$$m_{11} = \mathcal{E}\{XY\} = \int_{-\infty}^{\infty}\int_{-\infty}^{\infty} xyp_{X,Y}(x,y)\,\mathrm{d}x\mathrm{d}y\,. \tag{2.26}$$

Accordingly, the central product moment or the mean-adjusted product moment is defined by

$$\begin{aligned}\mu_{11} &= \mathrm{cov}(X,Y) = \mathcal{E}\{(X-\bar{X})(Y-\bar{Y})\}\\ &= \int_{-\infty}^{\infty}\int_{-\infty}^{\infty} (x-\bar{X})(y-\bar{Y})p_{X,Y}(x,y)\,\mathrm{d}x\mathrm{d}y\end{aligned} \tag{2.27}$$

and called the *covariance*.

In the special case where X and Y are independent, pdf $p_{X,Y}(x,y)$ is represented by the product as $p_{X,Y}(x,y) = p_X(x)p_Y(y)$, and the product moments become

$$m_{11} = \int_{-\infty}^{\infty} xp_X(x)\,\mathrm{d}x \int_{-\infty}^{\infty} yp_Y(y)\,\mathrm{d}y = \mathcal{E}\{X\}\mathcal{E}\{Y\}\,, \tag{2.28}$$

$$\mu_{11} = \int_{-\infty}^{\infty} (x - \bar{X}) p_X(x)\, dx \int_{-\infty}^{\infty} (y - \bar{Y}) p_Y(y)\, dy$$
$$= \mathcal{E}\{X - \bar{X}\}\mathcal{E}\{Y - \bar{Y}\} \,. \tag{2.29}$$

If two random variables X and Y are independent with properties (2.28) and (2.29), then it follows that they are also *uncorrelated*. However, the converse is generally not true, except in a few special cases. Note also that terminologically, if $\mathcal{E}\{XY\} = 0$, then the random variables X and Y are *orthogonal*.

It is often convenient to use normalized measures. The normalized mean-adjusted product moment is called the *correlation coefficient* or population Pearson correlation coefficient, and it is defined as

$$\rho_{X,Y} = \frac{\text{cov}(X, Y)}{\sigma_X \sigma_Y} \,, \tag{2.30}$$

where $\sigma_X > 0$ and $\sigma_Y > 0$ are standard deviations of variables X and Y.

As the degree of correlation between two random variables X and Y, the correlation coefficient ranges as $-1 \leqslant \rho_{X,Y} \leqslant 1$ due to the following limiting properties:

- If $X = Y$, the correlation ia maximum, $\rho_{X,Y} = 1$.
- If $X = -Y$, it is obvious that $\rho_{X,Y} = -1$.
- For uncorrelated X and Y, we have $\text{cov}(X, Y) = 0$ and hence $\rho_{X,Y} = 0$.

Product moments are fundamental when analyzing interactions between random variables. However, if two random variables are considered at two different points in some space or at two different points in time, it is necessary to use another function, called the *correlation function*.

2.1.3 Vector Random Variables

Let us now represent two random variable as column vectors

$$X = [\, X_1 \; X_2 \; \dots X_n \,]^T \in \mathbb{R}^n \,, \tag{2.31}$$

$$Y = [\, Y_1 \; Y_2 \; \dots Y_m \,]^T \in \mathbb{R}^m \,, \tag{2.32}$$

where entries are drawn from populations or variables of some random processes. Such vectors can be characterized by their *autocorrelation* and *cross-correlation*.

The *autocorrelation matrix* $\mathcal{R}_X$ and *autocovariance matrix* $\mathcal{C}_X$ of a vector X are defined as

$$\mathcal{R}_X = \mathcal{E}\{XX^T\}$$
$$= \begin{bmatrix} \mathcal{E}\{X_1^2\} & \mathcal{E}\{X_1X_2\} & \dots & \mathcal{E}\{X_1X_n\} \\ \mathcal{E}\{X_2X_1\} & \mathcal{E}\{X_2^2\} & \dots & \mathcal{E}\{X_2X_n\} \\ \vdots & \vdots & \ddots & \vdots \\ \mathcal{E}\{X_nX_1\} & \mathcal{E}\{X_nX_2\} & \dots & \mathcal{E}\{X_n^2\} \end{bmatrix} , \tag{2.33}$$

$$\mathcal{C}_X = \mathcal{E}\{(X - \bar{X})(X - \bar{X})^T\}$$
$$= \begin{bmatrix} \sigma_1^2 & \sigma_{12} & \dots & \sigma_{1n} \\ \sigma_{21} & \sigma_2^2 & \dots & \sigma_{2n} \\ \vdots & \vdots & \ddots & \vdots \\ \sigma_{n1} & \sigma_{n2} & \dots & \sigma_n^2 \end{bmatrix} , \tag{2.34}$$

where $\sigma_i^2 = \mathcal{E}\{(X_i - \bar{X}_i)^2\}$ is the variance of X_i and $\sigma_{ij} = \mathcal{E}\{(X_i - \bar{X}_i)(X_j - \bar{X}_j)\}$ is the covariance of X_i and X_j for $i, j \in [1, n]$. It follows that matrices $\mathcal{R}_X$ and C_X are square and symmetric. It can also be shown that for any vector $z \in \mathbb{R}^n$ the following values are nonnegative,

$$z^T \mathcal{R}_X z \geqslant 0 ,$$

$$z^T C_X z \geqslant 0 ,$$

and both $\mathcal{R}_X$ and C_X are thus *positive semidefinite*. Note that if the previous relationships are inequalities, matrices $\mathcal{R}_X$ and C_X are called *positive definite*.

Similarly, the *cross-correlation matrix* and *cross-covariance matrix* of random vectors X and Y are defined by, respectively,

$$\mathcal{R}_{XY} = \mathcal{E}\{XY^T\} = \begin{bmatrix} \mathcal{E}\{X_1Y_1\} & \mathcal{E}\{X_1Y_2\} & \dots & \mathcal{E}\{X_1Y_n\} \\ \mathcal{E}\{X_2Y_1\} & \mathcal{E}\{X_2Y_2\} & \dots & \mathcal{E}\{X_2Y_n\} \\ \vdots & \vdots & \ddots & \vdots \\ \mathcal{E}\{X_nY_1\} & \mathcal{E}\{X_nY_2\} & \dots & \mathcal{E}\{X_nY_n\} \end{bmatrix} , \tag{2.35}$$

$$C_{XY} = \mathcal{E}\{(X - \bar{X})(Y - \bar{Y})^T\} = \mathcal{E}\{XY^T\} - \bar{X}\bar{Y}^T . \tag{2.36}$$

Example 2.3 ***Vector Gaussian random variable.*** Suppose that the population of a real stochastic variable X is represented by the column vector X (2.31). The vector stochastic variable X is called Gaussian (Example 2.1) or normally distributed if its pdf is given by

$$p_X(x) = \frac{1}{\sqrt{(2\pi)^n |C_X|}} \exp\left[-\frac{1}{2}(x - \bar{X})^T C_X^{-1}(x - \bar{X})\right] \tag{2.37a}$$

$$= \mathcal{N}(\bar{X}, C_X) \tag{2.37b}$$

where C_X is the autocovariance (2.34), $|C_X|$ is the determinant of C_X, $\bar{X} = \mathcal{E}\{X\}$, and $x = [\, x_1 \;\dots\; x_n \,]^T$. Note that there is no closed form for cdf (2.2) of $\mathcal{N}(\bar{X}, C_X)$ and this function is calculated numerically.

In the bivariate case of two random variables X and Y, (2.37a) becomes

$$p_{X,Y}(x, y) = \frac{1}{2\pi\sigma_X\sigma_Y\sqrt{1 - \rho_{X,Y}^2}} \times \exp\left(-\frac{\sigma_X^2\tilde{x}^2 - 2\rho_{X,Y}\sigma_X\sigma_Y\tilde{x}\tilde{y} + \sigma_Y^2\tilde{y}^2}{2\sigma_X^2\sigma_Y^2(1 - \rho_{X,Y}^2)}\right) ,$$

where $\tilde{x} = x - \bar{X}_1, \tilde{y} = y - \bar{Y}_1$, and $\rho_{X,Y}$ is the correlation coefficient (2.30). If there is no correlation between X and Y and therefore $\rho_{X,Y} = 0$, then $p_{X,Y}(x, y)$ is represented by a product of two Gaussian pdfs as $p_{X,Y}(x, y) = p_X(x)p_Y(y)$. □

2.1.4 Conditional Probability: Bayes' Rule

We have already mentioned that random variables can be distributed jointly. They can also be distributed conditionally, when the probability of one variable depends on another already observed. To arrive at probabilistic relations associated with such variables, turn to [47] and consider two events A and B that have a joint probability $P(AB)$. We can now introduce the *conditional probability* $P(A|B)$ of the event A, assuming that B is observed. The conditional probability $P(A|B)$ can be defined as

$$P(A|B) = \frac{P(AB)}{P(B)} , \tag{2.38}$$

where the probability $P(B)$ of event B is called *marginal* probability because it is independent on other events. The conditional probability $P(A|B)$ has the following key properties [47, 91]:

$$P(A|B) \geqslant 0 ,$$

$$P(A|B) = 1 \quad \text{if} \quad B = A ,$$

$$P(A|B) \text{ is undefined if } P(B) = 0 .$$

Because $P(A|B)$ incorporates some already known information about the relation between two variables, it is also called an *a posteriori* probability, while $P(B)$ is also called an *a priori* probability.

By applying (2.38) to a collection of random variables $\{A_1, \dots, A_n\}$, we arrive at the important *chain rule*

$$P(A_1, \dots, A_n) = P(A_1|A_2, \dots, A_n)P(A_2, \dots, A_n) \tag{2.39}$$

that for three variables gives

$$P(A_1, A_2, A_3) = P(A_1|A_2, A_3)P(A_2|A_3)P(A_3) . \tag{2.40}$$

Note that the rule (2.39) plays a fundamental role in the Bayes theory.

Bayes' Theorem

Let us consider a general case of multiple events [145]. For a finite set $\{A_1, \dots, A_n\}$ of mutually exclusive events called a *finite partition* and another event B such that the events BA_i, $i \in [1, n]$, are mutually exclusive, we can refer to the *total probability theorem* and write

$$P(B) = P(BA_1) + \cdots + P(BA_n) .$$

Using (2.38), we can further modify this relation as

$$P(B) = P(B|A_1)P(A_1) + \cdots + P(B|A_n)P(A_n) , \tag{2.41}$$

because

$$P(BA_i) = P(B|A_i)P(A_i) . \tag{2.42}$$

On the other hand, since events A_i and B are interchangeable, we can write $P(BA_i) = P(A_i|B)P(B)$ and conclude that

$$P(A_i|B) = P(B|A_i)\frac{P(A_i)}{P(B)} , \tag{2.43}$$

$$\propto P(B|A_i)P(A_i) , \tag{2.44}$$

$$= cP(B|A_i)P(A_i) , \tag{2.45}$$

which is stated by *Bayes' theorem*. Note that Tomath Bayes was the first to use conditional probability and describe the probability of an event via prior knowledge of conditions related to the event.

The conditional probability $P(A_i|B)$ represented with (2.43) is the *likelihood* of event A_i given that event B is true. Likewise, $P(B|A_i)$ is the *likelihood* of event B given that event A_i is true. The *a priori* probability $P(B)$ is often omitted in (2.43), and the constant c is introduced (2.45) to normalize $P(A_i|B)$, to get the unit area.

Inserting (2.41) into (2.43) gives

$$\begin{aligned} P(A_i|B) &= \frac{P(B|A_i)P(A_i)}{P(B|A_1)P(A_1) + \cdots + P(B|A_n)P(A_n)} \\ &= \frac{P(B|A_i)P(A_i)}{\sum_{i=1}^{n} P(B|A_i)P(A_i)} , \end{aligned} \tag{2.46}$$

which generalizes Bayes' rule.

Conditional Probability Density

Considering two random variables X and Y instead of two events A and B and omitting the derivation, which can be found in [45][145][79], the relation (2.38) can be equivalently rewritten as

$$p_{X|Y}(x|y) = \frac{p_{XY}(x,y)}{p_Y(y)} , \tag{2.47}$$

where $p_{X|Y}(x|y)$ is the conditional pdf of X given Y, $p_{XY}(x,y)$ is the joint pdf of X and Y, and $p_Y(y)$ is called the *marginal* pdf of Y defined by

$$p_Y(y) = \int_x p_{XY}(x,y)\,\mathrm{d}x , \tag{2.48}$$

$$= \int_x p_{Y|X}(y|x)p_X(x)\,\mathrm{d}x = \mathcal{E}_X\{p_{Y|X}(y|x)\} . \tag{2.49}$$

For multiple random variables collected as $X = \{X_1, \ldots, X_n\}$ and $Y = \{Y_1, \ldots, Y_m\}$, (2.47) can be written as

$$p_{X|Y}(x_1, \ldots, x_n|y_1, \ldots, y_m) = \frac{p_{XY}(x_1, \ldots, x_n, y_1, \ldots, y_m)}{p_Y(y_1, \ldots, y_m)} \tag{2.50}$$

and the cdf found by integrating over X as

$$\begin{aligned} &F_{X|Y}(x_1, \ldots, x_n|y_1, \ldots, y_m) \\ &= \int_{-\infty}^{x_1} \ldots \int_{-\infty}^{x_n} \frac{p_{XY}(z_1, \ldots, z_n, y_1, \ldots, y_m)}{p_Y(y_1, \ldots, y_m)}\,\mathrm{d}z_n \ldots \mathrm{d}z_1 . \end{aligned} \tag{2.51}$$

In the case of three random variables X_1, X_2, and X_3, we accordingly have

$$p(x_1|x_2, x_3) = \frac{p(x_1, x_2, x_3)}{p(x_2, x_3)} ,$$

$$F(x_1|x_2, x_3) = \int_{-\infty}^{x_1} \frac{p(z, x_2, x_3)}{p(x_2, x_3)}\,\mathrm{d}z .$$

It is worth noting that the rule (2.48) allows excluding any primary variable from (2.50), and one can use (2.49) to exclude any conditional variable from (2.50). For example, given $p(x_1, x_2|y_1, y_2)$, we can find

$$p(x_1|y_1, y_2) = \int_{-\infty}^{\infty} p(x_1, x_2|y_1, y_2)\,\mathrm{d}x_2 , \tag{2.52}$$

$$p(x_1, x_2|y_2) = \int_{-\infty}^{\infty} p(x_1, x_2|y_1, y_2)p(y_1|y_2)\,\mathrm{d}y_1 . \tag{2.53}$$

Most generally, for a set $X = \{X_1, \dots, X_n\}$ one can write

$$p(x_1|x_2, \dots, x_n) = \frac{p_X(x_1, \dots, x_n)}{p(x_2, \dots, x_m)} \tag{2.54}$$

and, by the chain rule (2.39), obtain

$$p_X(x_1, x_2, \dots, x_n) = p(x_1|x_2, \dots, x_n)p(x_2|x_3, \dots, x_n) \\ \dots p(x_{n-1}|x_n)p(x_n) . \tag{2.55}$$

2.1.5 Transformation of Random Variables

Sometimes a deterministic transformation of one random variable X with known pdf $p_X(x)$ into another Y with unknown or desired pdf $p_Y(x)$ is required. For example, the pdf of the system input variable X is known, and the pdf of the output variable Y must be defined. It is also often required to transform two nonlinearly related random variables into two linearly related ones, or vice versa. An example can be found in tracking where measurements made in polar coordinates are to be represented in Cartesian coordinates.

Single-to-Single Variable Transformation

Let two random variables, X and Y, be related to each other as follows:

$$Y = g(X) , \tag{2.56}$$

$$X = h(Y) = g^{-1}(Y) , \tag{2.57}$$

where $g(\cdot)$ and $h(\cdot)$ are some known smooth functions. Suppose we know pdf $p_X(x)$ for X and would like to know pdf $p_Y(y)$ for Y. To find $p_Y(y)$, we can refer to the dependencies of Y on X (2.56) and X on Y (2.57) and assert that the probability that X lies within the interval $x + \mathrm{d}x$ is equal to the probability that Y lies within $y + \mathrm{d}y$. This can be formalized as

$$\int_x^{x+\mathrm{d}x} p_X(u)\,\mathrm{d}u = \begin{cases} \int_y^{y+\mathrm{d}y} p_Y(u)\,\mathrm{d}u & \text{if} \quad \mathrm{d}y > 0 \\ -\int_y^{y+\mathrm{d}y} p_Y(u)\,\mathrm{d}u & \text{if} \quad \mathrm{d}y < 0 \end{cases}$$

and further rewritten equivalently in the differential form as

$$p_X(x)\,\mathrm{d}x = p_Y(y)|\mathrm{d}y| .$$

Since we assume that $\mathrm{d}x > 0$, we arrive at the transformation rule

$$p_Y(y) = \frac{\mathrm{d}x}{|\mathrm{d}y|} p_X(x) = \left|\frac{\partial h(Y)}{\partial y}\right| p_X[h(Y)] . \tag{2.58}$$

Example 2.4 ***Nonlinearly coupled random variables.*** Gaussian random variable X with pdf (2.21) is related to another random variable Y as

$$Y = g(X) = X^3 , \quad X = h(Y) = Y^{1/3} .$$

The derivative $\partial X / \partial Y$ gives

$$\frac{\partial X}{\partial Y} = \frac{\partial h(Y)}{\partial Y} = \frac{1}{3} Y^{-\frac{2}{3}} > 0$$

and, by (2.58), pdf $p_Y(y)$ of Y becomes

$$p_Y(y) = \frac{1}{3y^{2/3}} \frac{1}{\sqrt{2\pi\sigma_X^2}} e^{-\frac{y^{2/3}}{2\sigma_X^2}}$$

that is not Gaussian. □

Transformation of Vector Random Variables

Given sets of random variables, $X = \{X_1, \dots, X_n\}$ represented with pdf $p_X(x_1, \dots, x_n)$ and $Y = \{Y_1, \dots, Y_n\}$ with $p_Y(y_1, \dots, y_n)$. Suppose the variables X_i and Y_i, $i \in [1, n]$, are related to each other as

$$X_i = h_i(Y_1, \dots, Y_n), \quad Y_i = g_i(X_1, \dots, X_n).$$

Then the $p_Y(y_1, \dots, y_n)$ of Y can be defined via $p_X(x_1, \dots, x_n)$ of X as

$$p_Y(y_1, \dots, y_n) = |J| p_X[h_1(y_1, \dots, y_n), \dots, h_n(y_1, \dots, y_n)], \tag{2.59}$$

where $|J|$ is the determinant of the Jacobian $J(\cdot)$ of the transformation,

$$J\left(\frac{x_1, \dots, x_n}{y_1, \dots, y_n}\right) = \begin{bmatrix} \frac{\partial h_1}{\partial y_1} & \cdots & \frac{\partial h_1}{\partial y_n} \\ \vdots & \ddots & \vdots \\ \frac{\partial h_n}{\partial y_1} & \cdots & \frac{\partial h_n}{\partial y_n} \end{bmatrix}. \tag{2.60}$$

Example 2.5 ***Cartesian-to-polar transformation of two random variables.*** Given two random variables X and Y represented in Cartesian coordinates with zero mean, $\bar{X} = \bar{Y} = 0$, equal variances $\sigma^2 = \sigma_X^2 = \sigma_Y^2$, and joint Gaussian pdf

$$p(x, y) = \frac{1}{2\pi\sigma^2} \exp\left(-\frac{x^2 + y^2}{2\sigma^2}\right). \tag{2.61}$$

We want to convert X and Y to other random variables r and φ, described in polar coordinates, in which case the variables are related to each other as

$$r = \sqrt{X^2 + Y^2}, \quad \varphi = \arctan\frac{Y}{X}, \quad |\varphi| \leqslant \pi,$$
$$X = r\cos\varphi, \quad Y = r\sin\varphi.$$

To provide a joint pdf $p(r, \varphi)$, define the determinant of the Jacobian of the transformation (2.60) as

$$\left|J\left(\frac{X, Y}{r, \varphi}\right)\right| = \begin{vmatrix} \cos\varphi & \sin\varphi \\ -r\sin\varphi & r\cos\varphi \end{vmatrix} = r,$$

and then follow (2.59) and obtain a joint pdf $p(r, \varphi)$ of r and φ,

$$\begin{aligned} p(r, \varphi) &= \frac{r}{2\pi\sigma^2} \exp\left[-\frac{(r\cos\varphi)^2 + (r\sin\varphi)^2}{2\sigma^2}\right] \\ &= \frac{r}{2\pi\sigma^2} \exp\left(-\frac{r^2}{2\sigma^2}\right). \end{aligned} \tag{2.62}$$

To obtain a marginal pdf of r, integrate out phase $|\varphi| \leqslant \pi \bmod 2\pi$,

$$p(r) = \frac{r}{\sigma^2} \exp\left(-\frac{r^2}{2\sigma^2}\right),$$

which is Rayleigh's pdf. By integrating out $r \geqslant 0$ from zero to infinity, arrive at the uniform distribution of phase $|\varphi| \leqslant \pi \bmod 2\pi$,

$$p(\varphi) = \begin{cases} \frac{1}{2\pi} , & -\pi \leqslant \varphi \leqslant \pi \\ 0 , & \text{otherwise} \end{cases}$$

Since the variables r and φ are independent of each other, we conclude that $p(r, \varphi) = p(r)p(\varphi)$. □

2.2 Stochastic Processes

Recall that a random variable X corresponds to some measurement outcome ξ. Since ξ exists at some time t as $\xi(t)$, the variable X is also a time function $X(t)$. The family of time functions $X(t)$ dependent on $\xi(t)$ is called a *stochastic process* or a *random process* $X(t) \triangleq X(t, \xi)$, where t and ξ are variables. As a collection of random variables, a stochastic process can be a *scalar stochastic process*, which is a set of random variables $X(t)$ in some coordinate space. It can also be a *vector stochastic process* $X(t)$, which is a collection of m random variables $X_i(t)$, $i \in [1, m]$, in some coordinate space.

The following forms of stochastic processes are distinguished:

- A stochastic process represented in continuous time with continuous values is called a *continuous stochastic process*.
- A *discrete stochastic process* is a process that is represented in continuous time with discrete values.
- A stochastic process represented in discrete time with continuous values is called a *stochastic sequence*.
- A *discrete stochastic sequence* is a process represented in discrete time with discrete values.

Using the concept of random variables, time-varying cdf $F(x, t) \triangleq F_X(x, t)$ and pdf $p(x, t) \triangleq p_X(x, t)$ of a scalar stochastic process process $X(t)$ can be represented as

$$F(x, t) = P\{X(t) \leqslant x\} , \tag{2.63}$$

$$p(x, t) = \frac{\partial F(x, t)}{\partial x} \quad \Leftrightarrow \quad F(x, t) = \int_{-\infty}^{x} p(u, t)\, du . \tag{2.64}$$

For a set of variables $x = \{x_1, \dots, x_n\}$ corresponding to a set of time instances $t = \{t_1, \dots, t_n\}$, we respectively have

$$F(x; t) = P\{X(t_1) \leqslant x_1, \dots, X(t_n) \leqslant x_n\} , \tag{2.65}$$

$$p(x; t) = \frac{\partial^n F(x_1, \dots, x_n; t_1, \dots, t_n)}{\partial x_1 \dots \partial x_n} , \tag{2.66}$$

$$F(x; t) = \int_{-\infty}^{x_1} \dots \int_{-\infty}^{x_n} p(u_1, \dots, u_n; t_1, \dots, t_n)\, du_n \dots du_1 . \tag{2.67}$$

A stochastic process can be ether *stationary* or *nonstationary*. There are the following types of stationary random processes:

- *Strictly stationary*, whose unconditional joint pdf does not change when shifted in time; that is, for all τ, it obeys

$$p(x_1, \dots, x_n; t_1 - \tau, \dots, t_n - \tau) = p(x_1, \dots, x_n; t_1, \dots, t_n) . \tag{2.68}$$

- *Wide-sense stationary*, whose mean and variance do not vary with respect to time.

- *Ergodic*, whose probabilistic properties deduced from a single random sample are the same as for the whole process.

It follows from these definitions that a strictly stationary random process is also a wide-sense random process, and an ergodic process is less common among other random processes. All other random processes are called *nonstationary*.

2.2.1 Correlation Function

The function that described the statistical correlation between random variables in some processes is called the *correlation function*. If the random variables represent the same quantity measured at two different points, then the correlation function is called the *autocorrelation function*. The correlation function of various random variables is called the *cross-correlation function*.

Autocorrelation Function

The *autocorrelation function* of a scalar random variable X measured at different time points t_1 and t_2 is defined as

$$\mathcal{R}_X(t_1, t_2) = \mathcal{E}\{X(t_1)X(t_2)\} \tag{2.69a}$$

$$= \int_{-\infty}^{\infty}\int_{-\infty}^{\infty} x_1 x_2 p_X(x_1, x_2; t_1, t_2)\, \mathrm{d}x_1 \mathrm{d}x_2 \,, \tag{2.69b}$$

where $p_X(x_1, x_2; t_1, t_2)$ is a joint time-varying pdf of $X(t_1)$ and $X(t_2)$.

For mean-adjusted processes, the correlation function is called an *autocovariance function* and is defined as

$$\mathcal{C}_X(t_1, t_2) = \mathcal{E}\{[X(t_1) - \bar{X}(t_1)][X(t_2) - \bar{X}(t_2)]\} \tag{2.70a}$$

$$= \int_{-\infty}^{\infty}\int_{-\infty}^{\infty} [x_1 - \bar{X}(t_1)][x_2 - \bar{X}(t_2)] p_X(x_1, x_2; t_1, t_2)\, \mathrm{d}x_1 \mathrm{d}x_2 \,. \tag{2.70b}$$

Both $\mathcal{R}_X(t_1, t_2)$ and $\mathcal{C}_X(t_1, t_2)$ tell us how much a variable $X(t_1)$ is coupled with its shifted version $X(t_2)$.

There exists a simple relation between the cross-correlation and autocorrelation functions assuming a complex random process $X(t)$,

$$\mathcal{R}_X(t_1, t_2) = \mathcal{E}\{X(t_1)X^*(t_2)\} \tag{2.71a}$$

$$= \mathcal{C}_X(t_1, t_2) + \bar{X}(t_1)\bar{X}^*(t_2) \,, \tag{2.71b}$$

where $X^*(t)$ is a complex conjugate of $X(t)$.

If the random process is stationary, its autocorrelation function does not depend on time, but depends on the time shift $\tau = t_2 - t_1$. For such processes, $R_X(t_1, t_2)$ is converted to

$$\mathcal{R}_X(\tau) = \mathcal{E}\{X(t)X(t + \tau)\} \tag{2.72a}$$

$$= \int_{-\infty}^{\infty}\int_{-\infty}^{\infty} x_1 x_2 p(x_1, x_2; \tau) dx_1 dx_2 \,. \tag{2.72b}$$

If the joint pdf $p(x_1, x_2; t_1, t_2)$ is not explicitly known and the stochastic process is supposed to be ergodic, then $\mathcal{R}_X(\tau)$ can be computed by averaging the product of the shifted variables as

$$\mathcal{R}_X(\tau) = \lim_{T\to\infty} \int_{-T}^{T} X(t)X(t+\tau)\,\mathrm{d}t\,, \tag{2.73}$$

and we notice that the rule (2.73) is common to experimental measurements of correlation. Similarly to (2.73), the covariance function $\mathcal{C}_X(t_1, t_2)$ (2.70b) can be measured for ergodic random processes.

Cross-Correlation Function

The correlation between two different random processes $X(t)$ and $Y(t)$ is described by the *cross-correlation function*. The relationship remains largely the same if we consider these variables at two different time instances and define the cross-correlation function as

$$\mathcal{R}_{XY}(t_1, t_2) = \mathcal{E}\{X(t_1)Y(t_2)\} \tag{2.74a}$$

$$= \int_{-\infty}^{\infty}\int_{-\infty}^{\infty} x_1 y_2 p(x_1, y_2; t_1, t_2)\,\mathrm{d}x_1\,\mathrm{d}y_2\,, \tag{2.74b}$$

and the *cross-covariance function* as

$$\mathcal{C}_{XY}(t_1, t_2) = \mathcal{E}\{[X(t_1) - \bar{X}(t_1)][Y(t_2) - \bar{Y}(t_2)]\} \tag{2.75a}$$

$$= \int_{-\infty}^{\infty}\int_{-\infty}^{\infty} [x_1 - \bar{X}(t_1)][y_2 - \bar{Y}(t_2)]p(x_1, y_2; t_1, t_2)\,\mathrm{d}x_1\,\mathrm{d}y_2\,. \tag{2.75b}$$

For stationary random processes, function (2.74a) can be computed as

$$\mathcal{R}_{XY}(\tau) = E\{X(t)Y(t+\tau)\} \tag{2.76a}$$

$$= \int_{-\infty}^{\infty}\int_{-\infty}^{\infty} x_1 y_2 p(x_1, y_2; \tau)\mathrm{d}x_1\,\mathrm{d}y_2 \tag{2.76b}$$

$$= \lim_{T\to\infty} \int_{-T}^{T} X(t)Y(t+\tau)\,\mathrm{d}t\,, \tag{2.76c}$$

and function (2.75a) modified accordingly.

Properties of Correlation Function

The following key properties of the autocorrelation and cross-correlation functions of two random processes $X(t)$ and $Y(t)$ are highlighted:

- The autocorrelation function is non-negative, $\mathcal{R}_X(t_1, t_2) \geqslant 0$, and has the property $\mathcal{R}_X(t_1, t_2) = \mathcal{R}_X^*(t_2, t_1)$.
- The cross-correlation function is a Hermitian function, $\mathcal{R}_{XY}(t_1, t_2) = \mathcal{R}_{YX}^*(t_2, t_1)$.
- The following Cauchy-Schwartz inequality holds,

$$|\mathcal{R}_{XY}(t_1, t_2)|^2 \leqslant \mathcal{E}\{|X(t_1)|^2\}\mathcal{E}\{|Y(t_2)|^2\} = \sigma_X^2(t_1)\sigma_Y^2(t_2)\,. \tag{2.77}$$

- For a stationary random process $X(t)$, the following properties apply:

$$\text{Symmetry}: \ \mathcal{R}_{XY}(\tau) = \mathcal{R}_{XY}(-\tau),$$

$$\text{Physical limit}: \ \lim_{\tau\to\infty} \mathcal{R}_{\xi}(\tau) = 0,$$

$$\text{Fourier transform}: \ \mathcal{F}\{\mathcal{R}_{XY}(\tau)\} \geqslant 0, \ \text{if exists.}$$

The latter property requires the study of a random process in the frequency domain, which we will do next.

2.2.2 Power Spectral Density

Spectral analysis of random processes in the frequency domain plays the same role as correlation analysis in the time domain. As in correlation analysis, two functions are recognized in the frequency domain: *power spectral density* (PSD) of a random process $X(t)$ and *cross power spectral density* (cross-PSD) of two random processes $X(t)$ and $Y(t)$.

Power Spectral Density

The PSD of a random process $X(t)$ is determined by the Fourier transform of its autocorrelation function. Because the Fourier transform requires a process that exists over all time, spectral analysis is applied to stationary processes that satisfy the Dirichlet conditions [165].

> ***Dirichlet conditions:*** *Any real-valued periodic function can be extended into the Fourier series if, over a period, a function 1) is absolutely integrable, 2) is finite, and 3) has a finite number of discontinuities.*

The Wiener-Khinchin theorem [145] states that the autocorrelation function $\mathcal{R}_X(\tau)$ of a wide-sense stationary random process $X(t)$ is related to its PSD $\mathcal{S}_X(\omega)$ by the Fourier transform pair as

$$\mathcal{S}_X(\omega) = \mathcal{F}\{\mathcal{R}_X(\tau)\} = \int_{-\infty}^{\infty} \mathcal{R}_X(\tau)e^{-j\omega\tau}\, d\tau \ , \tag{2.78}$$

$$\mathcal{R}_X(\tau) = \mathcal{F}^{-1}\{\mathcal{S}_X(\omega)\} = \frac{1}{2\pi}\int_{-\infty}^{\infty} \mathcal{S}_X(\omega)e^{j\omega\tau}\, d\omega \ , \tag{2.79}$$

and has the following fundamental properties:

- Since $\mathcal{R}_X(\tau)$ is conjugate symmetric, $\mathcal{R}_X(\tau) = \mathcal{R}_X^*(-\tau)$, then it follows that $\mathcal{S}_X(\omega)$ is a real function of ω of a real or complex stochastic process.
- If $X(t)$ is a real process, then $\mathcal{R}_X(\tau)$ is real and even, and $\mathcal{S}(\omega)$ is also real and even: $\mathcal{S}_X(\omega) = \mathcal{S}_X(-\omega)$. Otherwise, $\mathcal{S}_X(\omega)$ is not even.
- The PSD of a stationary process $X(t)$ is positive valued, $\mathcal{S}_X(\omega) \geqslant 0$.
- The variance σ_X^2 of a scalar $X(t)$ is provided by

$$\sigma_X^2 = \mathcal{R}_X(0) = \frac{1}{2\pi}\int_{-\infty}^{\infty} \mathcal{S}_X(\omega)\, d\omega \ , \tag{2.80}$$

- Given an LTI system with a frequency response $\mathcal{H}(j\omega)$, then $\mathcal{S}_X(\omega)$ of an input process $X(t)$ projects to $\mathcal{S}_Y(\omega)$ of an output process $Y(t)$ as

$$\mathcal{S}_Y(\omega) = |\mathcal{H}(j\omega)|^2 \mathcal{S}_X(\omega) . \tag{2.81}$$

Example 2.6 ***Gauss-Markov process.*** Consider a Gauss-Markov process $X(t)$, also known as the Ornstein-Uhlenbeck process [16], which the autocorrelation function exponentially decays with time as

$$\mathcal{R}_X(\tau) = \sigma_X^2 e^{-\beta|\tau|} ,$$

where σ_X^2 is the variance and $\beta^{-1} > 0$ is a time constant. The PSD of $X(t)$ is provided by (2.78) to be

$$\mathcal{S}_X(\omega) = \mathcal{F}\{\mathcal{R}_X(\tau)\} = \frac{\sigma_X^2}{\beta + j\omega} + \frac{\sigma_X^2}{\beta - j\omega} = \frac{2\sigma_X^2\beta}{\omega^2 + \beta^2} .$$

Figure 2.4 illustrates the autocorrelation function $\mathcal{R}_X(\tau)$ and PSD $\mathcal{S}_X(w)$ of this process.

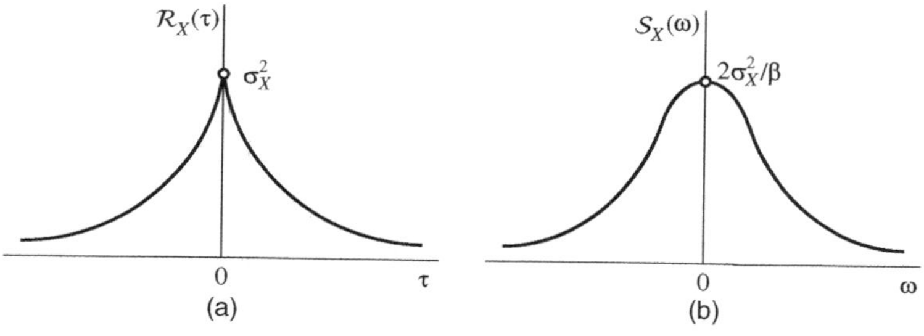

Figure 2.4 Autocorrelation function $\mathcal{R}_X(\tau)$ and PSD $\mathcal{S}_X(w)$ of a Gauss-Markov process.

As can be seen, both functions are real, positive-valued, and even. It is also seen that the correlation is maximum when $\tau = 0$, and the spectral content is concentrated around zero angular frequency ω. □

Cross Power Spectral Density

Like the PSD, the cross-PSD of two stationary random processes $X(t)$ and $Y(t)$ is defined by the Fourier transform pair as

$$\mathcal{S}_{XY}(j\omega) = \int_{-\infty}^{\infty} \mathcal{R}_{XY}(\tau) e^{-j\omega\tau} \, \mathrm{d}\tau , \tag{2.82}$$

$$\mathcal{R}_{XY}(\tau) = \frac{1}{2\pi} \int_{-\infty}^{\infty} \mathcal{S}_{XY}(j\omega) e^{j\omega\tau} \, \mathrm{d}\omega , \tag{2.83}$$

and we notice that the cross-PSD usually has complex values and differs from the PSD in the following properties:

- Since $\mathcal{R}_{XY}(\tau)$ and $\mathcal{R}_{YX}(\tau)$ are not necessarily even functions of τ, then it follows that $\mathcal{S}_{XY}(j\omega)$ and $\mathcal{S}_{YX}(j\omega)$ are not obligatorily real functions.
- Due to the Hermitian property $\mathcal{R}_{XY}(\tau) = \mathcal{R}_{YX}^*(\tau)$, functions $\mathcal{S}_{XY}(j\omega)$ and $\mathcal{S}_{YX}(j\omega)$ are complex conjugate of each other, $\mathcal{S}_{XY}(j\omega) = \mathcal{S}_{YX}^*(j\omega)$, and the sum of $\mathcal{S}_{XY}(j\omega)$ and $\mathcal{S}_{YX}(j\omega)$ is real.

If $Z(t)$ is the sum of two stationary random processes $X(t)$ and $Y(t)$, then the autocorrelation function $\mathcal{R}_Z(\tau)$ can be found as

$$\begin{aligned}\mathcal{R}_Z(\tau) &= E\{[X(t)+Y(t)][X(t+\tau)+Y(t+\tau)]\}\,,\\ &= \mathcal{R}_X(\tau)+\mathcal{R}_{XY}(\tau)+\mathcal{R}_{YX}(\tau)+\mathcal{R}_Y(\tau)\,.\end{aligned}$$

Therefore, the PSD of $Z(t)$ is generally given by

$$\mathcal{S}_Z(j\omega) = \mathcal{S}_X(\omega)+\mathcal{S}_{XY}(j\omega)+\mathcal{S}_{YX}(j\omega)+\mathcal{S}_Y(\omega)\,.$$

A useful normalized measure of the cross-PSD of two stationary processes $X(t)$ and $Y(t)$ is the *coherence* $\gamma^2_{X,Y}$ defined as

$$\gamma^2_{X,Y} = \frac{|\mathcal{S}_{X,Y}(j\omega)|}{\mathcal{S}_X(\omega)\mathcal{S}_Y(\omega)}\,, \tag{2.84}$$

which plays the role of the correlation coefficient (2.30) in the frequency domain. Maximum coherence is achieved when two processes are equal, and therefore $\gamma^2_{X,Y}=1$. On the other extreme, when two processes are uncorrelated, we have $\gamma^2_{X,Y}=0$, and hence the coherence varies in the interval $0 \leqslant \gamma^2_{X,Y} \leqslant 1$.

2.2.3 Gaussian Processes

As a collection of Gaussian variables, a Gaussian process is a stochastic process whose variables have a multivariate normal distribution. Because every finite linear combination of Gaussian variables is normally distributed, the Gaussian process plays an important role in modeling and state estimation as a useful and relatively simple mathematical idealization. It also helps in solving applied problems, since many physical processes after passing through narrowband paths acquire the property of Gaussianity. Some nonlinear problems can also be solved using the Gaussian approach [187].

Suppose that a Gaussian process is represented with a vector $X = [\,X_1\ \dots\ X_n\,]^T \in \mathbb{R}^n$ of Gaussian random variables corresponding to a set of time instances $\{t_1,\dots,t_n\}$ and that each variable is normally distributed with (2.37a). The pdf of this process is given by

$$p_X(x;t) = \frac{1}{\sqrt{(2\pi)^n|C_X|}}\exp\left[-\frac{1}{2}(x-\bar{X})^T C_X^{-1}(x-\bar{X})\right], \tag{2.85}$$

where $x=[\,x_1\ \dots\ x_n\,]^T \in \mathbb{R}^n$, $\bar{X}=\mathcal{E}\{X\}$, and the covariance C_X (2.34) is generally time-varying. The standard notation of Gaussian process is

$$X \sim \mathcal{N}(\bar{X}, C_X)\,.$$

If the Gaussian process is a collection of uncorrelated random variables and, therefore, $C_X(t_i,t_j)=0$ holds for $i \neq j$ and $C_X = \text{diag}(\sigma^2_{X_1}\ \dots\ \sigma^2_{X_n})$ is diagonal, then pdf (2.85) becomes a multiple product of the densities of each of the variables,

$$\begin{aligned}p_X(x;t) &= p_{X_1}(x_1;t_1)\dots p_{X_n}(x_n;t_n)\\ &= \frac{1}{\sqrt{(2\pi)^n\sigma^2_{X_1}\dots\sigma^2_{X_n}}}\exp\left[-\sum_{i=1}^{n}\frac{(x_i-\bar{X}_i)^2}{2\sigma^2_{X_i}}\right].\end{aligned} \tag{2.86}$$

The log-likelihood corresponding to (2.85) is

$$\ln\mathcal{L}(X|x;t) = -\frac{1}{2}\left[\ln|C_X| + (x-\bar{X})^T C_X^{-1}(x-\bar{X}) + n\ln(2\pi)\right] \tag{2.87}$$

and information entropy representing the average rate at which information is produced by a Gaussian process (2.85) is given by [2]

$$\begin{aligned} H[p_X(x;t)] &= -\int_{-\infty}^{\infty} \ln[p_X(x;t)]p_X(x;t)\,\mathrm{d}x \\ &= \frac{n}{2} + \frac{n}{2}\ln(2\pi) + \frac{1}{2}\ln|C_X| \,. \end{aligned} \tag{2.88}$$

Properties of Gaussian Processes

As a mathematical idealization of real physical processes, the Gaussian process exhibits several important properties that facilitate process analysis and state estimation. Researchers often choose to approximate data histograms with the normal law and use standard linear estimators, even if Gaussianity is clearly not observed. It is good if the errors are small. Otherwise, a more accurate approximation is required. In general, each random process requires an individual optimal estimator, unless its histogram can be approximated by the normal law to use standard solutions.

The following properties of the Gaussian process are recognized:

- In an exhaustive manner, the Gaussian process $X(t)$ is determined by the mean $\bar{X}(t)$ and the covariance matrix C_X, which is diagonal for uncorrelated Gaussian variables.
- Since the Gaussian variables are uncorrelated, it follows that they are also independent, and since they are independent, they are uncorrelated.
- The definitions of stationarity in the strict and wide sense are equivalent for Gaussian processes.
- The conditional pdf of the jointly Gaussian stochastic processes $X(t)$ and $Y(t)$ is also Gaussian. This follows from Bayes' rule (2.47), according to which

$$p_{X|Y}(x|y;t) = \frac{p_{XY}(x,y;t)}{p_Y(y;t)} \,.$$

- Linear transformation of a Gaussian process gives a Gaussian process; that is, the input Gaussian process $X(t)$ goes through the linear system to the output as $Y(t)$ in order to remain a Gaussian process.
- A linear operator can be found to convert a correlated Gaussian process to an uncorrelated Gaussian process and vice versa.

2.2.4 White Gaussian Noise

White Gaussian noise (WGN) occupies a special place among many other mathematical models of physical perturbations, As a stationary random process, WGN has the same intensity at all frequencies, and its PSD is thus constant in the frequency domain. Since the spectral intensity of any physical quantity decreases to zero with increasing frequency, it is said that WGN is an absolutely random process and as such does not exist in real life.

Continuous White Gaussian Noise

The WGN $w(t)$ is the most widely used form of white processes. Since noise is usually associated with zero mean, $E\{w(t)\} = 0$, WGN is often referred to as additive WGN (AWGN), which means that it can be added to any signal without introducing a bias. Since the PSD of the scalar $w(t)$ is constant, its autocorrelation function is delta-shaped and commonly written as

$$C_w(\tau) = \mathcal{R}_w(\tau) = \frac{N_0}{2}\delta(\tau) \,, \tag{2.89}$$

where $\delta(\tau)$ is the Dirac delta [165] and $N_0/2$ is some constant. It follows from (2.89) that the variance of WGN is infinite,

$$\sigma_w^2 = E\{w^2(t)\} = \mathcal{R}_w(0) = \frac{N_0}{2}\delta(0) = \infty \,,$$

and the Fourier transform (2.78) applied to (2.89) gives

$$\mathcal{S}_w(\omega) = \int_{-\infty}^{\infty} \frac{N_0}{2}\delta(\tau)e^{-j\omega\tau}\, \mathrm{d}\tau = \frac{N_0}{2} \,, \tag{2.90}$$

which means that the constant value $N_0/2$ in (2.89) has the meaning of a double-sided PSD $\mathcal{S}_w(\omega)$ of WGN, and N_0 is thus a one-sided PSD. Due to this property, the conditional pdf $p_X(x_k|x_l)$, $k > l$, of white noise is marginal,

$$p_X(x_k|x_l) = p_X(x_k) \,.$$

For zero mean vector WGN $w(t) = [\, w_1(t) \;\; w_2(t) \ldots w_n(t)\,]^T$, the covariance is defined by

$$\begin{aligned} C_w(\tau) = \mathcal{R}_w(\tau) &= E\{w(t)w^T(t+\tau)\} \\ &= \mathcal{S}_w\delta(\tau) = \frac{1}{2}\begin{bmatrix} N_{11} & \ldots & N_{1n} \\ \vdots & \ddots & \vdots \\ N_{n1} & \ldots & N_{nn} \end{bmatrix}\delta(\tau), \end{aligned} \tag{2.91}$$

where $\mathcal{S}_w$ is the PSD matrix of WGN, whose component $\frac{N_{ij}}{2}$, $\{i,j\} \in [1,n]$, is the PSD or cross-PSD of $w_i(t)$ and $w_j(t)$.

Discrete White Gaussian Noise

In discrete time index k, WGN is a discrete signal w_k, the samples of which are a sequence of uncorrelated random variables. Discrete WGN is defined as the average of the original continuous WGN $w(t)$ as

$$w_k = \frac{1}{\tau}\int_{t_k-\tau}^{t_k} w(t)\,\mathrm{d}t \,, \tag{2.92}$$

where $\tau = t_k - t_{k-1}$ is a proper time step.

Taking the expectation on the both sides of (2.92) gives zero,

$$E\{w_k\} = \frac{1}{\tau}\int_{t_k-\tau}^{t_k} E\{w(t)\}\,\mathrm{d}t = 0 \,,$$

and the variance of zero mean w_k can be found as

$$\begin{aligned} \sigma_{w_k}^2 &= E\{w_k^2\} \\ &= \frac{1}{\tau^2}\int_{t_k-\tau}^{t_k}\int_{t_k-\tau}^{t_k} E\{w(\theta_1)w(\theta_2)\}\,\mathrm{d}\theta_1\mathrm{d}\theta_2 \\ &= \frac{N_0}{2\tau^2}\int_{t_k-\tau}^{t_k}\int_{t_k-\tau}^{t_k} \delta(\theta_1-\theta_2)\,\mathrm{d}\theta_1\mathrm{d}\theta_2 \\ &= \frac{N_0}{2\tau^2}\int_{t_k-\tau}^{t_k} \mathrm{d}\theta_2 = \frac{N_0}{2\tau} = \frac{1}{\tau}\mathcal{S}_w \,, \end{aligned} \tag{2.93}$$

where $\mathcal{S}_w = \frac{N_0}{2}$ is the PSD (2.90) of WGN.

Accordingly, the pdf of a discrete WGN w_k can be written as

$$p_{w_k}(x) = \sqrt{\frac{\tau}{\pi N_0}} \exp\left(-\frac{\tau x^2}{N_0}\right)$$

and w_k denoted as $w_k \sim \mathcal{N}(0, \frac{N_0}{2\tau})$. Since a continuous WGN $w(t)$ can also be denoted as $w(t) \sim \mathcal{N}[0, \frac{N_0}{2}\delta(0)]$, the notations become equivalent when $\tau \to 0$.

For a discrete zero mean vector WGN $w_k = [\, w_{1k} \; \dots \; w_{nk} \,]^T$, whose components are given by (2.92) to have variance (2.93), the covariance R_w of w_k is defined following (2.93) as

$$R_w = E\{w_k w_k^T\} = \frac{1}{\tau}S_w = \frac{1}{2\tau}\begin{bmatrix} N_{11} & \dots & N_{1n} \\ \vdots & \ddots & \vdots \\ N_{n1} & \dots & N_{nn} \end{bmatrix}, \tag{2.94}$$

where S_w is the PSD matrix specified by (2.91). It follows from (2.91) and (2.94) that the covariance R_w of the discrete WGN evolves to the covariance $C_w(\tau) = \mathcal{R}_w(\tau)$ of the continuous WGN when $\tau \to 0$ as

$$\lim_{\tau\to 0} R_w = \lim_{\tau\to 0}\frac{1}{\tau}S_w \to S_w\delta(\tau) = C_w(\tau) = \mathcal{R}_w(\tau)\,.$$

Because R_w is defined for any $\tau \neq 0$, and $\mathcal{R}_w(\tau)$ is defined at $\tau = 0$, then it follows that there is no direct connection between R_w and $\mathcal{R}_w(\tau)$. Instead, $\mathcal{R}_w(\tau)$ can be viewed as a limited case of R_w when $\tau \to 0$.

2.2.5 Markov Processes

The Markov (or Markovian) process, which is also called *continuous-time Markov chain* or *continuous random walks*, is another idealization of real physical stochastic processes. Unlike white noise, the values of which do not correlate with each other at any two different time instances, correlation in the Markov process can be observed only between two nearest neighbors.

A stochastic process $X(t)$ is called Markov if its random variable $X(t)$, given $X(u)$ at $u < t$, does not depend on $X(s)$ since $s < u$. Thus, the following is an exhaustive property of all types of Markov processes.

> ***Markov process:*** *A random process is Markovian if on any finite time interval of time instances $t_1 < t_2 < \dots < t_n$ the conditional probability of a variable $X(t_n)$ given $X(t_1), X(t_2), \dots, X(t_{n-1})$ depends solely on $X(t_{n-1})$; that is,*
>
> $$P\{X(t_n) < x_n | X(t_1) = x_1, \dots, X(t_{n-1}) = x_{n-1}\} = P\{X(t_n) < x_n | X(t_{n-1}) = x_{n-1}\}\,. \tag{2.95}$$

The following probabilistic statement can be made about the future behavior of a Markov process: if the present state of a Markov process at t_j is known explicitly, then the future state at t_i, $i > j$, can be predicted without reference to any past state at t_k, $k < j$.

It follows from the previous definition that the multivariate pdf of a Markov process can be written as

$$p_X(x_1, \dots, x_n) = p_{X_1}(x_1)p_{X_2}(x_2|x_1)\dots p_{X_n}(x_n|x_{n-1}) = p_{X_1}(x_1)\prod_{i=1}^{n-1} p_{X_{i+1}}(x_{i+1}|x_i) \tag{2.96}$$

to be the chain rule (2.55) for Markov processes. In particular, for two Markovian variables one has

$$p_{X_1,X_2}(x_1, x_2) = p_{X_1}(x_1)p_{X_2}(x_2|x_1) .$$

An important example of Markov processes is the Wiener process, known in physics as Brownian motion.

Example 2.7 ***Wiener process.*** The Wiener process $W(t)$ appears when the white noise $w(t)$ passes through an ideal integrator as

$$W(t) = \int_0^t w(\theta)\,\mathrm{d}\theta \,. \tag{2.97}$$

The mean value of $W(t)$ is zero,

$$E\{W(t)\} = \int_0^t E\{w(\theta)\}\,\mathrm{d}\theta = 0 \,,$$

and the time-varying variance

$$\begin{aligned}\sigma_W^2(t) = E\{W^2(t)\} &= \int_0^t \int_0^t E\{w(\theta_1)w(\theta_2)\}\,\mathrm{d}\theta_1\mathrm{d}\theta_2 \\ &= \frac{N_0}{2}\int_0^t \int_0^t \delta(\theta_2 - \theta_1)\,\mathrm{d}\theta_1\mathrm{d}\theta_2 \\ &= \frac{N_0}{2}\int_0^t \mathrm{d}\theta = \frac{N_0}{2}t\end{aligned}$$

grows proportionally with time.

The autocorrelation function of $W(t)$ is given by

$$\begin{aligned}\mathcal{R}_W(t_1, t_2) = E\{W(t_1)W(t_2)\} &= \frac{N_0}{2}\int_0^{t_2} \int_0^{t_1} \delta(\theta_1 - \theta_2)\,\mathrm{d}\theta_1\mathrm{d}\theta_2 \\ &= \frac{N_0}{2}\begin{cases} t_2 \,, & t_2 < t_1 \\ t_1 \,, & t_1 < t_2 \end{cases}\end{aligned}$$

and also grows proportionally with time. □

When a Markov process is represented by a set of values at discrete time instances, it is called a *discrete-time Markov chain* or simply *Markov chain*. For a finite set of discrete variables $\{X_1, \dots, X_n\}$, specified at discrete time indexes $\{1, \dots, n\}$, the conditional probability (2.95) becomes

$$\begin{aligned}&P\{X_n < x_n|X_1 = x_1, \dots, X_{n-1} = x_{n-1}\} \\ &\quad = P\{X_n < x_n|X_{n-1} = x_{n-1}\} \,.\end{aligned} \tag{2.98}$$

Thus, the Markov chain is a stochastic sequence of possible events that obeys (2.98). A notable example of a Markov chain is the Poisson process [156, 191].

Property (2.98) can be written in pdf format as

$$p(x_n|x_1, \dots, x_{n-1}) = p(x_n|x_{n-1}) \,. \tag{2.99}$$

Since (2.99) defines the distribution of X_n via X_{n-1}, the conditional pdf $p(x_t|x_\theta)$ for $t > \theta$ is called the *transitional pdf* of the Markov process.

Example 2.8 ***Gauss-Markov sequence.*** Given (2.97), the WGN $w(t)$ can also be defined as

$$w(t) = \frac{\mathrm{d}}{\mathrm{d}t}W(t) \cong \frac{W(t) - W(t-\tau)}{\tau} ,$$

which leads to a first-order difference equation representing the Gauss-Markov sequence [79]

$$W_k = W_{k-1} + \tau w_k , \quad k = 0, 1, \dots , \tag{2.100}$$

where w_k is a white Gaussian sequence of variables, and the initial condition W_0 is Gaussian and does not depend on w_k.

The sequence (2.100) is clearly Markov, since W_k depends only on w_k provided W_{k-1}. Therefore, it is also called the Gauss-Markov colored *process noise* [185]. The transitional probability for this sequence can be written as

$$p(x_k|x_{k-1}) = p(x_k - x_{k-1}) .$$

With a zero initial condition $W_0 = 0$, the process noise (2.100) can be written in batch form as $W_k = \tau \sum_{i=1}^{k} w_i$ to have zero mean, $E\{W_k\} = 0$, and the variance

$$\sigma_W^2 = E\{W^2(t)\} = \tau^2 \sum_{i=1}^{k} E\{w_i^2\} = \tau k \frac{N_0}{2} = \tau k S_w ,$$

where $E\{w_k^2\}$ is defined by (2.93).

Note that if the noise in (2.100) is not gained with τ, as in the standard Gauss-Markov sequence [79], then the variance becomes $\sigma_W^2 = \frac{k}{\tau} S_w$. □

The transitional probability can be obtained for any two random variables X_k and X_m of the Markov process [79]. To show this, one can start with the rule (2.52) by rewriting it as

$$p(x_k|x_{k-2}) = \int_{-\infty}^{\infty} p(x_k, x_{k-1}|x_{k-2})\, \mathrm{d}x_{k-1} . \tag{2.101}$$

Now, according to the chain rule (2.96) and the Markov property (2.99), the integrand can be rewritten as

$$\begin{aligned} p(x_k, x_{k-1}|x_{k-2}) &= p(x_k|x_{k-1}, x_{k-2}) p(x_{k-1}|x_{k-2}) \\ &= p(x_k|x_{k-1}) p(x_{k-1}|x_{k-2}) \end{aligned}$$

and (2.101) transformed to the *Chapman-Kolmogorov equation*

$$p(x_k|x_{k-2}) = \int_{-\infty}^{\infty} p(x_k|x_{k-1}) p(x_{k-1}|x_{k-2})\, \mathrm{d}x_{k-1} . \tag{2.102}$$

Note that this equation also follows from the general probabilistic rule (2.53) and can be rewritten more generally for $m \geqslant k-2$ as

$$p(x_k|x_m) = \int_{-\infty}^{\infty} p(x_k|x_{k-1}) p(x_{k-1}|x_m)\, \mathrm{d}x_{k-1} . \tag{2.103}$$

The theory of Markov processes and chains establishes a special topic in the interpretation and estimation of real physical processes for a wide class of applications. The interested reader is referred to a number of fundamental and applied investigations discussed in [16, 45, 79, 145, 156, 191].

2.3 Stochastic Differential Equation

Dynamic physical processes can be both linear and nonlinear with respect to variables and perturbations. Many of them can be generalized by a multivariate differential equation in the form

$$\frac{\mathrm{d}}{\mathrm{d}t}X(t) = f(X, w, t)\,, \tag{2.104}$$

where $f(\cdot)$ is a nonlinear function of a general vector stochastic process $X(t)$ and noise $w(t)$. We encounter such a case in trajectory measurements where values and disturbances in Cartesian coordinates are nonlinearly related to values measured in polar coordinates (see Example 2.5).

2.3.1 Standard Stochastic Differential Equation

Since noise in usually less intensive than measured values, another form of (2.104) has found more applications,

$$\frac{\mathrm{d}}{\mathrm{d}t}X(t) = f(X, t) + g(X, t)w(t)\,, \tag{2.105}$$

$$\mathrm{d}X(t) = f(X, t)\,\mathrm{d}t + g(X, t)\,\mathrm{d}W(t)\,, \tag{2.106}$$

where $f(X, t)$ and $g(X, t)$ are some known nonlinear functions, $w(t)$ is some noise, and $W(t) = \int w(t)\,\mathrm{d}t$. A differential equation (2.105) or (2.106) can be thought of as SDE, because one or more its terms are random processes and the solution is also a random process. Since a typical SDE contains white noise $w(t)$ calculated by the derivative of the Wiener process $W(t)$, we will further refer to this case.

If the noise $w(t)$ in (2.105) is white Gaussian with zero mean $\mathcal{E}\{w(t)\} = 0$ and autocorrelation function $\mathcal{R}_w(\tau) = \frac{N_0}{2}\delta(\tau)$, and noise $W(t) = \int w(t)\,\mathrm{d}t$ in (2.106) is a Wiener process with zero mean $E\{W(t)\} = 0$ and autocorrelation function $\mathcal{R}_W(t_1, t_2) = \frac{N_0}{2}|t_2 - t_1|$, then (2.105) and (2.106) are called SDE if the following Lipschitz condition is satisfied.

> ***Lipschitz condition:*** *An equation (2.105) representing a scalar random process $X(t)$ is said to be SDE if zero mean noise $w(t)$ is white Gaussian with known variance and nonlinear functions $f(X, t)$ and $g(X, t)$ satisfy the Lipschitz condition*
>
> $$|f(x, t) - f(y, t)| + |g(x, t) - g(y, t)| \leqslant L|x - y| \tag{2.107}$$
>
> *for constant $L > 0$.*

The problem with integrating either (2.105) or (2.106) arises because the integrand, which has white noise properties, does not satisfy the Dirichlet condition, and thus the integral does not exist in the usual sense of Riemann and Lebesgue. However, solutions can be found if we use the Itô calculus and Stratonovich calculus.

2.3.2 Itô and Stratonovich Stochastic Calculus

Integrating SDE (2.106) from t_0 to t gives

$$X(t) = X(t_0) + \int_{t_0}^{t} f(X, t)\,\mathrm{d}t + \int_{t_0}^{t} g(X, t)\,\mathrm{d}W(t)\,, \tag{2.108}$$

where it is required to know the initial $X(t_0)$, and the first integral must satisfy the Dirichlet condition.

The second integral in (2.108), called the *stochastic integral*, can be defined in the Lebesgue sense as [140]

$$\int_{t_0}^{t} g(X,t)\,\mathrm{d}W(t) = \lim_{n\to\infty}\sum_{k=1}^{n} g(X,\bar{t}_k)[W(t_{k-1}) - W(t_k)]\,, \tag{2.109}$$

where the integration interval $[t_0, t]$ is divided into $n \to \infty$ subintervals as $t_0 < t_1 < \cdots < t_n = t$ and $\bar{t}_k \in [t_{k-1}, t_k]$. This calculus allows one to integrate the stochastic integral numerically if $\bar{t}_k$ is specified properly.

Itô proved that a stable solution to (2.109) can be found by assigning $\bar{t}_k = t_{k-1}$. The corresponding solution for (2.108) was named the Itô solution, and SDE (2.106) was called the Itô SDE [140]. Another calculus was proposed by Stratonovich [193], who suggested assigning $\bar{t}_k = \frac{1}{2}(t_{t-1} + t_k)$ at the midpoint of the interval. To distinguish the difference from Itô SDE, the Stratonovich SDE is often written as

$$\mathrm{d}X(t) = f(X,t)\,\mathrm{d}t + g(X,t) \circ \mathrm{d}W(t) \tag{2.110}$$

with a circle in the last term. The circle is also introduced into the stochastic integral as $\int_{t_0}^{t} g(X,t) \circ \mathrm{d}W(t)$ to indicate that the calculus (2.109) is in the Stratonovich sense. The analysis of Stratonovich's solution is more complicated, but Stratonovich's SDE can always be converted to Itô's SDE using a simple transformation rule [193]. Moreover, $g(X,t) = g(t)$ makes both solutions equivalent.

2.3.3 Diffusion Process Interpretation

Based on the rules of Itô and Stratonovich, the theory of SDE has been developed in great detail [16, 45, 125, 145, 156]. It has been shown that if the stochastic process $X(t)$ is represented with (2.105) or (2.106), then it is a Markovian process [45]. Moreover, the process described using (2.105) or (2.106) belongs to the class of *diffusion processes*, which are described using the *drift* $K_1(X,t)$ and *diffusion* $K_2(X,t)$ coefficients [193] defined as

$$K_1(x,t) = \lim_{\tau\to 0}\frac{1}{\tau}\mathcal{E}\{[X(t+\tau) - X(t)]|X = x\}\,, \tag{2.111}$$

$$K_2(x,t) = \lim_{\tau\to 0}\frac{1}{\tau}\mathcal{E}\{[X(t+\tau) - X(t)]^2|X = x\} \tag{2.112}$$

and associated with the first and second moments of the stochastic process $X(t)$. Depending on the choice of $\bar{t}_k$ in (2.109), the drift and diffusion coefficients can be defined in different senses. In the Stratonovich sense, the drift and diffusion coefficients become

$$K_1(X,t) = f(X,t) + \frac{N_0}{2} g(X,t)\frac{\partial g(X,t)}{\partial X}\,, \tag{2.113}$$

$$K_2(X,t) = \frac{N_0}{2} g^2(X,t)\,. \tag{2.114}$$

In the Itô sense, the second term vanishes on the right-hand side of (2.113).

In a more general case of a vector stochastic process $X(t)$, represented by a set of random subprocesses $\{X_1(t), \ldots, X_n(t)\}$, the ith stochastic process $X_i(t)$ can be described using SDE [193]

$$\mathrm{d}X_i(t) = f_i(X,t)\,\mathrm{d}t + \sum_{l=1}^{n} g_{il}(X,t)\,\mathrm{d}W_l(t)\,, \quad i \in [1,n]\,, \tag{2.115}$$

where the nonlinear functions $f_i(X,t)$ and $g_{il}(X,t)$ satisfy the Lipschitz condition (2.107) and $W_l(t)$, $l \in [1,n]$, are independent Wiener processes with zero mean $\mathcal{E}\{W_l(t)\} = 0$ and autocorrelation function

$$\mathcal{E}\{[W_l(t_1) - W_l(t_2)][W_j(t_1) - W_j(t_2)]\} = \frac{N_l}{2}|t_2 - t_1|\delta_{lj} ,$$

where $l, j \in [1,n]$, δ_{lj} is the Kronecker symbol, and $\frac{N_l}{2}$ is the PSD of $w_l(t)$.

Similarly to the scalar case, the vector stochastic process $X(t)$, described by an SDE (2.110), can be represented in the Stratonovich sense with the following drift coefficient $K_i(X,t)$ and diffusion coefficient $K_{ij}(X,t)$ [193],

$$K_i(X,t) = f_i(X,t) + \sum_{l=1}^{n}\sum_{j=1}^{n}\frac{N_l}{2}g_{jl}(X,t)\frac{\partial g_{il}(X,t)}{\partial X_j} , \tag{2.116}$$

$$K_{ij}(X,t) = \sum_{l=1}^{n}\frac{N_l}{2}g_{il}(X,t)g_{jl}(X,t) . \tag{2.117}$$

Given the drift and diffusion coefficients, SDE can be replaced by a probabilistic differential equation for the time-varying pdf $p(X,t)$ of $X(t)$. This equation is most often called the *Fokker-Planck-Kolmogorov* (FPK) *equation*, but it can also be found in the works of Einstein and Smoluchowski.

2.3.4 Fokker-Planck-Kolmogorov Equation

If the SDE (2.105) obeys the Lipschitz condition, then it can be replaced by the following FPK equation representing the dynamics of $X(t)$ in probabilistic terms as

$$\frac{\partial}{\partial t}p(x,t) = -\frac{\partial}{\partial x}[K_1(x,t)p(x,t)] + \frac{1}{2}\frac{\partial^2}{\partial x^2}[K_2(x,t)p(x,t)] . \tag{2.118}$$

This equation is also known as the *first* or *forward Kolmogorov equation*. Closed-form solutions of the partial differential equation (2.118) have been found so far for a few simple cases. However, the stationary case, assuming $\frac{\partial}{\partial t}p(x,t) = 0$ with $t \to \infty$, reduces it to a time-invariant probability flux

$$G(x) = K_1(x)p_{\rm st}(x) - \frac{1}{2}\frac{\partial}{\partial x}[K_2(x)p_{\rm st}(x)] = 0 , \tag{2.119}$$

which is equal to zero at all range points. Hence, the steady-state pdf $p_{\rm st}(x)$ can be defined as

$$p_{\rm st}(x) = \frac{c}{K_2(x)}\exp\left[2\int_{x_1}^{x}\frac{K_1(z)}{K_2(z)}\,\mathrm{d}z\right] \tag{2.120}$$

by integrating (2.119) from some point x_1 to x with the normalizing constant c. By virtue of this, in many cases, a closed-form solution is not required for (2.118), because pdf $p_{\rm st}(x)$ (2.120) contains all the statistical information about stochastic process $X(t)$ at $t \to \infty$.

The FPK equation (2.118) describes the forward dynamics of the stochastic process $X(t)$ in probabilistic terms. But if it is necessary to study the inverse dynamics, one can use the *second* or *backward Kolmogorov equation*

$$\frac{\partial}{\partial s}p(x,s) = -K_1(x,s)\frac{\partial}{\partial x}p(x,s) - \frac{1}{2}K_2(x,s)\frac{\partial^2}{\partial x^2}p(x,s) , \quad s \leqslant t , \tag{2.121}$$

specifically to learn the initial distribution of $X(t_0)$ at $t_0 < t$, provided that the distribution at t is already known.

2.3.5 Langevin Equation

A linear first-order SDE is known in physics as the *Langevin equation*, the physical nature of which can be found in Brownian motion. It also represents the first-order electric circuit driven by white noise $w(t)$. Although the Langevin equation is the simplest in the family of SDEs, it allows one to study the key properties of stochastic dynamics and plays a fundamental role in the theory of stochastic processes.

The Langevin equation can be written as

$$\frac{\mathrm{d}}{\mathrm{d}t}X(t) + \alpha X(t) = w(t)\,, \tag{2.122}$$

where α is a constant and $w(t)$ is white Gaussian noise with zero mean, $\mathcal{E}\{w(t)\} = 0$, and the autocorrelation function $\mathcal{R}_w(\tau) = \frac{N_0}{2}\delta(\tau)$. If we assume that the stochastic process $X(t)$ starts at time $t = 0$ as X_0, then the solution to (2.122) can be written as

$$X(t) = X_0 e^{-\alpha t} + \alpha e^{-\alpha t}\int_0^t e^{\alpha\theta} w(\theta)\,\mathrm{d}\theta\,, \tag{2.123}$$

which has the mean

$$\mathcal{E}\{X(t)\} = \bar{X}(t) = X_0 e^{-\alpha t} \tag{2.124}$$

and the variance

$$\begin{aligned}
\sigma_X^2(t) &= \mathcal{E}\{[X(t) - \bar{X}(t)]^2\} \\
&= \alpha^2 e^{-2\alpha t}\int_0^t\int_0^t e^{\alpha\theta_1} e^{\alpha\theta_2}\mathcal{E}\{w(\theta_1)w(\theta_2)\}\,\mathrm{d}\theta_1\,\mathrm{d}\theta_2 \\
&= \frac{\alpha^2 N_0}{2}e^{-2\alpha t}\int_0^t\int_0^t e^{\alpha\theta_1}e^{\alpha\theta_2}\delta(\theta_1 - \theta_2)\,\mathrm{d}\theta_1\,\mathrm{d}\theta_2 \\
&= \frac{\alpha^2 N_0}{2}e^{-2\alpha t}\int_0^t e^{2\alpha\theta}\,\mathrm{d}\theta \\
&= \frac{\alpha N_0}{4}(1 - e^{-2\alpha t})\,.
\end{aligned} \tag{2.125}$$

Since the Langevin equation is linear and the driving noise $w(t)$ is white Gaussian, it follows that $X(t)$ is also Gaussian with a nonstationary pdf

$$p(x,t) = \frac{1}{\sqrt{2\pi\sigma_X^2(t)}}\exp\left[-\frac{[x - \bar{X}(t)]^2}{2\sigma_X^2(t)}\right]\,, \tag{2.126}$$

where the mean $\bar{X}(t)$ is given by (2.124) and the variance $\sigma_X^2(t)$ by (2.125).

The drift coefficient (2.113) and diffusion coefficient (2.114) are defined for the Langevin equation (2.122) as

$$K_1(X,t) = -\alpha X(t)\,,$$

$$K_2 = \frac{\alpha^2 N_0}{2}\,,$$

and hence the FPK equation (2.118) becomes

$$\frac{\partial}{\partial t}p(x,t) = \alpha\frac{\partial}{\partial x}[xp(x,t)] + \frac{\alpha^2 N_0}{4}\frac{\partial^2}{\partial x^2}p(x,t)\,, \tag{2.127}$$

whose solution is a nonstationary Gaussian pdf (2.126) and the moments (2.124) and (2.125).

Langevin's equation is a nice illustration of how a stochastic process can be investigated in terms of a probability distribution, rather than by solving its SDE. Unfortunately, most high-order stochastic processes cannot be explored in this way due to the complexity, and the best way is to go into the state space and use state estimators, which we will discuss in the next chapter.

2.4 Summary

In this chapter we have presented the basics of probability and stochastic processes, which are essential to understand the theory of state estimation. Probability theory and methods developed for stochastic processes play a fundamental role in understanding the features of physical processes driven and corrupted by noise. They also enable the formulation of feature extraction requirements from noise processes and the development of optimal and robust estimators.

A random variable $X(\xi)$ of a physical process can be related to the corresponding measured element ξ in a linear or nonlinear manner. A collection of random variables is a random process. The probability $P\{X \leqslant x\}$ that X will occur below some constant x (event A) is calculated as the ratio of possible outcomes favoring event A to total possible outcomes. The corresponding function $F_X(x) = P\{X \leqslant x\} \geqslant 0$ is called cdf. In turn, pdf $p_X(x) \geqslant 0$ represents the concentration of X values around x and is equal to the derivative of $F_X(x)$ with respect to x, since cdf is the integral measure of pdf.

Each random process can be represented by a set of initial and central moments. The Gaussian process is the only one that is represented by a first-order raw moment (mean) and a second-order central moment (variance). Product moments represent the power of interaction between two different random variables X and Y. The normalized measure of interaction is called the correlation coefficient, which is in the ranges $-1 \leqslant \rho_{X,Y} \leqslant 1$. A stochastic process is associated with continuous time and stochastic sequence with discrete time.

The conditional probability $P(A|B)$ of the event A means that the event B is observed. Therefore, it is also called the *a posteriori* probability, and $P(A)$ and $P(B)$ are called the *a priori* probabilities. Bayes' theorem states that the joint probability $P(AB)$ can be represented as $P(AB) = P(A|B)P(B) = P(B|A)P(A)$, and there is always a certain rule for converting two correlated random variables into two uncorrelated random variables and vice versa.

The autocorrelation function establishes the degree of interaction between the values of a variable measured at two different time points, while the cross-correlation establishes the degree of interaction between the values of two different variables. According to the Wiener-Khinchin theorem, PSD and correlation function are related to each other by the Fourier transform.

Continuous white Gaussian noise has infinite variance, while its discrete counterpart has finite variance. A stochastic process is called Markov if its random variable $X(t)$, for a given $X(u)$ at $u < t$, does not depend on $X(s)$, since $s < u$.

The SDE can be solved either in the Itô sense or in the Stratonovich sense. It can also be viewed as a diffusion process and represented by the probabilistic FPK equation. The Langevin equation is a classical example of first-order stochastic processes associated with Brownian motion.

2.5 Problems

1 Two events A and B are mutually exclusive. Can they be uncorrelated and independent?

2 Two nodes N_1 and N_1 transmit the same message over the wireless network to a central station, which can only process the previously received one. Assuming a random delay in message delivery, what is the probability that 10 consecutive messages will belong to node N_1?

3 A network of 20 nodes contains 5 damaged ones. If we choose three nodes at random, what is the probability that at least one of these nodes is defective?

4 Show that if $P(B|A) = P(B)$, then 1) $P(AB) = P(A)P(B)$ and 2) $P(A|B) = P(A)$.

5 The binomial coefficients are computed by $\binom{n}{k} = \frac{n!}{k!(n-k)!}$. Show that 1) $\binom{n}{k} = \binom{n}{n-k}$ and 2) $\binom{n}{k} + \binom{n}{k+1} = \binom{n+1}{k+1}$.

6 Given events A, B, and B and using the chain rule, show that

$$P(AB) = P(A|B)P(B)\,,$$
$$P(AB|C) = P(A|BC)P(B|C)\,,$$
$$P(ABC) = P(A|BC)P(B|C)P(C)\,.$$

7 Given two independent identically distributed random variables $x \sim \mathcal{N}(0, \sigma^2)$ and $y \sim \mathcal{N}(0, \sigma^2)$ with zero mean, variance σ^2, and join pdf $p_{XY}(x, y) = p_X(x)p_Y(y)$, find the pdf of the following variables: 1) $z = \sqrt{x^2 + y^2}$, 2) $z = x^2 + y^2$, and 3) $z = x - y$.

8 The Bernoulli distribution of the discrete random variable $X \in \{0, 1\}$ is given by pmf

$$f(X; p) = p^X(1-p)^{1-X} \tag{2.128}$$

to represent random binary time delays in communication channels. Find the cdf for (2.128) and show that the mean is $\mathcal{E}\{X\} = p$ and the variance is $\sigma_X^2 = p(1-p)$.

9 A generalized normal distribution of a random variable X is given by the pdf

$$p_X(x) = \frac{\beta}{2\alpha\Gamma(1/\beta)} e^{-(|x-\mu|/\alpha)^\beta}\,, \tag{2.129}$$

where all are real, μ is location, $\alpha > 0$ is scale, and $\beta > 0$ is shape. Find the moments and cumulants of (2.129). Prove that the mean is $\mathcal{E}\{X\} = \mu$ and the variance is $\sigma_X^2 = \frac{\alpha^2\Gamma(3/\beta)}{\Gamma(3/\beta)}$. Find the values of β at which this distribution becomes Gaussian, almost rectangular, and heavy-tailed.

10 The random variable X representing the measurement noise has a Laplace distribution with pdf

$$p_X(x|\mu, b) = \frac{1}{2b}\exp\left(-\frac{|x-\mu|}{b}\right)\,, \tag{2.130}$$

where μ is the location parameter and $b > 0$ is the scale parameter. Find cdf, the mean $\bar{\mu}$, and the variance σ_L^2 of this variable.

11 Consider a set of N independent samples X_i, $i \in [1, N]$, each of which obeys the Laplace distribution (2.130) with variance σ_L^2. The maximum likelihood estimate of location is given by [13]

$$\hat{\mu} = \arg\min_{\mu} \sum_{i=1}^{N} \frac{1}{\sigma_L^2}|X_i - \mu| \tag{2.131}$$

and is called the *median* estimate. Redefine this estimate considering X_i as a state variable and explain the meaning of the estimate.

12 A measured random quantity X has a Cauchy distribution with pdf

$$p_X(x|\mu, b) = \frac{1}{\pi}\frac{b}{(x-\mu)^2 + b^2}\,, \tag{2.132}$$

where μ is the location parameter and $b > 0$ is the scale parameter. Prove that cdf is $F_X(x) = \frac{1}{\pi}\arctan\left(\frac{x-\mu}{b}\right) + \frac{1}{2}$ and that the mean and variance are not definable.

13 Consider a measurable random variable X_i represented by a Cauchy pdf (2.132) with $\mu = 0$ and $b = 1$. The set of measured variables passes through an electric circuit, where it is saturated by the power supply as

$$X_i = \begin{cases} X_i\,, & |X_i| < 5\,, \\ 5\,, & |X_i| \geqslant 5 \end{cases} .$$

Modify the Cauchy pdf (2.132) for saturated X_i and numerically compute the mean and variance.

14 The measurement of some scalar constant quantity is corrupted by the Gauss-Markov noise $v_k = \psi v_{k-1} + \xi_k$, where $0 < \psi < 1$ and ξ_k is the white Gaussian driving noise. Find the autocorrelation function and PSD of noise v_k. Describe the properties of v_k in two extreme cases: $\psi = 0$ and $\psi > 1$.

15 Given a discrete-time random process $x_k = \psi x_{k-1} + \xi_k$, where $0 < \psi < 1$ and ξ_k is some random driving force. Considering ξ_k as input and x_k as output, find the input-to-output transfer function $\mathcal{T}(z)$.

16 Explain the physical nature of the skewness γ_1 and kurtosis γ_2 and how these measures help to recognize the properties of a random variable X. Illustrate the analysis based on the Bernoulli, Gaussian, and generalized normal distributions and provide some practical examples.

17 The joint pdf of X and Y is given by

$$p_{X,Y}(x,y) = \begin{cases} x^2 + \frac{1}{3}xy\,, & 0 \leqslant x \leqslant 1, 0 \leqslant y \leqslant 2\,, \\ 0\,, & \text{otherwise} \end{cases}$$

Prove that the marginal pdfs are $p_X(x) = 2x^2 + \frac{2x}{3}$ and $p_Y(y) = \frac{1}{3}(1 + \frac{y}{2})$.

18 Two uncorrelated phase differences ϕ_1 and ϕ_2 are distributed with the conditional von Mises circular normal pdf

$$p(\phi|\gamma, \phi_0) = \frac{1}{2\pi I_0(\alpha)} e^{\alpha\cos(\phi-\phi_0)}\,, \tag{2.133}$$

where ϕ is a random phase mod 2π and ϕ_0 is its deterministic constituent, $I_0(\alpha)$ is a modified Bessel function of the first kind and zeroth order, and $\alpha(\gamma)$ is a parameter sensitive to the power *signal-to-noise ratio* (SNR) γ. Show that the phase difference $\Psi = \phi_2 - \phi_1$ with different SNRs $\gamma_1 \neq \gamma_2$ is conditionally distributed by

$$p(\Psi|\gamma_1, \gamma_2, \Psi_0) = \frac{1}{2\pi}\frac{I_0(r)}{I_0(\alpha_1)I_0(\alpha_2)} \tag{2.134}$$

and define the function $r(\Psi, \alpha_1, \alpha_2)$ for $I_0(r)$.

19 A stable system with random components is represented in discrete-time state space with the state equation $x_{k+1} = Fx_k + Bw_k$ and the observation equation $y_k = Hx_k + v_k$. Considering w_k as input and y_k as output, find the input-to-output transfer function $\mathcal{T}(z)$.

20 The Markov chain $X_n = X_{n-1}M$ with three states is specified using the transition matrix

$$M = \begin{bmatrix} 0.9 & 0.075 & 0.025 \\ 0.15 & 0.8 & 0.05 \\ 0.25 & 0.25 & 0.5 \end{bmatrix}.$$

Represent this chain as $X_n = X_0 L_n$, specify the transition matrix L_n, and find $\lim_{n\to\infty} L_n$.

21 Given a stationary process $x(t)$ with derivative $\dot{x}(t)$, show that for a given time t the random variables $x(t)$ and $\dot{x}(t)$ are orthogonal and uncorrelated.

22 The continuous-time three-state clock model is represented by SDE $\frac{\mathrm{d}}{\mathrm{d}t}x(t) = \begin{bmatrix} 0 & 1 & 0 \\ 0 & 0 & 1 \\ 0 & 0 & 0 \end{bmatrix} x(t) + w(t)$, where the zero mean noise $w(t) = [\ w_\varphi(t)\ \ w_{\dot\varphi}(t)\ \ w_{\ddot\varphi}(t)\ ]^T$ has the following components: $w_\varphi(t) \sim \mathcal{N}(0, S_\varphi \delta(\tau))$ is the phase noise, $w_\varphi(t) \sim \mathcal{N}(0, S_{\dot\varphi}\delta(\tau))$ is the frequency noise, and $w_\varphi(t) \sim \mathcal{N}(0, S_{\ddot\varphi}\delta(\tau))$ is the linear frequency drift noise. Show that the noise covariance for this model is given by

$$C_w(\tau) = \tau \begin{bmatrix} S_\varphi + \frac{S_{\dot\varphi}\tau^2}{3} + \frac{S_{\ddot\varphi}\tau^4}{20} & \frac{S_{\dot\varphi}\tau}{2} + \frac{S_{\ddot\varphi}\tau^3}{8} & \frac{S_{\ddot\varphi}\tau^2}{6} \\ \frac{S_{\dot\varphi}\tau}{2} + \frac{S_{\ddot\varphi}\tau^3}{8} & S_{\dot\varphi} + \frac{S_{\ddot\varphi}\tau^2}{3} & \frac{S_{\ddot\varphi}\tau}{2} \\ \frac{S_{\ddot\varphi}\tau^2}{6} & \frac{S_{\ddot\varphi}\tau}{2} & S_{\ddot\varphi} \end{bmatrix},$$

if we provide the integration from $t - \tau$ to t.

23 Two discrete random stationary processes w_k and v_k have autocorrelation functions

$$\mathcal{R}_w = \begin{bmatrix} 1 & 0 & \dots & 0 \\ 0 & 1 & \dots & 0 \\ \vdots & \vdots & \ddots & \vdots \\ 0 & 0 & \dots & 1 \end{bmatrix}, \quad \mathcal{R}_v = \begin{bmatrix} 1 & 1 & \dots & 1 \\ 1 & 1 & \dots & 1 \\ \vdots & \vdots & \ddots & \vdots \\ 1 & 1 & \dots & 1 \end{bmatrix}, \tag{2.135}$$

which are measured relative to each point on the horizon $[m, k]$ by shifting w_k and v_k by $\pm i$. What is the PSD of the first process w_k and the second process v_k?

24 Given an LTI system with the impulse response $h(t) = ae^{\alpha t}u(t)$, where $a > 0$, $\alpha > 0$, and $u(t)$ is the unit step function. In this system, the input is a stationary random process $x(t)$ with the autocorrelation function $R_x(\tau) = \frac{N}{2}\delta(t)$ applied at $t = 0$ and disconnected at $t = T$. Find the mean $\bar{y}(t)$ and the mean square value $\mathcal{E}\{y^2(t)\}$ of the output signal $y(t)$ and plot these functions.

3

State Estimation

Gauss's batch least squares is routinely used today, and it often gives accuracy that is superior to the best available EKF.

Fred Daum, [40], p. 65

Although SDEs provide accessible mathematical models and have become standard models in many areas of science and engineering, features extraction from high-order processes is usually much more complex than from first-order processes such as the Langevin equation. The state-space representation avoids many inconveniences by replacing high-order SDEs with first-order vector SDEs. Estimation performed using the state-space model solves simultaneously two problems: 1) filtering measurement noise and, in some cases, process noise and thus *filter*, and 2) solving state-space equations with respect to the process state and thus *state estimator*. For well-defined linear models, state estimation is usually organized using optimal, optimal unbiased, and unbiased estimators. However, any uncertainty in the model, interference, and/or errors in the noise description may require a norm-bounded state estimator to obtain better results. Methods of linear state estimation can be extended to nonlinear problems to obtain acceptable estimates in the case of smooth nonlinearities. Otherwise, special nonlinear estimators can be more successful in accuracy. In this chapter, we lay the foundations of state estimation and introduce the reader to the most widely used methods applied to linear and nonlinear stochastic systems and processes.

3.1 Lineal Stochastic Process in State Space

To introduce the state space approach, we consider a simple case of a single-input, single-output Kth order LTV system represented by the SDE

$$\sum_{n=0}^{K} a_n(t)\frac{\mathrm{d}^n}{\mathrm{d}t^n}z(t) = b(t)u(t) + c(t)w(t) , \tag{3.1}$$

where $a_n(t)$, $b(t)$, and $c(t)$ are known time-varying coefficients; $z(t)$ is the output; $u(t)$ is the input; and $w(t)$ is WGN. To represent (3.1) in state space, let us assign $z^{(n)}(t) = \frac{\mathrm{d}^n}{\mathrm{d}t^n}z(t)$, suppose that $a_K(t) = 1$, and rewrite (3.1) as

$$z^{(K)}(t) = -\sum_{n=0}^{K-1} a_n(t)z^{(n)}(t) + b(t)u(t) + c(t)w(t) . \tag{3.2}$$

Optimal and Robust State Estimation: Finite Impulse Response (FIR) and Kalman Approaches, First Edition.
Yuriy S. Shmaliy and Shunyi Zhao.

The nth-order derivative in (3.2) can be viewed as the nth system state and the state variables $x_1(t), \dots, x_K(t)$ assigned as

$$x_1(t) = z(t)\,,$$
$$x_2(t) = z'(t) = x_1'(t)\,,$$
$$\vdots$$
$$x_K(t) = z^{(K-1)}(t) = x_{K-1}'(t)\,,$$
$$-\sum_{i=1}^{K} a_{i-1}(t)x_i(t) + b(t)u(t) + c(t)w(t) = z^{(K)}(t) = x_K'(t)\,,$$

to be further combined into compact matrix forms

$$\begin{bmatrix} x_1' \\ \vdots \\ x_{K-1}' \\ x_K' \end{bmatrix} = \begin{bmatrix} 0 & 1 & \dots & 0 \\ \vdots & \vdots & \ddots & \vdots \\ 0 & 0 & & 1 \\ -a_0 & -a_1 & \dots & -a_{K-1} \end{bmatrix} \begin{bmatrix} x_1 \\ \vdots \\ x_{K-1} \\ x_K \end{bmatrix} + \begin{bmatrix} 0 \\ \vdots \\ 0 \\ b \end{bmatrix} u + \begin{bmatrix} 0 \\ \vdots \\ 0 \\ c \end{bmatrix} w\,,$$

$$z = \begin{bmatrix} 1 & 0 & \dots & 0 \end{bmatrix} \begin{bmatrix} x_1 & x_2 & \dots & x_K \end{bmatrix}^T$$

called *state-space equations*, where time t is omitted for simplicity. In a similar manner, a *multiple input multiple output* (MIMO) system can be represented in state space.

In applications, the system output $z(t)$ is typically measured in the presence of additive noise $v(t)$ and observed as $y(t) = z(t) + v(t)$. Accordingly, a continuous-time MIMO system can be represented in state space using the *state equation* and *observation equation*, respectively, as

$$x'(t) = A(t)x(t) + U(t)u(t) + L(t)w(t)\,, \tag{3.3}$$

$$y(t) = C(t)x(t) + D(t)u(t) + v(t)\,, \tag{3.4}$$

where $x(t) = [\,x_1 \dots x_K]^T \in \mathbb{R}^K$ is the *state vector*, $x'(t) = \frac{d}{dt}x(t)$, $y(t) = [\,y_1 \dots y_M]^T \in \mathbb{R}^M$ is the *observation vector*, $u(t) \in \mathbb{R}^L$ is the *input (control) signal vector*, $w(t) \sim \mathcal{N}(0, \mathcal{R}_w) \in \mathbb{R}^P$ is the *process noise*, and $v(t) \sim \mathcal{N}(0, \mathcal{R}_v) \in \mathbb{R}^M$ is the *observation* or *measurement noise*. Matrices $A(t) \in \mathbb{R}^{K\times K}$, $L(t) \in \mathbb{R}^{K\times P}$, $C(t) \in \mathbb{R}^{M\times K}$. $D(t) \in \mathbb{R}^{M\times L}$, and $U(t) \in \mathbb{R}^{K\times L}$ are generally time-varying. Note that the term $D(t)u(t)$ appears in (3.4) only if the order of $u(t)$ is the same as of the system, $L = K$. Otherwise, this term must be omitted. Also, if a system has no input, then both terms consisting $u(t)$ in (3.3) and (3.4) must be omitted. In what follows, we will refer to the following definition of the state space linear SDE.

> ***Linear SDE in state-space***: *The model in (3.3) and (3.4) is called linear if all its noise processes are Gaussian.*

Example 3.1 ***Second-order system driven by noise.*** A second-order LTI system driven by WGN $w(t)$ is represented by the following SDE

$$z''(t) + a_1 z'(t) + a_0 y(t) = a_0 w(t)\,,$$

where a_0 and a_1 are constant coefficients. For example, in an oscillatory system, a_0 is the squared angular fundamental frequency and a_1 is the angular bandwidth. The output $z(t)$ is measured in the presence of WGN $v(t)$ as $y(t) = z(t) + v(t)$. Referring to (3.3) and (3.4), the system can be represented in state space by the equations

$$x'(t) = Ax(t) + Bw(t)\,,$$
$$y(t) = Cx(t) + v(t)\,,$$

where matrices are specified as

$$A = \begin{bmatrix} 0 & 1 \\ -a_0 & -a_1 \end{bmatrix}, \quad B = \begin{bmatrix} 0 \\ a_0 \end{bmatrix}, \quad C = \begin{bmatrix} 1 & 0 \end{bmatrix}$$

and we notice that this system has no input. □

3.1.1 Continuous-Time Model

Solutions to (3.3) and (3.4) can be found in state space in two different ways for LTI and LTV systems.

Time-Invariant Case

When modeling LTI systems, all matrices are constant, and the solution to the state equation (3.3) can be found by multiplying both sides of (3.3) from the left-hand side with an integration factor e^{-At},

$$e^{-At}x'(t) - e^{-At}Ax(t) = e^{-At}Uu(t) + e^{-At}Lw(t) ,$$
$$[e^{-At}x(t)]' = e^{-At}Uu(t) + e^{-At}Lw(t) .$$

Further integration from t_0 to t gives

$$e^{-At}x(t) = e^{-At_0}x(t_0) + \int_{t_0}^{t} e^{-A\theta}Uu(\theta)\, d\theta + \int_{t_0}^{t} e^{-A\theta}Lw(\theta)\, d\theta \tag{3.5}$$

and the solution becomes

$$x(t) = \Phi(t, t_0)x(t_0) + \int_{t_0}^{t} \Phi(t, \theta)Uu(\theta)\, d\theta + \int_{t_0}^{t} \Phi(t, \theta)Lw(\theta)\, d\theta , \tag{3.6}$$

where the *state transition matrix* that projects the state from t_0 to t is

$$\Phi(t, t_0) = e^{A(t-t_0)} = e^{At}e^{-At_0} = \Phi(t)\Phi(-t_0) \tag{3.7}$$

and $\Phi(t) = e^{At}$ is called *matrix exponential.* Note that the stochastic integral in (3.6) can be computed either in the Itô sense or in the Stratonovich sense [193].

Substituting (3.6) into (3.4) transforms the observation equation to

$$y(t) = C\Phi(t, t_0)x(t_0) + C\int_{t_0}^{t} \Phi(t, \theta)[Uu(\theta) + Lw(\theta)]\, d\theta + v(t) , \tag{3.8}$$

where the matrix exponential $\Phi(t)$ has Maclaurin series

$$\Phi(t) = e^{At} = \sum_{i=0}^{\infty} \frac{t^i}{i!}A^i = I + At + \frac{A^2}{2!}t^2 + \ldots + \frac{A^i}{i!}t^i + \ldots , \tag{3.9}$$

in which $I = A^0$ is an identity matrix. The matrix exponential $\Phi(t)$ can also be transformed using the Cayley-Hamilton theorem [167], which states that $\Phi(t)$ can be represented with a finite series of length K as

$$\Phi(t) = e^{At} = \alpha_0 I + \alpha_1 A + \ldots + \alpha_{K-1}A^{K-1} , \tag{3.10}$$

if a constant α_i, $i \in [0, K-1]$, is specified properly.

Example 3.2 ***Second-order LTI system driven by noise.*** Consider the system discussed in Example 3.1. For $a_1 = 0$ and $a_0 = \omega_0^2$, matrix A becomes $A = \begin{bmatrix} 0 & 1 \\ -\omega_0^2 & 0 \end{bmatrix}$. By the Cayley-Hamilton theorem, the matrix exponential can be represented as $e^{At} = \alpha_1 I + \alpha_1 A$. To find α_0 and α_1, write the characteristic equation $\begin{vmatrix} -\lambda & 1 \\ -\omega_0^2 & -\lambda \end{vmatrix} = \lambda^2 + \omega_0^2 = 0$, and find the roots, $\lambda_1 = j\omega_0$ and $\lambda_2 = -j\omega_0$. The solution of the following equations

$$\alpha_0 + \alpha_1 \lambda_1 = e^{\lambda_1 t} ,$$
$$\alpha_0 + \alpha_1 \lambda_2 = e^{\lambda_2 t}$$

gives $\alpha_0 = \cos\omega_0 t$ and $\alpha_1 = \frac{1}{\omega_0}\sin\omega_0 t$, and $\Phi(t)$ can be represented by

$$\Phi(t) = e^{At} = \alpha_1 I + \alpha_1 A = \begin{bmatrix} \cos\omega_0 t & \frac{1}{\omega_0}\sin\omega_0 t \\ -\omega_0 \sin\omega_0 t & \cos\omega_0 t \end{bmatrix} .$$

The solutions (3.6) and (3.8) for this system finally become

$$x(t) = \Phi(t, t_0)x(t_0) + \int_{t_0}^{t} \Phi(t, \theta)Bw(\theta)\, d\theta , \tag{3.11}$$

$$y(t) = C\Phi(t, t_0)x(t_0) + C\int_{t_0}^{t} \Phi(t, \theta)Bw(\theta)\, d\theta + v(t) , \tag{3.12}$$

where $\Phi(t, t_0) = \Phi(t)\Phi(-t_0)$ is specified via (3.7). □

Time-Varying Case

For equations (3.3) and (3.4) of the LTV system with time-varying matrices, the proper integration factor cannot be found to obtain solutions like for (3.6) and (3.8) [167]. Instead, one can start with a homogenous *ordinary differential equation* (ODE) [28]

$$x'(t) = A(t)x(t) . \tag{3.13}$$

Matrix theory suggests that if $A(t)$ is continuous for $t \geqslant t_0$, then the solution to (3.13) for a known initial $x(t_0)$ can be written as

$$x(t) = Q(t)x(t_0) , \tag{3.14}$$

where the *fundamental matrix* $Q(t)$ that satisfies the differential equation

$$Q'(t) = A(t)Q(t) , \quad Q(t_0) = I , \tag{3.15}$$

is nonsingular for all t and is not unique.

To determine $Q(t)$, we can consider K initial state vectors $x_1(t_0), \dots, x_K(t_0)$, obtain from (3.13) K solutions $x_1(t), \dots, x_K(t)$, and specify the fundamental matrix as $Q(t) = [x_1(t)\; x_2(t) \dots x_K(t)]$. Since each particular solution satisfies (3.13), it follows that matrix $Q(t)$ defined in this way satisfies (3.14).

Provided $Q(t)$, the solution to (3.13) can be written as $x(t) = Q(t)s(t)$, where $Q(t)$ satisfies (3.15), and $s(t)$ is some unknown function of the same class. Referring to (3.15), equation (3.3) can then be transformed to

$$\begin{aligned} Q'(t)s(t) + Q(t)s'(t) &= A(t)Q(t)s(t) + U(t)u(t) + L(t)w(t) , \\ Q(t)s'(t) &= U(t)u(t) + L(t)w(t) \end{aligned} \tag{3.16}$$

and, from (3.16), we have

$$s'(t) = Q^{-1}(t)U(t)u(t) + Q^{-1}(t)L(t)w(t) ,$$

which, by known $s(t_0) = Q^{-1}(t_0)x(t_0)$, specifies function $s(t)$ as

$$s(t) = s(t_0) + \int_{t_0}^{t} Q^{-1}(\theta)U(\theta)u(\theta)\, d\theta + \int_{t_0}^{t} Q^{-1}(\theta)L(\theta)w(\theta)\, d\theta \,. \tag{3.17}$$

The solution to (3.3), by $x(t) = Q(t)s(t)$ and (3.17), finally becomes

$$x(t) = \Phi(t, t_0)x(t_0) + \int_{t_0}^{t} \Phi(t, \theta)[U(\theta)u(\theta) + L(\theta)w(\theta)]\, d\theta \,, \tag{3.18}$$

where the state transition matrix is given by

$$\Phi(t, t_0) = Q(t)Q^{-1}(t_0) \,. \tag{3.19}$$

Example 3.3 ***State transition matrix for LTV system [167].*** Consider the homogenous ODE $x'(t) = A(t)x(t)$ with $A(t) = \begin{bmatrix} 0 & t \\ 0 & 0 \end{bmatrix}$ and known initial conditions $x_1(0)$ and $x_2(0)$. Rewrite the ODE for two states as

$$x_1'(t) = tx_2(t) \quad \text{and} \quad x_2'(t) = 0 \,.$$

The solution of the second equation is $x_2(t) = x_2(0)$, and for the first equation we have

$$x_1(t) = \int_0^t \tau x_2(\tau) d\tau + x_1(0) = x_2(0) \int_0^t \theta\, d\theta + x_1(0) = \frac{t^2}{2} x_2(0) + x_1(0) \,.$$

We can now look at two different initial conditions. For example, for $x_1(0) = 0$ and $x_2(0) = 1$ define

$$x(0) = \begin{bmatrix} 0 \\ 1 \end{bmatrix} \quad \text{and} \quad x(t) = \begin{bmatrix} \frac{t^2}{2} \\ 1 \end{bmatrix}$$

and for $x_1(0) = 3$ and $x_2(0) = 1$ obtain

$$x(0) = \begin{bmatrix} 3 \\ 1 \end{bmatrix} \quad \text{and} \quad x(t) = \begin{bmatrix} \frac{t^2}{2} + 3 \\ 1 \end{bmatrix} .$$

Therefore, the fundamental matrix becomes $Q(t) = \begin{bmatrix} \frac{t^2}{2} & \frac{t^2}{2} + 3 \\ 1 & 1 \end{bmatrix}$ and the state transition matrix (3.19) is obtained as

$$\Phi(t, t_0) = Q(t)Q^{-1}(t_0) = \begin{bmatrix} \frac{t^2}{2} & \frac{t^2}{2} + 3 \\ 1 & 1 \end{bmatrix} \begin{bmatrix} -\frac{1}{3} & \frac{t_0^2}{6} + 1 \\ \frac{1}{3} & -\frac{t_0^2}{6} \end{bmatrix} = \begin{bmatrix} 1 & \frac{t^2 - t_0^2}{2} \\ 0 & 1 \end{bmatrix}$$

that formalizes the solution (3.18). □

Referring to (3.18), the observation equation can finally be rewritten as

$$y(t) = C(t)\Phi(t, t_0)x(t_0) + C(t) \int_{t_0}^{t} \Phi(t, \theta)[U(\theta)u(\theta) + L(\theta)w(\theta)]\, d\theta + v(t) \,. \tag{3.20}$$

It should be noted that the solutions (3.6) and (3.19) are formally equivalent. The main difference lies in the definitions of the state transition matrix: (3.7) for LTI systems and (3.19) for LTV systems.

3.1.2 Discrete-Time Model

The necessity of representing continuous-time state-space models in discrete time arises when numerical analysis is required or solutions are supposed to be obtained using digital blocks. In such cases, a solution to an SDE is considered at two discrete points $t_k = t$ and $t_{k-1} = t_0$ with a time step $\tau = t_k - t_{k-1}$, where k is a discrete time index. It is also supposed that a digital x_k represents accurately a discrete-time $x(t_k)$. Further, the FE or BE methods are used [26].

The FE method, also called the *standard Euler method* or Euler-Maruyama method, relates the numerically computed stochastic integrals to t_{k-1} and is associated with Itô calculus. The relevant discrete-time state-space model $x_k = Fx_{k-1} + w_{k-1}$, sometimes called a *prediction model*, is basic in control. A contradiction here is with the noise term w_{k-1} that exists even though the initial state x_{k-1} is supposed to be known and thus already affected by w_{k-1}.

The BE method, also known as an *implicit method*, relates all integrals to the current time t_k that yields $x_k = Fx_{k-1} + w_k$. This state equation is sometimes called a *real-time model* and is free of the contradiction inherent to the FE-based state equation.

The FE and BE methods are uniquely used to convert continuous-time differential equations to discrete time, and in the sequel we will use both of them. It is also worth noting that predictive state estimators fit better with the FE method and filters with the BE method.

Time-Invariant Case

Consider the solution (3.6) and the observation (3.4), substitute t with t_k, t_0 with t_{k-1}, and write

$$x(t_k) = e^{A\tau}x(t_{k-1}) + \int_{t_{k-1}}^{t_k} e^{A(t_k-\theta)}Uu(\theta)d\theta$$
$$+ \int_{t_{k-1}}^{t_k} e^{A(t_k-\theta)}Lw(\theta)d\theta \, ,$$
$$y(t_k) = Cx(t_k) + Du(t_k) + v(t_k) \, .$$

Because input $u(t)$ is supposed to be known, assume that it is also piecewise constant on each time step and put $u_k \cong u(t_k)$ outside the first integral. Next, substitute $x_k \cong x(t_k)$, $y_k \cong y(t_k)$, and $v_k \cong v(t_k)$, and write

$$x_k = Fx_{k-1} + Eu_k + w_k \, , \tag{3.21}$$

$$y_k = Hx_k + Du_k + v_k \, , \tag{3.22}$$

where $H = C$, $F = e^{A\tau}$, and

$$E = \int_{t_{k-1}}^{t_k} e^{A(t_k-\theta)}U \, d\theta = \int_0^{\tau} e^{A\theta}U \, d\theta \, , \tag{3.23}$$

$$w_k = \int_{t_{k-1}}^{t_k} e^{A(t_k-\theta)}Lw(\theta) \, d\theta \, . \tag{3.24}$$

The discrete process noise $w_k \sim \mathcal{N}(0, Q)$ given by (3.24) has zero mean, $\mathcal{E}\{w_k\} = 0$, and the covariance $Q = \mathcal{E}\{w_k w_k^T\}$ specified using the continuous WGN covariance (2.91) by

$$Q = \int_{t_{k-1}}^{t_k}\int_{t_{k-1}}^{t_k} e^{A(t_k-\theta_1)}L\mathcal{E}\{w(\theta_1)w^T(\theta_2)\}L^T e^{A(t_k-\theta_2)^T} \, d\theta_1 d\theta_2$$
$$= \int_{t_{k-1}}^{t_k}\int_{t_{k-1}}^{t_k} e^{A(t_k-\theta_1)}LS_w\delta(\theta_1-\theta_2)L^T e^{A(t_k-\theta_2)^T} \, d\theta_1 d\theta_2$$

$$
\begin{aligned}
&= \int_{t_{k-1}}^{t_k} e^{A(t_k-\theta)} L S_w L^T e^{A(t_k-\theta)^T} \, d\theta \\
&= \tau L S_w L^T + \gamma(\tau^i)|_{\gamma=0,L=I} \\
&= \tau S_w \, . \qquad (3.25)
\end{aligned}
$$

Observe that $\gamma(\tau^i)$ is a matrix of the same class as Q, which is a function of τ^i, $i > 1$. If τ is small enough, then this term can be set to zero. If also $L = I$ in (3.3), then $Q = \tau S_w$ establishes a rule of thumb relationship between the discrete and continuous process noise covariances.

The zero mean discrete observation noise $v_k = \frac{1}{\tau}\int_{t_{k-1}}^{t_k} v(t)\, dt \sim \mathcal{N}(0, R)$ (2.93) has the covariance $R = \mathcal{E}\{v_k v_k^T\}$ defined as

$$
\begin{aligned}
R &= \frac{1}{\tau^2}\int_{t_{k-1}}^{t_k}\int_{t_{k-1}}^{t_k} E\{v(t)v^T(t)\}\, dt dt \\
&= \frac{1}{2\tau}\begin{bmatrix} N_{11} & \dots & N_{1K} \\ \vdots & \ddots & \vdots \\ N_{K1} & \dots & N_{KK} \end{bmatrix} = \frac{1}{\tau} S_v , \qquad (3.26)
\end{aligned}
$$

where $N_{ij}/2$, $\{i,j\} \in [1,K]$, is a component of the PSD matrix S_v.

Example 3.4 ***Discrete LTI state-space model.*** Consider a continuous LTI system represented in state-space with

$$
\begin{aligned}
x'(t) &= Ax(t) + w(t) \, , \\
y(t) &= Cx(t) + v(t) \, ,
\end{aligned}
$$

where $A = \begin{bmatrix} 0 & 1 \\ 0 & 0 \end{bmatrix}$ and $C = [\, 1 \;\; 0 \,]$. A component $w_2(t)$ of WGN $w(t) = \begin{bmatrix} 0 \\ w_2(t) \end{bmatrix}$ has zero mean and the auto-covariance $C_2(\tau) = \frac{N_2}{2}\delta(\tau)$. The covariance function of $w(t)$ is thus given by $C_w(\tau) = \frac{N_2}{2}\begin{bmatrix} 0 & 0 \\ 0 & \delta(\tau) \end{bmatrix}$. The observation WGN $v(t)$ has zero mean and the covariance $C_v(\tau) = \frac{N_v}{2}\delta(\tau)$.

Following (3.21)–(3.25), the discrete state-space model becomes

$$
\begin{aligned}
x_k &= F x_{k-1} + w_k \, , \\
y_k &= H x_k + v_k \, ,
\end{aligned}
$$

where, by (3.9) with small τ, we have $F = \begin{bmatrix} 1 & \tau \\ 0 & 1 \end{bmatrix}$. The variance of noise v_k is given by $\sigma_v^2 = \frac{N_v}{2\tau}$ (2.94) and the covariance Q of w_k (3.25) is provided as follows. Consider $e^{A(t_k-\theta)} \cong I + A(t_k - \theta) = \begin{bmatrix} 1 & t_k - \theta \\ 0 & 1 \end{bmatrix}$, represent w_k as

$$
\begin{aligned}
w_k &= \int_{t_{k-1}}^{t_k} \Phi(t_k, \theta) w(\theta)\, d\theta = \int_{t_{k-1}}^{t_k} e^{A(t_k-\theta)} w(\theta)\, d\theta \\
&= \int_{t_{k-1}}^{t_k} \begin{bmatrix} 1 & t_k - \theta \\ 0 & 1 \end{bmatrix} w(\theta)\, d\theta = \int_{t_{k-1}}^{t_k} \begin{bmatrix} t_k - \theta \\ 1 \end{bmatrix} w_2(\theta)\, d\theta \, ,
\end{aligned}
$$

and transform the covariance $Q = \mathcal{E}\{w_k w_k^T\}$ as

$$
Q = \int_{t_{k-1}}^{t_k}\int_{t_{k-1}}^{t_k} \begin{bmatrix} t_k - \theta_1 \\ 1 \end{bmatrix} \mathcal{E}\{w_2(\theta_1) w_2^T(\theta_2)\} \begin{bmatrix} t_k - \theta_2 & 1 \end{bmatrix} \, d\theta_1 d\theta_2
$$

$$
\begin{aligned}
&= \frac{N_2}{2}\int_{t_{k-1}}^{t_k}\int_{t_{k-1}-\theta}^{t_k-\theta}\delta(\nu)\begin{bmatrix}(t_k-\theta)(t_k-\theta-\nu) & t_k-\theta\\ t_k-\theta-\nu & 1\end{bmatrix}\,\mathrm{d}\nu\,\mathrm{d}\theta\\
&= \frac{N_2}{2}\int_{t_{k-1}}^{t_k}\begin{bmatrix}(t_k-\theta)^2 & t_k-\theta\\ t_k-\theta & 1\end{bmatrix}\,\mathrm{d}\theta\\
&= \frac{N_2\tau}{2}\begin{bmatrix}\frac{\tau^2}{3} & \frac{\tau}{2}\\ \frac{\tau}{2} & 1\end{bmatrix},
\end{aligned}
$$

where we substituted $\theta = \theta_1$, $\theta_2 = \theta + \nu$, and $\tau = t_k - t_{k-1}$. Similarly, any other LTI process can be converted to discrete time. □

Time-Varying Case

For LTV systems, solutions in state space are given by (3.18) and (3.20). Substituting t with t_k and t_0 with t_{k-1} yields

$$
\begin{aligned}
x(t_k) &= \mathcal{Q}(t_k)\mathcal{Q}^{-1}(t_{k-1})x(t_{k-1}) + \mathcal{Q}(t_k)\int_{t_{k-1}}^{t_k}\mathcal{Q}^{-1}(\theta)U(\theta)u(\theta)\,\mathrm{d}\theta\\
&\quad + \mathcal{Q}(t_k)\int_{t_{k-1}}^{t_k}\mathcal{Q}^{-1}(\theta)L(\theta)w(\theta)\,\mathrm{d}\theta\,,\\
y(t_k) &= C(t_k)x(t_k) + D(t_k)u(t_k) + v(t_k)\,,
\end{aligned}
$$

where the fundamental matrix $\mathcal{Q}(t)$ must be assigned individually for each model. Reasoning along similar lines as for LTI systems, these equations can be represented in discrete-time as

$$x_k = F_k x_{k-1} + E_k u_k + w_k\,, \tag{3.27}$$

$$y_k = H_k x_k + D_k u_k + v_k\,, \tag{3.28}$$

where $H_k = C(t_k)$ and the time-varying matrices are given by

$$F_k = \mathcal{Q}(t_k)\mathcal{Q}^{-1}(t_{k-1})\,, \tag{3.29}$$

$$E_k = \mathcal{Q}(t_k)\int_{t_{k-1}}^{t_k}\mathcal{Q}^{-1}(\theta)U(\theta)\,\mathrm{d}\theta\,, \tag{3.30}$$

$$w_k = \mathcal{Q}(t_k)\int_{t_{k-1}}^{t_k}\mathcal{Q}^{-1}(\theta)L(\theta)w(\theta)\,\mathrm{d}\theta\,, \tag{3.31}$$

and noise $w_k \sim \mathcal{N}(0, Q_k)$ has zero mean, $\mathcal{E}\{w_k\} = 0$, and generally a time-varying covariance $Q_k = \mathcal{E}\{w_k w_k^T\}$. Noise $v_k \sim \mathcal{N}(0, R_k)$ also has zero mean, $\mathcal{E}\{v_k\} = 0$, and generally a time-varying covariance $R_k = \mathcal{E}\{v_k v_k^T\}$.

Example 3.5 ***Discrete-time LTV state-space model.*** Consider an LTV system (Example 3.3) represented in state space with

$$
\begin{aligned}
x'(t) &= A(t)x(t) + w(t)\,,\\
y(t) &= Cx(t) + v(t)\,,
\end{aligned}
$$

where $A(t) = \begin{bmatrix}0 & t\\ 0 & 0\end{bmatrix}$, $C = [\,1\;\;0\,]$, a component $w_2(t)$ of WGN $w(t) = \begin{bmatrix}0\\ w_2(t)\end{bmatrix}$ has zero mean and the auto-covariance $C_2(\tau) = \frac{N_2}{2}\delta(\tau)$, and the covariance function of $w(t)$ is $C_w(\tau) = \frac{N_2}{2}\begin{bmatrix}0 & 0\\ 0 & \delta(\tau)\end{bmatrix}$. The observation WGN $v(t)$ has zero mean and the covariance function $C_v(\tau) = \frac{N_v}{2}\delta(\tau)$.

The fundamental matrix was found in Example 3.3 to be $\mathcal{Q}(t) = \begin{bmatrix} \frac{t^2}{2} & \frac{t^2}{2}+3 \\ 1 & 1 \end{bmatrix}$ and the state transition matrix as $\Phi(t,t_0) = \mathcal{Q}(t)\mathcal{Q}^{-1}(t_0) = \begin{bmatrix} 1 & \frac{t^2-t_0^2}{2} \\ 0 & 1 \end{bmatrix}$. Accordingly, a discrete form of this model becomes

$$
\begin{aligned}
x_k &= F_k x_{k-1} + w_k \,, \\
y_k &= H x_k + v_k \,,
\end{aligned}
$$

where $H = C$, by (3.29) and $t_k = \tau k$, we have $F_k = \begin{bmatrix} 1 & \frac{\tau^2}{2}(2k-1) \\ 0 & 1 \end{bmatrix}$ and $\sigma_v^2 = \frac{N_v}{2\tau}$. Noise w_k is represented with

$$
w_k = \int_{t_{k-1}}^{t_k} \begin{bmatrix} t_k^2 - \theta^2 \\ 1 \end{bmatrix} w_2(\theta)\, \mathrm{d}\theta
$$

and its covariance $Q_k = \mathcal{E}\{w_k w_k^T\}$ can be defined as

$$
\begin{aligned}
Q_k &= \int_{t_{k-1}}^{t_k}\int_{t_{k-1}}^{t_k} \begin{bmatrix} t_k^2 - \theta_1^2 \\ 1 \end{bmatrix} \mathcal{E}\{w_2(\theta_1)w_2^T(\theta_2)\}[\ t_k^2-\theta_2^2 \ \ 1\ ]\, \mathrm{d}\theta_1 \mathrm{d}\theta_2 \\
&= \frac{N_2}{2}\int_{t_{k-1}}^{t_k}\int_{t_{k-1}-\theta}^{t_k-\theta} \delta(\nu) \begin{bmatrix} (t_k^2-\theta^2)(t_k^2-(\theta+\nu)^2) & t_k^2-\theta^2 \\ t_k^2-(\theta+\nu)^2 & 1 \end{bmatrix} \mathrm{d}\nu\, \mathrm{d}\theta \\
&= \frac{N_2}{2}\int_{t_{k-1}}^{t_k} \begin{bmatrix} (t_k^2-\theta^2)^2 & t_k^2-\theta^2 \\ t_k^2-\theta^2 & 1 \end{bmatrix} \mathrm{d}\theta \\
&= \frac{N_2\tau}{2} \begin{bmatrix} (k^4 - \frac{2}{3}k^2 + \frac{1}{5})\tau^4 & (k^2-\frac{1}{3})\tau^2 \\ (k^2-\frac{1}{3})\tau^2 & 1 \end{bmatrix} .
\end{aligned}
$$

In this example, the time-varying matrix $A(t)$ has became time-varying F_k in discrete time and the time-invariant system noise covariance $C_w(\tau)$ converted to the time-varying Q_k. □

3.2 Methods of Linear State Estimation

Provided adequate modeling of a real linear physical process in discrete-time state space, state estimation can be carried out using methods of optimal linear filtering based on the following state-space equations,

$$
x_k = F_k x_{k-1} + E_k u_k + B_k w_k \,, \tag{3.32}
$$

$$
y_k = H_k x_k + D_k u_k + v_k \,, \tag{3.33}
$$

where $x_k \in \mathbb{R}^K$ is the state vector, $u_k \in \mathbb{R}^L$ is the input (or control) signal vector, $y_k \in \mathbb{R}^P$ is the observation (or measurement) vector, $F_k \in \mathbb{R}^{K\times K}$ is the system (process) matrix, $H_k \in \mathbb{R}^{P\times K}$ is the observation matrix, $E_k \in \mathbb{R}^{K\times L}$ is the control (or input) matrix, $D_k \in \mathbb{R}^{P\times L}$ is the disturbance matrix, and $B_k \in \mathbb{R}^{K\times M}$ is the system (or process) noise matrix. The term with u_k is usually omitted in (3.33), assuming that the order of u_k is less than the order of y_k. At the other extreme, when the input disturbance appears to be severe, the noise components are often omitted in (3.32) and (3.33).

In many cases, system or process noise is assumed to be zero mean and white Gaussian $w_k \sim \mathcal{N}(0, Q_k) \in \mathbb{R}^M$ with known covariance $Q_k = \mathcal{E}\{w_k w_n^T\}\delta(k-n)$. Measurement or observation noise

$v_k \sim \mathcal{N}(0, R_k) \in \mathbb{R}^p$ is also often modeled as zero mean and white Gaussian with covariance $R_k = \mathcal{E}\{v_k v_n^T\}\delta(k-n)$. Moreover, many problems suggest that w_k and v_k can be viewed as physically uncorrelated and independent processes. However, there are other cases when w_k and v_k are correlated with each other and exhibit different kinds of coloredness.

Filters of two classes can be designed to provide state estimation based on (3.32) and (3.33). Estimators of the first class ignore the process dynamics described by the state equation (3.32). Instead, they require multiple measurements of the quantity in question to approach the true value through statistical averaging. Such estimators have batch forms, but recursive forms may also be available. An example is the LS estimator. Estimators of the second class require both state-space equations, so they are more flexible and universal. The recursive KF algorithm belongs to this class of estimators, but its batch form is highly inefficient in view of growing memory due to IIR. The batch FIR filter and LMF, both operating on finite horizons, are more efficient in this sense, but they cause computational problems when the batch is very large.

Regardless of the estimator structure, the notation $\hat{x}_{k|r}$ means an estimate of the state x_k at time index k, given observations of x_k up to and including at time index r. The state x_k to be estimated at the time index k is represented by the following standard variables:

- $\hat{x}_k^- \triangleq \hat{x}_{k|k-1}$ is the *a priori* (or prior) *state estimate* at k given observations up to and including at time index $k-1$;
- $\hat{x}_k \triangleq \hat{x}_{k|k}$ is the *a posteriori* (or posterior) *state estimate* at k given observations up to and including at k;
- The *a priori* estimation error is defined by

$$\varepsilon_k^- \triangleq \varepsilon_{k|k-1} = x_k - \hat{x}_k^- . \tag{3.34}$$

- The *a posteriori* estimation error is defined by

$$\varepsilon_k \triangleq \varepsilon_{k|k} = x_k - \hat{x}_k . \tag{3.35}$$

- The *a priori* error covariance is defined by

$$P_k^- \triangleq P_{k|k-1} = \mathcal{E}\{(x_k - \hat{x}_k^-)(x_k - \hat{x}_k^-)^T\} . \tag{3.36}$$

- The *a posteriori* error covariance is defined by

$$P_k \triangleq P_{k|k} = \mathcal{E}\{(x_k - \hat{x}_k)(x_k - \hat{x}_k)^T\} . \tag{3.37}$$

In what follows, we will refer to these definitions in the derivation and study of various types of state estimators.

3.2.1 Bayesian Estimator

Suppose there is some discrete stochastic process X_k, $k = 0, 1, \ldots$ measured as y_k and assigned a set of measurements $y_0^k = \{y_0, y_1, \ldots, y_k\}$. We may also assume that the *a posteriori* density $p(x_{k-1}|y_0^{k-1})$ is known at the past time index $k-1$. Of course, our interest is to obtain the *a posteriori* density $p(x_k|y_0^k)$ at k. Provided $p(x_k|y_0^k)$, the estimate $\hat{x}_k$ can be found as the first-order initial moment (2.10) and the estimation error as the second-order central moment (2.14). To this end, we use Bayes' rule (2.43) and obtain $p(x_k|y_0^k)$ as follows.

Consider the join density $p(x_k, y_k|y_0^{k-1})$ and represent it as

$$\begin{aligned} p(x_k, y_k|y_0^{k-1}) &= p(x_k|y_0^{k-1})p(y_k|x_k, y_0^{k-1}) \\ &= p(y_k|y_0^{k-1})p(x_k|y_k, y_0^{k-1}) . \end{aligned} \tag{3.38}$$

Note that y_k is independent of past measurements and write $p(y_k|x_k, y_0^{k-1}) = p(y_k|x_k)$. Also, the desirable *a posteriori* pdf can be written as $p(x_k|y_k, y_0^{k-1}) = p(x_k|y_0^k)$. Then use Bayesian inference (2.47) and obtain

$$p(x_k|y_0^k) = \frac{p(x_k|y_0^{k-1})p(y_k|x_k)}{p(y_k|y_0^{k-1})} = \alpha p(x_k|y_0^{k-1})p(y_k|x_k) , \tag{3.39}$$

where α is a normalizing constant, the conditional density $p(x_k|y_0^{k-1})$ represents the prior distribution of x_k, and $p(y_k|x_k)$ is the *likelihood function* of y_k given the outcome x_k. Note that the conditional pdf $p(x_k|y_0^{k-1})$ can be expressed via known $p(x_{k-1}|y_0^{k-1})$ using the rule (2.53) as

$$p(x_k|y_0^{k-1}) = \int_{-\infty}^{\infty} p(x_k|x_{k-1})p(x_{k-1}|y_0^{k-1})\, \mathrm{d}x_{k-1} . \tag{3.40}$$

Provided $p(x_k|y_0^k)$, the Bayesian estimate is obtained by (2.10) as

$$\hat{x}_k = \int_{-\infty}^{\infty} x_k p(x_k|y_0^k)\, \mathrm{d}x_k \tag{3.41}$$

and the estimation error is obtained by (2.14) as

$$P_k = \int_{-\infty}^{\infty} (x_k - \hat{x}_k)^2 p(x_k|y_0^k)\, \mathrm{d}x_k . \tag{3.42}$$

From the previous, it follows that the Bayesian estimate (3.41) is universal regardless of system models and noise distributions.

> ***Bayesian estimator****: Bayesian estimation can be universally applied to linear and nonlinear models with Gaussian and non-Gaussian noise.*

It is worth noting that for linear models the Bayesian approach leads to the optimal KF algorithm, which will be shown next.

Linear Model

Let us look at an important special case and show how the Bayesian approach can be applied to a linear Gaussian model (3.32) and (3.33) with $D_k = 0$. For clarity, consider a first-order stochastic process. The goal is to obtain the *a posteriori* density $p(x_k|y_0^k)$ in terms of the known *a posteriori* density $p(x_{k-1}|y_0^{k-1})$ and arrive at a computational algorithm.

To obtain the desired $p(x_k|y_0^k)$ using (3.39), we need to specify $p(x_k|y_0^{k-1})$ using (3.40) and define the likelihood function $p(y_k|x_k)$. Noting that all distributions are Gaussian, we can represent $p(x_{k-1}|y_0^{k-1})$ as

$$p(x_{k-1}|y_0^{k-1}) = \alpha_1 \exp\left[-\frac{(x_{k-1} - \hat{x}_{k-1})^2}{2P_{k-1}}\right] , \tag{3.43}$$

where $\hat{x}_{k-1}$ is the known *a posteriori* estimate of x_{k-1} and P_{k-1} is the variance of the estimation error. Hereinafter, we will use α_i as the normalizing coefficient.

Referring to (3.32), the conditional density $p(x_k|x_{k-1})$ can be written as

$$p(x_k|x_{k-1}) = \alpha_2 \exp\left[-\frac{(x_k - F_k x_{k-1} - E_k u_k)^2}{2\sigma_w^2}\right] , \tag{3.44}$$

where σ_w^2 is the variance of the process noise w_k.

Now, the *a priori* density $p(x_k|y_0^{k-1})$ can be transformed using (3.40) as

$$p(x_k|y_0^{k-1}) = \int_{-\infty}^{\infty} p(x_k|x_{k-1})p(x_{k-1}|y_0^{k-1})\, dx_{k-1} . \tag{3.45}$$

By substituting (3.43) and (3.44) and equating to unity the integral of the normally distributed x_{k-1}, (3.45) can be transformed to

$$p(x_k|y_0^{k-1}) = \alpha_3 \exp\left[-\frac{(x_k - F_k\hat{x}_{k-1} - E_k u_k)^2}{2(\sigma_w^2 + P_{k-1}F_k^2)}\right] . \tag{3.46}$$

Likewise, the likelihood function $p(y_k|x_k)$ can be written using (3.33) as

$$p(y_k|x_k) = \alpha_4 \exp\left[-\frac{(y_k - H_k x_k)^2}{2\sigma_v^2}\right] , \tag{3.47}$$

where σ_v^2 is the variance of the observation noise v_k. Then the required density $p(x_k|y_0^k)$ can be represented by (3.39) using (3.46) and (3.47) as

$$\begin{aligned} p(x_k|y_0^k) &= \alpha_5 \exp\left[-\frac{(x_k - F_k\hat{x}_{k-1} - E_k u_k)^2}{2(\sigma_w^2 + P_{k-1}F_k^2)} - \frac{(y_k - H_k x_k)^2}{2\sigma_v^2}\right] \\ &= \alpha_6 \exp\left[-\frac{(x_k - \hat{x}_k)^2}{2P_k}\right] , \end{aligned} \tag{3.48}$$

which holds if the following obvious relationship is satisfied

$$\frac{(x_k - F_k\hat{x}_{k-1} - E_k u_k)^2}{\sigma_w^2 + P_{k-1}F_k^2} + \frac{(y_k - H_k x_k)^2}{\sigma_v^2} = \frac{(x_k - \hat{x}_k)^2}{P_k} . \tag{3.49}$$

By equating to zero the free term and terms with x_k^2 and x_k, we arrive at several conditions

$$\frac{1}{P_k} = \frac{1}{\sigma_w^2 + P_{k-1}F_k^2} + \frac{H_k^2}{\sigma_v^2} , \tag{3.50}$$

$$K_k = \frac{P_k H_k}{\sigma_v^2} , \tag{3.51}$$

$$\hat{x}_k^- = F_k\hat{x}_{k-1} + E_k u_k , \tag{3.52}$$

$$\hat{x}_k = \hat{x}_k^- + K_k(y_k - H_k\hat{x}_k^-) , \tag{3.53}$$

which establish a recursive computational algorithm to compute the estimate $\hat{x}_k$ (3.53) and the estimation error P_k (3.50) starting with the initial $\hat{x}_0$ and P_0 for known σ_w^2 and σ_v^2. This is a one-state KF algorithm. What comes from its derivation is that the estimate (3.53) is obtained without using (3.41) and (3.42). Therefore, it is not a Bayesian estimator. Later, we will derive the K-state KF algorithm and discuss it in detail.

3.2.2 Maximum Likelihood Estimator

From (3.39) it follows that the *a posteriori* density $p(x_k|y_0^k)$ is given in terms of the *a priori* density $p(x_k|y_0^{k-1})$ and the likelihood function $p(y_k|x_k)$. It is also worth knowing that the likelihood function plays a special role in the estimation theory, and its use leads to the ML state estimator. For the first-order Gaussian stochastic process, the likelihood function is given by the expression (3.47). This function is maximized by minimizing the exponent power, and we see that the only solution following from $y_k - H_k x_k = 0$ is inefficient due to noise.

An efficient ML estimate can be obtained if we combine n measurement samples of the fully observed vector $x \triangleq x_k \in \mathbb{R}^K$ into an extended observation vector $Y \triangleq Y_k = [\, y_{1k}^T \cdots y_{nk}^T]^T \in \mathbb{R}^{nK}$

with a measurement noise $V \triangleq V_k = [v_{1k}^T \dots v_{nk}^T]^T \in \mathbb{R}^{nK}$ and write the observation equation as

$$Y = Mx + V \,, \tag{3.54}$$

where $M \triangleq M_k = [H_1^T \dots H_n^T]^T \in \mathbb{R}^{nK \times K}$ is the extended observation matrix. A relevant example can be found in sensor networks, where the desired quantity is simultaneously measured by n sensors. Using (3.54), the ML estimate of x can be found by maximizing the likelihood function $p(Y|x)$ as

$$\hat{x} = \arg\max_x \; p(Y|x) \,. \tag{3.55}$$

For Gaussian processes, the likelihood function becomes

$$p(Y|x) = \frac{1}{\sqrt{(2\pi)^n |R|}} \exp\left[-\frac{1}{2}(Y - Mx)^T R^{-1} (Y - Mx)\right] , \tag{3.56}$$

where $R \triangleq R_k \in \mathbb{R}^{nK \times nK}$ is the extended covariance matrix of the measurement noise V. Since maximizing (3.56) with (3.55) means minimizing the power of the exponent, the ML estimate can equivalently be obtained from

$$\hat{x} = \arg\min_x \; (Y - Mx)^T R^{-1} (Y - Mx) \,. \tag{3.57}$$

By equating the derivative with respect to x to zero,

$$\frac{\partial}{\partial x} (Y - Mx)^T R^{-1} (Y - Mx) = 0 \,,$$

using the identities (A.4) and (A.5), and assuming that $M^T R^{-1} M$ is nonsingular, the ML estimate can be found as

$$\hat{x} = (M^T R^{-1} M)^{-1} M^T R^{-1} Y \,, \tag{3.58}$$

which demonstrates several important properties when the number n of measurement samples grows without bound [125]:

- It converges to the true value of x.
- It is asymptotically optimal unbiased, normally distributed, and efficient.

> ***Maximum likelihood estimate*** *[125]: There is no other unbiased estimate whose covariance is smaller than the finite covariance of the ML estimate as a number of measurement samples grows without bounds.*

It follows from the previous that as n increases, the ML estimate (3.58) converges to the Bayesian estimate (3.41). Thus, the ML estimate has the property of unbiased optimality when n grows unboundedly, and is inefficient when n is small.

3.2.3 Least Squares Estimator

Another approach to estimate x from (3.54) is known as Gauss' *least squares* (LS) [54, 185] or LS estimator,[1] which is widely used in practice. In the LS method, the estimate $\hat{x}_k$ is chosen such that the sum of the squares of the residuals $Y - Mx$ minimizes the cost function $J = (Y - Mx)^T (Y - Mx)$ over all n to obtain

$$\hat{x} = \arg\min_x \; (Y - Mx)^T (Y - Mx) \,. \tag{3.59}$$

---

1 Carl Friedrich Gauss published his method of least squares in 1809 and claimed to have been in its possession since 1795. He showed that the arithmetic mean is the best estimate of the location parameter if the probability density is the (invented by him) normal distribution, now also known as Gaussian distribution.

Reasoning similarly to (3.57), the estimate can be found to be

$$\hat{x} = (M^T M)^{-1} M^T Y \; . \tag{3.60}$$

To improve (3.60), we can consider the weighted cost function $J = (Y - Mx)^T \bar{R}^{-1} (Y - Mx)$, in which $\bar{R}$ is a symmetric positive definite weight matrix. Minimizing of this cost yields the weighted LS (WLS) estimate

$$\hat{x} = (M^T \bar{R}^{-1} M)^{-1} M^T \bar{R}^{-1} Y \tag{3.61}$$

and we infer that the ML estimate (3.58) is a special case of WLS estimate (3.61), in which $\bar{R}$ is chosen to be the measurement noise covariance R.

3.2.4 Unbiased Estimator

If someone wants an unbiased estimator, whose average estimate is equal to the constant x, then the only performance criterion will be

$$\mathcal{E}\{\hat{x}\} = x \; . \tag{3.62}$$

To obtain an unbiased estimate, one can think of an estimate $\hat{x}$ as the product of some gain $\mathcal{H}$ and the measurement vector Y given by (3.54),

$$\hat{x} = \mathcal{H} Y = \mathcal{H}(Mx + V) \; .$$

Since V has zero mean, the unbiasedness condition (3.62) gives the unbiasedness constraint

$$\mathcal{H} M = I \; .$$

By multiplying the constraint by the identity $(M^T M)^{-1} M^T M$ from the right-hand side and discarding the nonzero M on both sides, we obtain $\mathcal{H} = (M^T M)^{-1} M^T$ and arrive at the unbiased estimate

$$\hat{x} = (M^T M)^{-1} M^T Y \; , \tag{3.63}$$

which is identical to the LS estimate (3.60). Since the product $\mathcal{H}Y$ is associated with discrete convolution, the unbiased estimator (3.63) can be thought of as an unbiased FIR filter.

Recursive Unbiased (LS) Estimator

We have already mentioned that batch estimators (3.63) and (3.60) can be computationally inefficient and cause a delay when the number of data points n is large. To compute the batch estimate recursively, one can start with $Y_i \triangleq Y_{ik}$, $i \in [1, n]$, associated with matrix $M_i \triangleq M_{ik}$, and rewrite (3.63) as

$$\begin{aligned} \hat{x}_i &= (M_i^T M_i)^{-1} M_i^T Y_i \\ &= G_i M_i^T Y_i \; , \end{aligned} \tag{3.64}$$

where $M_i = [H_1^T \dots H_i^T]^T$. Further, matrix $G_i = (M_i^T M_i)^{-1}$ can be represented using the forward and backward recursions as [179]

$$G_i = (H_i^T H_i + G_{i-1}^{-1})^{-1} \; , \tag{3.65}$$

$$G_{i-1} = (G_i^{-1} - H_i^T H_i)^{-1} \; , \tag{3.66}$$

and $M_i^T Y_i$ rewritten similarly as

$$M_i^T Y_i = H_i^T y_i + M_{i-1}^T Y_{i-1} .$$

Combining this recursive form with (3.64), we obtain

$$\hat{x}_i = G_i(H_i^T y_i + M_{i-1}^T Y_{i-1}) .$$

Next, substituting $M_{i-1}^T Y_{i-1} = G_{i-1}^{-1}\hat{x}_{k-1}$ taken from (3.64), combining with (3.66), and providing some transformations gives the recursive form

$$\hat{x}_i = \hat{x}_{i-1} + G_i H_i^T (y_i - H_i \hat{x}_{i-1}) , \tag{3.67}$$

in which G_i can be computed recursively by (3.65). To run the recursions, the variable i must range as $K + 1 \leqslant i \leqslant n$ for the inverse in (3.64) to exist and the initial values calculated at $i = K$ using the original batch forms.

Example 3.6 ***Moving average filter.*** Suppose that $M = [\ 1 \dots 1\ ]$, a constant quantity x is measured n times, and the measurements are combined into a vector $Y = [\ y_1 \dots y_n\ ]$. So we have $(M^T M)^{-1} = \frac{1}{n}$ and the batch estimate (3.63) is converted to *moving average estimate*

$$\hat{x} = \frac{1}{n} M^T Y = \frac{1}{n} \sum_{i=1}^{n} y_i .$$

> ***Moving average filter.*** *A moving average filter is optimal for a common problem, reducing random white noise while keeping the sharpest step response [189]. No other filter can produce noise less than simple averaging for Gaussian processes.*

The recursive computation of the moving average is provided by (3.67) as

$$\hat{x}_i = \hat{x}_{i-1} + \frac{1}{i}(y_i - \hat{x}_{i-1}) ,$$

if we increase i from $i = 2$ to $i = n$ for a given initial $\hat{x}_1$. The initial value can also be computed at $k = 2$ in batch form and then iteratively updated starting with $i = 3$. □

3.2.5 Kalman Filtering Algorithm

We now look again at the recursive algorithm in (3.50)–(3.53) associated with one-state linear models. The approach can easily be extended to the K-state linear model

$$x_k = F_k x_{k-1} + E_k u_k + B_k w_k , \tag{3.68}$$

$$y_k = H_k x_k + v_k , \tag{3.69}$$

where the noise vectors $w_k \sim \mathcal{N}(0, Q_k)$ and $v_k \sim \mathcal{N}(0, R_k)$ are uncorrelated. The corresponding algorithm was derived by Kalman in his seminal paper [84] and is now commonly known as the Kalman filter.[2]

The KF algorithm operates in two phases: 1) in the *prediction phase*, the *a priori* estimate and error covariance are predicted at k using measurements at $k - 1$, and 2) in the *update phase*, the *a priori* values are updated at k to the *a posteriori* estimate and error covariance using measurements at k. This strategy assumes two available algorithmic options: the *a posteriori* KF algorithm and the *a priori* KF algorithm. It is worth noting that the original KF algorithm is *a posteriori*.

---

2 Kalman derived his recursive filtering algorithm in 1960 by applying the Bayesian approach to linear processes with white Gaussian noise.

The *a posteriori* Kalman Filter

Consider a K-state space model as in (3.68) and (3.69). The Bayesian approach can be applied similarly to the one-state case, as was done by Kalman in [84], although there are several other ways to arrive at the KF algorithm. Let us show the simplest one.

The first thing to note is that the most reasonable prior estimate can be taken from (3.68) for the known estimate $\hat{x}_{k-1}$ and input u_k if we ignore the zero mean noise w_k. This gives

$$\hat{x}_k^- = F_k \hat{x}_{k-1} + E_k u_k \,, \tag{3.70}$$

which is also the prior estimate (3.52).

To update $\hat{x}_k^-$, we need to involve the observation (3.69). This can be done if we take into account the *measurement residual*

$$s_k = y_k - H_k \hat{x}_k^- \,, \tag{3.71}$$

which is the difference between data y_k and the predicted data $H_k \hat{x}_k^-$. The measurement residual covariance $S_k = S_k^T = \mathcal{E}\{s_k s_k^T\}$ can then be written as

$$\begin{aligned} S_k &= \mathcal{E}\{(H_k x_k + v_k - H_k \hat{x}_k^-)(\dots)^T\} \\ &= H_k P_k^- H_k^T + R_k \,. \end{aligned} \tag{3.72}$$

Since s_k is generally biased, because $\hat{x}_k^-$ is generally biased, the prior estimate can be updated as

$$\hat{x}_k = \hat{x}_k^- + K_k s_k \,, \tag{3.73}$$

where the matrix K_k is introduced to correct the bias in $\hat{x}_k^-$. Therefore, K_k plays the role of the *bias correction gain.*

As can be seen, the relations (3.70) and (3.73), which are reasonably extracted from the state-space model, are exactly the relations (3.52) and (3.53). Now notice that the estimate (3.73) will be optimal if we will find K_k such that the MSE is minimized. To do this, let us refer to (3.68) and (3.69) and define errors (3.34)–(3.37) involving (3.70)–(3.73).

The prior estimation error $\varepsilon_k^- = x_k - x_k^-$ (3.34) can be transformed as

$$\begin{aligned} \varepsilon_k^- &= F_k x_{k-1} + E_k u_k + B_k w_k - F_k \hat{x}_{k-1} - E_k u_k \\ &= F_k \varepsilon_{k-1} + B_k w_k \end{aligned} \tag{3.74}$$

and, for mutually uncorrelated ε_{k-1} and w_k, the prior error covariance (3.36) can be transformed to

$$P_k^- = F_k P_{k-1} F_k^T + B_k Q_k B_k^T \,. \tag{3.75}$$

Next, the estimation error $\varepsilon_k = x_k - \hat{x}_k$ (3.35) can be represented by

$$\begin{aligned} \varepsilon_k &= F_k x_{k-1} + E_k u_k + B_k w_k - F_k \hat{x}_{k-1} - E_k u_k - K_k (y_k - H_k \hat{x}_k^-) \\ &= F_k \varepsilon_{k-1} + B_k w_k - K_k (H_k F_k \varepsilon_{k-1} + H_k B_k w_k + v_k) \\ &= (I - K_k H_k)(F_k \varepsilon_{k-1} + B_k w_k) - K_k v_k \end{aligned} \tag{3.76}$$

and, for mutually uncorrelated ε_{k-1}, w_k, and v_k, the *a posteriori* error covariance (3.37) can be transformed to

$$\begin{aligned} P_k &= (I - K_k H_k)(F_k P_{k-1} F_k^T + B_k Q_k B_k^T)(I - K_k H_k)^T \\ &\quad + K_k R_k K_k^T \\ &= (I - K_k H_k) P_k^- (I - K_k H_k)^T + K_k R_k K_k^T \\ &= P_k^- - 2 K_k H_k P_k^- + K_k (H_k P_k^- H_k^T + R_k) K_k^T \,. \end{aligned} \tag{3.77}$$

What is left behind is to find the optimal bias correction gain K_k that minimizes MSE. This can be done by minimizing the trace of P_k, which is a convex function with a minimum corresponding to the optimal K. Using the matrix identities (A.3) and (A.6), the minimization of tr P_k by K_k can be carried out if we rewrite the equation $\frac{\partial}{\partial K_k}\text{tr } P_k = 0$ as

$$\frac{\partial}{\partial K_k}\text{tr } P_k = 2K_k S_k - 2P_k^- H_k^T = 0 .$$

This gives the optimal bias correction gain

$$K_k = P_k^- H_k^T S_k^{-1} . \tag{3.78}$$

The gain K_k (3.78) is known as the *Kalman gain*, and its substitution on the left-hand side of the last term in (3.77) transforms the error covariance P_k into

$$P_k = (I - K_k H_k) P_k^- . \tag{3.79}$$

Another useful form of P_k appears when we first refer to (3.78) and (3.72) and transform (3.77) as

$$\begin{aligned}
P_k &= P_k^- - 2K_k H_k P_k^- + K_k (H_k P_k^- H_k^T + R_k) K_k^T \\
&= P_k^- - 2P_k^- H_k^T S_k^{-1} H_k P_k^- + P_k^- H_k^T S_k^{-1} S_k S_k^{-1} H_k P_k^- \\
&= P_k^- - 2P_k^- H_k^T S_k^{-1} H_k P_k^- + P_k^- H_k^T S_k^{-1} H_k P_k^- \\
&= P_k^- - P_k^- H_k^T S_k^{-1} H_k P_k^- \\
&= P_k^- - P_k^- H_k^T (H_k P_k^- H_k^T + R_k)^{-1} H_k P_k^- .
\end{aligned}$$

Next, inverting the both sides of P_k and applying (A.7) gives [185]

$$\begin{aligned}
P_k^{-1} &= [P_k^- - P_k^- H_k^T (H_k P_k^- H_k^T + R_k)^{-1} H_k P_k^-]^{-1} \\
&= (P_k^-)^{-1} + H_k^T R_k^{-1} H_k
\end{aligned}$$

and another form of P_k appears as

$$P_k = [(P_k^-)^{-1} + H_k^T R_k^{-1} H_k]^{-1} . \tag{3.80}$$

Using (3.80), the Kalman gain (3.78) can also be modified as [185]

$$\begin{aligned}
K_k &= P_k^- H_k^T (H_k P_k^- H_k^T + R_k)^{-1} \\
&= (P_k P_k^{-1}) P_k^- H_k^T (H_k P_k^- H_k^T + R_k)^{-1} \\
&= P_k [(P_k^-)^{-1} + H_k^T R_k^{-1} H_k] P_k^- H_k^T (H_k P_k^- H_k^T + R_k)^{-1} \\
&= P_k (H_k^T + H_k^T R_k^{-1} H_k P_k^- H_k^T)(H_k P_k^- H_k^T + R_k)^{-1} \\
&= P_k H_k^T R_k^{-1} (R_k + H_k P_k^- H_k^T)(H_k P_k^- H_k^T + R_k)^{-1} \\
&= P_k H_k^T R_k^{-1} .
\end{aligned} \tag{3.81}$$

The previous derived standard KF algorithm gives an estimate at k utilizing data from zero to k. Therefore, it is also known as the *a posteriori* KF algorithm, the pseudocode of which is listed as Algorithm 1. The algorithm requires initial values $\hat{x}_0$ and P_0, as well as noise covariances Q_k and R_k to update the estimates starting at $k = 1$. Its operation is quite transparent, and recursions are easy to program, are fast to compute, and require little memory, making the KF algorithm suitable for many applications.

Algorithm 1: The *a posteriori* (Standard) KF Algorithm

Data: $y_k, u_k, \hat{x}_0, P_0, Q_k, R_k$
Result: $\hat{x}_k, P_k$
begin
  for $k = 1, 2, \cdots$ **do**
    $P_k^- = F_k P_{k-1} F_k^T + B_k Q_k B_k^T$;
    $S_k = H_k P_k^- H_k^T + R_k$;
    $K_k = P_k^- H_k^T S_k^{-1}$;
    $\hat{x}_k^- = F_k \hat{x}_{k-1} + E_k u_k$;
    $\hat{x}_k = \hat{x}_k^- + K_k(y_k - H_k \hat{x}_k^-)$;
    $P_k = (I - K_k H_k) P_k^-$;
  end for
end

The *a priori* Kalman Filter

Unlike the *a posteriori* KF, the *a priori* KF gives an estimate at $k+1$ using data y_k from k. The corresponding algorithm emerges from Algorithm 1 if we first substitute (3.79) in (3.74) and go to the *discrete difference Riccati equation* (DDRE)

$$\begin{aligned} P_{k+1}^- &= F_{k+1} P_k F_{k+1}^T + B_{k+1} Q_{k+1} B_{k+1}^T , \\ &= F_{k+1}(I - K_k H_k) P_k^- F_{k+1}^T + B_{k+1} Q_{k+1} B_{k+1}^T , \\ &= F_{k+1} P_k^- F_{k+1}^T - F_{k+1} P_k^- H_k^T (H_k P_k^- H_k^T + R_k)^{-1} H_k P_k^- F_{k+1}^T \\ &\quad + B_{k+1} Q_{k+1} B_{k+1}^T , \end{aligned}$$

which gives recursion for the covariance of the prior estimation error. Then the *a priori* estimate $\hat{x}_{k+1}^-$ can be obtained by the transformations

$$\begin{aligned} \hat{x}_k &= \hat{x}_k^- + K_k s_k , \\ F_{k+1}\hat{x}_k + E_{k+1} u_{k+1} &= F_{k+1}\hat{x}_k^- + E_{k+1} u_{k+1} + F_{k+1} K_k s_k , \\ \hat{x}_{k+1}^- &= F_{k+1}(\hat{x}_k^- + K_k s_k) + E_{k+1} u_{k+1} , \end{aligned}$$

and the *a priori* KF formalized with the pseudocode specified in Algorithm 2. As can be seen, this algorithm does not require data y_{k+1} to produce an estimate at $k+1$. However, it requires future values u_{k+1} and Q_{k+1} and is thus most suitable for LTI systems without input.

Optimality and Unbiasedness of the Kalman Filter

From the derivation of the Kalman gain (3.78) it follows that, by minimizing the MSE, the KF becomes optimal [36, 220]. In Chapter 4, the optimality of KF will be supported by the batch OFIR filter, whose recursions are exactly the KF recursions. On the other hand, the KF originates from the Bayesian approach. Therefore, many authors also call it *optimal unbiased*, and the corresponding proof given in [54] has been mentioned in many works [9, 58, 185]. However, in Chapter 4 it will be shown that the batch OUFIR filter has other recursions, which contradicts the proof given in [54].

To ensure that the KF is optimal, let us look at the explanation given in [54]. So let us consider the model (3.68) with $B_k = I$ and define an estimate $\hat{x}_k$ as [54]

$$\hat{x}_k = K_k' \hat{x}_k^- + K_k y_k , \tag{3.82}$$

Algorithm 2: The *a priori* KF Algorithm

Data: $y_k, u_k, \hat{x}_0^-, P_0^-, Q_k, R_k$

Result: $\hat{x}_k^-, P_k^-$

begin
 for $k = 0, 1, \cdots$ **do**
  $S_k = H_k P_k^- H_k^T + R_k$;
  $K_k = P_k^- H_k^T S_k^{-1}$;
  $\hat{x}_{k+1}^- = F_{k+1}[\hat{x}_k^- + K_k(y_k - H_k\hat{x}_k^-)] + E_{k+1}u_{k+1}$;
  $P_{k+1}^- = F_{k+1}(I - K_k H_{k+1})P_k^- F_{k+1}^T + B_{k+1}Q_{k+1}B_{k+1}^T$;
 end for
end

where $\hat{x}_k^- = F_k\hat{x}_{k-1}$ and the gain K_k' still needs to be determined. Next, let us examine the *a priori* error $\varepsilon_k^- = x_k - \hat{x}_k^-$ and the *a posteriori* error

$$\begin{aligned} \varepsilon_k &= x_k - \hat{x}_k \\ &= x_k - K_k'\hat{x}_k^- - K_k y_k \\ &= K_k'\varepsilon_k^- - (K_k' + K_k H_k - I)x_k - K_k v_k \ . \end{aligned}$$

Estimate (3.82) was stated in [54], p. 107, to be unbiased if $\mathcal{E}\{\varepsilon_k\} = 0$ and $\mathcal{E}\{\varepsilon_k^-\} = 0$. By satisfying these conditions using the previous relations, we obtain $K_k' = I - K_k H_k$ and the KF estimate becomes (3.73). This fact was used in [54] as proof that the KF is *optimal unbiased*.

Now, notice that the prior error ε_k^- is always biased in dynamic systems, so $\mathcal{E}\{\varepsilon_k^-\} \neq 0$. Otherwise, there would be no need to adjust the bias by K_k. Then we redefine the prior estimate as

$$\hat{x}_k^- = F_k\mathcal{E}\{x_{k-1}\} \tag{3.83}$$

and average the estimation error as

$$\begin{aligned} \mathcal{E}\{\varepsilon_k\} &= K_k'\mathcal{E}\{\varepsilon_k^-\} - (K_k' + K_k H_k - I)\mathcal{E}\{x_k\} \\ &= K_k'\mathcal{E}\{\hat{x}_k^-\} + (K_k H_k - I)\mathcal{E}\{x_k\} \ . \end{aligned}$$

The filter will be unbiased if $\mathcal{E}\{\varepsilon_k\}$ is equal to zero that gives

$$\begin{aligned} K_k'\mathcal{E}\{\hat{x}_k^-\} &= (I - K_k H_k)\mathcal{E}\{x_k\} \ , \\ K_k'\hat{x}_k^- &= (I - K_k H_k)F_k\mathcal{E}\{x_{k-1}\} \ . \end{aligned} \tag{3.84}$$

Substituting (3.84) into (3.82) gives the same estimate (3.73) with, however, different $\hat{x}_k^-$ defined by (3.83). The latter means that KF will be 1) *optimal*, if $\mathcal{E}\{x_{k-1}\}$ is *optimally biased*, and 2) *optimal unbiased*, if $\mathcal{E}\{x_{k-1}\}$ has no bias. Now note that the Kalman gain K_k makes an estimate optimal, and therefore there is always a small bias to be found in $\mathcal{E}\{x_{k-1}\}$, which contradicts the property of unbiased optimality and the claim that KF is unbiased.

Intrinsic Errors of Kalman Filtering

Looking at Algorithm 1 and assuming that the model matches the process, we notice that the required $\hat{x}_0$, P_0, Q_k, and R_k must be exactly known in order for KF to be optimal. However, due to the practical impossibility of doing this at every k and prior to running the filter, these values

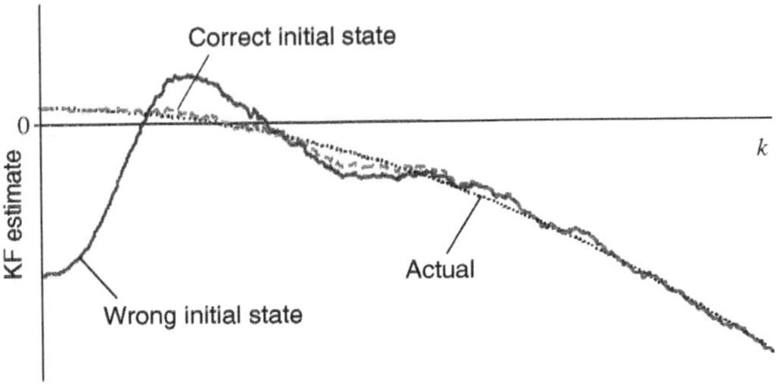

Figure 3.1 Typical errors in the KF caused by incorrectly specified initial conditions. Transients can last for a long time, especially in a harmonic state-space model.

are commonly estimated approximately, and the question arises about the practical optimality of KF. Accordingly, we can distinguish two intrinsic errors caused by incorrectly specified initial conditions and poorly known covariance matrices.

Figure 3.1 shows what happens to the KF estimate when $\hat{x}_0$ and P_0 are set incorrectly. With correct initial values, the KF estimate ranges close to the actual behavior over all k. Otherwise, incorrectly set initial values can cause large initial errors and long transients. Although the KF estimate approaches an actual state asymptotically, it can take a long time. Such errors also occur when the process undergoes temporary changes associated, for example, with jumps in velocity, phase, and frequency. The estimator inability to track temporary uncertain behaviors causes transients that can last for a long time, especially in harmonic models [179].

Now, suppose the error covariances are not known exactly and introduce the worst-case scaling factor β as Q/β^2 or $\beta^2 R$. Figure 3.2 shows the typical root MSEs (RMSEs) produced by KF as functions of β, and we notice that $\beta < 1$ makes the estimate noisy and $\beta > 1$ makes it biased with respect to the optimal case of $\beta = 1$. Although the most favorable case of $\beta^2 Q$ and $\beta^2 R$ assumes that the effects will be mutually exclusive, in practice this rarely happens due to the different nature of the process and measurement noise. We add that the intrinsic errors considered earlier appear at the KF output regardless of the model and differ only in values.

3.2.6 Backward Kalman Filter

Some applications require estimating the past state of a process or retrieving information about the initial state and error covariance. For example, multi-pass filtering (forward-backward-forward-...) [231] updates the initial values using a backward filter, which improves accuracy.

The backward *a posteriori* KF can be obtained similarly to the standard KF if we represent the state equation (3.68) backward in time as

$$x_{k-1} = F_k^{-1}(x_k - E_k u_k - B_k w_k)$$

and then, reasoning along similar lines as for the *a posteriori* KF, define the *a priori* state estimate at $k-1$,

$$\tilde{x}_{k-1}^- = F_k^{-1}(\tilde{x}_k - E_k u_k) \,,$$

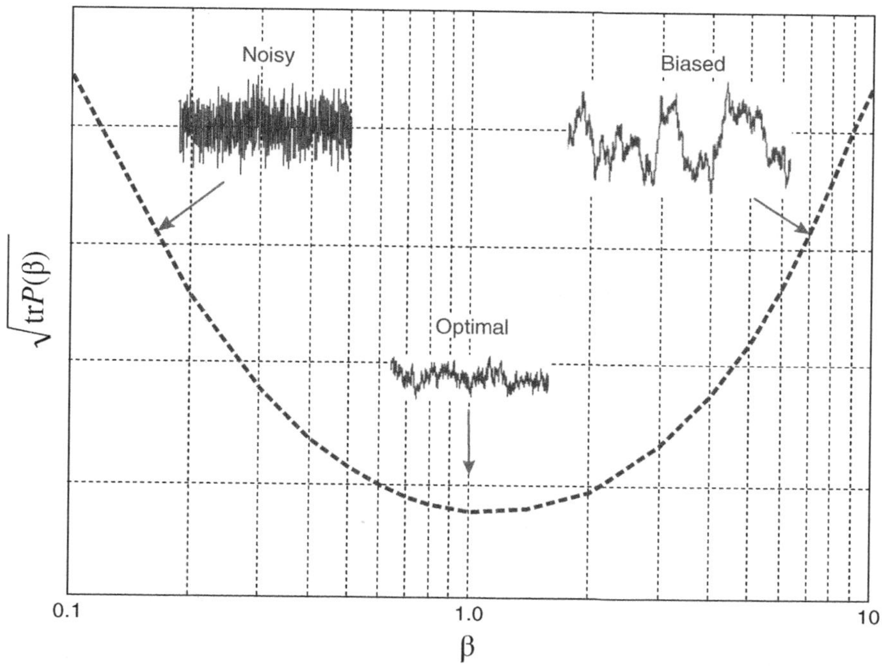

Figure 3.2 Effect of errors in noise covariances, Q/β^2 and $\beta^2 R$, on the RMSEs produced by KF in the worst case: $\beta = 1$ corresponds to the optimal estimate, $\beta < 1$ makes the estimate more noisy, and $\beta > 1$ makes it more biased.

the measurement residual $s_{k-1} = y_{k-1} - H_{k-1}\tilde{x}^-_{k-1}$, the measurement residual covariance

$$S_{k-1} = H_{k-1}P^-_{k-1}H^T_{k-1} + R_{k-1} ,$$

the backward estimate

$$\tilde{x}_{k-1} = \tilde{x}^-_{k-1} + K_{k-1}s_{k-1} ,$$

the *a priori* error covariance

$$P^-_{k-1} = F^{-1}_k(P_k + B_kQ_kB^T_k)F^{-T}_k ,$$

and the *a posteriori* error covariance

$$P_{k-1} = P^-_{k-1} - 2K_{k-1}H_{k-1}P^-_{k-1} + K_{k-1}S_{k-1}K^T_{k-1} .$$

Further minimizing P_{k-1} by K_{k-1} gives the optimal gain

$$K_{k-1} = P^-_{k-1}H^T_{k-1}S^{-1}_{k-1}$$

and the modified error covariance

$$P_{k-1} = (I - K_{k-1}H_{k-1})P^-_{k-1} .$$

Finally, the pseudocode of the backward *a posteriori* KF, which updates the estimates from k to 0, becomes as listed in Algorithm 3. The algorithm requires initial values $\hat{x}_k$ and P_k at k and updates the estimates back in time from $k-1$ to zero to obtain estimates of the initial state $\tilde{x}_0$ and error covariance P_0.

Algorithm 3: Backward *a posteriori* KF Algorithm

Data: $y_k, u_k, \hat{x}_k, P_k, Q_k, R_k$
Result: $\hat{x}_0, P_0$
begin
  for $n = k-1, k-2, \cdots, 0$ **do**
    $P_n^- = F_{n+1}^{-1}(P_{n+1} + B_{n+1}Q_{n+1}B_{n+1}^T)F_{n+1}^{-T}$;
    $S_n = H_n P_n^- H_n^T + R_n$;
    $K_n = P_n^- H_n^T S_n^{-1}$;
    $\tilde{x}_n^- = F_{n+1}^{-1}(\tilde{x}_{n+1} - E_{n+1}u_{n+1})$;
    $\tilde{x}_n = \tilde{x}_n^- + K_n(y_n - H_n\tilde{x}_n^-)$;
    $P_n = (I - K_n H_n)P_n^-$;
  end for
end

3.2.7 Alternative Forms of the Kalman Filter

Although the KF Algorithm 1 is the most widely used, other available forms of KF may be more efficient in some applications. Collections of such algorithms can be found in [9, 58, 185] and some other works. To illustrate the flexibility and versatility of KF in various forms, next we present two versions known as the *information filter* and *alternate KF form*. Noticing that these modifications were shown in [185] for the FE-based model, we develop them here for the BE-based model. Some other versions of the KF can be found in the section "Problems".

Information Filter

When minimal information loss is required instead of minimal MSE, the information matrix $\mathcal{I}_k$, which is the inverse of the covariance matrix P_k, can be used. Since P_k is symmetric and positive definite, it follows that the inverse of P_k is unique and KF can be modified accordingly.

By introducing $\mathcal{I}_k = P_k^{-1}$ and $\mathcal{I}_k^- = (P_k^-)^{-1}$, the error covariance (3.80) can be transformed to

$$\mathcal{I}_k = \mathcal{I}_k^- + H_k^T R_k^{-1} H_k \ , \tag{3.85}$$

where, by (3.75) and $\bar{Q}_k = B_k Q_k B_k^T$, matrix $\mathcal{I}_k^-$ becomes

$$\mathcal{I}_k^- = [F_k \mathcal{I}_{k-1}^{-1} F_k^T + \bar{Q}_k]^{-1} \ . \tag{3.86}$$

Using the Woodbury identity (A.7), one can further rewrite (3.86) as

$$\mathcal{I}_k^- = \bar{Q}_k^{-1} - \bar{Q}_k^{-1} F_k(\mathcal{I}_{k-1} + F_k^T \bar{Q}_k^{-1} F_k)^{-1} F_k^T \bar{Q}_k^{-1}$$

and arrive at the information KF [185]. Given y_k, $\hat{x}_0$, $\mathcal{I}_0$, Q_k, and R_k, the algorithm predicts and updates the estimates as

$$\mathcal{I}_k^- = \bar{Q}_k^{-1} - \bar{Q}_k^{-1} F_k(\mathcal{I}_{k-1} + F_k^T \bar{Q}_k^{-1} F_k)^{-1} F_k^T \bar{Q}_k^{-1} \ , \tag{3.87}$$

$$\mathcal{I}_k = \mathcal{I}_k^- + H_k^T R_k^{-1} H_k \ , \tag{3.88}$$

$$K_k = \mathcal{I}_k^{-1} H_k^T R_k^{-1} \ , \tag{3.89}$$

$$\hat{x}_k^- = F_k \hat{x}_{k-1} + E_k u_k \ , \tag{3.90}$$

$$\hat{x}_k = \hat{x}_k^- + K_k(y_k - H_k \hat{x}_n^-) \ . \tag{3.91}$$

It is worth noting that the information KF is mathematically equivalent to the standard KF for linear models. However, its computational load is greater because more inversions are required in the covariance propagation.

Alternate Form of the KF

Another form of KF, called *alternate KF form*, has been proposed in [185] to felicitate obtaining optimal smoothing. The advantage is that this form is compatible with the game theory-based H_∞ filter [185]. The modification is provided by inverting the residual covariance (3.77) through the information matrix using (A.7). KF recursions modified in this way for a BE-based model lead to the algorithm

$$\hat{x}_k^- = F_k \hat{x}_{k-1} + E_k u_k \,,$$
$$K_k = P_k^- (I + H_k^T R_k^{-1} H_k P_k^-)^{-1} H_k^T R_k^{-1} \,,$$
$$\hat{x}_k = \hat{x}_k^- + K_k (y_k - H_k \hat{x}_k^-) \,,$$
$$P_{k+1}^- = F_k P_k^- (I + H_k^T R_k^{-1} H_k P_k^-)^{-1} F_k^T + B_k Q_k B_k^T \,.$$

It follows from [185] that matrix $P_k^-(I + H_k^T R_k^{-1} H_k P_k^-)^{-1}$ is equal to P_k. Therefore, this algorithm can be written in a more compact form. Given $y_k, \hat{x}_0, P_0, Q_k$, and R_k, the initial prior error covariance is computed by $P_1^- = A_1 P_0 A_1^T + B_1 Q_1 B_1^T$ and the update equations for $k = 1, 2 \ldots$ become

$$P_k = P_k^- (I + H_k^T R_k^{-1} H_k P_k^-)^{-1} \,, \tag{3.92}$$

$$K_k = P_k H_k^T R_k^{-1} \,, \tag{3.93}$$

$$\hat{x}_k^- = F_k \hat{x}_{k-1} + E_k u_k \,, \tag{3.94}$$

$$\hat{x}_k = \hat{x}_k^- + K_k (y_k - H_k \hat{x}_k^-) \,, \tag{3.95}$$

$$P_{k+1}^- = F_k P_k F_k^T + B_k Q_k B_k^T \,. \tag{3.96}$$

The proof of the equivalence of matrices (3.92) and (3.77) is postponed to the section “Problems”.

3.2.8 General Kalman Filter

Looking into the KF derivation procedure, it can be concluded that the assumptions regarding the whiteness and uncorrelatedness of Gaussian noise are too strict and that the algorithm is thus “ideal,” since white noise does not exist in real life. More realistically, one can think of system noise and measurement noise as color processes and approximate them by a Gauss-Markov process driven by white noise.

To make sure that such modifications are in demand in real life, consider a few examples. When noise passes through bandlimited and narrowband channels, it inevitably becomes colored and correlated. Any person traveling along a certain trajectory follows a control signal generated by the brain. Since this signal cannot be accurately modeled, deviations from its mean appear as system colored noise. A visual camera catches and tracks objects asymptotically, which makes the measurement noise colored.

In Fig. 3.3a, we give an example of the *colored measurement noise* (CMN) noise, and its simulation as the Gauss-Markov process is illustrated in Fig. 3.3b. An example (Fig. 3.3a) shows the signal strength colored noise discovered in the GPS receiver [210]. The noise here was

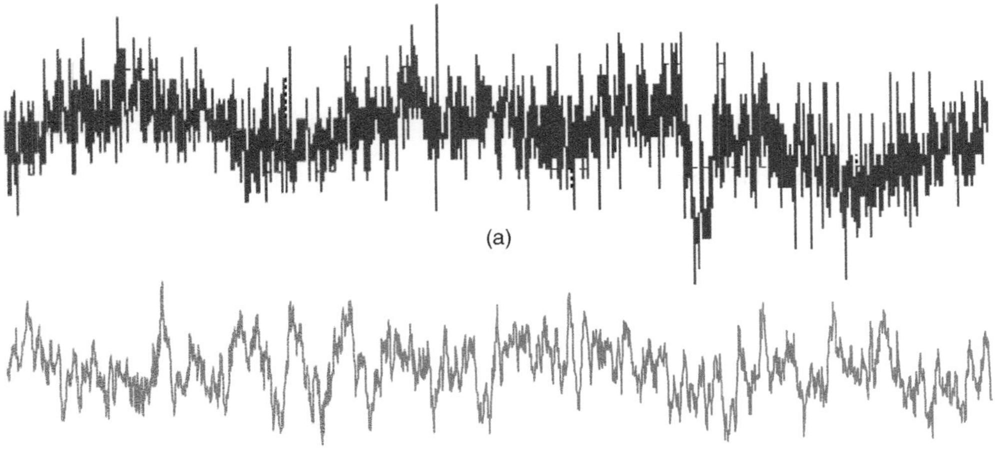

Figure 3.3 Examples of CMN in electronic channels: (a) signal strength CMN in a GPS receiver Based on [210] and (b) Gauss-Markov noise simulated by $v_n = 0.9v_{n-1} + \xi_n$ with $\xi_n \sim \mathcal{N}(0, 16)$.

extracted by removing the antenna gain attenuation from the signal strength measurements and then normalizing to zero relative to the mean. The simulated colored noise shown in Fig. 3.3b comes from the Gauss-Markov sequence $x_n = 0.9x_{n-1} + w_n$ with white Gaussian driving noise $w_n \sim \mathcal{N}(0, 16)$. Even a quick glance at the drawing (Fig. 3.3) allows us to think that both processes most likely belong to the same class and, therefore, colored noise can be of Markov origin.

Now that we understand the need to modify the KF for colored and correlated noise, we can do it from a more general point of view than for the standard KF and come up with a universal linear algorithm called *general Kalman filter* (GKF) [185], which can have various forms. Evidence for the better performance of the GKF, even a minor improvement, can be found in many works, since white noise is unrealistic.

In many cases, the dynamics of CMN v_k and *colored process noise* (CPN) w_k can be viewed as Gauss-Markov processes, which complement the state-space model as

$$x_k = F_k x_{k-1} + E_k u_k + B_k w_k \,, \tag{3.97}$$

$$w_k = \Theta_k w_{k-1} + \mu_k \,, \tag{3.98}$$

$$v_k = \Psi_k v_{k-1} + \xi_k \,, \tag{3.99}$$

$$y_k = H_k x_k + v_k \,, \tag{3.100}$$

where $Q_{\mu k} = E\{\mu_k \mu_k^T\}$ and $R_{\xi k} = E\{\xi_k \xi_k^T\}$ are the covariances of white Gaussian noise vectors $\mu_k \sim \mathcal{N}(0, Q_{\mu k})$ and $\xi_k \sim \mathcal{N}(0, R_{\xi k})$ and cross-covariances $L_k = E\{\mu_k \xi_k^T\}$ and $L_k^T = E\{\xi_k \mu_k^T\}$ represent the time-correlation between μ_k and ξ_k. Equations (3.98) and (3.99) suggest that the coloredness factors Θ_k and Ψ_k should be chosen so that w_k and v_k remain stationary. Since the model (3.97)–(3.100) is still linear with white Gaussian μ_k and ξ_k, the Kalman approach can be applied and the GKF derived in the same way as KF.

Two approaches have been developed for applying KF to (3.97)–(3.100). The first *augmented states* approach was proposed by A. E. Bryson *et al.* in [24, 25] and is the most straightforward. It proposes to combine the dynamic equations (3.97)–(3.99) into a single augmented state equation and represent the model as

$$\begin{bmatrix} x_k \\ w_k \\ v_k \end{bmatrix} = \begin{bmatrix} F_k & B_k\Theta_k & 0 \\ 0 & \Theta_k & 0 \\ 0 & 0 & \Psi_k \end{bmatrix} \begin{bmatrix} x_{k-1} \\ w_{k-1} \\ v_{k-1} \end{bmatrix} + \begin{bmatrix} E_k \\ 0 \\ 0 \end{bmatrix} u_k + \begin{bmatrix} B_k & 0 \\ I & 0 \\ 0 & I \end{bmatrix} \begin{bmatrix} \mu_k \\ \xi_k \end{bmatrix} , \tag{3.101}$$

$$y_k = \begin{bmatrix} H_k & 0 & I \end{bmatrix} \begin{bmatrix} x_k \\ w_k \\ v_k \end{bmatrix} + 0 , \tag{3.102}$$

where $E\left\{ \begin{bmatrix} \mu_k \\ \xi_k \end{bmatrix} \begin{bmatrix} \mu_k^T & \xi_k^T \end{bmatrix} \right\} = \begin{bmatrix} Q_{\mu k} & L_k \\ L_k^T & R_{\xi k} \end{bmatrix}$. Obviously, the augmented equations (3.101) and (3.102) can be easily simplified to address only CPN, CMN, or noise correlation problems. An easily seen drawback is that the observation equation (3.102) has zero measurement noise. This makes it impossible to apply some KF versions that require inversion of the covariance matrix R_k, such as information KF and alternate KF form. Therefore, the KF is said to be *ill conditioned* for state augmentation [25].

The second approach, which avoids the state augmentation, was proposed and developed by A. E. Bryson *et al.* in [24, 25] and is now known as Bryson's algorithm. This algorithm was later improved in [147] to what is now referred to as the Petovello algorithm. The idea was to use *measurement differencing* to convert the model with CMN to a standard model with new matrices, but with white noise. The same idea was adopted in [182] for systems with CPN using *state differencing*. Next, we show several GKF algorithms for time-correlated and colored noise sources.

Time-Correlated Noise

If noise has a very short correlation time compared to time intervals of interest, it is typically considered as white and its coloredness can be neglected. Otherwise, system may experience a great noise interference [58].

When the noise sources are white, $w_k = \mu_k$ and $v_k = \xi_k$, but time correlated with $L_k \neq 0$, equations (3.98) and (3.99) can be ignored, and (3.101) and (3.102) return to their original forms (3.32) and (3.33) with time-correlated w_k and v_k. There are two ways to apply KF for time-correlated w_k and v_k. We can either de-correlate w_k and v_k or derive a new bias correction gain (Kalman gain) to guarantee the filter optimality.

Noise de-correlation. To de-correlate w_n and v_n and thereby make it possible to apply KF, one can use the Lagrange method [15] and combine (3.32) with the term $\Lambda_k(y_k - H_k x_k - v_k)$, where y_k is a data vector and the Lagrange multiplier Λ_k is yet to be determined. This transforms the state equation to

$$\begin{aligned} x_k &= F_k x_{k-1} + E_k u_k + B_k w_k + \Lambda_k(y_k - H_k x_k - v_k) \\ &= \bar{F}_k x_{k-1} + \bar{u}_k + \bar{w}_k , \end{aligned} \tag{3.103}$$

where we assign

$$\bar{F}_k = (I - \Lambda_k H_k) F_k , \tag{3.104}$$

$$\bar{u}_k = (I - \Lambda_k H_k) E_k u_k + \Lambda_k y_k , \tag{3.105}$$

$$\bar{w}_k = (I - \Lambda_k H_k) B_k w_k - \Lambda_k v_k , \tag{3.106}$$

and white noise $\bar{w}_k \sim \mathcal{N}(0, \bar{Q}_k) \in \mathbb{R}^K$ has the covariance $\bar{Q}_k = E\{\bar{w}_k \bar{w}_k^T\}$,

$$\bar{Q}_k = E\{[(I - \Lambda_k H_k) B_k w_k - \Lambda_k v_k][\ldots]^T\}$$

$$= (I - \Lambda_k H_k) B_k Q_k B_k^T (I - \Lambda_k H_k)^T + \Lambda_k R_k \Lambda_k^T$$
$$- (I - \Lambda_k H_k) B_k L_k \Lambda_k^T - \Lambda_k L_k^T B_k^T (I - \Lambda_k H_k)^T \,. \tag{3.107}$$

To find a matrix Λ_k such that $\bar{w}_n$ and v_n become uncorrelated, the natural statement $E\{\bar{w}_k v_k^T\} = 0$ leads to

$$E\{\bar{w}_k v_k^T\} = E\{[(I - \Lambda_k H_k) B_k w_k - \Lambda_k v_k] v_k^T\}$$
$$= (I - \Lambda_k H_k) B_k L_k - \Lambda_k R_k = 0 \tag{3.108}$$

and the Lagrange multiplier can be found as

$$\Lambda_k = L_k (H_k B_k L_k + R_k)^{-1} \,. \tag{3.109}$$

Finally, replacing in (3.107) the product $\Lambda_k R_k$ taken from (3.108) removes R_k, and (3.107) becomes

$$\bar{Q}_k = (I - \Lambda_k H_k) B_k Q_k B_k^T (I - \Lambda_k H_k)^T$$
$$- \Lambda_k L_k^T B_k^T (I - \Lambda_k H_k)^T \,. \tag{3.110}$$

Given y_n, $\hat{x}_0$, P_0, $\bar{Q}_n$, R_n, and L_k, the GKF estimates can be updated using Algorithm 1, in which Q_k should be substituted with $\bar{Q}_k$ (3.110) using Λ_k given by (3.109) and $E_k u_k$ with $\bar{u}_k$ (3.105). We thus have

$$P_k^- = \bar{F}_k P_{k-1} \bar{F}_k^T + \bar{Q}_k \,, \tag{3.111}$$

$$S_k = H_k P_k^- H_k^T + R_k \,, \tag{3.112}$$

$$K_k = P_k^- H_k^T S_k^{-1} \,, \tag{3.113}$$

$$\hat{x}_k^- = \bar{F}_k \hat{x}_{k-1} + \bar{u}_k \,, \tag{3.114}$$

$$\hat{x}_k = \hat{x}_k^- + K_k (y_k - H_k \hat{x}_k^-) \,, \tag{3.115}$$

$$P_k = (I - K_k H_k) P_k^- \,, \tag{3.116}$$

where $\bar{F}_k$ is given by (3.104).

New Kalman gain. Another possibility to apply KF to models with time-correlated w_n and v_n implies obtaining a new Kalman gain K_k. In this case, first it is necessary to change the error covariance (3.76) for correlated w_n and v_n as

$$P_k = (I - K_k H_k) P_k^- (I - K_k H_k)^T + K_k R_k K_k^T$$
$$- (I - K_k H_k) B_k L_k K_k^T - K_k L_k^T B_k^T (I - K_k H_k)^T$$
$$= P_k^- - 2K_k (H_k P_k^- + L_k^T B_k^T)$$
$$+ K_k (S_k + H_k B_k L_k + L_k^T B_k^T H_k^T) K_k^T \,, \tag{3.117}$$

where P_k^- is given by (3.74) and S_k by (3.77). Using (A.3) and (A.6), the derivative of the trace of P_k with respect to K_k can be found as

$$\frac{\partial}{\partial K_k} \text{tr}\; P_k = 2K_k (S_k + H_k B_k L_k + L_k^T B_k^T H_k^T) - 2(P_k^- H_k^T + B_k L_k) \,.$$

By equating the derivative to zero, $\frac{\partial}{\partial K_k} \text{tr}\; P_k = 0$, we can find the optimal Kalman gain as

$$K_k = (P_k^- H_k^T + B_k L_k)(S_k + H_k B_k L_k + L_k^T B_k^T H_k^T)^{-1} \tag{3.118}$$

and notice that it becomes the original gain (3.78) for uncorrelated noise, $L_k = 0$. Finally, (3.118) simplifies the error covariance (3.117) to

$$P_k = (I - K_k H_k)P_k^- - 2K_k L_k^T B_k^T + B_k L_k K_k^T \tag{3.119}$$

$$= (I - K_k H_k)P_k^- - K_k L_k^T B_k^T . \tag{3.120}$$

Given y_k, $\hat{x}_0$, P_0, Q_k, R_k, and L_k, the GKF algorithm for time-correlated w_k and v_k becomes

$$P_k^- = F_k P_{k-1} F_k^T + B_k Q_k B_k^T , \tag{3.120}$$

$$S_k = H_k P_k^- H_k^T + R_k , \tag{3.121}$$

$$K_k = (P_k^- H_k^T + B_k L_k)(S_k + H_k B_k L_k + L_k^T B_k^T H_k^T)^{-1} , \tag{3.122}$$

$$\hat{x}_k^- = F_k \hat{x}_{k-1} + E_k u_k , \tag{3.123}$$

$$\hat{x}_n = \hat{x}_k^- + K_k(y_k - H_k \hat{x}_k^-) , \tag{3.124}$$

$$P_k = (I - K_k H_k)P_k^- - K_k L_k^T B_k^T . \tag{3.125}$$

As can be seen, when the correlation is removed by $L_k = 0$, the algorithm (3.120)–(3.125) becomes the standard KF Algorithm 1. It is also worth noting that the algorithms (3.111)–(3.116) and (3.120)–(3.125) produce equal estimates and, thus, they are identical with no obvious advantages over each other. We will see shortly that the corresponding algorithms for CMN are also identical.

Colored Measurement Noise

The suggestion that CMN may be of Markov origin was originally made by Bryson *et al.* in [24] for continuous-time and in [25] for discrete-time tracking systems. Two fundamental approaches have also been proposed in [25] for obtaining GKF for CMN: *state augmentation* and *measurement differencing*. The first approach involves reassigning the state vector so that the white sources are combined in an augmented system noise vector. This allows using KF directly, but makes it ill-conditioned, as shown earlier. Even so, state augmentation is considered to be the main solution to the problem caused by CMN [54, 58, 185].

For CMN, the state-space equations can be written as

$$x_k = F_k x_{k-1} + E_k u_k + B_k w_k , \tag{3.126}$$

$$v_k = \Psi_k v_{k-1} + \xi_k , \tag{3.127}$$

$$y_k = H_k x_k + v_k , \tag{3.128}$$

where ξ_k and w_k are zero mean white Gaussian with the covariances $Q_k = E\{w_k w_k^T\}$ and $R_{\xi k} = E\{\xi_k \xi_k^T\}$ and the property $E\{w_k \xi_n^T\} = 0$ for all n and k. Augmenting the state vector gives

$$\begin{bmatrix} x_k \\ v_k \end{bmatrix} = \begin{bmatrix} F_k & 0 \\ 0 & \Psi_k \end{bmatrix} \begin{bmatrix} x_{k-1} \\ v_{k-1} \end{bmatrix} + \begin{bmatrix} E_k \\ 0 \end{bmatrix} u_k + \begin{bmatrix} B_k & 0 \\ 0 & I \end{bmatrix} \begin{bmatrix} w_k \\ \xi_k \end{bmatrix} ,$$

$$y_k = \begin{bmatrix} H_k & I \end{bmatrix} \begin{bmatrix} x_k \\ v_k \end{bmatrix} + 0 ,$$

where, as in (3.101), zero observation noise makes KF ill-conditioned.

To avoid increasing the state vector, one can apply the measurement differencing approach [25] and define a new observation z_n as

$$\begin{aligned} z_k &= y_k - \Psi_k y_{k-1} , \\ &= H_k x_k + v_k - \Psi_k H_{k-1} x_{k-1} - \Psi_k v_{k-1} . \end{aligned} \tag{3.129}$$

By substituting x_{k-1} taken from (3.126) and v_{k-1} taken from (3.127), the observation equation (3.129) becomes

$$\bar{z}_k = z_k - \bar{E}_k u_k = \bar{H}_k x_k + \bar{v}_k , \tag{3.130}$$

where we introduced

$$\bar{H}_k = H_k - \Gamma_k , \tag{3.131}$$

$$\bar{E}_k = \Gamma_k E_k , \tag{3.132}$$

$$\bar{v}_k = \Gamma_k B_k w_k + \xi_k , \tag{3.133}$$

$\Gamma_k = \Psi_k H_{k-1} F_k^{-1}$, and white noise $\bar{v}_k$ with the properties

$$E\{\bar{v}_k \bar{v}_k^T\} = \Gamma_k \Phi_k + R_k , \tag{3.134}$$

$$E\{\bar{v}_k w_k^T\} = \Gamma_k B_k Q_k , \tag{3.135}$$

$$E\{w_k \bar{v}_k^T\} = Q_k B_k^T \Gamma_k^T , \tag{3.136}$$

where

$$\Phi_k = B_k Q_k B_k^T \Gamma_k^T . \tag{3.137}$$

Now the modified model (3.126) and (3.130) contains white and time-correlated noise sources w_k and $\bar{v}_k$, for which the prior estimate $\hat{x}_k^-$ is defined by (3.70) and the prior error covariance P_k^- by (3.74).

For (3.130), the measurement residual can be written as

$$\begin{aligned} s_k &= \bar{z}_k - \bar{H}_k \hat{x}_k^- \\ &= \bar{H}_k (F_k x_{k-1} + E_k u_k + B_k w_k) + \bar{v}_k - \bar{H}_k (F_k \hat{x}_{k-1} + E_k u_k) \\ &= \bar{H}_k F_k \epsilon_{n-1} + \bar{H}_k B_k w_k + \bar{v}_k \end{aligned} \tag{3.138}$$

and the innovation covariance $S_k = E\{s_k s_k^T\}$ can be transformed as

$$\begin{aligned} S_n &= \bar{H}_k F_k P_{k-1} F_k^T \bar{H}_k^T + \bar{H}_k B_k Q_k B_k^T \bar{H}_k^T \\ &\quad + \Gamma_k \Phi_k + R_k + \bar{H}_k \Phi_k + \Phi_k^T \bar{H}_k^T \\ &= \bar{H}_k P_k^- \bar{H}_k^T + R_k + H_k \Phi_k + \Phi_k^T \bar{H}_k^T . \end{aligned} \tag{3.139}$$

The recursive GKF estimate can thus be written as

$$\begin{aligned} \hat{x}_k &= \hat{x}_k^- + K_k s_k \\ &= \hat{x}_k^- + K_k (\bar{z}_k - \bar{H}_k \hat{x}_k^-) , \end{aligned} \tag{3.140}$$

where the Kalman gain K_k should be found for either time-correlated or de-correlated w_k and $\bar{v}_k$. Thereby, one can come up with two different, although mathematically equivalent, algorithms as discussed next.

Correlated noise. For time-correlated w_k and $\bar{v}_k$, KF can be applied if we find the Kalman gain K_k by minimizing the trace of the error covariance P_k. Representing the estimation error $\epsilon_k = x_k - \hat{x}_k$ with

$$\epsilon_k = (I - K_k\bar{H}_k)(F_k\epsilon_{k-1} + B_k w_k) - K_k\bar{v}_k \tag{3.141}$$

and noticing that ϵ_{k-1} does not correlate w_k and $\bar{v}_k$, we obtain the error covariance $P_k = E\{\epsilon_k\epsilon_k^T\}$ as

$$\begin{aligned}
P_k &= (I - K_k\bar{H}_k)P_k^-(I - K_k\bar{H}_k)^T \\
&\quad + K_k(\Gamma_k\Phi_k + R_k)K_k^T - (I - K_k\bar{H}_k)\Phi_k K_k^T \\
&\quad - K_k\Phi_k^T(I - K_k\bar{H}_k)^T
\end{aligned} \tag{3.142a}$$

$$\begin{aligned}
&= (I - K_k D_k)P_k^-(I - K_k\bar{H}_k)^T \\
&\quad + K_k[(H_k - \bar{H}_k)\Phi_k + R_k]K_k^T - (I - K_k\bar{H}_k)\Phi_k K_k^T \\
&\quad - K_k\Phi_k^T(I - K_k\bar{H}_k)^T \\
&= (I - K_k\bar{H}_k)P_k^-(I - K_k\bar{H}_k)^T + K_k(H_k\Phi_k + R_k)K_k^T \\
&\quad - \Phi_k K_k^T - K_k\Phi_k^T(I - K_k\bar{H}_k)^T \\
&= (I - K_k\bar{H}_k)P_k^-(I - K_k\bar{H}_k)^T + K_k R_k K_k^T \\
&\quad - K_k\Phi_k^T(I - K_k\bar{H}_k)^T - (I - K_k H_k)\Phi_k K_k^T \\
&= P_k^- - (P_k^-\bar{H}_k^T + \Phi_k)K_k^T - K_k(P_k^-\bar{H}_k^T + \Phi_k)^T \\
&\quad + K_k S_k K_k^T ,
\end{aligned} \tag{3.142b}$$

where P_k^- is given by (3.120) and S_k by (3.139).

Now the optimal Kalman gain K_k can be found by minimizing $\text{tr}(P_k)$ using (A.3), (A.4), and (A.6). Equating to zero the derivative of $\text{tr}(P_k)$ with respect to K_k gives

$$\frac{\partial\, \text{tr}P_k}{\partial K_k} = -2(P_k^-\bar{H}_k^T + \Phi_k) + 2K_k S_k = 0 \tag{3.143}$$

and we obtain

$$K_k = (P_k^-\bar{H}_k^T + \Phi_k)S_k^{-1} . \tag{3.144}$$

Finally, replacing the first component in the last term in (3.142b) with (3.144) gives

$$P_k = P_k^- - K_k(\bar{H}_k P_k^- + \Phi_k^T) . \tag{3.145}$$

Given $y_k, \hat{x}_0, P_0, Q_k, R_k$, and Ψ_k, the GKF equations become [183]

$$P_k^- = F_k P_{k-1} F_k^T + B_k Q_k B_k^T , \tag{3.146}$$

$$S_k = \bar{H}_k P_k^- \bar{H}_k^T + R_k + H_k\Phi_k + \Phi_k^T\bar{H}_k^T , \tag{3.147}$$

$$K_k = (P_k^-\bar{H}_k^T + \Phi_k)S_k^{-1} , \tag{3.148}$$

$$\hat{x}_k^- = F_k\hat{x}_{k-1} + E_k u_k , \tag{3.149}$$

$$\hat{x}_k = \hat{x}_k^- + K_k(y_k - \Psi_k y_{k-1} - \bar{E}_k u_k - \bar{H}_k\hat{x}_k^-) , \tag{3.150}$$

$$P_k = (I - K_k\bar{H}_k)P_k^- - K_k\Phi_k^T \tag{3.151}$$

and we notice that $\Psi_k = 0$ causes $\Gamma_k = 0$, $\bar{H}_k = H_k$, $\bar{E}_k = 0$, and $\Phi_k = 0$, and this algorithm converts to the standard KF Algorithm 1.

De-correlated noise. Alternatively, efforts can be made to de-correlate w_n and $\bar{v}_n$ by repeating steps (3.103)–(3.110) as shown next. Similarly to (3.103), rewrite the state equation as

$$\begin{aligned} x_k &= F_k x_{k-1} + E_k u_k + B_k w_k + \Lambda_k(\bar{z}_k - \bar{H}_k x_k - \bar{v}_k) \\ &= \bar{F}_k x_{k-1} + \bar{u}_k + \bar{w}_k , \end{aligned} \tag{3.152}$$

where the following variables are assigned

$$\bar{F}_k = (I - \Lambda_k \bar{H}_k) F_k , \tag{3.153}$$

$$\bar{u}_k = (I - \Lambda_k \bar{H}_k) E_k u_k + \Lambda_k \bar{z}_k , \tag{3.154}$$

$$\bar{w}_k = (I - \Lambda_k \bar{H}_k) B_k w_k - \Lambda_k \bar{v}_k , \tag{3.155}$$

and require that noise $\bar{w}_k$ be white with the covariance

$$\begin{aligned} \bar{Q}_k &= E\{\bar{w}_k \bar{w}_k^T\} = E\{[(I - \Lambda_k \bar{H}_k) B_k w_k - \Lambda_k \bar{v}_k][\dots]^T\} \\ &= E\{[(I - \Lambda_k \bar{H}_k - \Lambda_k \Gamma_k) B_k w_k - \Lambda_k \xi_k][\dots]^T\} \\ &= E\{[(I - \Lambda_k H_k) B_k w_k - \Lambda_k \xi_k][\dots]^T\} \\ &= (I - \Lambda_k H_k) B_k Q_k B_k^T (I - \Lambda_k H_k)^T + \Lambda_k R_k \Lambda_k^T . \end{aligned} \tag{3.156}$$

Matrix Λ_k that makes ζ_k and $\bar{v}_k$ uncorrelated can be found by representing $E\{\zeta_k \bar{v}_k^T\} = 0$ with

$$\begin{aligned} E\{\zeta_k \bar{v}_k^T\} &= E\{[(I - \Lambda_k \bar{H}_k) B_k w_k - \Lambda_k \bar{v}_k](\Gamma_k B_k w_k + \xi_k)^T\} , \\ &= E\{[(I - \Lambda_k H_k) B_k w_k - \Lambda_k \xi_k](\Gamma_k B_k w_k + \xi_k)^T\} , \\ &= (I - \Lambda_k H_k) \Phi_k - \Lambda_k R_k = 0 \end{aligned} \tag{3.157}$$

that gives the following Lagrange multiplier

$$\Lambda_k = \Phi_k (H_k \Phi_k + R_k)^{-1} . \tag{3.158}$$

Substituting $\Lambda_k R_k$ taken from (3.157) into the last term in (3.156) removes R_k and (3.156) takes the form

$$\bar{Q}_k = (I - \Lambda_k H_k) B_k Q_k B_k^T (I - \Lambda_k \bar{H}_k)^T . \tag{3.159}$$

Given y_k, $\hat{x}_0$, P_0, Q_k, R_k, and Ψ_k, the GKF equations become [183]

$$\bar{P}_k^- = \bar{F}_k P_{k-1} \bar{F}_k^T + \bar{Q}_k , \tag{3.160}$$

$$\bar{S}_k = \bar{H}_k \bar{P}_k^- \bar{H}_k^T + \Gamma_k \Phi_k + R_k , \tag{3.161}$$

$$\bar{K}_k = \bar{P}_k^- \bar{H}_k^T \bar{S}_k^{-1} , \tag{3.162}$$

$$\hat{\bar{x}}_k^- = \bar{F}_k \hat{x}_{k-1} + \bar{u}_k , \tag{3.163}$$

$$\hat{x}_k = \hat{\bar{x}}_k^- + \bar{K}_k (y_k - \Psi_k y_{k-1} - \bar{E}_k u_k - \bar{H}_k \hat{\bar{x}}_k^-) , \tag{3.164}$$

$$P_k = (I - \bar{K}_k \bar{H}_k) \bar{P}_k^- . \tag{3.165}$$

Again we notice that $\Psi_k = 0$ makes $\Gamma_k = 0$, $\Phi_k = 0$, $\bar{F}_k = A_k$, $\bar{H}_k = H_k$, $\bar{Q}_k = Q_k$, and $\bar{E}_k = 0$, and this algorithm transforms to KF Algorithm 1. Next, we will show that algorithms (3.146)–(3.151) and (3.160)–(3.165) give identical estimates and therefore are equivalent.

Equivalence of GKF Algorithms for CMN

The equivalence of the Bryson and Petovello algorithms for CMN was shown in [33], assuming time-invariant noise covariances that do not distinguish between FE- and BE-based models. We will now show that the GKF algorithms (3.146)–(3.151) and (3.160)–(3.165) are also equivalent [183]. Note that the algorithms can be said to be equivalent in the MSE sense if they give the same estimates for the same input; that is, for the given $\hat{x}_0$ and P_0, the outputs $\hat{x}_n$ and P_n in both algorithms are identical.

First, we represent the prior estimate $\hat{\bar{x}}_k^-$ (3.163) via the prior estimate $\hat{x}_k^-$ (3.149) and the prior error covariance $\bar{P}_k^-$ (3.160) via P_k^- (3.146) as

$$\begin{aligned}\hat{\bar{x}}_k^- &= \bar{F}_k\hat{x}_{k-1} + \Lambda_k\bar{z}_k\\ &= M_n\hat{x}_k^- + \Lambda_k\bar{z}_k\,,\end{aligned}\tag{3.166}$$

$$\begin{aligned}\bar{P}_k^- &= \bar{F}_k P_{k-1}\bar{F}_k^T + \bar{Q}_k\\ &= M_k P_k^- M_k^T - \Lambda_k \Phi_k^T M_k^T\,,\end{aligned}\tag{3.167}$$

where $M_k = I - \Lambda_k\bar{H}_k$. We now equate estimates (3.150) and (3.164),

$$\hat{x}_k^- + K_k(\bar{z}_k - \bar{H}_k\hat{x}_k^-) = \hat{\bar{x}}_k^- + \bar{K}_k(\bar{z}_k - \bar{H}_k\hat{\bar{x}}_k^-)\,,$$

substitute (3.166), combine with (3.153), transform as

$$\begin{aligned}K_n(z_k - \bar{H}_k\hat{x}_k^-) &= \Lambda_k(z_k - \bar{H}_k\hat{x}_k^-) + \bar{K}_k[(I - \bar{H}_k\Lambda_k)z_k\\ &\quad -(I - \bar{H}_k\Lambda_k)\bar{H}_k\hat{x}_k^-)]\,,\\ (K_k - \Lambda_k)(z_k - \bar{H}_k\hat{x}_k^-) &= \bar{K}_k(I - \bar{H}_k\Lambda_k)(z_k - \bar{H}_k\hat{x}_k^-)\,,\end{aligned}$$

and skip the nonzero equal terms from the both sides. This relates the gains K_k and $\bar{K}_k$ as

$$K_k = \bar{K}_k(I - \bar{H}_k\Lambda_k) + \Lambda_k\,,\tag{3.168}$$

which is the main condition for both estimates to be identical.

If also (3.168) makes the error covariances identical, then the algorithms can be said to be equivalent. To make sure that this is the case, we equate (3.151) and (3.165) as

$$\begin{aligned}(I - K_k\bar{H}_k)P_k^- - K_k\Phi_k^T &= (I - \bar{K}_k\bar{H}_k)\bar{P}_k^-\,,\\ P_k^- - K_k(\bar{H}_kP_k^- + \Phi_k^T) &= (I - \bar{K}_k\bar{H}_k)\bar{P}_k^-\,,\end{aligned}\tag{3.169}$$

substitute (3.168) and (3.167), rewrite (3.169) as

$$\begin{aligned}&P_k^- - [\bar{K}_k(I - \bar{H}_k\Lambda_k) + \Lambda_k](\bar{H}_kP_k^- + \Phi_k^T)\\ &\quad = (I - \bar{K}_k\bar{H}_k)(M_kP_k^-M_k^T - \Lambda_k\Phi_k^TM_k^T)\,,\end{aligned}$$

introduce $Y_k = (I - M_k^T)$, and arrive at

$$\begin{aligned}(I - \bar{K}_k\bar{H}_k)M_kP_k^-Y_k &= \bar{K}_k(\bar{H}_k\Lambda_k\Phi_k^TM_k^T + \Phi_k^T\\ &\quad - \bar{H}_k\Lambda_k\Phi_k^T) + \Lambda_k\Phi_k^TY_k\,.\end{aligned}$$

By rearranging the terms, this relationship can be transformed into

$$(I - \bar{K}_k\bar{H}_k)(M_kP_k^- - \Lambda_k\Phi_k^T)Y_k = \bar{K}_k\Phi_k^T$$

$$\begin{aligned}&(I - \bar{K}_k\bar{H}_k)(M_kP_k^- - \Lambda_k\Phi_k^T) - (I - \bar{K}_k\bar{H}_k)\bar{P}_k^-\\ &\quad = \bar{K}_k\Phi_k^T\end{aligned}$$

$$
\begin{aligned}
(I - \bar{K}_k\bar{H}_k)(M_kP_k^- - \Lambda_k\Phi_k^T)M_k^T - (I - \bar{K}_k\bar{H}_k)\bar{P}_k^-M_k^T &= \bar{K}_k\Phi_k^TM_k^T \\
(I - \bar{K}_k\bar{H}_k)\bar{P}_k^- - (I - \bar{K}_k\bar{H}_k)\bar{P}_k^-M_k^T &= \bar{K}_k\Phi_k^TM_k^T \\
(I - \bar{K}_k\bar{H}_k)\bar{P}_k^-Y_k &= \bar{K}_k\Phi_k^TM_k^T
\end{aligned}
\tag{3.170}
$$

and, by substituting $\bar{K}_k = \bar{P}_k^-\bar{H}_k^T\bar{S}_k^{-1}$, represented with

$$
(I - \bar{P}_k^-\bar{H}_k^T\bar{S}_k^{-1}\bar{H}_k)\bar{P}_k^-(I - M_k^T) = \bar{P}_k^-\bar{H}_k^T\bar{S}_k^{-1}\Phi_k^TM_k^T ,
$$

which, after tedious manipulations with matrices, becomes

$$
\begin{aligned}
\Lambda_k^T &= \bar{S}_k^{-1}(\Phi_k^T + \bar{H}_k\bar{P}_k^-\bar{H}_k^T\Lambda_k^T - \Phi_k^T\bar{H}_k^T\Lambda_k^T) , \\
\bar{S}_k\Lambda_k^T &= \Phi_k^T + \bar{H}_k\bar{P}_k^-\bar{H}_k^T\Lambda_k^T - \Phi_k^T\bar{H}_k^T\Lambda_k^T .
\end{aligned}
$$

By substituting $\bar{H}_k\bar{P}_k^-\bar{H}_k^T$ taken from $\bar{S}_k = \bar{H}_k\bar{P}_k^-\bar{H}_k^T + \Gamma_k\Phi_k + R_k$, we finally end up with

$$
\begin{aligned}
0 &= \Phi_k^T - \Gamma_k\Phi_k\Lambda_k^T - R_k\Lambda_k^T - \Phi_k^T\bar{H}_k^T\Lambda_k^T , \\
H_k\Phi_k &= \Phi_k^T\Gamma_k^T + \bar{H}_k\Phi_k .
\end{aligned}
\tag{3.171}
$$

Because $\Gamma_k\Phi_k$ is symmetric, $\Phi_k^T\Gamma_k^T = \Gamma_k\Phi_k$, and (3.131) means $\bar{H}_k = H_k - \Gamma_k$, then it follows that (3.171) is the equality $\bar{H}_k\Phi_k = \bar{H}_k\Phi_k$, and we conclude that the algorithms (3.146)–(3.151) and (3.160)–(3.165) are equivalent. It can also be shown that they do not demonstrate significant advantages over each other.

Colored Process Noise

Process noise becomes colored for various reasons, often when the internal control mechanism is not reflected in the state equation and thus becomes the error component. For example, the IEEE Standard 1139-2008 [75] states that, in addition to white noise, the oscillator phase PSD has four independent CPNs with slopes f^{-1}, f^{-2}, f^{-3}, and f^{-4}, where f is the Fourier frequency. Unlike CMN, which needs to be filtered out, slow CPN behavior can be tracked, as in localization and navigation. But when solving the identification problem, both the CMN and the CPN are commonly removed. This important feature distinguishes GKFs designed for CMN and CPN.

A linear process with CPN can be represented in state space in the standard formulation of the Gauss-Markov model with the equations

$$
x_k = F_kx_{k-1} + E_ku_k + B_kw_k , \tag{3.172}
$$

$$
w_k = \Theta_kw_{k-1} + \mu_k , \tag{3.173}
$$

$$
y_k = H_kx_k + v_k . \tag{3.174}
$$

where $x_k \in \mathbb{R}^K$, $y_k \in \mathbb{R}^M$, $F_k \in \mathbb{R}^{K\times K}$, and $H_k \in \mathbb{R}^{M\times K}$. To obtain GKF for CPN, we assign $w_k \in \mathbb{R}^K$, $B_k \in \mathbb{R}^{K\times K}$, and $\Theta_k \in \mathbb{R}^{K\times K}$, and think that F_k, B_k, and Θ_k are nonsingular. We also suppose that the noise vectors $\mu_k \sim \mathcal{N}(0, Q_k) \in \mathbb{R}^K$ and $v_k \sim \mathcal{N}(0, R_k) \in \mathbb{R}^M$ are uncorrelated with the covariances $E\{\mu_k\mu_k^T\} = Q_k$ and $E\{v_kv_k^T\} = R_k$, and choose the matrix Θ_k so that CPN w_k remains stationary.

The augmented state-space model for CPN becomes

$$\begin{bmatrix} x_k \\ w_k \end{bmatrix} = \begin{bmatrix} F_k & B_k\Theta_k \\ 0 & \Theta_k \end{bmatrix} \begin{bmatrix} x_{k-1} \\ w_{k-1} \end{bmatrix} + \begin{bmatrix} E_k \\ 0 \end{bmatrix} u_k + \begin{bmatrix} B_k \\ I \end{bmatrix} \mu_k ,$$

$$y_k = \begin{bmatrix} H_k & 0 \end{bmatrix} \begin{bmatrix} x_k \\ w_k \end{bmatrix} + v_k ,$$

and we see that, unlike (3.101), this does not make the KF ill-conditioned. On the other hand, colored noise w_k repeated in the state x_k can cause extra errors in the KF output under certain conditions.

To avoid the state augmentation, we now reason similarly to the measurement differencing approach [25] and derive the GKF for CPN based on the model (3.172)–(3.174) using the *state differencing* approach [182].

Using the state differencing method, the new state χ_k can be written as

$$\chi_k = x_k - \Pi_k x_{k-1} \tag{3.175a}$$

$$\begin{aligned} &= F_k x_{k-1} + E_k u_k + B_k w_k - \Pi_k x_{k-1} \\ &= \tilde{F}_k \chi_{k-1} + \tilde{u}_k + \tilde{w}_k - \tilde{F}_k \chi_{k-1} - \tilde{u}_k - \tilde{w}_k \\ &\quad + (F_k - \Pi_k) x_{k-1} + E_k u_k + B_k w_k , \end{aligned} \tag{3.175b}$$

where Π_k, $\tilde{F}_k$, $\tilde{u}_k$, and $\tilde{w}_k$ are still to be determined. The KF can be applied if we rewrite the new state equation as

$$\chi_k = \tilde{F}_k \chi_{k-1} + \tilde{u}_k + \tilde{w}_k \tag{3.176}$$

and find the conditions under which the noise $\tilde{w}_k$ will be zero mean and white Gaussian, $\tilde{w}_k \sim \mathcal{N}(0, \tilde{Q}_k) \in \mathbb{R}^K$. By referring to (3.176), unknown variables can be retrieved by putting to zero the remainder in (3.175b),

$$-\tilde{F}_k \chi_{k-1} - \tilde{u}_k - \tilde{w}_k + (F_k - \Pi_k) x_{k-1} + E_k u_k + B_k w_k = 0 . \tag{3.177}$$

Then we replace χ_{k-1} taken from (3.175a) into (3.177), combine with $x_{k-2} = F_{k-1}^{-1}(x_{k-1} - E_{k-1}u_{k-1} - B_{k-1}w_{k-1})$ taken from (3.172), and write

$$\begin{aligned} &(F_k - \Pi_k - \tilde{F}_k + \tilde{F}_k \Pi_{k-1} F_{k-1}^{-1}) x_{k-1} \\ &= \tilde{u}_k + \tilde{w}_k - E_k u_k + \tilde{F}_k \Pi_{k-1} F_{k-1}^{-1} u_{k-1} - B_k \mu_k - \\ &\quad - (B_k \Theta_k - \tilde{F}_k \Pi_{k-1} F_{k-1}^{-1} B_{k-1}) w_{k-1} . \end{aligned} \tag{3.178}$$

Although $x_0 = 0$ may hold for some applications, the case $x_{k-1} = 0$ for arbitrary k is isolated, and a solution to (3.178) can thus be found with simultaneous solution of four equations

$$\tilde{F}_k = F_k - \Pi_k + \tilde{F}_k \Pi_{k-1} F_{k-1}^{-1} , \tag{3.179}$$

$$\tilde{u}_k = E_k u_k - \tilde{F}_k \Pi_{k-1} F_{k-1}^{-1} u_{k-1} , \tag{3.180}$$

$$B_k \Theta_k = \tilde{F}_k \Pi_{k-1} F_{k-1}^{-1} B_{k-1} , \tag{3.181}$$

$$\tilde{w}_k = B_k \mu_k , \tag{3.182}$$

which suggest that the modified control signal $\tilde{u}_k$ is given by (3.180), and for nonsingular F_k, $\tilde{F}_k$, and Θ_k (3.181) gives

$$\Pi_n = \tilde{F}_{k+1}^{-1} \bar{\Theta}_{k+1} F_k , \tag{3.183}$$

where $\bar{\Theta}_k = B_k \Theta_k B_{k-1}^{-1}$ is a weighted version of Θ_k. Next, substituting (3.183) into (3.179) and providing some transformations give the backward and forward recursions for $\tilde{F}_k$,

$$\tilde{F}_k = F_k - \tilde{F}_{k+1}^{-1} \bar{\Theta}_{k+1} F_k + \bar{\Theta}_k \,, \tag{3.184}$$

$$\tilde{F}_{k-1} = F_{k-1} - \tilde{F}_k^{-1} \bar{\Theta}_k F_{k-1} + \bar{\Theta}_{k-1} \,,$$

$$\tilde{F}_k^{-1} \bar{\Theta}_k F_{k-1} = F_{k-1} - \tilde{F}_{k-1} + \bar{\Theta}_{k-1} \,,$$

$$\tilde{F}_k^{-1} \bar{\Theta}_k = I - (\tilde{F}_{k-1} - \bar{\Theta}_{k-1}) F_{k-1}^{-1} \,,$$

$$\tilde{F}_k = \bar{\Theta}_k [I - (\tilde{F}_{k-1} - \bar{\Theta}_{k-1}) F_{k-1}^{-1}]^{-1} \,. \tag{3.185}$$

It can be seen that $\tilde{F}_k$ becomes F_k for white noise if $\Pi_k = 0$ and $\bar{\Theta}_k = 0$, which follows from the backward recursion (3.184), but is not obvious from the forward recursion (3.185).

What follows is that the choice of the initial $\tilde{F}_0$ is critical to run (3.185). Indeed, if we assume that $\tilde{F}_0 = F_0$, then (3.185) gives $\tilde{F}_1 = \bar{\Theta}_1 F_0 \bar{\Theta}_0^{-1}$, which may go far beyond the proper matrix. But for the diagonal $\bar{\Theta}_1$ with equal components $\tilde{F}_0 = F_0$ is the only solution. On the other hand, $\tilde{F}_0 = \bar{\Theta}_0$ gives $\tilde{F}_1 = \bar{\Theta}_1$, which makes no sense. One way to specify $\tilde{F}_0$ is to assume that the process is time-invariant up to $k = 0$, convert (3.184) to the *nonsymmetric algebraic Riccati equation* (NARE) [107] or *quadratic matrix equation*

$$\tilde{F}^2 - \tilde{F}(F + \bar{\Theta}) + \bar{\Theta} F = 0 \,, \tag{3.186}$$

and solve (3.186) for $\tilde{F} = \tilde{F}_0$. However, the solution to (3.186) is generally not unique [107], and efforts must be made to choose the proper one.

So the modified state equation (3.176) becomes

$$\chi_k = \tilde{F}_k \chi_{k-1} + \tilde{u}_k + B_k \mu_k \tag{3.187}$$

with white Gaussian μ_k and matrix $\tilde{F}_k$, which is obtained recursively using (3.185) for the properly chosen fit $\tilde{F}_0$ of (3.186). To specify the initial χ_0, we consider (3.175a) at $k = 0$ and write $\chi_0 = x_0 - \Pi_0 x_{-1}$. Since x_{-1} is not available, we set $x_{-1} = x_0$ and take $\chi_0 = (I - \Pi_0)x_0 = (I - \tilde{F}_1^{-1} \bar{\Theta}_1 F_0)x_0$ as a reasonable initial state to run (3.187).

Finally, extracting $x_k = \chi_k + \Pi_k x_{k-1}$ from (3.175a), substituting it into the observation equation (3.174), and rearranging the terms, we obtain the modified observation equation

$$\tilde{y}_k = H_k \chi_k + v_k \tag{3.188}$$

with respect to $\tilde{y}_k = y_k - H_k \Pi_k x_{k-1}$, in which x_{k-1} can be replaced by the estimate $\hat{x}_{k-1}$.

Given y_k, $\hat{x}_0$, P_0, Q_k, R_k, and Θ_k, the GKF equations for CPN become

$$P_k^- = \tilde{F}_k P_{k-1} \tilde{F}_k^T + B_k Q_k B_k^T \,, \tag{3.189}$$

$$S_k = H_k P_k^- H_k^T + R_k \,, \tag{3.190}$$

$$K_k = P_k^- H_k^T S_k^{-1} \,, \tag{3.191}$$

$$\hat{\chi}_k^- = \tilde{F}_k \hat{\chi}_{k-1} + \tilde{u}_k \,, \tag{3.192}$$

$$\hat{\chi}_k = \hat{\chi}_k^- + K_k (y_k - H_k \tilde{F}_{k+1}^{-1} B_{k+1} \Theta_{k+1} B_k^{-1} F_k \hat{x}_{k-1} - H_k \hat{\chi}_k^-) \,, \tag{3.193}$$

$$\hat{x}_k = \hat{\chi}_k + \tilde{F}_{k+1}^{-1} B_{k+1} \Theta_{k+1} B_k^{-1} F_k \hat{x}_{k-1} \,, \tag{3.194}$$

$$P_k = (I - K_k H_k) P_k^- \,. \tag{3.195}$$

A specific feature of this algorithm is that the future values of B_k and Θ_k are required at $k+1$, which is obviously not a problem for LTI systems. It is also seen that $\Theta_k = 0$ results in $\tilde{F}_k = F_k$ and $\tilde{u}_k = E_k u_k$, and the algorithm becomes the standard KF Algorithm 1.

3.2.9 Kalman-Bucy Filter

The digital world today requires fast and accurate digital state estimators. But when the best sampling results in significant loss of information due to limitations in the operation frequency of digital devices, then there will be no choice but to design and implement filters physically in continuous time.

The optimal filter for continuous-time stochastic linear systems was obtained in [85] by Kalman and Bucy and is called the *Kalman-Bucy filter* (KBF). Obtaining KBF can be achieved in a standard way by minimizing MSE. Another way is to convert KF to continuous time as shown next.

Consider a stochastic LTV process represented in continuous-time state space by the following equations

$$x'(t) = A(t)x(t) + U(t)u(t) + w(t) , \tag{3.196}$$

$$y(t) = C(t)x(t) + v(t) , \tag{3.197}$$

where the initial $x(0)$ is supposed to be known.

For a sufficiently short time interval $\tau = t_k - t_{k-1}$, we can assume that all matrices and input signal are piecewise constant when $t_{k-1} \leqslant t \leqslant t_k$ and take $A(t) \cong A(t_k)$, $U(t) \cong U(t_k)$, $C(t) \cong C(t_k)$, and $u(t) \cong u(t_k)$. Following (3.6) and assuming that τ is small, we can then approximate $e^{A(t_k)\tau}$ with $e^{A(t_k)\tau} \cong I + A(t_k)\tau$ and write the solution to (3.196) as

$$x_k = e^{A(t_k)\tau} x_{k-1} + \int_{t_{k-1}}^{t_k} e^{A(t_k)(t_k-\theta)} \mathrm{d}\theta \; U(t_k)u(t_k) + \int_{t_{k-1}}^{t_k} e^{A(t_k)(t_k-\theta)} w(\theta)\mathrm{d}\theta .$$

Thus, the discrete analog of model (3.196) and (3.197) is

$$x_k = F_k x_{k-1} + E_k u_k + B_k w_k , \tag{3.198}$$

$$y_k = H_k x_k + v_k , \tag{3.199}$$

where we take $H_k = C(t_k)$,

$$F_k = e^{A(t_k)\tau} \cong I + A(t_k)\tau$$
$$E_k \cong \left[I + A(t_k)\frac{\tau}{2}\right] \tau U(t_k) \cong \tau U(t_k) ,$$
$$B_k \cong \tau I ,$$

refer to (3.25) and (3.26), and define the noise covariances as $Q_k = E\{w_k w_k^T\} = \tau S_w$ and $R_k = E\{v_k v_k^T\} = \frac{1}{\tau} S_v$, where S_w is the PSD of $w(t)$ and S_v is the PSD of $v(t)$. Ensured correspondence between continuous-time model (3.196) and (3.197) and discrete-time model (3.198) and (3.199) for small τ, the KBF can be obtained as follows.

We first notice that the continuous time estimate does not distinguish between the prior and posterior estimation errors due to $\tau = 0$. Therefore, the optimal Kalman gain K_k (3.78) for small τ can be transformed as

$$K_k = P_k^- H_k^T (H_k P_k^- H_k^T + R_k)^{-1} ,$$

$$= P(t_k)C^T(t_k)\left[C(t_k)P(t_k)C^T(t_k) + \frac{S_v}{\tau}\right]^{-1} ,$$

$$= P(t)C^T(t)S_v^{-1}\tau = K(t)\tau , \tag{3.200}$$

$$K(t) = P(t)C^T(t)S_v^{-1} , \tag{3.201}$$

where $K(t)$ is the optimal Kalman gain in continuous time.

The Kalman estimate (3.72) can now be transformed for small τ as

$$\hat{x}_k = F_k\hat{x}_{k-1} + E_k u_k + K_k(y_k - H_k F_k \hat{x}_{k-1} - H_k E_k u_k) ,$$
$$\frac{\hat{x}(t_k) - \hat{x}(t_{k-1})}{\tau} = A(t_k)\hat{x}(t_{k-1}) + U(t_k)u(t_k)$$
$$+ K(t_k)[y(t_k) - C(t_k)\hat{x}(t_{k-1}) - \tau C(t_k)U(t_k)] ,$$
$$\hat{x}'(t) = A(t)\hat{x}(t) + U(t)u(t) + K(t)[y(t) - C(t)\hat{x}(t)] , \tag{3.202}$$

where $K(t)$ is given by (3.201).

Reasoning along similar lines as for (3.201), error covariance P_k (3.79) combined with P_k^-, defined as (3.75), can also be transformed for small τ as

$$P_k = (I - K_k H_k)(F_k P_{k-1} F_k^T + Q_k) ,$$
$$P(t_k) = [I - \tau K(t_k)C(t_k)][I + \tau A(t_k)]P(t_{k-1})[I + \tau A(t_k)]^T$$
$$+ S_w\tau - K(t_k)C(t_k)S_w\tau^2 ,$$
$$= P(t_{k-1}) + P(t_{k-1})\tau A^T(t_k)$$
$$+ \tau[A(t_k) - K(t_k)C(t_k)]P(t_{k-1}) + S_w\tau$$
$$\frac{P(t_k) - P(t_{k-1})}{\tau} = P(t_{k-1})A^T(t_k) + A(t_k)P(t_{k-1})$$
$$- K(t_k)C(t_k)P(t_{k-1}) + S_w .$$
$$P'(t) = P(t)A^T(t) + A(t)P(t) - K(t)C(t)P(t) + S_w , \tag{3.203}$$

where (3.203) is the *Riccati differential equation* (RDE) (A.25).

Thus, the KBF can be represented by two differential equations

$$\hat{x}'(t) = A(t)\hat{x}(t) + U(t)u(t)$$
$$+ P(t)C^T(t)S_v^{-1}[y(t) - C(t)\hat{x}(t)] , \tag{3.204}$$

$$P'(t) = P(t)A^T(t) + A(t)P(t)$$
$$- P(t)C^T(t)S_v^{-1}C(t)P(t) + S_w , \tag{3.205}$$

given the initial state $\hat{x}(0)$ and error covariance $P(0)$.

It is worth noting that although less attention is paid to the RBP in the modern digital world, its hardware implementation is required when the main spectral information is located above the operation frequency of the digital device.

3.3 Linear Recursive Smoothing

When better noise reduction is required than using optimal filtering, we think of *optimal smoothing*. The term *smoothing* comes from antiquity, and solutions to the smoothing problem can be found in the works of Gauss, Kolmogorov, and Wiener. Both optimal filtering and optimal smoothing minimize the MSE using data taken from the past up to the current time index k. What distinguishes the two approaches is that optimal filtering gives an estimate in the current time index k, while optimal smoothing refers the estimate to the past point $k-q$ with a lag $q>0$. Therefore, it is common to say that $\hat{x}_{k|k}$ is the output of an *optimal filter* and $\hat{x}_{k-q|k}$ is the output of a q-lag *optimal smoother*. Smoothing can also be organized at the current point k by involving q future data. In this case, it is said that the estimate $\hat{x}_{k|k+q}$ is produced by a q-lag *smoothing filter*.

The following smoothing problems can be solved:

- *Fixed-lag smoothing* gives an estimate at $m<k-q<k$ with a fixed lag q over a horizon $[m,k]$, where m can be zero.
- *Fixed-interval smoothing* gives an estimate at any point in a fixed interval $[m,k]$ with a lag q ranging from $q=1$ to $q=m$.
- *Fixed-point smoothing* gives an estimate at a fixed point $m<v<k$ in $[m,k]$ with variable lag $q=k-v$, where m can be zero.

The opposite problem to *smoothing* is called *prediction*. The prediction refers to an estimate of the future point with some step $p>0$, and it is said that $\hat{x}_{k+p|k}$ is the p-step predictive estimate and $\hat{x}_{k|k-p}$ is the p-step predictive filtering estimate.

Several two-pass (forward-backward) smoothers have been developed using Kalman recursions [9, 83]. The idea behind each such solution is to provide forward filtering using KF and then arrange the backward pass to obtain a q-lag smoothing estimate at $k-q$. Next we will look at a few of the most widely used recursive smoothers.

3.3.1 Rauch-Tung-Striebel Algorithm

In the Rauch-Tung-Striebel (RTS) smoother developed for fixed intervals [154], the forward state estimates $\hat{x}_k$ and $\hat{x}_k^-$ and the error covariances P_k and P_k^- are taken at each k from KF, and then the q-lag smoothed state estimate $\tilde{x}_{k-q|k}$ and error covariance $\tilde{P}_{k-q|k}$ are computed at $r=k-q<k$ backward using backward recursions.

If we think of a smoothing estimate $\tilde{x}_r \triangleq \tilde{x}_{r|k}$ as an available filtering estimate $\hat{x}_r$ adjusted for the predicted residual $\tilde{x}_{r+1}-\hat{x}_{r+1}^-$ by the gain $\tilde{K}_r$ as

$$\tilde{x}_r = \hat{x}_r + \tilde{K}_r(\tilde{x}_{r+1} - \hat{x}_{r+1}^-) \,, \tag{3.206}$$

then the RTS smoother derivation can be provided as follows.

Define the smoother error $\tilde{\varepsilon}_r = x_r - \tilde{x}_r$ by

$$\begin{aligned}\tilde{\varepsilon}_r &= x_r - \hat{x}_r - \tilde{K}_r(\tilde{x}_{r+1} - \hat{x}_{r+1}^-)\\ &= \hat{\varepsilon}_r - \tilde{K}_r(-x_{r+1} + \tilde{x}_{r+1} + x_{r+1} - \hat{x}_{r+1}^-)\\ &= \hat{\varepsilon}_r + \tilde{K}_r(\tilde{\varepsilon}_{r+1} - \hat{\varepsilon}_{r+1}^-) \,,\end{aligned} \tag{3.207}$$

which gives the error covariance

$$\tilde{P}_r = P_r + \tilde{K}_r(\tilde{P}_{r+1} - P_{r+1}^-)\tilde{K}_r^T \,. \tag{3.208}$$

To find the optimal smoother gain $\tilde{K}_r$, substitute $\hat{\varepsilon}_{r+1}^- = F_{r+1}\hat{\varepsilon}_r$, represent (3.107) as

$$\begin{aligned}\tilde{\varepsilon}_r &= \hat{\varepsilon}_r + \tilde{K}_r(\tilde{\varepsilon}_{r+1} - F_{r+1}\hat{\varepsilon}_r)\\ &= (I - \tilde{K}_rF_{r+1})\hat{\varepsilon}_r + \tilde{K}_r\tilde{\varepsilon}_{r+1} \,,\end{aligned}$$

and arrive at another form of the error covariance

$$\tilde{P}_r = (I - \tilde{K}_rF_{r+1})P_r(I - \tilde{K}_rF_{r+1})^T + \tilde{K}_r\tilde{P}_{r+1}\tilde{K}_r^T \,.$$

Now, apply the derivative with respect to $\tilde{K}_r$ to the trace of $\tilde{P}_r$, consider $\tilde{P}_{r+1}$ as the process noise covariance Q_{r+1}, and obtain $\tilde{K}_r = P_r F_{r+1}^T (P_{r+1}^-)^{-1}$. The RTS smoothing algorithm can then be formalized with the following steps,

$$\tilde{K}_r = P_r F_{r+1}^T (P_{r+1}^-)^{-1} , \tag{3.209}$$

$$\tilde{x}_r = \hat{x}_r + \tilde{K}_r (\tilde{x}_{r+1} - \hat{x}_{r+1}^-) , \tag{3.210}$$

$$\tilde{P}_r = P_r + \tilde{K}_r (\tilde{P}_{r+1} - P_{r+1}^-) \tilde{K}_r^T , \tag{3.211}$$

where the required KF estimates $\hat{x}_k$, $\hat{x}_k^-$, P_k, and P_k^- must be available (saved) from the filtering procedure. A more complete analysis of RTS smoother can be found in [83, 185].

3.3.2 Bryson-Frazier Algorithm

Another widely used modified Bryson-Frazier (MBF) smoother, developed by Bierman [20] for fixed intervals, also uses data stored from the KF forward pass. In this algorithm, the backward pass for $\tilde{\Lambda}_k = 0$ and $\hat{\lambda}_k = 0$ is organized using the following recursions,

$$\tilde{\Lambda}_r = H_r^T S_r^{-1} H_r + \Pi_r^T \hat{\Lambda}_r \Pi_r , \tag{3.212}$$

$$\hat{\Lambda}_{r-1} = F_r^T \tilde{\Lambda}_r F_r , \tag{3.213}$$

$$\tilde{\lambda}_r = -H_r^T S_r^{-1} y_r + \Pi_r^T \hat{\lambda}_r , \tag{3.214}$$

$$\hat{\lambda}_{r-1} = F_r^T \tilde{\lambda}_r , \tag{3.215}$$

where $\Pi_r = I - K_r H_r$ and other definitions are taken from Algorithm 4. The MBF q-lag smoothing estimate and error covariance are then found as

$$\tilde{x}_r = \hat{x}_r - P_r \hat{\lambda}_r , \tag{3.216}$$

$$\tilde{P}_r = P_r - P_r \hat{\Lambda}_r P_r . \tag{3.217}$$

It can be seen that the inversion of the covariance matrix is not required for this algorithm, which is an important advantage over the RTS algorithm.

3.3.3 Two-Filter (Forward-Backward) Smoothing

Another solution to the smoothing problem can be proposed if we consider data on $[m, \dots, r, \dots, k]$, where m can be zero, provide forward filtering on $[m, r]$ and backward filtering on $[k, r]$, and then combine both estimates at r. The resulting two-filter (or forward-backward) smoother is suitable for the fixed interval problem [83], but can also be applied to other smoothing problems. Furthermore, both recursive and batch structures can be designed for two-filter smoothing by fusing either filtering or predictive estimates.

A two-filter smoother can be designed if we assume that the forward filtering estimate is obtained at r over $[m, r]$ as $\hat{x}_r^{\mathrm{f}}$ and the backward filtering estimate is obtained at r over $[k, r]$ as $\hat{x}_r^{\mathrm{b}}$. The smoothing estimate $\tilde{x}_r$ can then be obtained by combining both estimates as

$$\tilde{x}_r = K_r^{\mathrm{f}} \hat{x}_r^{\mathrm{f}} + K_r^{\mathrm{b}} \hat{x}_r^{\mathrm{b}} , \tag{3.218}$$

where the gains K_r^{f} and K_r^{b} are required such that the MSE is minimized for the optimal $\tilde{x}_r$. Since K_r^{f} and K_r^{b} should not cause a regular error (bias), they can be linked as $K_r^{\mathrm{f}} + K_r^{\mathrm{b}} = I$. Then (3.218) can be represented as

$$\tilde{x}_r = K_r^{\mathrm{f}}\hat{x}_r^{\mathrm{f}} + (I - K_r^{\mathrm{f}})\hat{x}_r^{\mathrm{b}} . \tag{3.219}$$

By defining the forward filtering error as $\varepsilon_r^{\mathrm{f}} = x_r - \hat{x}_r^{\mathrm{f}}$, the backward one as $\varepsilon_r^{\mathrm{b}} = x_r - \hat{x}_r^{\mathrm{b}}$, and the smoother error as $\tilde{\varepsilon}_r = x_r - \tilde{x}_r$, we next represent the error covariance $\tilde{P}_r = \mathcal{E}\{\tilde{\varepsilon}_r\tilde{\varepsilon}_r^T\}$ as

$$\begin{aligned}
\tilde{P}_r &= \mathcal{E}\{[x_r - K_r^{\mathrm{f}}\hat{x}_r^{\mathrm{f}} - (I - K_r^{\mathrm{f}})\hat{x}_r^{\mathrm{b}}][\dots]^T\} \\
&= \mathcal{E}\{(\varepsilon_r^{\mathrm{b}} - K_r^{\mathrm{f}}\hat{x}_r^{\mathrm{f}} + K_r^{\mathrm{f}}\hat{x}_r^{\mathrm{b}})(\dots)^T\} \\
&= \mathcal{E}\{(\varepsilon_r^{\mathrm{b}} + K_r^{\mathrm{f}}x_r - K_r^{\mathrm{f}}\hat{x}_r^{\mathrm{f}} - K_r^{\mathrm{f}}x_r + K_r^{\mathrm{f}}\hat{x}_r^{\mathrm{b}})(\dots)^T\} \\
&= \mathcal{E}\{(\varepsilon_r^{\mathrm{b}} + K_r^{\mathrm{f}}\varepsilon_r^{\mathrm{f}} - K_r^{\mathrm{f}}\varepsilon_r^{\mathrm{b}})(\dots)^T\} \\
&= P_r^{\mathrm{b}} + K_r^{\mathrm{f}}(P_r^{\mathrm{f}} + P_r^{\mathrm{b}})K_r^{\mathrm{f}^T} - 2K_r^{\mathrm{f}}P_r^{\mathrm{b}} ,
\end{aligned}$$

equate to zero the derivative applied to the trace of $\tilde{P}_r$ with respect to K_r^{f}, and obtain the optimal gains as

$$K_r^{\mathrm{f}} = P_r^{\mathrm{b}}(P_r^{\mathrm{f}} + P_r^{\mathrm{b}})^{-1} , \tag{3.220}$$

$$K_r^{\mathrm{b}} = P_r^{\mathrm{f}}(P_r^{\mathrm{f}} + P_r^{\mathrm{b}})^{-1} . \tag{3.221}$$

Using (3.220) and (3.221), the error covariance $\tilde{P}_r$ can finally be transformed to

$$\tilde{P}_r = \left[(P_r^{\mathrm{f}})^{-1} + (P_r^{\mathrm{b}})^{-1}\right]^{-1} , \tag{3.222}$$

and we see that the information $\tilde{\mathcal{I}}_r = \tilde{P}_r^{-1}$ about the smoothing estimate (3.219) is additively combined with the information $\tilde{\mathcal{I}}_r^{\mathrm{f}} = (\tilde{P}_r^{\mathrm{f}})^{-1}$ at the output of the forward filter and $\tilde{\mathcal{I}}_r^{\mathrm{b}} = (\tilde{P}_r^{\mathrm{b}})^{-1}$ at the output of the backward filter,

$$\tilde{\mathcal{I}}_r = \tilde{\mathcal{I}}_r^{\mathrm{f}} + \tilde{\mathcal{I}}_r^{\mathrm{b}} . \tag{3.223}$$

The proof of (3.222) can be found in [83, 185] and is postponed to "Problems".

Thus, two-filter (forward-backward) smoothing is provided by (3.218) with optimal gains (3.220) and (3.221) and error covariance (3.222). It also follows that the information identity (3.223) is fundamental to two-filter optimal smoothing and must be obeyed regardless of structure (batch or recursive). Practical algorithms of these smoothers are given in [185], and more details about the two-filter smoothing problem can be found in [83].

3.4 Nonlinear Models and Estimators

So far, we have looked at linear models and estimators. Since many systems are nonlinear in nature and therefore have nonlinear dynamics, their mathematical representations require nonlinear ODEs [80] and algebraic equations. More specifically, we will assume that the process (or system) is nonlinear and that measurement can also be provided using nonlinear sensors. For white Gaussian noise, the exact solution of the nonlinear problem in continuous time was found by Stratonovich [193] and Kushner [92] in the form of a *stochastic partial differential equation* (SPDE). The state probability density associated with the SPDE is given by the Stratonovich-Kushner equation, which belongs to the FPK family of equations (2.118), and KBF is its particular solution. Because the Stratonovich-Kuchner equation has no general solution and is therefore impractical in discrete time, much more attention has been drawn to approximate solutions developed during decades to provide a state estimate with sufficient accuracy.

In discrete-time state-space, the nonlinear state and observation equations can be written, respectively, as

$$x_k = f_k(x_{k-1}, u_k, w_k) , \tag{3.224}$$

$$y_k = h_k(x_k, v_k) , \tag{3.225}$$

where f_k and h_k are some nonlinear time-varying functions, u_k is an input, and w_k and v_k are white Gaussian with known statistics. If the noise is of low intensity or acts additively, the model (3.224) and (3.225) is often written as

$$x_k = f_k(x_{k-1}, u_k) + w_k , \tag{3.226}$$

$$y_k = h_k(x_k) + v_k . \tag{3.227}$$

Provided the posterior distribution $p(x_k|y_0^k)$ of the state x_k by (3.39), the Bayesian estimate $\hat{x}_k$ for nonlinear models can be found as (3.41) with error covariance (3.42). For Gauss-Markov processes, $p(x_k|y_0^k)$ can be specified by solving the FPK equation (2.118) in the stationary mode as (2.120). This gives fairly accurate estimates, but in practice difficulties arise when f_k and/or h_k are not smooth enough and the noise is large.

To find approximate solutions, Cox in [38] and others extended KF to nonlinear models in the first-order approximation, and Athans, Wishner, and Bertolini provided it in [12] in the second-order approximation. However, it was later told [178, 185] that nothing definite can be said about the second-order approximation, and the first-order extended KF (EKF) was recommended as the main tool. Yet another versions of EKF have been proposed, such as the divided difference filter [136] and quadrature KF [10], and we notice that several solutions beyond the EKF can be found in [40].

Referring to particularly poor performance of the EKF when the system is highly nonlinear and noise large [49], Julier, Uhlmann, and Durrant-Whyte developed the unscented KF (UKF) [81, 82, 199] to pick samples around the mean, propagate through the nonlinearity, and then recover the mean and covariance. Although the UKF often outperforms the EKF and can be improved by using high-order statistics, it loses out in many applications to another approach known as *particle filtering*.

The idea behind the particle filter (PF) approach [44] is to use sequential Monte Carlo (MC) methods [116] and solve the filtering problem by computing the posterior distributions of the states of some Markov process, given some noisy and partial observations. Given enough time to generate a large number of particles, the PF usually provides better accuracy than other nonlinear estimators. Otherwise, it can suffer from two effects called *sample degeneracy* and *sample impoverishment*, which cause divergence.

Next we will discuss the most efficient nonlinear state estimators. Our focus will be on algorithmic solutions, advantages, and drawbacks.

3.4.1 Extended Kalman Filter

The EKF is viewed in state estimation as a nonlinear version of KF that linearizes the mean and covariance between two neighboring discrete points. For smooth rather than harsh nonlinearities, EKF has become the *de facto* standard technique used in many areas of system engineering.

For the model in (3.226) and (3.227), the first-order EKF (EKF-1) and the second-order EKF (EKF-2) can be obtained using the Taylor series as shown next. Suppose that both $f_k(x_{k-1})$ and

$h_k(x_k)$ are smooth functions and approximate them on a time step between two neighboring discrete points using the second-order Taylor series. For simplicity, omit the known input u_k.

The nonlinear function $f_k(x_{k-1})$ can be expanded around the estimate $\hat{x}_{k-1}$ and $h_k(x_k)$ around the prior estimate $\hat{x}_k^-$ as [178]

$$f_k(x_{k-1}) \cong f_k(\hat{x}_{k-1}) + \dot{F}_k \varepsilon_{k-1} + \frac{1}{2}\alpha_k , \tag{3.228}$$

$$h_k(x_k) \cong h_k(\hat{x}_k^-) + \dot{H}_k \varepsilon_k^- + \frac{1}{2}\beta_k , \tag{3.229}$$

where the Jacobian matrices (A.20) are

$$\dot{F}_k = \left.\frac{\partial f_k}{\partial x}\right|_{x=\hat{x}_{k-1}} , \tag{3.230}$$

$$\dot{H}_k = \left.\frac{\partial h_k}{\partial x}\right|_{x=\hat{x}_k^-} , \tag{3.231}$$

$\varepsilon_k^- = x_k - \hat{x}_k^-$ is the prior estimation error, and $\varepsilon_k = x_k - \hat{x}_k$ is the estimation error. The second-order terms are represented by [15]

$$\alpha_k = \sum_{i=1}^{K} e_i^K \varepsilon_{k-1}^T \ddot{F}_{ik} \varepsilon_{k-1} , \tag{3.232}$$

$$\beta_k = \sum_{j=1}^{M} e_j^M \varepsilon_k^{-T} \ddot{H}_{jk} \varepsilon_k^- , \tag{3.233}$$

where the Hessian matrices (A.21) are

$$\ddot{F}_{ik} = \left.\frac{\partial^2 f_{ik}}{\partial x^2}\right|_{x=\hat{x}_{k-1}} , \tag{3.234}$$

$$\ddot{H}_{jk} = \left.\frac{\partial^2 h_{jk}}{\partial x^2}\right|_{x=\hat{x}_k^-} , \tag{3.235}$$

and f_{ik}, $i \in [1, K]$, and h_{jk}, $j \in [1, M]$, are the ith and jth components, respectively, of vectors $f_k(x_{k-1})$ and $h_k(x_k)$. Also, $e_i^K \in \mathbb{R}^K$ and $e_j^M \in \mathbb{R}^M$ are Cartesian basis vectors with the ith and jth components equal to unity and all others equal to zero.

Referring to (3.228) and (3.229), the nonlinear model (3.226) and (3.227) can now be approximated with

$$x_k = \dot{F}_k x_{k-1} + \eta_k + w_k , \tag{3.236}$$

$$\tilde{y}_k = \dot{H}_k x_k + v_k , \tag{3.237}$$

where the modified observation vector is $\tilde{y}_k = y_k - \psi_k$, also

$$\eta_k = f_k(\hat{x}_{k-1}) - \dot{F}_k \hat{x}_{k-1} + \frac{1}{2}\alpha_k , \tag{3.238}$$

$$\psi_k = h_k(\hat{x}_k^-) - \dot{H}_k \hat{x}_k^- + \frac{1}{2}\beta_k , \tag{3.239}$$

and other matrices are specified earlier. As can be seen, the second-order terms α_k and β_k affect only η_k and ψ_k. If α_k and β_k contribute insignificantly to the estimates [185], they are omitted in the EKF-1. Otherwise, the second-order EKF-2 is used.

The EKF algorithm can now be summarized with the recursions

$$P_k^- = \dot{F}_k P_{k-1} \dot{F}_k^T + Q_k , \tag{3.240}$$

$$S_k = \dot{H}_k P_k^- \dot{H}_k^T + R_k \,, \tag{3.241}$$

$$K_k = P_k^- \dot{H}_k^T S_k^{-1} \,, \tag{3.242}$$

$$\hat{x}_k = f_k(\hat{x}_{k-1}) + K_k\{y_k - \psi_k - h_k[f_k(\hat{x}_{k-1})]\} \,, \tag{3.243}$$

$$P_k = (I - K_k \dot{H}_k) P_k^- \,. \tag{3.244}$$

Now it is worth mentioning again that the EKF usually works very accurately when both nonlinearities are weak. However, if this condition is not met and the nonlinearity is severe, another approach using the "unscented transformation" typically gives better estimates. The corresponding modification of the KF [81] is considered next.

3.4.2 Unscented Kalman Filter

Referring to typically large errors and possible divergence of the KF with sharp nonlinearities [49], Julier, Uhlrnann, and Durrant-Whyte have developed another approach to nonlinear filtering, following the intuition that it is easier to approximate a probability distribution than it is to approximate an arbitrary nonlinear function or transformation [82]. The approach originally proposed by Uhlrnann in [199] was called the *unscented transformation* (UT) and resulted in the unscented (or Uhlrnann) KF (UKF) [81, 82]. The UKF has been further deeply investigated and used extensively in nonlinear filtering practice. Next we will succinctly introduce UT and UKF, referring the reader to [185] and many other books, tutorials, and papers discussing the theory and applications of UT and UKF for further learning.

The Unscented Transformation

Consider a nonlinear function $f_k(x_{k-1})$. The problem with its linear approximation (3.210) is that the matrix $\dot{F}_k$ projects x_{k-1} into the next time index k in a linear way, which may not be precise and even cause divergence. To avoid the linearization and not use matrix $\dot{F}_k$, we can project through $f_k(x_{k-1})$ the statistics of x_{k-1} and obtain at k the mean and approximate covariance. First of all, we are interested in the projection of the mean and the deviations from the mean by the standard deviation.

Referring to [82], we can consider a K-state process and assign $2K+1$ weighted samples of *sigma points* $\chi_{k-1}^{(i)}$ at $k-1$ for $i \in [0, K]$ as

$$\begin{aligned}
\chi_{k-1}^{(0)} &= \bar{x}_{k-1} \,, \quad i = 0 \,, \\
\chi_{k-1}^{(i)} &= \bar{x}_{k-1} + \left(\sqrt{(K+\kappa)P_{k-1}}\right)_i \,, \quad i \in [1, K] \,, \\
\chi_{k-1}^{(i+K)} &= \bar{x}_{k-1} - \left(\sqrt{(K+\kappa)P_{k-1}}\right)_i \,, \quad i \in [1, K] \,,
\end{aligned} \tag{3.245}$$

where $A = \sqrt{P}$, and A_i means the ith column of A if $P = AA^T$. Otherwise, if $P = A^T A$, A_i means the ith raw of A. Note that, to find the matrix square root, the Cholesky decomposition method can be used [59]. A tuning factor κ can be either positive or negative, although such that $K + \kappa \neq 0$. This factor is introduced to affect the fourth and higher-order sample moments of the sigma points.

While propagating through $f_k(x_{k-1})$, the sigma points (3.245) are projected at k as

$$\bar{x}_k^{(i)} = f_k(\chi_{k-1}^{(i)}) \,, \tag{3.246}$$

and one may compute the predicted mean $\bar{x}_k$ by the weighed sum of $\bar{x}_k^{(i)}$ as

$$\bar{x}_k = \sum_{i=0}^{2K} W_i \bar{x}_k^{(i)}, \tag{3.247}$$

where the weighted function W_i is specified in [82] as $W_0 = \kappa/(K+\kappa)$ for $i = 0$ and as $W_i = 0.5/(K+\kappa)$ for $i \neq 0$. But it can also be chosen in another way, for example, as $W_0 = 1$ for $i = 0$ and $W_0 = 1/2K$ for $i \neq 0$ [185]. The advantage of the UT approach is that the weights W_i can be chosen in such a way that this also affects higher-order statistics.

Provided $\bar{x}_k$ by (3.247), the error covariance can be predicted as

$$P_k^{(x)} = \sum_{i=0}^{2K} W_i [\bar{x}_k^{(i)} - \bar{x}_k][\bar{x}_k^{(i)} - \bar{x}_k]^T . \tag{3.248}$$

Referring to (3.246) and $h_k(x_k)$ in (3.227), the predicted observations can then be formed as

$$\bar{y}_k^{(i)} = h_k(\bar{x}_k^{(i)}), \tag{3.249}$$

and further averaged to obtain

$$\bar{y}_k = \sum_{i=0}^{2K} W_i \bar{y}_k^{(i)} \tag{3.250}$$

with the prediction error covariance

$$P_k^{(y)} = \sum_{i=0}^{2K} W_i [\bar{y}_k^{(i)} - \bar{y}_k][\bar{y}_k^{(i)} - \bar{y}_k]^T . \tag{3.251}$$

The cross-covariance between $\bar{x}_k$ and $\bar{y}_k$ can be found as

$$P_k^{(xy)} = \sum_{i=0}^{2K} W_i [\bar{x}_k^{(i)} - \bar{x}_k][\bar{y}_k^{(i)} - \bar{y}_k]^T . \tag{3.252}$$

It follows that the UT just described provides a set of statistics at the outputs of the nonlinear blocks for an optimal solution to the filtering problem. The undoubted advantage over the Taylor series (3.228) and (3.229) is that the Jacobian matrices are not used, and projections are obtained directly through nonlinear functions.

UKF Algorithm

So the ultimate goal of the UKF algorithm is to get rid of all matrices of the EKF algorithm (3.240)–(3.244) using the previous introduced projections. Let us show how to do it. Since (3.248) predicts the error covariance, we can approximate P_k^- with $P_k^{(x)}$, and since (3.251) predicts the innovation covariance (3.223), we can approximate S_k with $P_k^{(y)}$. Then the product $P_k^- \dot{H}_k^T$ in (3.242) can be transformed to $P_k^{(xy)}$ defined by (3.252) as

$$\begin{aligned} P_k^- \dot{H}_k^T &= P_k^{(x)} \dot{H}_k^T \\ &= \sum_{i=0}^{2K} W_i [\bar{x}_k^{(i)} - \bar{x}_k][\bar{x}_k^{(i)} - \bar{x}_k]^T \dot{H}_k^T \\ &= \sum_{i=0}^{2K} W_i [\bar{x}_k^{(i)} - \bar{x}_k][\dot{H}_k \bar{x}_k^{(i)} - \dot{H}_k \bar{x}_k]^T \end{aligned}$$

$$
\begin{aligned}
&= \sum_{i=0}^{2K} W_i [\bar{x}_k^{(i)} - \bar{x}_k][\bar{y}_k^{(i)} - \bar{y}_k]^T \\
&= P_k^{(xy)}
\end{aligned}
\tag{3.253}
$$

and the gain K_k (3.242) represented as $K_k = P_k^{(xy)} P_k^{(y)^{-1}}$. Finally, (3.244) can be transformed as

$$
\begin{aligned}
P_k &= (I - K_k \dot{H}_k) P_k^- \\
&= P_k^- - K_k \dot{H}_k P_k^- \\
&= P_k^- - K_k P_k^{(xy)^T} \\
&= P_k^- - K_k P_k^{(y)^T} K_k^T .
\end{aligned}
\tag{3.254}
$$

It can be seen from the previous that there is now a complete set of predictions to get rid of matrices in EKF, and the UKF algorithm is thus the following

$$K_k = P_k^{(xy)} P_k^{(y)^{-1}} , \tag{3.255}$$

$$\hat{x}_k = \bar{x}_k + K_k (y_k - \bar{y}_k) , \tag{3.256}$$

$$P_k = P_k^- - K_k P_k^{(y)} K_k^T , \tag{3.257}$$

where $P_k^{(xy)}$ is computed by (3.252), $P_k^- = P_k^{(x)}$ by (3.248), $P_k^{(y)}$ by (3.251), $\bar{x}_k$ by (3.247), and $\bar{y}_k$ by (3.250).

Many studies have confirmed that the UKF in (3.255)–(3.257) is more accurate than EKF when the nonlinearity is not smooth. Moreover, using the additional degree of freedom κ in the weights W_i, the UKF can be made even more accurate. It should be noted, however, that the same statistics can be predicted by UT for different distributions, which is a serious limitation on accuracy. Next, we will consider the "particle filtering" approach that does not have this drawback.

3.4.3 Particle Filtering

Along with the linear optimal KF, the Bayesian approach has resulted in two important suboptimal nonlinear solutions: *point mass filter* (PMF) [8] and *particle filter* (PF) [62]. The PMF computes posterior density recursively over a deterministic state-space grid. It can be applied to any nonlinear and non-Gaussian model, but it has a serious limitation: complexity. The PF performs sequential MC estimation by representing the posterior densities with samples or particles, similarly to PMF, and is generally more successful in accuracy than EKF and UKF for highly nonlinear systems with large Gaussian and non-Gaussian noise.

To introduce the PF approach, we will assume that a nonlinear system is well represented by a first-order Markov model and is observed over a nonlinear equation with (3.206) and (3.207) ignoring the input u_k. We will be interested in the best estimate for $x_{1:k} = \{x_1\ x_2\ \dots\ x_k\}$, given measurements $y_{1:k} = \{y_1\ y_2\ \dots\ y_k\}$. More specifically, we will try to obtain a filtering estimate from $p(x_k|x_{1:k})$. Within the Bayesian framework, the required estimate and the estimation error can be extracted from the conditional posterior probability density $p(x_{1:k}|y_{1:k})$. This can be done, albeit approximately, using the classical "bootstrap" PF considered next.

Bootstrap

The idea behind PF is simple and can be described in two steps, referring to the *bootstrap* proposed by Gordon, Salmond, and Smith [62].

Prediction. Draw N samples[3] (or particles) $x_{k-1}^{(i)} \sim p(x_{k-1}|y_{1:k-1})$ and $w_k^{(i)} \sim p(w_k)$. Pass these samples through the system model (3.206) to obtain N prior samples $x_k^{-(i)} = f_k(x_{k-1}^{(i)}, w_k^{(i)})$. Since prediction may not be accurate, update the result next.

Update. Using y_k, evaluate the likelihood of each prior sample as $p(y_k|x_k^{-(i)})$, normalize by $\Sigma = \sum_{i=1}^{N} p(y_k|x_k^{-(i)})$, and obtain a normalized weight $\omega_i = p(y_k|x_k^{-(i)})/\Sigma$, which is a discrete pmf over $x_k^{-(i)}$. From ω_i, find the estimate of x_k and evaluate errors. Optionally, resample N times from ω_i to generate samples $\tilde{x}_k^{(i)}$ such that $P\{\tilde{x}_k^{(j)} = x_k^{-(i)}\} = \omega_i$ holds for any j.

If we generate a huge number of the particles at each k, then the bootstrap will be as accurate as the best hypothetical filter. This, however, may not be suitable for many real-time applications. But even if the computation time is not an issue, incorrectly set initial values cause errors in the bootstrap to grow at a higher rate than in the KF. Referring to these specifics, next we will consider a more elaborated and complete theory of PF.

Particle Filter Theory

Within the Bayesian framework, we are interested in the posterior pmf $p(x_{1:k}|y_{1:k})$ to obtain an estimate of $x_{1:k}$ and evaluate the estimation errors. Thus, we may suppose that the past $p(x_{1:k-1}|y_{1:k-1})$ is known, although in practice it may be too brave.

Assuming that $p(x_{1:k}|y_{1:k})$ is analytically inaccessible, we can find its discrete approximation at N points as a function of $i \in [1, N]$. Using the sifting property of Dirac delta, we can also write

$$p(x_{1:k}|y_{1:k}) = \int p(x_{1:k}^{(i)}|y_{1:k})\delta(x_{1:k} - x_{1:k}^{(i)})\, \mathrm{d}x_{1:k}^{(i)} \tag{3.258}$$

and go to the discrete form by drawing samples $x_{1:k}^{(i)}$ from $p(x_{1:k}^{(i)}|y_{1:k})$, which we think is not easy. But we can introduce the *importance density* $q(x_{1:k}|y_{1:k})$, from which $x_{1:k}^{(i)}$ can easily be drawn. Of course, the importance density must be of the same class as $p(x_{1:k}|y_{1:k})$, so it is pmf. At this point, we can rewrite (3.258) as

$$p(x_{1:k}|y_{1:k}) = \int \frac{p(x_{1:k}^{(i)}|y_{1:k})}{q(x_{1:k}^{(i)}|y_{1:k})}\delta(x_{1:k} - x_{1:k}^{(i)}) \times q(x_{1:k}^{(i)}|y_{1:k})\, \mathrm{d}x_{1:k}^{(i)} \tag{3.259a}$$

$$= \int g(x_{1:k}^{(i)})q(x_{1:k}^{(i)}|y_{1:k})\, \mathrm{d}x_{1:k}^{(i)}\,, \tag{3.259b}$$

where $g(x_{1:k}^{(i)})$ is defined by comparing (3.259a) and (3.259b). Since (3.259b) *de facto* is the expectation of $g(x_{1:k}^{(i)})$, we then apply the *sequential importance sampling* (SIS), which is the MC method of integration, and rewrite (3.259b) in discrete form as

$$p(x_{1:k}|y_{1:k}) \approx \sum_{i=1}^{N} \frac{p(x_{1:k}^{(i)}|y_{1:k})}{q(x_{1:k}^{(i)}|y_{1:k})}\delta(x_{1:k} - x_{1:k}^{(i)}) \tag{3.260a}$$

$$= \sum_{i=1}^{N} \omega_k^{(i)}\delta(x_{1:k} - x_{1:k}^{(i)})\,, \tag{3.260b}$$

where the *importance weight*

$$\omega_k^{(i)} \propto \frac{p(x_{1:k}^{(i)}|y_{1:k})}{q(x_{1:k}^{(i)}|y_{1:k})} \tag{3.261}$$

---

3 Notation $x_k^{(i)} \backsim p(x_k)$ means drawing N samples $x_k^{(i)}$, $i \in [1, N]$, from $p(x_k)$.

corresponds to each of the generated particles $x_{1:k}^{(i)}$. Note that (3.261) still needs to be normalized for $\sum_{i=1}^{N} \omega_k^{(i)} = 1$ in order to be a density. Therefore, the sign $\propto$ is used.

To formalize the recursive form $\omega_{k-1}^{(i)} \to \omega_k^{(i)}$ for (3.261), we apply Bayes' rule to $p(x_{1:k}, y_k|y_{1:k-1})$ and write

$$p(x_{1:k}|y_{1:k}) = \frac{p(y_k|x_{1:k}, y_{1:k-1})p(x_{1:k}|y_{1:k-1})}{p(y_{1:k-1})} . \tag{3.262}$$

Using $p(x_{1:k}|y_{1:k-1}) = p(x_{1:k-1}|y_{1:k-1})p(x_k|x_{1:k-1}, y_{1:k-1})$ and the Markov property, we next represent (3.262) as

$$p(x_{1:k}|y_{1:k}) \propto p(y_k|x_k)p(x_k|x_{k-1})p(x_{1:k-1}|y_{1:k-1}) . \tag{3.263}$$

Since the importance density is of the same class as $p(x_{1:k}|y_{1:k})$, we follow [43, 155], apply Bayes' rule to $q(x_{1:k}|y_{1:k}) = q(x_k, x_{1:k-1}|y_{1:k})$, and choose the importance density to factorize of the form

$$\begin{aligned} q(x_{1:k}|y_{1:k}) &= q(x_k|x_{1:k-1}, y_{1:k})q(x_{1:k-1}|y_{1:k}) \\ &= q(x_k|x_{1:k-1}, y_{1:k})q(x_{1:k-1}|y_{1:k-1}) \\ &= q(x_k|x_{k-1}, y_k)q(x_{1:k-1}|y_{1:k-1}) . \end{aligned} \tag{3.264}$$

By combining (3.261), (3.263), and (3.264), the recursion for (3.261) can finally be found as

$$\omega_k^{(i)} \propto \omega_{k-1}^{(i)} \frac{p(y_k|x_k^{(i)})p(x_k^{(i)}|x_{k-1}^{(i)})}{q(x_k^{(i)}|x_{1:k-1}^{(i)}, y_{1:k})} . \tag{3.265}$$

If we now substitute (3.265) in (3.260b), then we can use the density $p(x_{1:k}|y_{1:k})$ to find the Bayesian estimate

$$\hat{x}_{1:k|k} = \int x_{1:k} p(x_{1:k}|y_{1:k}) \, dx_{1:k} \tag{3.266}$$

and the estimation error covariance

$$P_{1:k|k} = \int (x_{1:k} - \hat{x}_{1:k|k})(x_{1:k} - \hat{x}_{1:k|k})^T p(x_{1:k}|y_{1:k}) \, dx_{1:k} . \tag{3.267}$$

It should be noted that (3.266) and (3.267) are valid for both smoothing and filtering. In the case of filtering, the weight (3.265) ignores the past history and becomes

$$\omega_k^{(i)} \propto \omega_{k-1}^{(i)} \frac{p(y_k|x_k^{(i)})p(x_k^{(i)}|x_{k-1}^{(i)})}{q(x_k^{(i)}|x_{k-1}^{(i)}, y_k)} , \tag{3.268}$$

the posterior pmf (3.260b) transforms to

$$p(x_k|y_{1:k}) = \sum_{i=1}^{N} \omega_k^{(i)} \delta(x_k - x_k^{(i)}) , \tag{3.269}$$

and the filtering estimates can be found as

$$\hat{x}_{k|k} = \int x_k p(x_k|y_{1:k}) \, dx_k \tag{3.270a}$$

$$\approx \sum_{i=1}^{N} x_k^{(i)} \omega_k^{(i)} , \tag{3.270b}$$

$$P_{k|k} = \int (x_k - \hat{x}_{k|k})(x_k - \hat{x}_{k|k})^T p(x_k|y_{1:k}) \, dx_k \tag{3.271a}$$

$$\approx \sum_{i=1}^{N}(x_k^{(i)} - \hat{x}_{k|k})(x_k^{(i)} - \hat{x}_{k|k})^T \omega_k^{(i)} . \tag{3.271b}$$

Although PF is quite transparent and simple algorithmically, it has a serious drawback associated with the divergence caused by two effects called *sample degeneracy* and *sample impoverishment* [43, 62, 155]. Both these effects are associated with "bad" distribution of particles. The PF divergence usually occurs at low noise levels and is aggravated by incorrect initial values. There is a simple remedy for the divergency: set $N \to \infty$ to make both (3.248) and (3.251) true densities. However, this causes computational problems in real-time implementations.

These disadvantages can be effectively overcome by improving the diversity of samples in hybrid structures [114]. Examples are the PF/UKF [201], PF/KF [43], and PF combined with the UFIR filter [143]. Nevertheless, none of the known solutions is protected from divergency when the noise is very low and N is small. The following issues need to be addressed to improve the performance.

Degeneracy. Obviously, the best choice of importance density is the posterior density itself. Otherwise, the unconditional variance of the importance weights increases with time for the observations $y_{1:k}$, interpreted as random variables [43]. This unavoidable phenomenon, known as *sample degeneracy* [43], has the following practical appearance: after a few iterations, all but one of the normalized weights are very close to zero. Thus, a large computational load will be required to update the particles, whose contribution to the final estimate is almost zero.

Importance density. The wrong choice of importance density may crucially affect the estimate. Therefore, efforts should be made to justify acceptable solutions aimed at minimizing the variance of the importance weights.

Resampling. Resampling of particles is often required to reduce the effects of degeneracy. The idea behind resampling is to skip particles that have small weights and multiply ones with large weights. Although the resampling step mitigates the degeneracy, it poses another problem, which is discussed next.

Sample impoverishment. As a consequence of resampling, particles with large weights are statistically selected many times. It results in a loss of divergency between the particles, since the resulting sample will contain many repeated points. This problem is called *sample impoverishment*, and its effect can be especially destroying when the noise is low and the number of particles in insufficient.

3.5 Robust State Estimation

Robustness is required of an estimator when model errors occur due to mismodeling, noise environments are uncertain, and a system (process) undergoes unpredictable temporary changes such as jumps in velocity, phase, frequency, etc. Because the KF is poorly protected against such factors, efforts should be made whenever necessary to "robustify" the performance.

When a system is modeled with errors in poorly known noise environments, then matrices F_k, E_k, B_k, H_k, D_k, Q_k, and R_k in (3.32) and (3.33) can be complicated by uncertain additions ΔF_k, ΔE_k, ΔB_k, ΔH_k, ΔD_k, ΔQ_k, and ΔR_k, and the state-space equations written as

$$x_k = (F_k + \Delta F_k)x_{k-1} + (E_k + \Delta E_k)u_k + (B_k + \Delta B_k)w_k , \tag{3.272}$$

$$y_k = (H_k + \Delta H_k)x_k + (D_k + \Delta D_k)u_k + v_k , \tag{3.273}$$

where the covariance Q_k of noise w_k and R_k of v_k may also have uncertain parts resulting in $Q_k + \Delta Q_k$ and $R_k + \Delta R_k$.

Each uncertainty ΔY of a system can be completely or partially unknown, being either deterministic or stochastic [35]. But even if ΔY is uncertain, the maximum values of its components are usually available in applications. This allows restricting the estimation error covariance, as was done by Toda and Patel in [197] for nonzero ΔF, ΔH, ΔQ, and ΔR. Minimizing errors for a set of maximized uncertain parameters, all of each have bounded norms, creates the foundation for *robust filtering* [65]. The robust state estimation problem also arises when all matrices are certain, but noise is not Gaussian, heavy-tailed, or Gaussian with outliers.

The minimax approach is most popular in the design of robust estimators [87, 123, 124, 135, 207] under the previous assumptions. Shown that a saddle-point property holds, the minimax estimator designed to have worst-case optimal performance is referred to as *minimax robust* [160]. Huber showed in [73] that such an estimator can be treated as the ML estimator for the least favorable member of the class [160]. Referring to [73], several approaches have been developed over the decades to obtain various kinds of minimax state estimators, such as the robust KF and H_∞ filter.

3.5.1 Robustified Kalman Filter

After leaving the comfortable Gaussian environment, the KF output degrades, and efforts must be made to improve performance. In many cases, this is easier to achieve by modifying the KF algorithm rather than deriving a new filter. For measurement noise with uncertain distribution (supposedly heavy-tailed), Masreliez and Martin robustified KF [124] to protect against outliers in the innovation residual. To reduce errors caused by such an uncertainty, they introduced a symmetric vector influence function $\Psi(\nu_k)$, where $\nu_k = T_k(y_k - H_k x_k^-)$ is the weighted residual and T_k is a transformation matrix of a certain type. For $u_k = 0$, $B_k = I$, and assuming all uncertain increments in (3.272) and (3.273) are zero, these innovations have resulted in the following robustified KF algorithm [124]

$$P_k^- = F_k P_{k-1} F_k^T + Q_k \,, \tag{3.274}$$

$$\hat{x}_k^- = F_k \hat{x}_{k-1} \,, \tag{3.275}$$

$$\nu_k = T_k(y_k - H_k \hat{x}_k^-) \,, \tag{3.276}$$

$$\hat{x}_k = \hat{x}_k^- + P_k^- H_k^T T_k^T \Psi(\nu_k) \,, \tag{3.277}$$

$$P_k = P_k^- - P_k^- H_k^T T_k^T T_k H_k P_k^- E_0\{\psi_p'(\nu)\} \,, \tag{3.278}$$

where a component $\psi_p(\nu)$ of $\Psi(\nu)$ is an odd symmetric scalar influence function corresponding to the minimax estimate and $\psi_p'(\nu) = d\psi_p(\nu)/d\nu$. To saturate the outliers efficiently, function $\psi_p(\nu)$ was written as (yet another function $\psi_\epsilon(\nu)$ was suggested in [124])

$$\psi_p(\nu) = \begin{cases} \frac{1}{sy_p}\tan\left(\frac{\nu}{2sy_p}\right) \,, & |\nu| \leqslant y_p \\ \frac{1}{sy_p}\tan\left(\frac{1}{2s}\right)\operatorname{sgn}(\nu) \,, & |\nu| > y_p \end{cases} \,,$$

where $0 < p < 1$ is a selected value, $s = s(p)$ is such that $\inf_{0<p<1} s(p) = \frac{1}{\pi}$ [121], and y_p is the saturation point. The factor $E_0\{\psi_p'(\nu)\}$ is completely defined by the selected p-value as

$$E_0\{\psi'(\nu)\} = \frac{1}{s^2 y_p^2}\left[1 - p\left(1 + \tan^2\frac{1}{2s}\right)\right]$$

and more detail can be found in [124] and [121].

The advantage of this algorithm is that the error covariance P_k for selected p-point is always less than its value when the components of the transformed residual vector ν_k are independently and identically distributed with the least favorable marginal distribution [87]. As mentioned in [87], this worst-case covariance is the optimum covariance for the case where the transformed residuals have this least favorable property. However, this estimate does not provide a saddle point for the MSE due to the constraints of the linear model, as is discussed in [124]. A similar problem was latter solved by Schick and Mitter in [160].

3.5.2 Robust Kalman Filter

An advanced approach to state estimation under uncertain matrices was developed by Khargonekar, Petersen, and Zhou in [89]. It was assumed that $\Delta \Upsilon$ is coupled with the bounded uncertainty Ω as $H_1 \Omega E$, where $\Omega^T \Omega \leqslant I$ and H_1 and E are some known matrices. Then the error minimization problem is solved for the maximized norm-bounded uncertain block. Based on this approach, many robust estimators have been developed for time-varying and time-invariant systems, assuming different operation conditions and environments [42, 55, 57, 113, 129, 215]. Next we give a robust KF (RKF) scheme obtained by Xie, Soh, and de Souza in [215] and generalized in [111] for uncertain models with known uncorrelated white noise sources.

Consider a discrete-time state-space model

$$x_{k+1} = F_\delta x_k + w_k \,, \tag{3.279}$$

$$y_k = H_\delta x_k + v_k \tag{3.280}$$

and unite the uncertain matrices $F_\delta = F + \Delta F_k$ and $H_\delta = H + \Delta H_k$ in a parameter uncertainties vector as

$$\begin{bmatrix} F_\delta \\ H_\delta \end{bmatrix} = \begin{bmatrix} F \\ H \end{bmatrix} + \begin{bmatrix} H_1 \\ H_2 \end{bmatrix} \Omega_k E \,, \tag{3.281}$$

where Ω_k is an unknown uncertain matrix such that $\Omega_k^T \Omega_k \leqslant I$ for all k and H_1, H_2, and E are known matrices of appropriate dimensions. Noise vectors w_k and v_k are supposed to be zero mean Gaussian and uncorrelated with the covariances $E\{w_k w_k^T\} = Q$ and $E\{v_k v_k^T\} = R$. It is also supposed that the system is quadratically stable.

> ***Quadratic stability of a linear system*** *[142]: A system (3.261) is said to be quadratically stable if there exists a symmetric positive definite matrix P such that*
>
> $$(F + \Delta F_k)^T P (F + \Delta F_k) - P < 0 \tag{3.282}$$
>
> *for all admissible uncertainties ΔF_k.*

We can look for a stable KF estimate in the form of [215]

$$\hat{x}_{k+1} = \hat{F}_k \hat{x}_k + \hat{K}_k y_k \,, \quad \hat{x}_0 = 0 \,, \tag{3.283}$$

where $\hat{F}_k$ and $\hat{K}_k$ are to be determined in a way such that the estimation error $\varepsilon_k = x_k - \hat{x}_k$ is asymptotically stable and the ε_k dynamics satisfies

$$\mathcal{E}\{\varepsilon_k^T \varepsilon_k\} \leqslant \text{tr}\, \mathcal{P}_k \tag{3.284}$$

for a symmetric nonnegative definite matrix $\mathcal{P}_k$ being an optimized upper bound of the error covariance to provide a guaranteed cost by solving the minimax problem

$$\mathcal{J} = \min_{\epsilon_k} \max_{\Delta F_k, \Delta H_k} \mathcal{P}_k(\epsilon_k, \Delta F_k, \Delta H_k) .$$

Now, since matrices $\hat{F}_k$ and $\hat{K}_k$ are specified, it is necessary to link the estimate (3.283) with the standard KF having no uncertainties.

Because error ϵ_k depends on the system state, we can introduce an augmented vector $\xi_k = \begin{bmatrix} \epsilon_k \\ \hat{x}_k \end{bmatrix}$ and represent it recursively as

$$\xi_{k+1} = (\bar{F}_k + \bar{D}_k \Delta_k \bar{E})\xi_k + \bar{G}_k \eta_k , \tag{3.285}$$

where $\xi_0 = \begin{bmatrix} x_0 \\ 0 \end{bmatrix}$, $\eta_k = \begin{bmatrix} w_k \\ v_k \end{bmatrix}$, $\bar{Q} = \mathcal{E}\{\eta_k \eta_k^T\}$, and

$$\bar{F}_k = \begin{bmatrix} F - \hat{K}_k H & F - \hat{F}_k - \hat{K}_k H \\ \hat{K}_k H & \hat{F}_k + \hat{K}_k H \end{bmatrix}, \quad \bar{G}_k = \begin{bmatrix} I & -\hat{K}_k \\ 0 & \hat{K}_k \end{bmatrix},$$
$$\bar{D}_k = \begin{bmatrix} H_1 - \hat{K}_k H_2 \\ \hat{K}_k H_2 \end{bmatrix}, \quad \bar{E} = \begin{bmatrix} E & E \end{bmatrix}, \quad \bar{Q} = \begin{bmatrix} Q & 0 \\ 0 & R \end{bmatrix}. \tag{3.286}$$

For given Δ_k, the covariance matrix $\Sigma_k = \mathcal{E}\{\xi_k \xi_k^T\}$ of ξ_k satisfies the following Lyapunov equation,

$$(\bar{F}_k + \bar{D}_k \Delta_k \bar{E})\Sigma_k(\bar{F}_k + \bar{D}_k \Delta_k \bar{E})^T - \Sigma_{k+1} + \bar{G}_k \bar{Q} \bar{G}_k^T = 0 . \tag{3.287}$$

Since the uncertainty Δ_k is unknown, we are interested in finding an upper bound $\bar{\Sigma}$ of Σ_k for all admissible Δ_k. By replacing Σ_k with $\bar{\Sigma}$, we arrive at the Lyapunov inequality

$$(\bar{F}_k + \bar{D}_k \Delta_k \bar{E})\bar{\Sigma}_k(\bar{F}_k + \bar{D}_k \Delta_k \bar{E})^T - \bar{\Sigma}_{k+1} + \bar{G}_k \bar{Q} \bar{G}_k^T \leqslant 0 \tag{3.288}$$

and notice that $\Sigma_k \leqslant \bar{\Sigma}_k$ holds for all Δ_k, provided that Σ_k for optimal estimates stays at a saddle.

It has been shown [111] that the filter (3.283) can be said to be robust if for some scaling factor $\varsigma_k > 0$ there exists a bounded, positive definite, and symmetric $\tilde{\Sigma}_k \geqslant 0$ that satisfies the DDRE

$$\begin{aligned} \tilde{\Sigma}_{k+1} &= \bar{F}_k \tilde{\Sigma}_k \bar{F}_k^T + \bar{F}_k \tilde{\Sigma}_k \bar{E}^T (\varsigma_k^{-1} I - \bar{E} \tilde{\Sigma}_k \bar{E}^T) \bar{E} \tilde{\Sigma}_k \bar{F}_k^T \\ &\quad + \varsigma_k^{-1} \bar{D}_k \bar{D}_k^T + \bar{G}_k \bar{Q} \bar{G}_k^T \end{aligned} \tag{3.289}$$

such that $I - \varsigma_k \bar{E} \tilde{\Sigma}_k \bar{E}^T > 0$ for $\tilde{\Sigma}_0 = \bar{\Sigma}_0 = \begin{bmatrix} P_{x_0} & 0 \\ 0 & 0 \end{bmatrix}$.

Without going into other details, which can be found in [215, 233], we summarize the RKF estimate proposed in [111] with

$$\hat{x}_{k+1} = (F + \tilde{F}_k)\hat{x}_k + \hat{K}_k[y_k - (H - \tilde{H}_k)\hat{x}_k] , \quad \hat{x}_0 = 0 , \tag{3.290}$$

where the necessary matrices with uncertain components are computed by

$$\begin{aligned} \tilde{F}_k &= \varsigma_k F \mathcal{P}_k F^T (I - \varsigma_k E \mathcal{P}_k E^T)^{-1} E , \\ \tilde{H}_k &= \varsigma_k H \mathcal{P}_k E^T (I - \varsigma_k E \mathcal{P}_k E^T)^{-1} E , \\ R_\gamma &= R + \varsigma_k^{-1} H_2 H_2^T , \\ Y_k^{-1} &= \mathcal{P}_k^{-1} - \varsigma_k E^T E , \\ \hat{K}_k &= (F Y_k H^T + \varsigma_k^{-1} H_1 H_2^T)(R_\varsigma + H Y_k H^T)^{-1} , \end{aligned} \tag{3.291}$$

and the optimized covariance bound $\mathcal{P}_k > 0$ of the filtering error ε_k is specified by solving the DDRE

$$\mathcal{P}_{k+1} = FY_kF^T - (FY_kH^T + \varsigma_k^{-1}H_1H_2^T)(R_\varsigma + HY_kH^T)^{-1} \times (FY_kH^T + \varsigma_k^{-1}H_1H_2^T)^T + \varsigma_k^{-1}H_1H_1^T + Q\,, \quad (3.292)$$

subject to

$$Y_k^{-1} = \mathcal{P}_k^{-1} - \varsigma_k E^T E > 0$$

over a horizon of N data points in order for $\mathcal{P}_k$ to be positive definite.

Let us now assume that the model in (3.289) and (3.290) has no uncertainties. By $H_1 = 0, H_2 = 0$, and $E = 0$, we thus have $\tilde{F}_k = 0, \tilde{H}_k = 0, R_\gamma = R, Y_k = \mathcal{P}_k$. Since the DDRE gives a recursion for the prior estimation error [185], we substitute $\mathcal{P}_k = P_k^-$, transform gain (3.273) to

$$\hat{\mathcal{K}}_k = FP_k^-H^T(R + HP_k^-H^T)^{-1} = FK_k\,, \quad (3.293)$$

where K_k is the Kalman gain (3.79), and rewrite (3.292) as

$$P_{k+1}^- = FP_k^-F^T - FP_k^-H^T(R + HP_k^-H^T)^{-1}HP_k^-F^T + Q = F(P_k^- - K_kHP_k^-)F^T + Q\,, \quad (3.294)$$

which also requires substituting $\hat{x}_k$ with $\hat{x}_k^-$ in (3.290) to obtain the prior estimate

$$\hat{x}_{k+1}^- = F[\hat{x}_k^- + K_k(y_k - H\hat{x}_k^-)]\,. \quad (3.295)$$

It now follows that the RKF (3.290)–(3.292) is the *a priori* RKF [111] and that without model uncertainties it reduces to the *a priori* KF (Algorithm 2). Note that the *a posteriori* RKF, which reduces to the *a posteriori* KF (Algorithm 1) without model uncertainties, can be found in [111].

3.5.3 H_∞ Filtering

Although the H_∞ approach was originally introduced and long after that has been under investigation in control theory [111][34], its application to filtering has also attracted researchers owing to several useful features. Unlike KF, the H_∞ filter does not require any statistical information, while guaranteeing a prescribed level of noise reduction and robustness to model errors and temporary uncertainties.

To present the H_∞ filtering approach, we consider an LTI system represented in state space with equations

$$x_{k+1} = Fx_k + w_k\,, \quad (3.296)$$

$$y_k = Hx_k + v_k\,, \quad (3.297)$$

where w_k and v_k are some error vectors, even deterministic, each with bounded energy. We wish to have a minimax estimator that for the maximized errors w_k and v_k minimizes the estimation error $\varepsilon_k = x_k - \hat{x}_k$. Since the solution for Gaussian w_k and v_k is KF, and arbitrary w_k and v_k do not violate the estimator structure, the recursive estimate has the form

$$\hat{x}_{k+1} = F\hat{x}_k + \mathcal{K}_k(y_k - H\hat{x}_k)\,, \quad (3.298)$$

where the unknown gain $\mathcal{K}_k$ is still to be determined.

The objective is to minimize γ in a way such that for the estimates obtained on a horizon $[1, N]$ the following relation is satisfied

$$\sum_{k=1}^{N} \|\varepsilon_k\|^2 < \gamma^2 \left[\|\varepsilon_0\|^2_{P_{x_0}^{-1}} + \sum_{k=0}^{N-1} (\|w_k\|^2 + \|v_k\|^2) \right] ,$$

which leads to the following cost function of the *a priori* H_∞ filter

$$J = \sum_{k=1}^{N} \|\varepsilon_k\|^2 - \gamma^2 \left[\varepsilon_0^T P_{x_0}^{-1} \varepsilon_0 + \sum_{k=0}^{N-1} (\|w_k\|^2 + \|v_k\|^2) \right] < 0 , \tag{3.299}$$

where P_{x_0} is a positive definite matrix representing the uncertainty in the initial state, and the squared ℓ_2-norm $\|x\|^2 \triangleq \|x\|_2^2$ is depicted as $\|x\|^2 = x^T x$ or $\|x\|_P^2 = x^T P x$. Thus, the following minimax problem should be solved,

$$\mathcal{J} = \min_{\varepsilon_k} \max_{\varepsilon_0, w_k, v_k} \mathcal{P}_k(\varepsilon_k, \varepsilon_0, w_k, v_k) .$$

It was shown in [111] that for a cost (3.299) with $\gamma > 0$ there exists an *a priori* H_∞ filter over the horizon $[1, N]$ if and only if there exists a positive definite solution $\mathcal{P}_k = \mathcal{P}_k^T > 0$ for $\mathcal{P}_0 = P_{x_0} > 0$ to the following DDRE,

$$\begin{aligned} \mathcal{P}_{k+1} &= F\mathcal{P}_k F^T - \mathcal{K}_k H \mathcal{P}_k F^T + \mathcal{Q} \\ &\quad + \gamma^{-2} \mathcal{P}_{k+1} (I + \gamma^{-2} \mathcal{P}_{k+1})^{-1} \mathcal{P}_{k+1} , \end{aligned} \tag{3.300}$$

where the filter gain is given by

$$\mathcal{K}_k = F\mathcal{P}_k H^T (\mathcal{R} + H\mathcal{P}_k H^T)^{-1} , \tag{3.301}$$

$\mathcal{Q}$ is the error matrix of w_k, and $\mathcal{R}$ is the error matrix of v_k. Note that $\mathcal{Q}$ and $\mathcal{R}$ have different meanings than the noise covariances Q and R in KF. It can be shown that when w_k and v_k are both zero mean Gaussian and uncorrelated with known covariances Q and R, then substituting $\mathcal{Q} = Q$, $\mathcal{R} = R$, $\mathcal{P}_k = P_k^-$, and $\mathcal{K}_k = FK_k$, and neglecting the terms with large γ transform the *a priori* H_∞ filter to the *a priori* KF (Algorithm 2). Let us add that the *a posteriori* H_∞ filter, which for Gaussian models becomes the *a posteriori* KF (Algorithm 1), is also available from [111].

3.5.4 Game Theory H_∞ Filter

Another approach to H_∞ filtering was developed using the game theory. An existence of the game-theoretic solutions for minimax robust linear filters and predictors was examined by Martin and Mintz in [122]. Shortly thereafter, Verdú in [207] presented a generalization of Huber's approach to the minimax robust Bayesian statistical estimation problem of location with respect to the least favorable prior distributions. The minimax H_∞ problem, solved on FH, considers the game cost function as a ratio of the filter error norm, which must be minimized, and the sum of the initial error norm and bounded error covariance norms, which must be maximized [14, 163, 190]. The corresponding recursive H_∞ filtering algorithm was obtained by Simon in [185], and it becomes KF without uncertainties.

In the standard formulation of game theory applied to model (3.278) and (3.279), the cost function for H_∞ filtering is defined as follows:

$$\bar{J} = \frac{\sum_{k=0}^{N-1} \|\varepsilon_k\|^2_{\bar{S}_k}}{\|\varepsilon_0\|^2_{\mathcal{P}_0^{-1}} + \sum_{k=0}^{N-1} \left(\|w_k\|^2_{\mathcal{Q}_k^{-1}} + \|v_k\|^2_{\mathcal{R}_k^{-1}} \right)} , \tag{3.302}$$

in order to find an estimate that minimizes $\bar{J}$. Similarly to the H_∞ filter, (3.302) can be modified to

$$J = -\frac{1}{\theta}\|\varepsilon_0\|^2_{\mathcal{P}_0^{-1}} + \sum_{k=0}^{N-1}\left[\|\varepsilon_k\|^2_{\bar{S}_k} - \frac{1}{\theta}\left(\|w_k\|^2_{\mathcal{Q}_k^{-1}} + \|v_k\|^2_{\mathcal{R}_k^{-1}}\right)\right] < 1\,, \tag{3.303}$$

and the H_∞ filtering estimate can be found by solving the minimax problem

$$\mathcal{J} = \min_{\varepsilon_k}\max_{\varepsilon_0, w_k, v_k} \mathcal{P}_k(\varepsilon_k, \varepsilon_0, w_k, v_k)\,.$$

For the FE-based model (3.296) and (3.297), the *a priori* game theory-based H_∞ filter derived in [185] using the cost (3.303) is represented with the following recursions

$$K_k = \mathcal{P}_k^-[I - \theta\bar{S}_k\mathcal{P}_k^- + H_k^T\mathcal{R}_k^{-1}H_k\mathcal{P}_k^-]^{-1}H_k^T\mathcal{R}_k^{-1}\,, \tag{3.304}$$

$$\hat{x}_{k+1}^- = F_k[\hat{x}_k^- + K_k(y_k - H_k\hat{x}_k^-)]\,, \tag{3.305}$$

$$\mathcal{P}_{k+1}^- = F_k\mathcal{P}_k^-[I - \theta\bar{S}_k\mathcal{P}_k^- + H_k^T\mathcal{R}_k^{-1}H_k\mathcal{P}_k^-]^{-1}F_k^T + \mathcal{Q}_k\,, \tag{3.306}$$

where matrix $\bar{S}_n$ is constrained by a positive definite matrix

$$(\mathcal{P}_n^-)^{-1} - \theta\bar{S}_n + H^T\mathcal{R}^{-1}H > 0\,.$$

It is the user choice to assign $\bar{S}_n$, which is introduced to weight the prior estimation error. If the goal is to weight all error components equally [115], one can set $\bar{S}_n = I$. Because the cost $\bar{J}$ represents the squared norm error-to-error ratio and is guaranteed in the *a priori* H_∞ filter to be $J < 1/\theta$, the scalar bound $\theta > 0$ must be small enough.

To organize *a posteriori* filtering that matches the BE-based model (3.32) and (3.33) with $u_k = 0$ and $B_k = I$, we can consider (3.286)–(3.288), replace F_k with F_{k+1} and $\mathcal{Q}_k$ with $\mathcal{Q}_{k+1}$, note that $\hat{x}_{k+1}^- = F_{k+1}\hat{x}_k$, refer to the derivation procedure given in (3.96), and arrive at the *a posteriori* H_∞ filtering algorithm [180]

$$\mathcal{P}_k = \mathcal{P}_k^-(I - \theta\bar{S}_k\mathcal{P}_k^- + H_k^T\mathcal{R}_k^{-1}H_k\mathcal{P}_k^-)^{-1}\,, \tag{3.307}$$

$$K_k = \mathcal{P}_kH_k^T\mathcal{R}_k^{-1}\,, \tag{3.308}$$

$$\hat{x}_k = F_k\hat{x}_{k-1} + K_k(y_k - H_kF_k\hat{x}_{k-1})\,, \tag{3.309}$$

$$\mathcal{P}_{k+1}^- = F_k\mathcal{P}_kF_k^T + \mathcal{Q}_k\,, \tag{3.310}$$

which for given $\mathcal{P}_0$ can be initiated with $\mathcal{P}_1^- = F_1\mathcal{P}_0F_1^T + \mathcal{Q}_1$.

For Gaussian noise and no disturbances, $\theta = 0$ makes this filter an alternate KF (3.92)–(3.96). Otherwise, a properly set small $\theta > 0$ improves the H_∞ filter accuracy. It is worth noting that when θ approaches the boundary specified by the constraint, the H_∞ filter goes to instability. This means that in order to achieve the best possible effect, θ must be carefully selected and its value optimized.

3.6 Summary

As a process of filtering out erroneous data with measurement noise and simultaneously solving state-space equations for unknown state variables at a given time instant, the state estimation theory is a powerful tool for many branches of engineering and science. To provide state estimation using computational resources, we need to know how to represent a dynamic process in the continuous-time state space and then go correctly to the discrete-time state space. Since a stochastic integral can be computed in various senses, the transition from continuous to discrete time is organized using either the FE method, associated with Itô calculus, or the BE method. The FE-based state model predicts future state and is therefore fundamental in state feedback control. The BE-based state model is used to solve various signal processing problems associated with filtering, smoothing, prediction, and identification.

The Bayesian approach is based on the use of Bayesian inference and is used to develop Bayesian estimators for linear and nonlinear state space models with noise having any distribution. For linear models, the Bayesian approach results in the optimal KF. Using the likelihood function, it is possible to obtain an ML estimator whose output converges to the Bayesian estimate as the number of measurement samples grows without bounds.

In the Gauss approach, the LS estimate is chosen so that the sum of the squares of the measurement residuals minimizes the mean square residual. The weighted LS approach considers the ML estimator as a particular case for Gaussian processes. The unbiased estimator only satisfies the unbiasedness condition and gives an estimate identical to LS.

The linear recursive KF algorithm can be *a posteriori* or *a priori*. The KF algorithm can be represented in various forms depending on the process, applications, and required outputs. However, the KF algorithm is poorly protected from initial errors, model errors, errors in noise statistics, and temporary uncertainties. The GKF is a modification of KF for time-correlated and colored Gaussian noise. The KBF is an optimal state estimator in continuous time. For nonlinear state-space models, the most widely used estimators are EKF, UKF, and PF.

Robust state estimation is required when the model is uncertain and/or operates under disturbances. The minimax approach is most popular when developing robust estimators. This implies minimizing estimation errors for a set of maximized uncertain parameters. Shown that a saddle-point holds, the minimax estimator provides the worst-case optimal performance and is therefore minimax robust.

3.7 Problems

1 Find conditions under which the FE-based and BE-based state-space models become identical. Which of these models provides the most accurate matched estimate?

2 The FE-based state space model is basic in feedback state control, and the BE-based state space models is used for signal processing when prediction is not required. Why is this so? Give a reasonable explanation.

3 Use the Bayes inference formula

$$p(x_k|y_0^k) = \frac{p(x_k|y_0^{k-1})p(y_k|x_k)}{p(y_k|y_0^{k-1})}$$

and show that the ML estimate coincides with the most probable Bayesian estimate given a uniform prior distribution on the parameters.

4 Consider the LS estimation problem of a constant quantity Q

$$\hat{Q} = \arg\min_{Q} \ (Y - MQ)^T(Y - MQ) \ ,$$

where $Y - MQ$ is the measurement residual, Y is a vector of multiple measurements of Q, and M is a proper matrix. Derive the LS estimator of Q and relate it to the UFIR estimator.

5 Given an oscillatory system of the second order. Following Example 3.2, derive the matrix exponential for the system matrix $A = \begin{bmatrix} 0 & 1 \\ -\omega_0^2 & -2\delta \end{bmatrix}$, where 2δ is an angular bandwidth.

6 **Solved problem: Matrix identity.** Matrix L_k is given by (3.92). Show that this matrix is identical to the error covariance matrix P_k given by (3.79). Equate (3.92) and (3.79), substitute (3.78) and (3.77), and write

$$\begin{aligned} P_k^-(I + H_k^T R_k^{-1} H_k P_k^-)^{-1} &= P_k^- - K_k H_k P_k^- \,, \\ P_k^-(I + H_k^T R_k^{-1} H_k P_k^-)^{-1} &= P_k^- - P_k^- H_k^T (H_k P_k^- H_k^T + R_k)^{-1} H_k P_k^- \,, \\ (I + H_k^T R_k^{-1} H_k P_k^-)^{-1} &= I - H_k^T (H_k P_k^- H_k^T + R_k)^{-1} H_k P_k^- \,. \end{aligned}$$

By the Woodbury matrix identity (A.7),

$$(H_k P_k^- H_k^T + R_k)^{-1} = R_k^{-1} - R_k^{-1} H_k (I + P_k^- H_k^T R_k^{-1} H_k)^{-1} P_k^- H_k^T R_k^- \,,$$

transform the previous relation as

$$\begin{aligned} (I + H_k^T R_k^{-1} H_k P_k^-)^{-1} &= I - H_k^T [R_k^{-1} - R_k^{-1} H_k (I + P_k^- H_k^T R_k^{-1} H_k)^{-1} \\ &\quad \times P_k^- H_k^T R_k^-] H_k P_k^- \,. \\ &= I - H_k^T R_k^{-1} H_k P_k^- + H_k^T R_k^{-1} H_k \\ &\quad \times (I + P_k^- H_k^T R_k^{-1} H_k)^{-1} P_k^- H_k^T R_k^- H_k P_k^- \,. \end{aligned}$$

Because $H_k^T R_k^{-1} H_k$, P_k^-, and $I - P_k^- H_k^T R_k^{-1} H_k$ are symmetric, rearrange the matrix products as

$$\begin{aligned} (I + P_k^- H_k^T R_k^{-1} H_k)^{-1} &= I - P_k^- H_k^T R_k^{-1} H_k + P_k^- H_k^T R_k^{-1} H_k \\ &\quad \times (I + P_k^- H_k^T R_k^{-1} H_k)^{-1} P_k^- H_k^T R_k^- H_k \,, \end{aligned}$$

introduce $D = P_k^- H_k^T R_k^{-1} H_k$, rewrite as $(I + D)^{-1} = I - D + D(I + D)^{-1} D$, substitute $I = (I + D)(I + D)^{-1}$, provide the transformations, and end up with an identity $I = I$, which proves that L_k given by (3.92) and P_k given by (3.79) are identical.

7 A linear discrete-time state-space model is augmented with the Gauss-Markov CPN $w_k = \Theta_k w_{k-1} + \mu_k$, where $\mu_k \sim \mathcal{N}(0, Q_\mu)$, and CMN $v_k = \Psi_k v_{k-1} + \xi_k$, where $\xi_k \sim \mathcal{N}(0, R_\xi)$. In what case is an estimator required to filter out only the CMN? What problem is solved by filtering out both CPN and CMN? Is there an application that only requires CPN filtering?

8 **Solved problem: Steady state KF.** Consider the KF Algorithm 1. Set $k \to \infty$ and suppose that at a steady state $P_k \cong P_{k-1} = P$ and $P_k^- \cong P_{k-1}^- = P^-$. Then transform the *a priori* error covariance as

$$\begin{aligned} P_k^- &= F_k P_{k-1} F_k^T + Q_k \\ &= F_k (P_{k-1}^- - K_{k-1} H_{k-1} P_{k-1}^-) F_k^T + Q_k \,, \\ &= F_k P_{k-1}^- F_k^T - F_k P_{k-1}^- H_{k-1}^T (H_{k-1} P_{k-1}^- H_{k-1}^T + R_{k-1})^{-1} \\ &\quad \times H_{k-1} P_{k-1}^- F_k^T + Q_k \end{aligned}$$

and arrive at the *discrete algebraic Riccati equation* (DARE)

$$P^- = FP^-F^T - FP^-H^T(H^-P^-H^T + R)^{-1}HP^-F^T + Q \,.$$

Next, represent the Kalman gain as $K = P^-H^T(HP^-H^T + R)^{-1}$ and end up with the steady state KF algorithm:

$$\begin{aligned} P^- &= (\text{obtain by solving the DARE}) \\ K &= P^-H^T(HP^-H^T + R)^{-1} \,, \\ \hat{x}_k &= F\hat{x}_{k-1} + K(y_k - HF\hat{x}_{k-1}) \,. \end{aligned}$$

9 The recursive LS estimate is given by

$$\hat{x}_k = \hat{x}_{k-1} + G_kH^T(y_k - H\hat{x}_{k-1}) \,,$$

where matrix G_k is computed recursively as $G_k = (H^TH + G_{k-1}^{-1})^{-1}$. How does G_k change with the increase in the number of measurement k? How is the recursive LS estimate related to UFIR estimate?

10 A constant quantity Q is measured in the presence of Laplace noise and is estimated using the median estimator

$$\hat{Q} = \arg\min_{Q} \sum_{i=1}^{N} |y_i - Q| \,.$$

Modify this estimator under the assumption that Q is not constant and has K states.

11 Under the Cauchy noise, a constant quantity Q is estimated using the myriad estimator [13]

$$\hat{Q} = \arg\min_{Q} \sum_{i=1}^{N} \log\{\gamma^2 + (y_i - Q)^2\} \,,$$

where γ is the linearity parameter. Modify the myriad estimator for a dynamic quantity Q represented in state space by K states.

12 **Solved problem: Proof of (3.222).** Consider the two-filter smoother and its error covariance $\tilde{P}_r = P_r^{\mathrm{b}} + K_r^{\mathrm{f}}(P_r^{\mathrm{f}} + P_r^{\mathrm{b}})K_r^{\mathrm{f}^T} - 2K_r^{\mathrm{f}}P_r^{\mathrm{b}}$, substitute (3.220) and (3.221), and represent formally as

$$\begin{aligned} P &= B + B(A+B)^{-1}B - 2B(A+B)^{-1}B \\ &= B - B(A+B)^{-1}B \,, \end{aligned}$$

where A and B are positive definite, symmetric, and invertible matrices. Using (A.10), write $(A+B)^{-1} = B^{-1} - B^{-1}(I + AB^{-1})^{-1}AB^{-1}$ and transform the previous matrix equation to

$$\begin{aligned} P &= B - B(A+B)^{-1}B \,, \\ &= B - B[B^{-1} - B^{-1}(I + AB^{-1})^{-1}AB^{-1}]B \,, \\ &= (I + AB^{-1})^{-1}A \,, \\ &= (A^{-1} + B^{-1})^{-1} \end{aligned}$$

that completes the proof of (3.222).

13 The second-order expansions of nonlinear functions are given by (3.228) and (3.229). Derive coefficients α_k (3.232) and β_k (3.233) using Hessian matrices (3.234) and (3.235). Find an explanation to the practical observation that the second-order EKF either decreases or increases the estimation error, and nothing definite can be said about its performance.

14 The UKF has the following limitation: the same statistics can be predicted for different distributions. Suggest a feasible solution to improve performance using the high-order statistics. Will it perform better than the second-order EKF?

15 The PF can dramatically improve estimation performance when the model has sever nonlinearity. Why is it impossible to fully guarantee the PF stability? What are the most effective ways to avoid the divergency in the PF? Why is there no guarantee that the PF will not diverge when the number of particles is insufficient?

16 The robust KF approach for uncertain models suggests representing the uncertain system matrix F_δ and observation matrix H_δ using an unknown matrix Ω_k as (3.281),

$$\begin{bmatrix} F_\delta \\ H_\delta \end{bmatrix} = \begin{bmatrix} F \\ H \end{bmatrix} + \begin{bmatrix} H_1 \\ H_2 \end{bmatrix} \Omega_k E ,$$

where F and H are known and the newly introduced auxiliary matrices H_1, H_2, and E are supposed to be known as well. Give examples when such a model is 1) advantageous, 2) can hardly be applied, and 3) is not feasible. Consider both the LTI and LTV systems.

17 In the presence of unknown norm-bounded noise, the H_∞ filtering estimate is given by (3.298) with the bias correction gain (3.301) and error covariance computed recursively by (3.300), where the optimal tuning factor γ_{opt} guarantees the best filtering performance and robustness. How to specify γ_{opt}? Can we find γ_{opt} analytically? What will happen if we set 1) $\gamma < \gamma_{\text{opt}}$ and 2) $\gamma > \gamma_{\text{opt}}$?

18 Solve the problems listed in the previous item, considering instead the tuning factor θ of the game theory H_∞ filter, recursively represented by (3.307)–(3.310).

4

Optimal FIR and Limited Memory Filtering

The user of the system naturally tries to minimize the inaccuracies caused by…noise —by filtering.

Brian D. O. Anderson, John B. Moore [9], p. 2.

The *state estimator* that relates the output to the current time point is also called a *filter*. As such, it can be designed to have either FIR or IIR. An FIR filter requires data from an FH $[m, k]$ of N points, from $m = k - N + 1$ to k, so the length of its impulse response is limited by N points. In contrast, transients in IIR filters last indefinitely due to internal feedback and decay over time. This important specific feature predetermines two critical properties of FIR estimators: 1) no feedback is required, and thus round-off errors are not compounded at the output by summed iterations and 2) inherent BIBO stability. Known bottlenecks are that 1) the batch form can cause a computational problem when N is large and 2) even iterative FIR algorithms work about N times slower than KF. Note that the latter drawback can be overcome using parallel computing. But what makes the FIR estimator attractive in the modern world of powerful computers is that it can process full block error matrices, while efficient recursions are available only for diagonal such matrices associated with white noise. This means that batch FIR filters are basically more accurate state estimators than recursive schemes.

Convolution is a general operator for linear systems. Consequently, the convolution-based batch form is common to linear state estimators. All the basic properties of linear state estimators can be extracted from batch form, but not all of them are available from recursions. An example is the KF, whose optimality follows from the OFIR filter, but whose unbiased optimality is not supported by the OUFIR filter.

The advantage of the FIR approach is illustrated in Fig. 4.1. Suppose that the disturbance f_k appears in the model at three time points as (I), (II), and (III) (Fig. 4.1a), and note that KF having IIR requires all data, from 0 to k, while the FIR filter requires data from $[m, k]$. Disturbance (I) acts close to k and causes the same errors ε_k in both filters (Fig. 4.1b). The effect of disturbance (II) is weakened on both filters, but the FIR filter sees only half of the disturbance (II) and thus produces fewer errors. Finally, disturbance (III) acts outside $[m, k]$; the FIR filter does not see it, while the KF still responds. The general idea behind this example was formulated by Schweppe [161] as an old estimate updated in discrete time index k not over all data on $[0, k]$ but over a horizon $[m, k]$ of N most recent observations. It was later rephrased by Maybeck [126] that it is preferable to rerun the growing memory filter over the data horizon for each k.

Optimal and Robust State Estimation: Finite Impulse Response (FIR) and Kalman Approaches, First Edition.
Yuriy S. Shmaliy and Shunyi Zhao.

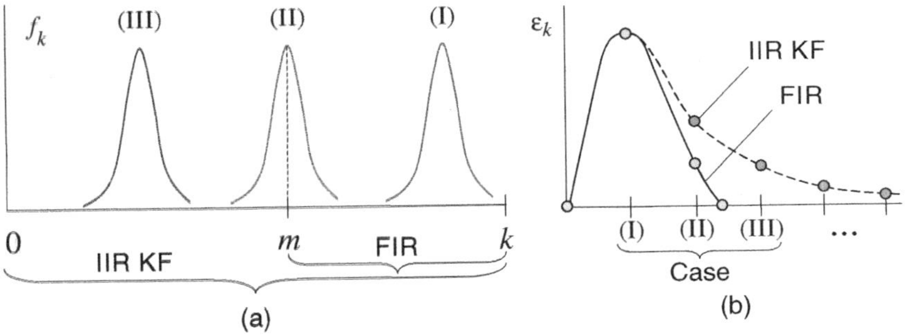

Figure 4.1 Effect of the disturbance f_k, which appears at three different time instances as (I), (II), and (III), on the estimation errors ε_k produced by the KF and FIR filter: (a) disturbances and (b) estimation errors.

In this chapter, we present the OFIR filter and LMF theory in both convolution-based batch form and iterative form using recursions. We will show that the OFIR filter is the most general of the other available linear optimal state estimators and that the optimal Kalman recursions (Algorithm 4) serve equivalently to the OFIR filter on finite and infinite horizons.

4.1 Extended State-Space Model

Consider an LTV system represented in discrete-time state space with the following state and observation equations, respectively,

$$x_k = F_k x_{k-1} + E_k u_k + B_k w_k \,, \tag{4.1}$$

$$y_k = H_k x_k + v_k \,, \tag{4.2}$$

where $x_k \in \mathbb{R}^K$ is the state vector, $u_k \in \mathbb{R}^L$ is the input vector, $y_k \in \mathbb{R}^P$ is the observation vector, $w_k \sim \mathcal{N}(0, Q_k) \in \mathbb{R}^M$ is the process noise, $v_k \sim \mathcal{N}(0, R_k) \in \mathbb{R}^P$ is the observation noise. Assume that $F_k \in \mathbb{R}^{K\times K}$, $H_k \in \mathbb{R}^{P\times K}$, $E_k \in \mathbb{R}^{K\times L}$, and $B_k \in \mathbb{R}^{K\times M}$ are known matrices.

The model in (4.1) and (4.2) cannot be used directly in FIR filtering and requires an extension on $[m, k]$. This can be done if we first rewrite (4.1) using the backward-in-time solutions as

$$x_k = F_k x_{k-1} + E_k u_k + B_k w_k \,, \tag{4.3a}$$

$$x_{k-1} = F_{k-1} x_{k-2} + E_{k-1} u_{k-1} + B_{k-1} w_{k-1} \,, \tag{4.3b}$$

$$\vdots$$

$$x_{m+2} = F_{m+2} x_{m+1} + E_{m+2} u_{m+2} + B_{m+2} w_{m+2} \,, \tag{4.3c}$$

$$x_{m+1} = F_{m+1} x_m + E_{m+1} u_{m+1} + B_{m+1} w_{m+1} \,, \tag{4.3d}$$

$$x_m = x_m + E_m u_m + B_m w_m \,, \tag{4.3e}$$

where the initial state x_m is supposed to be known and hence $u_m = 0$ and $w_m = 0$ in (4.3e). Then substituting (4.3d) into (4.3c) to modify (4.3c) for the initial state x_m and doing so until (4.3b) and (4.3a) are also modified for x_m allow extending (4.1) on $[m, k]$. By introducing the extended vectors

$$X_{m,k} = [\, x_m^T \; x_{m+1}^T \; \cdots \; x_k^T \,]^T \in \mathbb{R}^{NK} \,, \tag{4.4}$$

$$U_{m,k} = [\, u_m^T \; u_{m+1}^T \; \cdots \; u_k^T \,]^T \in \mathbb{R}^{NL} \,, \tag{4.5}$$

$$W_{m,k} = [\, w_m^T \; w_{m+1}^T \; \cdots \; w_k^T \,]^T \in \mathbb{R}^{NM} \,, \tag{4.6}$$

and referring to (4.3a)–(4.3e), the extended state equation can be written as

$$X_{m,k} = F_{m,k}x_m + S_{m,k}U_{m,k} + D_{m,k}W_{m,k} \,, \tag{4.7}$$

where the extended matrices are

$$F_{m,k} = \begin{bmatrix} I & F_{m+1}^T & \dots & (\mathcal{F}_{k-1}^{m+1})^T & (\mathcal{F}_k^{m+1})^T \end{bmatrix}^T \in \mathbb{R}^{NK\times K} \,, \tag{4.8}$$

$$S_{m,k} = \begin{bmatrix} E_m & 0 & \dots & 0 & 0 \\ F_{m+1}E_m & E_{m+1} & \dots & 0 & 0 \\ \vdots & \vdots & \ddots & \vdots & \vdots \\ \mathcal{F}_{k-1}^{m+1}E_m & \mathcal{F}_{k-1}^{m+2}E_{m+1} & \dots & E_{k-1} & 0 \\ \mathcal{F}_k^{m+1}E_m & \mathcal{F}_k^{m+2}E_{m+1} & \dots & F_kE_{k-1} & E_k \end{bmatrix} \in \mathbb{R}^{NK\times NL} \,, \tag{4.9}$$

$$\mathcal{F}_r^g = \begin{cases} F_rF_{r-1}\dots F_g \,, & g < r+1 \,, \\ I \,, & g = r+1 \\ 0 \,, & g > r+1 \end{cases} \,, \tag{4.10}$$

and matrix $D_{m,k} \in \mathbb{R}^{NK\times NM}$ can be written in the same manner as (4.9), if we substitute E_i with B_i for $i \in [m,k]$.

Similarly, the observation equation (4.2) can be written as

$$y_k = H_kx_k + v_k \,, \tag{4.11a}$$

$$y_{k-1} = H_{k-1}x_{k-1} + v_{k-1} \,, \tag{4.11b}$$

$$\vdots$$

$$y_m = H_mx_m + v_m \,. \tag{4.11c}$$

By substituting x_k, x_{k-1}, ... , x_m taken from (4.3a)–(4.3e) into (4.11a)–(4.11c) and assigning two vectors

$$Y_{m,k} = [\, y_m^T \; y_{m+1}^T \; \dots \; y_k^T \,]^T \in \mathbb{R}^{NP} \,, \tag{4.12}$$

$$V_{m,k} = [\, v_m^T \; v_{m+1}^T \; \dots \; v_k^T \,]^T \in \mathbb{R}^{NP} \,, \tag{4.13}$$

we arrive at the extended observation equation

$$Y_{m,k} = H_{m,k}x_m + L_{m,k}U_{m,k} + G_{m,k}W_{m,k} + V_{m,k} \,, \tag{4.14}$$

in which the extended matrices are

$$H_{m,k} = \bar{H}_{m,k}F_{m,k} \in \mathbb{R}^{NP\times K} \,, \tag{4.15}$$

$$L_{m,k} = \bar{H}_{m,k}S_{m,k} \in \mathbb{R}^{NP\times NL} \,, \tag{4.16}$$

$$G_{m,k} = \bar{H}_{m,k}D_{m,k} \in \mathbb{R}^{NP\times NM} \,, \tag{4.17}$$

and matrix $\bar{H}_{m,k} = \text{diag}\left(H_m \; H_{m+1} \; \dots \; H_k\right)$ is diagonal.

The extended state-space equations (4.7) and (4.14) can be used to derive all kinds of linear convolution-based batch state estimators (filters, smoothers, and predictors) for a given cost function. An FIR filter will require an FH $[m,k]$ of N points, and an IIR filter can be designed keeping $m = 0$ for each k. This approach also gives an idea of the batch KF, which, however, does not make practical sense in view of growing memory and increasing dimensions of all vectors and matrices.

4.2 The *a posteriori* Optimal FIR Filter

The generalized structure of a linear *a posteriori* OFIR filter operating on a horizon $[m, k]$ is shown in Fig. 4.2.

Measurement data received as y_k in the presence of white Gaussian noise go to the input along with the input (control) signal u_k. Using the extended vectors $Y_{m,k}$ and $U_{m,k}$, the output estimate $\hat{x}_k$ can be written in batch form as

$$\hat{x}_k \triangleq \hat{x}_{k|k} = \mathcal{H}^{\mathrm{h}}_{m,k} Y_{m,k} + \mathcal{H}^{\mathrm{f}}_{m,k} U_{m,k} \,, \tag{4.18}$$

where $\mathcal{H}^{\mathrm{h}}_{m,k}$ is the *homogenous gain* and $\mathcal{H}^{\mathrm{f}}_{m,k}$ is the *forced gain*, to be specified in the MSE sense. Because the initial state $\hat{x}_m$ and error covariance P_m are taken at the initial point of the horizon $[m, k]$, they become variables of gains $\mathcal{H}^{\mathrm{h}}_{m,k}$ and $\mathcal{H}^{\mathrm{f}}_{m,k}$. Below we will obtain these gains, investigate filtering errors, represent an estimate (4.18) with an iterative algorithm using recursions, and link to the *a posteriori* KF.

4.2.1 Batch Estimate and Error Covariance

Given $\hat{x}_k$ by (4.18), the estimation error $\varepsilon_k = x_k - \hat{x}_k$ can be determined if we represent x_k by the last raw vector in $X_{m,k}$ given by (4.7) as

$$x_k = \mathcal{F}^{m+1}_k x_m + \bar{S}_{m,k} U_{m,k} + \bar{D}_{m,k} W_{m,k} \,, \tag{4.19}$$

where $\bar{S}_{m,k}$ is the last raw vectors in $S_{m,k}$ and so is $\bar{D}_{m,k}$ in $D_{m,k}$.

Applying the unbiasedness condition

$$\mathcal{E}\{\hat{x}_k\} = \mathcal{E}\{x_k\} \tag{4.20}$$

to (4.18) and (4.19) gives two unbiasedness constraints,

$$\mathcal{F}^{m+1}_k = \mathcal{H}^{\mathrm{h}}_{m,k} H_{m,k} \,, \tag{4.21}$$

$$\mathcal{H}^{\mathrm{f}}_{m,k} = \bar{S}_{m,k} - \mathcal{H}^{\mathrm{h}}_{m,k} L_{m,k} \,. \tag{4.22}$$

To find the optimal gain $\mathcal{H}^{\mathrm{h}}_{m,k}$, the standard way is to minimize the MSE $\mathcal{E}\{\varepsilon_k^T \varepsilon_k\}$, which is the trace of the error covariance $\mathcal{E}\{\varepsilon_k \varepsilon_k^T\}$, as

$$\mathcal{H}^{\mathrm{h}}_{m,k} = \underset{\mathcal{H}^{\mathrm{h}}_{m,k}}{\arg\min}\ \mathcal{E}\{\varepsilon_k^T \varepsilon_k\} \,.$$

The optimal gain is obtained by taking the derivative of $\mathcal{E}\{\varepsilon_k^T \varepsilon_k\}$ or tr $\mathcal{E}\{\varepsilon_k \varepsilon_k^T\}$ with respect to $\mathcal{H}^{\mathrm{h}}_{m,k}$ and equating it to zero. This gives $\mathcal{H}^{\mathrm{h}}_{m,k}$ and leads to an important finding: the estimate (4.18) will be optimal if the estimation error ε_k demonstrates orthogonality on $[m, k]$ to both inputs, Y_k and U_k. That can be achieved by fulfilling the orthogonality condition [176]

$$\mathcal{E}\{(x_k - \mathcal{H}^{\mathrm{h}}_{m,k} Y_{m,k} - \mathcal{H}^{\mathrm{f}}_{m,k} U_{m,k}) Y^T_{m,k}\} = 0 \,. \tag{4.23}$$

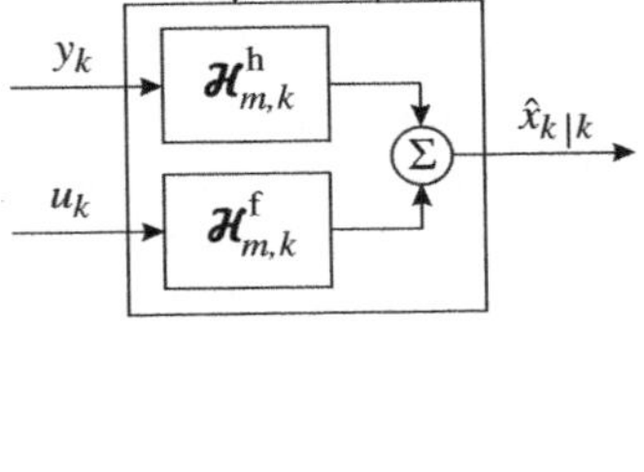

Figure 4.2 Generalized structure of a linear *a posteriori* OFIR filter.

If we now substitute in (4.23) $Y_{m,k}$ with (4.14), introduce

$$\chi_m = \mathcal{E}\{x_m x_m^T\} , \quad \Psi_{m,k} = \mathcal{E}\{U_{m,k}U_{m,k}^T\} ,$$
$$\mathcal{Q}_{m,k} = \mathcal{E}\{W_{m,k}W_{m,k}^T\} , \quad \mathcal{R}_{m,k} = \mathcal{E}\{V_{m,k}V_{m,k}^T\} ,$$

and take into account that x_m, $U_{m,k}$, $W_{m,k}$, and $V_{m,k}$ are mutually independent and uncorrelated, then (4.23) can be transformed to

$$\begin{aligned}&\mathcal{B}_{m,k}\chi_m H_{m,k}^T + \mathcal{W}_{m,k}\mathcal{Q}_{m,k}G_{m,k}^T - \mathcal{V}_{m,k}\mathcal{R}_{m,k}\\ &= (\mathcal{H}_{m,k}^{\mathrm{h}}L_{m,k} - \bar{S}_{m,k} + \mathcal{H}_{m,k}^{\mathrm{f}})\Psi_{m,k}L_{m,k}^T ,\end{aligned} \tag{4.24}$$

where the error residual matrices

$$\mathcal{B}_{m,k} = \mathcal{F}_k^{m+1} - \mathcal{H}_{m,k}^{\mathrm{h}}H_{m,k} , \tag{4.25}$$
$$\mathcal{W}_{m,k} = \bar{D}_{m,k} - \mathcal{H}_{m,k}^{\mathrm{h}}G_{m,k} , \tag{4.26}$$
$$\mathcal{V}_{m,k} = \mathcal{H}_{m,k}^{\mathrm{h}} \tag{4.27}$$

represent, respectively, the *bias error residual*, *system error residual*, and *measurement error residual*. Obviously, matrices (4.25)–(4.27) are fully responsible for optimal cancelation of regular (bias) and random errors. We add that the random initial state x_m can be represented as $x_m = \hat{x}_m + \varepsilon_m$, and then χ_m in (4.24) replaced with

$$\chi_m = E\{x_m x_m^T\} = E\{(\hat{x}_m + \varepsilon_m)(\hat{x}_m + \varepsilon_m)^T\} = \hat{\chi}_m + P_m ,$$

where $\hat{\chi}_m = \hat{x}_m\hat{x}_m^T$ and $P_m = E\{\varepsilon_m\varepsilon_m^T\}$ is the initial error covariance.

The optimal gains $\mathcal{H}_{m,k}^{\mathrm{h}}$ and $\mathcal{H}_{m,k}^{\mathrm{f}}$ for the *a posteriori* OFIR filter can now be found as follows. First, recall that the homogenous impulse response of the LTV system is obtained for zero input. Then set $\Psi_{m,k} = 0$ to (4.24) and obtain $\mathcal{H}_{m,k}^{\mathrm{h}}$ as

$$\mathcal{H}_{m,k}^{\mathrm{h}} = \mathcal{H}_{m,k}^{1\mathrm{h}}\mathcal{H}_{m,k}^{2\mathrm{h}} \tag{4.28a}$$
$$= (\mathcal{F}_k^{m+1}\chi_m H_{m,k}^T + \mathcal{Z}_1)(\mathcal{Z}_\chi + \mathcal{Z}_2 + \mathcal{R}_{m,k})^{-1} , \tag{4.28b}$$
$$= (\hat{\mathcal{H}}_{m,k}^{\mathrm{h}}\mathcal{Z}_\chi + \mathcal{Z}_1)(\mathcal{Z}_\chi + \mathcal{Z}_2 + \mathcal{R}_{m,k})^{-1} , \tag{4.28c}$$

where $\mathcal{H}_{m,k}^{1\mathrm{h}}$ and $\mathcal{H}_{m,k}^{2\mathrm{h}}$ are the first and second product components in (4.28b),

$$\mathcal{Z}_\chi = H_{m,k}\chi_m H_{m,k}^T , \quad \mathcal{Z}_1 = \bar{D}_{m,k}\mathcal{Q}_{m,k}G_{m,k}^T ,$$
$$\mathcal{Z}_2 = G_{m,k}\mathcal{Q}_{m,k}G_{m,k}^T ,$$

and the homogeneous gain $\hat{\mathcal{H}}_{m,k}^{\mathrm{h}}$ of the UFIR filter is given by [179]

$$\hat{\mathcal{H}}_{m,k}^{\mathrm{h}} = \mathcal{F}_k^{m+1}(H_{m,k}^T H_{m,k})^{-1}H_{m,k}^T \tag{4.29a}$$
$$= (C_{m,k}^T C_{m,k})^{-1}C_{m,k}^T , \tag{4.29b}$$

where $C_{m,k} = H_{m,k}(\mathcal{F}_k^{m+1})^{-1}$. Note that (4.28c) is obtained by multiplying the first χ_m in (4.28b) from the left-hand side with an identity $(H_{m,k}^T H_{m,k})^{-1}H_{m,k}^T H_{m,k}$.

What follows from the previous is that by neglecting noise we have $\mathcal{H}_{m,k}^{\mathrm{h}} = \hat{\mathcal{H}}_{m,k}$, and the OFIR filter for deterministic models becomes the UFIR filter. Another observation can be made regarding the initial state x_m. It was shown in [173, 176] that if we smooth the OFIR estimate back in time to the first horizon point, then x_m can be determined by solving the DARE, thereby removing the requirement of the initial state. Yet another and even easier way to find x_m exists if we use the ML. This means that the initial state is not an unsolved problem for OFIR filtering.

The forced impulse response of an LTV system is defined for zero initial conditions. Substituting (4.28b) into (4.24), its left-hand side becomes identically zero, and the remaining part gives the forced gain $\mathcal{H}^{\mathrm{f}}_{m,k} = \bar{S}_{m,k} - \mathcal{H}^{\mathrm{h}}_{m,k} L_{m,k}$. Now note that this relationship is due to the unbiasedness constraint (4.22), which is thus inherently build into the OFIR filter structure.

The batch *a posteriori* OFIR filter (4.18) is thus given by

$$\hat{x}_k = \hat{x}^{\mathrm{h}}_k + \hat{x}^{\mathrm{f}}_k \,, \tag{4.30a}$$

$$= \mathcal{H}_{m,k} Y_{m,k} + (\bar{S}_{m,k} - \mathcal{H}_{m,k} L_{m,k}) U_{m,k} \,, \tag{4.30b}$$

where $Y_{m,k}$ and $U_{m,k}$ contain real data collected on $[m,k]$, $\hat{x}^{\mathrm{h}}_k$ is the first component in (4.30b), $\hat{x}^{\mathrm{f}}_k$ is the second component in (4.30b), and $\mathcal{H}_{m,k} \triangleq \mathcal{H}^{\mathrm{h}}_{m,k}$ can be said to be the *fundamental OFIR filter gain*. Since the constraint (4.22) is embedded into (4.30b) by design, it follows that the input U_k is tracked by the OFIR filter (4.30b) exactly. Note that the *a posteriori* OFIR filter without input was originally designed in [176].

The batch error covariance P_k for the *a posteriori* OFIR filter (4.30b) can be easily found if we use (4.19) and (4.30b) and do some transformations. This gives

$$\begin{aligned} P_k &= \mathcal{E}\{\varepsilon_k \varepsilon_k^T\} \\ &= \mathcal{B}_{m,k} \mathcal{X}_m \mathcal{B}^T_{m,k} + \mathcal{W}_{m,k} \mathcal{Q}_{m,k} \mathcal{W}^T_{m,k} \\ &\quad + \mathcal{V}_{m,k} \mathcal{R}_{m,k} \mathcal{V}^T_{m,k} \,, \end{aligned} \tag{4.31}$$

where the error residual matrices are specified by (4.25)–(4.27). It should be noted right away that the error covariance (4.31) does not depend on input u_k due to the built-in constraint (4.22). The first term in (4.31) containing $\mathcal{X}_m$ represents the bias error, and the two last terms with $\mathcal{Q}_{m,k}$ and $\mathcal{R}_{m,k}$ represent the random errors. Both these errors are optimally balanced by matrices (4.25)–(4.27), and the estimate (4.30b) is thus truly optimal.

4.2.2 Recursive Forms

It follows from (4.31) that the OFIR estimate (4.30b) is *a posteriori* optimal. It is also seen that the batch form is computationally time-consuming, especially when $N \gg 1$, due to large dimensions of all extended vectors and matrices. An efficient computation of (4.30b) can be done if we find recursions for $\hat{x}^{\mathrm{h}}_k$ and $\hat{x}^{\mathrm{f}}_k$ and combine them into an iterative algorithm as stated by the following theorem.

Theorem 4.1 Given a batch OFIR filtering estimate (4.30b). Its iterative computation on $[m,k]$ for given $\hat{x}_m$ and P_m is provided by Kalman recursions (Algorithm 4) if we change the auxiliary time-index i from $m+1$ to k and take output when $i = k$.

Algorithm 4: The *a posteriory* Optimal Kalman Recursions

1 $\hat{x}^-_k = F_k \hat{x}_{k-1} + E_k u_k$ (*a priori* state estimate);
2 $P^-_k = F_k P_{k-1} F^T_k + B_k Q_k B^T_k$ (*a priori* error covariance);
3 $z_k = y_k - H_k \hat{x}^-_k$ (measurement residual);
4 $S_k = H_k P^-_k H^T_k + R_k$ (innovation covariance);
5 $K_k = P^-_k H^T_k S^{-1}_k$ (Kalman gain);
6 $\hat{x}_k = \hat{x}^-_k + K_k z_k$ (*a posteriori* state estimate);
7 $P_k = (I - K_k H_k) P^-_k$ (*a posteriori* error covariance);

Proof: Consider $\mathcal{H}^{1h}_{m,k}$ defined by (4.28b). Decompose the following matrices as $\mathcal{F}^{m+1}_k = F_k\mathcal{F}^{m+1}_{k-1}$,

$$\bar{D}_{m,k} = [F_k\bar{D}_{m,k-1} \;\; B_k] \,,$$

$$G_{m,k} = \begin{bmatrix} G_{m,k-1} & 0 \\ H_kF_k\bar{D}_{m,k-1} & H_kB_k \end{bmatrix} ,$$

take into account that $\mathcal{Q}_{m,k} = \text{diag}(\mathcal{Q}_{m,k-1} \; Q_k)$ and

$$F_k\bar{D}_{m,k-1}\mathcal{Q}_{m,k-1}\bar{D}^T_{m,k-1}F^T_k + B_kQ_kB^T_k = \bar{D}_{m,k}\mathcal{Q}_{m,k}\bar{D}^T_{m,k} \,, \tag{4.32}$$

and represent $\mathcal{H}^{1h}_{m,k}$ recursively for $k \geq m+1$ as

$$\begin{aligned} \mathcal{H}^{1h}_{m,k} &= \mathcal{F}^{m+1}_k \chi_m H^T_{m,k} + \bar{D}_{m,k}\mathcal{Q}_{m,k}G^T_{m,k} \\ &= [F_k\mathcal{H}^{1h}_{m,k-1} \;\; M_kH^T_k] \,, \end{aligned} \tag{4.33}$$

where $M_k = \mathcal{F}^{m+1}_k\chi_m\mathcal{F}^{m+1^T}_k + \bar{D}_{m,k}\mathcal{Q}_{m,k}\bar{D}^T_{m,k}$.

To derive a recursive form for $\mathcal{H}^{2h}_{m,k}$ (4.28b), first transform $Z_\chi + Z_2$ as

$$\begin{aligned} Z_\chi + Z_2 &= H_{m,k}\chi_mH^T_{m,k} + G_{m,k}\mathcal{Q}_{m,k}G^T_{m,k} \\ &= \begin{bmatrix} \mathcal{H}^{2h^{-1}}_{m,k-1} - \mathcal{R}_{m,k-1} & \tilde{\mathcal{H}}^{2h^T}_{m,k-1} \\ \tilde{\mathcal{H}}^{2h}_{m,k-1} & H_kM_kH^T_k \end{bmatrix} , \end{aligned}$$

where $\tilde{\mathcal{H}}^{2h}_{m,k-1} = H_kF_k\mathcal{H}^{1h}_{m,k-1}$ and M_k is given earlier. Next, refer to $\mathcal{R}_{m,k} = \text{diag}\,(\mathcal{R}_{m,k-1} \; R_k)$ and represent $Z_\chi + Z_2 + \mathcal{R}_{m,k}$ as

$$Z_\chi + Z_2 + \mathcal{R}_{m,k} = \begin{bmatrix} \mathcal{H}^{2h^{-1}}_{m,k-1} & \tilde{\mathcal{H}}^{2h^T}_{m,k-1} \\ \tilde{\mathcal{H}}^{2h}_{m,k-1} & H_kM_kH^T_k + R_k \end{bmatrix} .$$

To simplify inverting this matrix, separate it into the two block matrices $\bar{Z}_{m,k} = \text{diag}\,(\mathcal{H}^{2h^{-1}}_{m,k-1} \; R_k)$ and

$$\tilde{Z}_{m,k} = \begin{bmatrix} 0 & \tilde{\mathcal{H}}^{2h^T}_{m,k-1} \\ \tilde{\mathcal{H}}^{2h}_{m,k-1} & H_kM_kH^T_k \end{bmatrix} ,$$

and decompose $\mathcal{H}^{2h}_{m,k}$ as

$$\begin{aligned} \mathcal{H}^{2h}_{m,k} &= (\bar{Z}_{m,k} + \tilde{Z}_{m,k})^{-1} \\ &= \bar{Z}^{-1}_{m,k}(I + \tilde{Z}_{m,k}\bar{Z}^{-1}_{m,k})^{-1} \\ &= \bar{Z}^{-1}_{m,k}\hat{Z}^{-1}_{m,k} \,, \end{aligned} \tag{4.34}$$

where $\hat{Z}_{m,k} = I + \tilde{Z}_{m,k}\bar{Z}^{-1}_{m,k}$. Now, using the Schur complement (A.18), represent $\hat{Z}^{-1}_{m,k}$ as

$$\hat{Z}^{-1}_{m,k} = \begin{bmatrix} \mathcal{Z}_{11} & \mathcal{Z}_{12} \\ \mathcal{Z}_{21} & \mathcal{Z}_{22} \end{bmatrix} , \tag{4.35}$$

where $\mathcal{Z}_{11} = I + \bar{\Omega}_k$, $\mathcal{Z}_{12} = -\tilde{\mathcal{H}}^{2h^T}_{m,k-1}R^{-1}_k\Omega^{-1}_k$, $\mathcal{Z}_{21} = -\Omega^{-1}_k\tilde{\mathcal{H}}^{2h}_{m,k-1}\mathcal{H}^{2h}_{m,k-1}$, $\mathcal{Z}_{22} = \Omega^{-1}_k$, $\bar{\Omega}_k = \tilde{\mathcal{H}}^{2h^T}_{m,k-1} R^{-1}_k\Omega^{-1}_k\tilde{\mathcal{H}}^{2h}_{m,k-1}\mathcal{H}^{2h}_{m,k-1}$, and

$$\Omega_k = I + H_k\Lambda^-_kH^T_kR^{-1}_k \,, \tag{4.36}$$

$$\Lambda^-_k = M_k - F_k\mathcal{H}^{h}_{m,k-1}\mathcal{H}^{1h^T}_{m,k-1}F^T_k \,. \tag{4.37}$$

Substitute (4.35) into (4.34), combine with (4.33), and arrive at the recursion for $\mathcal{H}_{m,k}$,

$$\mathcal{H}_{m,k} = [(F_k - K_k H_k F_k)\mathcal{H}^{\mathrm{h}}_{m,k-1} \quad K_k] \,, \tag{4.38}$$

where

$$K_k = \Lambda_k^- H_k^T (H_k \Lambda_k^- H_k^T + R_k)^{-1} \tag{4.39}$$

and Λ_k^- is defined by (4.37). Now, substituting (4.38) into $\hat{x}_k^{\mathrm{h}}$ specified by (4.30b) yields the recursive estimate

$$\hat{x}_k^{\mathrm{h}} = F_k \hat{x}_{k-1}^{\mathrm{h}} + K_k(y_k - H_k F_k \hat{x}_{k-1}^{\mathrm{h}}) \,. \tag{4.40}$$

Next, consider $\hat{x}_k^{\mathrm{f}}$ defined by (4.30b), refer to $\bar{S}_{m,k} = [F_k \bar{S}_{m,k-1} \quad E_k]$ and

$$L_{m,k} = \begin{bmatrix} L_{m,k-1} & 0 \\ H_k F_k \bar{S}_{m,k-1} & H_k E_k \end{bmatrix} ,$$

and come up with another recursion

$$\hat{x}_k^{\mathrm{f}} = (I - K_k H_k) F_k \hat{x}_{k-1}^{\mathrm{f}} + (I - K_k H_k) E_k u_k \,. \tag{4.41}$$

Combining (4.40) and (4.41) into (4.30a) finally gives the *a posteriori* KF estimate

$$\hat{x}_k = \hat{x}_k^- + K_k(y_k - H_k \hat{x}_k^-) \,, \tag{4.42}$$

where $\hat{x}_k^- = F_k \hat{x}_{k-1} + E_k u_k$ and K_k given by (4.39) is the iteratively updated bias correction gain of the *a posteriori* OFIR filter.

What is left behind is to find a recursion for the batch Λ_k^- (4.37). To do this, substitute $M_k = M_k^T$ given in (4.33), $\mathcal{H}^{1\mathrm{h}}_{m,k}$ by (4.33), and $\mathcal{H}_{m,k}$ by (4.38); provide the transformations; and arrive at the DDRE

$$\begin{aligned} \Lambda_k^- =& F_k \Lambda_{k-1}^- F_k^T + B_k Q_k B_k^T - F_k \Lambda_{k-1}^- H_{k-1}^T \\ & \times (H_{k-1} \Lambda_{k-1}^- H_{k-1}^T + R_{k-1})^{-1} H_{k-1} \Lambda_{k-1}^- F_k^T \,, \end{aligned}$$

which, if we involve (4.39), can be further represented as

$$\Lambda_k^- = F_k \Lambda_{k-1}^- F_k^T + B_k Q_k B_k^T - F_k K_{k-1} H_{k-1} \Lambda_{k-1}^- F_k^T \,. \tag{4.43}$$

Now combine the first and the last matrix products in (4.43), assign

$$\Lambda_k = (I - K_k H_k) \Lambda_k^- \,, \tag{4.44}$$

and transform (4.43) to

$$\Lambda_k^- = F_k \Lambda_{k-1} F_k^T + B_k Q_k B_k^T \,. \tag{4.45}$$

Finally, recall that DDRE requires the matrix Λ_k^- to be symmetric and positive definite. Also observe that (4.39), (4.42), (4.44), and (4.45) are exactly the relations listed in Algorithm 4. Then rename $\Lambda_k^- = P_k^-$ and $\Lambda_k = P_k$ and complete the proof. □

Iterative OFIR algorithm: *Iterative computation of the batch OFIR filtering estimate (4.30b) is provided on $[m, k]$ by Kalman recursions listed in Algorithm 4.*

The pseudocode of the *a posteriori* iterative OFIR filtering algorithm is listed as Algorithm 5.

Algorithm 5: Iterative *a posteriori* OFIR Filtering Algorithm

Data: $y_k, u_k, \hat{x}_m, P_m, Q_k, R_k, N$
begin
  for $k = 1, 2, \cdots$ **do**
    $m = k - N + 1$ if $k > N - 1$ and $m = 0$ otherwise;
    for $i = m + 1, m + 2, \cdots, k$ **do**
      Algorithm 1: $\hat{x}_i, P_i$
    end for
  end for
  Result: $\hat{x}_k, P_k$
end

Given $\hat{x}_m$ and P_m, this algorithm iteratively updates values from $i = m + 1$ to $i = k$ using optimal recursions (Algorithm 4) and obtains $\hat{x}_k$ and P_k when $i = k$. It is worth noting that although the OFIR filter improves the estimate with each new iteration, the number of iterations can be limited by N_{opt} of the UFIR filter [179]. Thus, the OFIR filter can be more robust to uncertainties than KF, provided that the initial $\hat{x}_m$ and P_m are set correctly.

A special case of Algorithm 5 is the full-horizon OFIR Algorithm 6, which operates on $[0, k]$ and, therefore, is the standard *a posteriori* KF algorithm.

Algorithm 6: Iterative *a posteriori* Full Horizon OFIR Filtering Algorithm, Which Is the KF Algorithm

Data: $y_k, u_k, \hat{x}_0, P_0, Q_k, R_k$
begin
  for $k = 1, 2, \cdots$ **do**
    Algorithm 1
  end for
  Result: $\hat{x}_k, P_k$
end

This means that KF can be viewed as a full horizon OFIR filter operating on $[0, k]$, and thus we have further evidence that KF is optimal.

4.2.3 System Identification

Another benefit of the FIR approach is that a system can be optimally identified at k if we apply the orthogonality condition (4.22) to a noiseless state $x_k = \mathcal{F}_k^{m+1} x_m + \bar{S}_{m,k} U_{m,k}$. This means that we would like to remove both the system noise and the measurement noise and to estimate optimally the inherent system state [173]. Without going into details, we note that in this case the orthogonality condition (4.25) is transformed into (4.26) with the same error residual matrices $\mathcal{B}_{m,k}$ and $\mathcal{V}_{m,k}$ and another system error residual matrix $\mathcal{W}_{m,k} = -\mathcal{H}_{m,k} G_{m,k}$. Then the *a posteriori* estimate (4.30b) can be used to provide system identification with the error covariance (4.31).

4.3 The *a posteriori* Optimal Unbiased FIR Filter

In some cases, requiring an initial state $\hat{x}_m$ and error covariance P_m may be too daring, and efficient methods will be needed to get around the problem. We have already mentioned that since data on $[m, k-1]$ and a past estimate $\hat{x}_{k-1}$ are available, the initial values can be obtained using backward filtering algorithms. There is also a more radical way to embed the unbiasedness constraint (4.23) in the *a posteriori* OFIR filter and go to the *a posteriori* optimal unbiased (OUFIR) filter.

There are two ways to instill unbiasedness in filters. We can simply remove the term with the constraint (4.21) in the OFIR filter error covariance to obtain the OUFIR-I filter gain. Since this is akin to a modification of the OFIR filter for zero initial conditions, the resulting batch can be expected to be computed using optimal Kalman recursions. We can also fully embed unbiasedness by minimizing the trace of the error covariance subject to (4.21) using the Lagrange multiplier approach. This results in an OUFIR-II filter that is truly optimal unbiased. It is worth noting that the property of unbiased optimality is achieved in the OUFIR-II filter with other recursive forms than Kalman's recursions. Next we will consider both the OUFIR-I and OUFIR-II filtering algorithms.

The generalized structure of a linear *a posteriori* OUFIR filter operating on the horizon $[m, k]$ is shown in Fig. 4.3.

Measurement data y_k come to the input together with the input (control) signal u_k. The output estimate $\hat{x}_k$ can be written in batch form as

$$\hat{x}_k = \dot{\mathcal{H}}^{\mathrm{h}}_{m,k} Y_{m,k} + \dot{\mathcal{H}}^{\mathrm{f}}_{m,k} U_{m,k} , \tag{4.46}$$

where $\dot{\mathcal{H}}^{\mathrm{h}}_{m,k}$ and $\dot{\mathcal{H}}^{\mathrm{f}}_{m,k}$ are the gains of the *a posteriori* OUFIR filter, specified in the MSE sense and subject to the unbiasedness constraint that removes the initial values from the variables. Next, we will discuss two possible types of OUFIR filters.

4.3.1 Batch OUFIR-I Estimate and Error Covariance

For the batch *a posteriori* OUFIR-I filtering estimate provided by (4.46), the unbiasedness test (4.20) gives two constraints

$$\mathcal{F}^{m+1}_k = \dot{\mathcal{H}}_{m,k} H_{m,k} , \tag{4.47}$$

$$\dot{\mathcal{H}}^{\mathrm{f}}_{m,k} = \bar{S}_{m,k} - \dot{\mathcal{H}}_{m,k} L_{m,k} , \tag{4.48}$$

where $\dot{\mathcal{H}}_{m,k} \triangleq \dot{\mathcal{H}}^{\mathrm{h}}_{m,k}$. Since we wish to avoid the part of the OFIR filter error covariance associated with (4.47), we apply the orthogonality condition to (4.46) as

$$\mathcal{E}\{(x_k - \dot{\mathcal{H}}_{m,k} Y_{m,k} - \dot{\mathcal{H}}^{\mathrm{f}}_{m,k} U_{m,k}) Y^T_{m,k}\} = 0 \tag{4.49}$$

and remove some terms using (4.47). Further, (4.49) can be transformed to

$$\begin{aligned} &\dot{\mathcal{W}}_{m,k} \mathcal{Q}_{m,k} G^T_{m,k} - \dot{\mathcal{V}}_{m,k} \mathcal{R}_{m,k} \\ &\quad = (\dot{\mathcal{H}}_{m,k} L_{m,k} - \bar{S}_{m,k} + \dot{\mathcal{H}}^{\mathrm{f}}_{m,k}) \Psi_{m,k} L^T_{m,k} , \end{aligned} \tag{4.50}$$

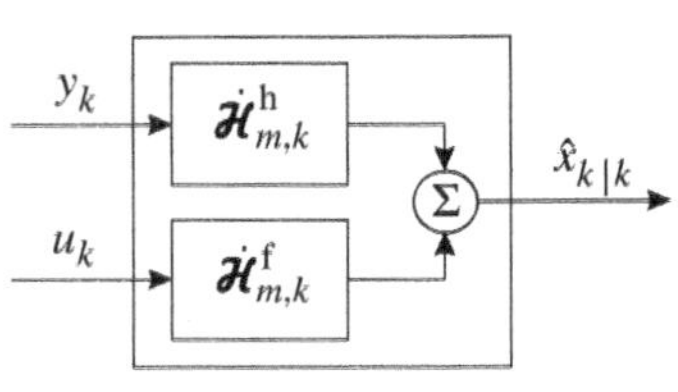

Figure 4.3 Generalized structure of a linear *a posteriori* OUFIR filter.

where the *error residual* matrices are specified as

$$\dot{\mathcal{W}}_{m,k} = \bar{D}_{m,k} - \dot{\mathcal{H}}_{m,k} G_{m,k} , \tag{4.51}$$

$$\dot{\mathcal{V}}_{m,k} = \dot{\mathcal{H}}_{m,k} . \tag{4.52}$$

Equality (4.50) has two solutions: (4.48) and the filter gain

$$\dot{\mathcal{H}}^{(1)}_{m,k} = \bar{D}_{m,k} \mathcal{Q}_{m,k} G^T_{m,k} \Omega^{-1}_{m,k} , \tag{4.53}$$

where $\Omega_{m,k} = G_{m,k}\mathcal{Q}_{m,k}G^T_{m,k} + \mathcal{R}_{m,k}$. Therefore, the batch OUFIR-I filtering estimate (4.46) becomes

$$\hat{x}_k = \hat{x}^{\rm h}_k + \hat{x}^{\rm f}_k \tag{4.54a}$$

$$= \dot{\mathcal{H}}^{(1)}_{m,k} Y_{m,k} + (\bar{S}_{m,k} - \dot{\mathcal{H}}^{(1)}_{m,k} L_{m,k}) U_{m,k} , \tag{4.54b}$$

where $\hat{x}^{\rm h}_k$ and $\hat{x}^{\rm f}_k$ are the relevant terms in (4.54b) and $Y_{m,k}$ and $U_{m,k}$ are vectors of real data. Similarly to (4.31), the error covariance of the OUFIR-I filter can be found to be

$$P_k = \dot{\mathcal{W}}_{m,k}\mathcal{Q}_{m,k}\dot{\mathcal{W}}^T_{m,k} + \dot{\mathcal{V}}_{m,k}\mathcal{R}_{m,k}\dot{\mathcal{V}}^T_{m,k} . \tag{4.55}$$

As can be seen, (4.55) has lost the bias term and is determined only by the noise covariance matrices. But this does not mean that the filter is really optimal unbiased. It rather means that by removing the term associated with the constraint (4.47), we simply set the initial values to zero. If we think in this way, then $\chi_m = 0$ in the optimal gain (4.28c) makes it equal to (4.5), and the OUFIR-I filter thus can be viewed as a special case of the OFIR filter for zero initial values.

4.3.2 Recursive Forms for OUFIR-I Filter

Recursions for (4.53) can be found similarly to (4.30b), if we find recursive forms for (4.48) and (4.54b), which is stated by the following theorem.

Theorem 4.2 Given the a posteriori OUFIR-I filtering estimate (4.54b), its iterative computation on $[m, k]$ is provided using Kalman recursions (Algorithm 4) by changing the iterative variable i from $m + 2$ to k and taking the output when $i = k$. Initial values are obtained at $m + 1$ by (4.54b) and (4.55) as

$$\begin{aligned}\hat{x}_{m+1} &= \dot{\mathcal{H}}^{(1)}_{m,m+1} Y_{m,m+1} \\ &\quad + (\bar{S}_{m,m+1} - \dot{\mathcal{H}}^{(1)}_{m,m+1}) U_{m,m+1} ,\end{aligned} \tag{4.56}$$

$$\begin{aligned}P_{m+1} &= \dot{\mathcal{W}}_{m,m+1}\mathcal{Q}_{m,m+1}\dot{\mathcal{W}}^T_{m,m+1} \\ &\quad + \dot{\mathcal{V}}_{m,m+1}\mathcal{R}_{m,m+1}\dot{\mathcal{V}}^T_{m,m+1} ,\end{aligned} \tag{4.57}$$

where $\dot{\mathcal{H}}^{(1)}_{m,m+1}$ is given by (4.53).

Proof: Represent (4.53) as $\dot{\mathcal{H}}_{m,k} = \dot{\mathcal{H}}^{\rm 1h}_{m,k}\dot{\mathcal{H}}^{\rm 2h}_{m,k}$. To find a recursion for $\dot{\mathcal{H}}^{\rm 1h}_{m,k} = \bar{D}_{m,k}\mathcal{Q}_{m,k}G^T_{m,k}$, refer to (4.33) and write

$$\dot{\mathcal{H}}^{\rm 1h}_{m,k} = [F_k \dot{\mathcal{H}}^{\rm 1h}_{m,k-1} \quad \dot{M}_k H^T_k] , \tag{4.58}$$

where $\dot{M}_k = \bar{D}_{m,k}\mathcal{Q}_{m,k}\bar{D}^T_{m,k}$. Decompose $\dot{\mathcal{H}}^{\rm 2h}_{m,k} = \Omega^{-1}_{m,k}$ similarly to (4.34) and obtain

$$\dot{\mathcal{H}}_{m,k} = [(F_k - \dot{K}_k H_k F_k)\dot{\mathcal{H}}_{m,k-1} \quad \dot{K}_k] , \tag{4.59}$$

where $\dot{K}_k = \dot{\Lambda}_k^- H_k^T (H_k \dot{\Lambda}_k^- H_k^T + R_k)^{-1}$ and

$$\dot{\Lambda}_k^- = \dot{M}_k - F_k \dot{\mathcal{H}}_{m,k-1} \dot{\mathcal{H}}_{m,k-1}^{1h^T} F_k^T \,. \tag{4.60}$$

Substitute (4.58) into (4.54b), refer to (4.41), and obtain

$$\hat{x}_i = \hat{x}_i^- + \dot{K}_i (y_i - H_i \hat{x}_i^-) \,, \tag{4.61}$$

where $\hat{x}_i^- = F_i \hat{x}_{i-1} + E_i u_i$ and $\dot{K}_i$ is given in (4.59). Since $\dot{K}_i$ depends on $\dot{\Lambda}_i^-$ specified by (4.60), find a recursion for $\dot{\Lambda}_i^-$. To do this, refer to a similarity between $\dot{\Lambda}_i^-$ and Λ_i^- and write

$$\dot{\Lambda}_i^- = F_i \dot{\Lambda}_{i-1} F_i^T + B_i Q_i B_i^T \,,$$
$$\dot{\Lambda}_i = (I - \dot{K}_i H_i) \dot{\Lambda}_i^- \,.$$

Following theorem 4.1, rename $\dot{\Lambda}_i^- = P_i^-$, $\dot{\Lambda}_i = P_i$, and $\dot{K}_i = K_i$ and arrive at the *a posteriori* optimal recursions (Algorithm 4). To initialize iterations, compute $\hat{x}_{m+1}$ and P_{m+1} in short batch forms on $[m, m+1]$ by (4.54b) and (4.55) as (4.56) and (4.57) and complete the proof. □

Theorem 4.2 suggests that an iterative *a posteriori* OUFIR-I filtering algorithm, represented with the pseudocode in Algorithm 7, can be used to avoid requiring initial values.

Algorithm 7: Iterative *a posteriori* OUFIR-I Filtering Algorithm

```
Data: y_k, u_k, Q_k, R_k, N
begin
    for k = 1, 2, ··· do
        m = k − N + 1 if k > N − 1 and m = 0 otherwise;
        Compute x̂_{m+1} by (4.56) and P_{m+1} by (4.57) ;
        for i = m + 2, m + 3, ···, k do
            Algorithm 1
        end for
    end for
    Result: x̂_k, P_k
end
```

Unlike the *a posteriori* OFIR Algorithm 5, here the initial $\hat{x}_{m+1}$ and P_{m+1} are self-computed on $[m, m+1]$ in short batch forms.

4.3.3 Batch OUFIR-II Estimate and Error Covariance

Our next goal is to make the OFIR filter truly optimal unbiased. Then let us go back to (4.46) and embed the constraint (4.47) to obtain the *a posteriori* OUFIR-II filter. To do this, we start with the error covariance (4.55) and subject it to (4.47) using the Lagrange matrix multiplier Λ. Referring to (4.51) and (4.52), we then write the cost function as

$$J = \arg\min_{\dot{\mathcal{H}}_{m,k}, \Lambda} \operatorname{tr} [(\bar{\mathcal{D}}_{m,k} - \dot{\mathcal{H}}_{m,k} \mathcal{G}_{m,k}) \mathcal{Q}_{m,k} (\bar{\mathcal{D}}_{m,k} - \dot{\mathcal{H}}_{m,k} \mathcal{G}_{m,k})^T$$
$$+ \dot{\mathcal{H}}_{m,k} \mathcal{R}_{m,k} \dot{\mathcal{H}}_{m,k}^T + \Lambda (I - \dot{\mathcal{H}}_{m,k} \mathcal{C}_{m,k})] \,. \tag{4.62}$$

Applying the derivatives with respect to $\dot{\mathcal{H}}_{m,k}$ and Λ, (4.62) is equivalently split into two equations

$$0 = \frac{\partial}{\partial \dot{\mathcal{H}}_{m,k}} \text{tr}\ [(\bar{D}_{m,k} - \dot{\mathcal{H}}_{m,k}G_{m,k})\mathcal{Q}_{m,k}(\bar{D}_{m,k} - \dot{\mathcal{H}}_{m,k}G_{m,k})^T$$
$$+ \dot{\mathcal{H}}_{m,k}\mathcal{R}_{m,k}\dot{\mathcal{H}}^T_{m,k} + \Lambda(I - \dot{\mathcal{H}}_{m,k}C_{m,k})]\ , \tag{4.63}$$

$$0 = \frac{\partial}{\partial \Lambda} \text{tr}\ [\Lambda(I - \dot{\mathcal{H}}_{m,k}C_{m,k})] = I - \dot{\mathcal{H}}_{m,k}C_{m,k}\ , \tag{4.64}$$

and a simple transformation of (4.63) gives

$$\Lambda^T C^T_{m,k} = 2\dot{\mathcal{H}}_{m,k}\mathcal{R}_{m,k} + 2\dot{\mathcal{H}}_{m,k}G_{m,k}\mathcal{Q}_{m,k}G^T_{m,k} - 2\bar{D}_{m,k}\mathcal{Q}_{m,k}G^T_{m,k}\ ,$$
$$C_{m,k}\Lambda = 2(G_{m,k}\mathcal{Q}_{m,k}G^T_{m,k} + \mathcal{R}_{m,k})\dot{\mathcal{H}}^T_{m,k} - 2G_{m,k}\mathcal{Q}_{m,k}\bar{D}^T_{m,k}$$
$$= 2\Omega_{m,k}\dot{\mathcal{H}}^T_{m,k} - 2G_{m,k}\mathcal{Q}_{m,k}\bar{D}^T_{m,k}\ , \tag{4.65}$$

where matrix

$$\Omega_{m,k} = G_{m,k}\mathcal{Q}_{m,k}G^T_{m,k} + \mathcal{R}_{m,k} \tag{4.66}$$

is symmetric and positive definite.

Multiplying both parts of (4.65) by $C^T_{m,k}\Omega^{-1}_{m,k}$ from the left-hand side and referring to (4.64), we obtain

$$C^T_{m,k}\Omega^{-1}_{m,k}C_{m,k}\Lambda = 2C^T_{m,k}\Omega^{-1}_{m,k}\Omega_{m,k}\dot{\mathcal{H}}^T_{m,k} - 2C^T_{m,k}\Omega^{-1}_{m,k}G_{m,k}\mathcal{Q}_{m,k}\bar{D}^T_{m,k}\ ,$$
$$C^T_{m,k}\Omega^{-1}_{m,k}C_{m,k}\Lambda = 2C^T_{m,k}\dot{\mathcal{H}}^T_{m,k} - 2C^T_{m,k}\Omega^{-1}_{m,k}G_{m,k}\mathcal{Q}_{m,k}\bar{D}^T_{m,k}\ ,$$
$$C^T_{m,k}\Omega^{-1}_{m,k}C_{m,k}\Lambda = 2I - 2C^T_{m,k}\Omega^{-1}_{m,k}G_{m,k}\mathcal{Q}_{m,k}\bar{D}^T_{m,k} \tag{4.67}$$

that gives the Lagrange multiplier

$$\Lambda = 2(C^T_{m,k}\Omega^{-1}_{m,k}C_{m,k})^{-1}(I - C^T_{m,k}\Omega^{-1}_{m,k}G_{m,k}\mathcal{Q}_{m,k}\bar{D}^T_{m,k})\ . \tag{4.68}$$

Now we look at (4.65) again, obtain

$$\dot{\mathcal{H}}_{m,k} = \frac{1}{2}\Lambda^T C^T_{m,k}\Omega^{-1}_{m,k} + \bar{D}_{m,k}\mathcal{Q}_{m,k}G^T_{m,k}\Omega^{-1}_{m,k}\ ,$$

substitute Λ defined by (4.68), end up with the gain $\dot{\mathcal{H}}^{(2)}_{m,k}$ of the OUFIR-II filter,

$$\dot{\mathcal{H}}^{(2)}_{m,k} = (C^T_{m,k}\Omega^{-1}_{m,k}C_{m,k})^{-1}C^T_{m,k}\Omega^{-1}_{m,k} + \bar{D}_{m,k}\mathcal{Q}_{m,k}G^T_{m,k}\Omega^{-1}_{m,k}$$
$$\times [I - C_{m,k}(C^T_{m,k}\Omega^{-1}_{m,k}C_{m,k})^{-1}C^T_{m,k}\Omega^{-1}_{m,k}] \tag{4.69a}$$
$$= (C^T_{m,k}\Omega^{-1}_{m,k}C_{m,k})^{-1}C^T_{m,k}\Omega^{-1}_{m,k} +$$
$$\times \dot{\mathcal{H}}^{(1)}_{m,k}[I - C_{m,k}(C^T_{m,k}\Omega^{-1}_{m,k}C_{m,k})^{-1}C^T_{m,k}\Omega^{-1}_{m,k}]\ , \tag{4.69a}$$

and conclude that it differs significantly from (4.53).

Thus, the OUFIR-II filtering estimate is obtained with

$$\hat{x}_k = \dot{\mathcal{H}}^{(2)}_{m,k}Y_{m,k} + (\bar{S}_{m,k} - \dot{\mathcal{H}}^{(2)}_{m,k}L_{m,k})U_{m,k}\ , \tag{4.70}$$

and the error covariance is determined by (4.55) if we use the gain (4.69a). Noticing that this estimate is also unaffected by either the initial state x_m or input u_m, we finally conclude that the unbiasedness is fully embedded in (4.69a), and the *a posteriori* OUFIR-II filter is indeed optimal unbiased. Therefore, in what follows we will refer to it as the *OUFIR filter*.

4.3.4 Recursion Forms for OUFIR-II Filter

The theory of batch OUFIR filtering was developed in [222], where recursive forms were also shown. An iterative OUFIR filtering algorithm using recursions is formulated in the following theorem, and other details can be found in [222].

Theorem 4.3 Given the batch OUFIR filtering estimate (4.70) obtained using the gain (4.69a), its iterative computation is provided by Algorithm 8.

Algorithm 8: Iterative *a posteriori* OUFIR Filtering Algorithm

Data: y_k, Q_k, R_k

begin
- **for** $k = N-1 : \infty$ **do**
  - $m = k - N + 1,\ r = \max\{m + K, m + 2\},\ \alpha = r - 1$;
  - {Auxiliary Variables};
  - $\bar{F}_{\alpha-1} = \bar{B}_{m,\alpha-1} Q_{m,\alpha-1} H^T_{m,\alpha-1}$;
  - $\bar{F}_{\alpha} = \bar{B}_{m,\alpha} Q_{m,\alpha} H^T_{m,\alpha}$;
  - $O_{\alpha} = \bar{B}_{m,\alpha} Q_{m,\alpha} \bar{B}^T_{m,\alpha}$;
  - {Initial Values};
  - $P_\alpha = U_\alpha - F_\alpha \bar{F}_{\alpha-1} Z_{w+v,\alpha-1} \bar{F}^T_{\alpha-1} F^T_\alpha$;
  - $N_\alpha = (H^T_{m,\alpha} Z^{-1}_{w+v,\alpha} H_{m,\alpha})^{-1}$;
  - $G_\alpha = \mathcal{F}^{m+1}_\alpha - F_\alpha \bar{F}_{\alpha-1} Z^{-1}_{w+v,\alpha-1} H_{m,\alpha}$;
  - $\hat{x}_\alpha = (\mathcal{F}^{m+1}_\alpha N_\alpha H^T_{m,\alpha} + \bar{F}_\alpha - \bar{F}_\alpha Z^{-1}_{w+v,\alpha} H_{m,\alpha} N_\alpha H^T_{m,\alpha}) Z^{-1}_{w+v,\alpha} Y_{m,\alpha}$;
  - **for** $l = r : k$ **do**
    - $P_l = F_l P_{l-1} F^T_l + B_l Q_l B^T_l - F_l P_{l-1} H^T_{l-1} S^{-1}_{l-1} H_{l-1} P_{l-1} F^T_l$;
    - $S_l = H_l P_l H^T_l + R_l$;
    - $G_l = F_l (I - K_{l-1} H_{l-1}) G_{l-1}$;
    - $K_l = P_l H^T_l S^{-1}_l$;
    - $\bar{K}_l = G^T_l H^T_l (H_l P_l H^T_l + R_l)^{-1}$;
    - $N_l = (N^{-1}_{l-1} + \bar{K}_l H_l G_l)^{-1}$;
    - $\tilde{K}_l = (I - K_l H_l) G_l N_l \bar{K}_l$;
    - $\hat{x}_l = F_l \hat{x}_{l-1} + (K_l + \tilde{K}_l)(y_l - H_l F_l \hat{x}_{l-1})$;
  - **end for**
- **end for**
- **Result**: $\hat{x}_k$, P_k

end

Proof: Introduce an auxiliary time index l and define auxiliary variables $\hat{x}^{[1]}_l$, $\hat{x}^{[2]}_l$ and $\hat{x}^{[3]}_l$ as

$$\hat{x}^{[1]}_l = \mathcal{F}^{m+1}_l N_l H^T_{m,l} \Delta^{-1}_{w+v,m} Y_{m,l} , \tag{4.71}$$

$$\hat{x}^{[2]}_l = \bar{F}_l \Delta^{-1}_{w+v,l} Y_{m,l} , \tag{4.72}$$

$$\hat{x}^{[3]}_l = \bar{F}_l \Delta^{-1}_{w+v,l} H_{m,l} N_l H^T_{m,l} \Delta^{-1}_{w+v,l} Y_{m,l} , \tag{4.73}$$

where $\bar{F}_l = \bar{B}_{m,l} Q_{m,l} H^T_{m,l}$ and $N_l = (H^T_{m,l} \Delta^{-1}_{w+v,l} H_{m,l})^{-1}$. Rewrite the batch OUFIR filtering estimate (4.69b) as

$$\hat{x}_l = \hat{x}^{[1]}_l + \hat{x}^{[2]}_l - \hat{x}^{[3]}_l , \tag{4.74}$$

where $\hat{x}_l = \hat{x}_n$ when $l = n$. Use a decompositions of $G_{m,l}$ given before (4.32) and represent $\Delta_{w+v,l}$ with

$$\Delta_{w+v,l} = \Delta'_{w+v,l} + \Delta''_{w+v,l} \,, \tag{4.75}$$

where $\Delta'_{w+v,l} = \text{diag}(R_l, \Delta_{w+v,l-1})$ is nonsingular and

$$\Delta''_{w+v,l} = \begin{bmatrix} H_l O_l H_l^T & \bar{F}_{l-1}^T F_l^T H_l^T \\ H_l F_l \bar{F}_{l-1} & 0 \end{bmatrix} ,$$

where the positive definite symmetric matrix O_l is given by

$$\begin{aligned} O_l &= B_l Q_l B_l^T + F_l \bar{B}_{m,l-1} Q_{m,l-1} \bar{B}_{m,l-1}^T F_l^T \\ &= \bar{B}_{m,l} Q_{m,l} \bar{B}_{m,l}^T \,. \end{aligned} \tag{4.76}$$

Since $\Delta'_{w+v,l}$ is invertible, represent the inverse of $\Delta_{w+v,l}$ as $\Delta_{w+v,l}^{-1} = \Delta'^{-1}_{w+v,l} \Omega_l^{-1}$, where Ω_l is partitioned into a block form using block matrices $\Omega_{l11} = I + H_l O_l H_l^T R_l^{-1}$, $\Omega_{l12} = H_l F_l \bar{F}_{l-1} \Delta_{w+v,l-1}^{-1}$, $\Omega_{l21} = \bar{F}_{l-1}^T F_l^T H_l^T R_l^{-1}$, and $\Omega_{l22} = I$. Using the Schur complement of Ω_{l11}, compute Ω_l^{-1} by

$$\Omega_l^{-1} = \left[\Omega_l^T[1] \;\; \Omega_l^T[2]\right]^T \,, \tag{4.77}$$

where matrices $\Omega_l[1] = [\bar{\Omega}_{l11}^{-1} \;\; -\bar{\Omega}_{l11}^{-1} \Omega_{l12}]$ and $\Omega_l[2] = [-\Omega_{l21} \bar{\Omega}_{l11}^{-1} \;\; I + \Omega_{l21} \bar{\Omega}_{l11}^{-1} \Omega_{l12}]$ are written using a matrix $\bar{\Omega}_{l11} = I + H_l P_l H_l^T R_l^{-1}$, in which

$$P_l = O_l - F_l \bar{F}_{l-1} \Delta_{w+v,l-1}^{-1} \bar{F}_{l-1}^T F_l^T \,. \tag{4.78}$$

Now consider (4.71), (4.72), and (4.73) and notice that matrices $H_{m,l}^T \Delta_{w+v,l}^{-1}$, $\bar{F}_l \Delta_{w+v,l}^{-1}$, and N_l require decompositions. Then combine $\Delta_{w+v,l}^{-1}$ and $H_{m,l}$, make some rearrangements, and write

$$H_{m,l}^T \Delta_{w+v,l}^{-1} = \left[\bar{K}_l \;\; H_{m,l-1}^T \Delta_{w+v,l-1}^{-1} - \bar{K}_l \Omega_{l12}\right] \,, \tag{4.79}$$

where $\bar{K}_l = G_l^T H_l^T R_l^{-1} \bar{\Omega}_{l11}^{-1}$ and

$$G_l = \mathcal{F}_l^{m+1} - F_l \bar{F}_{l-1} \Delta_{w+v,l-1}^{-1} H_{m,l-1} \,. \tag{4.80}$$

Next consider $\bar{F}_l \Delta_{w+v,l}^{-1}$, where $\bar{F}_l$ can be computed recursively as $\bar{F}_l = [O_l H_l^T \;\; F_l \bar{F}_{l-1}]$, and obtain

$$\bar{F}_l \Delta_{w+v,l}^{-1} = \left[K_l \;\; F_l \bar{F}_{l-1} \Delta_{w+v,l-1}^{-1} - K_l \Omega_{l12}\right] \,, \tag{4.81}$$

where

$$K_l = P_l H_l^T R_l^{-1} \bar{\Omega}_{l11}^{-1} \,. \tag{4.82}$$

Substitute (4.79) into N_l, rearrange the terms to have $N_l^{-1} = Z_{l-1}^{-1} + \bar{K}_l H_l G_l$, and come up with

$$N_l = N_{l-1} - N_{l-1}(I + \bar{K}_l H_l G_l N_{l-1})^{-1} \bar{K}_l H_l G_l N_{l-1} \,. \tag{4.83}$$

Substitution of (4.78) into (4.71) using (4.83) gives the decomposition

$$\hat{x}_l^{[1]} = F_l \hat{x}_{l-1}^{[1]} - \mathcal{F}_l^{m+1} N_l \bar{K}_l \hat{x}_{l-1}^{\epsilon} + \mathcal{F}_l^{m+1} N_l \bar{K}_l \bar{y}_l \,, \tag{4.84}$$

where $\hat{x}_{l-1}^{\epsilon} = H_l F_l (\hat{x}_{l-1}^{[1]} - \hat{x}_{l-1}^{[3]})$ and $\bar{y}_l = y_l - H_l F_l \bar{F}_{l-1} \Delta_{w+v,l-1}^{-1} Y_{m,l-1}$. Then combine (4.82) and (4.72) and obtain

$$\hat{x}_l^{[2]} = F_l \hat{x}_{l-1}^{[2]} + K_l \bar{y}_l \,. \tag{4.85}$$

Reasoning along similar lines, transform $\hat{x}_l^{[3]}$ as

$$\begin{aligned}\hat{x}_l^{[3]} =& F_l\hat{x}_{l-1}^{[3]} - F_l\bar{F}_{l-1}\Delta_{w+v,l-1}^{-1}H_{m,l-1}N_l\bar{K}_l\hat{x}_{l-1}^{\epsilon} \\ &+ K_l\hat{x}_{l-1}^{\epsilon} - K_lH_lG_lN_l\bar{K}_l\hat{x}_{l-1}^{\epsilon} \\ &+ F_l\bar{F}_{l-1}\Delta_{w+v,l-1}^{-1}H_{m,l-1}N_l\bar{K}_l\bar{y}_l \\ &+ K_lH_lG_lN_l\bar{K}_l\bar{y}_l \ . \end{aligned} \tag{4.86}$$

At this point, combine (4.86), (4.85), (4.84), and (4.74) and arrive at the recursion for the *a posteriori* OUFIR filtering estimate

$$\hat{x}_l = F_l\hat{x}_{l-1} + (K_l + \tilde{K}_l)(y_l - H_lF_l\hat{x}_{l-1}) \ , \tag{4.87}$$

where $\bar{y}_l - \hat{x}_{l-1}^{\epsilon} = y_l - H_lF_l\hat{x}_{l-1}$ is the measurement residual and

$$\tilde{K}_l = (I - K_lH_l)G_lN_lG_l^TH_l^T(H_lP_lH_l^T + R_l)^{-1} \tag{4.88}$$

is an addition to the bias correction gain K_l.

Go back to (4.82) and (4.88) and note that gains K_l and $\tilde{K}_l$ depend on matrices P_l, G_l, and N_l, where the iteration for N_l is given by (4.83), which also requires P_l and G_l. Using (4.78), obtain the recursion for the error covariance P_l,

$$P_l = F_lP_{l-1}F_l^T + B_lQ_lB_l^T - F_lK_{l-1}H_{l-1}P_{l-1}F_l^T \ .$$

Finally, substitute (4.81) into (4.80), make some rearrangements, and come up with the recursive form $G_l = F_l(I - K_{l-1}H_{l-1})G_{l-1}$. Specify the initial values by computing P_l, G_l, N_l, and $\hat{x}_l$ using their definitions at $l = \alpha$ and complete the proof. □

What immediately catches attention is that the bias correction gain (line 21) is combined with two gains as $K_l + \tilde{K}_l$ and is thus not the Kalman gain. Unlike the KF, Algorithm 8 self-determines the initial matrices (lines 4–10). It then updates estimates iteratively (lines 12–17) at each time index l, and the final estimates go to the output when $l = k$. Accordingly, Algorithm 8 demonstrates all the advantages of FIR filtering: better robustness against temporary model uncertainties, lower sensitivity to noise, and smaller round-off errors. But since the estimation errors decrease with each time step, the OUFIR filter can also operate in one iteration cycle like KF.

It is also worth noting that the additional gain $\tilde{K}_l$ is due to the embedded unbiasedness constraint. As the horizon length N increases, this gain decreases to zero, and Algorithm 8 transforms to the KF Algorithm when $N \gg N_{\text{opt}}$. Consequently, an essential difference between the outputs of the KF and OUFIR filter can be observed only at relatively short horizons, $N < N_{\text{opt}}$.

4.4 Maximum Likelihood FIR Estimator

Another useful linear approach to state estimation is to use the likelihood function and obtain an ML FIR estimate. The FIR ML estimator can be derived to have two different forms. We can obtain the ML estimate directly at k or, alternatively, first find the ML estimate at the start point m and then project it onto k. Both of these solutions are discussed next.

4.4.1 ML-I FIR Filtering Estimate

Consider the state-space model

$$x_k = F_k x_{k-1} + B_k w_k \,, \tag{4.89}$$

$$y_k = H_k x_k + v_k \,. \tag{4.90}$$

Using the extended equations (4.7) and (4.14) with $U_{m,k} = 0$, the *a posteriori* ML-I FIR filtering estimate can be found over data taken from $[m, k]$ by maximizing the likelihood $p(Y_{m,k}|x_k)$ of x_k as

$$\hat{x}_{k|k} = \arg\max_{x_k} \; p(Y_{m,k}|x_k) \,. \tag{4.91}$$

To solve the maximization problem (4.91), we define x_k by (4.19) for zero input $u_k = 0$ as

$$x_k = \mathcal{F}_k^{m+1} x_m + \bar{D}_{m,k} W_{m,k} \,.$$

By moving $\bar{D}_{m,k} W_{m,k}$ to the left-hand side and multiplying both sides by $(\mathcal{F}_k^{m+1})^{-1}$, we determine the initial state as

$$x_m = (\mathcal{F}_k^{m+1})^{-1} x_k - (\mathcal{F}_k^{m+1})^{-1} \bar{D}_{m,k} W_{m,k} \,. \tag{4.92}$$

Further substituting (4.92) into (4.14) and rearranging the terms gives the measurement residual

$$Y_{m,k} - C_{m,k} x_k = \mathcal{N}_{m,k} \,, \tag{4.93}$$

where $C_{m,k} = H_{m,k}(\mathcal{F}_k^{m+1})^{-1}$ and

$$\mathcal{N}_{m,k} = (G_{m,k} - C_{m,k}\bar{D}_{m,k}) W_{m,k} + V_{m,k} \tag{4.94}$$

is a random noise component.

Now, the likelihood of x_k can be defined by the multidimensional Gaussian distribution as

$$p(Y_{m,k}|x_k) \propto \exp\left\{ -\frac{1}{2}(Y_{m,k} - C_{m,k}x_k)^T \Sigma_{m,n}^{-1}(\ldots) \right\} \,, \tag{4.95}$$

where $(\ldots)$ means the term that is equal to the relevant preceding term and $\Sigma_{m,k}$ is given by

$$\Sigma_{m,k} = \mathcal{E}\{\mathcal{N}_{m,k}\mathcal{N}_{m,k}^T\} \tag{4.96a}$$

$$= (G_{m,k} - C_{m,k}\bar{D}_{m,k})\mathcal{Q}_{m,k}(\ldots)^T + \mathcal{R}_{m,k} \,. \tag{4.96b}$$

Using (4.95), the maximization problem (4.91) can now be solved by minimizing the quadratic form as

$$\hat{x}_{k|k} = \arg\min_{x_k} \; \left\{ -\frac{1}{2}(Y_{m,k} - C_{m,k}x_k)^T \Sigma_{m,k}^{-1}(\ldots) \right\} \,. \tag{4.97}$$

By taking the derivative of the right-hand side of (4.97) with respect to x_k, equating it to zero, and then solving the obtained equation, we finally arrive at the *a posteriori* ML-I FIR filtering estimate [223],

$$\hat{x}_{k|k} = (C_{m,k}^T \Sigma_{m,k}^{-1} C_{m,k})^{-1} C_{m,k}^T \Sigma_{m,k}^{-1} Y_{m,k} \tag{4.98a}$$

$$= \mathcal{H}_{m,k}^{\mathrm{ML1}} Y_{m,k} \,, \tag{6.98b}$$

where $\mathcal{H}_{m,k}^{\mathrm{ML1}}$ is the gain of the ML-I FIR filter. In what follows we will show that the ML form (4.98a) unifies all bias-constrained optimal and unbiased FIR filters.

4.4.2 Equivalence of ML-I FIR and OUFIR Filters

It can be seen that the ML-I filter (4.98a) becomes the UFIR filter (4.29b) if we put $\Sigma_{m,k} = I$. On the other hand, it was shown in [223] that embedding unbiasedness in an OFIR filter [222] makes it an OUFIR filter (4.69a). To show that (4.98a) unifies these filters as well, next we will prove that (4.98a) is equivalent to the OUFIR estimate (4.69a) with zero input.

Theorem 4.4 Given model (4.89) and (4.90), then the ML-I FIR filter (4.98a) is equivalent to the OUFIR-I filter (4.69a).

Proof: Consider (4.96b) and represent as

$$\begin{aligned}\Sigma_{m,k} &= \underbrace{G_{m,k}\mathcal{Q}_{m,k}G^T_{m,k} + \mathcal{R}_{m,k}}_{\Omega_{m,k}} - \underbrace{\left(C_{m,k}\bar{D}_{m,k}\mathcal{Q}_{m,k}G^T_{m,k} + \Psi_{m,k}\right)}_{\bar{\Delta}_{m,k}} \\ &= \Omega_{m,k} - \bar{\Delta}_{m,k}\,, \end{aligned} \tag{4.99}$$

where $\Psi_{m,k} = G_{m,k}\mathcal{Q}_{m,k}\bar{D}^T_{m,k}C^T_{m,k} - C_{m,k}\bar{D}_{m,k}\mathcal{Q}_{m,k}\bar{D}^T_{m,k}C^T_{m,k}$. Next, decompose the gain $\mathcal{H}^{\mathrm{ML}}_{m,k}$ (4.98a) as

$$\begin{aligned}\mathcal{H}^{\mathrm{ML1}}_{m,k} &= \left[C^T_{m,k}(\Omega_{m,k} - \bar{\Delta}_{m,k})^{-1}C_{m,k}\right]^{-1} C^T_{m,k} \\ &\quad \times \left(\Omega_{m,k} - \bar{\Delta}_{m,k}\right)^{-1} . \end{aligned} \tag{4.100}$$

By applying the matrix inversion lemma (A.11) to $(\Omega_{m,k} - \bar{\Delta}_{m,k})^{-1}$ obtain

$$\mathcal{H}^{\mathrm{ML1}}_{m,k} = \mathcal{H}^{(1)}_{m,k} + \mathcal{H}^{(2)}_{m,k} \tag{4.101}$$

and, by applying (A.11) again, obtain

$$\begin{aligned}\mathcal{H}^{(1)}_{m,k} &= \left[C^T_{m,k}(\Omega_{m,k} - \bar{\Delta}_{m,k})^{-1}C_{m,k}\right]^{-1} C^T_{m,k}\Omega^{-1}_{m,k}\,, \\ &= (\mathcal{X}_{m,k} + \mathcal{Y}_{m,k})^{-1}C^T_{m,k}\Omega^{-1}_{m,k} \\ &= \mathcal{X}^{-1}_{m,k}C^T_{m,k}\Omega^{-1}_{m,k} - (\mathcal{X}_{m,k} + \mathcal{Y}_{m,k})^{-1} \\ &\quad \times \mathcal{Y}_{m,k}\mathcal{X}^{-1}_{m,k}H^T_{m,k}\Omega^{-1}_{m,k}\,, \end{aligned} \tag{4.102}$$

$$\begin{aligned}\mathcal{H}^{(2)}_{m,k} &= \mathcal{X}^{-1}_{m,k}C^T_{m,k}(\Omega_{m,k} - \bar{\Delta}_{m,k})^{-1}\bar{\Delta}_{m,k}\Omega^{-1}_{m,k} \\ &\quad - (\mathcal{X}_{m,k} + \mathcal{Y}_{m,k})^{-1}\mathcal{Y}_{m,k}\mathcal{X}^{-1}_{m,n}C^T_{m,k} \\ &\quad \times (\Omega_{m,k} - \bar{\Delta}_{m,k})^{-1}\bar{\Delta}_{m,k}\Omega^{-1}_{m,k}\,, \end{aligned} \tag{4.103}$$

where $\mathcal{X}_{m,k} = C^T_{m,k}\Omega^{-1}_{m,k}C_{m,k}$ and

$$\mathcal{Y}_{m,k} = C^T_{m,k}(\Omega_{m,k} - \bar{\Delta}_{m,k})^{-1}\bar{\Delta}_{m,k}\Omega^{-1}_{m,k}C_{m,k}\,. \tag{4.104}$$

Now combine (4.102) and (4.103) and represent (4.101) as

$$\mathcal{H}^{\mathrm{ML1}}_{m,k} = \mathcal{H}^{(1,1)}_{m,k} + \mathcal{H}^{(a)}_{m,k} - \mathcal{H}^{(b)}_{m,k}\,, \tag{4.105}$$

where $\mathcal{H}^{(1,1)}_{m,k} = \mathcal{X}^{-1}_{m,k}C^T_{m,k}\Omega^{-1}_{m,k}$ and

$$\begin{aligned}\mathcal{H}^{(a)}_{m,k} &= \mathcal{X}^{-1}_{m,k}C^T_{m,k}(\Omega_{m,k} - \bar{\Delta}_{m,k})^{-1}\bar{\Delta}_{m,k}\Omega^{-1}_{m,k}\,, \\ \mathcal{H}^{(b)}_{m,k} &= (\mathcal{X}_{m,k} + \mathcal{Y}_{m,k})^{-1}\mathcal{Y}_{m,k}\mathcal{X}^{-1}_{m,k}C^T_{m,k}(\Omega_{m,k} - \bar{\Delta}_{m,k})^{-1}\,. \end{aligned}$$

Use $\bar{\Delta}_{m,k}$ taken from (4.99) and represent $\mathcal{H}^{(a)}_{m,k}$ as

$$\begin{aligned}\mathcal{H}^{(a)}_{m,k} &= \mathcal{X}^{-1}_{m,k}(\mathcal{X}_{m,k} + \mathcal{Y}_{m,k})\mathcal{F}_{m,k} \\ &= \mathcal{F}_{m,k} + \mathcal{F}'_{m,k} ,\end{aligned} \tag{4.106}$$

where $\mathcal{F}_{m,k} = (\bar{D}_{m,k}\mathcal{Q}_{m,k}G^T_{m,k} + \bar{\Psi}_{m,k})\Omega^{-1}_{m,k}$,

$$\begin{aligned}\bar{\Psi}_{m,k} &= [C^T_{m,k}(C_{m,k}C^T_{m,k})^{-1}G_{m,k} - \bar{D}_{m,k}] \\ &\quad \times \mathcal{Q}_{m,k}\bar{D}_{m,k}{}^T C^T_{m,k} , \\ \mathcal{F}'_{m,k} &= \mathcal{X}^{-1}_{m,k}\mathcal{Y}_{m,k}\mathcal{F}_{m,k} .\end{aligned}$$

Replace the second matrix $\bar{\Delta}_{m,k}$ on the right-hand side of (4.104) by its form taken from (4.99), and go to

$$\begin{aligned}\mathcal{Y}_{m,k} &= C^T_{m,k}(\Omega_{m,k} - \bar{\Delta}_{m,k})^{-1}C_{m,k} \\ &\quad \times (\bar{D}_{m,k}\mathcal{Q}_{m,k}G^T_{m,k} + \bar{\Psi}_{m,k})\Omega^{-1}_{m,k}C_{m,k} , \\ &= (\mathcal{X}_{m,k} + \mathcal{Y}_{m,k})\mathcal{F}_{m,k}C_{m,k} .\end{aligned} \tag{4.107}$$

Then substituting $\mathcal{Y}_{m,k}$ in $\mathcal{F}'_{m,k}$ with (4.107) gives

$$\mathcal{F}'_{m,k} = (\mathcal{F}_{m,k} + \mathcal{F}'_{m,k})\bar{\Delta}_{m,k}\Omega^{-1}_{m,k}$$

that can be transformed to

$$\begin{aligned}\mathcal{F}'_{m,k} &= \mathcal{F}_{m,k}\bar{\Delta}_{m,k}\Omega^{-1}_{m,k}\left(I - \bar{\Delta}_{m,k}\Omega^{-1}_{m,k}\right)^{-1} \\ &= \mathcal{F}_{m,k}\Xi_{m,k}\left(I - \Xi_{m,k}\right)^{-1} ,\end{aligned} \tag{4.108}$$

where $\Xi_{m,k} \triangleq \bar{\Delta}_{m,k}\Omega^{-1}_{m,k} = \tilde{H}_{m,k}\mathcal{F}_{m,k}$ and $\mathcal{H}^{(a)}_{m,k}$ becomes

$$\mathcal{H}^{(a)}_{m,k} = \mathcal{F}_{m,k} + \mathcal{F}_{m,k}\Xi_{m,k}\left(I - \Xi_{m,k}\right)^{-1} . \tag{4.109}$$

Consider (4.105) again and transform $\mathcal{H}^{(b)}_{m,k}$ to

$$\begin{aligned}\mathcal{H}^{(b)}_{m,k} &= \mathcal{F}_{m,k}\bar{\mathcal{X}}_{m,k}\Omega^{-1}_{m,k} \\ &\quad + \underbrace{\mathcal{F}_{m,k}\bar{\mathcal{X}}_{m,k}(\Omega_{m,k} - \bar{\Delta}_{m,k})^{-1}}_{\mathcal{H}^{(b)}_{m,k}}\bar{\Delta}_{m,k}\Omega^{-1}_{m,k} \\ &= \mathcal{F}_{m,k}\bar{\mathcal{X}}_{m,k}\Omega^{-1}_{m,k} + \Upsilon_{m,k} ,\end{aligned} \tag{4.110}$$

where $\bar{\mathcal{X}}_{m,k} = C_{m,k}\mathcal{X}^{-1}_{m,k}C^T_{m,k}$ and $\Upsilon_{m,k} = \mathcal{H}^{(b)}_{m,k}\Xi_{m,k}$. Next, replace $\mathcal{H}^{(b)}_{m,k}$ with (4.110) in the last relationship and obtain

$$\Upsilon_{m,k} = \mathcal{F}_{m,k}\bar{\mathcal{X}}_{m,k}\Omega^{-1}_{m,k}\Xi_{m,k}\left(I - \Xi_{m,k}\right)^{-1} . \tag{4.111}$$

At this point, combining (4.109), (4.110), (4.111), and (4.105) yields

$$\begin{aligned}\mathcal{H}^{\mathrm{ML1}}_{m,k} &= \mathcal{X}^{-1}_{m,k}C^T_{m,k}\Omega^{-1}_{m,k} + \mathcal{F}_{m,k}\left(I - \bar{\mathcal{X}}_{m,k}\Omega^{-1}_{m,k}\right) \\ &\quad + \mathcal{F}_{m,k}\left(I - \bar{\mathcal{X}}_{m,k}\Omega^{-1}_{m,k}\right)\Xi_{m,k}\left(I - \Xi_{m,k}\right)^{-1} .\end{aligned} \tag{4.112}$$

Note that the second term in (4.112) is identically zero, since

$$
\begin{aligned}
(I - \bar{\chi}_{m,k}\Omega_{m,k}^{-1})\Xi_{m,k} &= [I - C_{m,k}(C_{m,k}^T\Omega_{m,k}^{-1}C_{m,k})^{-1}C_{m,k}^T\Omega_{m,k}^{-1}] \\
&\quad \times C_{m,k}\mathcal{F}_{m,k} \\
&= [C_{m,k} - C_{m,k}(C_{m,k}^T\Omega_{m,k}^{-1}C_{m,k})^{-1}C_{m,k}^T\Omega_{m,k}^{-1}C_{m,k}] \\
&\quad \times \mathcal{F}_{m,k} \\
&= (C_{m,k} - C_{m,k})\mathcal{F}_{m,k} = 0 .
\end{aligned}
$$

The first term in (4.112) can be transformed as

$$
\begin{aligned}
\mathcal{F}_{m,k}(I - \bar{\chi}_{m,k}\Omega_{m,k}^{-1}) &= (\bar{D}_{m,k}\mathcal{Q}_{m,k}G_{m,k}^T + \bar{\Psi}_{m,k})\Omega_{m,k}^{-1}(I - \bar{\chi}_{m,k}\Omega_{m,k}^{-1}) \\
&= \bar{D}_{m,k}\mathcal{Q}_{m,k}G_{m,k}^T\Omega_{m,k}^{-1}(I - \bar{\chi}_{m,k}\Omega_{m,k}^{-1}) + \\
&\quad + \bar{\Psi}_{m,k}\Omega_{m,k}^{-1}(I - \bar{\chi}_{m,k}\Omega_{m,k}^{-1}) ,
\end{aligned}
$$

where, for the previous reasons, the last term is also identically zero

$$
\begin{aligned}
\bar{\Psi}_{m,k}\Omega_{m,k}^{-1}(I - \bar{\chi}_{m,k}\Omega_{m,k}^{-1}) &= [C_{m,k}^T(C_{m,k}C_{m,k}^T)^{-1}G_{m,k} - \bar{D}_{m,k}] \\
&\quad \times \mathcal{Q}_{m,k}\bar{D}_{m,k}^{\ T}C_{m,k}^T\Omega_{m,k}^{-1}(I - \bar{\chi}_{m,k}\Omega_{m,k}^{-1}) = 0 .
\end{aligned}
$$

Finally, the gain (4.112) can be written in two equivalent forms:

$$
\begin{aligned}
\mathcal{H}_{m,k}^{\text{ML1}} &= (C_{m,k}^T\Omega_{m,k}^{-1}C_{m,k})^{-1}C_{m,k}^T\Omega_{m,k}^{-1} + D_{m,k}\mathcal{Q}_{m,k}G_{m,k}^T\Omega_{m,k}^{-1} \\
&\quad \times [I - C_{m,k}(C_{m,k}^T\Omega_{m,k}^{-1}C_{m,k})^{-1}C_{m,k}^T\Omega_{m,k}^{-1}] \quad (4.113a) \\
&= D_{m,k}\mathcal{Q}_{m,k}G_{m,k}^T\Omega_{m,k}^{-1} + (I - D_{m,k}\mathcal{Q}_{m,k}G_{m,k}^T\Omega_{m,k}^{-1}C_{m,k}) \\
&\quad \times (C_{m,k}^T\Omega_{m,k}^{-1}C_{m,k})^{-1}C_{m,k}^T\Omega_{m,k}^{-1} . \quad (4.113b)
\end{aligned}
$$

All that follows from this derivation is that the gain $\mathcal{H}_{m,k}^{\text{ML1}}$ given by (4.113a) or (4.113b) is identical to the OUFIR filter gain given by (4.69a) or (4.69b). Thus, these estimators are equivalent, which completes the proof. □

4.4.3 ML-II FIR Filtering Estimate

As we mentioned earlier, the FIR approach allows us to find an ML estimate of the initial state at m and then project it onto k, which leads to the ML-II FIR estimator. Referring to (4.89) and (4.90), the ML estimate of x_m obtained over $[m, k]$ can be found by maximizing the likelihood $p(Y_{m,k}|x_m)$ as

$$
\tilde{x}_{m|k} = \arg\max_{x_m} \; p(Y_{m,k}|x_m) . \quad (4.114)
$$

From (4.14) we have

$$
Y_{m,k} - H_{m,k}x_k = \mathcal{N}_{m,k} , \quad (4.115)
$$

$$
\mathcal{N}_{m,k} = G_{m,k}W_{m,k} + V_{m,k} \quad (4.116)
$$

and the likelihood of x_m becomes

$$
p(Y_{m,k}|x_m) \propto \exp\left\{-\frac{1}{2}(Y_{m,k} - H_{m,k}x_m)^T\Omega_{m,k}^{-1}(\dots)\right\} , \quad (4.117)
$$

where $\Omega_{m,k}$ is given by (4.66). The maximization problem (4.114) can now be solved by minimizing the quadratic form as

$$
\tilde{x}_{m|k} = \arg\min_{x_m} \; \left\{-\frac{1}{2}(Y_{m,k} - H_{m,k}x_m)^T\Omega_{m,k}^{-1}(\dots)\right\} , \quad (4.118)
$$

which gives the *a posteriori* ML FIR filtering estimate of the initial state x_m,

$$\tilde{x}_{m|k} = (H_{m,k}^T \Omega_{m,k}^{-1} H_{m,k})^{-1} H_{m,k}^T \Omega_{m,k}^{-1} Y_{m,k} \,. \tag{4.119}$$

Since an estimate (4.119) is obtained at m over data available on $[m, k]$, it can be used to specify the initial state in FIR filtering algorithms. But we have another goal here, and project (4.119) onto k, keeping in mind the unbiasedness of ML estimators. Thus, we arrive at the ML-II FIR filtering estimate

$$\begin{aligned}\hat{x}_{k|k} &= \mathcal{F}_k^{m+1} \hat{x}_{m|k} \\ &= \mathcal{F}_k^{m+1} (H_{m,k}^T \Omega_{m,k}^{-1} H_{m,k})^{-1} H_{m,k}^T \Omega_{m,k}^{-1} Y_{m,k} \\ &= (C_{m,k}^T \Omega_{m,k}^{-1} C_{m,k})^{-1} C_{m,k}^T \Omega_{m,k}^{-1} Y_{m,k} \\ &= \mathcal{H}_{m,k}^{\text{ML2}} Y_{m,k} \,,\end{aligned} \tag{4.120}$$

where $\mathcal{H}_{m,k}^{\text{ML2}}$ is the gain of the ML-II FIR filter. What can be seen is that the gain $\mathcal{H}_{m,k}^{\text{ML1}}$ defined by (4.98a) unifies the gain $\mathcal{H}_{m,k}^{\text{ML2}}$ defined by (4.120) if we replace $\Sigma_{m,k}$ with $\Omega_{m,k}$. It is also worth noting that the batch RH MVF obtained in [106] has the same form (4.120).

4.4.4 Properties of ML FIR State Estimators

It is now appropriate to summarize the basic properties of ML FIR state estimators (filters). First, we notice that for Gaussian processes the ML estimator gives an MVU estimate [88], and thus the ML FIR and OUFIR estimates are strongly related. These are other important properties:

- *Equivalence:* The ML-I FIR estimator and OUFIR-II estimator give equivalent estimates. Other bias-constrained estimators, such as the ML-II FIR filter and OUFIR-I filter, can be considered as approximations.
- *Recursions:* Recursive forms for the bias-constrained FIR state estimators (Algorithm 8) are not Kalman recursions (Algorithm 4).
- *Unbiasedness:* A state estimator is unbiased if it satisfies the unbiasedness condition $\mathcal{E}\{\hat{x}_k\} = \mathcal{E}\{x_k\}$ for all k. Since the ML FIR and OUFIR filters satisfy this condition, they have the property of unbiased optimality, which is not the case of the OFIR and OFIR-I filters. The KF does not serve bias-constrained FIR filters and is therefore not optimal unbiased.
- *Deadbeat property:* A state estimator has the deadbeat property if for any noiseless model it satisfies $\hat{x}_{k|k} = x_k$. To see if an estimator has this property, first notice that for noiseless models the ML FIR estimator becomes the UFIR estimator (4.29b) when $\Sigma_{m,k} = I$. Next, suppose the noise is low and approximate $\Sigma_{m,k}^{-1}$ with $\Sigma_{m,k}^{-1} = (I + \tilde{\Sigma}_{m,k})^{-1} \cong I - \tilde{\Sigma}_{m,k}$, where $\tilde{\Sigma}_{m,k}$ is a small addition due to noise. Neglecting $\tilde{\Sigma}_{m,k}$ for deterministic systems, we arrive at (4.29b) and conclude that the ML FIR filter is a deadbeat estimator.
- *Error covariance:* Following [223], it can be shown that the error covariance of the ML FIR filter is equal to the error covariance (4.55) of the *a posteriori* OUFIR filter,

$$P_k^{\text{ML}} = \dot{\mathcal{W}}_{m,k} \mathcal{Q}_{m,k} \dot{\mathcal{W}}_{m,k}^T + \dot{\mathcal{V}}_{m,k} \mathcal{R}_{m,k} \dot{\mathcal{V}}_{m,k}^T \,. \tag{4.121}$$

The general conclusion that can be made is that the ML FIR form (4.98a) is the most versatile and, as such, unifies all bias-constrained OUFIR and UFIR estimators. It is also worth noting again that optimal Kalman recursions (Algorithm 4) do not serve batch FIR forms that have the property of unbiased optimality (ses Algorithm 8).

4.5 The *a priori* FIR Filters

An *a priori* state estimate is required in at least two cases where 1) data y_k are temporary or permanently unavailable (or delayed) at k and 2) the *a posteriori* estimate $\hat{x}_k$ turns out to be rougher than its predicted or prior value [106, 111]. The *a priori* FIR filter gives an estimate at $k+1$ using data from $[m, k]$. Therefore, it is akin to a predictor. However, since it still requires u_{k+1} and w_{k+1}, it is not a strong predictor. Meanwhile, the RH FIR filter [106] produces an estimate at k over all variables taken from $[k-N, k-1]$ and thus is a one-step predictive filter.

4.5.1 The *a priori* Optimal FIR Filter

The *a priori* OFIR filtering estimate can be found if we consider a prior estimate and represent it at $k+1$ over $[m, k]$ as

$$\begin{aligned}
\hat{x}^-_{k+1} &= F_{k+1}\hat{x}_k + E_{k+1}u_{k+1} , \\
&= F_{k+1}(\mathcal{H}^{\mathrm{h}}_{m,k}Y_{m,k} + \mathcal{H}^{\mathrm{f}}_{m,k}U_{m,k}) + E_{k+1}u_{k+1} , \\
&= F_{k+1}\mathcal{H}^{\mathrm{h}}_{m,k}(H_{m,k}x_m + L_{m,k}U_{m,k} + G_{m,k}W_{m,k} + V_{m,k}) \\
&\quad + F_{k+1}\mathcal{H}^{\mathrm{f}}_{m,k}U_{m,k} + E_{k+1}u_{k+1} ,
\end{aligned} \tag{4.122}$$

where $\mathcal{H}^{\mathrm{h}}_{m,k}$ is given by (4.28b) and $\mathcal{H}^{\mathrm{f}}_{m,k}$ by (4.22). Similarly, the state can be represented at $k+1$ as

$$\begin{aligned}
x_{k+1} &= F_{k+1}x_k + E_{k+1}u_{k+1} + B_{k+1}w_{k+1} , \\
&= F_{k+1}(\mathcal{F}^{m+1}_k x_m + \bar{S}_{m,k}U_{m,k} + \bar{D}_{m,k}W_{m,k}) \\
&\quad + E_{k+1}u_{k+1} + B_{k+1}w_{k+1} .
\end{aligned} \tag{4.123}$$

Now applying the unbiasedness condition $\mathcal{E}\{\hat{x}^-_{k+1}\} = \mathcal{E}\{x_{k+1}\}$ to (4.122) and (4.123) gives the unbiasedness constraints (4.21) and (4.22), which are valid for the *a posteriori* OFIR filter.

It can be shown that the orthogonality condition $\mathcal{E}\{\varepsilon^-_{k+1}Y^T_{m,k}\} = 0$ gives $\mathcal{H}^{\mathrm{h}}_{m,k}$ defined by (4.28b) and $\mathcal{H}^{\mathrm{f}}_{m,k}$ defined by (4.22). Accordingly, the batch error covariance $P^-_{k+1} = \mathcal{E}\{\varepsilon^-_{k+1}\varepsilon^{-T}_{k+1}\}$ of the *a priori* OFIR filter becomes

$$\begin{aligned}
P^-_{k+1} &= F_{k+1}(\mathcal{B}_{m,k}\mathcal{X}_m\mathcal{B}^T_{m,k} + \mathcal{W}_{m,k}\mathcal{Q}_{m,k}\mathcal{W}^T_{m,k} \\
&\quad + \mathcal{V}_{m,k}\mathcal{R}_{m,k}\mathcal{V}^T_{m,k})F^T_{k+1} + B_{k+1}Q_{k+1}B^T_{k+1} \\
&= F_{k+1}P_kF^T_{k+1} + B_{k+1}Q_{k+1}B^T_{k+1} ,
\end{aligned} \tag{4.124}$$

where P_k is the error covariance of the *a posteriori* OFIR filter.

Since the *a priori* OFIR filter does not utilize data from $k+1$, it is obviously less accurate than the *a posteriori* OFIR filter. It can also be shown that batch forms (4.122) and (4.124) can be computed using the *a priori* KF Algorithm 2, given the initial values at m. The justification of this claim is postponed to "Problems," and the reader is encouraged to find a proof.

4.5.2 The *a priori* Optimal Unbiased FIR Filter

We can now modify the *a priori* OFIR filter to be the *a priori* OUFIR-I filter, in which case the estimate is defined by

$$\hat{x}^-_{k+1} = F_{k+1}(\dot{\mathcal{H}}^{\mathrm{h}}_{m,k}Y_{m,k} + \dot{\mathcal{H}}^{\mathrm{f}}_{m,k}U_{m,k}) + E_{k+1}w_{k+1} , \tag{4.125}$$

where $\dot{\mathcal{H}}^{\mathrm{h}}_{m,k}$ is given by (4.53) and $\dot{\mathcal{H}}^{\mathrm{f}}_{m,k}$ by (4.48). After removing the term with the initial values using (4.47), the error covariance takes the form

$$P^-_{k+1} = F_{k+1}(\dot{\mathcal{W}}_{m,k}\mathcal{Q}_{m,k}\dot{\mathcal{W}}^T_{m,k} + \dot{\mathcal{V}}^-_{m,k}\mathcal{R}_{m,k}\dot{\mathcal{V}}^T_{m,k})F^T_{k+1}$$
$$+ B_{k+1}Q_{k+1}B^T_{k+1} , \tag{4.126}$$

where $\dot{\mathcal{W}}_{m,k}$ is given by (4.51) and $\dot{\mathcal{V}}_{m,k}$ by (4.52).

It can be shown that recursions for (4.125) and (4.126) are provided by the *a priori* KF Algorithm 2. This does not mean, however, that the *a priori* KF is optimal unbiased, because the OUFIR-I filter is nothing more than the OFIR filter applied for zero initial conditions. The derivation of recursions for (4.125) and (4.126) is postponed to "Problems."

4.6 Limited Memory Filtering

The idea of limited memory filtering came about in attempts to instill KF more robustness. Analyzing errors in the Kalman, ML, MVU (for Gaussian processes), and LS estimators, Jazwinski stated in [79] that the *LMF is the only device for preventing divergence in the presence of unbounded perturbations* in control signal. The general idea of the LMF was formulated by Schweppe [161] as an old estimate updated not over all data but over most resent observations. Latter, Bruckstein and Kailath have confirmed in [23] that the LMF is the best linear estimator of the process (signal) of interest given noisy observations. It was also shown in [101] that the KF-based LMFs are equivalent. A historical note on the theory of LMF can be found in [179].

The LMF works similarly to the FIR filter but requires an initial state x_{m-1} beyond $[m, k]$ as an input variable. This means that the LMF has IIR, because the best x_{m-1} is provided by the estimate $\hat{x}_{m-1}$, and the LMF structure thus implies state feedback.

Batch LMF can be obtained if we extend the model in (4.1) and (4.2) on $[m, k]$ relative to x_{m-1} as

$$X_{m,k} = \mathcal{A}_{m,k}x_{m-1} + S_{m,k}U_{m,k} + D_{m,k}W_{m,k} , \tag{4.127}$$

$$Y_{m,k} = \mathcal{C}_{m,k}x_{m-1} + L_{m,k}U_{m,k} + G_{m,k}W_{m,k} + V_{m,k} , \tag{4.128}$$

where $m = 1$ stands for batch KF. Extended vectors and matrices used to define (4.127) and (4.128) can be found in the discussion of (4.7) and (4.14), with the exception of matrices $\mathcal{A}_{m,k}$ and $\mathcal{C}_{m,k}$ defined by

$$\mathcal{A}_{m,k} = [F^T_m, \cdots , (\mathcal{F}^m_{k-1})^T, (\mathcal{F}^m_k)^T]^T , \tag{4.129}$$

$$\mathcal{C}_{m,k} = \bar{C}_{m,k}\mathcal{A}_{m,k} . \tag{4.130}$$

Since the model in (4.127) and (4.128) involves x_{m-1}, it does not require $u_m = 0$ and $w_m = 0$, unlike (4.7) and (4.14). This feature makes a difference between FIR filters and LMF.

4.6.1 Batch Limited Memory Filter

The generalized structure of the LMF is shown in Fig. 4.4.

Unlike FIR filtering, here the initial state x_{m-1} is taken beyond the averaging horizon $[m, k]$. Therefore, x_{m-1} cannot be found among the LMF gain variables and serves as an input along with the data y_k and the control signal u_k. Accordingly, the batch LMF estimate can be represented in the discrete convolution-based form as

$$\hat{x}_k = \bar{\mathcal{H}}^{\mathrm{h}}_{m,k}Y_{m,k} + \bar{\mathcal{H}}^{\mathrm{f}}_{m,k}U_{m,k} + \bar{\mathcal{H}}^{\mathrm{s}}_{m,k}x_{m-1} , \tag{4.131}$$

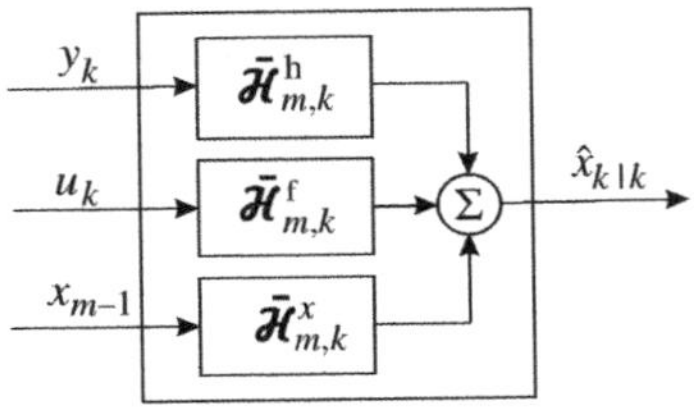

Figure 4.4 Generalized structure of the LMF.

where gains $\bar{\mathcal{H}}^{\mathrm{h}}_{m,k}$, $\bar{\mathcal{H}}^{\mathrm{f}}_{m,k}$, and $\bar{\mathcal{H}}^{\mathrm{s}}_{m,k}$ minimize the MSE for three orthogonal inputs: y_k, u_k, and x_{m-1}. The state x_k can be specified by the last raw vector in (4.127) as

$$x_k = \mathcal{F}^m_k x_{m-1} + \bar{S}_{m,k} U_{m,k} + \bar{D}_{m,k} W_{m,k} \,, \tag{4.132}$$

and then the unbiasedness condition $E\{\hat{x}_k\} = E\{x_k\}$ applied to (4.131) and (4.132) gives two constraints:

$$\bar{\mathcal{H}}^{\mathrm{f}}_{m,k} = \bar{S}_{m,k} - \bar{\mathcal{H}}^{\mathrm{h}}_{m,k} L_{m,k} \,, \tag{4.133}$$

$$\bar{\mathcal{H}}^{\mathrm{s}}_{m,k} = \mathcal{F}^m_k - \bar{\mathcal{H}}^{\mathrm{h}}_{m,k} C_{m,k} \,. \tag{4.134}$$

By applying the orthogonality condition

$$E\{(x_k - \bar{\mathcal{H}}^{\mathrm{h}}_{m,k} Y_{m,k} - \bar{\mathcal{H}}^{\mathrm{f}}_{m,k} U_{m,k} - \bar{\mathcal{H}}^{\mathrm{s}}_{m,k} \mathbf{x}_{m-1}) Y^T_{m,k}\} = 0 \tag{4.135}$$

and providing the averaging, we obtain

$$\begin{aligned} &\bar{\mathcal{B}}_{m,k} \mathcal{X}_{m-1} C^T_{m,k} + \bar{\mathcal{W}}_{m,k} \mathcal{Q}_{m,k} G^T_{m,k} - \bar{\mathcal{V}}_{m,k} \mathcal{R}_{m,k} \\ &\quad = (\bar{\mathcal{H}}^{\mathrm{h}}_{m,k} L_{m,k} - \bar{S}_{m,k} + \bar{\mathcal{H}}^{\mathrm{f}}_{m,k}) \Psi_{m,k} L^T_{m,k} , \end{aligned} \tag{4.136}$$

where the error residual matrices are defined as

$$\bar{\mathcal{B}}_{m,k} = \mathcal{F}^m_k - \bar{\mathcal{H}}^{\mathrm{h}}_{m,k} C_{m,k} - \bar{\mathcal{H}}^{\mathrm{s}}_{m,k} \,, \tag{4.137}$$

$$\bar{\mathcal{W}}_{m,k} = \bar{D}_{m,k} - \bar{\mathcal{H}}^{\mathrm{h}}_{m,k} G_{m,k} \,, \tag{4.138}$$

$$\bar{\mathcal{V}}_{m,k} = \bar{\mathcal{H}}^{\mathrm{h}}_{m,k} \,. \tag{4.139}$$

By setting $U_{m,k} = 0$ and $\mathcal{X}_{m-1} = 0$, solutions to (4.136) appear as constraints (4.133) and (4.134) and give the fundamental (homogeneous) gain $\bar{\mathcal{H}}_{m,k} \triangleq \bar{\mathcal{H}}^{\mathrm{h}}_{m,k}$,

$$\bar{\mathcal{H}}_{m,k} = \bar{D}_{m,k} \mathcal{Q}_{m,k} G^T_{m,k} \Omega^{-1}_{m,k} \,, \tag{4.140}$$

where $\Omega_{m,k}$ is specified by (4.66). Accordingly, the batch LMF becomes

$$\hat{x}_k = \hat{x}^{\mathrm{h}}_k + \hat{x}^{\mathrm{f}}_k + \hat{x}^{\mathrm{s}}_k \tag{4.141a}$$

$$\begin{aligned} &= \bar{\mathcal{H}}_{m,k} Y_{m,k} + (\bar{S}_{m,k} - \bar{\mathcal{H}}_{m,k} L_{m,k}) U_{m,k} \\ &\quad + (\mathcal{F}^m_k - \bar{\mathcal{H}}_{m,k} C_{m,k}) x_{m-1} \,, \end{aligned} \tag{4.141b}$$

where $\hat{x}^{\mathrm{h}}_k$, $\hat{x}^{\mathrm{f}}_k$, and $\hat{x}^{\mathrm{s}}_k$ are relevant terms in (4.141b). For $\bar{\mathcal{H}}_{m,k}$ defined by (4.140), the left-hand side of (4.136) becomes identically zero. Therefore, the forced gain $\bar{\mathcal{H}}^{\mathrm{f}}_{m,k}$ is defined by (4.133).

What can be inferred now is that the LMF gain (4.140) is equivalent to the OUFIR-I filter gain (4.53). Since the latter is a special case of the OFIR filter for zero initial values, the LMF is also optimal. However, the LMF is not a special case of the OFIR filter with removed initial conditions. It is strictly optimal, because its estimate (4.141b) is affected by the initial state x_{m-1} that will be seen in the LMF error covariance. Thus, we conclude that for $m = 1$ the LMF is nothing more and nothing less than the convolution-based *batch optimal KF*.

> ***OFIR filter vs. LMF:*** *The key difference between the OFIR filter and LMF, both working on* $[m, k]$ *, is that the former requires initial values at* m *, while the latter at* $m-1$ *. Consequently,* x_m *is a variable of the OFIR filter gain* $\mathcal{H}_{m,k}$ *, while* x_{m-1} *is an input of the LMF.*

Similarly to (4.31), the error covariance of the LMF can be written as

$$P_k = \bar{\mathcal{B}}_{m,k}\mathcal{X}_{m-1}\bar{\mathcal{B}}^T_{m,k} + \bar{\mathcal{W}}_{m,k}\mathcal{Q}_{m,k}\bar{\mathcal{W}}^T_{m,k} + \bar{\mathcal{V}}_{m,k}\mathcal{R}_{m,k}\bar{\mathcal{V}}^T_{m,k} , \tag{4.142}$$

where the error residual matrices are specified by (4.137)–(4.139). Like the OFIR filter, the LMF optimally balances between the bias errors associated with the initial state χ_{m-1} and random errors caused by the system and measurement noise. Thus, the LMF is an *optimal estimator*.

4.6.2 Iterative LMF Algorithm Using Recursions

Recursions for the batch LMF (4.141b) can be found if we take into account a similarity of $\bar{\mathcal{H}}_{m,k}$ defined by (4.140) and $\mathcal{H}_{m,k}$ defined by (4.28b).

Theorem 4.5 Given a batch LMF estimate (4.141b). Its iterative computation on $[m, k]$ is provided by optimal Kalman recursions (Algorithm 4) for initial $\hat{x}_{m-1}$ and P_{m-1} by changing a time index i from m to k and taking the output when $i = k$.

Proof: The proof of this theorem can be found in "Problems." □

It follows from the previous that, like the batch OFIR filtering estimate, the batch LMF estimate can also be computed using the Kalman recursions listed in Algorithm 4. Furthermore, the best LMF is the *limited memory KF* (LMKF), since a batch LMF is equivalently converted to an iterative LMKF represented by Algorithm 9.

Algorithm 9: Limited Memory KF

Data: $y_k, u_k, \hat{x}_{m-1}, P_{m-1}, Q_k, R_k, N$
begin
  for $k = 1, 2, \cdots$ **do**
    $m = k - N + 1$ if $k > N - 1$ and $m = 0$ otherwise;
    for $i = m, m+1, \cdots, k$ **do**
      Algorithm 4
    end for
  end for
  Result: $\hat{x}_k, P_k$
end

Thus, (4.141b) is the batch LMKF.

While LMKF is more robust due to the ability to operate with most resent data, it requires initial values at $m-1$ for each horizon $[m, k]$. Otherwise, large bias errors may appear in the output, especially on a short horizon. This flaw explains why the LMKF is not used as widely as the KF.

It is also noticeable that the transitions from batch LMKF to batch KF and from iterative LMKF Algorithm 9 to recursive KF Algorithm are quite simple: for $m = 1$, LMKF becomes a batch KF operating on $[1, k]$ for a given x_0.

4.7 Continuous-Time Optimal FIR Filter

Continuous time filtering techniques are always on the table when digital computing resources do not provide sufficient accuracy in real time due to operating frequency limitations. In such cases, filters are implemented physically rather than using digital hardware and software.

To find a solution for continuous-time OFIR filtering, we consider the following state-space model

$$x'(t) = A(t)x(t) + U(t)u(t) + w(t) , \tag{4.143}$$

$$y(t) = C(t)x(t) + v(t) , \tag{4.144}$$

where noise $w(t) \sim \mathcal{N}(0, \mathcal{R}_w)$ has the covariance $\mathcal{R}_w(\tau) = \mathcal{S}_w\delta(\tau)$ and $v(t) \sim \mathcal{N}(0, \mathcal{R}_v)$ has $\mathcal{R}_v(\tau) = \mathcal{S}_v\delta(\tau)$. For successful FIR filtering, it is assumed that the initial conditions are set at point $t - T$, where T is the length of the horizon $[t - T, t]$, and we will also assume that $x(t - T)$, $w(t)$, and $v(t)$ are random and mutually uncorrelated.

In the time domain, two operators can be used to find the OFIR filtering estimate based on (4.143) and (4.144). The convolution-based operator provides the optimal impulse response, and the differential equation operator leads to the KBF form. In presenting both of these solutions, next we will mainly follow the results obtained by Kwon *et al.* in [66, 100, 104].

4.7.1 Optimal Impulse Response

Following the discrete-time approach, the continuous-time FIR filtering estimate $\hat{x}(t) \triangleq \hat{x}(t|t)$ can be defined as

$$\begin{aligned}\hat{x}(t) &= \hat{x}^{\mathrm{h}}(t) + \hat{x}^{\mathrm{f}}(t) \\ &= \int_{t-T}^{t} H^{\mathrm{h}}(t, \nu)y(\nu)\, \mathrm{d}\nu + \int_{t-T}^{t} H^{\mathrm{f}}(t, \nu)u(\nu)\, \mathrm{d}\nu ,\end{aligned} \tag{4.145}$$

where $H^{\mathrm{h}}(t, \nu) \triangleq H^{\mathrm{h}}(t, \nu; T)$ is the homogenous impulse response matrix and $H^{\mathrm{f}}(t, \nu) \triangleq H^{\mathrm{f}}(t, \nu; T)$ is the forced impulse response matrix, both time-varying, dependent on T, and equal to zero beyond a horizon $[t - T, t]$.

The solution to (4.143) for initial values at ν is given by (3.18) as

$$x(t) = \Phi(t, \nu)x(\nu) + \int_{\nu}^{t} \Phi(t, \theta)[U(\theta)u(\theta) + L(\theta)w(\theta)]\, \mathrm{d}\theta , \tag{4.146}$$

where a time variable ν ranges as $t - T \leqslant \nu \leqslant t$. With respect to $x(\nu)$, (4.146) can further be expressed as

$$x(\nu) = \Phi(\nu, t)x(t) - \int_{\nu}^{t} \Phi(\nu, \theta)[U(\theta)u(\theta) + L(\theta)w(\theta)]\, \mathrm{d}\theta \tag{4.147}$$

and then $y(\nu) = C(\nu)x(\nu) + v(\nu)$ required by (4.145) becomes

$$y(\nu) = C(\nu)\left[\Phi(\nu, t)x(t) - \int_{\nu}^{t} \Phi(\nu, \theta)[U(\theta)u(\theta) + L(\theta)w(\theta)]\, \mathrm{d}\theta\right] + v(\nu) .$$

To express $H^{\mathrm{f}}(t,\nu)$ via $H^{\mathrm{h}}(t,\nu)$, it was proposed in [66, 104] to use $H^{\mathrm{h}}(t,\tau)$ for supposedly known $y(t)$ and $u(t)$ in the solution

$$
\begin{aligned}
y(\nu) + C(\nu)\int_{\nu}^{t} \Phi(\nu,\theta)U(\theta)u(\theta)\,\mathrm{d}\theta \\
= C(\nu)\left[\Phi(\nu,t)x(t) - \int_{\nu}^{t} \Phi(\nu,\theta)L(\theta)w(\theta)\,\mathrm{d}\theta\right] + v(\nu)
\end{aligned}
$$

and represent the estimate as [100]

$$
\begin{aligned}
\hat{x}(t) &= \int_{t-T}^{t} H^{\mathrm{h}}(t,\nu)\left[y(\nu) + C(\nu)\int_{\nu}^{t} \Phi(\nu,\theta)U(\theta)u(\theta)\,\mathrm{d}\theta\right]\mathrm{d}\nu \\
&= \hat{x}^{\mathrm{h}}(t) + \int_{t-T}^{t} H^{\mathrm{h}}(t,\nu)C(\nu)\left[\int_{\nu}^{t} \Phi(\nu,\theta)U(\theta)u(\theta)\,\mathrm{d}\theta\right]\mathrm{d}\nu\,, \\
&= \hat{x}^{\mathrm{h}}(t) + \int_{\nu}^{t}\left[\int_{t-T}^{\theta} H^{\mathrm{h}}(t,\nu)C(\nu)\Phi(\nu,\theta)U(\theta)\mathrm{d}\nu\right]u(\theta)\mathrm{d}\theta \\
&= \hat{x}^{\mathrm{h}}(t) + \int_{\nu}^{t} H^{\mathrm{f}}(t,\theta)u(\theta)\mathrm{d}\theta = \hat{x}^{\mathrm{h}}(t) + \int_{t-T}^{t} H^{\mathrm{f}}(t,\nu)u(\nu)\mathrm{d}\nu\,, \\
&= \hat{x}^{\mathrm{h}}(t) + \hat{x}^{\mathrm{f}}(t)\,,
\end{aligned}
\tag{4.148}
$$

where the forced impulse response matrix is defined by

$$
H^{\mathrm{f}}(t,s) = \int_{t-T}^{s} H^{\mathrm{h}}(t,\nu)C(\nu)\Phi(\nu,s)U(s)\,\mathrm{d}\nu\,, \quad t-T \leqslant s \leqslant t\,. \tag{4.149}
$$

Referring to the obtained $H^{\mathrm{f}}(t,s)$ by (4.149), the homogenous impulse response matrix $H^{\mathrm{h}}(t,\nu)$ can be found for zero input, $u(t) = 0$, if to consider the orthogonality condition

$$
\int_{t-T}^{t} [x(t) - \hat{x}(t)]y^{T}(s)\,\mathrm{d}s = 0
$$

and transform it to the Fredholm integral equation of the second kind [100] with respect to the optimal matrix $H^{\mathrm{h}}(t,s)$; that is,

$$
H^{\mathrm{h}}(t,s) = P_{x}(t,s)C^{T}(x)S_{v}^{-1} - \int_{t-T}^{t} H^{\mathrm{h}}(t,v)C(v)P_{x}(v,s)C^{T}(s)S_{v}^{-1}\,\mathrm{d}v\,, \tag{4.150}
$$

where

$$
P(t,s) = \mathcal{E}\{x(t)x^{T}(s)\} = \Phi(t,s)P_{x}(s,s)\,,
$$

$\Phi(t,s)$ is the state transition matrix, and $P_{x}(t,t)$ is provided for given $P_{x}(0{,}0)$ by the RDE [100]

$$
\frac{\mathrm{d}}{\mathrm{d}t}P(t,t) = A(t)P(t,t) + P(t,t)A^{T}(t) + S_{w}\,.
$$

Equation (4.150) can be solved numerically for $H^{\mathrm{h}}(t,\nu)$ and then $H^{\mathrm{f}}(t,\nu)$ computed via (4.149) that finally gives the OFIR estimate (4.145). To avoid solving the integral equation (4.150), one can find the optimal $H^{\mathrm{h}}(t,s)$ by solving two differential equations, as suggested in [100]. Thus, the optimal impulse response can be computed as the product of the inverted and direct solutions of the differential equations in a final form that resembles the optimal gain (4.28b) of a discrete-time

OFIR filter. On the other hand, if we start from zero and compute all integrals with a growing horizon $T = t$, then the impulse response $H^{\mathrm{h}}(t, \nu)$, $0 \leqslant \nu \leqslant t$, will correspond to KBF.

4.7.2 Differential Equation Form

Let us now look at equations (4.143) and (4.144) again. Assuming that the initial conditions are set to zero, we conclude that the optimal filter for this model is KBF (3.204) and (3.205). If we assume that the initial conditions are set at $t - T$, then KBF will still work, provided that the initial conditions at $t - T$ are set correctly. To find $\hat{x}(t - T)$ and $P(t - T)$, one can run another KBF in parallel, take its estimates at $t - T$, and set these estimates as the initial conditions for the main KBF. But this means that the continuous-time OFIR filter *de facto* is the KBF operating on $[t - T, t]$. The last conclusion will become more obvious if we recall that for any linear stochastic continuous-time state-space model there is a unique KBF regardless of the initial time and the initial conditions.

Thus, the continuous-time OFIR filter is a KBF operating on the horizon $[t - T, t]$ for the given initial conditions $\hat{x}(t - T)$ and $P(t - T)$,

$$\begin{aligned}\hat{x}'(t) &= A(t)\hat{x}(t) + U(t)u(t) \\ &\quad + P(t)C^T(t)S_v^{-1}[y(t) - C(t)\hat{x}(t)] , \quad [t - T, t] ,\end{aligned} \tag{4.151}$$

$$\begin{aligned}P'(t) &= P(t)A^T(t) + A(t)P(t) \\ &\quad - P(t)C^T(t)S_v^{-1}C(t)P(t) + S_w , \quad [t - T, t] .\end{aligned} \tag{4.152}$$

The main problem with (4.151) and (4.152) is the practical inability of providing exact values of $\hat{x}(t - T)$ and $P(t - T)$. In fact, there is no other filter that can estimate the initial values at $t - T$ better than KBF. Thus, we can estimate the initial values only approximately, and if the approximation is less accurate than the KBF estimate obtained at $t - T$ from zero, then it follows that an OFIR filter operating on $[t - T, t]$ will have no advantage over KBF operating on $[0, t]$.

We now arrive at two important findings, illustrated earlier in Fig. 4.1. An OFIR filter operating on $[t - T, t]$ cannot outperform KBF operating on $[0, t]$ if the model is explicitly specified without errors. However, if some uncertainty has a noticeable effect on the KBF, then the OFIR filter may be more accurate even for approximately set initial conditions at $t - T$. Thus, the OFIR filter is more robust than the KBF.

4.8 Extended *a posteriori* OFIR Filtering

Extended discrete-time state-space equations are generally not available for nonlinear models in view of the complexity caused by multiple enclosed nonlinear functions. Therefore, extended OFIR filtering is commonly organized by converting a nonlinear state space model to a linear one using the Taylor series and then modifying the linear algorithms.

Consider the nonlinear state-space equations

$$x_k = f_k(x_{k-1}) + w_k , \tag{4.153}$$

$$y_k = h_k(x_k) + v_k , \tag{4.154}$$

where w_k and v_k are zero mean white Gaussian noise components. Following the derivation of EKF given in Chapter 3, equations (4.153) and (4.154) can be approximated using the second-order

Taylor series expansion as

$$x_k = \dot{F}_k x_{k-1} + \eta_k + w_k \,, \tag{4.155}$$

$$\tilde{y}_k = \dot{H}_k x_k + v_k \,, \tag{4.156}$$

where $\tilde{y}_k = y_k - \psi_k$ is the modified observation vector,

$$\eta_k = f_k(\hat{x}_{k-1}) - \dot{F}_k \hat{x}_{k-1} + \frac{1}{2}\alpha_k \,, \tag{4.157}$$

$$\psi_k = h_k(\hat{x}_k^-) - \dot{H}_k \hat{x}_k^- + \frac{1}{2}\beta_k \,, \tag{4.158}$$

$\alpha_k = \sum_{i=1}^{K} e_i^K \varepsilon_{k-1}^T \ddot{F}_{ik} \varepsilon_{k-1}$, $\beta_k = \sum_{j=1}^{M} e_j^M \varepsilon_k^{-T} \ddot{H}_{jk} \varepsilon_k^-$, and $e_i^K \in \mathbb{R}^K$ and $e_j^M \in \mathbb{R}^M$ are Cartesian basis vectors for which the ith and jth components are equal to one, and all the others are equal to zero. The nonlinear functions are represented as

$$f_k(x_{k-1}) \cong f_k(\hat{x}_{k-1}) + \dot{F}_k \varepsilon_{k-1} + \frac{1}{2}\alpha_k \,,$$

$$h_k(x_n) \cong h_k(\hat{x}_k^-) + \dot{H}_k \varepsilon_n^- + \frac{1}{2}\beta_k \,,$$

where $\dot{F}_k = \left.\frac{\partial f_k}{\partial x}\right|_{x=\hat{x}_{k-1}}$ and $\dot{H}_k = \left.\frac{\partial h_k}{\partial x}\right|_{x=\hat{x}_k^-}$ are Jacobian matrices and $\ddot{F}_{ik} = \left.\frac{\partial^2 f_{ik}}{\partial x^2}\right|_{x=\hat{x}_{k-1}}$ and $\ddot{H}_{jk} = \left.\frac{\partial^2 h_{jk}}{\partial x^2}\right|_{x=\hat{x}_k^-}$ are Hessian matrices.

Based on the transformations presented previously, the extended *a posteriori* OFIR filtering algorithms of the first-order (EOFIR-I) and of the second-order (EOFIR-II) can be developed.

The pseudocode of the EOFIR filtering algorithm, developed using EKF recursions, is listed in Algorithm 10, where the matrix ψ_k is given by (4.158) through matrix functions defined by the Taylor series expansion.

Algorithm 10: The *a posteriori* EOFIR Filtering Algorithm

Data: $y_k, \hat{x}_m, P_m, Q_k, R_k, N$
Result: $\hat{x}_k$

```
begin
    for k = 1, 2, ··· do
        m = k − N + 1 if k > N − 1 and m = 0 otherwise ;
        for l = m + 1 : k do
            P_l^- = F_l P_{l-1} F_l^T + Q_l ;
            S_l = H_l P_l^- H_l^T + R_l ;
            K_l = P_l^- H_l^T S_l^{-1} ;
            x̄_l = f_l(x̄_{l-1}) + K_l{y_l − ψ_l − h_l[f_l(x̄_{l-1})]} ;
            P_l = (I − K_l H_l) P_l^- ;
        end for
        x̂_k = x̄_k ;
    end for
end
```

(In lines 5–9, F_l and H_l are printed with dots: $\dot{F}_l$, $\dot{H}_l$.)

Like the EKF, the EOFIR algorithm uses the original nonlinear functions $f_k(x_{k-1})$ and $h_k(x_k)$ to update estimates, and we notice that this code is universal for both *a posteriori* EOFIR-I and

EOFIR-II filters with the following specifics. For EOFIR-I, the second-order terms β_k and α_k in the matrix ψ_k are omitted. Therefore, all the basic properties of this filter are due to the Jacobian matrices $\dot{F}_k$ and $\dot{H}_k$. For EOFIR-II, the full mathematical form associated with the second-order Taylor series expansion and Hessian matrices $\ddot{F}_k$ and $\ddot{H}_k$ is used.

In discussing EKF, we have already mentioned that the second-order extended filters do not demonstrate clear advantages over the first-order filters [153, 185]. Therefore, EKF-I and EOFIR-I filters can be found much more frequently in applications.

4.9 Properties of FIR State Estimators

We can now do a comparative analysis of different types of FIR state estimators (filters) and highlight the differences. Many of the properties of FIR filters follow directly from their batch forms, but not all are available from recursions. Therefore, we first summarize the batch and recursive forms, as shown in Fig. 4.5, and then will draw some important conclusions.

It was previously shown that the OFIR filter is the most general and universal optimal FIR estimator for linear stochastic models. Indeed, because of the built-in unbiasedness, it becomes an OUFIR estimator, which is equivalent to the ML FIR estimator. On the other hand, if we ignore zero mean noise, the OFIR filter becomes an UFIR filter. However, neither the OUFIR filter can be converted back to the OFIR filter due to the lost information about bias errors, nor the UFIR filter can be converted to the OFIR and OUFIR filters due to the lack of information about the bias and random errors. Another optimal structure is LMF, which has an IIR caused by the state feedback. The peculiarity of the LMF is that it represents the batch KF when the horizon grows without limits.

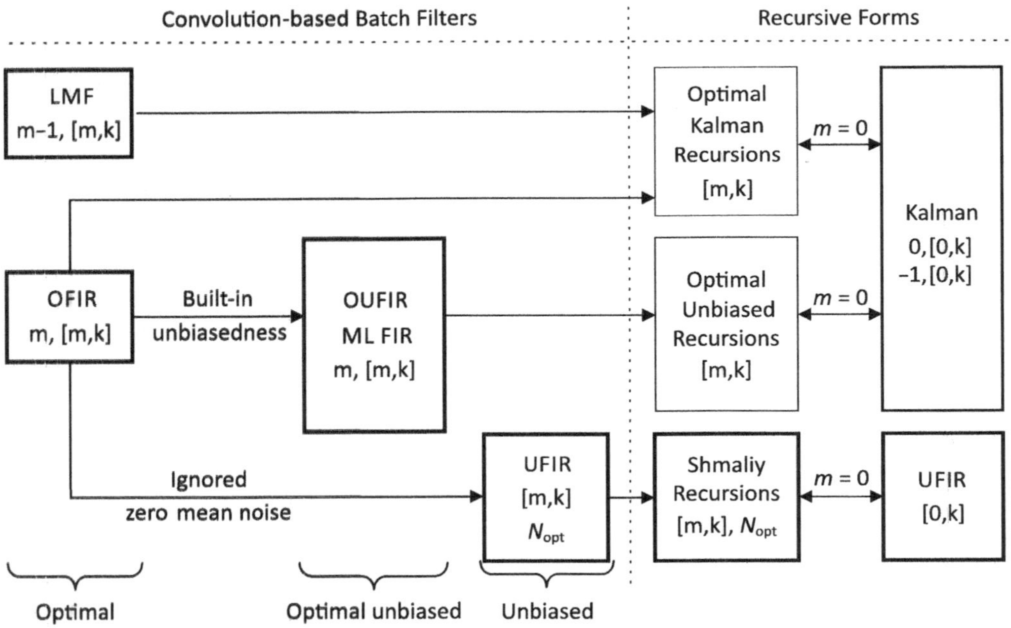

Figure 4.5 Batch linear state estimators (filters) and recursive forms: "$n, [m,k]$" means that the initial values are required at n for data taken from $[m,k]$.

The batch OFIR and LMF estimates are computed by Kalman recursions (Algorithm 4), which proves the KF optimality. Consequently, the growing memory full horizon OFIR batch represents on $[0, k \to \infty]$ a batch KF, and the best LMF operating on $[m, k]$ is LMKF. A bias-constrained ML FIR batch cannot be computed using Kalman recursions, and optimal unbiased recursions (Algorithm 8) must be used. This may be somewhat surprising, because KF is stated by many authors as being optimal unbiased. But we have already shown evidence of KF's inability to handle optimal unbiased FIR batches. In turn, the UFIR filter completely ignores zero mean noise and initial values and is computed iteratively using recursions derived by Shmaliy in [175].

Referring to the previous, next we will consider the most critical properties of these state estimators.

Deterministic Models

OFIR filter: When $Q_k = 0$ and $R_k = 0$, the OFIR filter becomes a UFIR filter since the optimal gain $\mathcal{H}_{m,k}$ (4.28b) is converted to an unbiased gain $\hat{\mathcal{H}}_{m,k}$ (4.29b) that only satisfies the unbiasedness constraint. Thus, the OFIR filter can be applied to both deterministic and stochastic systems.

OUFIR filters: When $Q_k = 0$, the OUFIR-I filter gain (4.53) becomes identically zero and cannot be applied. For the same reason, the second component in the OUFIR-II filter gain (4.69b) also becomes identically zero. If we further assume that $R_k = 0$, then matrix $\Omega_{m,k}$ becomes singular, and the OUFIR-II filter can no longer process data.

Resume: The OFIR filter can be applied to arbitrary linear systems, and the OUFIR filter can be applied only to stochastic systems.

Maximum Likelihood

OFIR filter: The unbiasedness constraint built into (4.28b) removes the initial state requirement, and the OFIR filter becomes an OUFIR-II filter, which belongs to the class of ML estimators.

OUFIR filter: Because of the built-in unbiasedness, the OUFIR-II filter belongs to the class of ML estimators by design.

Resume: The OUFIR-II filter is an ML estimator. The OFIR filter becomes an ML estimator by embedding the unbiasedness constraint.

BIBO Stability

OFIR filter: Suppose that data $Y_{m,k}$ and input $U_{m,k}$ are both bounded, $\|Y_{m,k}\| < \infty$ and $\|U_{m,k}\| < \infty$. The OFIR estimate will also be bounded, and the OFIR filter will be BIBO stable if $\mathcal{H}_{m,k}$ (4.28b) is bounded. Then consider three limiting cases:

- If the initial state and noise are norm-bounded, $\|x_m\| < \infty$, $\|\mathcal{Q}_{m,k}\| < \infty$, and $\|\mathcal{R}_{m,k}\| < \infty$, then it follows that $\|\mathcal{H}_{m,k}\| < \infty$.
- For arbitrary x_m but bounded noise, (4.28b) in the limiting case can be written as

  $$\mathcal{H}_{m,k} \approx \mathcal{F}_k^{m+1} \chi_m H_{m,k}^T (H_{m,k} \chi_m H_{m,k}^T)^{-1} .$$

  Now multiply the first χ_m from the left-hand side by the identity matrix $(H_{m,k}^T H_{m,k})^{-1} H_{m,k}^T H_{m,k}$; then drop the identity on the right-hand side and obtain

  $$\mathcal{H}_{m,k} \approx \mathcal{F}_k^{m+1} (H_{m,k}^T H_{m,k})^{-1} H_{m,k}^T ,$$

  which means that $\|\mathcal{H}_{m,k}\| < \infty$ holds for stable systems.
- When $\|x_m\| < \infty$ and 1) $\|\mathcal{Q}_{m,k}\| < \infty$ and $\mathcal{R}_{m,k}$ is not bounded or 2) $\|\mathcal{R}_{m,k}\| < \infty$ and $\mathcal{Q}_{m,k}$ is not bounded, then we still have $\|\mathcal{H}_{m,k}\| < \infty$ for stable systems.

Thus, the OFIR filter is BIBO stable for stable systems even if either x_m, $\mathcal{Q}_{m,k}$, or $\mathcal{R}_{m,k}$ is not bounded.

OUFIR filter: For $\|Y_{m,k}\| < \infty$ and $\|U_{m,k}\| < \infty$, the OUFIR filter is BIBO stable, because (4.69b) is bounded for stable systems given any $\mathcal{Q}_{m,k}$ and $\mathcal{R}_{m,k}$ [110].

LMF: For $\|Y_{m,k}\|<\infty$, $\|U_{m,k}\|<\infty$, and $\|x_{m-1}\|<\infty$, the LMF is BIBO stable, because $\bar{\mathcal{H}}_{m,k}$ is bounded for stable systems with any $\mathcal{Q}_{m,k}$, and $\mathcal{R}_{m,k}$ [110]. However, this filter will not be BIBO stable if the initial state x_{m-1} is unbounded.

Resume: The OFIR and OUFIR filters are BIBO stable. The LMF is BIBO stable only if the initial state is bounded.

Deadbeat Property

OFIR filter: For $U_{m,k}=0$, $\mathcal{Q}_{m,k}=0$, and $\mathcal{R}_{m,k}=0$, the OFIR estimate becomes $\hat{x}_k=\mathcal{H}_{m,k}H_{m,k}x_m$, the model (4.19) is transformed to $x_k=\mathcal{F}_k^{m+1}x_m$, and the OFIR filter becomes a UFIR filter (4.29b). Accordingly, we have $\hat{x}_k=x_k$ and conclude that the OFIR filter is a *deadbeat* filter [106].

OUFIR filter: The OUFIR filter has the deadbeat property by design, because of the built-in unbiasedness.

LMF: By $U_{m,k}=0$, $\mathcal{Q}_{m,k}=0$, and $\mathcal{R}_{m,k}=0$, the LMF estimate (4.141b) and model (4.132) become identical, and we have $\hat{x}_k=x_k$. Thus, the LMF also has the deadbeat property.

Resume: The OFIR and OUFIR filters as well as the LMF are deadbeat state estimators.

Accuracy

If a model has no uncertainties and the noise statistics and initial values are known exactly, then there is no need to use FIR or LMF state estimators [176]. In this ideal case, the full horizon OFIR filter and KF are equivalent and the best linear state estimators.

Robustness

In terms of robustness, the state estimators can be compared as follows:

- The iterative OUFIR Algorithm 8 can be more robust than the iterative OFIR Algorithm 5 only if the initial values in each iteration cycle are self-computed more accurately than those set for the OFIR filter.
- The LMF can be more robust than KF if its initial values in each iteration cycle are set more precisely than the values obtained with KF.
- Since the UFIR filter does not require initial values, noise covariances, or other tuning factors on a given horizon $[m,k]$ of N_{opt} points, it is the most robust of all linear state estimators.

To illustrate the properties of state estimators, we will look at a few typical examples next.

Example 4.1 ■ ***GPS-based navigation of a maneuvering vehicle.*** The coordinates x, m and y, m of a rapidly maneuvering vehicle are measured every second, $\tau=1$ s, using a GPS navigation unit. Vehicle dynamics along the coordinate x is represented in state space with a two-state polynomial tracking model

$$x_k=Fx_{k-1}+Bw_k\,,$$
$$y_k=Hx_k+v_k\,,$$

where matrices are defined as

$$F=\begin{bmatrix}1 & \tau\\ 0 & 1\end{bmatrix},\quad H=\begin{bmatrix}1 & 0\end{bmatrix},\quad B=\begin{bmatrix}\tau/2\\ 1\end{bmatrix}.$$

It is assumed that the process noise w_k affects only the second state with the standard deviation of $\sigma_w=2$ m/s. The GPS data noise v_k has the standard deviation of $\sigma_v=3.75$ m, and the optimal horizon is measured as $N_{\text{opt}}=5$.

A part of the vehicle trajectory with a single fast maneuver along the coordinate x is shown in Fig. 4.6. Since the velocity jumps during one second and then returns, the processes are considered as step and impulse responses.

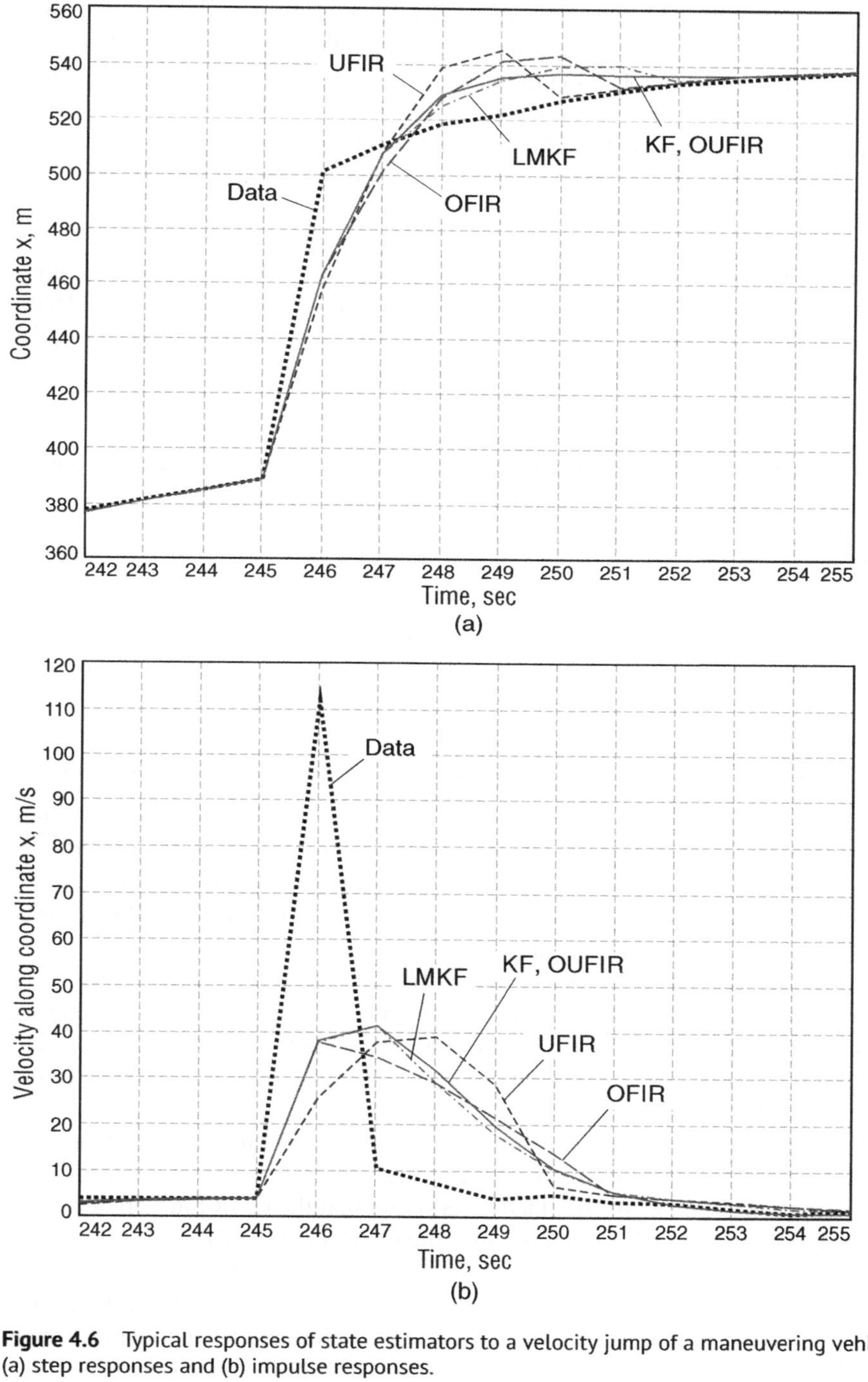

Figure 4.6 Typical responses of state estimators to a velocity jump of a maneuvering vehicle: (a) step responses and (b) impulse responses.

It can be seen that the UFIR filter has the shortest transients with a duration of $N_{opt} = 5$ points. The OFIR filter operating on a horizon of $N = 2$ points and initialized with an UFIR filter gives the correct estimate at the next point following the transient of the UFIR filter. Since the OUFIR filter is not successful in accurate self-determining the initial values on short horizons, we run it on $N = 3N_{opt}$ points. Accordingly, the OUFIR filter and KF exhibit the longest decaying responses. The LMF requires initial values at $m - 1$, so its transients end at the next points following the OFIR filter's transients. □

Example 4.1 clearly demonstrates that all linear state estimators respond consistently to temporary uncertainties and that the discrepancies are mainly due to different impulse responses. Here, we have evidence of typical effects inherent to linear FIR state estimators: *UFIR filter produces the largest excursions and shortest transients, while optimal filters produce the shortest excursions and longest transients.*

Example 4.2 ■ ***Filtering under uncertainties.*** Consider the polynomial tracking state-space model

$$x_k = F_k x_{k-1} + w_k ,$$
$$y_k = Hx_k + v_k ,$$

where $H = [\,1\ 0\,]$ and the uncertain system matrix is defined as

$$F_k = \begin{bmatrix} 1 & (1 + d_k a)\tau \\ 0 & 1 \end{bmatrix} ,$$

in which $\tau = 1$ s, $d_k = \begin{bmatrix} 5 , & 160 \leqslant k \leqslant 200 \\ 0 , & \text{otherwise} \end{bmatrix}$, and a is a correcting coefficient. The white noise w_k has the variance $\sigma_1^2 = 10^{-4}$ in the first state and $\sigma_2^2 = 4 \times 10^{-6}/\text{s}^2$ in the second state. The variance of white measurement noise is $\sigma_v^2 = 0.0225$, and the process is generated for $a = 1$ starting with the initial state components $x_{10} = 1$ and $x_{20} = 0.01$. Then two filtering modes are considered:

Case-I ($a = 1$): The system matrix F_k is known at every point k and filters fit the model exactly.

Case-II ($a = 0$): Filters do not fit the model when $d_k \neq 0$.

For the OFIR filter, horizons inside and outside of region $160 \leqslant k \leqslant 200$ are specified with $N = 14$ and $N = 29$, respectively. Under ideal conditions of exactly known model and noise, the filtering errors as sketched in Fig. 4.7 for *Case-I* and *Case-II* , where the estimation error bounds (EBs) are depicted in the 3σ sense. It can be seen that in both cases the OFIR filter and KF give similar errors with discrepancies caused by the FH lengths.

Filtering was then repeated assuming more realistic operation conditions, when the noise statistics are not known exactly, by replacing the system noise variances in the algorithms with $\sigma_1^2 = 0.5^2 \times 10^{-4}$ and $\sigma_2^2 = 0.25^2 \times 4 \times 10^{-6}/\text{s}^2$. Figure 4.8 shows that the OFIR filter produces the smallest excursions and shortest transients and is thus more robust than the KF. □

Example 4.1 and example 4.2 provide more evidence for the higher robustness of FIR structures, supporting the key theoretical finding previously made: 1) OFIR filter and KF are identical state estimators under ideal operation conditions and 2) otherwise, the OFIR filter is more robust.

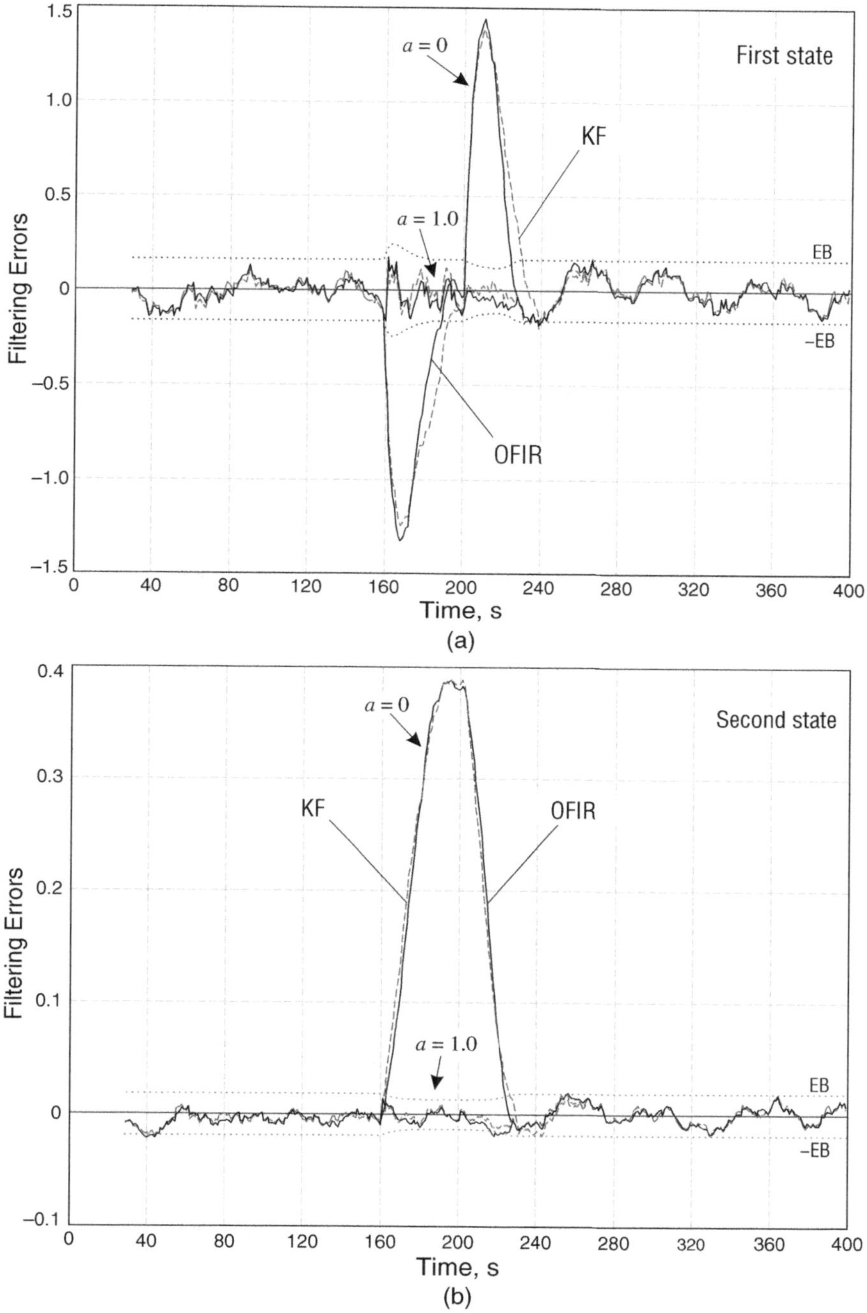

Figure 4.7 Typical errors produced by the OFIR filter and KF under ideal conditions when $a = 1$ and $a = 0$: (a) first state and (b) second state.

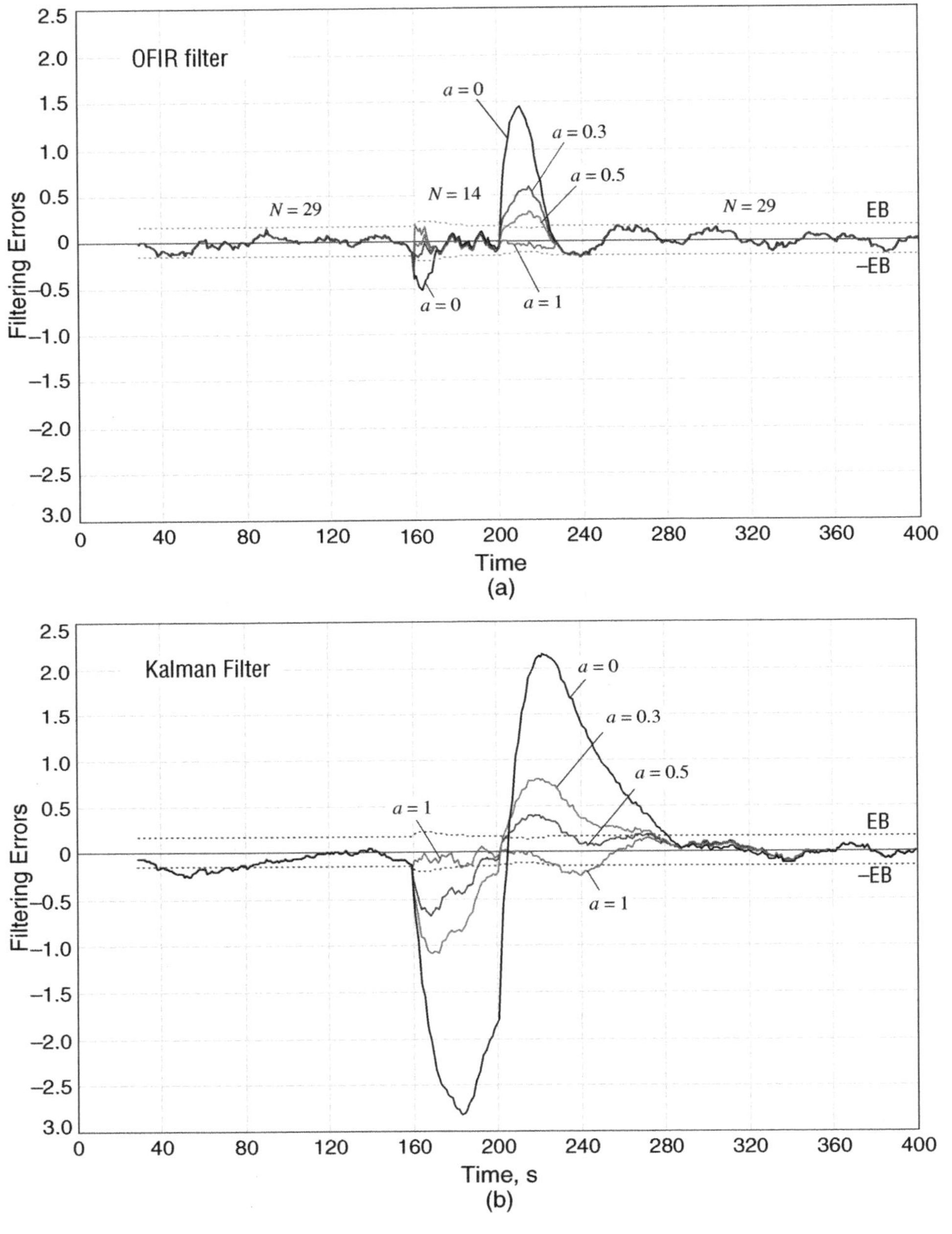

Figure 4.8 Typical estimation errors in the first state produced by the OFIR filter and KF under errors in noise statistics: (a) OFIR filter and (b) KF.

4.10 Summary

As a mathematical tool for state estimation in stochastic processes represented by linear equations in state space, optimal filtering solves many problem arising in statistical signal processing, stochastic control, and related fields. In continuous time, KBF is the solution. In discrete time, KF is the perfect solution because of its optimality, simplicity, fast computation, and small memory requirements, although what folks call the "KF" is not a filter but a recursive computational algorithm that computes batch OFIR and LMF estimates.

There are three main linear FIR state estimators: 1) *optimal*, which is an OFIR filter, 2) *optimal unbiased*, which is an OUFIR filter or ML FIR filter, and 3) *unbiased*, which is a UFIR filter. In continuous time, the OFIR filter is a KBF operating on finite horizons. In discrete time, the batch OFIR filtering estimate can be computed using Kalman recursions, OUFIR using other (Kalman-like) recursions, and UFIR using Shmaliy recursions.

The concepts of *a priori* filtering and *a posteriori* filtering play an important role in discrete time. They allow us to use a prior estimate instead of a posterior estimate when data are temporary unavailable. It is worth noting that the difference between the *a priori* and *a posteriori* state estimates is indistinguishable in continuous time due to the zero time step.

The OFIR filter is most general for optimal state estimation in stochastic linear systems. If we embed unbiasedness, this filter is converted to a bias constrained OUFIR filter, which is equivalent to the ML FIR state estimator. At the other extreme, if we ignore zero mean noise, the OFIR filter becomes a robust UFIR filter that does not require initial values and noise covariances. Another batch optimal filter, called LMF, requires initial values outside the averaging horizon. It is worth noting that the best LMF is the LMKF, and the LMF operating on unlimited horizons is the batch KF.

The OFIR filter can be applied to stochastic and deterministic processes, while the bias-constrained optimal filters can be applied only to stochastic processes. By embedding unbiasedness, the OFIR filter becomes an ML FIR estimator. All FIR filters are BIBO stable for any initial state, while the LMF is BIBO stable only for bounded initial state. Moreover, all FIR filters and LMF are deadbeat state estimators.

If the model and Gaussian noise are known exactly, then there will be no need to use FIR or LMF state estimators. In this ideal case, the full horizon OFIR filter and KF are equivalent and are the best state estimators. Otherwise, the FIR and LMF structures will be more robust, especially to temporary uncertainties and errors in noise statistics. This property of FIR structures is seen as a major advantage over IIR state estimators.

4.11 Problems

1 Two discrete-time systems are represented in state space with equations $x_{k+1} = Fx_k + w_k$ and $y_k = [\,1\ 0\,]x_k + v_k$, where $x_k = [\,x_{1k}\ x_{2k}\,]^T$ is the two-state vector. The first system has a matrix $F = \begin{bmatrix} 1 & \tau \\ 0 & 1 \end{bmatrix}$, where $\tau = t_{k+1} - t_k$, and the second system has $F = \begin{bmatrix} 1 & 0 \\ 0 & 1 \end{bmatrix}$. Represent the states x_{1k} and x_{2k} analytically and make a conclusion about the system feasibility.

2 A second-order LTI system is represented in continuous time with equations $x'(t) = Ax(t) + Bw(t)$ and $y(t) = Cx(t) + v(t)$, where matrices are specified as

$$A = \begin{bmatrix} 0 & 1 \\ -1 & -2 \end{bmatrix}, \quad B = \begin{bmatrix} 0 \\ 1 \end{bmatrix}, \quad C = \begin{bmatrix} 1 & 0 \end{bmatrix}. \tag{4.159}$$

Represent this model in discrete time using the FE and BE methods.

3 Show that the error covariance (4.121) of an ML FIR filter is equal to the error covariance (4.55) of an OUFIR filter.

4 Given the matrix identity $(H^T\Omega^{-1}H)^{-1}H^T\Omega^{-1}H = I$, find the conditions under which the following identity holds,

$$H(H^T\Omega^{-1}H)^{-1}H^T\Omega^{-1} = I\,.$$

5 The block covariance matrix $\mathcal{Q}_{m,k}$ of a colored noise has one of the following structures

$$\mathcal{Q}_{m,k} = Q\begin{bmatrix} 1 & 1 & \dots & 1 \\ 1 & 1 & \dots & 1 \\ \vdots & \vdots & \ddots & \vdots \\ 1 & 1 & \dots & 1 \end{bmatrix}, \quad \mathcal{Q}_{m,k} = Q\begin{bmatrix} 1 & 1 & \dots & 0 & 0 \\ 1 & 1 & \dots & 0 & 0 \\ \vdots & \vdots & \ddots & \vdots & \vdots \\ 0 & 0 & \dots & 1 & 1 \\ 0 & 0 & \dots & 1 & 1 \end{bmatrix}.$$

Find recursive forms for these matrices and compare with the recursive representation $\mathcal{Q}_{m,k} = \begin{bmatrix} \mathcal{Q}_{m,k-1} & 0 \\ 0 & Q \end{bmatrix}$ of the diagonal covariance matrix $\mathcal{Q}_{m,k} = \text{diag}(\, Q\; Q \dots Q\,)$ of white noise.

6 The batch *a priori* OFIR filter is given by (4.107) and the error covariance by (4.109). Show that these batches can be computed using the *a priori* KF Algorithm 2 for given initial values at m.

7 Given the batch *a priori* OUFIR filter (4.110) and error covariance (4.111), show that these batch forms can be computed with the *a priori* KF Algorithm 2 for the initial values specified at m.

8 The ML FIR state estimator unifies different kinds of bias-constrained optimal FIR estimators in the canonic ML form (4.98a),

$$\hat{x}_{k|k} = (C_{m,k}^T \Sigma_{m,k}^{-1} C_{m,k})^{-1} C_{m,k}^T \Sigma_{m,k}^{-1} Y_{m,k}\,,$$

that does not require the initial conditions. Why does the recursive OUFIR filtering Algorithm 8, which is equivalent to the ML FIR algorithm, requires initial values? Does this contradict the property if unbiased optimality?

9 Explain why Kalman recursions serve both the OFIR filter and the LMF having IIR?

10 **Solved Problem: Proof of Theorem 4.4**. Consider the batch LMF (4.141b). To arrive at an iterative computational algorithm using Kalman recursions (Algorithm 4) for given $\hat{x}_{m-1}$ and P_{m-1}, start with (4.28b), set $\chi_m = 0$, and replace $C_{m,k}$ with $C_{m,k}$ and $\mathcal{H}_{m,k}$ with $\bar{\mathcal{H}}_{m,k}$.

Using $C_{m,k} = [C^T_{m,k-1} \quad (H_k \mathcal{F}^m_k)^T]^T$ and following the proof of theorem 4.1, represent $\bar{\mathcal{H}}_{m,k}$ recursively as

$$\bar{\mathcal{H}}_{m,k} = [(I - K_k H_k) F_k \bar{\mathcal{H}}_{m,k-1} \quad K_k] \, ,$$

where the Kalman gain is given by

$$K_k = P_k^- H_k^T (H_k P_k^- H_k^T + R_k)^{-1}$$

and P_k^- is updated using the DDRE as

$$P_k^- = F_k (I - K_{k-1} H_{k-1}) P_{k-1}^- F_k^T + B_k Q_k B_k^T \, .$$

The two first terms on the right-hand side of (4.141a) give

$$\begin{aligned}\hat{x}_k^{\text{hf}} &= \bar{\mathcal{H}}_{m,k} Y_{m,k} + (\bar{S}_{m,k} - \bar{\mathcal{H}}_{m,k} L_{m,k}) U_{m,k} \\ &= F_k \hat{x}_{k-1}^{\text{hf}} + E_k u_k \\ &\quad + K_k [y_k - H_k (F_k \hat{x}_{k-1}^{\text{hf}} + E_k u_k)] \, .\end{aligned}$$

Likewise, represent the remaining third term in (4.141a) as $\hat{x}_k^{\text{s}} = (I - K_k H_k) F_k \hat{x}_{k-1}^{\text{s}}$. Next, combine $\hat{x}_k^{\text{s}}$ with $\hat{x}_k^{\text{hf}}$ and end up with a recursion

$$\hat{x}_k = \hat{x}_k^- + K_k (y_k - H_k \hat{x}_k^-)$$

that completes the proof of theorem 4.4. We can now develop an iterative algorithm for batch LMF.

11 A quasi periodic process is approximated and measured as

$$y_k = a_{0k} + a_{1k} \cos\left(\frac{\pi}{12} k + \phi_1\right) + a_{2k} \cos\left(\frac{2\pi}{12} k + \phi_2\right) + v_k \, ,$$

where $v_k \sim \mathcal{N}(0, \sigma_v^2)$. Consider a_{0k}, a_{1k}, a_{2k}, ϕ_1, and ϕ_2 as state variables and represent the process with the state and observation equations. Simulate the process for some σ_v^2 and system noise $w_k \sim \mathcal{N}(0, \sigma_w^2)$, apply the KF and OFIR algorithms, and compare the estimation errors.

12 The harmonic model is given in state space as

$$\begin{aligned} x_k &= \begin{bmatrix} 0.96596 + \delta_k & 0.25869 \\ -0.25869 & 0.96596 \end{bmatrix} x_{k-1} + \begin{bmatrix} 0.1 \\ 0.1 \end{bmatrix} u_k + \begin{bmatrix} 1 \\ 1 \end{bmatrix} w_k \, , \\ y_k &= [\, 1 \quad 0 \,] x_k + v_k \, , \end{aligned}$$

with the initial state $x_k^T = [\, 1 \quad 1 \,]$ and noise variances $\sigma_w^2 = 1$ and $\sigma_v^2 = 1$. The disturbance $\delta_k = 0.02$ occurs when $100 \leqslant k \leqslant 110$. Simulate this process and estimate x_k using the OFIR filter and KF. Show plots for each of the states and compare the estimation errors.

13 Consider the state-space model $x_k = F x_{k-1} + w_k$ and $y_k = H x_k + v_k$ with $w_k \sim \mathcal{N}(0, Q)$ and $v_k \sim \mathcal{N}(0, R)$, where Q and R are known. To initialize the iterative OUFIR filtering algorithm on $[m, k]$ the initial state and error covariance are required at m. Solve this problem by developing a hybrid OUFIR/UFIR algorithm.

14 A process is represented with the state equation $x_k = F x_{k-1} + w_k$. Data arrive with a delay on n_k discrete points, and the observation is given by $y_k = H x_{k-n_k} + v_k$, where w_k and v_k are white Gaussian noise vectors. Convert the observation to another without delay and modify the iterative OFIR and OUFIR filtering algorithms accordingly.

15 A quantity described with the state equation $x_k = Fx_{k-1} + B_k w_k$ is multiply measured via a WSN with n sensors. The ith, $i \in [1, n]$, observation and total observation are given, respectively, by the equations

$$y_k^{(i)} = H_k^{(i)} x_k + v_k^{(i)} ,$$
$$y_k = H_k x_k + v_k ,$$

where $y_k = [\, y_k^{(1)^T} , \dots , y_k^{(n)^T} \,]^T$ is the total observation vector and $H_k = [\, H_k^{(1)^T} , \dots , H_k^{(n)^T} \,]$ is the WSN matrix. Develop the distributed OFIR filter with consensus on measurements for this model.

16 A measurement of a distant quantity Q is transmitted with time-stamped delayed and missing data. The state space model is given by

$$x_k = Fx_{k-1} + w_k ,$$
$$\tilde{y}_k = HFx_{k-1} ,$$
$$y_k = \kappa_k Hx_{k-n_k} + (1 - \kappa_k)\tilde{y}_k + v_k ,$$

where w_k and v_k are white Gaussian. When data arrive successfully, the missing data scalar κ_k becomes $\kappa_k = 1$; otherwise, it becomes $\kappa_k = 0$. Transform the state-space model into another without latency and modify the OFIR filtering algorithm accordingly.

17 Measurements arrive at a receiver with randomly delayed data. The state-space model is given by

$$x_k = F_k x_{k-1} + B_k w_k ,$$
$$y_k = (1 - \gamma_k)Hx_k + \gamma_k x_{k-1} + v_k ,$$

where the delay factor $\gamma_k = 0$ indicates that there is no latency and $\gamma_k = 1$ otherwise. Vectors w_k and v_k are white Gaussian, and the factor γ_k has a binary Bernoulli distribution with a given probability for each k. Transform this model to another without latency and modify the OFIR filtering algorithm accordingly.

18 A control system is represented in discrete-time state space with the block diagram shown in Fig. 4.9.

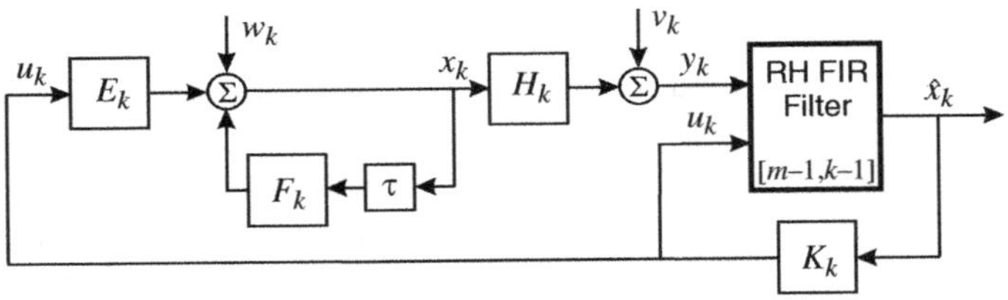

Figure 4.9 Discrete-time control system with an RH FIR filter.

State estimation is provided with an RH FIR filter that gives an estimate $\hat{x}_k$ over data on $[m-1, k-1]$. The filter output is used in state feedback control. Represent this system with state-space equations.

19 A digital linear system is represented with equations $x_{k+1} = Fx_k + Eu_k + Bw_k$ and $y_k = Hx_k + v_k$, where $w_k \sim \mathcal{N}(0, Q)$ and $v_k \sim \mathcal{N}(0, R)$. Due to the state feedback $u_k = Ky_k$, the state-space model becomes

$$\begin{aligned} x_{k+1} &= (F + EKH)x_k + \xi_k , \\ y_k &= Hx_k + v_k , \end{aligned} \tag{4.160}$$

where $\xi_k = Bw_k + EKv_k$ and, therefore, Gaussian ξ_k and v_k are time-correlated. De-correlate ξ_k and v_k and design a batch OFIR filter and iterative algorithm.

20 Consider the control system shown in Fig. 4.9 and modify it using an ML FIR filter instead of an RH FIR filter. Why can the output of the ML FIR filter not be used directly in state feedback?

21 A wireless multisensor system is represented in state space with

$$\begin{aligned} x_{k+1} &= (F_k + A_k\alpha_k)x_k + w_k , \\ z_k^{(i)} &= (H_k^{(i)} + C_k^{(i)}\beta_k^{(i)})x_k + v_k^{(i)} , \\ y_k^{(i)} &= \gamma_k^{(i)} z_k^{(i)} + (1 - \gamma_k^{(i)})\mathcal{I}^{(i)} , \end{aligned} \tag{4.161}$$

where $y_k^{(i)}$, $i \in [1, n]$, is the Tobit observation model, n is the number of sensors, α_k and $\beta_k^{(i)}$ are multiplicative noise sequences, and $\mathcal{I}^{(i)}$ is the sensor threshold such that the sensor probability is $\gamma_k^{(i)} = 1$ when $z_k^{(i)} > \mathcal{I}^{(i)}$ and $\gamma_k^{(i)} = 0$ otherwise. Derive a federated ML FIR predictor for this model.

22 To mitigate the requirement of the initial state, optimal FIR filtering is organized over the data available on $[m, k]$ in three steps. First, $\hat{x}_k$ is provided for a roughly set x_m, then a backward OFIR filter is used to estimate $\hat{x}_m$, and finally the OFIR filter is run again for an improved $\hat{x}_m$. When does this approach give the best accuracy for 1) Gaussian noise, 3) temporary model uncertainty, or 2) heavy tailed noise?

23 Hybrid navigation systems can be designed using different methods and state estimators. Using probabilistic weights, develop fusion algorithms by combining 1) UFIR and OFIR filters, 2) UFIR and ML FIR filters, and 3) OFIR and ML FIR filters.

5

Optimal FIR Smoothing

It seems important to note that there are an infinity of smoothing algorithms.

John B. Moore [130], p. 163

5.1 Introduction

Smoothing is a better tool for denoising at some past data point. It is generally not a real-time but rather a post-processing technique. Since the computational complexity of smoothing algorithms is usually not an issue in practice, their usefulness is appreciated in many areas of signal processing. Smoothers are also useful for obtaining pseudo ground truths, which are later used to tune filtering algorithms, avoiding expensive reference sources. Various approaches to signal smoothing have been developed over the decades [63, 127]. In recursive state estimation, improvement is obtained by processing backward the available KF estimates. An example is the fixed-interval RTS smoother (3.209)–(3.211) [154]. With the batch approach, smoothing is organized at a certain point on the horizon $[m, k]$ with a delay-lag q in such a way that $m < k - q < k$. An example is the Savitzky–Golay smoothing filter [158], which was obtained using the LS approach and in which an estimate is related to the middle of the averaging horizon, although not in state space. A more general UFIR smoother [174] provides a q-lag smoothing estimate in one step, although much less attention has been paid to OFIR smoothing.

In this chapter, we will discuss OFIR smoothing techniques and problems by looking at and using various combinations of forward and backward models and filtering schemes.

5.2 Smoothing Problem

The first attempts to develop an optimal FIR smoother were made in [102]. The solution was found by modifying the one-step predictive MVF [97], the computational complexity $\mathcal{O}(N^2)$ of which was reduced to $\mathcal{O}(N)$ due to the improvement of computation sequences and the use of the system matrix property. Around the same time, another order-recursive FIR smoother based on the Cholesky factorization was obtained in [219]. Further progress in RH FIR smoothing was achieved by W. H. Kwon and his followers. The fixed-lag bias-constrained minimum variance FIR smoother was derived in discrete time in [94, 96] and in continuous time in [93]. In [95], it was argued that

Optimal and Robust State Estimation: Finite Impulse Response (FIR) and Kalman Approaches, First Edition.
Yuriy S. Shmaliy and Shunyi Zhao.

the size of the half-horizon lag, which is set for the SG smoothing filter, is not generally the best in the MSE sense. This observation was later confirmed in [174], and we will discuss it in detail in the next chapter. The fixed-lag RH ML FIR smoother was derived in [4] to have the same structure as other bias-constrained RH FIR smoothers. The interested reader is referred to [224] and Chapter 4, which show that the ML FIR canonical form (1.19) unifies all other types of optimal FIR structures with embedded unbiasedness. It is worth noting that the OFIR smoothing, filtering, and prediction problems can be solved universally using the p-shift OFIR filter [176], which provides q-lag smoothing by $q = -p > 0$, filtering by $p = 0$, and p-step prediction by $p > 0$.

To immediately comprehend the difference between smoothing, filtering, and prediction in terms of noise reduction, we provide an example using the UFIR approach. Although the outputs of the p-shift OFIR and UFIR filters are different, their similarity when $p \approx 0$ gives an idea about the main effect.

Example 5.1 ***UFIR smoothing, filtering, and prediction.*** Consider the 1-degree state-space polynomial model in (4.1) and (4.2) with $F = \begin{bmatrix} 1 & \tau \\ 0 & 1 \end{bmatrix}$, $H = [\,1 \;\; 0\,]$, $u_k = 0$, $\tau = 1$ s, $w_k = 0$, and $v_k \sim \mathcal{N}(0, \sigma_v^2)$. The NPG of the 1-degree p-shift UFIR filter [174] is defined for the first state as the ratio of the output noise variance $\sigma_{\hat{x}}^2$ to the input noise variance σ_v^2 as [174]

$$\text{NPG} = \frac{\sigma_{\hat{x}}^2}{\sigma_v^2} = \frac{2(2N-1)(N-1) + 12p(N-1+p)}{N(N^2-1)} .$$

The NPG representing the filter precision is illustrated in Fig. 5.1 as a function of p. It can be seen that the smoothing estimate $\hat{x}_{k-q|k}$, $N-1 < q = -p < 0$, is more precise than the filtering estimate $\hat{x}_{k|k}$, and that the latter is more precise than the predicted estimate $\hat{x}_{k+p|k}$. Note that UFIR smoothing guarantees the same errors for $q = 0$ and $q = N-1$ on $[m, k]$, and more detail can be found in the next chapter.

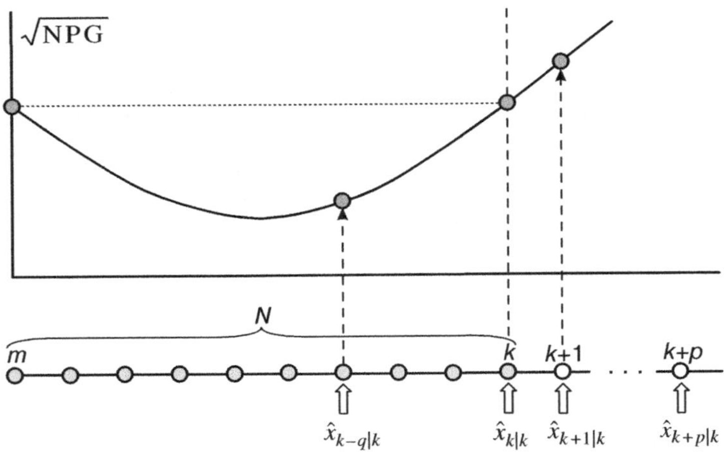

Figure 5.1 NPG of the 1-degree polynomial UFIR smoother $\hat{x}_{k-q|k}$, filter $\hat{x}_{k|k}$, and predictor $\hat{x}_{k+p|k}$. Based on [175] and [176]. □

Referring to this example, folks may wonder why not forget about filtering and always use smoothing. The problem is that the biased q-lag smoothing estimate may not always be sufficiently accurate at the current time point, and future data may not be available for smoothing filtering.

Four OFIR smoothing algorithms can be developed using forward and backward models as well as forward and backward OFIR filtering:

- Forward filter/forward model (FFFM)
- Backward filter/backward model (BFBM)
- Forward filter/backward model (FFBM)
- Backward filter/forward model (BFFM)

Other possible schemes include two-filter FIR smoothing and ML FIR smoothing. We will go over all these algorithms next, noting that reasonably simple recursive forms are generally unavailable for batch FIR smoothers, even for LTI processes.

5.3 Forward Filter/Forward Model q-lag OFIR Smoothing

Batch FFFM q-lag OFIR smoothing can be organized if we process data optimally forward on $[m, k]$ and relate the estimate to $k - q$ using a forward model, as shown in Fig. 5.2. The approach was developed in [173, 176] as a p-shift OFIR filter that becomes the q-lag smoother by $q = -p > 0$. The corresponding FFFM smoother is designed using a state-space model extended forward from m to k, processing data forward on $[m, k]$, and relating the estimate to $k - q$. The gain for this smoother can be determined using the orthogonality condition and employing the state model extended from m to $k - q$, as will be shown next.

5.3.1 Batch Smoothing Estimate

Consider an LTV stochastic process represented in state space with equations

$$x_k = F_k x_{k-1} + E_k u_k + B_k w_k \,, \tag{5.1}$$

$$y_k = H_k x_k + v_k \tag{5.2}$$

and extend it on $[m, k]$ as

$$X_{m,k} = F_{m,k} x_m + S_{m,k} U_{m,k} + D_{m,k} W_{m,k} \,, \tag{5.3}$$

$$Y_{m,k} = H_{m,k} x_m + L_{m,k} U_{m,k} + G_{m,k} W_{m,k} + V_{m,k} \,, \tag{5.4}$$

where the extended vectors and matrices are given after (4.7) and (4.14).

Following [173, 176], the FFFM q-lag OFIR smoothing estimate can be defined as

$$\tilde{x}_{k-q} \triangleq \tilde{x}_{k-q|k} = \mathcal{H}^{\mathrm{h}(q)}_{m,k} Y_{m,k} + \mathcal{H}^{\mathrm{f}(q)}_{m,k} U_{m,k} \,, \tag{5.5}$$

where $\mathcal{H}^{(q)}_{m,k} \triangleq \mathcal{H}^{\mathrm{h}(q)}_{m,k}$ is the q-variant homogenous gain and $\mathcal{H}^{\mathrm{f}(q)}_{m,k}$ is the q-variant forced gain. It follows from Fig. 5.2 that data taken from $[m, k]$ are processed forward using the forward model to produce a smoothing estimate at $k - q$. The forward state model for this estimate can be defined at

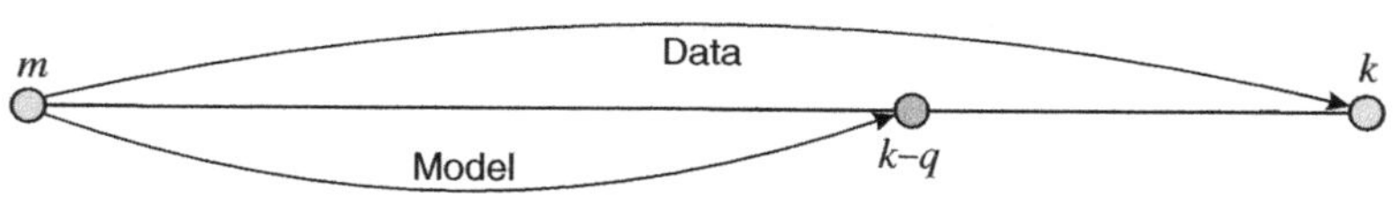

Figure 5.2 Forward filter/forward model q-lag OFIR smoothing strategy.

$k-q$ by the $(N-q)$th raw vector of (5.3) as [176]

$$x_{k-q} = \mathcal{F}_{k-q}^{m+1} x_m + \bar{S}_{m,k}^{(N-q)} U_{m,k} + \bar{D}_{m,k}^{(N-q)} W_{m,k} \,, \tag{5.6}$$

where $\bar{S}_{m,k}^{(N-q)}$ is the $(N-q)$th raw vector in $S_{m,k}$ given by (4.9) and so is $\bar{D}_{m,k}^{(N-q)}$ in $D_{m,k}$.

The unbiasedness condition $\mathcal{E}\{\tilde{x}_{k-q}\} = \mathcal{E}\{x_{k-q}\}$ applied to (5.5) and (5.6) gives two unbiasedness constraints

$$\mathcal{F}_{k-q}^{m+1} = \mathcal{H}_{m,k}^{(q)} H_{m,k} \,, \tag{5.7}$$

$$\mathcal{H}_{m,k}^{\mathrm{f}(q)} = \bar{S}_{m,k}^{(N-q)} - \mathcal{H}_{m,k}^{(q)} L_{m,k} \,, \tag{5.8}$$

and the smoothing error $\varepsilon_{k-q} \triangleq \varepsilon_{k-q|k} = x_{k-q} - \tilde{x}_{k-q}$ can be transformed to

$$\begin{aligned}
\varepsilon_{k-q} &= \mathcal{F}_{k-q}^{m+1} x_m + \bar{S}_{m,k}^{(N-q)} U_{m,k} + \bar{D}_{m,k}^{(N-q)} W_{m,k} \\
&\quad - \mathcal{H}_{m,k}^{(q)} Y_{m,k} - \mathcal{H}_{m,k}^{\mathrm{f}(q)} U_{m,k} \\
&= (\mathcal{F}_{k-q}^{m+1} - \mathcal{H}_{m,k}^{(q)} H_{m,k}) x_m \\
&\quad + (\bar{S}_{m,k}^{(N-q)} - \mathcal{H}_{m,k}^{(q)} L_{m,k} - \mathcal{H}_{m,k}^{\mathrm{f}(q)}) U_{m,k} \\
&\quad + (\bar{D}_{m,k}^{(N-q)} - \mathcal{H}_{m,k}^{(q)} G_{m,k}) W_{m,k} - \mathcal{H}_{m,k}^{(q)} V_{m,k} \,.
\end{aligned}$$

Now, the orthogonality condition $\mathcal{E}\{\varepsilon_{k-q} Y_{m,k}^T\} = 0$ gives

$$\begin{aligned}
&\mathcal{B}_{m,k}^{(q)} \mathcal{X}_m H_{m,k}^T + \mathcal{W}_{m,k}^{(q)} \mathcal{Q}_{m,k} G_{m,k}^T - \mathcal{V}_{m,k}^{(q)} \mathcal{R}_{m,k} \\
&= (\mathcal{H}_{m,k}^{\mathrm{h}(q)} L_{m,k} - \bar{S}_{m,k}^{(N-q)} + \mathcal{H}_{m,k}^{\mathrm{f}(q)}) \Psi_{m,k} L_{m,k}^T \,,
\end{aligned} \tag{5.9}$$

where the q-variant error residual matrices,

$$\mathcal{B}_{m,k}^{(q)} = \mathcal{F}_{k-q}^{m+1} - \mathcal{H}_{m,k}^{(q)} H_{m,k} \,, \tag{5.10}$$

$$\mathcal{W}_{m,k}^{(q)} = \bar{D}_{m,k}^{(N-q)} - \mathcal{H}_{m,k}^{(q)} G_{m,k} \,, \tag{5.11}$$

$$\mathcal{V}_{m,k}^{(q)} = \mathcal{H}_{m,k}^{(q)} \,, \tag{5.12}$$

represent, respectively, the *bias error residual*, *system error residual*, and *measurement error residual*. Note that, as in OFIR filtering, matrices (5.10)–(5.12) are fully responsible for optimal cancellation of regular bias and random errors in the FFFM q-lag OFIR smoothing estimate at $k-q$.

For zero input $U_{m,k} = 0$, relation (5.9) gives the fundamental gain

$$\begin{aligned}
\mathcal{H}_{m,k}^{(q)} &= (\mathcal{F}_{k-q}^{m+1} \mathcal{X}_m H_{m,k}^T + \bar{D}_{m,k}^{(N-q)} \mathcal{Q}_{m,k} G_{m,k}^T) \\
&\quad \times (H_{m,k} \mathcal{X}_m H_{m,k}^T + G_{m,k} \mathcal{Q}_{m,k} G_{m,k}^T + \mathcal{R}_{m,k})^{-1} \,,
\end{aligned} \tag{5.13}$$

where a matrix $\bar{D}_{m,k}^{(N-q)}$ can be represented with

$$\bar{D}_{m,k}^{(N-q)} = [\ \mathcal{F}_{k-q}^{m+1} B_m \quad \mathcal{F}_{k-q}^{m+2} B_{m+1} \ \cdots \ \mathcal{F}_{k-q}^{k} B_{k-1} \quad \mathcal{F}_{k-q}^{k+1} B_k\ ] \,. \tag{5.14}$$

Multiplying the first $\mathcal{X}_m$ in (5.13) from the left-hand side with an identity $(H_{m,k}^T H_{m,k})^{-1} H_{m,k}^T H_{m,k}$

and referring to (4.28c), we transform (5.13) to

$$\mathcal{H}_{m,k}^{(q)} = (\mathcal{F}_{k-q}^{m+1}\chi_m H_{m,k}^T + \mathcal{Z}_1^{(q)})(\mathcal{Z}_\chi + \mathcal{Z}_2 + \mathcal{R}_{m,k})^{-1} \tag{5.15a}$$

$$= (\hat{\mathcal{H}}_{m,k}^{(q)}\mathcal{Z}_\chi + \mathcal{Z}_1^{(q)})(\mathcal{Z}_\chi + \mathcal{Z}_2 + \mathcal{R}_{m,k})^{-1}\,, \tag{5.15b}$$

where matrices $\mathcal{Z}_\chi$, $\mathcal{Z}_1^{(q)}$, and $\mathcal{Z}_2$ are defined by comparing (5.15a) and (5.13), and the q-lag UFIR smoother gain $\hat{\mathcal{H}}_{m,k}^{(q)}$ is given by

$$\hat{\mathcal{H}}_{m,k}^{(q)} = \mathcal{F}_{k-q}^{m+1}(H_{m,k}^T H_{m,k})^{-1}H_{m,k}^T \tag{5.16a}$$

$$= (\mathcal{F}_k^{k-q+1})^{-1}(C_{m,k}^T C_{m,k})^{-1}C_{m,k}^T \tag{5.16b}$$

$$= (\mathcal{F}_k^{k-q+1})^{-1}\hat{\mathcal{H}}_{m,k}\,, \tag{5.16c}$$

where $\hat{\mathcal{H}}_{m,k} = (C_{m,k}^T C_{m,k})^{-1}C_{m,k}^T$ is the UFIR filter gain.

Another useful form of $\hat{\mathcal{H}}_{m,k}^{(q)}$ can be shown if we assign $\mathcal{H}_{m,k}^{2\text{h}} = (\mathcal{Z}_\chi + \mathcal{Z}_2 + \mathcal{R}_{m,k})^{-1}$ and transform (5.15b) as

$$\begin{aligned}\mathcal{H}_{m,k}^{(q)} &= (\hat{\mathcal{H}}_{m,k}\mathcal{Z}_\chi + \mathcal{Z}_1 + \hat{\mathcal{H}}_{m,k}^{(q)}\mathcal{Z}_\chi - \hat{\mathcal{H}}_{m,k}\mathcal{Z}_\chi + \mathcal{Z}_1^{(q)} - \mathcal{Z}_1)\mathcal{H}_{m,k}^{2\text{h}}\\ &= (\hat{\mathcal{H}}_{m,k}\mathcal{Z}_\chi + \mathcal{Z}_1)\mathcal{H}_{m,k}^{2\text{h}} + [(\hat{\mathcal{H}}_{m,k}^{(q)} - \hat{\mathcal{H}}_{m,k})\mathcal{Z}_\chi + \mathcal{Z}_1^{(q)} - \mathcal{Z}_1]\mathcal{H}_{m,k}^{2\text{h}}\\ &= \mathcal{H}_{m,k} + [((\mathcal{F}_k^{k-q+1})^{-1} - I)\hat{\mathcal{H}}_{m,k}\mathcal{Z}_\chi + \mathcal{Z}_1^{(q)} - \mathcal{Z}_1]\mathcal{H}_{m,k}^{2\text{h}}\,.\end{aligned}$$

Rewriting (5.15b) for $q = 0$ as $\mathcal{H}_{m,k} = (\hat{\mathcal{H}}_{m,k}\mathcal{Z}_\chi + \mathcal{Z}_1)\mathcal{H}_{m,k}^{2\text{h}}$, multiplying both sides from the left-hand side with an identity $(\hat{\mathcal{H}}_{m,k}\mathcal{Z}_\chi + \mathcal{Z}_1)(\hat{\mathcal{H}}_{m,k}\mathcal{Z}_\chi + \mathcal{Z}_1)^T[\dots]^{-1}$, where $[\dots]$ depicts the preceding product, and discarding nonzero $\hat{\mathcal{H}}_{m,k}\mathcal{Z}_\chi + \mathcal{Z}_1$ on both sides, we obtain

$$\mathcal{H}_{m,k}^{2\text{h}} = (\hat{\mathcal{H}}_{m,k}\mathcal{Z}_\chi + \mathcal{Z}_1)^T[\dots]^{-1}\mathcal{H}_{m,k}\,. \tag{5.17}$$

We next represent $\mathcal{H}_{m,k}^{(q)}$ as

$$\begin{aligned}\mathcal{H}_{m,k}^{(q)} &= \{I + [((\mathcal{F}_k^{k-q+1})^{-1} - I)\hat{\mathcal{H}}_{m,k}\mathcal{Z}_\chi + \mathcal{Z}_1^{(q)} - \mathcal{Z}_1]\bar{\mathcal{Z}}\}\mathcal{H}_{m,k}\\ &= (\hat{\mathcal{H}}_{m,k}^{(q)}\mathcal{Z}_\chi + \mathcal{Z}_1^{(q)})\bar{\mathcal{Z}}\mathcal{H}_{m,k}\,,\end{aligned} \tag{5.18}$$

where $\bar{\mathcal{Z}} = (\hat{\mathcal{H}}_{m,k}\mathcal{Z}_\chi + \mathcal{Z}_1)^T[(\hat{\mathcal{H}}_{m,k}\mathcal{Z}_\chi + \mathcal{Z}_1)(\hat{\mathcal{H}}_{m,k}\mathcal{Z}_\chi + \mathcal{Z}_1)^T]^{-1}$ and both $\hat{\mathcal{H}}_{m,k}^{(q)}$ and $\mathcal{H}_{m,k}$ have available recursive forms. It can be seen that $q = 0$ makes (5.18) an identity, $\mathcal{H}_{m,k}^{(q)} = \mathcal{H}_{m,k}$. Assuming $u_k = 0$, we finally obtain a q-lag OFIR smoothing estimate through the OFIR filtering estimate $\hat{x}_k$ as

$$\begin{aligned}\tilde{x}_{k-q} &= \mathcal{H}_{m,k}^{(q)}Y_{m,k}\\ &= (\hat{\mathcal{H}}_{m,k}^{(q)}\mathcal{Z}_\chi + \mathcal{Z}_1^{(q)})\bar{\mathcal{Z}}\hat{x}_k\,.\end{aligned} \tag{5.19}$$

Referring to (5.8), the FFFM q-lag OFIR smoother finally becomes

$$\tilde{x}_{k-q} = \mathcal{H}_{m,k}^{(q)}Y_{m,k} + (\bar{S}_{m,k}^{(N-q)} - \mathcal{H}_{m,k}^{(q)}L_{m,k})U_{m,k}\,, \tag{5.20}$$

where the gain $\mathcal{H}^{(q)}_{m,k}$ exists in various batch forms (5.13), (5.15b), and (5.18). Now it is worth noting the distinctive advantage of the batch smoother (5.20) over the recursive RTS smoother: *it can work with full block noise and error covariance matrices, which provides greater accuracy for nonwhite noise*. Also note that although the noise coloredness is not supported by the stochastic model in (5.1) and (5.2), it is supported by the H_2 OFIR state estimator (Chapter 8), which has the same gain for LTI processes as the OFIR state estimator. Next, we will consider the error covariance of the FFFM OFIR smoother.

5.3.2 Error Covariance

Having determined the smoothing error $\varepsilon_{k-q} = x_{k-q} - \tilde{x}_{k-q}$, the error covariance P_{k-q} of the FFFM q-lag OFIR smoother can be represented similarly to the OFIR filter as

$$\begin{aligned} P_{k-q} &= \mathcal{E}\{\varepsilon_{k-q}\varepsilon^T_{k-q}\} \\ &= \mathcal{B}^{(q)}_{m,k}\mathcal{X}_m\mathcal{B}^{(q)^T}_{m,k} + \mathcal{W}^{(q)}_{m,k}\mathcal{Q}_{m,k}\mathcal{W}^{(q)^T}_{m,k} \\ &\quad + \mathcal{V}^{(q)}_{m,k}\mathcal{R}_{m,k}\mathcal{V}^{(q)^T}_{m,k} , \end{aligned} \tag{5.21}$$

where the error residual matrices are given by (5.10)–(5.12).

To comprehend the effect of q on error covariance (5.21), it is necessary to study how q affects matrices (5.10)–(5.12). Since the superscript $N-q$ in matrix $\bar{D}^{(N-q)}_{m,k}$ denotes the number of nonzero components and matrix $\mathcal{F}^{m+1}_{k-q}$ in stable systems acquires smaller values with q, it follows that the gain $\mathcal{H}^{\mathrm{h}(q)}_{m,k}$ decreases with increasing q, as does the last term in (5.21) due to $\mathcal{V}^{(q)}_{m,k} = \mathcal{H}^{\mathrm{h}(q)}_{m,k}$. This confirms that smoothing removes measurement noise better than filtering. However, the same cannot be said for bias errors and system noise. Indeed, even though the components of q-varying matrices decrease in (5.10) and (5.12) with increasing q, the output effect turns out to be either less than in (5.12) or even negative, and bias errors are not reduced as much as measurement noise. It also confirms that *smoothing is just a noise reduction technique*.

Example 5.2 ***FFFM OFIR smoothing.*** Consider a two-state polynomial model $x_k = Fx_{k-1} + Bw_k$ and $y_k = Hx_k + v_k$, where $F = \begin{bmatrix} 1 & \tau \\ 0 & 1 \end{bmatrix}$, $B = \begin{bmatrix} \tau \\ 1 \end{bmatrix}$, $H = [\,1 \;\; 0\,]$, $\tau = 0.1$ s, and $w_k \sim \mathcal{N}(0,2)$. Observation is provided in the presence of two types of measurement noise v_k: 1) $v_k \sim \mathcal{N}(0,10)$ that corresponds to $N_{\mathrm{opt}} = 16$ for the UFIR filter and 2) strongly colored $v_k = \varphi v_{k-1} + \xi_k$ with $\varphi = 0.95$ and $\xi_k \sim \mathcal{N}(0,10)$ (Fig. 3.3c).

The theoretical (solid) and measured (dashed-dotted) RMSEs averaged over 10^4 MC runs are sketched in Fig. 5.3a.

In the first case of white noise, the block covariance matrix $\mathcal{R}_{m,k}$ for v_k is diagonal, which is also required by RTS smoothing. As we can see, here the RTS smoother produces the smallest and the UFIR smoother the largest RMSEs across all delay-lags (Fig. 5.3a).

The situation changes dramatically with a strong CMN, when matrix $\mathcal{R}_{m,k}$ on the horizon $[m, k]$ of 16 points becomes

$$\mathcal{R}_{m,k} = \begin{bmatrix} 102.58 & 97.44 & 92.56 & \dots & 50.02 \\ 97.44 & 102.58 & 97.44 & \dots & 52.65 \\ 92.56 & 97.44 & 102.58 & \dots & 55.42 \\ \vdots & \vdots & \vdots & \ddots & \vdots \\ 50.02 & 52.65 & 55.42 & \dots & 102.58 \end{bmatrix},$$

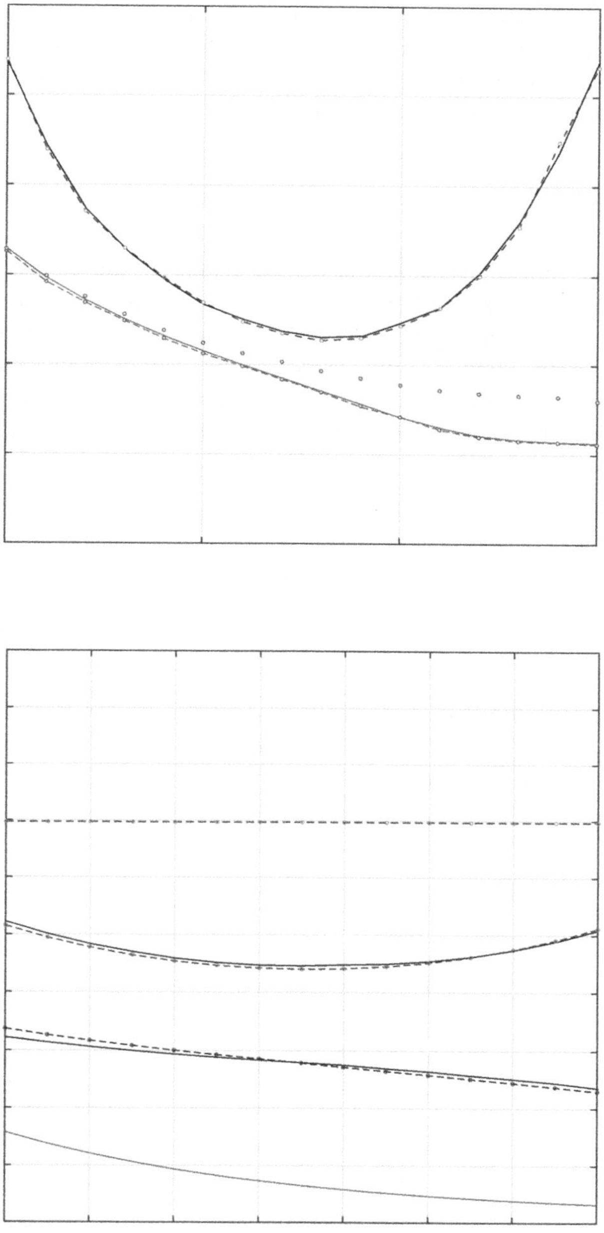

Figure 5.3 RMSE produced by the FFFM OFIR, UFIR, and RTS smoothers: (a) white measurement noise and (b) CMN with $\varphi = 0.95$. Solid lines are theoretical, and dashed-dotted lines represent numerical RMSEs averaged over 10^4 MC runs.

where discarding entries outside the main diagonal can lead to large smoothing errors. We watch for this effect in the recursive RTS smoother, which turns out to be much less accurate than the FFFM OFIR smoother operating with full matrix $\mathcal{R}_{m,k}$ (Fig. 5.3b). Note that the theoretical RMSE boundary shown here for RTS smoothing corresponds to white noise. □

5.4 Backward OFIR Filtering

Filtering that is organized in the opposite direction is sometimes called smoothing, although it does not match the smoothing strategy and is nothing more than backward filtering. But because we will soon be using backward filtering when designing BFFM and BFBM OFIR smoothers, we will cover it in this chapter. It is of practical importance that the initial values estimated at m by a backward filter can be used in forward optimal filtering, thereby removing the requirement of the initial state.

The backward OFIR filter can be obtained similarly to the forward OFIR filter if we extend the state-space model on $[m, k]$ back in time.

5.4.1 Backward State-Space Model

To extend the model in (5.1) on $[m, k]$ back in time, we first invert it as

$$x_{k-1} = F_k^{-1} x_k - F_k^{-1} E_k u_k - F_k^{-1} B_k w_k \tag{5.22}$$

and then project (5.22) to x_m.

By introducing the backward extended vectors and depicting them with left arrows over as $\overleftarrow{X}_{k,m} = [\, x_k^T \; x_{k-1}^T \; \dots \; x_m^T \,]^T$, $\overleftarrow{U}_{k,m} = [\, u_k^T \; u_{k-1}^T \; \dots \; u_m^T \,]^T$, and $\overleftarrow{W}_{k,m} = [\, w_k^T \; w_{k-1}^T \; \dots \; w_m^T \,]^T$, we then combine the projections into the extended linear state equation

$$\overleftarrow{X}_{k,m} = \overleftarrow{F}_{k,m} x_k - \overleftarrow{S}_{k,m} \overleftarrow{U}_{k,m} - \overleftarrow{D}_{k,m} \overleftarrow{W}_{k,m} \,, \tag{5.23}$$

where, using the product matrix

$$\mathcal{X}_r^g = \begin{cases} (F_r F_{r-1} \dots F_g)^{-1} & , \; r \geq g \\ 0 & , \; r < g \end{cases} ,$$

we define the extended matrices as

$$\begin{aligned} \overleftarrow{F}_{k,m} &= \left[\, (\mathcal{F}_k^{k+1})^{-T} \; (\mathcal{F}_k^{k})^{-T} \; \dots \; (\mathcal{F}_k^{m+2})^{-T} \; (\mathcal{F}_k^{m+1})^{-T} \,\right]^T \\ &= \left[\, I \; \{\mathcal{X}_k^k\}^T \; \dots \; \{\mathcal{X}_k^{m+2}\}^T \; \{\mathcal{X}_k^{m+1}\}^T \,\right]^T , \end{aligned} \tag{5.24}$$

$$\overleftarrow{S}_{k,m} = \begin{bmatrix} 0 & 0 & \dots & 0 & 0 \\ \mathcal{X}_k^k E_k & 0 & \dots & 0 & 0 \\ \mathcal{X}_k^{k-1} E_k & \mathcal{X}_{k-1}^{k-1} E_{k-1} & \dots & 0 & 0 \\ \vdots & \vdots & \ddots & \vdots & \vdots \\ \mathcal{X}_k^{m+2} E_k & \mathcal{X}_{k-1}^{m+2} E_{k-1} & \dots & 0 & 0 \\ \mathcal{X}_k^{m+1} E_k & \mathcal{X}_{k-1}^{m+1} E_{k-1} & \dots & \mathcal{X}_{m+1}^{m+1} E_{m+1} & 0 \end{bmatrix} , \tag{5.25}$$

and write matrix $\overleftarrow{D}_{k,m}$ in the same manner as (5.25) by substituting E_i with B_i for $i \in [k, m]$.

In a similar manner, we introduce the backward extended vectors $\overleftarrow{Y}_{k,m} = [\, y_k^T \; y_{k-1}^T \; \dots \; y_m^T \,]^T$ and $\overleftarrow{V}_{k,m} = [\, v_k^T \; v_{k-1}^T \; \dots \; v_m^T \,]^T$ and go to the backward extended observation equation

$$\overleftarrow{Y}_{k,m} = \overleftarrow{H}_{k,m} x_k - \overleftarrow{L}_{k,m} \overleftarrow{U}_{k,m} - \overleftarrow{G}_{k,m} \overleftarrow{W}_{k,m} + \overleftarrow{V}_{k,m} \,, \tag{5.26}$$

where $\overleftarrow{H}_{k,m} = \bar{H}_{k,m}\overleftarrow{F}_{k,m}$, $\overleftarrow{L}_{k,m} = \bar{H}_{k,m}\overleftarrow{S}_{k,m}$, $\overleftarrow{G}_{k,m} = \bar{H}_{k,m}\overleftarrow{D}_{k,m}$, and $\bar{H}_{k,m} = \text{diag}\,(H_k \;\; H_{k-1} \;\; \dots \;\; H_m)$ is a diagonal matrix.

5.4.2 Batch Estimate

Based on model (5.23) and (5.26), the batch backward OFIR filter is stated by the following theorem.

Theorem 5.1 Given a LTV digital system represented in state space with equations (5.23) and (5.26), then the batch backward a posteriori OFIR filter is given by

$$\tilde{x}_m = \overleftarrow{\mathcal{H}}_{k,m}\overleftarrow{Y}_{k,m} + (\overleftarrow{\mathcal{H}}_{k,m}\overleftarrow{L}_{k,m} - \bar{\overleftarrow{S}}_{k,m})\overleftarrow{U}_{k,m}\,, \tag{5.27}$$

where the fundamental (homogeneous) gain $\overleftarrow{\mathcal{H}}_{k,m} \triangleq \overleftarrow{\mathcal{H}}^{\text{h}}_{k,m}$ is defined as

$$\begin{aligned}\overleftarrow{\mathcal{H}}_{k,m} &= (\mathcal{X}_k^{m+1}\chi_k\overleftarrow{H}^T_{k,m} + \bar{\overleftarrow{D}}_{k,m}\mathcal{Q}_{k,m}\overleftarrow{G}^T_{k,m})\\ &\quad\times(\overleftarrow{H}_{k,m}\chi_k\overleftarrow{H}^T_{k,m} + \overleftarrow{G}_{k,m}\mathcal{Q}_{k,m}\overleftarrow{G}^T_{k,m} + \mathcal{R}_{k,m})^{-1}\,,\end{aligned} \tag{5.28}$$

$\chi_k = \mathcal{E}\{x_k x_k^T\}$, $\overleftarrow{\mathcal{Q}}_{k,m} = \mathcal{E}\{\overleftarrow{W}_{k,m}\overleftarrow{W}^T_{k,m}\}$, $\overleftarrow{\mathcal{R}}_{k,m} = \mathcal{E}\{\overleftarrow{V}_{k,m}\overleftarrow{V}^T_{k,m}\}$, $\bar{\overleftarrow{S}}_{k,m}$ is the last row vector in $\overleftarrow{S}_{k,m}$ and so is $\bar{\overleftarrow{D}}_{k,m}$ in $\overleftarrow{D}_{k,m}$.

Proof: Define the backward OFIR filtering estimate $\tilde{x}_m$ in the batch form as

$$\tilde{x}_m = \overleftarrow{\mathcal{H}}_{k,m}\overleftarrow{Y}_{k,m} + \overleftarrow{\mathcal{H}}^{\text{f}}_{k,m}\overleftarrow{U}_{k,m}\,, \tag{5.29}$$

where the fundamental gain $\overleftarrow{\mathcal{H}}_{k,m}$ is determined for zero input $\overleftarrow{U}_{k,m} = 0$ and the forced gain $\overleftarrow{\mathcal{H}}^{\text{f}}_{k,m}$ for zero initial state $x_k = 0$.

Extract the last Nth row vector from (5.23) to be the state model at m,

$$x_m = \mathcal{X}_k^{m+1}x_k - \bar{\overleftarrow{S}}_{k,m}\overleftarrow{U}_{k,m} - \bar{\overleftarrow{D}}_{k,m}\overleftarrow{W}_{k,m}\,, \tag{5.30}$$

where the last row vectors in $\overleftarrow{S}_{k,m}$ and $\overleftarrow{D}_{k,m}$ are defined by

$$\bar{\overleftarrow{S}}_{k,m} = [\mathcal{X}_k^{m+1}E_k \;\; \mathcal{X}_{k-1}^{m+1}E_{k-1} \;\; \cdots \;\; \mathcal{X}_{m+1}^{m+1}E_{m+1} \;\; 0]\,, \tag{5.31}$$

$$\bar{\overleftarrow{D}}_{k,m} = [\mathcal{X}_k^{m+1}B_k \;\; \mathcal{X}_{k-1}^{m+1}B_{k-1} \;\; \cdots \;\; \mathcal{X}_{m+1}^{m+1}B_{m+1} \;\; 0]\,. \tag{5.32}$$

Find the unbiasedness constraints produced by the unbiasedness condition $E\{\hat{x}_m\} = E\{x_m\}$ applied to (5.29) and (5.30),

$$\mathcal{X}_k^{m+1} = \overleftarrow{\mathcal{H}}^{\text{h}}_{k,m}\overleftarrow{H}_{k,m}\,, \tag{5.33}$$

$$\overleftarrow{\mathcal{H}}^{\text{f}}_{k,m} = \overleftarrow{\mathcal{H}}^{\text{h}}_{k,m}\overleftarrow{L}_{k,m} - \bar{\overleftarrow{S}}_{k,m}\,. \tag{5.34}$$

To come up with the backward OFIR estimate, define the estimation error as $\varepsilon_m = x_m - \tilde{x}_m$ and apply the orthogonality condition on $[m, k]$ as

$$\mathcal{E}\{(x_m - \overleftarrow{\mathcal{H}}^{\text{h}}_{k,m}\overleftarrow{Y}_{k,m} - \overleftarrow{\mathcal{H}}^{\text{f}}_{k,m}\overleftarrow{U}_{k,m})\overleftarrow{Y}^T_{k,m}\} = 0\,. \tag{5.35}$$

Next, substitute $\overleftarrow{Y}_{k,m}$ with (5.26), and introduce the mean square initial state $\chi_k = \mathcal{E}\{x_k x_k^T\}$ and the auxiliary matrices $\Psi_{k,m} = \mathcal{E}\{\overleftarrow{U}_{k,m}\overleftarrow{U}^T_{k,m}\}$, $\mathcal{Q}_{k,m} = \mathcal{E}\{\overleftarrow{W}_{k,m}\overleftarrow{W}^T_{k,m}\}$, and $\mathcal{R}_{k,m} = \mathcal{E}\{\overleftarrow{V}_{k,m}\overleftarrow{V}^T_{k,m}\}$.

Take into account that x_k, $\overleftarrow{U}_{k,m}$, $\overleftarrow{W}_{k,m}$, and $\overleftarrow{V}_{k,m}$ are mutually independent and uncorrelated, and transform (5.35) to

$$\overleftarrow{\mathcal{B}}_{k,m}\chi_k\overleftarrow{H}^T_{k,m} + \overleftarrow{\mathcal{W}}_{k,m}\mathcal{Q}_{k,m}\overleftarrow{G}^T_{k,m} - \overleftarrow{\mathcal{V}}_{k,m}\mathcal{R}_{k,m}$$
$$= (\overleftarrow{\mathcal{H}}^{h}_{k,m}\overleftarrow{L}_{k,m} - \overleftarrow{\bar{S}}_{k,m} - \overleftarrow{\mathcal{H}}^{f}_{k,m})\Psi_{k,m}\overleftarrow{L}^T_{k,m}\,, \tag{5.36}$$

where the error residual matrices

$$\overleftarrow{\mathcal{B}}_{k,m} = \mathcal{X}^{m+1}_k - \overleftarrow{\mathcal{H}}^{h}_{k,m}H_{k,m}\,, \tag{5.37}$$

$$\overleftarrow{\mathcal{W}}_{k,m} = \overleftarrow{\bar{D}}_{k,m} - \overleftarrow{\mathcal{H}}^{h}_{k,m}\overleftarrow{G}_{k,m}\,, \tag{5.38}$$

$$\overleftarrow{\mathcal{V}}_{k,m} = \overleftarrow{\mathcal{H}}^{h}_{k,m} \tag{5.39}$$

represent, respectively, the bias error, system error, and measurement error.

Put $\Psi_{k,m} = 0$ in (5.36), provide the transformations, and derive the fundamental gain $\overleftarrow{\mathcal{H}}_{k,m}$ (5.28). Next, set $\chi_k = 0$, which makes the left-hand side in (5.28) identically zero, and find the forced gain $\overleftarrow{\mathcal{H}}^{f}_{k,m}$, which is given by (5.34). Finally, substitute (5.34) into (5.29), end up with (5.27), and complete the proof. □

The error covariance $P_m = \mathcal{E}\{(x_m - \tilde{x}_m)(x_m - \tilde{x}_m)^T\}$ for x_m given by (5.30) and $\tilde{x}_m$ by (5.29) can be found by providing the averaging and referring to (5.37)–(5.39) that finally yields

$$P_m = \overleftarrow{\mathcal{B}}_{k,m}\chi_k\overleftarrow{\mathcal{B}}^T_{k,m} + \overleftarrow{\mathcal{W}}_{k,m}\mathcal{Q}_{k,m}\overleftarrow{\mathcal{W}}^T_{k,m}$$
$$+ \overleftarrow{\mathcal{V}}_{k,m}\mathcal{R}_{k,m}\overleftarrow{\mathcal{V}}^T_{k,m}\,. \tag{5.40}$$

As in the case of forward OFIR filtering, the error residual matrices (5.37)–(5.39) are fully responsible in (5.40) for the optimal cancellation of the regular bias and random errors in the backward estimate $\tilde{x}_m$. Moreover, for diagonal noise covariance matrices, the batch estimate (5.27) and error covariance (5.40) can be represented recursively, as shown next.

5.4.3 Recursive Estimate and Error Covariance

The following theorem introduces recursive forms for the batch backward *a posteriori* OFIR filter (5.30).

Theorem 5.2 Given a batch backward a posteriori OFIR filter (5.27) with the fundamentals gain (5.28) and known x_k and P_k, the recursive forms for this filter are the following,

$$P^-_{k-1} = F^{-1}_k(P_k + B_kQ_kB^T_k)F^{-T}_k\,, \tag{5.41}$$

$$\tilde{x}^-_{k-1} = F^{-1}_k(\tilde{x}_k - E_ku_k)\,, \tag{5.42}$$

$$S_{k-1} = H_{k-1}P^-_{k-1}H^T_{k-1} + R_{k-1}\,, \tag{5.43}$$

$$K_{k-1} = P^-_{k-1}H^T_{k-1}S^{-1}_{k-1}\,, \tag{5.44}$$

$$\tilde{x}_{k-1} = \tilde{x}^-_{k-1} + K_{k-1}(y_{k-1} - H_{k-1}\tilde{x}^-_{k-1})\,, \tag{5.45}$$

$$P_{k-1} = (I - K_{k-1}H_{k-1})P^-_{k-1}\,, \tag{5.46}$$

where the estimates are updated iteratively on $[m, k]$ back-in-time from $k-1$ to m for given initial x_k and P_k.

Proof: Recursions for (5.27) can be found similarly to the OFIR filter if we derive recursions for $\overleftarrow{\mathcal{H}}^{1h}_{k,m}$, which is represented with the first parentheses in (5.28), and $\overleftarrow{\mathcal{H}}^{2h}_{k,m}$, which is represented with the inverse of the second parentheses in (5.28).

To find a recursive form for $\overleftarrow{\mathcal{H}}^{1h}_{k,m}$, start with

$$\overleftarrow{\mathcal{H}}^{1h}_{k,m} = \mathcal{X}_k^{m+1} \chi_k \overleftarrow{H}^T_{k,m} + \overleftarrow{\bar{D}}_{k,m} \mathcal{Q}_{k,m} \overleftarrow{G}^T_{k,m} , \tag{5.47}$$

decompose the following matrices as

$$\begin{aligned}
\overleftarrow{H}^T_{k,m} &= [\ \overleftarrow{H}^T_{k,m+1} \quad \mathcal{X}_k^{m+1^T} H_m^T\ ] ,\\
\overleftarrow{\bar{D}}_{k,m} &= F^{-1}_{m+1}[\ \overleftarrow{\bar{D}}_{k,m+1} + \overleftarrow{\bar{B}}_{k,m+1} \quad 0\ ] ,\\
\mathcal{Q}_{k,m} &= \mathrm{diag}(\ \mathcal{Q}_{k,m+1} \quad Q_m\ ) ,\\
\overleftarrow{G}^T_{k,m} &= \begin{bmatrix} \overleftarrow{D}^T_{k,m+1} \overleftarrow{H}^T_{k,m+1} & (\overleftarrow{\bar{D}}_{k,m+1} + \overleftarrow{\bar{B}}_{k,m+1})^T F^{-T}_{m+1} H^T_m \\ 0 & 0 \end{bmatrix} ,
\end{aligned}$$

$$\overleftarrow{\bar{B}}_{k,m+1} = [0 \quad 0 \quad \cdots \quad 0 \quad B_{m+1}] ,$$

substitute the decompositions into $\overleftarrow{\mathcal{H}}^{1h}_{k,m}$, provide the transformations, and arrive at

$$\begin{aligned}
\overleftarrow{\mathcal{H}}^{1h}_{k,m} &= [\ \mathcal{X}_k^{m+1} \chi_k \overleftarrow{H}^T_{k,m+1} \qquad \mathcal{X}_k^{m+1} \chi_k \mathcal{X}_k^{m+1^T} H^T_m\ ]\\
&\quad + [\ \Xi_1 \quad \Xi_2\ ] ,
\end{aligned}$$

where $\Xi_1 = F^{-1}_{m+1}(\overleftarrow{\bar{D}}_{k,m+1} + \overleftarrow{\bar{B}}_{k,m+1}) \mathcal{Q}_{k,m+1} \overleftarrow{D}^T_{k,m+1} \overleftarrow{H}^T_{k,m+1}$ and

$$\begin{aligned}
\Xi_2 &= F^{-1}_{m+1}(\overleftarrow{\bar{D}}_{k,m+1} + \overleftarrow{\bar{B}}_{k,m+1}) \mathcal{Q}_{k,m+1} \overleftarrow{\bar{D}}^T_{k,m+1}\\
&\quad \times (\overleftarrow{\bar{D}}_{k,m+1} + \overleftarrow{\bar{B}}_{k,m+1})^T F^{-T}_{m+1} H^T_m .
\end{aligned}$$

Observe that the last components in $\overleftarrow{\bar{D}}_{k,m+1}$ and $\overleftarrow{D}^T_{k,m+1}$ are identically zero, $\overleftarrow{\bar{B}}_{k,m+1} \mathcal{Q}_{k,m+1} \overleftarrow{D}^T_{k,m+1} = 0$ and $\overleftarrow{\bar{B}}_{k,m+1} \mathcal{Q}_{k,m+1} \overleftarrow{\bar{D}}^T_{k,m+1} = 0$, refer to a decomposition

$$\begin{aligned}
\overleftarrow{\bar{D}}_{k,m} \mathcal{Q}_{k,m} \overleftarrow{\bar{D}}_{k,m} &= F^{-1}_{m+1} \overleftarrow{\bar{D}}_{k,m+1} \mathcal{Q}_{k,m+1} \overleftarrow{\bar{D}}^T_{k,m+1} F^{-T}_{m+1}\\
&\quad + F^{-1}_{m+1} B_{m+1} Q_{m+1} B^T_{m+1} F^{-T}_{m+1} ,
\end{aligned} \tag{5.48}$$

and represent $\overleftarrow{\mathcal{H}}^{1h}_{k,m}$ recursively as

$$\overleftarrow{\mathcal{H}}^{1h}_{k,m} = [\ F^{-1}_{m+1} \overleftarrow{\mathcal{H}}^{1h}_{k,m+1} \qquad \overleftarrow{M}_m H^T_m\ ] , \tag{5.49}$$

where

$$\overleftarrow{M}_m = \mathcal{X}_k^{m+1} \chi_k \mathcal{X}_k^{m+1^T} + \overleftarrow{\bar{D}}_{k,m} \mathcal{Q}_{k,m} \overleftarrow{\bar{D}}_{k,m} . \tag{5.50}$$

To find a recursion for $\overleftarrow{\mathcal{H}}^{2h}_{k,m}$, consider

$$\overleftarrow{\mathcal{H}}^{2h}_{k,m} = (\overleftarrow{H}_{k,m} \chi_k \overleftarrow{H}^T_{k,m} + \overleftarrow{G}_{k,m} \mathcal{Q}_{k,m} \overleftarrow{G}^T_{k,m} + \mathcal{R}_{k,m})^{-1} , \tag{5.51}$$

refer to the above-provided decompositions, and represent the sum of the first two matrix products as

$$Z_\chi + Z_1 = \begin{bmatrix} \Pi_{k,m+1} & (H_m F_{m+1}^{-1} \overleftarrow{\mathcal{H}}^{1h}_{k,m+1})^T \\ H_m F_{m+1}^{-1} \overleftarrow{\mathcal{H}}^{1h}_{k,m+1} & H_m \bar{\overleftarrow{M}}_m H_m^T \end{bmatrix} , \tag{5.52}$$

where $\Pi_{k,m+1} = \bar{\overleftarrow{H}}_{k,m+1} \mathcal{X}_k \bar{\overleftarrow{H}}^T_{k,m+1} + \bar{\overleftarrow{G}}_{k,m+1} \mathcal{Q}_{k,m+1} \bar{\overleftarrow{G}}^T_{k,m+1}$. From (5.51) find

$$\Pi_{k,m+1} = (\overleftarrow{\mathcal{H}}^{2h}_{k,m+1})^{-1} - \mathcal{R}_{k,m+1} ,$$

introduce $\tilde{\overleftarrow{\mathcal{H}}}^{2h}_{k,m+1} = H_m F_{m+1}^{-1} \overleftarrow{\mathcal{H}}^{1h}_{k,m+1}$, and rewrite (5.52) as

$$Z_\chi + Z_1 = \begin{bmatrix} (\overleftarrow{\mathcal{H}}^{2h}_{k,m+1})^{-1} - \mathcal{R}_{k,m+1} & (\tilde{\overleftarrow{\mathcal{H}}}^{2h}_{k,m+1})^T \\ \tilde{\overleftarrow{\mathcal{H}}}^{2h}_{k,m+1} & H_m \bar{\overleftarrow{M}}_m H_m^T \end{bmatrix} . \tag{5.53}$$

Next, refer to $\mathcal{R}_{k,m} = \mathrm{diag}(\ \mathcal{R}_{k,m+1} \quad R_m\ )$, combine with (5.53), and obtain

$$\overleftarrow{\mathcal{H}}^{2h^{-1}}_{k,m} = \begin{bmatrix} \overleftarrow{\mathcal{H}}^{2h^{-1}}_{k,m+1} & \tilde{\overleftarrow{\mathcal{H}}}^{2h^T}_{k,m+1} \\ \tilde{\overleftarrow{\mathcal{H}}}^{2h}_{k,m+1} & H_m \bar{\overleftarrow{M}}_m H_m^T + R_m \end{bmatrix} = \bar{Z}_{k,m} + \tilde{Z}_{k,m} , \tag{5.54}$$

where the block matrices are given by

$$\bar{Z}_{k,m} = \mathrm{diag}(\overleftarrow{\mathcal{H}}^{2h^{-1}}_{k,m+1} \quad R_m) ,$$

$$\tilde{Z}_{k,m} = \begin{bmatrix} 0 & \tilde{\overleftarrow{\mathcal{H}}}^{2h^T}_{k,m+1} \\ \tilde{\overleftarrow{\mathcal{H}}}^{2h}_{k,m+1} & H_m \bar{\overleftarrow{M}}_m H_m^T \end{bmatrix} .$$

Now, decompose $\overleftarrow{\mathcal{H}}^{2h}_{k,m} = (\bar{Z}_{k,m} + \tilde{Z}_{k,m})^{-1}$ as

$$\begin{aligned} \overleftarrow{\mathcal{H}}^{2h}_{k,m} &= \bar{Z}^{-1}_{k,m}(I + \tilde{Z}_{k,m}\bar{Z}^{-1}_{k,m})^{-1} \\ &= \bar{Z}^{-1}_{k,m}\hat{Z}^{-1}_{k,m} , \end{aligned} \tag{5.55}$$

where $\hat{Z}_{k,m} = I + \tilde{Z}_{k,m}\bar{Z}^{-1}_{k,m}$. Using the Schur complement, represent $\hat{Z}^{-1}_{k,m}$ as

$$\hat{Z}^{-1}_{k,m} = \begin{bmatrix} \mathcal{Z}_{11} & \mathcal{Z}_{12} \\ \mathcal{Z}_{21} & \mathcal{Z}_{22} \end{bmatrix} , \tag{5.56}$$

where the components are given by

$$\mathcal{Z}_{11} = I + \tilde{\overleftarrow{\mathcal{H}}}^{2h^T}_{k,m+1} R_m^{-1} \Omega_m^{-1} \tilde{\overleftarrow{\mathcal{H}}}^{2h}_{k,m+1} \overleftarrow{\mathcal{H}}^{2h}_{k,m+1} ,$$

$$\mathcal{Z}_{12} = -\tilde{\overleftarrow{\mathcal{H}}}^{2h^T}_{k,m+1} R_m^{-1} \Omega_m^{-1} ,$$

$$\mathcal{Z}_{21} = -\Omega_m^{-1} \tilde{\overleftarrow{\mathcal{H}}}^{2h}_{k,m+1} \overleftarrow{\mathcal{H}}^{2h}_{k,m+1} ,$$

$$\mathcal{Z}_{22} = \Omega_m^{-1} ,$$

and the following matrices are used,

$$\Omega_m = I + H_m \Lambda_m^- H_m^T R_m^{-1} , \tag{5.57}$$

$$\Lambda_m^- = \bar{\overleftarrow{M}}_m - F_{m+1}^{-1} \overleftarrow{\mathcal{H}}^{h}_{k,m+1} \overleftarrow{\mathcal{H}}^{1h^T}_{k,m+1} F_m^{-T} . \tag{5.58}$$

Provided (5.56) and $\bar{Z}_{k,m}^{-1}$, the recursive form for $\overleftarrow{\mathcal{H}}^{2h}_{k,m}$ is given by (5.55). To find a recursive form for $\overleftarrow{\mathcal{H}}_{k,m} = \overleftarrow{\mathcal{H}}^{1h}_{k,m}\overleftarrow{\mathcal{H}}^{2h}_{k,m}$, combine (5.49) with (5.55), and transform this product to

$$\overleftarrow{\mathcal{H}}_{k,m} = [\ \Theta_{k,m+1} \quad \Theta_m\ ]\,, \tag{5.59}$$

where matrices are specified as $\Theta_{k,m+1} = F_{m+1}^{-1}\overleftarrow{\mathcal{H}}_{k,m+1}Z_{11} + \overleftarrow{M}_m H_m^T R_m^{-1} Z_{12}$ and $\Theta_m = F_{m+1}^{-1}\overleftarrow{\mathcal{H}}_{k,m+1} Z_{12} + \overleftarrow{M}_m H_m^T R_m^{-1} Z_{22}$. Because Θ_m applied to the last mth component of $\overleftarrow{Y}_{k,m}$ plays the role of the bias correction gain, assign $\Theta_m = K_m$ and come up with

$$\begin{aligned} K_m &= (\overleftarrow{M}_m - F_{m+1}^{-1}\overleftarrow{\mathcal{H}}_{k,m+1}\overleftarrow{\mathcal{H}}^{1h^T}_{k,m+1}F_{m+1}^{-T})H_m^T R_m^{-1}\Omega_m^{-1} \\ &= \Lambda_m^- H_m^T R_m^{-1}(I + H_m\Lambda_m^- H_m^T R_m^{-1})^{-1} \\ &= \Lambda_m^- H_m^T (H_m \Lambda_m^- H_m^T + R_m)^{-1}\,. \end{aligned} \tag{5.60}$$

Represent matrix $\Theta_{k,m+1}$ as

$$\Theta_{k,m+1} = (I - K_m H_m)F_{m+1}^{-1}\overleftarrow{\mathcal{H}}_{k,m+1}$$

and arrive at the recursion for $\overleftarrow{\mathcal{H}}_{k,m}$,

$$\overleftarrow{\mathcal{H}}_{k,m} = \left[(I - K_m H_m)F_{m+1}^{-1}\overleftarrow{\mathcal{H}}_{k,m+1} \quad K_m\right]. \tag{5.61}$$

Now set $u_k = 0$ and use (5.61). This gives a recursion for the homogenous estimate $\tilde{x}_m^{\mathrm{h}}$,

$$\begin{aligned} \tilde{x}_m^{\mathrm{h}} &= \overleftarrow{\mathcal{H}}_{k,m}\overleftarrow{Y}_{k,m} \\ &= \left[(I - K_m H_m)F_{m+1}^{-1}\overleftarrow{\mathcal{H}}_{k,m+1} \quad K_m\right]\begin{bmatrix}\overleftarrow{Y}_{k,m+1} \\ y_m\end{bmatrix} \\ &= F_{m+1}^{-1}\tilde{x}_{m+1}^{\mathrm{h}} + K_m(y_m - H_m F_{m+1}^{-1}\tilde{x}_{m+1}^{\mathrm{h}})\,, \end{aligned} \tag{5.62}$$

where the bias correction gain K_m is given by (5.60).

A recursion for Λ_m^- required by K_m can be found if we substitute $\overleftarrow{M}_m$ (5.50) into (5.58), refer to (5.60), and transform Λ_m^- to

$$\begin{aligned} \Lambda_m^- &= F_m^{-1}(\overleftarrow{M}_{m+1} - \overleftarrow{\mathcal{H}}^{h}_{k,m+1}\overleftarrow{\mathcal{H}}^{1h^T}_{k,m+1} + B_{m+1}Q_{m+1}B_{m+1}^T)F_{m+1}^{-T}\,, \\ &= F_m^{-1}[(I - K_{m+1}H_{m+1})\Lambda_{m+1}^- + B_{m+1}Q_{m+1}B_{m+1}^T]F_{m+1}^{-T}, \end{aligned}$$

which can be shown to be the DARE. Then introduce

$$\Lambda_m = (I - K_m H_m)\Lambda_m^- \tag{5.63}$$

and rewrite Λ_m^- as

$$\Lambda_m^- = F_m^{-1}(\Lambda_{m+1} + B_{m+1}Q_{m+1}B_{m+1}^T)F_{m+1}^{-T}\,. \tag{5.64}$$

Now consider the forced estimate $\tilde{x}_m^{\mathrm{f}}$ in (5.29), decompose matrices as

$$\begin{aligned} \overleftarrow{S}_{k,m} &= F_{m+1}^{-1}[\ \overleftarrow{S}_{k,m+1} + \overleftarrow{\bar{E}}_{k,m+1} \quad 0\ ]\,, \\ \overleftarrow{L}_{k,m} &= \begin{bmatrix} \overleftarrow{\bar{H}}_{k,m+1}\overleftarrow{S}_{k,m+1} & 0 \\ H_m F_{m+1}^{-1}(\overleftarrow{S}_{k,m+1} + \overleftarrow{\bar{E}}_{k,m+1}) & 0 \end{bmatrix}, \\ \overleftarrow{\bar{E}}_{k,m+1} &= [0 \ \ 0 \ \ \cdots \ \ 0 \ \ E_{m+1}]\,, \end{aligned}$$

provide some transformations, and arrive at the recursion

$$\tilde{x}^{\mathrm{f}}_m = (I - K_m H_m) F^{-1}_{m+1} (\tilde{x}^{\mathrm{f}}_{m+1} - E_{m+1} u_{m+1}) \,. \tag{5.65}$$

Finally, combine (5.62) with (5.65) and end up with the recursive form for the backward OFIR estimate,

$$\tilde{x}_m = \tilde{x}^-_m + K_m (y_m - H_m \tilde{x}^-_m) \,, \tag{5.66}$$

where the prior estimate is given by

$$\tilde{x}^-_m = F^{-1}_{m+1} (\tilde{x}_{m+1} - E_{m+1} u_{m+1}) \,. \tag{5.67}$$

Rename $\Lambda^-_m = P^-_m$ and $\Lambda_m = P_m$, substitute $m+1 = k$ and $m = k-1$, arrive at the algorithm (5.41)–(5.46), and complete the proof. □

All that follows from this proof is that the backward OFIR filter recursive forms (5.41)–(5.46) are exactly the backward KF recursions (Algorithm 3). The estimate (5.45) is updated iteratively on $[m, k]$ back-in-time from $k-1$ to m, and the error covariance (5.40) is computed iteratively using (5.41) and (5.46). More detail about backward OFIR filtering can be found in [230], and the q-lag BFBM OFIR smoother involving backward filtering will be considered next.

5.5 Backward Filter/Backward Model g-lag OFIR Smoother

Smoothing is commonly thought of as an improvement of a filtering estimate $\hat{x}_k$ at past with lag q as $\tilde{x}_{k-q}$ in a back-in-time procedure. In this regard, the BFBM structure seems to be the most straightforward, although not unique. In contrast to the forward-in-time extended linear model (5.8) and (5.9), the initial state in the backward-in-time extended model (k-to-m) is an available estimate $\hat{x}_k$, and the final state is x_m. The BFBM estimation strategy is illustrated in Fig. 5.4 with two time-reversal procedures implying processing the data backward over $[m, k]$ and producing the estimate at $k - q = m + g$. It should be noted that the BFBM scheme is not equivalent to the FFFM scheme in view of the predictive state equation (5.22).

5.5.1 Batch Smoothing Estimate

The g-lag BFBM OFIR smoothing estimate can be defined as

$$\tilde{x}_{m+g} \triangleq \tilde{x}_{m+g|m} = \overleftarrow{H}^{\mathrm{h}(g)}_{k,m} \overleftarrow{Y}_{k,m} + \overleftarrow{H}^{\mathrm{f}(g)}_{k,m} \overleftarrow{U}_{k,m} \,, \tag{5.68}$$

where the discrete step g is related to the delay-lag q as $g = N - 1 - q$, $0 < \{q, g\} < N - 1$, and $\overleftarrow{H}^{(q)}_{k,m} \triangleq \overleftarrow{H}^{\mathrm{h}(q)}_{k,m}$ and $\overleftarrow{H}^{\mathrm{f}(q)}_{k,m}$ are the backward gains. The state model extracted at point $m+g$ from (5.23) as the $(N-g)$th raw vector is defined as follows:

$$x_{m+g} = \mathcal{F}^{m+g+1}_k x_k - \overleftarrow{\bar{S}}^{(g)}_{k,m} \overleftarrow{U}_{k,m} - \overleftarrow{\bar{D}}^{(g)}_{k,m} \overleftarrow{W}_{k,m} \,, \tag{5.69}$$

where $\overleftarrow{\bar{S}}^{(g)}_{k,m}$ is the $(N-g)$th raw vector in $\overleftarrow{S}_{k,m}$ given by (5.25) and so is $\overleftarrow{\bar{D}}^{(g)}_{k,m}$ in $\overleftarrow{D}_{k,m}$.

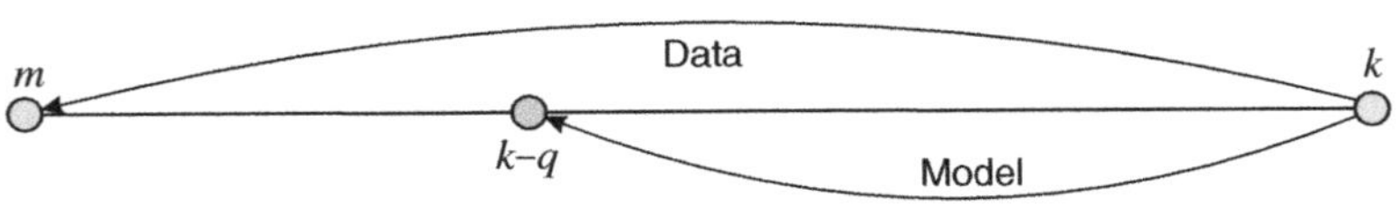

Figure 5.4 Backward filter/backward model g-lag OFIR smoothing strategy to produce an estimate at $k - q = m + g$.

The unbiasedness condition $\mathcal{E}\{\tilde{x}_{m+g}\} = \mathcal{E}\{x_{m+g}\}$ applied to (5.68) and (5.69) gives two unbiasedness constraints

$$\mathcal{X}_k^{m+g+1} = \overleftarrow{\mathcal{H}}_{k,m}^{(g)} \overleftarrow{H}_{k,m} , \tag{5.70}$$

$$\overleftarrow{\mathcal{H}}_{k,m}^{f(g)} = \overleftarrow{\mathcal{H}}_{k,m}^{(g)} \overleftarrow{L}_{k,m} - \overleftarrow{\bar{S}}_{k,m}^{(g)} \tag{5.71}$$

and the estimation error $\varepsilon_{m+g} = x_{m+g} - \tilde{x}_{m+g}$ is transformed to

$$\begin{aligned}
\varepsilon_{m+g} &= \mathcal{X}_k^{m+g+1} x_k - \overleftarrow{\bar{S}}_{k,m}^{(g)} \overleftarrow{U}_{k,m} - \overleftarrow{\bar{D}}_{k,m}^{(g)} \overleftarrow{W}_{k,m} \\
&\quad - \overleftarrow{\mathcal{H}}_{k,m}^{(g)} \overleftarrow{Y}_{k,m} - \overleftarrow{\mathcal{H}}_{k,m}^{f(g)} \overleftarrow{U}_{k,m} \\
&= (\mathcal{X}_k^{m+g+1} - \overleftarrow{\mathcal{H}}_{k,m}^{(g)} \overleftarrow{H}_{k,m}) x_k \\
&\quad - (\overleftarrow{\bar{S}}_{k,m}^{(g)} - \overleftarrow{\mathcal{H}}_{k,m}^{(g)} \overleftarrow{L}_{k,m} + \overleftarrow{\mathcal{H}}_{k,m}^{f(g)}) \overleftarrow{U}_{k,m} \\
&\quad - (\overleftarrow{\bar{D}}_{k,m}^{(g)} - \overleftarrow{\mathcal{H}}_{k,m}^{(g)} \overleftarrow{G}_{k,m}) \overleftarrow{W}_{k,m} - \overleftarrow{\mathcal{H}}_{k,m}^{(g)} \overleftarrow{V}_{k,m} .
\end{aligned}$$

The orthogonality condition $\mathcal{E}\{\varepsilon_{m+g} \overleftarrow{Y}_{k,m}^T\} = 0$ gives

$$\begin{aligned}
&\overleftarrow{\mathcal{B}}_{k,m}^{(g)} \chi_k \overleftarrow{H}_{k,m}^T + \overleftarrow{\mathcal{W}}_{k,m}^{(g)} \overleftarrow{\mathcal{Q}}_{k,m} \overleftarrow{G}_{k,m}^T - \overleftarrow{\mathcal{V}}_{k,m}^{(g)} \overleftarrow{\mathcal{R}}_{k,m} \\
&\quad = -(\overleftarrow{\bar{S}}_{k,m}^{(g)} - \overleftarrow{\mathcal{H}}_{k,m}^{h(g)} \overleftarrow{L}_{k,m} + \overleftarrow{\mathcal{H}}_{k,m}^{f(g)}) \overleftarrow{\Psi}_{k,m} \overleftarrow{L}_{k,m}^T ,
\end{aligned} \tag{5.72}$$

where $\overleftarrow{\mathcal{Q}}_{k,m} = \mathrm{diag}[\ Q_k\ Q_{k-1}\ \dots\ Q_m\ ]$, $\overleftarrow{\mathcal{R}}_{k,m} = \mathrm{diag}[\ R_k\ R_{k-1}\ \dots\ R_m\ ]$, $\overleftarrow{\Psi}_{k,m} = \mathrm{diag}[\ u_k\ u_{k-1}\ \dots\ u_m\ ]$, and the g-variant error residual matrices are

$$\overleftarrow{\mathcal{B}}_{k,m}^{(g)} = \mathcal{X}_k^{m+g+1} - \overleftarrow{\mathcal{H}}_{k,m}^{(g)} \overleftarrow{H}_{k,m} , \tag{5.73}$$

$$\overleftarrow{\mathcal{W}}_{k,m}^{(g)} = \overleftarrow{\bar{D}}_{k,m}^{(g)} - \overleftarrow{\mathcal{H}}_{k,m}^{(g)} \overleftarrow{G}_{k,m} , \tag{5.74}$$

$$\overleftarrow{\mathcal{V}}_{k,m}^{(g)} = \overleftarrow{\mathcal{H}}_{k,m}^{(g)} . \tag{5.75}$$

For zero input $\overleftarrow{U}_{k,m} = 0$, relation (5.9) gives the homogenous gain

$$\begin{aligned}
\overleftarrow{\mathcal{H}}_{k,m}^{(g)} &= (\mathcal{X}_k^{m+g+1} \chi_k \overleftarrow{H}_{k,m}^T + \overleftarrow{\bar{D}}_{k,m}^{(g)} \overleftarrow{\mathcal{Q}}_{k,m} \overleftarrow{G}_{k,m}^T) \\
&\quad \times (\overleftarrow{H}_{k,m} \chi_k \overleftarrow{H}_{k,m}^T + \overleftarrow{G}_{k,m} \overleftarrow{\mathcal{Q}}_{k,m} \overleftarrow{G}_{k,m}^T + \overleftarrow{\mathcal{R}}_{k,m})^{-1} \\
&= [\mathcal{X}_k^{m+g+1} (\overleftarrow{H}_{k,m}^T \overleftarrow{H}_{k,m})^{-1} \overleftarrow{H}_{k,m}^T \overleftarrow{H}_{k,m} \chi_k \overleftarrow{H}_{k,m}^T + \overleftarrow{\mathcal{Z}}_1^{(g)}] \\
&\quad \times (\overleftarrow{\mathcal{Z}}_\chi + \overleftarrow{\mathcal{Z}}_2 + \overleftarrow{\mathcal{R}}_{k,m})^{-1} \\
&= (\hat{\overleftarrow{\mathcal{H}}}_{k,m}^{h(g)} \overleftarrow{\mathcal{Z}}_\chi + \overleftarrow{\mathcal{Z}}_1^{(g)})(\overleftarrow{\mathcal{Z}}_\chi + \overleftarrow{\mathcal{Z}}_2 + \overleftarrow{\mathcal{R}}_{k,m})^{-1} ,
\end{aligned} \tag{5.76}$$

where $\overleftarrow{\mathcal{Z}}_\chi$, $\overleftarrow{\mathcal{Z}}_1^{(g)}$, and $\overleftarrow{\mathcal{Z}}_2$ follow directly from (5.76) and the backward g-lag UFIR smoother gain $\hat{\overleftarrow{\mathcal{H}}}_{k,m}^{(g)}$ appears from the constraint (5.70) as

$$\hat{\overleftarrow{\mathcal{H}}}_{k,m}^{(g)} = \mathcal{X}_k^{m+g+1} (\overleftarrow{H}_{k,m}^T \overleftarrow{H}_{k,m})^{-1} \overleftarrow{H}_{k,m}^T \tag{5.77a}$$

$$= \mathcal{F}_{m+g}^{m+1} (\mathcal{F}_k^{m+1})^{-1} (\overleftarrow{H}_{k,m}^T \overleftarrow{H}_{k,m})^{-1} \overleftarrow{H}_{k,m}^T \tag{5.77b}$$

$$= \mathcal{F}_{m+g}^{m+1} \hat{\overleftarrow{\mathcal{H}}}_{k,m} . \tag{5.77c}$$

Note that the backward UFIR filter homogeneous gain $\overleftarrow{\mathcal{H}}_{k,m}$ used in (5.77c) is specified by (5.77a) for $g = 0$ as

$$\overleftarrow{\mathcal{H}}_{k,m} = \mathcal{X}_k^{m+1}(\overleftarrow{\bar{H}}_{k,m}^T \overleftarrow{\bar{H}}_{k,m})^{-1}\overleftarrow{\bar{H}}_{k,m}^T \,. \tag{5.78}$$

The q-lag batch BFBM OFIR smoother can finally be represented with

$$\tilde{x}_{m+g} = \overleftarrow{\mathcal{H}}_{k,m}^{(g)} \overleftarrow{Y}_{k,m} + (\overleftarrow{\bar{S}}_{k,m}^{(g)} + \overleftarrow{\mathcal{H}}_{k,m}^{(g)} \overleftarrow{\bar{L}}_{k,m})\overleftarrow{U}_{k,m} \tag{5.79}$$

to produce an estimate at $k - q = m + g$ for given $\hat{x}_k$.

5.5.2 Error Covariance

The error covariance P_{m+g} for the q-lag BFBM OFIR smoother can be obtained by reasoning along the same lines as for the OFIR filter,

$$\begin{aligned} P_{m+g} &= \mathcal{E}\{\varepsilon_{m+g}\varepsilon_{m+g}^T\} \\ &= \overleftarrow{\mathcal{B}}_{k,m}^{(g)} \mathcal{X}_k \overleftarrow{\mathcal{B}}_{k,m}^{(g)T} + \overleftarrow{\mathcal{W}}_{k,m}^{(g)} \overleftarrow{\mathcal{Q}}_{k,m} \overleftarrow{\mathcal{W}}_{k,m}^{(g)T} \\ &\quad + \overleftarrow{\mathcal{V}}_{k,m}^{(g)} \overleftarrow{\mathcal{R}}_{k,m} \overleftarrow{\mathcal{V}}_{k,m}^{(g)T} \,, \end{aligned} \tag{5.80}$$

where the error residual matrices are given by (5.73)–(5.75). The derivation of (5.80) is postponed to "Problems." It follows from the previous that an increase in the variable g leads to an increase in the components of matrices $\mathcal{F}_{m+g}^{m+1}$ and $\overleftarrow{\bar{D}}_{k,m}^{(g)}$, which, in turn, causes an increase in the gain $\mathcal{H}_{k,m}^{(g)}$. Consequently, the components of the error residual matrix (5.75) increase with increasing g or, equivalently, decrease with increasing q. This means more efficient reduction of measurement noise.

As in the case of FFFM smoothing, the effect of g on matrices (5.73) and (5.74) is less pronounced. It is also worth noting that, strictly speaking, the errors produced by the FFFM and BFBM OFIR smoothers at $k - q$ are not equal. This inference follows from the fact that the back-in-time model (5.22) used to obtain the BFBM smoothing estimate is predictive, but the model (5.1) used to derive the FFFM smoother is not. In this respect, FFFM smoothing can be considered more accurate.

Example 5.3 ***BFBM OFIR smoothing.*** Consider the system presented in Example 5.2 and solve the smoothing problem under the same conditions using the BFBM OFIR smoother. As benchmarks, use the UFIR and RTS smoothers. The theoretical (solid) and numerical (dashed-dotted) RMSEs averaged over 10^4 MC runs are shown in Fig. 5.5 as functions q.

The BFBM OFIR smoother works in the opposite direction of the FFFM OFIR smoother (Fig. 5.3), and we can see that when the noise is white, the RTS smoother performs better (Fig. 5.5a). However, a strong CMN changes the picture in such a way that, as with FFFM, the BFBM OFIR smoother performs better, while the RTS gives worse estimates (Fig. 5.5b). □

5.6 Forward Filter/Backward Model q-Lag OFIR Smoother

The q-lag FFBM OFIR smoothing solution can be found by following the strategy illustrated in Fig. 5.6.

An appropriate algorithm can be obtained using the p-shift forward OFIR filter in which the MSE is minimized at $k - q$ with $q = -p$ for the back-in-time projected state model.

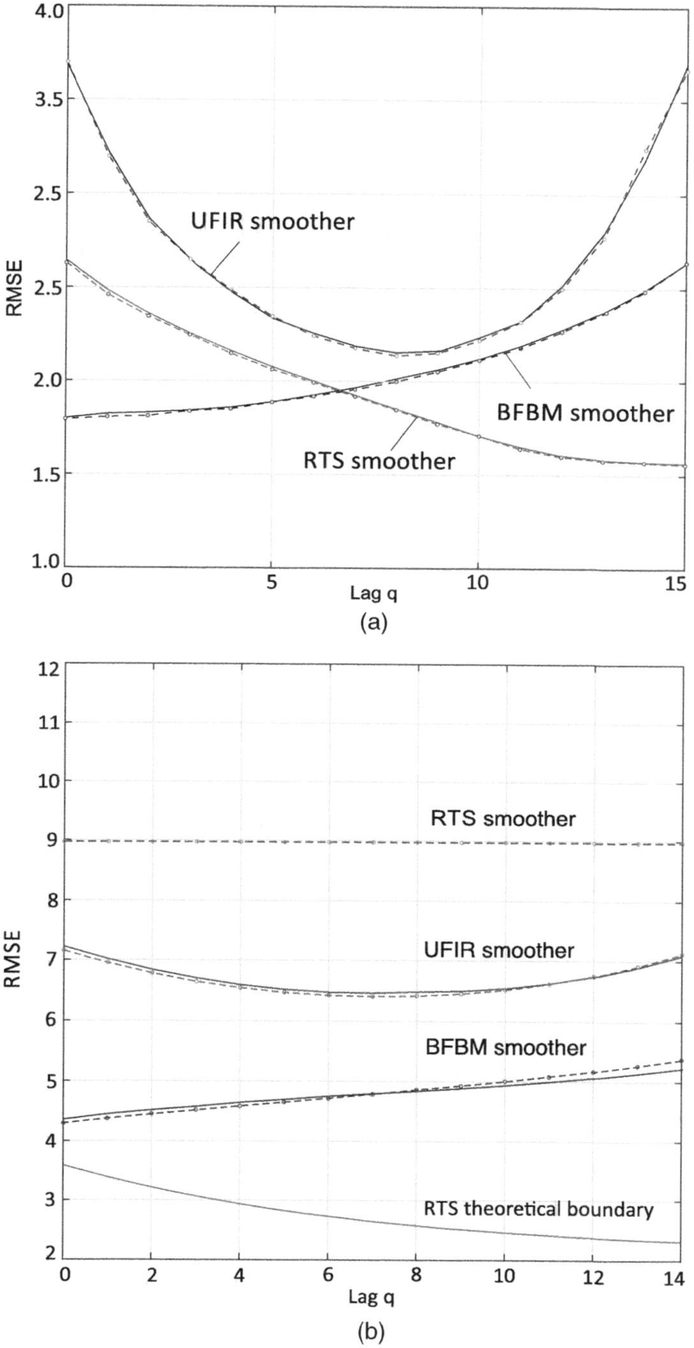

Figure 5.5 RMSEs produced by the BFBM OFIR, UFIR, and RTS smoothers: (a) white measurement noise and (b) CMN with $\varphi = 0.95$. Solid lines are theoretical, and dashed-dotted lines represent simulated RMSEs averaged over 10^4 MC runs.

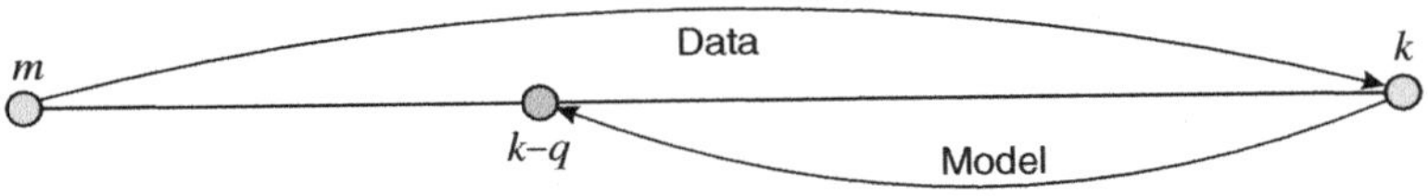

Figure 5.6 Forward filter/backward model q-lag OFIR smoothing strategy.

5.6.1 Batch Smoothing Estimate

To derive a q-lag FFBM OFIR smoother, we consider the backward model (5.69), in which we replace $m+g$ with $k-q$ and write

$$x_{k-q} = \mathcal{X}_k^{k-q+1} x_k - \overleftarrow{S}^{(g)}_{k,m} \overleftarrow{U}_{k,m} - \overleftarrow{\bar{D}}^{(g)}_{k,m} \overleftarrow{W}_{k,m} \,.$$

We then use the forward estimate (5.5),

$$\tilde{x}_{k-q} \triangleq \tilde{x}_{k-q|k} = \mathcal{H}^{(q)}_{m,k} Y_{m,k} + \mathcal{H}^{f(q)}_{m,k} U_{m,k} \,,$$

and apply the unbiasedness conditions $\mathcal{E}\{\tilde{x}_{k-q}\} = \mathcal{E}\{x_{k-q}\}$. This gives two unbiasedness constraints

$$\mathcal{F}^{m+1}_{k-q} = \mathcal{H}^{(q)}_{m,k} H_{m,k} \,, \tag{5.81}$$

$$\mathcal{H}^{f(q)}_{m,k} = \overleftarrow{\tilde{S}}^{(q)}_{m,k} - \mathcal{H}^{(q)}_{m,k} L_{m,k} \,, \tag{5.82}$$

where matrix $\overleftarrow{\tilde{S}}^{(q)}_{m,k} = \mathcal{X}_k^{k-q+1} \bar{S}_{m,k} - \overleftarrow{S}^{(g)}_{m,k}$ is composed as

$$\overleftarrow{\tilde{S}}^{(q)}_{m,k} = \Big[\, \mathcal{F}^{m+1}_{k-q} E_m \quad \mathcal{F}^{m+2}_{k-q} E_{m+1} \quad \dots \quad E_{k-q}$$
$$-F^{-1}_{k-q+1} E_{k-q+1} \quad \dots \quad -\mathcal{X}_k^{k-q+1} E_k \Big]$$

and matrix $\overleftarrow{\tilde{D}}^{(q)}_{m,k}$ can be obtained by substituting E_k with B_k in $\overleftarrow{\tilde{S}}^{(q)}_{m,k}$.

Now the estimation error $\varepsilon_{k-q} = x_{k-q} - \tilde{x}_{k-q}$ can be transformed to

$$\begin{aligned}
\varepsilon_{k-q} &= (\mathcal{F}^{m+1}_{k-q} - \mathcal{H}^{(q)}_{m,k} H_{m,k}) x_m \\
&\quad + (\overleftarrow{\tilde{S}}^{(q)}_{m,k} - \mathcal{H}^{(q)}_{m,k} L_{m,k} - \mathcal{H}^{f(q)}_{m,k}) U_{m,k} \\
&\quad + (\overleftarrow{\tilde{D}}^{(q)}_{m,k} - \mathcal{H}^{(q)}_{m,k} G_{m,k}) W_{m,k} - \mathcal{H}^{(q)}_{m,k} V_{m,k}
\end{aligned}$$

and then the orthogonality condition $\mathcal{E}\{\varepsilon_{k-q} \overleftarrow{Y}^{T}_{m,k}\} = 0$ yields

$$\begin{aligned}
&\mathcal{B}^{(q)}_{m,k} \mathcal{X}_m H^T_{m,k} + \mathcal{W}^{(q)}_{m,k} \mathcal{Q}_{m,k} G^T_{m,k} - \mathcal{V}^{(q)}_{m,k} \mathcal{R}_{m,k} \\
&\quad = (\mathcal{H}^{h(q)}_{m,k} L_{m,k} - \overleftarrow{\tilde{S}}^{(q)}_{m,k} + \mathcal{H}^{f(q)}_{m,k}) \Psi_{m,k} L^T_{m,k} \,,
\end{aligned} \tag{5.83}$$

where matrix $\mathcal{B}^{(q)}_{m,k}$ is defined by (5.10), $\mathcal{V}^{(q)}_{m,k}$ by (5.12), and $\mathcal{W}^{(q)}_{m,k}$ is

$$\mathcal{W}^{(q)}_{m,k} = \overleftarrow{\tilde{D}}^{(q)}_{m,k} - \mathcal{H}^{(q)}_{m,k} G_{m,k} \,. \tag{5.84}$$

At this point, the fundamental gain can be found as

$$\begin{aligned}
\mathcal{H}^{(q)}_{m,k} &= (\mathcal{F}^{m+1}_{k-q} \mathcal{X}_m H^T_{m,k} + \overleftarrow{\tilde{D}}^{(q)}_{m,k} \mathcal{Q}_{m,k} G^T_{m,k}) \\
&\quad \times (H_{m,k} \mathcal{X}_m H^T_{m,k} + G_{m,k} \mathcal{Q}_{m,k} G^T_{m,k} + \mathcal{R}_{m,k})^{-1}
\end{aligned} \tag{5.85}$$

and substituted into the forced gain (5.82). The q-lag FFBM OFIR smoothing estimate finally becomes

$$\tilde{x}_{k-q} = \mathcal{H}^{(q)}_{m,k} Y_{m,k} + (\overleftarrow{S}^{(q)}_{m,k} - \mathcal{H}^{(q)}_{m,k} L_{m,k}) U_{m,k} , \tag{5.86}$$

and we notice that other forms of $\mathcal{H}^{(q)}_{m,k}$ can be adopted from the FFFM OFIR filter, if we substitute $\bar{D}^{(q)}_{m,k}$ with $\overleftarrow{D}^{(q)}_{m,k}$ and $\bar{S}^{(q)}_{m,k}$ with $\overleftarrow{S}^{(q)}_{m,k}$.

5.6.2 Error Covariance

In the same way, as for the OFIR smoothers discussed earlier, we obtain the error covariance $P_{k-q} = \mathcal{E}\{\varepsilon_{k-q}\varepsilon^T_{k-q}\}$ of the FFBM OFIR smoother as

$$P_{k-q} = \mathcal{B}^{(q)}_{m,k} \mathcal{X}_m \mathcal{B}^{(q)^T}_{m,k} + \mathcal{W}^{(q)}_{m,k} \mathcal{Q}_{m,k} \mathcal{W}^{(q)^T}_{m,k} + \mathcal{V}^{(q)}_{m,k} \mathcal{R}_{m,k} \mathcal{V}^{(q)^T}_{m,k} , \tag{5.87}$$

where the error residual matrices are given by (5.10), (5.12), and (5.84). We also notice that since this smoother is derived using a back-in-time predictive state model, it has properties similar to the BFBM OFIR smoother.

Example 5.4 ***FFBM OFIR smoothing.*** Given a two-state polynomial model considered in Example 5.2, solve the smoothing problem using the FFBM OFIR smoother and the UFIR and RTS smoothers as benchmarks. The RMSEs produced by the smoothers are shown in Fig. 5.7 as functions of lag q. Since the difference between the FFBM and FFFM structures lies solely in the models used, the results shown in Fig. 5.7 are similar to those in Fig. 5.3. Note that the slight difference between the FFFM and FFBM outputs is due to the noise samples being processed in the forward and backward directions. □

5.7 Backward Filter/Forward Model q-Lag OFIR Smoother

Another solution, which belongs to the family of OFIR smoothers, appears by combining backward filtering with a forward model and minimizing MSE at $k-q$ in what can be said to be the q-lag BFFM OFIR smoother. The BFFM strategy is illustrated in Fig. 5.8 to show that it is the anti-counterpart of the FFBM strategy shown in Fig. 5.8.

Indeed, in contrast to the FFBM case, here the smoothing estimate appears at $k-q$ by using the backward extended state-space model and minimizing errors using the forward state model.

5.7.1 Batch Smoothing Estimate

Without going deep into the details of the derivation, which are similar to the previously derived smoothers, next we list the main steps leading to the batch q-lag BFFM OFIR smoother.

For the backward g-lag OFIR smoothing estimate (5.27) and forward state model (5.6), the unbiasedness condition $\mathcal{E}\{\tilde{x}_{m+g}\} = \mathcal{E}\{x_{m+g}\}$ gives two unbiasedness constraints

$$(\mathcal{F}^{k-q+1}_k)^{-1} = \overleftarrow{H}^{(g)}_{k,m} \overleftarrow{H}_{k,m} , \tag{5.88}$$

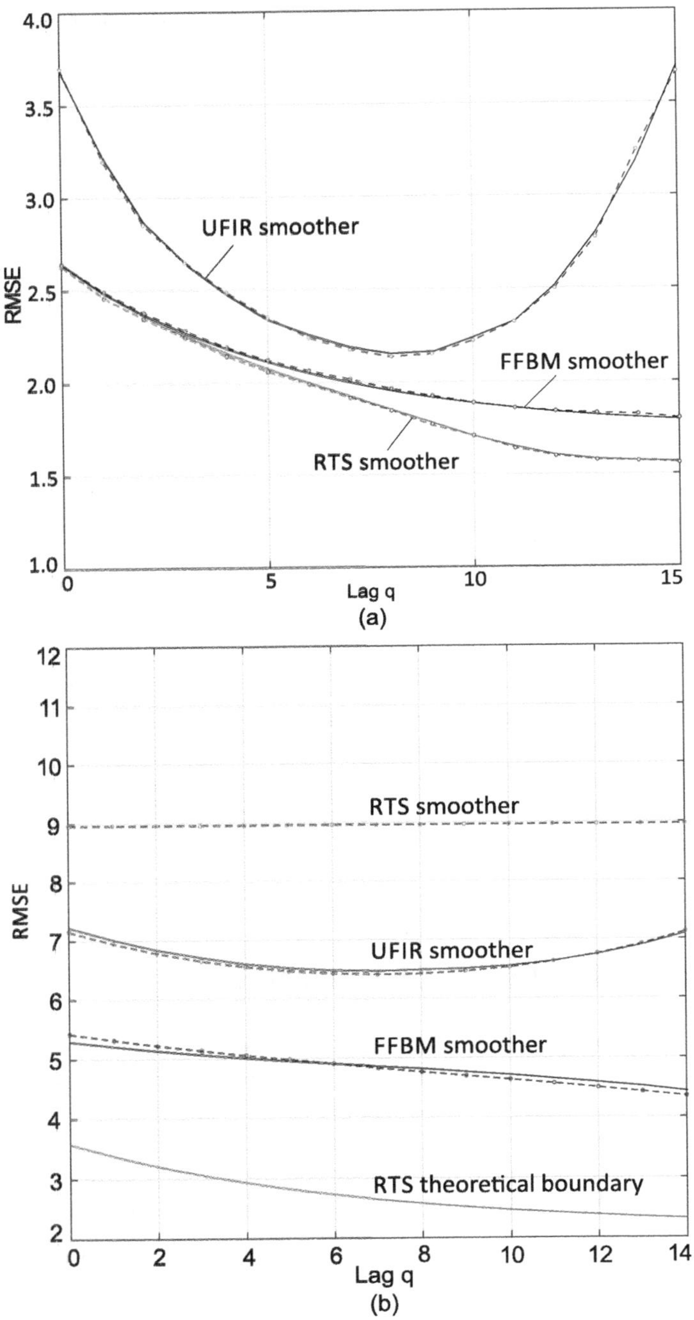

Figure 5.7 RMSE produced by the FFBM OFIR, UFIR, and RTS smoothers: (a) white measurement noise and (b) CMN with $\varphi = 0.95$. Solid lines are theoretical, and dashed-dotted lines represent simulated RMSEs averaged over 10^4 MC runs.

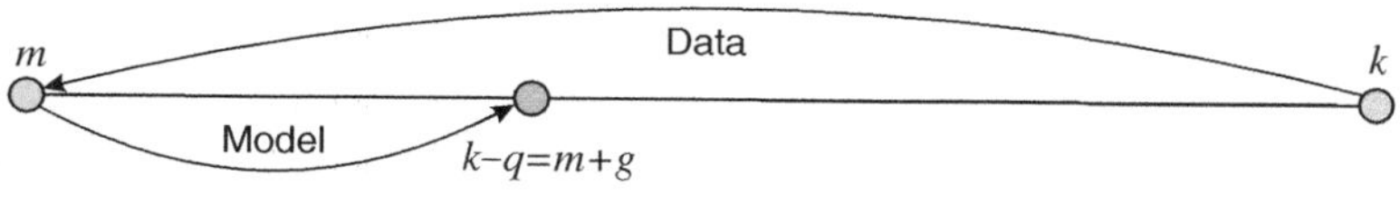

Figure 5.8 Backward filter/forward model q-lag OFIR smoothing strategy.

$$\overleftarrow{\mathcal{H}}^{\mathrm{f}(g)}_{k,m} = \bar{S}^{(g)}_{k,m} + \overleftarrow{\mathcal{H}}^{(g)}_{k,m}(\overleftarrow{L}_{k,m} - \overleftarrow{H}_{k,m}\overleftarrow{\bar{S}}_{k,m}) \tag{5.89}$$

and the estimation error $\varepsilon_{m+g} = x_{m+g} - \tilde{x}_{m+g}$ becomes

$$\begin{aligned}\varepsilon_{m+g} &= \overleftarrow{\mathcal{B}}^{(g)}_{k,m}x_m + \overleftarrow{\mathcal{W}}^{(g)}_{k,m}\overleftarrow{W}_{k,m} - \overleftarrow{\mathcal{V}}^{(g)}_{k,m}\overleftarrow{V}_{k,m} \\ &\quad + [\overleftarrow{\bar{S}}^{(g)}_{k,m} + \overleftarrow{H}^{(g)}_{k,m}(\overleftarrow{L}_{k,m} - \overleftarrow{H}_{k,m}\overleftarrow{\bar{S}}_{k,m}) - \overleftarrow{\mathcal{H}}^{\mathrm{f}(g)}_{k,m}]\overleftarrow{U}_{k,m} ,\end{aligned}$$

where the error residual matrices are given by

$$\overleftarrow{\mathcal{B}}^{(g)}_{k,m} = \mathcal{F}^{m+1}_{m+g} - \overleftarrow{H}^{(g)}_{k,m}\overleftarrow{H}_{k,m}\mathcal{F}^{m+1}_{k} , \tag{5.90}$$

$$\overleftarrow{\mathcal{W}}^{(g)}_{k,m} = \overleftarrow{\bar{D}}^{(g)}_{k,m} + \overleftarrow{\mathcal{H}}^{(g)}_{k,m}(\overleftarrow{G}_{k,m} - \overleftarrow{H}_{k,m}\overleftarrow{\bar{D}}_{k,m}) , \tag{5.91}$$

$$\overleftarrow{\mathcal{V}}^{(g)}_{k,m} = \overleftarrow{\mathcal{H}}^{(g)}_{k,m} . \tag{5.92}$$

The orthogonality condition $\mathcal{E}\{\varepsilon_{m+g}\overleftarrow{Y}^{T}_{k,m}\} = 0$ produces

$$\begin{aligned}&\overleftarrow{\mathcal{B}}^{(g)}_{k,m}\chi_k\overleftarrow{H}^{T}_{k,m} - \overleftarrow{\mathcal{W}}^{(g)}_{k,m}\overleftarrow{\mathcal{Q}}_{k,m}\overleftarrow{G}^{T}_{k,m} - \overleftarrow{\mathcal{V}}^{(g)}_{k,m}\overleftarrow{\mathcal{R}}_{k,m} \\ &= (\overleftarrow{\bar{S}}^{(g)}_{k,m} + \overleftarrow{\mathcal{H}}^{(g)}_{k,m}\overleftarrow{L}_{k,m} - \overleftarrow{\mathcal{H}}^{\mathrm{f}(g)}_{k,m})\overleftarrow{\Psi}_{k,m}\overleftarrow{L}^{T}_{k,m} ,\end{aligned} \tag{5.93}$$

where $\overleftarrow{\mathcal{Q}}_{k,m} = \mathrm{diag}[\ Q_k\ Q_{k-1}\ \dots\ Q_m\ ]$, $\overleftarrow{\mathcal{R}}_{k,m} = \mathrm{diag}[\ R_k\ R_{k-1}\ \dots\ R_m\ ]$, and $\overleftarrow{\Psi}_{k,m} = \mathrm{diag}[\ u_k\ u_{k-1}\ \dots\ u_m\ ]$. For zero input $\overleftarrow{U}_{k,m} = 0$, relation (5.93) gives the fundamental gain

$$\begin{aligned}\overleftarrow{\mathcal{H}}^{(g)}_{k,m} &= (\mathcal{X}^{m+g+1}_{k}\chi_k\overleftarrow{H}^{T}_{k,m} - \overleftarrow{\bar{D}}^{(g)}_{k,m}\overleftarrow{\mathcal{Q}}_{k,m}\overleftarrow{G}^{T}_{k,m}) \\ &\quad \times(\overleftarrow{H}_{k,m}\chi_k\overleftarrow{H}^{T}_{k,m} + \overleftarrow{G}_{k,m}\overleftarrow{\mathcal{Q}}_{k,m}\overleftarrow{G}^{T}_{k,m} + \overleftarrow{\mathcal{R}}_{k,m})^{-1} \\ &= [\mathcal{X}^{m+g+1}_{k}(\overleftarrow{H}^{T}_{k,m}\overleftarrow{H}_{k,m})^{-1}\overleftarrow{H}^{T}_{k,m}\overleftarrow{H}_{k,m}\chi_k\overleftarrow{H}^{T}_{k,m} - \overleftarrow{\mathcal{Z}}^{(g)}_{1}] \\ &\quad \times(\overleftarrow{\mathcal{Z}}_{\chi} + \overleftarrow{\mathcal{Z}}_{2} + \overleftarrow{\mathcal{R}}_{k,m})^{-1} \\ &= (\overleftarrow{\hat{\mathcal{H}}}^{(g)}_{k,m}\overleftarrow{\mathcal{Z}}_{\chi} - \overleftarrow{\mathcal{Z}}^{(g)}_{1})(\overleftarrow{\mathcal{Z}}_{\chi} + \overleftarrow{\mathcal{Z}}_{2} + \overleftarrow{\mathcal{R}}_{k,m})^{-1} ,\end{aligned} \tag{5.94}$$

where matrices $\overleftarrow{\mathcal{Z}}_{\chi}$, $\overleftarrow{\mathcal{Z}}^{(g)}_{1}$, and $\overleftarrow{\mathcal{Z}}_{2}$ are defined by comparing the equalities in (5.94). Note that the gain $\overleftarrow{\hat{\mathcal{H}}}^{(g)}_{k,m}$ for the backward g-lag UFIR smoother appears from the unbiasedness constraint (5.88) as

$$\overleftarrow{\hat{\mathcal{H}}}^{(g)}_{k,m} = \mathcal{X}^{m+g+1}_{k}(\overleftarrow{H}^{T}_{k,m}\overleftarrow{H}_{k,m})^{-1}\overleftarrow{H}^{T}_{k,m} \tag{5.95a}$$

$$= \mathcal{F}^{m+1}_{m+g}(\mathcal{F}^{m+1}_{k})^{-1}(\overleftarrow{H}^{T}_{k,m}\overleftarrow{H}_{k,m})^{-1}\overleftarrow{H}^{T}_{k,m} \tag{5.95b}$$

$$= \mathcal{F}^{m+1}_{m+g}\overleftarrow{\hat{\mathcal{H}}}_{k,m} \tag{5.95c}$$

and, if we set $g = 0$, becomes the backward UFIR filter gain

$$\overleftarrow{\hat{\mathcal{H}}}_{k,m} = \mathcal{X}_k^{m+1}(\overleftarrow{H}_{k,m}^T \overleftarrow{H}_{k,m})^{-1}\overleftarrow{H}_{k,m}^T \,. \tag{5.96}$$

The batch q-lag BFFM OFIR smoothing estimate finally becomes

$$\tilde{x}_{m+g} = \overleftarrow{\mathcal{H}}_{k,m}^{(g)}\overleftarrow{Y}_{k,m} + (\overleftarrow{S}_{k,m}^{(g)} + \overleftarrow{\mathcal{H}}_{k,m}^{(g)}\overleftarrow{L}_{k,m})\overleftarrow{U}_{k,m} \tag{5.97}$$

with the gain $\overleftarrow{\mathcal{H}}_{k,m}^{(g)}$ defined by (5.94).

5.7.2 Error Covariance

Provided the estimate (5.97), the error covariance $P_{m+g} = \mathcal{E}\{\varepsilon_{m+g}\varepsilon_{m+g}^T\}$ of the q-lag BFFM OFIR smoother can be written as

$$\begin{aligned} P_{m+g} &= \mathcal{E}\{\varepsilon_{m+g}\varepsilon_{m+g}^T\} \\ &= \overleftarrow{\mathcal{B}}_{k,m}^{(g)}\mathcal{X}_k\overleftarrow{\mathcal{B}}_{k,m}^{(g)T} + \overleftarrow{\mathcal{W}}_{k,m}^{(g)}\overleftarrow{\mathcal{Q}}_{k,m}\overleftarrow{\mathcal{W}}_{k,m}^{(g)T} \\ &\quad + \overleftarrow{\mathcal{V}}_{k,m}^{(g)}\overleftarrow{\mathcal{R}}_{k,m}\overleftarrow{\mathcal{V}}_{k,m}^{(g)T} \,, \end{aligned} \tag{5.98}$$

where the error residual matrices are specified by (5.90)–(5.92). Again, we notice that the measurement noise is more efficiently suppressed here, while bias errors and system noise undergo more complex corrections by smaller orders of magnitudes.

Example 5.5 ***BFFM OFIR smoothing.*** Given a two-state polynomial model (Example 5.2), solve the smoothing problem using the BFFM OFIR, UFIR, and RTS smoothers. The RMSEs produced by these smoothers are shown in Fig. 5.9. Note the similarity of the results shown in Fig. 5.9 and Fig. 5.5, because the difference between the BFBM and BFFM structures is only due to the models used. Note that the slight difference between the BFFM and BFBM outputs is insignificant as it only relates to opposite directions of noise samples processing. □

5.8 Two-Filter q-lag OFIR Smoother

A simple and quite straightforward smoothing technique is known as fixed-interval *two-filter smoothing* [83]. Unlike the FFFM, FFBM, BFFM, and BFBM OFIR schemes discussed earlier, the two-filter approach can be easily implemented recursively using forward and backward optimal filters. Originally developed using KF, as discussed in Chapter 3, the two-filter smoothing method can be extended in the same way to OFIR filtering.

The q-lag forward-backward (FB) OFIR smoothing strategy is illustrated in Fig. 5.10.

It deals with data on a fixed interval $[m, \dots, k-q, \dots, k]$ and implies that the forward OFIR filter operates on the horizon $[m, k-q]$ to obtain a forward filtering estimate $\hat{x}_{k-q}^{\mathrm{f}}$, and the backward OFIR filter operates on $[k-q, k]$ to obtain a backward filtering estimate $\hat{x}_{k-q}^{\mathrm{b}}$.

Given $\hat{x}_{k-q}^{\mathrm{f}}$ and $\hat{x}_{k-q}^{\mathrm{b}}$, the FB OFIR smoother can be designed by fusing the estimates as

$$\tilde{x}_{k-q} = K_{k-q}^{\mathrm{f}}\hat{x}_{k-q}^{\mathrm{f}} + K_{k-q}^{\mathrm{b}}\hat{x}_{k-q}^{\mathrm{b}} \,, \tag{5.99}$$

where the gains K_{k-q}^{f} and K_{k-q}^{b} are chosen such that the MSE is minimized in $\tilde{x}_{k-q}$. To avoid regular bias errors, the forward and backward gains are related as $K_{k-q}^{\mathrm{f}} + K_{k-q}^{\mathrm{b}} = I$, and (5.99) thus becomes

$$\tilde{x}_{k-q} = K_{k-q}^{\mathrm{f}}\hat{x}_{k-q}^{\mathrm{f}} + (I - K_{k-q}^{\mathrm{f}})\hat{x}_{k-q}^{\mathrm{b}} \,. \tag{5.100}$$

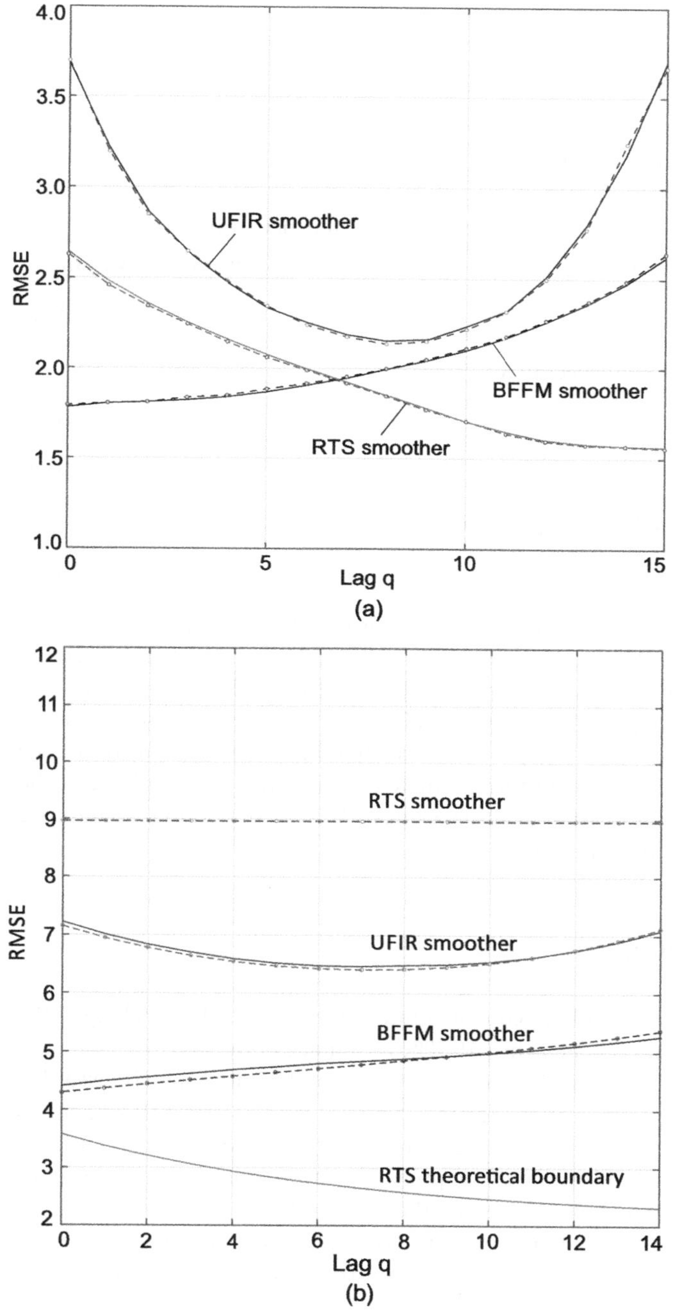

Figure 5.9 RMSE produced by the FFBM OFIR, UFIR, and RTS smoothers: (a) white measurement noise and (b) CMN with $\varphi = 0.95$. Solid lines are theoretical, and dashed-dotted lines represent simulated RMSEs averaged over 10^4 MC runs.

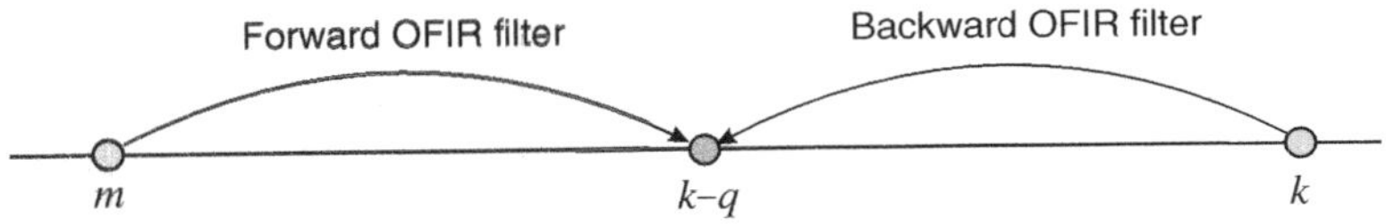

Figure 5.10 Two-filter q-lag FB OFIR smoothing strategy.

For the forward estimation error defined as $\varepsilon^{\mathrm{f}}_{k-q} = x_{k-q} - \hat{x}^{\mathrm{f}}_{k-q}$ and backward error as $\varepsilon^{\mathrm{b}}_{k-q} = x_{k-q} - \hat{x}^{\mathrm{b}}_{k-q}$, the FB OFIR smoothing error $\tilde{\varepsilon}_{k-q} = x_{k-q} - \tilde{x}_{k-q}$ corresponds to the error covariance

$$\tilde{P}_{k-q} = P^{\mathrm{b}}_{k-q} + K^{\mathrm{f}}_{k-q}(P^{\mathrm{f}}_{k-q} + P^{\mathrm{b}}_{k-q})K^{\mathrm{f}^T}_{k-q} - 2K^{\mathrm{f}}_{k-q}P^{\mathrm{b}}_{k-q} \,. \tag{5.101}$$

The derivative applied to $\mathrm{tr}(\tilde{P}_{k-q})$ with respect to K^{f}_{k-q} gives

$$K^{\mathrm{f}}_{k-q} = P^{\mathrm{b}}_{k-q}(P^{\mathrm{f}}_{k-q} + P^{\mathrm{b}}_{k-q})^{-1} \,, \tag{5.102}$$

$$K^{\mathrm{b}}_{k-q} = P^{\mathrm{f}}_{k-q}(P^{\mathrm{f}}_{k-q} + P^{\mathrm{b}}_{k-q})^{-1} \tag{5.103}$$

and, accordingly, (5.101) is transformed to a simple form

$$\tilde{\mathcal{I}}_{k-q} = \tilde{\mathcal{I}}^{\mathrm{f}}_{k-q} + \tilde{\mathcal{I}}^{\mathrm{b}}_{k-q} \,, \tag{5.104}$$

where $\tilde{\mathcal{I}}_{k-q} = \tilde{P}^{-1}_{k-q}$, $\tilde{\mathcal{I}}^{\mathrm{f}}_{k-q} = (\tilde{P}^{\mathrm{f}}_{k-q})^{-1}$, and $\tilde{\mathcal{I}}^{\mathrm{b}}_{k-q} = (\tilde{P}^{\mathrm{b}}_{k-q})^{-1}$ are information matrices.

It is easy to show that the FB OFIR smoothing estimate obtained by two OFIR filters acting forward and backward on the horizons $[m, k-q]$ and $[k-g, k]$, respectively, is less noisy than any of the filtering estimates. Indeed, if we suppose that the initial values at m and k are set properly, then it follows that $\tilde{P}^{\mathrm{f}}_{k-q} \cong \tilde{P}^{\mathrm{b}}_{k-q} \cong P$ holds for any lag $0 < q < N-1$ and that the smoother error covariance becomes $\tilde{P}_{k-q} \cong 0.5P$. Otherwise, the best smoothing estimate can be found in the middle of the averaging horizon. Thus, in the best case of accurate initial values, FB OFIR smoothing can improve the noise reduction by a factor of $\sqrt{2}$ over all lags.

5.9 q-Lag ML FIR Smoothing

In Chapter 4, we have already shown that all biased-constrained optimal FIR filters belong to the class of ML estimators and have the canonical form 1.19. The advantage of the ML FIR estimator, which is equivalent to the OUFIR estimator, is that the requirement of the initial state is discarded by the property of *unbiased optimality*. Thus, it is clear that different kinds of ML FIR smoothers having the FFFM, FFBM, BFFM, and BFBM structures can also be designed using forward and backward models and OUFIR filters. To illustrate the approach, next we look at q-lag FFFM ML FIR smoothing and postpone derivation of other ML smoothers to "Problems."

The q-lag FFFM ML FIR smoother, hereinafter referred to as the q-lag ML FIR smoother, can be obtained by taking into account the ML estimate (4.91) at $k-q$. To do this, we start with the model

$$x_k = F_k x_{k-1} + B_k w_k \,, \tag{5.105}$$

$$y_k = H_k x_k + v_k \,, \tag{5.106}$$

and extend it on $[m, k]$ in the traditional forms of

$$X_{m,k} = F_{m,k}x_m + D_{m,k}W_{m,k} , \tag{5.107}$$

$$Y_{m,k} = H_{m,k}x_m + G_{m,k}W_{m,k} + V_{m,k} , \tag{5.108}$$

for which the definitions of all extended vectors and matrices can be found after (4.7) and (4.14).

Using (5.6), the state x_{k-q} can be defined at $k-q$ for $u_k = 0$ as

$$x_{k-q} = \mathcal{F}_{k-q}^{m+1}x_m + \bar{D}_{m,k}^{(N-q)}W_{m,k} , \tag{5.109}$$

where matrix $\bar{D}_{m,k}^{(N-q)}$ is given by (5.14). By reorganizing the terms in (5.109), we represent the initial state x_m as

$$x_m = (\mathcal{F}_{k-q}^{m+1})^{-1}x_{k-q} - (\mathcal{F}_{k-q}^{m+1})^{-1}\bar{D}_{m,k}^{(N-q)}W_{m,k} \tag{5.110}$$

and find the q-lag ML FIR estimate and estimation error in batch forms as in the following.

5.9.1 Batch q-lag ML FIR Estimate

Substituting (5.110) into (5.108), we separate the regular and random terms as

$$Y_{m,k} - C_{m,k}^{(q)}x_{k-q} = \mathcal{N}_{m,k} , \tag{5.111}$$

where $C_{m,k}^{(q)} = H_{m,k}(\mathcal{F}_{k-q}^{m+1})^{-1}$ and the random term

$$\mathcal{N}_{m,k} = H_{m,k}(D_{m,k} - C_{m,k}^{(q)}\bar{D}_{m,k}^{(N-q)})W_{m,k} + V_{m,k} \tag{5.112}$$

combines the extended noise vectors $W_{m,k}$ and $V_{m,k}$.

For Gaussian w_k and v_k, the likelihood of x_{k-q} can be written as

$$p(Y_{m,k}|x_{k-q}) \propto \exp\left\{-\frac{1}{2}\left(Y_{m,k} - C_{m,k}^{(q)}x_{k-q}\right)^T \Sigma_{m,k}^{-1}(\dots)\right\} , \tag{5.113}$$

where $\Sigma_{m,k} = \mathcal{E}\{\mathcal{N}_{m,k}\mathcal{N}_{m,k}^T\}$, and the q-lag ML FIR estimate $\tilde{x}_{k-q}$ can be found by minimizing the quadratic form as

$$\tilde{x}_{k-q} = \arg\min_{x_{k-q}} \left\{-\frac{1}{2}\left(Y_{m,k} - C_{m,k}^{(q)}x_{k-q}\right)^T \Sigma_{m,k}^{-1}(\dots)\right\} . \tag{5.114}$$

By equating to zero the derivative

$$\frac{\partial}{\partial x_{k-q}}(Y_{m,k} - C_{m,k}^{(q)}x_{k-q})^T \Sigma_{m,k}^{-1}(\dots)$$
$$= C_{m,k}^{(q)^T}\Sigma_{m,k}^{-1}(Y_{m,k} - C_{m,k}^{(q)^T}x_{k-q}) = 0 ,$$

we obtain the q-lag ML FIR smoothing estimate in the canonical ML form

$$\begin{aligned}\hat{x}_{k-q} &= (C_{m,k}^{(q)^T}\Sigma_{m,k}^{-1}C_{m,k}^{(q)})^{-1}C_{m,k}^{(q)^T}\Sigma_{m,k}^{-1}Y_{m,k} \\ &= \mathcal{H}_{m,k}^{\mathrm{ML}(q)}Y_{m,k} .\end{aligned} \tag{5.115}$$

This gives the q-lag ML FIR smoother gain

$$\mathcal{H}^{\mathrm{ML}(q)}_{m,k} = (\mathcal{C}^{(q)^T}_{m,k} \Sigma^{-1}_{m,k} \mathcal{C}^{(q)}_{m,k})^{-1} \mathcal{C}^{(q)^T}_{m,k} \Sigma^{-1}_{m,k} , \tag{5.116}$$

which, by $q = 0$, becomes the ML FIR filter gain (4.98b). We then conclude that (5.115) can be applied universally to solving problems of filtering and smoothing using ML FIR structures. The error covariance for this smoothing is discussed next.

5.9.2 Error Covariance

Since the ML FIR filter and OUFIR-I filter are equivalent, it can be assumed that the error covariances of the q-lag ML FIR and q-lag OUFIR-I smoothers are also equivalent. Obtaining the error covariance of the q-lag ML FIR smoother can be done in a similar manner, resulting in the following standard form

$$P_{k-q} = \dot{\mathcal{W}}^{(q)}_{m,k} \mathcal{Q}_{m,k} \dot{\mathcal{W}}^{(q)^T}_{m,k} + \dot{\mathcal{V}}^{(q)}_{m,k} \mathcal{R}_{m,k} \dot{\mathcal{V}}^{(q)^T}_{m,k} , \tag{5.117}$$

where the q-varying error residual matrices are given by

$$\dot{\mathcal{W}}^{(q)}_{m,k} = \bar{\mathcal{D}}^{(N-q)}_{m,k} - \mathcal{H}^{\mathrm{ML}(q)}_{m,k} \mathcal{G}_{m,k} , \tag{5.118}$$

$$\dot{\mathcal{V}}^{(q)}_{m,k} = \mathcal{H}^{\mathrm{ML}(q)}_{m,k} . \tag{5.119}$$

It is worth noting again that in (5.117) there is no term representing bias errors, and thus q-lag ML FIR smoothing (5.115) has the property of unbiased optimality.

Error analysis in the q-lag ML FIR smoother is rather tedious and even complicated, since the error covariance (5.117) computed at $k-q$ over $[m, k]$ is multifactorial. Even so, some observations can be made.

The noise reduction with (5.116) is mainly due to the matrix $\mathcal{C}^{(q)}_{m,k}$, which depends on the product matrix $\mathcal{F}^{m+1}_{k-q}$, which changes from $\mathcal{F}^{m+1}_{k}$ when $q = 0$ to I when $q = N-1$. For example, if $F = \begin{bmatrix} 1 & 0.5 \\ 0 & 1 \end{bmatrix}$ and $N = 10$, then the product matrix changes from $\mathcal{F}^{m+1}_{k} = \begin{bmatrix} 1 & 4.5 \\ 0 & 1 \end{bmatrix}$ when $q = 0$ to $\mathcal{F}^{m+1}_{m} = I$ when $q = N-1$. Accordingly, with increasing q, the components in $\mathcal{C}^{(q)}_{m,k}$ grow, and the gain (5.116) decreases. This reduces the second matrix product in (5.117) via (5.119), which means that smoothing provides an effective reduction in measurement noise.

5.10 Summary

When the best possible noise reduction is required, smoothing does it with maximum efficiency. The peculiarity is that smoothing associated the estimate with some past point and, therefore, is not a real-time method. Smoothing solutions can be found mainly in post-processing algorithms, although some real-time estimators also incorporate smoothing procedures where the lag q does not introduce significant bias errors. Since post-processing is usually not strictly time-bound, batch FIR smoothing algorithms that operate with full block noise and error matrices are preferable in terms of accuracy over recursive schemes that are available when such matrices are diagonal. Other important observations related to FIR smoothing follow.

Unlike recursive KF-based smoothers, FIR smoothers operate on a horizon of N points and produce an estimate at any point on it. Therefore, they are typically derived as q-lag FIR smoothers, where the lag $q > 0$ is not allowed to be greater than $N-2$. For a given N_{opt}, RMSE of the UFIR smoothing estimate is a symmetric function on $[m, k]$. In contrast, the RMSE of the OFIR smoothing estimate decreases monotonously with increasing q.

There are four basic OFIR smoothing algorithms, which are available using forward and backward filters and models: FFFM, FFBM, BFFM, and BFBM. The FFFM and FFBM OFIR smoothers reduce the RMSE back in time, from k to m, while the BFFM and BFBM smoothers do it in an opposite direction, from m to k. The discrepancies between the FFFM and FFBM OFIR smoothing estimates and the BFFM and BFBM smoothing estimates are due to opposite directions of noise reduction. They are usually minor.

Backward OFIR filtering, organized on $[m, k]$ from k to m, is sometimes associated with smoothing, although this is not consistent with smoothing strategies. For white Gaussian noise, the batch backward OFIR filtering estimate is computed iteratively using backward Kalman recursions. It is noteworthy that even for white Gaussian noise, recursive forms are not available for q-lag OFIR smoothing, and their derivation remains a challenging problem.

In the two-filter OFIR smoothing approach, the forward and backward OFIR filtering estimates are statistically fused on $[m, k]$. For white Gaussian noise, such smoothing can be easily developed using forward and backward Kalman recursions to improve the noise reduction by a factor of $\sqrt{2}$ over $[m, k]$. An effectiveness of this approach can be increased by fusing the FFFM (or FFBM) and BFFM (or BFBM) OFIR smoothing estimates to achieve maximum noise reduction.

Another available denoising technique is known as q-lag ML FIR smoothing, which exhibits the property of unbiased optimality. Various kinds of ML FIR smoothers can be designed as FFFM, FFBM, BFFM, and BFBM structures using forward and backward models and OUFIR estimates. Unbiased optimality can also be achieved by subjecting the FFFM, FFBM, BFFM, and BFBM OFIR smoothers to the unbiasedness constraint at $k - q$. Finally, it is worth noting again that recursive forms are still unavailable for batch FIR smoothers, even when noise is white Gaussian.

5.11 Problems

1 The q-lag FFFM OFIR smoother is given in batch form by (5.20). Suppose the input is zero, $u_k = 0$, and find recursions for the homogeneous estimate. Then find recursions for the forced estimate.

2 Find recursions for the error covariance (5.21) of the q-lag FFFM OFIR smoother and determine the optimal lag q_{opt} that minimizes MSE.

3 A constant scalar quantity Q is observed on $[m, k]$ as $Y_{m,k} = H_N Q + V_{m,k}$, where the vectors $Y_{m,k}$ and $V_{m,k}$ and the matrix H_N are defined as

$$Y_{m,k} = \begin{bmatrix} y_m \\ y_{m+1} \\ \vdots \\ y_k \end{bmatrix}, H_N = \begin{bmatrix} 1 \\ 1 \\ \vdots \\ 1 \end{bmatrix}, V_{m,k} = \begin{bmatrix} v_m \\ v_{m+1} \\ \vdots \\ v_k \end{bmatrix}.$$

The zero mean white Gaussian measurement noise $v_k \sim \mathcal{N}(0, \sigma_v^2)$ has the variance σ_v^2. Obtain the q-lag FFFM OFIR smoothing estimate for $\sigma_v^2 = 0$ and show that for all lags q on $[m, k]$ the estimation error is the same.

4 It follows from Fig. 5.3b that the Gauss-Markov CMN $v_k = \varphi v_{k-1} + \xi_k$, where ξ_k is white Gaussian, makes the RTS smoother much less accurate than FFFM OFIR and UFIR smoothers. Confirm this inference analytically by comparing the error covariance (5.21) of the FFFM OFIR smoother and (3.211) of the RTS smoother.

5 Consider the forward model (5.6) and the forward FIR estimate (5.5) at $k-q$. Subject the orthogonality condition $\mathcal{E}\{\varepsilon_{k-q} Y_{m,k}^T\} = 0$ to the unbiasedness constraint $\mathcal{F}_{k-q}^{m+1} = \mathcal{H}_{m,k}^{(q)} H_{m,k}$ given by (5.7), and obtain a q-lag FFFM OUFIR smoother. Find the error covariance for this smoother.

6 A periodically varying stochastic process is given in discrete-time state space with the equations

$$x_k = \begin{bmatrix} \cos\frac{\pi}{64} & \sin\frac{\pi}{64} \\ -\sin\frac{\pi}{64} & \cos\frac{\pi}{64} \end{bmatrix} x_{k-1} + \begin{bmatrix} 1 \\ 1 \end{bmatrix} w_k \,,$$

$$y_k = \begin{bmatrix} 1 & 0 \end{bmatrix} x_k + v_k \,,$$

where the zero mean white Gaussian $w_k =\sim \mathcal{N}(0, \sigma_w^2)$ and $v_k =\sim \mathcal{N}(0, \sigma_v^2)$ have the variances $\sigma_w^2 = 1$ and $\sigma_v^2 = 100$. The optimal horizon for the UFIR filter is measured as $N_{\text{opt}} = 111$. The process starts with $x_{10} = 100$ and $x_{20} = 10^{-2}$. Apply q-lag FFFM OFIR smoothing and examine the RMSE as a function of lag q.

7 Consider the q-lag OFIR smoothing problem and rewrite model (5.6) from $k-q$ to k as

$$x_k = \mathcal{F}_k^{k-q+1} x_{k-q} + \bar{\bar{S}}_{k-q,k}^{(N)} U_{k-q,k} + \bar{D}_{k-q,k}^{(N)} W_{k-q,k} \,.$$

Represent x_{k-q} via x_k and derive another q-lag FFFM OFIR smoother. Analyze the difference with the smoother represented by (5.20).

8 Using the state model discussed in the previous above item and converted from x_{k-q} to x_k, obtain another q-lag BFFM OFIR smoother.

9 Consider the backward batch OFIR filter (5.28) stated by theorem 5.1. Follow the derivation of the forward OFIR filter and validate the recursive forms stated by theorem 5.2.

10 Given the batch q-lag BFBM OFIR smoothing estimate (5.80), provide the derivation of the error covariance (5.81).

11 Consider the error covariance (5.21) of the FFFM OFIR smoother, (5.81) of the BFBM OFIR smoother, (5.88) of the FFBM OFIR smoother, and (5.99) of the BFFM OFIR smoother. Show that the errors produced by the FFFM and FFBM OFIR smoothers and the BFFM and BFBM OFIR smoothers are statistically equivalent.

12 Consider the limiting case of the error covariances (5.21), (5.81), (5.88), and (5.99) for zero process and measurement noise and show that the corresponding OFIR smoothers are converted into a unique UFIR smoother.

13 Consider the backward model (5.70) and the backward FIR estimate (5.69) at $m+g$. Subject the orthogonality condition $\mathcal{E}\{\varepsilon_{m+g} \overleftarrow{Y}_{k,m}^T\} = 0$ to the unbiasedness constraint $\mathcal{X}_k^{m+g+1} =$

$\overleftarrow{H}^{(g)}_{k,m}\overleftarrow{H}_{k,m}$ given by (5.71) and derive a q-lag BFBM OUFIR smoother. Find the error covariance for this smoother.

14 Consider the back-in-time state model (5.70) and rewrite it from $m+g$ to m as

$$x_m = \mathcal{F}^{m+1}_{m+g} x_{m+g} - \overleftarrow{S}^{(g)}_{m+g,m}\overleftarrow{U}_{m+g,m} - \overleftarrow{D}^{(g)}_{m+g,m}\overleftarrow{W}_{m+g,m} \,.$$

Based upon this model, represent x_{m+g} via x_m and derive another q-lag BFBM OFIR smoother. Analyze the properties of this smoother.

15 Using the state model given in the previous item and converted from x_{k+g} to x_m, derive another q-lag FFBM OFIR smoother.

16 Consider the backward model (5.70) and the forward estimate (5.5) at $k-q$. Subject the orthogonality condition $\mathcal{E}\{\varepsilon_{k-q}\overleftarrow{Y}^{T}_{k,m}\} = 0$ to the unbiasedness constraint $\mathcal{F}^{m+1}_{k-q} = \mathcal{H}^{(q)}_{m,k}H_{m,k}$ and derive a q-lag FFBM OUFIR smoother. Find the error covariance for this smoother.

17 A stochastic process is represented with the state-space equations

$$x_k = \begin{bmatrix} 1 & 0.1 \\ 0 & 1 \end{bmatrix} x_{k-1} + \begin{bmatrix} 0 \\ 1 \end{bmatrix} w_k \,,$$

$$y_k = \begin{bmatrix} 1 & 0 \end{bmatrix} x_k + v_k \,,$$

where the zero mean white Gaussian $w_k =\sim \mathcal{N}(0,\sigma_w^2)$ and $v_k =\sim \mathcal{N}(0,\sigma_v^2)$ have the variances $\sigma_w^2 = 2/\text{s}$ and $\sigma_v^2 = 10$. The process starts with $x_{10} = 0$ and $x_{20} = 0$. Apply q-lag ML FIR smoothing and two-filter FIR smoothing on $[m,k]$ of $N = 30$ points and examine the RMSEs as functions of lag q. Analyze the accuracy of these solutions.

18 Considering the cases of LTI and LTV systems, find recursive forms for the batch q-lag BFBM OFIR smoothing estimate given by (5.80) with zero input.

19 Provide the derivation for the error covariance (5.88) of the q-lag BFFM OFIR smoother.

20 A batch q-lag FFBM OFIR smoothing estimate is given by (5.87). Find recursive forms for this estimate. Consider LTI and LTV systems with and without an input signal.

21 Consider the forward model (5.6) and the backward FIR estimate (5.28) at $m+g$. Apply the orthogonality condition $\mathcal{E}\{\varepsilon_{m+g}\overleftarrow{Y}^{T}_{k,m}\} = 0$ subject to the unbiasedness constraint $(\mathcal{F}^{k-q+1}_{k})^{-1} = \overleftarrow{H}^{(g)}_{k,m}\overleftarrow{H}_{k,m}$ and derive a q-lag BFFM OUFIR smoother. Obtain the error covariance for this smoother.

6

Unbiased FIR State Estimation

The arithmetic mean is the most probable value.

Carl F. Gauss [53], p. 244

It is usually taken for granted that the right method for determining the constants is the method of least squares.

Karl Pearson [146], p. 266

6.1 Introduction

Optimal state estimation requires information about noise and initial values, which is not always available. The requirement of initial values is canceled in the optimal unbiased or ML estimators, which still require accurate noise information. Obviously, these estimators can produce large errors if the noise statistics are inaccurate or the noise is far from Gaussian. In another extreme method of state estimation that gives UFIR estimates corresponding to the LS, the observation mean is tracked only under the zero mean noise assumption. Designed to satisfy only the unbiasedness constraint, such an estimator discards all other requirements and in many cases justifies suboptimality by being more robust. The great thing about the UFIR estimator is that, unlike OFIR, OUFIR, and ML FIR, it only needs an optimal horizon of N_{opt} points to minimize MSE. It is worth noting that determining N_{opt} requires much less effort than noise statistics. Given N_{opt}, the UFIR state estimator, which has no other tuning factors, appears to be the most robust in the family of linear state estimators.

In this chapter, we discuss different kinds of UFIR state estimators, mainly filters and smoothers and, to a lesser extent, predictors.

6.2 The *a posteriori* UFIR Filter

Let us consider an LTV system represented in discrete-time state space with the following state and observation equations, respectively,

$$x_k = F_k x_{k-1} + E_k u_k + B_k w_k \,, \tag{6.1}$$

$$y_k = H_k x_k + v_k \,. \tag{6.2}$$

Optimal and Robust State Estimation: Finite Impulse Response (FIR) and Kalman Approaches, First Edition.
Yuriy S. Shmaliy and Shunyi Zhao.

These equations can be extended on $[m, k]$ as

$$X_{m,k} = F_{m,k}x_m + S_{m,k}U_{m,k} + D_{m,k}W_{m,k} , \tag{6.3}$$

$$Y_{m,k} = H_{m,k}x_m + L_{m,k}U_{m,k} + G_{m,k}W_{m,k} + V_{m,k} , \tag{6.4}$$

using the definitions of vectors and matrices given for (4.7) and (4.14).

6.2.1 Batch Form

The unbiased filtering problem can be solved if our goal is to satisfy only the unbiasedness condition

$$\mathcal{E}\{\hat{x}_{k|k}\} = \mathcal{E}\{x_k\} , \tag{6.5}$$

and then minimize errors by choosing the optimal horizon of N_{opt} points.

For the FIR estimate defined as

$$\hat{x}_k \triangleq \hat{x}_{k|k} = \mathcal{H}_{m,k}Y_{m,k} + \mathcal{H}^{\text{f}}_{m,k}U_{m,k} , \tag{6.6}$$

and the state model represented with the Nth row vector in (6.3),

$$x_k = \mathcal{F}^{m+1}_k x_m + \bar{S}_{m,k}U_{m,k} + \bar{D}_{m,k}W_{m,k} , \tag{6.7}$$

the condition (6.5) gives two unbiasedness constraints,

$$\mathcal{F}^{m+1}_k = \mathcal{H}_{m,k}H_{m,k} , \tag{6.8}$$

$$\mathcal{H}^{\text{f}}_{m,k} = \bar{S}_{m,k} - \mathcal{H}_{m,k}L_{m,k} . \tag{6.9}$$

By multiplying both sides of (6.8) from the right-hand side by the matrix identity $(H^T_{m,k}H_{m,k})^{-1}H^T_{m,k}$ $H_{m,k}$ and discarding the nonzero $H_{m,k}$ on both sides, we find the fundamental gain $\hat{H}_{m,k}$ of the UFIR filter

$$\hat{\mathcal{H}}_{m,k} = \mathcal{F}^{m+1}_k (H^T_{m,k}H_{m,k})^{-1}H^T_{m,k} \tag{6.10a}$$

$$= (C^T_{m,k}C_{m,k})^{-1}C^T_{m,k} , \tag{6.10b}$$

where $C_{m,k} = H_{m,k}(\mathcal{F}^{m+1}_k)^{-1}$. Then, referring to the forced gain $\hat{H}^{\text{f}}_{m,k}$ given by (6.9), the *a posteriori* UFIR filtering estimate can be written as

$$\hat{x}_k = \hat{\mathcal{H}}_{m,k}Y_{m,k} + (\bar{S}_{m,k} - \hat{\mathcal{H}}_{m,k}L_{m,k})U_{m,k} . \tag{6.11}$$

As can be seen, the gain $\hat{H}_{m,k}$ (6.10a) does not contain any information about noise and initial values, which means that the UFIR filter does not have the inherent disadvantages of the KF and OFIR filter.

The error covariance for the *a posteriori* UFIR filter (6.11) can be written similarly to the OFIR filter (4.31) as

$$P_k = \mathcal{W}_{m,k}\mathcal{Q}_{m,k}\mathcal{W}^T_{m,k} + \mathcal{V}_{m,k}\mathcal{R}_{m,k}\mathcal{V}^T_{m,k} , \tag{6.12}$$

where the error residual matrices are given by

$$\mathcal{W}_{m,k} = \bar{D}_{m,k} - \hat{\mathcal{H}}_{m,k}G_{m,k} , \tag{6.13}$$

$$\mathcal{V}_{m,k} = \hat{\mathcal{H}}_{m,k} . \tag{6.14}$$

It now follows that (6.12) has the same structure as (4.55) of the *a posteriori* OUFIR filter with, however, another gain $\hat{\mathcal{H}}_{m,k}$ in (6.13) and (6.14).

Because $\hat{\mathcal{H}}_{m,k}$ given by (6.10a) does not require noise covariances, the UFIR filter is not optimal and therefore generally gives larger errors than the OUFIR filter. In practice, however, this flaw can be ignored if a higher robustness is required, as in many applications, especially industrial.

Generalized Noise Power Gain

An important indicator of the effectiveness of FIR filtering is the NPG introduced by Trench in [198]. The NPG is the ratio of the output noise variance σ^2_{out} to the input noise variance σ^2_{in}, which is akin to the noise figure in wireless communications. For white Gaussian noise, the NPG is equal to the sum of the squared coefficients of the FIR filter impulse response h_k,

$$\text{NPG} = \frac{\sigma^2_{\text{out}}}{\sigma^2_{\text{in}}} = \sum_{k=0}^{N-1} h_k^2 \,, \tag{6.15}$$

which is the squared norm of h_k.

In state space, the homogeneous gain $\hat{\mathcal{H}}_{m,k}$ represents the coefficients of the FIR filter impulse response. Therefore, the product $\mathcal{G}_k = \hat{\mathcal{H}}_{m,k}\hat{\mathcal{H}}^T_{m,k}$ plays the role of a *generalized NPG* (GNPG) [179]. Referring to (6.10a) and (6.10b), GNPG can be written in the following equivalent forms:

$$\mathcal{G}_k = \hat{\mathcal{H}}_{m,k}\hat{\mathcal{H}}^T_{m,k} \tag{6.16a}$$

$$= \mathcal{F}_k^{m+1}(H^T_{m,k}H_{m,k})^{-1}\mathcal{F}_k^{m+1^T} \tag{6.16b}$$

$$= (C^T_{m,k}C_{m,k})^{-1} \,. \tag{6.16c}$$

It follows that GNPG is a symmetric square matrix $\mathcal{G}_k = \mathcal{G}_k^T \in \mathcal{R}^{K\times K}$, where the main diagonal components represent the NPGs for the system states, and the remaining components the cross NPGs. The main property of $\mathcal{G}_k$ is that its trace decreases with increasing horizon length, which provides effective noise reduction. On the other hand, an increase in N causes an increase in bias errors, and therefore $\mathcal{G}_k$ must be optimally set by choosing an optimal horizon length.

6.2.2 Iterative Algorithm Using Recursions

Recursions for the batch *a posteriori* UFIR filtering estimate (6.11) can be found by decomposing (6.11) as $\hat{x}_k = \hat{x}_k^{\text{h}} + \hat{x}_k^{\text{f}}$, where

$$\hat{x}_k^{\text{h}} = \hat{\mathcal{H}}_{m,k}Y_{m,k} \,, \tag{6.17}$$

$$\hat{x}_k^{\text{f}} = (\bar{S}_{m,k} - \hat{\mathcal{H}}_{m,k}L_{m,k})U_{m,k} \,, \tag{6.18}$$

$\hat{x}_k^{\text{h}}$ is the homogeneous estimate and $\hat{x}_k^{\text{f}}$ is the forced estimate.

Using (6.16b), the estimate x_k^{h} can be transformed to

$$\begin{aligned} x_k^{\text{h}} &= \mathcal{F}_k^{m+1}(H^T_{m,k}H_{m,k})^{-1}H^T_{m,k}Y_{m,k} \\ &= \mathcal{F}_k^{m+1}(H^T_{m,k}H_{m,k})^{-1}(\mathcal{F}_k^{m+1})^T(\mathcal{F}_k^{m+1})^{-T}H^T_{m,k}Y_{m,k} \\ &= \mathcal{G}_k(\mathcal{F}_k^{m+1})^{-T}H^T_{m,k}Y_{m,k} \,, \end{aligned} \tag{6.19}$$

and, by applying the decomposition

$$H^T_{m,k} = [H^T_{m,k-1} \quad (\mathcal{F}_k^{m+1})^T H_k^T] \,, \tag{6.20}$$

the recursion for the GNPG (6.16b) can be obtained if we transform $\mathcal{G}_k^{-1}$ as

$$\mathcal{G}_k^{-1} = (\mathcal{F}_k^{m+1})^{-T}H^T_{m,k}H_{m,k}(\mathcal{F}_k^{m+1})^{-1}$$

$$= (\mathcal{F}_k^{m+1})^{-T} \begin{bmatrix} H_{m,k-1}^T & (\mathcal{F}_k^{m+1})^T H_k^T \end{bmatrix} \begin{bmatrix} H_{m,k-1} \\ H_k \mathcal{F}_k^{m+1} \end{bmatrix} (\mathcal{F}_k^{m+1})^{-1}$$
$$= H_k^T H_k + (F_k \mathcal{G}_{k-1} F_k^T)^{-1} .$$

This gives the following forward and backward recursive forms:

$$\mathcal{G}_k = [H_k^T H_k + (F_k \mathcal{G}_{k-1} F_k^T)^{-1}]^{-1} , \tag{6.21}$$

$$\mathcal{G}_{k-1} = F_k^{-1} (\mathcal{G}_k^{-1} - H_k^T H_k)^{-1} F_k^{-T} . \tag{6.22}$$

Using (6.20) and (6.22) and extracting $H_{m,k-1}^T = (\mathcal{F}_{k-1}^{m+1})^T \mathcal{G}_{k-1}^{-1} \hat{\mathcal{H}}_{m,k-1}^{\mathrm{h}}$ from (6.10a), we next transform the product $H_{m,k}^T Y_{m,k}$ as

$$H_{m,k}^T Y_{m,k} = \begin{bmatrix} H_{m,k-1}^T & (\mathcal{F}_k^{m+1})^T H_k^T \end{bmatrix} \begin{bmatrix} Y_{m,k-1} \\ y_k \end{bmatrix}$$
$$= H_{m,k-1}^T Y_{m,k-1} + (\mathcal{F}_k^{m+1})^T H_k^T y_k$$
$$= (\mathcal{F}_{k-1}^{m+1})^T \mathcal{G}_{k-1}^{-1} \hat{\mathcal{H}}_{m,k-1} Y_{m,k-1} + (\mathcal{F}_k^{m+1})^T H_k^T y_k$$
$$= (\mathcal{F}_{k-1}^{m+1})^T \mathcal{G}_{k-1}^{-1} \hat{x}_{k-1}^{\mathrm{h}} + (\mathcal{F}_k^{m+1})^T H_k^T y_k$$
$$= (\mathcal{F}_k^{m+1})^T (\mathcal{G}_k^{-1} - H_k^T H_k) F_k \hat{x}_{k-1}^{\mathrm{h}} + (\mathcal{F}_k^{m+1})^T H_k^T y_k . \tag{6.23}$$

By combining (6.21) and (6.23), we finally obtain the following recursion for the homogeneous estimate

$$x_k^{\mathrm{h}} = \mathcal{G}_k [(\mathcal{G}_k^{-1} - H_k^T H_k) F_k \hat{x}_{k-1}^{\mathrm{h}} + H_k^T y_k]$$
$$= (I - \mathcal{G}_k H_k^T H_k) F_k \hat{x}_{k-1}^{\mathrm{h}} + \mathcal{G}_k H_k^T y_k$$
$$= F_k \hat{x}_{k-1}^{\mathrm{h}} + \hat{K}_k (y_k - H_k F_k \hat{x}_{k-1}^{\mathrm{h}}) , \tag{6.24}$$

where $\hat{K}_k = \mathcal{G}_k H_k^T$ is the bias correction gain of the UFIR filter.

To find the recursion for the forced estimate (6.18), we first show that

$$\bar{S}_{m,k} U_{m,k} = \begin{bmatrix} F_k \bar{S}_{m,k-1} & E_k \end{bmatrix} \begin{bmatrix} U_{m,k-1} \\ u_k \end{bmatrix}$$
$$= F_k \bar{S}_{m,k-1} U_{m,k-1} + E_k u_k . \tag{6.25}$$

Then, by extracting $\hat{\mathcal{H}}_{m,k}$ from (6.19), $\hat{\mathcal{H}}_{m,k} L_{m,k} U_{m,k}$ can be transformed as

$$\hat{\mathcal{H}}_{m,k} L_{m,k} U_{m,k} = \mathcal{G}_k (\mathcal{F}_k^{m+1})^{-T} \begin{bmatrix} H_{m,k-1}^T & (\mathcal{F}_k^{m+1})^T H_k^T \end{bmatrix}$$
$$\times \begin{bmatrix} L_{m,k-1} & 0 \\ H_k F_k \bar{S}_{m,k-1} & H_k E_k \end{bmatrix} \begin{bmatrix} U_{m,k-1} \\ u_k \end{bmatrix}$$
$$= \mathcal{G}_k [(\mathcal{F}_k^{m+1})^{-T} H_{m,k-1}^T L_{m,k-1}$$
$$+ H_k^T H_k F_k \bar{S}_{m,k-1}] U_{m,k-1} + \mathcal{G}_k H_k^T H_k E_k u_k . \tag{6.26}$$

Now, by combining (6.25) and (6.26) and substituting (6.22), the forced estimate (6.18) can be represented with the recursion

$$\hat{x}_k^{\mathrm{f}} = (I - \mathcal{G}_k H_k^T H_k)(F_k \hat{x}_{k-1}^{\mathrm{f}} + E_k u_k) , \tag{6.27}$$

which, together with (6.24), gives the recursions for the *a priori* and *a posteriori* UFIR filtering estimates, respectively,

$$\hat{x}_k^- = F_k \hat{x}_{k-1} + E_k u_k \,, \tag{6.28}$$

$$\hat{x}_k = \hat{x}_k^- + \mathcal{G}_k H_k^T (y_k - H_k \hat{x}_k^-) \,. \tag{6.29}$$

The recursive computation of the *a posteriori* UFIR filtering estimate (6.11) can now be summarized with the following theorem.

Theorem 6.1 Given the batch a posteriori UFIR filtering estimate (6.11), its iterative computation on $[m, k]$ is provided by Algorithm 11 starting with $\mathcal{G}_s$ and $\hat{x}_s$ computed in short batch forms (6.16c) and (6.11) on $[m, s]$ for $s = m + K - 1$, by varying an auxiliary time-index i from $s + 1$ to k, and taking the output when $i = k$.

Algorithm 11: Iterative *a posteriori* UFIR Filtering Algorithm

Data: y_k, u_k

begin
  for $k = N - 1, N, \cdots$ **do**
    $m = k - N + 1, s = k - N + K$;
    $\mathcal{G}_s = (C_{m,s}^T C_{m,s})^{-1}$;
    $\hat{x}_s = \mathcal{G}_s C_{m,s}^T (Y_{m,s} - L_{m,s} U_{m,s}) + \bar{S}_{m,s} U_{m,s}$;
    for $i = s + 1 : k$ **do**
      $\check{x}_i^- = F_i \check{x}_{i-1} + E_i u_i$;
      $\mathcal{G}_i = [H_i^T H_i + (F_i \mathcal{G}_{i-1} F_i^T)^{-1}]^{-1}$;
      $K_i = \mathcal{G}_i H_i^T$;
      $\check{x}_i = \check{x}_i^- + K_i (y_i - H_i \check{x}_i^-)$;
    end for
    $\hat{x}_k = \check{x}_k$;
  end for
  Result: $\hat{x}_k$
end

Proof: The derivation of the recursive forms for (6.11) is provided by (6.17)–(6.29). □

Note that the need to compute the initial $\mathcal{G}_s$ and $\hat{x}_s$ in short batch forms arises from the fact that the inverse in (6.16c) does not exist on shorter horizons.

Modified GNPG

It was shown in [50, 195] that the robustness of the UFIR filter can be improved by scaling the GNPG (6.21) with the weight γ_k as

$$\mathcal{G}_k = \gamma_k [H_k^T H_k + (F_k \mathcal{G}_{k-1} F_k^T)^{-1}]^{-1} \,.$$

The weight γ_k is defined by

$$\gamma_k = \frac{1}{\lfloor N/2 \rfloor} \sum_{i=k_0}^{k} \sqrt[K]{\eta_i / \eta_{i-1}} \,,$$

where $k_0 = k - \lfloor N/2 \rfloor + 1$, $\lfloor N/2 \rfloor$ is the integer part of $N/2$, and K is the number of the states. The RMS deviation η_k of the estimate is computed using the innovation residual as

$$\eta_k = \sqrt{\frac{1}{\kappa}(y_k - H_k\hat{x}_k)^T(y_k - H_k\hat{x}_k)}\,,$$

where κ means the dimension of the target motion. With this adaptation of the GNPG to the operation conditions, the UFIR filter can be approximately twice as accurate as the KF when tracking maneuvering targets [50].

6.2.3 Recursive Error Covariance

Although the UFIR filter does not require error covariance P_k to obtain an estimate, P_k may be required to evaluate the quality of the estimate on a given horizon. Since the batch form (6.12) is not always suitable for fast estimation, next we present the recursive forms obtained in [226].

Consider P_k given by (6.12) and represent with

$$P_k = P_k^{(1)} - P_k^{(2)} - P_k^{(3)} + P_k^{(4)} + P_k^{(5)}\,, \tag{6.30}$$

where the components are defined as

$$P_k^{(1)} = \bar{D}_{m,k}\mathcal{Q}_{m,k}\bar{D}_{m,k}^T\,, \quad P_k^{(2)} = \bar{D}_{m,k}\mathcal{Q}_{m,k}G_{m,k}^T\hat{\mathcal{H}}_{m,k}^T\,,$$
$$P_k^{(3)} = \hat{\mathcal{H}}_{m,k}^{h}G_{m,k}\mathcal{Q}_{m,k}\bar{D}_{m,k}^T\,, \quad P_k^{(4)} = \hat{\mathcal{H}}_{m,k}G_{m,k}\mathcal{Q}_{m,k}G_{m,k}^T\hat{\mathcal{H}}_{m,k}^T\,,$$
$$P_k^{(5)} = \hat{\mathcal{H}}_{m,k}\mathcal{R}_{m,k}\hat{\mathcal{H}}_{m,k}^T\,.$$

Using the decomposition $\bar{D}_{m,k} = [F_k\bar{D}_{m,k-1} \quad B_k]$, we represent matrix $P_k^{(1)}$ recursively as

$$P_k^{(1)} = F_kP_{k-1}^{(1)}F_k^T + B_kQ_kB_k^T\,. \tag{6.31}$$

Then, referring to (6.20), (6.22), (6.31), and

$$G_{m,k} = \begin{bmatrix} G_{m,k-1} & 0 \\ H_kF_k\bar{D}_{m,k-1} & H_kB_k \end{bmatrix}\,, \tag{6.32}$$
$$\mathcal{H}_{m,k} = \mathcal{G}_k(\mathcal{F}_k^{m+1})^{-T}H_{m,k}^T\,,$$

and substituting $H_{m,k-1}(\mathcal{F}_{k-1}^{m+1})^{-1}$ with $\mathcal{H}_{m,k-1}^T\mathcal{G}_{k-1}^{-1}$, we represent matrices $P_k^{(2)}$ and $P_k^{(3)}$ recursively with

$$\begin{aligned} P_k^{(2)} &= F_kP_{k-1}^{(2)}F_k^T - F_kP_{k-1}^{(2)}F_k^TH_k^TH_k\mathcal{G}_k \\ &\quad + F_kP_{k-1}^{(1)}F_k^TH_k^TH_k\mathcal{G}_k + B_kQ_kB_k^TH_k^TH_k\mathcal{G}_k\,. \end{aligned} \tag{6.33}$$

$$\begin{aligned} P_k^{(3)} &= F_kP_{k-1}^{(3)}F_k^T - \mathcal{G}_kH_k^TH_kF_kP_{k-1}^{(3)}F_k^T \\ &\quad + \mathcal{G}_kH_k^TH_kF_kP_{k-1}^{(1)}F_k^T + \mathcal{G}_kH_k^TH_kB_kQ_kB_k^T\,. \end{aligned} \tag{6.34}$$

Similarly, the recursions for $P_k^{(4)}$ and $P_k^{(4)}$ can be obtained as

$$\begin{aligned} P_k^{(4)} &= F_kP_{k-1}^{(4)}F_k^T - F_kP_{k-1}^{(4)}F_k^TH_k^TH_k\mathcal{G}_k - \mathcal{G}_kH_k^TH_kF_kP_{k-1}^{(4)}F_k^T \\ &\quad + \mathcal{G}_kH_k^TH_kF_kP_{k-1}^{(4)}F_k^TH_k^TH_k\mathcal{G}_k + \mathcal{G}_kH_k^TH_kF_kP_{k-1}^{(2)}F_k^T \\ &\quad F_kP_{k-1}^{(3)}F_k^TH_k^TH_kG_k - \mathcal{G}_kH_k^TH_kF_kP_{k-1}^{(3)}F_k^TH_k^TH_k\mathcal{G}_k \\ &\quad - \mathcal{G}_kH_k^TH_kF_kP_{k-1}^{(2)}F_k^TH_k^TH_k\mathcal{G}_k + \mathcal{G}_kH_k^TH_kF_kP_{k-1}^{(2)}F_k^T \\ &\quad + \mathcal{G}_kH_k^TH_k(F_kP_{k-1}^{(1)}F_k^T + B_kQ_kB_k^T)H_k^TH_k\mathcal{G}_k\,, \end{aligned} \tag{6.35}$$

$$P_k^{(5)} = F_k P_{k-1}^{(5)} F_k^T - F_k P_{k-1}^{(5)} F_k^T H_k^T H_k \mathcal{G}_k - \mathcal{G}_k H_k^T H_k F_k P_{k-1}^{(5)} F_k^T$$
$$+ \mathcal{G}_k H_k^T H_k F_k P_{k-1}^{(5)} F_k^T H_k^T H_k \mathcal{G}_k + \mathcal{G}_k H_k^T R_k H_k \mathcal{G}_k \,. \tag{6.36}$$

By combining (6.31), (6.34), (6.35), and (6.36) with (6.30), we finally arrive at the recursive form for the error covariance P_k of the *a posteriori* UFIR filter

$$P_k = (I - \mathcal{G}_k H_k^T H_k)(F_k P_{k-1} F_k^T + B_k Q_k B_k^T)(I - \mathcal{G}_k H_k^T H_k)^T$$
$$+ \mathcal{G}_k H_k^T R_k H_k \mathcal{G}_k \,. \tag{6.37}$$

It is worth noting that (6.37) is equivalent to the error covariance of the *a posteriori* KF (3.77), in which the Kalman gain K_k must be replaced with the bias correction gain $\mathcal{G}_k H_k^T$ of the UFIR filter. The notable difference is that the Kalman gain minimizes the MSE and is thus optimal, whereas the $\mathcal{G}_k H_k^T$ derived from the unbiased constraint (6.17) makes the UFIR estimate truly unbiased, but noisier, since no effort has been made so far to minimize error covariance. The minimum MSE can be reached in the UFIR filter output at an optimal horizon of N_{opt} points, which will be discussed next.

6.2.4 Optimal Averaging Horizon

Unlike KF, which has IIR, the UFIR filter operates with data on the horizon $[m, k]$ of N points, which is sometimes called the *FIR filter memory*. To minimize MSE, N should be optimally chosen as N_{opt}. Otherwise, if $N < N_{\text{opt}}$, noise reduction will be ineffective, and, for $N > N_{\text{opt}}$, bias errors will prevail as shown in Fig. 6.1.

Since the UFIR filter is not selective, there is no exact relationship between filtering order and memory. Optimizing the memory of a UFIR filter in state space requires finding the derivative of the trace $\text{tr}(P_k)$ of the error covariance with respect to N, which is problematic, especially for LTV systems. In some cases, N can be found heuristically for real data. If a reference model or the

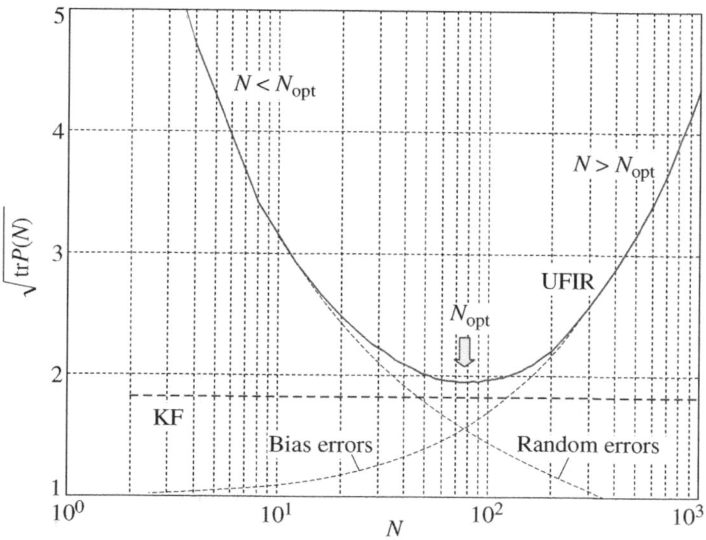

Figure 6.1 The RMSE produced by the UFIR filter as a function of N. An optimal balance is achieved when $N = N_{\text{opt}}$, where the UFIR estimate is still less accurate than the KF estimate.

ground truth is available at the test stage, then N_{opt} can be found by minimizing $\text{tr}(P_k)$. For polynomial models, N_{opt} was found in [169] analytically through higher order states, which has explicit limitations. If the ground truth is not available, as in many applications, then N_{opt} can be measured via the derivative of the trace of the measurement residual covariance, as shown in [153]. Next we will discuss several such cases based on the time-invariant state-space model

$$x_k = Fx_{k-1} + Bw_k \,, \tag{6.38}$$

$$y_k = Hx_k + v_k \,. \tag{6.39}$$

All methods will be separated into two classes depending on the operation conditions: with and without the ground truth.

Available Ground Truth

When state x_k is available through the ground truth measurement, which means that P_k is also available, N_{opt} can be found by referring to Fig. 6.1 and solving on $[0, k]$ the following optimization problem

$$N_{\text{opt}} = \arg\min_k (\text{tr}P_k) + 1 \,. \tag{6.40}$$

There are several approaches to finding N_{opt} using (6.40).

- When $\text{tr}(P_k)$ in (6.40) reaches a minimum with $N_{\text{opt}} = k + 1$, we can assume that $\hat{P} \triangleq \hat{P}(N_{\text{opt}}) = P_{k+1} = P_k$ and represent the error covariance (6.37) by the *discrete Lyapunov equation* (A.34)

  $$\hat{P} - \Upsilon\hat{P}\Upsilon^T = \Psi \,, \tag{6.41}$$

  where $\Upsilon = (I - \mathcal{G}H^T H)F$ is required to be stable and matrix $\Psi = (I - \mathcal{G}H^T H)R(I - \mathcal{G}H^T H)^T + \mathcal{G}H^T QH\mathcal{G}$ is symmetric and positive definite. Note that the bias correction gain $\mathcal{G}H^T$ in (6.41) is related to GNPG $\mathcal{G}$, which decreases as the reciprocal of N. The solution to (6.41) is given by the infinite sum [83]

  $$\hat{P}(N) = \sum_{i=0}^{\infty} \Upsilon^i \Psi (\Upsilon^i)^T \,, \tag{6.42}$$

  whose convergence depends on the model. The trace of (6.42) can be plotted, and then N_{opt} can be determined at the point of minimum. For some models, N_{opt} can be found analytically using (6.42) [169].
- For P_k given by (6.37), the optimization problem (6.40) can also be solved numerically with respect to N_{opt} by increasing k until $\text{tr}(P_k)$ reaches a minimum. However, for some models, there may be ambiguities associated with multiple minima. Therefore, to make the batch estimator as fast as possible, N_{opt} should be determined by the first minimum [153]. It is worth noting that the solution (6.42) agrees with the fact that when the model is deterministic, $R = 0$ and $Q = 0$, and the filter is a perfect match, then we have $N_{\text{opt}} = \infty$ and $P = 0$.
- From Fig. 6.1 and many other investigations, it can be concluded that the difference between the KF and UFIR estimates vanishes on larger horizons, $N_{\text{opt}} \gg 1$ [179]. Furthermore, the UFIR filter is low sensitive to N; that is, this filter produces acceptable errors when N changes up to 30% around N_{opt} [179]. Thus, we can roughly characterize errors in the UFIR estimate at k in terms of the error covariance of the KF. If the product $H_k^T H_k$ is invertible, then replacing the Kalman gain K_k with the bias correction gain $\mathcal{G}_k H_k^T$ in the error covariance of the KF as $P_k = (I - \mathcal{G}_k H_k^T H_k)P_k^-$ and then providing simple transformations give

  $$\mathcal{G}_k^{-1} \cong H_k^T H_k P_k^- (P_k^- - P_k)^{-1} \,, \tag{6.43}$$

where P_k^- is prior error covariance of the KF. For the given (6.43), the backward recursion (6.22) can be used to compute the GNPG back in time as

$$\mathcal{G}_{k-1}^{-1} = F_k^T(\mathcal{G}_k^{-1} - H_k^T H_k)F_k \,. \tag{6.44}$$

Because an increase in N always results in a decrease in the trace of $\mathcal{G}_k$ [179], then it follows that N_{opt} can be estimated by calculating the number of steps backward until $\mathcal{G}_{n-1}$ finally becomes singular. Since $\mathcal{G}_n$ is singular when $N < K$, where K is the number of the states, the optimal horizon can be measured as $N_{\text{opt}} \cong k + K$. This approach can serve when $k > N_{\text{opt}}$ or when there are enough data points to find N_{opt}. Its advantage is that N_{opt} can be found even for LTV systems.

The previously considered methods make it possible to obtain N_{opt} by minimizing $\text{tr}(P_k)$. However, if the ground truth is unavailable and the noise covariances are not well known, then N_{opt} measured in this ways may be incorrect. If this is the case, then another approach based on the measurement residual rather than the error covariance may yield better results.

Unavailable Ground Truth

In cases when the ground truth is not available, N_{opt} can be estimated using the available measurement residual covariance S_k [153]. To justify the approach, we represent S_k as

$$\begin{aligned} S_k &= \mathcal{E}\{(y_k - H\hat{x}_k^-)(y_k - H\hat{x}_k^-)^T\} \\ S_k &= \mathcal{E}\{(H\varepsilon_k^- + v_k)(H\varepsilon_k^- + v_k)^T\} \\ &= HP_k^- H^T + R + H\mathcal{E}\{\varepsilon_k^- v_k^T\} + \mathcal{E}\{v_k \varepsilon_k^{-T}\}H^T \,, \end{aligned} \tag{6.45}$$

where the two last terms strongly depend on N. Now, three limiting cases can be considered:

- When $k = N - 1 = K - 1$, the UFIR filter degree is equal to the number of the data points, noise reduction is not provided, and the error $H\varepsilon_k^-$ with the sign reversed becomes almost equal to the measurement noise v_k. Replacing $H\varepsilon_k^-$ with $-v_k$ makes (6.45) close to zero

$$S_{K-1} = HP_{K-1}^- H^T + R - R - R \cong 0 \,. \tag{6.46}$$

- At the optimal point $k_{\text{opt}} = N_{\text{opt}} - 1$, the last two terms in (6.45) vanish by the orthogonality condition, the measurement noise prevails over the estimation error, $R \gg HP_{K-1}^- H^T$, and the residual becomes

$$S_{N_{\text{opt}}-1} = HP_{N_{\text{opt}}-1}^- H^T + R \gtrsim R \,. \tag{6.47}$$

- When $k \gg N_{\text{opt}}$, the estimation error is mainly associated with increasing bias errors. Accordingly, the first term in (6.45) becomes dominant, and S_k approaches it,

$$S_{k \gg N_{\text{opt}}} \to HP_{k \gg N_{\text{opt}}}^- H^T \,. \tag{6.48}$$

The transition between the S_k values can be learned if we take into account (6.46), (6.47), and (6.48). Indeed, when $K \leqslant N \leqslant N_{\text{opt}}$, the bias errors are practically insignificant. They grow with increasing N and approach the standard deviation in the estimate at N_{opt} when the filter becomes slightly inconsistent with the system due to process noise.

Since $\text{tr}(P_k^-)$ decreases monotonously with increasing N due to noise reduction, $\text{tr}(S_k)$ is also a monotonic function when $N < N_{\text{opt}}$. It grows monotonously from $\text{tr}(S_{K-1})$ given by (6.46) to $\text{tr}(S_{N_{\text{opt}}-1})$ given by (6.47) and passes through N_{opt} with a minimum rate when $\text{tr}(P_k)$ is minimum. It matters that the rate of $\text{tr}(S_k)$ around N_{opt} cannot be zero due to increasing bias errors. And this is irrespective of the model, as the UFIR filter reduces white noise variance as the reciprocal of N, and the effect of bias is still small when $N \leqslant N_{\text{opt}}$.

Another behavior of $\text{tr}(S_k)$ can be observed for $N > N_{\text{opt}}$, when the bias errors grow not equally in different models. Although $\text{tr}(P_k)$ is always convex on N in stable filters, bias errors affecting its ascending right side can grow either monotonously in polynomial models or eventually oscillating in harmonic models. The latter means that $\text{tr}(S_k)$ can reach its value (6.48) with oscillations, even having a constant average rate.

It follows from the previous that the derivative of $\text{tr}(S_k)$ with respect to N can pass through multiple minima and that the first minimum gives the required value of N_{opt}. The approach developed in [153] suggests that N_{opt} can be determined in the absence of the ground truth by solving the following minimization problem

$$N_{\text{opt}} \cong \arg\min_{k} \left(\frac{\partial}{\partial k}\text{tr}S_k\right) + 1\,, \tag{6.49}$$

where minimization should be ensured by increasing k, starting from $K-1$, until the first minimum is reached. To avoid ambiguity when solving the problem (6.49), the number of points should be large enough. Moreover, $\text{tr}(S_k)$ may require smoothing before applying the derivative.

Let us now look at the S_k properties in more detail and find a stronger justification for (6.49). Since the difference between the *a priori* and *a posteriori* UFIR estimates does not affect the dependence of $\text{tr}(S_k)$ on N, we can replace $\hat{x}_k^-$ with $\hat{x}_k$, ε_k^- with ε_k, and P_k^- with P_k. We also observe that, since $\mathcal{E}\{\hat{x}_{k-1}v_k^T\}=0$, $\mathcal{E}\{x_{k-1}v_k^T\}=0$, $\mathcal{E}\{w_k v_k^T\}=0$, and $\hat{x}_k$ can be replaced by the UFIR estimate (6.29), the last two terms in (6.45) can be represented as

$$\begin{aligned}
\mathcal{E}\{\varepsilon_k^- v_k^T\} &\cong \mathcal{E}\{\varepsilon_k v_k^T\}\\
&= \mathcal{E}\{(x_k-\hat{x}_k)v_k^T\} = -\mathcal{E}\{\hat{x}_k v_k^T\}\\
&= -\mathcal{E}\{[F\hat{x}_{k-1}+\mathcal{G}_k H^T(y_k - HF\hat{x}_{k-1})]v_k^T\\
&= -\mathcal{G}_k H^T\mathcal{E}\{y_k v_k^T\} = -\mathcal{G}_k H^T\mathcal{E}\{(Hx_k+v_k)v_k^T\}\\
&= -\mathcal{G}_k H^T\mathcal{E}\{Hx_k v_k^T\} - \mathcal{G}_k H^T R\\
&= -\mathcal{G}_k H^T\mathcal{E}\{H(Fx_{k-1}+w_k)v_k^T\} - \mathcal{G}_k H^T R\\
&= -\mathcal{G}_k H^T R\,,\\
\mathcal{E}\{v_k\varepsilon_k^T\} &= -RH\mathcal{G}_k\,.
\end{aligned}$$

This transforms (6.45) to

$$S_k \cong HP_kH^T + R - \mathcal{G}_k H^T R - RH\mathcal{G}_k\,. \tag{6.50}$$

Again we see the same picture. By $N = K$, the bias correction gain in the UFIR filter is close to unity, $\mathcal{G}_{N-1}H^T \cong I$, and (6.50) gives a value close to zero, as in (6.46). When $N = N_{\text{opt}}$, P_k and $\mathcal{G}_k H^T$ are small enough and (6.50) transforms into (6.47). Finally, for $N \gg N_{\text{opt}}$, P_k becomes large and $\mathcal{G}_k H^T$ small, which leads to (6.48).

Although we have already discussed many details, it is still necessary to prove that the slope of $\text{tr}(S_k)$ is always positive up to and around N_{opt} and that this function is convex on N. To show this, represent the derivative of S_k as $S_k' = S_k - S_{k-1}$, assuming a unit time-step for simplicity. For (6.50) at the point $k = N_{\text{opt}} - 1$ we have $P_k \cong P_{k-1}$ and, therefore, S_k' can be written as

$$S_k' \cong (\mathcal{G}_{k-1}-\mathcal{G}_k)H^T R + RH(\mathcal{G}_{k-1}-\mathcal{G}_k)\,.$$

Since the GNPG $\mathcal{G}_k$ decreases with increasing N, then we have $\mathcal{G}_{k-1}-\mathcal{G}_k > 0$, and it follows that $S_k' > 0$ and so is the derivative of $\text{tr}(S_k)$,

$$\left.\frac{\mathrm{d}}{\mathrm{d}k}\text{tr}(S_k)\right|_{k=N_{\text{opt}}-1} > 0\,.$$

Thus, $\text{tr}(S_k)$ passes through $N_{\text{opt}} - 1$ with minimum positive slope. Moreover, since bias errors force S_k to approach P_k as N increases, function $\frac{d}{dk}\text{tr}(S_k)$ is reminiscent of P_k. Finally, further minimizing $\frac{d}{dk}\text{tr}(S_k)$ with N yields N_{opt}, which can be called the key property of S_k.

Example 6.1 ***Optimal horizon for two-state polynomial model [153].*** Consider a two-state polynomial model (6.38) and (6.39) with $F = \begin{bmatrix} 1 & \tau \\ 0 & 1 \end{bmatrix}$, $H = [1\ \ 0]$, $B = [\tau/2\ \ 1]^T$, $w_k \sim \mathcal{N}(0, \sigma_w^2)$, $v_k \sim \mathcal{N}(0, \sigma_v^2)$, $Q = \sigma_w^2$, $R = \sigma_v^2$, $\tau = 0.1$ s, $\sigma_w = 0.01/\text{s}$, and $\sigma_v = 0.1$.

Functions $\sqrt{\text{tr}(P_k)}$, $\sqrt{\text{tr}(S_k)}$, and $\sqrt{\frac{d}{dk}\text{tr}(S_k)}$ computed analytically and numerically are shown in Fig. 6.2. As can be seen, $N_{\text{opt}} = 34$ is measured either at the minimum of $\sqrt{\text{tr}(P_k)}$ or at the minimum of $\sqrt{\frac{d}{dk}\text{tr}(S_k)}$. For this model, N_{opt} can be calculated using the empirical relationship found in [153] using the noise variances. It also gives

$$N_{\text{opt}} \cong \sqrt{\frac{12\sigma_v}{\tau\sigma_w}} \cong 34 \,. \tag{6.51}$$

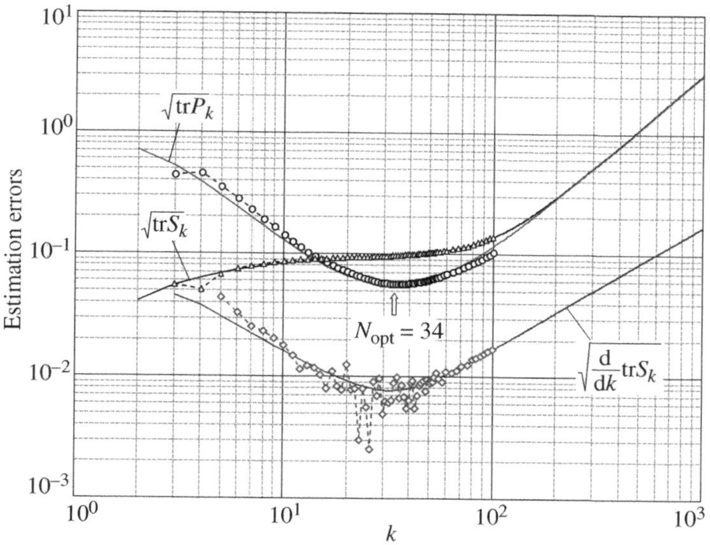

Figure 6.2 Determining N_{opt} for a UFIR filter applied to two-state polynomial model using functions $\sqrt{\text{tr}(P_k)}$, $\sqrt{\text{tr}(S_k)}$, and $\sqrt{\frac{d}{dk}\text{tr}(S_k)}$. Solid lines are theoretical, and dotted lines are experimental.

A example of measuring N_{opt} for a two-state harmonic model can be found in [153]. □

Example 6.2 ***Filtering a harmonic model [177].*** Given a continuous-time two-state harmonic model $\dot{x}(t) = Ax(t) + w(t)$ of a conservative oscillator having the system matrix $A = \begin{bmatrix} 0 & 1 \\ -\alpha^2\omega_0^2 & 0 \end{bmatrix}$, where ω_0^2 is the fundamental angular frequency, $\alpha = 1 + \Delta$, and Δ is an uncertain factor. An oscillator is observed in discrete time as $y_k = Hx_k + v_k$ with $H = [1\ \ 0]$. The stationary noise $w(t) = [0\ \ w_2(t)]^T$ is zero mean white Gaussian with the covariance

$R_w(t-\theta) = \mathcal{E}\{w(t)w^T(\theta)\} = S_w\delta(t-\theta)$, where $S_w = \begin{bmatrix} 0 & 0 \\ 0 & S_{w_2} \end{bmatrix}$ and S_{w_2} is the uniform double-sided PSD of $w_2(t)$. The measurement noise is $v_k \sim \mathcal{N}(0, \sigma_v^2)$.

The continuous-time state model is converted to $x_k = Fx_{k-1} + w_k$, where the system matrix is

$$F = \begin{bmatrix} \cos \alpha\omega_0\tau & \frac{1}{\alpha\omega_0} \sin \alpha\omega_0\tau \\ -\alpha\omega_0 \sin \alpha\omega_0\tau & \cos \alpha\omega_0\tau \end{bmatrix}$$

and zero mean noise $w_k = [\, w_{1k} \;\; w_{2k}]^T$ has the covariance

$$R = S_{w_2} \begin{bmatrix} \frac{\tau}{2\alpha^2\omega_0^2}\left(1 - \frac{\sin 2\alpha\omega_0\tau}{2\alpha\omega_0\tau}\right) & \frac{\tau}{4\alpha^2\omega_0^2}(1 - \cos 2\alpha\omega_0\tau) \\ \frac{\tau}{4\alpha^2\omega_0^2}(1 - \cos 2\alpha\omega_0\tau) & \frac{\tau}{2}\left(1 + \frac{\sin 2\alpha\omega_0\tau}{2\alpha\omega_0\tau}\right) \end{bmatrix} .$$

The KF and UFIR filter are run for $\omega_0 = 0.1$ rad/s, $\sigma_{w_2}^2 = 10^{-4}$, $\sigma_v^2 = 10$, $N_{\text{opt}} = 34$, and $\Delta = 1$ when $100 \leqslant k \leqslant 120$ and $\Delta = 0$ otherwise. Filtering errors are shown in Fig. 6.3, where the typical effects inherent to harmonic models are observed. Influenced by the uncertainty $\Delta = 1$ acting over a short period of time [100,120], KF produces long and oscillating transients, while the UFIR filter returns monotonously to normal filtering after N_{opt} points.

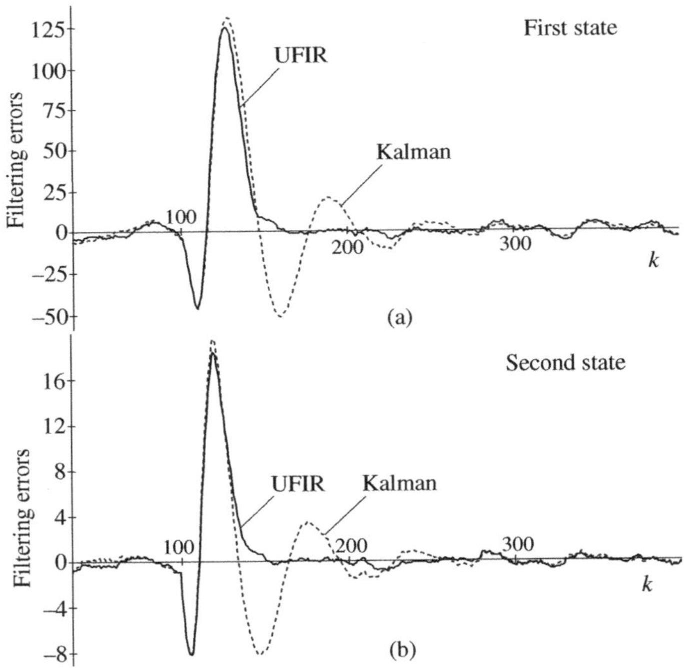

Figure 6.3 Typical filtering errors produced by the UFIR filter and KF for a harmonic model with a temporary uncertainty in the time span [100,119]: (a) first state and (b) second state. □

6.3 Backward *a posteriori* UFIR Filter

The backward UFIR filter can be derived similarly to the forward UFIR filter if we refer to (5.23) and (5.26) and write the extended model as

$$\overleftarrow{X}_{k,m} = \overleftarrow{F}_{k,m}x_k - \overleftarrow{S}_{k,m}\overleftarrow{U}_{k,m} - \overleftarrow{D}_{k,m}\overleftarrow{W}_{k,m} , \tag{6.52}$$

$$\overleftarrow{Y}_{k,m} = \overleftarrow{H}_{k,m}x_k - \overleftarrow{L}_{k,m}\overleftarrow{U}_{k,m} - \overleftarrow{G}_{k,m}\overleftarrow{W}_{k,m} + \overleftarrow{V}_{k,m} , \tag{6.53}$$

for which the definitions of all vectors and matrices can be found after (5.23) and (5.26). Batch and recursive forms for this filter can be obtained as shown next.

6.3.1 Batch Form

The batch backward UFIR estimate $\tilde{x}_m$ can be defined as

$$\begin{aligned}\tilde{x}_m &= \tilde{x}_m^{\mathrm{h}} + \tilde{x}_m^{\mathrm{f}} \\ &= \overleftarrow{\hat{\mathcal{H}}}{}^{\mathrm{h}}_{k,m}\overleftarrow{Y}_{k,m} + \overleftarrow{\hat{\mathcal{H}}}{}^{\mathrm{f}}_{k,m}\overleftarrow{U}_{k,m} ,\end{aligned} \tag{6.54}$$

where $\overleftarrow{Y}_{k,m}$ is given by (6.53). The homogenous gain $\overleftarrow{\hat{\mathcal{H}}}_{k,m} \triangleq \overleftarrow{\hat{\mathcal{H}}}{}^{\mathrm{h}}_{k,m}$ and the forced gain $\overleftarrow{\hat{\mathcal{H}}}{}^{\mathrm{f}}_{k,m}$ can be determined by satisfying the unbiasedness condition $E\{\hat{x}_m\} = E\{x_m\}$ applied to (6.54) and the model

$$x_m = \mathcal{F}_k^{m+1}x_k - \overleftarrow{\bar{S}}_{k,m}\overleftarrow{U}_{k,m} - \overleftarrow{\bar{D}}_{k,m}\overleftarrow{W}_{k,m} , \tag{6.55}$$

which is the last Nth row vector in (6.52). This gives two unbiasedness constraints

$$\mathcal{F}_k^{m+1} = \overleftarrow{\hat{\mathcal{H}}}_{k,m}\overleftarrow{H}_{k,m} , \tag{6.56}$$

$$\overleftarrow{\hat{\mathcal{H}}}{}^{\mathrm{f}}_{k,m} = \overleftarrow{\hat{\mathcal{H}}}_{k,m}\overleftarrow{L}_{k,m} - \overleftarrow{\bar{S}}_{k,m} , \tag{6.57}$$

and the first constraint (6.56) yields the fundamental gain

$$\overleftarrow{\hat{\mathcal{H}}}_{k,m} = \mathcal{F}_k^{m+1}(\overleftarrow{H}{}^T_{k,m}\overleftarrow{H}_{k,m})^{-1}\overleftarrow{H}{}^T_{k,m} . \tag{6.58}$$

For $\overleftarrow{\hat{\mathcal{H}}}_{k,m}$ obtained by (6.58), the forced gain is defined by (6.57), and we notice that the same result appears when the noise is neglected in the backward OFIR estimate (5.28). The backward *a posteriori* UFIR filtering estimate is thus given in a batch form by (6.54) with the gains (6.58) and (6.57).

The error covariance P_m for the backward UFIR filter can be written as

$$P_m = \overleftarrow{\mathcal{W}}_{k,m}\mathcal{Q}_{k,m}\overleftarrow{\mathcal{W}}{}^T_{k,m} + \overleftarrow{\mathcal{V}}_{k,m}\mathcal{R}_{k,m}\overleftarrow{\mathcal{V}}{}^T_{k,m} , \tag{6.59}$$

where the error residual matrices are specified by

$$\overleftarrow{\mathcal{W}}_{k,m} = \overleftarrow{\bar{D}}_{k,m} - \overleftarrow{\hat{\mathcal{H}}}_{k,m}\overleftarrow{G}_{k,m} , \tag{6.60}$$

$$\overleftarrow{\mathcal{V}}_{k,m} = \overleftarrow{\hat{\mathcal{H}}}_{k,m} . \tag{6.61}$$

Note that, as in the forward UFIR filter, matrices (6.60) and (6.61) provide an optimal balance between bias and random errors in the backward UFIR filtering estimate $\tilde{x}_m$ if we optimally set the averaging horizon of N_{opt} points, as shown in Fig. 6.1 and Fig. 6.2.

6.3.2 Recursions and Iterative Algorithm

The batch estimate (6.54) can be computed iteratively on $[m, k]$ if we find recursions for $\tilde{x}_m^{\mathrm{h}}$ and $\tilde{x}_m^{\mathrm{f}}$.

By introducing the backward GNPG

$$\begin{aligned}\overleftarrow{\mathcal{G}}_m &= \hat{\overleftarrow{\mathcal{H}}}_{k,m}\hat{\overleftarrow{\mathcal{H}}}^T_{k,m} \\ &= \mathcal{X}_k^{m+1}(\overleftarrow{\bar{H}}^T_{k,m}\overleftarrow{\bar{H}}_{k,m})^{-1}\mathcal{X}_k^{m+1^T} ,\end{aligned} \tag{6.62}$$

we write the homogeneous estimate as

$$\tilde{x}_m^{\mathrm{h}} = \hat{\overleftarrow{\mathcal{H}}}_{k,m}\overleftarrow{Y}_{k,m} \tag{6.63a}$$

$$= \overleftarrow{\mathcal{G}}_m(\mathcal{X}_k^{m+1})^{-T}\overleftarrow{\bar{H}}^T_{k,m}\overleftarrow{Y}_{k,m} , \tag{6.63b}$$

and represent recursively the inverse of the GNPG (6.62) by

$$\begin{aligned}\overleftarrow{\mathcal{G}}_m^{-1} &= (\mathcal{X}_k^{m+1})^{-T}\overleftarrow{\bar{H}}^T_{k,m}\overleftarrow{\bar{H}}_{k,m}\mathcal{X}_k^{m+1^T}(\mathcal{X}_k^{m+1})^{-1} \\ &= (\mathcal{X}_k^{m+1})^{-T}\begin{bmatrix}\overleftarrow{\bar{H}}^T_{k,m+1} & \mathcal{X}_k^{m+1^T}H_m^T\end{bmatrix}\begin{bmatrix}\overleftarrow{\bar{H}}_{k,m+1} \\ H_m\mathcal{X}_k^{m+1}\end{bmatrix}\mathcal{X}_k^{m+1^T}(\mathcal{X}_k^{m+1})^{-1} \\ &= H_m^T H_m + (F_{m+1}^{-1}\overleftarrow{\mathcal{G}}_{m+1}F_{m+1}^{-T})^{-1} .\end{aligned} \tag{6.64}$$

This gives two recursive forms

$$\overleftarrow{\mathcal{G}}_m = [H_m^T H_m + (F_{m+1}^{-1}\overleftarrow{\mathcal{G}}_{m+1}F_{m+1}^{-T})^{-1}]^{-1} , \tag{6.65}$$

$$\overleftarrow{\mathcal{G}}_{m+1} = F_{m+1}(\overleftarrow{\mathcal{G}}_m^{-1} - H_m^T H_m)^{-1}F_{m+1}^T . \tag{6.66}$$

Referring to $\overleftarrow{\bar{H}}^T_{k,m} = \mathcal{X}_k^{m+1^T}\overleftarrow{\mathcal{G}}_m^{-1}\hat{\overleftarrow{\mathcal{H}}}_{k,m}$ taken from (6.63a) and (6.63b), we next represent the product $\overleftarrow{\bar{H}}^T_{k,m}\overleftarrow{Y}_{k,m}$ as

$$\begin{aligned}\overleftarrow{\bar{H}}^T_{k,m}\overleftarrow{Y}_{k,m} &= \begin{bmatrix}\overleftarrow{\bar{H}}^T_{k,m+1} & \mathcal{X}_k^{m+1^T}H_m^T\end{bmatrix}\begin{bmatrix}\overleftarrow{Y}_{k,m+1} \\ y_m\end{bmatrix} , \\ &= \overleftarrow{\bar{H}}^T_{k,m+1}\overleftarrow{Y}_{k,m+1} + \mathcal{X}_k^{m+1^T}H_m^T y_m \\ &= \mathcal{X}_k^{m+2^T}\overleftarrow{\mathcal{G}}_{m+1}^{-1}\hat{\overleftarrow{\mathcal{H}}}_{k,m+1}\overleftarrow{Y}_{k,m+1} + \mathcal{X}_k^{m+1^T}H_m^T y_m \\ &= \mathcal{X}_k^{m+2^T}\overleftarrow{\mathcal{G}}_{m+1}^{-1}\tilde{x}_{m+1}^{\mathrm{h}} + \mathcal{X}_k^{m+1^T}H_m^T y_m .\end{aligned} \tag{6.67}$$

By combining (6.65), (6.66), and (6.67), a recursion for $\tilde{x}_m^{\mathrm{h}}$ can now be written as

$$\tilde{x}_m^{\mathrm{h}} = F_{m+1}^{-1}\tilde{x}_{m+1}^{\mathrm{h}} + \overleftarrow{\mathcal{G}}_m H_m^T(y_m - H_m F_{m+1}^{-1}\tilde{x}_{m+1}^{\mathrm{h}}) , \tag{6.68}$$

where $\overleftarrow{\mathcal{G}}_m H_m^T$ is the bias correction gain of the UFIR filter.

To derive a recursion for the forced gain $\tilde{x}_m^{\mathrm{f}}$, we use the decompositions

$$\overleftarrow{S}_{k,m} = F_{m+1}^{-1}[\ \overleftarrow{S}_{k,m+1} + \overleftarrow{E}_{k,m+1} \quad 0\ ] ,$$

$$\overleftarrow{G}^T_{k,m} = \begin{bmatrix} \overleftarrow{G}^T_{k,m+1} & (\overleftarrow{S}_{k,m+1} + \overleftarrow{\bar{E}}_{k,m+1})^T F^{-T}_{m+1} H^T_m \\ 0 & 0 \end{bmatrix} ,$$

$$\overleftarrow{\bar{E}}_{k,m+1} = [0 \;\; 0 \;\; \cdots \;\; 0 \;\; E_{m+1}] ,$$

follow the derivation of the recursive forms for the backward OFIR filter, provide routine transformations, and arrive at

$$\begin{aligned} \tilde{x}^{\mathrm{f}}_m &= (\overleftarrow{S}_{k,m} - \overleftarrow{\hat{H}}_{k,m} \overleftarrow{L}_{k,m}) \overleftarrow{U}_{k,m} , \\ &= (I - \overleftarrow{\mathcal{G}}_m H^T_m H_m) F^{-1}_{m+1} (\tilde{x}^{\mathrm{f}}_{m+1} + E_{m+1} u_{m+1}) . \end{aligned} \tag{6.69}$$

A simple combination of (6.68) and (6.69) finally gives

$$\begin{aligned} \tilde{x}_m &= \tilde{x}^{\mathrm{h}}_m + \tilde{x}^{\mathrm{f}}_m \\ &= \tilde{x}^-_m + \overleftarrow{\mathcal{G}}_m H^T_m (y_m - H_m \tilde{x}^-_m) , \end{aligned} \tag{6.70}$$

where the prior estimate is specified by

$$\tilde{x}^-_m = F^{-1}_{m+1} (\tilde{x}_{m+1} + E_{m+1} u_{m+1}) . \tag{6.71}$$

The backward iterative *a posteriori* UFIR filtering algorithm can now be generalized with the pseudocode listed as Algorithm 12. The algorithm starts with the initial $\overleftarrow{\mathcal{G}}_s$ and $\tilde{x}_s$ computed at s in short batch forms (6.62) and (6.54). The iterative computation is performed back in time on the horizon of N points so that the filter gives estimates from zero to $m = k - N + 1$.

Algorithm 12: Backward *a posteriori* UFIR Filtering Algorithm

---

Data: y_k, u_k

begin
  for $l = N-1, N, \cdots, k$ **do**
    $n = k - l, s = n + N - K, r = n + N - 1$;
    $\overleftarrow{\mathcal{G}}_s = \mathcal{X}^{s+1}_r (\overleftarrow{H}^T_{r,s} \overleftarrow{H}_{r,s})^{-1} \mathcal{X}^{s+1^T}_r$;
    $\check{x}_s = \overleftarrow{\mathcal{G}}_s (\mathcal{X}^{s+1}_r)^{-T} \overleftarrow{H}^T_{r,s} (\overleftarrow{Y}_{r,s} - \overleftarrow{L}_{r,s} \overleftarrow{U}_{r,s}) + \overleftarrow{S}_{r,s} \overleftarrow{U}_{r,s}$;
    for $i = 1 : N - K$ **do**
      $g = s - i + 1$;
      $\check{x}^-_{g-1} = F^{-1}_g (\check{x}_g + E_g u_g)$;
      $\overleftarrow{\mathcal{G}}_{g-1} = (H^T_{g-1} H_{g-1} + F^T_g \overleftarrow{\mathcal{G}}^{-1}_g F_g)^{-1}$;
      $K_{g-1} = \overleftarrow{\mathcal{G}}_{g-1} H^T_{g-1}$;
      $\check{x}_{g-1} = \check{x}^-_{g-1} + K_{g-1} (y_{g-1} - H_{g-1} \check{x}^-_{g-1})$;
    end for
    $\hat{x}_n = \check{x}_n$;
  end for
  Result: $\hat{x}_n$
end

Example 6.3 ***Forward and backward filtering estimates.*** Consider a moving vehicle whose trajectory, measured each second, $\tau = 1$ s, by a GPS navigation unit, is used as ground truth. To estimate the vehicle state, use the two-state polynomial model (6.38) and (6.39) with matrices specified in Example 6.1. The scalar noise components $w_k \sim \mathcal{N}(0, \sigma_w^2)$ and $v_k \sim \mathcal{N}(0, \sigma_v^2)$ are specified with $\sigma_w = 2/\text{s}$, and $\sigma_v = 3.75$, and the optimal horizon for the UFIR filter is measured as $N_{\text{opt}} = 5$.

Estimates of the vehicle coordinate x, m obtained with the forward and backward KF and UFIR filter in the presence of a temporary uncertainty are shown in Fig. 6.4. As can be seen, beyond the uncertainty, all filters give consistent and hardly distinguishable estimates. The uncertainty that occurs as a step function at $k = 789$ affects the estimates in such a way that UFIR filters produce larger excursions and shorter transients, while KFs produce smaller excursions and longer transients.

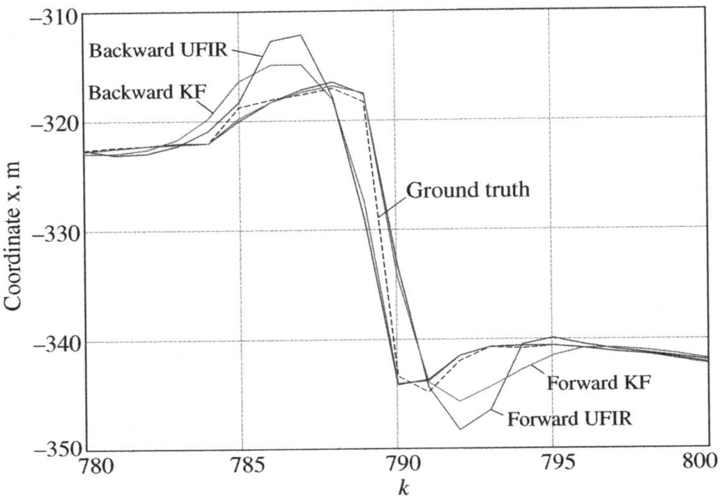

Figure 6.4 Estimates of the coordinate x , m of a moving vehicle obtained using forward and backward KFs and UFIR filters. The GPS-based ground truth is dashed.

□

It is worth noting that the typical differences between KFs and UFIR filters illustrated in Fig. 6.4 are recognized as fundamental [179]. Another thing to mention is that forward and backward filters act in opposite directions, so their responses appear antisymmetric.

6.3.3 Recursive Error Covariance

The recursive form for the error covariance (6.59) of the backward *a posteriori* UFIR filter can be found similarly to the forward UFIR filter. To this end, we first represent (6.59) using (6.60) and (6.61) as

$$\begin{aligned}
P_m &= (\overleftarrow{\bar{D}}_{k,m} - \overleftarrow{\hat{\mathcal{H}}}_{k,m}\overleftarrow{G}_{k,m})\mathcal{Q}_{k,m}(\overleftarrow{\bar{D}}_{k,m} - \overleftarrow{\hat{\mathcal{H}}}_{k,m}\overleftarrow{G}_{k,m})^T \\
&\quad + \overleftarrow{\hat{\mathcal{H}}}_{k,m}\mathcal{R}_{k,m}\overleftarrow{\hat{\mathcal{H}}}^T_{k,m} \\
&= \overleftarrow{\bar{D}}_{k,m}\mathcal{Q}_{k,m}\overleftarrow{\bar{D}}^T_{k,m} - 2\overleftarrow{\bar{D}}_{k,m}\mathcal{Q}_{k,m}\overleftarrow{G}^T_{k,m}\overleftarrow{\hat{\mathcal{H}}}^T_{k,m} \\
&\quad + \overleftarrow{\hat{\mathcal{H}}}_{k,m}\overleftarrow{G}_{k,m}\mathcal{Q}_{k,m}\overleftarrow{G}^T_{k,m}\overleftarrow{\hat{\mathcal{H}}}^T_{k,m} + \overleftarrow{\hat{\mathcal{H}}}_{k,m}\mathcal{R}_{k,m}\overleftarrow{\hat{\mathcal{H}}}^T_{k,m} .
\end{aligned}$$

We then find recursions for each of the matrix products in the previous relationship, combine the recursive forms obtained, and finally come up with

$$P_m = (I - \overleftarrow{\mathcal{G}}_m H_m^T H_m) F_{m+1}^{-1} (P_{m+1} + B_{m+1} Q_{m+1} B_{m+1}^T) \\ \times F_{m+1}^{-T} (I - \overleftarrow{\mathcal{G}}_m H_m^T H_m)^T + \overleftarrow{\mathcal{G}}_m H_m^T R_m H_m \overleftarrow{\mathcal{G}}_m \,. \tag{6.72}$$

It can now be shown that there is no significant difference between the error covariances of the forward and backward UFIR filters. Since these filters process the same data, but in opposite directions, it follows that the errors on the given horizon $[m, k]$ are statistically equal.

6.4 The q-lag UFIR Smoother

In postprocessing and when filtering stationary and quasi stationary signals, smoothing may be the best choice because it provides better noise reduction. Various types of smoothers can be designed using the UFIR approach, although many of them appear to be equivalent as opposed to OFIR smoothing. Here we will first derive the q-lag FFFM UFIR smoother and then show that all other UFIR smoothing structures of this type are equivalent due to their ability to ignore noise.

Let us look again at the q-lag FFFM FIR smoothing strategy illustrated in Fig. 5.2. The corresponding UFIR smoother can be designed to satisfy the unbiasedness condition

$$\mathcal{E}\{\tilde{x}_{k-q|k}\} = \mathcal{E}\{x_{k-q}\} \,, \tag{6.73}$$

where the q-lag estimate can be defined as

$$\tilde{x}_{k-q} \triangleq \tilde{x}_{k-q|k} = \hat{\mathcal{H}}_{m,k}^{(q)} Y_{m,k} + \hat{\mathcal{H}}_{m,k}^{\mathrm{f}(q)} U_{m,k} \,, \tag{6.74}$$

and the state model represented by the $(N-q)$th row vector in (6.3) as

$$x_{k-q} = \mathcal{F}_{k-q}^{m+1} x_m + \bar{S}_{m,k}^{(N-q)} U_{m,k} + \bar{D}_{m,k}^{(N-q)} W_{m,k} \,, \tag{6.75}$$

where $\bar{S}_{m,k}^{(N-q)}$ is the $(N-q)$th row vector in (4.9) and so is $\bar{D}_{m,k}^{(N-q)}$ in $\bar{D}_{m,k}$.

6.4.1 Batch and Recursive Forms

Condition (6.73) applied to (6.74) and (6.75) gives two unbiasedness constraints,

$$\mathcal{F}_{k-q}^{m+1} = \hat{\mathcal{H}}_{m,k}^{(q)} H_{m,k} \,, \tag{6.76}$$

$$\hat{\mathcal{H}}_{m,k}^{\mathrm{f}(q)} = \bar{S}_{m,k}^{(N-q)} - \hat{\mathcal{H}}_{m,k}^{(q)} L_{m,k} \,, \tag{6.77}$$

and after simple manipulations the first one in (6.76) gives a fundamental UFIR smoother gain $\hat{\mathcal{H}}_{m,k}^{(q)}$,

$$\hat{\mathcal{H}}_{m,k}^{(q)} = \mathcal{F}_{k-q}^{m+1} (H_{m,k}^T H_{m,k})^{-1} H_{m,k}^T \,. \tag{6.78}$$

Referring to $\mathcal{F}_{k-q}^{m+1} = (\mathcal{F}_k^{k-q+1})^{-1} \mathcal{F}_k^{m+1}$, we next transform (6.78) to

$$\hat{\mathcal{H}}_{m,k}^{(q)} = (\mathcal{F}_k^{k-q+1})^{-1} \mathcal{F}_k^{m+1} (H_{m,k}^T H_{m,k})^{-1} H_{m,k}^T \\ = (\mathcal{F}_k^{k-q+1})^{-1} \hat{\mathcal{H}}_{m,k} \,, \tag{6.79}$$

where $\hat{\mathcal{H}}_{m,k}$ is the homogeneous gain (6.10a) of the UFIR filter, and express the q-lag homogeneous UFIR smoothing estimate in terms of the filtering estimate $\hat{x}_k$ as

$$\begin{aligned}\tilde{x}^{\mathrm{h}}_{k-q} &= (\mathcal{F}_k^{k-q+1})^{-1}\hat{\mathcal{H}}_{m,k}Y_{m,k} \\ &= (\mathcal{F}_k^{k-q+1})^{-1}\hat{x}_k \end{aligned} \tag{6.80}$$

that does not require recursion; that is, the recursively computed filtering estimate $\hat{x}_k$ is projected into $k-q$ in one step by the matrix $(\mathcal{F}_k^{k-q+1})^{-1}$.

For the forced estimate, the recursive form

$$\begin{aligned}\tilde{x}^{\mathrm{f}}_{k-q} &= \hat{\mathcal{H}}^{\mathrm{f}(q)}_{m,k}U_{m,k} \\ &= \bar{S}^{(N-q)}_{m,k}U_{m,k} - \hat{\mathcal{H}}^{(q)}_{m,k}L_{m,k}U_{m,k}\end{aligned} \tag{6.81}$$

appears if we first represent the last row vector $\bar{S}_{m,k}$ of the matrix $S_{m,k}$ given by (4.9) as

$$\bar{S}_{m,k} = \mathcal{F}_k^{k-q+1}\bar{S}^{(N-q)}_{m,k} + \tilde{S}^{(q)}_{m,k} ,$$

where matrix $\tilde{S}^{(q)}_{m,k}$ becomes matrix $\tilde{D}^{(q)}_{m,k}$ if we replace B_k by E_k. Also $\bar{S}^{(N-q)}_{m,k}$ can be written as

$$\bar{S}^{(N-q)}_{m,k} = (\mathcal{F}_k^{k-q+1})^{-1}(\bar{S}_{m,k} - \tilde{S}^{(q)}_{m,k}) ,$$

and then the subsequent modification of (6.81) gives

$$\begin{aligned}\tilde{x}^{\mathrm{f}}_{k-q} &= (\mathcal{F}_k^{k-q+1})^{-1}(\bar{S}_{m,k} - \hat{\mathcal{H}}_{m,k}L_{m,k} - \tilde{S}^{(q)}_{m,k})U_{m,k} \\ &= (\mathcal{F}_k^{k-q+1})^{-1}(\hat{x}^{\mathrm{f}}_k - \tilde{S}^{(q)}_{m,k}U_{m,k}) .\end{aligned}$$

Next, using the decomposition $\tilde{S}^{(q)}_{m,k} = [\ F_k\tilde{S}^{(q)}_{m,k-1} \quad E_k\ ]$, we transform the forced estimate to the form

$$\tilde{x}^{\mathrm{f}}_{k-q} = (\mathcal{F}_k^{k-q+1})^{-1}(\hat{x}^{\mathrm{f}}_k - \check{x}^{\mathrm{f}}_{k-q}) , \tag{6.82}$$

where the q-varying product correction term $\check{x}^{\mathrm{f}}_{k-q} = \tilde{S}^{(q)}_{m,k}U_{m,k}$ is computed recursively as

$$\begin{aligned}\check{x}^{\mathrm{f}}_{k-q} &= \begin{bmatrix} F_k\tilde{S}^{(q)}_{m,k-1} & E_k \end{bmatrix}\begin{bmatrix} U_{m,k-1} \\ u_k \end{bmatrix} \\ &= F_k\check{x}^{\mathrm{f}}_{k-q-1} + E_k u_k .\end{aligned} \tag{6.83}$$

By combining (6.80) and (6.82), we finally arrive at the recursion

$$\begin{aligned}\tilde{x}_{k-q} &= \tilde{x}^{\mathrm{h}}_{k-q} + \tilde{x}^{\mathrm{f}}_{k-q} \\ &= (\mathcal{F}_k^{k-q+1})^{-1}(\hat{x}_k - \check{x}^{\mathrm{f}}_{k-q}) ,\end{aligned} \tag{6.84}$$

using which, the recursive q-lag FFFM *a posteriori* UFIR smoothing algorithm can be designed as follows. Reorganize (6.83) as

$$\check{x}^{\mathrm{f}}_{k-q-1} = F_k^{-1}(\check{x}^{\mathrm{f}}_{k-q} - E_k u_k) ,$$

set $q=0$, assign $\check{x}^{\mathrm{f}}_k = \hat{x}^{\mathrm{f}}_k$, and compute for $q = 1, 2, \ldots$ until this recursion gives $\check{x}^{\mathrm{f}}_{k-q}$. Given $\check{x}^{\mathrm{f}}_k$ and $\check{x}^{\mathrm{f}}_{k-q}$, the smoothing estimate is obtained by (6.84). It is worth noting that in the particular case of an autonomous system, $u_k = 0$, the q-lag *a posteriori* UFIR smoothing estimate is computed using a simple projection (6.80).

6.4.2 Error Covariance

In batch form, the q-varying error covariance P_{k-q} of the FFFM UFIR smoother is determined by (6.12), although with the renewed matrices,

$$P_{k-q} = \mathcal{W}^{(q)}_{m,k}\mathcal{Q}_{m,k}\mathcal{W}^{(q)^T}_{m,k} + \mathcal{V}^{(q)}_{m,k}\mathcal{R}_{m,k}\mathcal{V}^{(q)^T}_{m,k} , \tag{6.85}$$

where the error residual matrices are given by

$$\mathcal{W}^{(q)}_{m,k} = \bar{D}^{(N-q)}_{m,k} - \hat{\mathcal{H}}^{(q)}_{m,k}G_{m,k} , \tag{6.86}$$

$$\mathcal{V}^{(q)}_{m,k} = \hat{\mathcal{H}}^{(q)}_{m,k} . \tag{6.87}$$

To find a recursive form for (6.85), we write it in the form

$$P_{k-q} = (\mathcal{F}^{k-q+1}_{k})^{-1}(P_k + \tilde{P}_q)(\mathcal{F}^{k-q+1}_{k})^{-T} , \tag{6.88}$$

where P_k is the error covariance (6.37) of the UFIR filter, and represent the q-varying amendment $\tilde{P}_q$ as

$$\begin{aligned}\tilde{P}_q &= -\bar{D}_{m,k}\mathcal{Q}_{m,k}\tilde{D}^{(q)^T}_{m,k} - \tilde{D}^{(q)}_{m,k}\mathcal{Q}_{m,k}\bar{D}^{T}_{m,k} + \tilde{D}^{(q)}_{m,k}\mathcal{Q}_{m,k}G^T_{m,k}\hat{\mathcal{H}}^T_{m,k} \\ &\quad + \hat{\mathcal{H}}_{m,k}G_{m,k}\mathcal{Q}_{m,k}\tilde{D}^{(q)^T}_{m,k} + \tilde{D}^{(q)}_{m,k}\mathcal{Q}_{m,k}\tilde{D}^{(q)^T}_{m,k} \\ &= -\mathcal{Z}^{(1)}_q - \mathcal{Z}^{(2)}_q + \mathcal{Z}^{(3)}_q + \mathcal{Z}^{(4)}_q + \mathcal{Z}^{(5)}_q .\end{aligned} \tag{6.89}$$

It can be seen that $\tilde{P}_q$ naturally becomes zero at $q = 0$ due to $\tilde{D}^{(0)}_{m,k} = 0$. Furthermore, the structure of the matrix $\tilde{D}^{(q)}_{m,k}$ suggests that $\mathcal{Z}^{(1)}_q = \mathcal{Z}^{(2)}_q = \mathcal{Z}^{(5)}_q$ and also $\mathcal{Z}^{(4)}_q = \mathcal{Z}^{(3)^T}_q$.

By considering several cases of $\mathcal{Z}^{(1)}_q$ for $q = 1,2,3$,

$$\begin{aligned}\mathcal{Z}^{(1)}_1 &= B_kQ_kB^T_k , \\ \mathcal{Z}^{(1)}_2 &= \mathcal{Z}^{(1)}_1 + F_kB_{k-1}Q_{k-1}B^T_{k-1}F^T_k , \\ \mathcal{Z}^{(1)}_3 &= \mathcal{Z}^{(2)}_1 + F_kF_{k-1}B_{k-2}Q_{k-2}B^T_{k-2}F^T_{k-1}F^T_k ,\end{aligned}$$

and reasoning deductively, we obtain the following recursion for $\mathcal{Z}^{(1)}_q$,

$$\mathcal{Z}^{(1)}_q = \mathcal{Z}^{(1)}_{q-1} + M_q(\mathcal{F}^{k-q+2}_k)^{-1} , \tag{6.90}$$

where the matrix M_q is still in batch form as

$$M_q = \mathcal{F}^{k-q+2}_k B_{k-q+1}Q_{k-q+1}B^T_{k-q+1} . \tag{6.91}$$

Similarly, we represent $\mathcal{Z}^{(3)}_q$ in special cases as

$$\begin{aligned}\mathcal{Z}^{(3)}_1 &= B_kQ_kB^T_kH^T_kH_k\mathcal{G}_k , \\ \mathcal{Z}^{(3)}_2 &= \mathcal{Z}^{(3)}_1 + F_kB_{k-1}Q_{k-1}B^T_{k-1}(H^T_{k-1}H_{k-1}F^{-1}_k + F^T_kH^T_{k-1}H_k)\mathcal{G}_k , \\ \mathcal{Z}^{(3)}_3 &= \mathcal{Z}^{(3)}_2 + F_kF_{k-1}B_{k-2}Q_{k-2}B^T_{k-2}(H^T_{k-2}H_{k-2}F^{-1}_{k-1}F^{-1}_k \\ &\quad + F^T_{k-1}H^T_{k-2}H_{k-1}F^{-1}_k + F^T_{k-1}F^T_kH^T_{k-2}H_k)\mathcal{G}_k ,\end{aligned}$$

and then replace the sum in the parentheses with

$$L_q = \sum_{i=1}^{q}\mathcal{F}^{k-q+2^T}_{k-q+i}H^T_{k-q+i}H_{k-q+i}(\mathcal{F}^{k-q+i+1}_k)^{-1} , \tag{6.92}$$

which gives the following recursion for $\mathcal{Z}_q^{(3)}$,

$$\mathcal{Z}_q^{(3)} = \mathcal{Z}_{q-1}^{(3)} + M_q L_q \mathcal{G}_k \,. \tag{6.93}$$

By combining (6.90) and (6.93) with (6.89), we finally obtain the recursive form for the error covariance as

$$\tilde{P}_q = \tilde{P}_{q-1} - M_q \mathcal{F}_k^{k-q+2^T} + M_q L_q \mathcal{G}_k + \mathcal{G}_k L_q^T M_q^T \,, \tag{6.94}$$

which should be computed starting with $q = 2$ for the initial value

$$\tilde{P}_1 = -B_k Q_k B_k^T + B_k Q_k B_k^T H_k^T H_k \mathcal{G}_k + \mathcal{G}_k H_k^T H_k B_k Q_k B_k^T \tag{6.95}$$

using the matrix M_q given by (6.91) and L_q by (6.92). Noticing that the recursive form for the batch matrix L_q given by (6.92) is not available in this procedure, we postpone to "Problems" the alternative derivation of recursive forms for (6.85).

Time-Invariant Case

For LTI systems, the matrix L_q, given in batch form as (6.92), can easily be represented recursively as

$$L_q = L_{q-1} F^{-1} + F^{q-1^T} H^T H$$

and the q-lag FFFM UFIR smoothing algorithm can be modified accordingly. Computing the initial $\tilde{P}_1$ by (6.95) and knowing the matrix $L_1 = H^T H$, one can update the estimates for $q = 2,3...$ as

$$M_q = F^{q-1} B Q B^T \,, \tag{6.96}$$

$$L_q = L_{q-1} F^{-1} + F^{q-1^T} H^T H \,, \tag{6.97}$$

$$\tilde{P}_q = \tilde{P}_{q-1} - M_q F^{q-1^T} + M_q L_q \mathcal{G} + \mathcal{G} L_q^T M_q^T \,, \tag{6.98}$$

$$P_{k-q} = F^{q^{-1}} (P_k + \tilde{P}_q) F^{q^{-T}} \,. \tag{6.99}$$

It should be noted that the computation of the error covariance P_{k-q} using (6.99) does not have to be necessarily included in the iterative cycle and can be performed only once after q reaches the required lag-value.

6.4.3 Equivalence of UFIR Smoothers

Other types of q-lag UFIR smoothers can be obtained if we follow the FFBM, BFFM, and BFBM strategies discussed in Chapter 5. To show the equivalence of these smoothers, we will first look at the FFBM UFIR smoother and then draw an important conclusion.

The q-lag FFBM UFIR smoother can be obtained similarly to the FFFM UFIR smoother in several steps. Referring to the backward state-space model in (5.23) and (5.26), we define the state at $k-q$) as

$$x_{k-q} = \mathcal{X}_k^{k-q+1} x_k - \overleftarrow{\bar{S}}_{k,m}^{(q)} \overleftarrow{U}_{k,m} - \overleftarrow{\bar{D}}_{k,m}^{(q)} \overleftarrow{W}_{k,m} \tag{6.100}$$

and the forward UFIR smoothing estimate as

$$\tilde{x}_{k-q} = \hat{H}_{m,k}^{(q)} Y_{m,k} + \hat{H}_{m,k}^{f(q)} U_{m,k} \,. \tag{6.101}$$

From (6.100) at the point $q = N-1$, we can also obtain

$$x_m = \mathcal{X}_k^{m+1} x_k - \overleftarrow{\bar{S}}_{k,m} \overleftarrow{U}_{k,m} - \overleftarrow{\bar{D}}_{k,m} \overleftarrow{W}_{k,m} \,, \tag{6.102}$$

where $\overleftarrow{\bar{S}}_{k,m}$ is the last row vector in (5.41) and so is $\overleftarrow{\bar{D}}_{k,m}$ in $\overleftarrow{D}_{k,m}$.

The unbiasedness condition $\mathcal{E}\{\tilde{x}_{k-q}\} = \mathcal{E}\{x_{k-q}\}$ applied to (6.100) and (6.101) gives

$$\hat{H}^{(q)}_{m,k}H_{m,k}x_m - \mathcal{X}^{k-q+1}_k x_k = -\overleftarrow{S}^{(q)}_{k,m}\overleftarrow{U}_{k,m} - (\hat{H}^{(q)}_{m,k}L_{m,k} + \hat{H}^{f(q)}_{m,k})U_{m,k} .$$

Taking into account that $\overleftarrow{S}^{(q)}_{k,m}\overleftarrow{U}_{k,m} = \overleftarrow{S}^{(q)}_{m,k}U_{m,k}$ and replacing x_k extracted from (6.102) with $\mathcal{E}\{\overleftarrow{W}_{k,m}\} = 0$, the previous relationship can be split into two unbiasedness constraints

$$\hat{H}^{(q)}_{m,k}H_{m,k} = \mathcal{F}^{m+1}_{k-q} , \tag{6.103}$$

$$\hat{H}^{f(q)}_{m,k} = \mathcal{X}^{k-q+1}_k \bar{S}_{m,k} - \overleftarrow{S}^{(q)}_{m,k} - \hat{H}^{(q)}_{m,k}L_{m,k} . \tag{6.104}$$

What now follows is that the first constraint (6.103) is exactly the constraint (6.76) for the FFFM UFIR smoother. We then observe that for $q \geq 1$ we have $\mathcal{X}^{k-q+1}_k = (\mathcal{F}^{k-q+1}_k)^{-1}$ and thus $\overleftarrow{S}^{(q)}_{m,k}$ can be transformed as

$$\begin{aligned}
\overleftarrow{S}^{(q)}_{m,k} &= \begin{bmatrix} 0 & \dots & 0 & F^{-1}_{k-q+1}E_{k-q+1} & \dots & \mathcal{X}^{k-q+1}_{k-1}E_{k-1} & \mathcal{X}^{k-q+1}_k E_k \end{bmatrix} \\
&= \mathcal{X}^{k-q+1}_k \begin{bmatrix} 0 & \dots & 0 & (\mathcal{X}^{k-q+2}_k)^{-1}E_{k-q+1} & \dots & F_kE_{k-1} & E_k \end{bmatrix} \\
&= (\mathcal{F}^{k-q+1}_k)^{-1}\tilde{S}^{(q)}_{m,k} .
\end{aligned}$$

Finally, we transform the second constraint (6.104) to

$$\hat{H}^{f(q)}_{m,k} = (\mathcal{F}^{k-q+1}_k)^{-1}(\bar{S}_{m,k} - \tilde{S}^{(q)}_{m,k} - \hat{H}_{m,k}L_{m,k}) \tag{6.105}$$

and conclude that this is constraint (6.77) for FFFM UFIR smoothing.

Now the following important conclusion can be drawn. Because the FFBM and FFFM UFIR smoothers obey the same unbiasedness constraints in (6.76) and (6.77), then it follows that these smoothers are equivalent. And this is not surprising, since all UFIR structures obeying only the unbiasedness constraint ignore noise. By virtue of that, the forward and backward models become identical at $k - q$, and thus the FFFM and FMBM UFIR smoothers are equivalent. In addition, since the forward and backward UFIR filters are equivalent for the same reason, then it follows that the BFFM and BFBM UFIR smoothers are also equivalent, and an important finding follows.

***Equivalence of UFIR smoothers**: Satisfied only the unbiasedness condition, q-lag FFFM, FFBM, BFFM, and BFBM UFIR smoothers are equivalent.*

In view of the previous definition, the q-lag FFFM UFIR smoother can be used universally as a UFIR smoother in both batch and iterative forms. Other UFIR smoothing structures such as FFBM, BFFM, and BFBM have rather theoretical meaning.

6.5 State Estimation Using Polynomial Models

There is a wide class of systems and processes whose states change slowly over time and, thus, can be represented by degree polynomials. Examples can be found in target tracking, networking, and biomedical applications. Signal envelopes in narrowband wireless communication channels, remote wireless control, and remote sensing are also slowly changing.

The theory of UFIR state estimation developed for discrete time-invariant polynomial models [172] implies that the process can be represented on $[m, k]$ using the following state-space equations

$$x_k = Fx_{k-1} + Bw_k \,, \tag{6.106}$$

$$y_k = x_k + v_k \,, \tag{6.107}$$

where w_k and v_k are zero mean noise vectors. The $i \geq 1$ power of the system matrix $F \in \mathbb{R}^{K\times K}$ has a specific structure

$$F^i = \begin{bmatrix} 1 & \tau i & \frac{(\tau i)^2}{2} & \dots & \frac{(\tau i)^{K-1}}{(K-1)!} \\ 0 & 1 & \tau i & \dots & \frac{(\tau i)^{K-2}}{(K-2)!} \\ 0 & 0 & 1 & \dots & \frac{(\tau i)^{K-3}}{(K-3)!} \\ \vdots & \vdots & \vdots & \ddots & \vdots \\ 0 & 0 & 0 & \dots & 1 \end{bmatrix} , \tag{6.108}$$

which means that, by $i = 1$, each row in F is represented with the descending ith degree, $i \in [0, K-1]$, Taylor or Maclaurin series, where K is the number of the states.

6.5.1 Problems Solved with UFIR Structures

The UFIR approach applied to the model in (6.106) and (6.107) to satisfy the unbiasedness condition gives a unique ith degree polynomial impulse response function $h_k^{(i)}(p)$ for each of the states separately, where $p \gtrless 0$ is a discrete time shift relative to the current point k. Function $h_k^{(i)}(p)$ has many useful properties, which make the UFIR estimate near optimal. Depending on p, the following types of $h_k^{(i)}(p)$ can be recognized. When $p = 0$, function $h_k^{(i)}$ is used to obtain UFIR filtering. When $p < 0$, function $h_k^{(i)}(-q)$ serves to obtain $(q = -p)$-lag UFIR smoothing filtering. Finally, when $p > 0$, function $h_k^{(i)}(p)$ is used to obtain p-step UFIR predictive filtering.

Accordingly, the following problems can be solved by applying UFIR structures to a polynomial signal s_k measured as y_k in the presence of zero mean additive noise:

- **Filtering** provides an estimate at k based on data taken from $[m, k]$,

$$\hat{s}_{k|k} = \sum_{n=0}^{N-1} h_n^{(i)} y_{k-n} \,. \tag{6.109}$$

- **Smoothing filtering** provides a q-lag, $q > 0$, smoothing estimate at k based on data taken from $[m+q, k+q]$,

$$\tilde{s}_{k|k+q} = \sum_{n=-q}^{N-1-q} h_n^{(i)}(-q) y_{k-n} \,. \tag{6.110}$$

- **Predictive filtering** provides a p-step, $p > 0$, predictive estimate at k based on data taken from $[m-p, k-p]$,

$$\tilde{s}_{k|k-p} = \sum_{n=p}^{N-1+p} h_n^{(i)}(p) y_{k-n} \,. \tag{6.111}$$

Note that one-step predictive UFIR filtering, $p = 1$, was originally developed for polynomial models in [71]. In state space, it is known as RH FIR filtering [106] used in state-feedback control and MPC.

- **Smoothing** provides a q-lag smoothing estimate at $k-q$, $q>0$, based on data taken from $[m,k]$,

$$\tilde{s}_{k-q|k} = \sum_{n=0}^{N-1} \tilde{h}_n^{(i)}(-q) y_{k-n} . \tag{6.112}$$

- **Prediction** provides a p-step, $p>0$, prediction at $k+p$ based on data taken from $[m,k]$,

$$\tilde{s}_{k+p|k} = \sum_{n=0}^{N-1} \tilde{h}_n^{(i)}(p) y_{k-n} . \tag{6.113}$$

In the previous definitions of the state estimation problems, the functions $h_n^{(i)}(p)$ and $\tilde{h}_n^{(i)}(p)$ are not equal for $p \neq 0$, but can be transformed into each other. More detail can be found in [181].

6.5.2 The *p*-shift UFIR Filter

Looking at the details of the UFIR strategy, we notice that the approach that ignores zero mean noise allows us to solve universally the filtering problem (6.109), the smoothing filtering problem (6.110), and the predictive filtering problem (6.111) by obtaining a p-shift estimate [176]. The UFIR approach also assumes that a shift to the past can be achieved at point k using data taken from $[m+p, k+p]$ with a positive smoother lag $q=-p$, a shift to the future at point k using data taken from $[m-p, k-p]$ with a positive prediction step $p>0$, and that $p=0$ means filtering.

Thus, the p-shift UFIR filtering estimate can be defined as

$$\begin{aligned} \hat{x}_{k|k-p} &= \hat{\mathcal{H}}_N(p) Y_{k-p,m-p} \\ &= \sum_{n=p}^{N-1+p} \mathcal{H}_n(p) y_{k-n} , \end{aligned} \tag{6.114}$$

where the components of the gain $\hat{\mathcal{H}}_N(p) = [\ \mathcal{H}_0(p)\ \ \mathcal{H}_1(p)\ \ \dots\ \ \mathcal{H}_{N-1}(p)\ ]$ are diagonal matrices specified by the matrix

$$\mathcal{H}_n(p) = \text{diag}(\ h_n^{(K-1)}(p)\ h_n^{(K-2)}(p)\ \dots\ h_n^{(0)}(p)\ ) ,$$

whose components, in turn, are the values of the function $h_n^{(i)}(p)$. The unbiasedness condition applied to (6.114) gives the unbiasedness constraint

$$F^{N-1+p} = \hat{\mathcal{H}}_N(p) H_N(p) , \tag{6.115}$$

where $H_N(p) = [\ (F^{N-1+p})^T\ \ (F^{N-2+p})^T\ \dots\ (F^{1+p})^T\ \ (F^p)^T\ ]^T$.

For the lth system state, the p-shift UFIR filtering estimate is defined as

$$\hat{x}_{l(k+p)} = \hat{\mathcal{H}}_{Nl}(p) H_{Nl}(p) x_m \tag{6.116}$$

and the constraint (6.115) is transformed to

$$(F^{N-1+p})_l = \hat{\mathcal{H}}_{Nl}(p) H_{Nl}(p) , \tag{6.117}$$

where $(F)_l$ means the lth row in F and the remaining lth rows are given by

$$\hat{\mathcal{H}}_{Nl}(p) = [\ h_0^{(K-l)}(p)\ h_1^{(K-l)}(p)\ \dots\ h_{N-1}^{(K-l)}(p)\ ] , \tag{6.118}$$

$$H_{Nl} = [\ (F_l^{N-1+p})^T\ \ (F_l^{N-2+p})^T\ \dots\ (F_l^{1+p})^T\ \ (F_l^p)^T\ ]^T . \tag{6.119}$$

It is worth noting that the linear matrix equation (6.117) can be solved analytically for the ith degree polynomial impulse response $h_k^{(i)}(p)$. This gives the p-varying function

$$h_k^{(i)}(p) = \sum_{j=0}^{i} a_{ji}(p)k^j , \tag{6.120}$$

where $i \in [0, K-1]$, $k \in [p, N-1+p]$, and the coefficient $a_{ji}(p)$ is defined by

$$a_{ji}(p) = (-1)^j \frac{M_{(j+1)1}^{(i)}(p)}{|\Lambda_i(p)|} , \tag{6.121}$$

where the determinant $|\Lambda_i(p)|$ of the p-varying Hankel matrix $\Lambda_i(p) = \Theta_i^T(p)\Theta_i(p)$ is specified via the Vandermonde matrix $\Theta_i(p)$ as

$$\Lambda_i(p) = \begin{bmatrix} c_0(p) & c_1(p) & \dots & c_i(p) \\ c_1(p) & c_2(p) & \dots & c_{i+1}(p) \\ \vdots & \vdots & \dots & \vdots \\ c_i(p) & c_{i+1}(p) & \dots & c_{2i}(p) \end{bmatrix} \tag{6.122}$$

and $M_{(j+1)1}^{(i)}(p)$ is the minor of $\Lambda_i(p)$. The uth component c_u, $u \in [0,2i]$, of matrix (6.122) is the power series

$$c_u(p) = \sum_{i=p}^{N-1+p} i^u = \frac{1}{u+1}[B_{u+1}(N+p) - B_{u+1}(p)] , \tag{6.123}$$

where $B_n(x)$ is the Bernoulli polynomial.

The coefficients of several low-degree polynomials $h_k^{(i)}(p)$, which are most widely used in practice, are given in Table 6.1. Using this table, or (6.121) for higher-degree systems, one can obtain an analytic form for $h_k^{(i)}(p)$ as a function of p, where $p = 0$ is for UFIR filtering, $p = -q < 0$ for UFIR smoothing filtering, and $p > 0$ for UFIR predictive filtering.

Properties of p-shift UFIR Filters

The most important properties of the impulse response function $h_k^{(i)}(p)$ are listed in Table 6.2, which also summarizes some of the critical findings [181]. If we set $p = 0$, we can use this table to examine

Table 6.1 Coefficients $a_{ji}(p)$ of Low-Degree Functions $h_k^{(i)}(p)$.

$h_k^{(i)}(p)$	Coefficients
Uniform:	$a_{00} = \frac{1}{N}$
Ramp:	$a_{01} = \frac{2(2N-1)(N-1)+12p(N-1+p)}{N(N^2-1)}$
	$a_{11} = -\frac{6(N-1+2p)}{N(N^2-1)}$
Quadratic:	$a_{02} = 3\frac{3N^4-12N^3+17N^2-12N+4+12(N-1)(2N^2-5N+2)p}{N(N^4-5N^2+4)}$ $+3\frac{12(7N^2-15N+7)p^2+120(N-1)p^3+60p^4}{N(N^4-5N^2+4)}$
	$a_{12} = -18\frac{2N^3-7N^2+7N-2+2(7N^2-15N+7)p+30(N-1)p^2+20p^3}{N(N^4-5N^2+4)}$
	$a_{22} = 30\frac{N^2-3N+2+6(N-1)p+6p^2}{N(N^4-5N^2+4)}$

Table 6.2 Main Properties of $h_k^{(i)}(p)$.

Property	
Region of existence:	$p \leqslant k \leqslant N-1+p$
z-transform at $\omega = 0$:	$H_i(z=1,p) = 1$ (UFIR filter is an LP filter)
Unit area:	$\sum_{k=p}^{N-1+p} h_k^{(i)}(p) = 1$
Energy (NPG):	$\sum_{k=p}^{N-1+p} h_k^{(i)^2}(p) = h_0^{(i)}(p) = a_{0i}(p)$
Value at zero:	$h_0^{(i)}(p) > 0$
Zero moments:	$\sum_{k=p}^{N-1+p} h_k^{(i)}(p)k^u = 0\,,\ 1 \leqslant u \leqslant i$
	$\sum_{k=p}^{N-1+p} h_k^{(l)}(p)h_k^{(i)}(p)k^u = 0\,,\ 1 \leqslant u \leqslant \lvert i-l \rvert$
Orthogonality:	$\sum_{k=0}^{N-1} \frac{2k}{N(N-1)} h_k^{(l)} h_k^{(i)} = \frac{i+1}{N(N-1)} \prod_{i=0}^{r} \frac{N-1-r}{N+r} \delta_{ki}\,,$ $\sum_{k=p}^{N-1+p} \rho(k,p) h_k^{(l)} h_k^{(i)} = d_i^2(k,p)\delta_{li}\,,$ $l, i \in [0, K-1]$
Unbiasedness:	$\int_0^{2\pi} \mathcal{T}^{(i)}(e^{j\omega T})\mathrm{d}(\omega T) = \int_0^{2\pi} \lvert\mathcal{T}^{(i)}(e^{j\omega T})\rvert^2 \mathrm{d}(\omega T)$
	$\sum_{n=0}^{N-1} \lvert\mathcal{T}^{(i)}(n)\rvert^2 = \sum_{n=0}^{N-1} \mathcal{T}^{(i)}(n)$

the properties of the UFIR filter. We can also characterize the UFIR filter in terms of system theory. Indeed, since the transfer function $\mathcal{T}^{(i)}(z)$ of the ith degree UFIR filter is equal to unity at zero frequency, $z = 1$, then it follows that this filter is essentially a *low-pass* (LP) filter. Moreover, if we analyze other types of FIR structures in a similar way, we can come to the following general conclusion.

> ***State estimator in the transform domain****: All optimal, optimal unbiased, and unbiased state estimators are essentially LP structures.*

Since the sum of the values of $h_k^{(i)}(p)$ is equal to unity, it follows that the UFIR filter is strictly stable. More specifically, it is a BIBO stable filter due to the FIR. The sum of the squared values of $h_k^{(i)}(p)$ represents the filter NPG, which is equal to the energy of the function $h_k^{(i)}$. The important thing is that NPG is equal to the function $h_0^{(i)}$ at zero, which, in turn, is equal to the zero-degree coefficient $a_{0i}(p)$. This means that the denoising properties of the UFIR filter can be fully explored using the value NPG $= a_{0i}(p)$. It is also worth noting that the family of functions $h_k^{(l)}$ and $h_k^{(i)}$, $\{l, i\} \in [0, K-1]$, establish an orthogonal basis, and thus high-degree impulse responses can be computed through low-degree impulse responses using a recurrence relation.

The fact that all moments of the function $h_k^{(i)}(p)$ are equal to zero means nothing more and nothing less than the UFIR filter is strictly unbiased by design. The unbiasedness of FIR filters can also be checked by equating the area of the transfer function and the area of the squared magnitude frequency response. The same test for unbiasedness in the *discrete Fourier transform* (DSP) domain is ensured by equating the sum of the DSP values and the sum of the squared magnitude values. At the end of this chapter, when we will consider practical implementations, we will take a closer look at the properties of UFIR filters in the transform domain.

6.5.3 Filtering of Polynomial Models

The UFIR filter can be viewed as a special case of the p-shift UFIR filter when $p = 0$. This transforms the impulse response function (6.120) to

$$h_k^{(i)} = \sum_{j=0}^{i} a_{ji} k^j \,, \quad i \in [0, K-1], \quad k \in [0, N-1] \,, \tag{6.124}$$

where the coefficient a_{ji} is defined by (6.121) as

$$a_{ji} = (-1)^j \frac{M_{(j+1)1}^{(i)}}{|\Lambda_i|} \,. \tag{6.125}$$

The constant (zero-degree, $i = 0$) FIR function $h_k^{(0)} = a_{00} = \frac{1}{N}$ is used for simple averaging. The ramp (first-degree, $i = 1$) FIR function

$$h_k^{(1)} = a_{01} + a_{11} k = \frac{2(2N-1) - 6k}{N(N+1)} \tag{6.126}$$

is applicable for linear signals. The quadratic (second-degree, $i = 2$) FIR function

$$\begin{aligned} h_k^{(2)} &= a_{02} + a_{12} k + a_{22} k^2 \\ &= \frac{3(3N^2 - 3N + 2) - 18(2N-1) - 30}{N(N+1)(N+2)} \end{aligned} \tag{6.127}$$

is applicable for quadratically changing signals, etc.

There is also a recurrence relation [131]

$$\begin{aligned} h_k^{(i)} &= 2\frac{i^2(2N-1) - k(4i^2-1)}{i(2i-1)(N+i)} h_k^{(i-1)} \\ &\quad - \frac{(2i+1)(N-i)}{(2i-1)(N+i)} h_k^{(i-2)} \end{aligned} \tag{6.128}$$

that can be used when $i \geq 1$ and $h_k^{(-i)} = 0$ to compute $h_k^{(i)}$ of any degree in terms of the lower-degree functions. Several low-degree functions $h_k^{(i)}$ computed using (6.128) are shown in Fig. 6.5.

The NPG of a UFIR filter, which is defined as $\text{NPG} = a_{0i} = h_0^{(i)}$, suggests that the best noise reduction associated with the lowest NPG is obtained by simple averaging (case $i = 0$ in Fig. 6.5). An increase in the filter degree leads to an increase in the filter output noise, and the following statements can be made.

> ***Noise reduction with polynomial filters****: 1) A zero-degree filter (simple averaging) is optimal in the sense of minimum produced noise; 2) An increase in the filter degree or, which is the same, the number of states leads to an increase in random errors.*

Therefore, because of the better noise reduction, the low-degree UFIR state estimators are most widely used. In Fig. 6.5, we see an increase in $\text{NPG} = h_0^{(i)}$ caused by an increase in the degree i at $k = 0$.

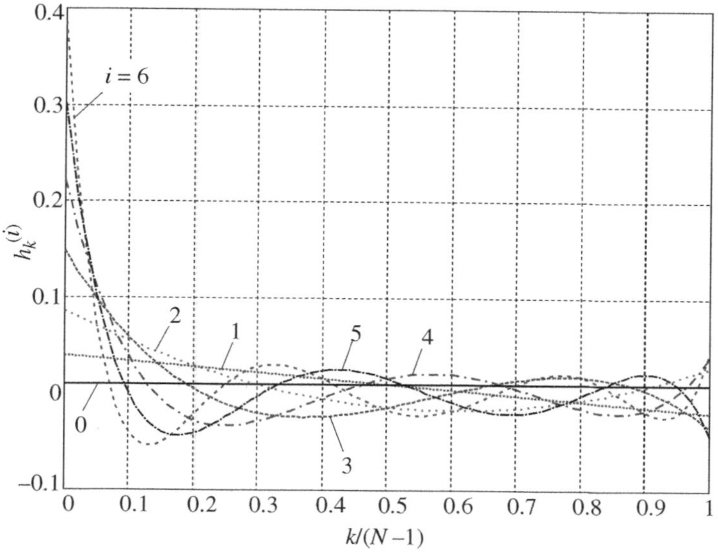

Figure 6.5 Low-degree polynomial FIR functions $h_k^{(i)}$.

6.5.4 Discrete Shmaliy Moments

In [131], the class of the ith-degree polynomial FIR functions $h_k^{(i)}$, defined by (6.124) and having the previously listed fundamental properties, was tested using the orthogonality condition

$$\sum_{k=0}^{N-1} \rho(k,N) h_k^{(l)} h_k^{(i)} = d_i^2(N) \delta_{li} , \tag{6.129}$$

where δ_{li} is the Kronecker symbol and $\{l, i\} \in [0, K-1]$. It was found that the set of functions $\{h_k^{(l)} h_k^{(i)}\}$ for $\{l, i\} \geq 1$ and $l \neq i$ is orthogonal on $[0, N-1]$ with the square of the weighted norm $d_i^2(N)$ of $h_k^{(i)}$ given by

$$d_i^2(N) = \frac{i+1}{N(N-1)} \prod_{g=0}^{i} \frac{N-1-g}{N+g} = \frac{(i+1)(N-i-1)_i}{N(N)_{i+1}} , \tag{6.130}$$

where $(a)_0 = 1$ and $(a)_i = a(a+1)\dots(a+i-1)$ for $i \geq 1$ is the Pochhammer symbol. The non-negative weight $\rho(k, N)$ in (6.129) is the ramp probability density function

$$\rho(k,N) = \frac{2k}{N(N-1)} \geq 0 . \tag{6.131}$$

An applied significance of this property for signal analysis is that, due to the orthogonality, higher-degree functions $h_k^{(i)}$ can be computed in terms of lower-degree functions using a recurrence relation (6.128).

Since the functions $h_k^{(i)}$ have the properties of discrete orthogonal polynomials (DOP), they were named in [61, 162] *discrete Shmaliy moments* (DSM) and investigated in detail. It should be noted that, due to the embedded unbiasedness, DSM belong to the one-parameter family of DOP, while the classical Meixner and Krawtchouk polynomials belong to the two-parameter family, and the most general Hahn polynomials belong to the three-parameter family of DOP [131]. This makes the DSM more suitable for unbiased analysis and reconstruction if signals than the classical DOP and Tchebyshev polynomials. Note that DSM are also generalized by Hanh's polynomials along with the classical DOP.

6.5.5 Smoothing Filtering and Smoothing

Both the q-lag UFIR smoothing filtering problem (6.110) and the UFIR smoothing problem (6.112) can be solved universally using (6.120) and (6.121) if we assign $p = -q$, $q > 0$.

Smoothing Filtering

A feature of UFIR smoothing filtering is that the function $h_0^{(i)}(q)$ exists on the horizon $-q \leqslant k \leqslant N - 1 - q$, and otherwise it is equal to zero, while the lag q is limited to $0 < q < N - 1$. Zero-degree UFIR smoothing filtering is still provided by simple averaging. First-degree UFIR smoothing filtering can be obtained using the q-varying ramp response function

$$h_k^{(1)}(q) = h_k^{(1)} - 12q\frac{N-1-q-k}{N(N^2-1)}\,, \tag{6.132}$$

where $h_k^{(1)}$ is the UFIR filter ramp impulse response (6.126). What can be observed is that better noise reduction is accompanied in (6.132) by loss of stability. Indeed, when N approaches unity, the second term in (6.132) grows indefinitely, which also follows from NPG of the first-degree given by

$$a_{01}(q) = a_{01} - 12q\frac{N-1-q}{N(N^2-1)} \tag{6.133}$$

$$\cong \frac{4}{N}\left(1 - \frac{3}{N}q\right)\,, N \gg 1, q \ll N\,. \tag{6.134}$$

Approximation (6.134) valid for large N and $q \ll N$ clearly shows that increasing the horizon length N leads to a decrease in NPG, which improves noise reduction. Likewise, increasing the lag q results in better noise reduction.

The features discussed earlier are inherent to UFIR smoothing filters of any degree, although with the important specifics illustrated in Fig. 6.6. This figure shows that all smoothing filters of

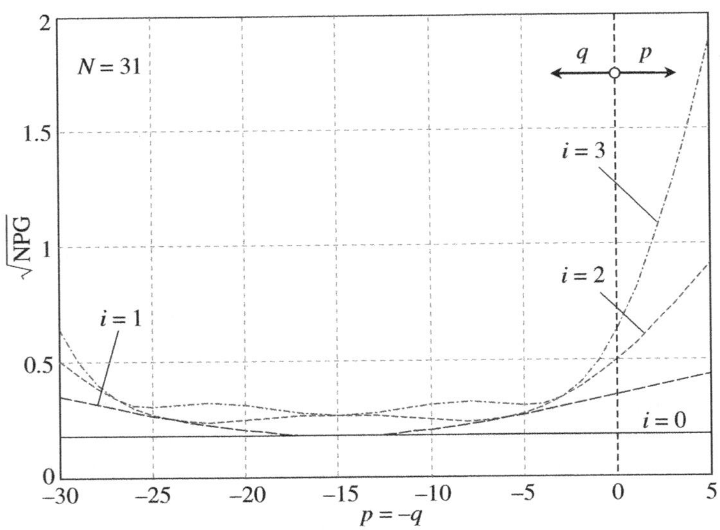

Figure 6.6 The q-varying NPG of a UFIR smoothing filter for several low-degrees.

degree $i > 0$ provide better noise reduction as q increases, starting from zero. However, NPG can have multiple minima, and therefore the optimal lag q_{opt} does not necessarily correspond to the middle of the averaging horizon $[m, k]$. For odd degrees, q_{opt} can be found exactly in the middle of $[m, k]$, while for even degrees at other points. For more information see [174].

Smoothing

The smoothing problem defined by (6.112) and discussed in state space in Chapter 5 can be solved if we introduce a gain $\tilde{h}_k^{(i)}(q)$ as

$$\tilde{h}_k^{(i)}(q) = h_{k-q}^{(i)}(q) , \tag{6.135}$$

which exists on $[0, N-1]$ with the same major properties as $h_k^{(i)}(q)$. Indeed, the ramp UFIR smoother can be designed using the FIR function

$$\tilde{h}_k^{(1)}(q) = h_k^{(1)} - 6q\frac{N-1-2k}{N(N^2-1)} . \tag{6.136}$$

The NPG for this smoother is defined by

$$\text{NPG}(q) = h_q^{(1)} - 6q\frac{N-1-q}{N(N^2-1)} \tag{6.137}$$

$$= a_{01} - 12q\frac{N-1-q}{N(N^2-1)} . \tag{6.138}$$

As expected, NPG (6.138) is exactly the same as (6.133) of the UFIR smoothing filter, since noise reduction is provided by both structures with equal efficiency. Note that similar conclusions can be drawn for other UFIR smoothing structures applied to polynomial models.

6.5.6 Generalized Savitzky-Golay Filter

A special case of the UFIR smoothing filter was originally shown in [158] and is now called the Savitzky-Golay (SG) filter. The convolution-based smoothed estimate appears at the output of the SG filter with a lag $q = \frac{N-1}{2}$ in the middle of the averaging horizon as

$$\tilde{s}_{k|k+\frac{N-1}{2}} = \sum_{n=-\frac{N-1}{2}}^{\frac{N-1}{2}} \psi_n y_{k-n} , \tag{6.139}$$

where the set of N convolution coefficients ψ_n is determined by the linear LS method to fit with typically low-degree polynomial processes. Since the coefficients ψ_n can be taken directly from the FIR function $h_n^{(i)}(-q)$, the SG filter is a special case of (6.110) with the following restrictions:

- The horizon length N must be odd; otherwise, a fractional number appears in the sum limits.
- The fixed-lag is set as $q = \frac{N-1}{2}$, while some applications require different lags and the optimal lag may not be equal to this value.

It then follows that the UFIR smoothing filter (6.110), developed for arbitrary $N > 1$ and lags $0 < q < N-1$, generalizes the SG filter (6.139) in the particular case of odd N and $q = \frac{N-1}{2}$. Also note that the lag in (6.110) can be optimized for even-degree polynomials as shown in [174].

6.5.7 Predictive Filtering and Prediction

The predictive filtering problem (6.111) can be solved directly using the p-shift FIR function (6.120) if we set $p > 0$. Like filtering and smoothing, zero-degree UFIR predictive filtering is provided by simple averaging. The first-degree predictive filter can be designed using a p-varying ramp function

$$h_k^{(1)}(p) = h_k^{(1)} + 12p\frac{N-1+p-k}{N(N^2-1)}, \tag{6.140}$$

which makes a difference with smoothing filtering. The NPG of the predictive filter is determined by

$$a_{01}(p) = a_{01} + 12p\frac{N-1+p}{N(N^2-1)} \tag{6.141}$$

$$\cong \frac{4}{N}\left(1+\frac{3}{N}p\right), N \gg 1, p \ll N, \tag{6.142}$$

and we note an important feature: increasing the prediction step p leads to an increase in NPG, and denoising becomes less efficient. Looking at the region around $p = 0$ in Fig. 6.5 and referring to (6.138) and (6.142), we come to the obvious conclusion that *prediction is less precise than filtering* and *filtering is less precise than smoothing.*

The prediction problem (6.113) can be solved similarly to the smoothing problem by introducing the gain

$$\tilde{h}_k^{(l)}(p) = h_{k+p}^{(l)}(p), \tag{6.143}$$

which exists on $[0, N-1]$ with the same main properties as $\tilde{h}_k^{(l)}(q)$. The ramp UFIR predictor can be designed using the FIR function

$$\tilde{h}_k^{(1)}(p) = h_k^{(1)} + 6p\frac{N-1-2k}{N(N^2-1)} \tag{6.144}$$

and its efficiency can be estimated using the NPG (6.141). Note that similar conclusions can be drawn for other UFIR predictors corresponding to polynomial models.

Example 6.4 ***Smoothing, filtering, and prediction errors [177].*** Consider the dynamics of an object described by the following continuous-time state model $\dot{x}(t) = Ax(t) + w(t)$ with $A = \begin{bmatrix} 0 & 1 \\ 0 & 0 \end{bmatrix}$. Convert this model in discrete time to $x_k = Fx_{k-1} + w_k$, where $F = \begin{bmatrix} 1 & \tau \\ 0 & 1 \end{bmatrix}$ and $\tau = 0.1$ s. The noise $w_k \sim \mathcal{N}(0, R)$ has the covariance $R = \begin{bmatrix} S_{w_2}\tau + \frac{\tau^3}{3}S_{w_3} & \frac{\tau^2}{2}S_{w_3} \\ \frac{\tau^2}{2}S_{w_3} & \tau S_{w_3} \end{bmatrix}$, where S_{w_2} and S_{w_3} are the uniform double-sides PSDs of the second and third state noise, respectively. The object is observed as $y_k = Hx_k + v_k$, where $H = [\,1 \;\; 0\,]$ and $v_k \sim \mathcal{N}(0, \sigma_v^2)$.

To compare p-varying estimation errors, we set $\sigma_{w_2}^2 = 10^{-1}/\text{s}$, $\sigma_{w_3}^2 = 10^{-3}/\text{s}^2$, and $\sigma_v^2 = 1.33$ and run the p-shift OFIR and UFIR filters. The estimation errors are shown in Fig. 6.7 for $N = 10$ and $N = 20$. It can be seen that the minimum errors correspond to the lag $q = N/2$ and grow at any deviation from this point. □

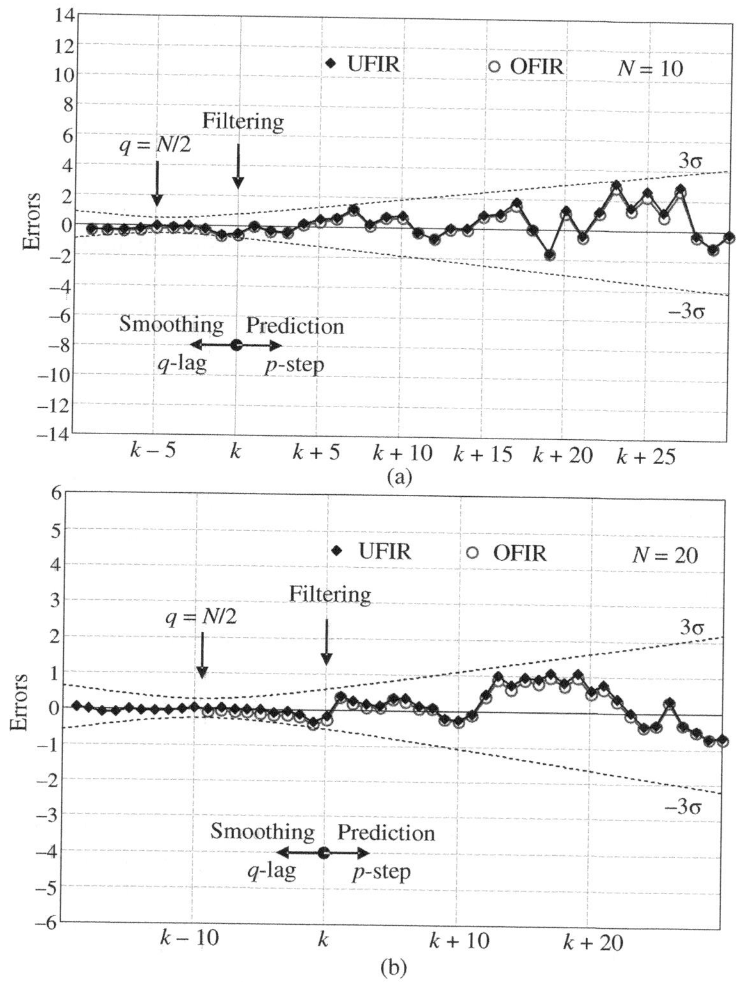

Figure 6.7 Typical smoothing, filtering, and prediction errors produced by OFIR and UFIR filters for a two-state tracking model: (a) $N = 10$ and (b) $N = 20$.

6.6 UFIR State Estimation Under Colored Noise

Like KF, the UFIR filter can also be generalized for Gauss-Markov colored noise, if we take into account that unbiased averaging ignores zero mean noise. Accordingly, if we convert a model with colored noise to another with white noise and then ignore the white noise sources, the UFIR filter can be used directly. In Chapter 3, we generalized KF for CMN and CPN. In what follows, we will look at the appropriate modifications to the UFIR filter and focus on what makes it better in the first place: the ability to filter out more realistic nonwhite noise. Although such modifications require tuning factors, and therefore the filter may be more vulnerable and less robust, the main effect is usually positive.

6.6.1 Colored Measurement Noise

We consider the following state-space model with Gauss-Markov CMN, which was used in Chapter 3 to design the GKF,

$$x_k = F_k x_{k-1} + E_k u_k + B_k w_k \,, \tag{6.145}$$

$$v_k = \Psi_k v_{k-1} + \xi_k \,, \tag{6.146}$$

$$y_k = H_k x_k + v_k \,, \tag{6.147}$$

where $w_k \sim \mathcal{N}(0, Q_k)$ and $\xi_k \sim \mathcal{N}(0, R_{\xi k})$ have the covariances $Q_k = E\{w_k w_k^T\}$ and $R_{\xi k} = E\{\xi_k \xi_k^T\}$. The coloredness factor Ψ_k is chosen such that the Gauss-Markov noise v_k is always stationary, as required. Using measurement differencing as $z_k = y_k - \Psi_k y_{k-1}$, we write a new observation as

$$\bar{z}_k = z_k - \bar{E}_k u_k = \bar{H}_k x_k + \bar{v}_k \,, \tag{6.148}$$

where $\bar{H}_k = H_k - \Gamma_k$ is the new observation matrix, the auxiliary matrices are defined as $\bar{E}_k = \Gamma_k E_k$ and $\Gamma_k = \Psi_k H_{k-1} F_k^{-1}$, and the noise

$$\bar{v}_k = \Gamma_k B_k w_k + \xi_k \tag{6.149}$$

is zero mean white Gaussian with the properties $E\{\bar{v}_k \bar{v}_k^T\} = \Gamma_k \Phi_k + R_k$, $E\{\bar{v}_k w_k^T\} = \Gamma_k B_k Q_k$, and $E\{w_k \bar{v}_k^T\} = Q_k B_k^T \Gamma_k^T$, where $\Phi_k = B_k Q_k B_k^T \Gamma_k^T$.

It can be seen that the modified state-space model in (6.145) and (6.148) contains white and time-correlated noise sources w_k and $\bar{v}_k$. Unlike KF, the UFIR filter does not require any information about noise, except for the zero mean assumption. Therefore, both w_k and $\bar{v}_k$ can be ignored, and thus the UFIR filter is unique for both correlated and de-correlated w_k and $\bar{v}_k$. However, the UFIR filter cannot ignore CMN v_k, which is biased on a finite horizon $[m, k]$.

The pseudocode of the *a posteriori* UFIR filtering algorithm for CMN is listed as Algorithm 13. To initialize iterations avoiding singularities, the algorithm requires a short measurement vector $Y_{m,s} = [\, y_m^T \ \dots \ y_s^T \,]^T$ and an auxiliary block matrix

$$C_{m,s} = \begin{bmatrix} \bar{H}_m (F_s \dots F_{m+1})^{-1} \\ \vdots \\ \bar{H}_{s-1} F_s^{-1} \\ \bar{H}_s \end{bmatrix} .$$

It can be seen that for $\Psi_n = 0$ this algorithm becomes the standard UFIR filtering algorithm. More details about the UFIR filter developed for CMN can be found in [183].

The error covariance for the UFIR filter is given by [179]

$$\begin{aligned} P_k &= (I - \mathcal{G}_k \bar{H}_k^T \bar{H}_k) P_k^- (I - \mathcal{G}_k \bar{H}_k^T \bar{H}_k)^T + \mathcal{G}_k \bar{H}_k^T \\ &\quad \times (\Gamma_k \Phi_k + R_k) \bar{H}_k \mathcal{G}_k - 2(I - \mathcal{G}_k \bar{H}_k^T \bar{H}_k) \Phi_k \bar{H}_k \mathcal{G}_k \\ &= P_k^- - 2(P_k^- \bar{H}_k^T + \Phi_k) \bar{H}_k \mathcal{G}_k + \mathcal{G}_k \bar{H}_k^T S_k \bar{H}_k \mathcal{G}_k \\ &= P_k^- - (2P_k^- \bar{H}_k^T + 2\Phi_k + \mathcal{G}_k \bar{H}_k^T S_k) \bar{H}_k \mathcal{G}_k \,, \end{aligned} \tag{6.150}$$

Algorithm 13: UFIR Filtering Algorithm for CMN

Data: y_k, N
Result: $\hat{x}_k$
begin
  for $k = N-1, N, \cdots$ **do**
    $m = k - N + 1\,, s = k - N + K$
    $\mathcal{G}_s = (C^T_{m,s} C_{m,s})^{-1}$
    $\bar{x}_s = \mathcal{G}_s C^T_{m,s} Y_{m,s}$
    for $l = s+1 : k$ **do**
      $\bar{H}_l = H_l - \Psi_l H_{l-1} F_l^{-1}$
      $\bar{z}_l = y_l - \Psi_l y_{l-1}$
      $\mathcal{G}_l = [\bar{H}_l^T \bar{H}_l + (F_l \mathcal{G}_{l-1} F_l^T)^{-1}]^{-1}$
      $K_l = \mathcal{G}_l \bar{H}_l^T$
      $\bar{x}_l^- = F_l \bar{x}_{l-1}$
      $\bar{x}_l = \bar{x}_l^- + K_l(\bar{z}_l - \bar{H}_l \bar{x}_l^-)$
    end for
    $\hat{x}_k = \bar{x}_k$
  end for
end

where the matrices $\bar{H}_k$, Ψ_k, and Γ_k are defined earlier and the reader should remember that the GNPG matrix $\mathcal{G}_k$ is symmetric.

Typical RMSEs produced by the KF and UFIR algorithms versus ψ are shown in Fig. 6.8 [183], where we recognize several basic features. It can be seen that the KF and UFIR filter modified for CMN give fewer errors than the original ones. It should also be noted that GKF performs better

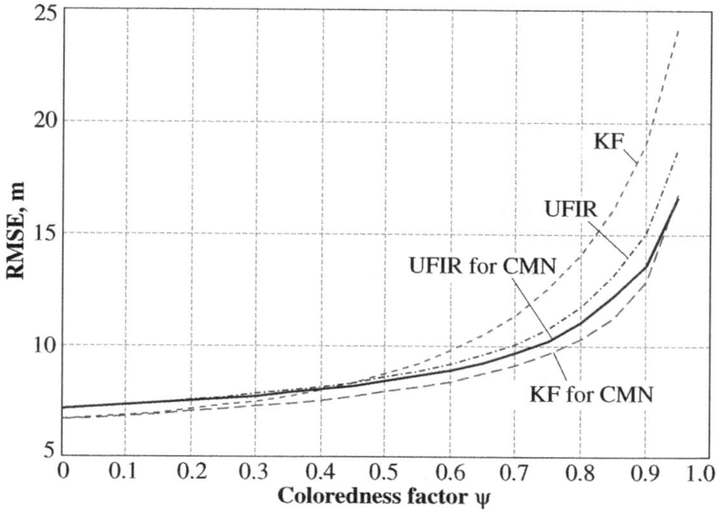

Figure 6.8 Typical RMSEs produced by KF and UFIR filter for a two-state model versus the coloredness factor ψ [183].

when $0 < \psi < 0.95$, and that the general UFIR filter is more accurate when $0.95 < \psi < 1$. This means that a more robust UFIR filter may be a better choice when measurement noise is heavily colored.

6.6.2 Colored Process Noise

Unlike the CMN, which always needs to be filtered out, the CPN, or at least its slow spectral components, can be tracked to avoid losing information about the process behavior. Let us show how to deal with CMN based on the following the state-space model

$$x_k = F_k x_{k-1} + E_k u_k + B_k w_k \,, \tag{6.151}$$

$$w_k = \Theta_k w_{k-1} + \mu_k \,, \tag{6.152}$$

$$y_k = H_k x_k + v_k \,, \tag{6.153}$$

where matrices $F_k \in \mathbb{R}^{K\times K}$, $B_k \in \mathbb{R}^{K\times K}$, and $\Theta_k \in \mathbb{R}^{K\times K}$ are nonsingular, $H_k \in \mathbb{R}^{M\times K}$, and $w_k \in \mathbb{R}^K$ is the Gauss-Markov CPN. Noise vectors $\mu_k \sim \mathcal{N}(0, Q_k) \in \mathbb{R}^K$ and $v_k \sim \mathcal{N}(0, R_k) \in \mathbb{R}^M$ are mutually uncorrelated with the covariances $E\{\mu_k \mu_k^T\} = Q_k$ and $E\{v_k v_k^T\} = R_k$. The coloredness factor matrix Θ_k is chosen such that the CPN w_k is always stationary.

Using state differencing (3.175a), a new state equation can be written as

$$\chi_k = \tilde{F}_k \chi_{k-1} + \tilde{u}_k + B_k \mu_k \,, \tag{6.154}$$

where μ_k is white Gaussian, $\tilde{u}_k = E_k u_k - \tilde{F}_k \Pi_{k-1} F_{k-1}^{-1} u_{k-1}$, $\Pi_k = \tilde{F}_{k+1}^{-1} \bar{\Theta}_{k+1} F_k$, and $\tilde{F}_k$ is defined by solving for initial $\tilde{F} = \bar{\Theta}$ the NARE $\tilde{F}^2 - \tilde{F}(F + \bar{\Theta}) + \bar{\Theta} F = 0$, where $\bar{\Theta} = B\Theta B^{-1}$.

Using (3.188), we write the modified observation equation as

$$\tilde{y}_k = y_k - H_k \Pi_k x_{k-1} = H_k \chi_k + v_k \,, \tag{6.155}$$

where x_{k-1} can be substituted with the available past estimate $\hat{x}_{k-1}$.

The pseudocode of the UFIR algorithm developed for CPN is listed as Algorithm 14. To initialize iterations, Algorithm 14 employs a short data vector $Y_{m,s} = [\, y_m^T \ \cdots \ y_s^T \,]^T$ and an auxiliary matrix

$$C_{m,s} = \begin{bmatrix} H_m(\tilde{F}_s \dots \tilde{F}_{m+1})^{-1} \\ \vdots \\ H_{s-1}\tilde{F}_s^{-1} \\ H_s \end{bmatrix}.$$

Note that, by setting $\bar{\Theta}_k = 0$ and $\tilde{F}_k = F_k$, Algorithm 14 becomes the standard iterative UFIR filtering algorithm.

The error covariance of the UFIR filter modified for CPN can be found if we notice that $\varepsilon_k = x_k - \hat{x}_k = \epsilon_k + \Pi_k \epsilon_{k-1}$, where $\epsilon_k = \chi_k - \hat{\chi}_k$. Since this estimate is subject to the constraint $\tilde{F}_k \Pi_{k-1} F_{k-1}^{-1} \bar{\Theta}_k^{-1} = I$ [182], the error ϵ_k for $K_k = \mathcal{G}_k H_k^T$ can be transformed to

$$\begin{aligned} \epsilon_k &= \tilde{F}_k \chi_{k-1} + \mu_k - \tilde{F}_k \hat{\chi}_{k-1} - K_k(z_k - H_k \tilde{F}_k \hat{\chi}_{k-1}) \\ &= (I - K_k H_k)\tilde{F}_k \epsilon_{k-1} + (I - K_k H_k)\mu_k - K_k v_k \,, \end{aligned}$$

and the corresponding error covariance $\bar{P}_k = E\{\epsilon_k \epsilon_k^T\}$ found to be

$$\begin{aligned} \bar{P}_k &= (I - K_k H_k)(\tilde{F}_k \bar{P}_{k-1} \tilde{F}_k^T + B_k Q_k B_k^T) \\ &\quad \times (I - K_k H_k)^T + K_k R_k K_k^T \,. \end{aligned}$$

Algorithm 14: UFIR Filtering Algorithm for CPN

Data: y_k, N, Θ_k
Result: $\hat{x}_k$

begin
  for $k = N-1, N, \cdots$ **do**
    $m = k - N + 1\,, s = k - N + K\,;$
    $\mathcal{G}_s = (C_{m,s}^T C_{m,s})^{-1}\,;$
    $\bar{x}_s = \mathcal{G}_s C_{m,s}^T Y_{m,s}\,;$
    for $l = s+1 : k$ **do**
      $z_l = y_l - H_l \tilde{F}_{l+1}^{-1} \bar{\Theta}_{l+1} F_l \hat{\mathbf{x}}_{l-1}\,;$
      $\mathcal{G}_l = [H_l^T H_l + (\tilde{F}_l \mathcal{G}_{l-1} \tilde{F}_l^T)^{-1}]^{-1}\,;$
      $K_l = \mathcal{G}_l H_l^T\,;$
      $\bar{\chi}_l^- = \tilde{F}_l \bar{\chi}_{l-1}\,;$
      $\bar{\chi}_l = \bar{\chi}_l^- + K_l(z_l - H_l \bar{\chi}_l^-)\,;$
    end for
    $\hat{x}_k = \bar{\chi}_k + \tilde{F}_{k+1}^{-1} \bar{\Theta}_{k+1} F_k \hat{x}_{k-1}\,;$
  end for
end

This finally gives

$$P_k = \bar{P}_k + \Pi_k P_{k-1} \Pi_k^T\ .$$

Typical RMSEs produced by the modified and original filters for a two-state polynomial model with CPN are shown in Fig. 6.9 as functions of the scalar coloredness factor θ [182]. The filtering effect here is reminiscent of the effect shown in Fig. 6.6 for CMN, and we notice that the accuracy of

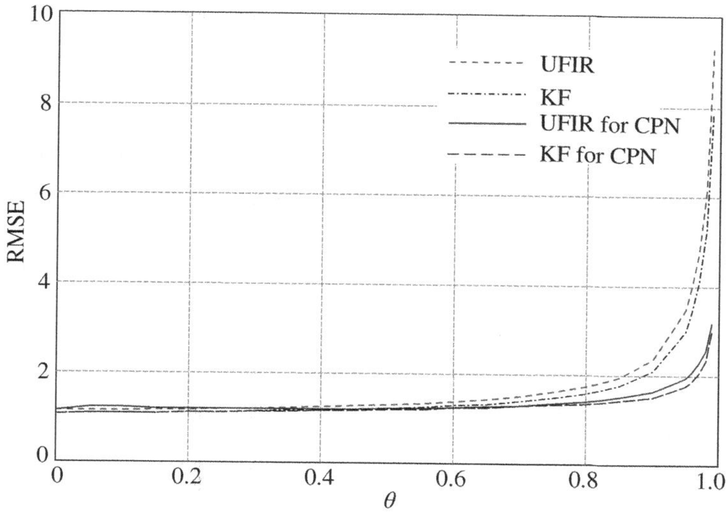

Figure 6.9 Typical RMSEs produced by the two-state UFIR filter, KF, and modified KF and UFIR filter in the presence of CPN as functions of the coloredness factor θ [182].

the original filters has been improved. However, a significant improvement in performance is recognized only when the coloredness factor is relatively large, $\theta > 0.5$. Otherwise, the discrepancies between the filter outputs are not significant.

Considering the previous modifications of the UFIR filter, we finally conclude that the filtering effect in the presence of CMN and/or CPN is noticeable only with strong coloration.

6.7 Extended UFIR Filtering

Representation of physical processes and approximation of systems using linear models does not always fit with practical needs. Looking at the nonlinear model in (3.226) and (3.227) and analyzing the Taylor series approach that results in the extended KF algorithms, we conclude that UFIR filtering can also be adapted to nonlinear behaviors [178], as will be shown next.

Given a nonlinear state-space model

$$x_k = f_k(x_{k-1}) + w_k \,, \tag{6.156}$$

$$y_k = h_k(x_k) + v_k \,, \tag{6.157}$$

where w_k and v_k are mutually uncorrelated, zero mean, and not obligatorily Gaussian additive noise vectors. In Chapter 3, when derived EKF, it was shown that (6.156) and (6.157) can be approximated using the second-order Taylor series expansion as

$$x_k = \dot{F}_k x_{k-1} + \eta_k + w_k \,, \tag{6.158}$$

$$\tilde{y}_k = \dot{H}_k x_k + v_k \,, \tag{6.159}$$

where $\tilde{y}_k = y_k - \psi_k$ is the modified observation vector and η_k and ψ_k represent the components resulting from the linearization,

$$\eta_k = f_k(\hat{x}_{k-1}) - \dot{F}_k \hat{x}_{k-1} + \frac{1}{2}\alpha_k \,, \tag{6.160}$$

$$\psi_k = h_k(\hat{x}_k^-) - \dot{H}_k \hat{x}_k^- + \frac{1}{2}\beta_k \,, \tag{6.161}$$

in which $\alpha_k = \sum_{i=1}^{K} e_i^K \varepsilon_{k-1}^T \ddot{F}_{ik} \varepsilon_{k-1}$, $\beta_k = \sum_{j=1}^{M} e_j^M \varepsilon_k^{-T} \ddot{H}_{jk} \varepsilon_k^-$, and $e_i^K \in \mathbb{R}^K$ and $e_j^M \in \mathbb{R}^M$ are Cartesian basis vectors with the ith and jth components unity, and all others are zeros. The nonlinear functions are represented by

$$f_k(x_{k-1}) \cong f_k(\hat{x}_{k-1}) + \dot{F}_k \varepsilon_{k-1} + \frac{1}{2}\alpha_k \,,$$

$$h_k(x_k) \cong h_k(\hat{x}_k^-) + \dot{H}_k \varepsilon_k^- + \frac{1}{2}\beta_k \,,$$

where $\dot{F}_k = \left.\frac{\partial f_k}{\partial x}\right|_{x=\hat{x}_{k-1}}$ and $\dot{H}_k = \left.\frac{\partial h_k}{\partial x}\right|_{x=\hat{x}_k^-}$ are Jacobian matrices and $\ddot{F}_{ik} = \left.\frac{\partial^2 f_{ik}}{\partial x^2}\right|_{x=\hat{x}_{k-1}}$ and $\ddot{H}_{jk} = \left.\frac{\partial^2 h_{jk}}{\partial x^2}\right|_{x=\hat{x}_k^-}$ are Hessian matrices.

Based on (6.158) and (6.159), we can now develop the first- and second-order extended UFIR filtering algorithms. Similarly to the EKF-1 and EKF-2 algorithms, we will refer to the extended UFIR algorithms as EFIR-1 and EFIR-2.

6.7.1 First-Order Extended UFIR Filter

Let us look at the model in (6.158) and (6.159) again and assume that the mutually uncorrelated w_k and v_k are zero mean and white Gaussian. The EFIR-1 (first-order) filtering algorithm can be designed using the following recursions [178]

$$\hat{x}_k = f_k(\hat{x}_{k-1}) + \mathcal{G}_k \dot{H}_k^T [y_k - h_k(f_k(\hat{x}_{k-1}))] \,, \tag{6.162}$$

$$\mathcal{G}_k = [\dot{H}_k^T \dot{H}_k + (\dot{F}_k \mathcal{G}_{k-1} \dot{F}_k^T)^{-1}]^{-1} \,. \tag{6.163}$$

The pseudocode of the EFIR-1 filtering algorithm is listed as Algorithm 15. As in the EKF-1 algorithm, here the prior estimate $\bar{x}_l^-$ is obtained using a nonlinear projection $f_l(\bar{x}_{l-1})$, and then $\bar{x}_l^-$ is projected onto the observation as $h_l(\bar{x}_l^-)$. Also, the system and observation matrices $\dot{F}_k$ and $\dot{H}_k$ are Jacobian. The error covariance for this filter can be computed using (6.37) and the Jacobian matrices $\dot{F}_k$ and $\dot{H}_k$.

Algorithm 15: First-Order EFIR-1 Filtering Algorithm

Data: y_k, N
Result: $\hat{x}_k$
1 **begin**
2   **for** $k = N-1, N, \cdots$ **do**
3     $m = k - N + 1\,, s = k - N + K$;
4     $G_s = (C_{m,s}^T C_{m,s})^{-1}$;
5     $\bar{x}_s = G_s C_{m,s}^T Y_{m,s}$;
6   **for** $l = s+1 : k$ **do**
7       $G_l = [\dot{H}_l^T \dot{H}_l + (\dot{F}_l G_{l-1} \dot{F}_l^T)^{-1}]^{-1}$;
8       $\bar{x}_l^- = f_l(\bar{x}_{l-1})$;
9       $\bar{x}_l = \bar{x}_l^- + G_l H_l^T [y_l - h_l(\bar{x}_l^-)]$;
10   **end for**
11     $\hat{x}_k = \bar{x}_k$;
12   **end for**
13 **end**

The EFIR-1 filtering algorithm developed in this way turns out to be simple and in most cases truly efficient. However, it should be noted that the initial state $\bar{x}_s$, computed linearly in line 5 of Algorithm 15, may be too rough when the nonlinearity is strong. If so, then $\bar{x}_s$ can be computed using multiple projections as $\bar{x}_{m+1} = f_{m+1}(\bar{x}_m)$, $\bar{x}_{m+2} = f_{m+2}(\bar{x}_{m+1})$, ..., $\bar{x}_s = f_s(\bar{x}_{s-1})$. Otherwise, an auxiliary EKF-1 algorithm can be used to obtain $\bar{x}_s$.

6.7.2 Second-Order Extended UFIR Filter

The derivation of the second-order EFIR-2 filter is more complex, and its application may not be very useful because it loses the ability to ignore noise covariances. To come up with the EFIR-2 algorithm, we will mainly follow the results obtained in [178].

We first define the prior estimate by

$$\hat{x}_k^- = f_k(\hat{x}_{k-1}) + \frac{1}{2}\bar{\alpha}_k \,, \tag{6.164}$$

for which, by the cyclic property of the trace operator, the expectation of α_n can be found as

$$\mathcal{E}\{\alpha_k\} = \bar{\alpha}_k = \sum_{l=1}^{K} e_l^K \operatorname{tr}\left\{\ddot{F}_{lk} P_{k-1}\right\} . \tag{6.165}$$

We can show that for $\hat{x}_k^-$ given by (6.164), the expectation of the prior error $\mathcal{E}\{\epsilon_k^-\} = \mathcal{E}\{x_k - \hat{x}_k^-\}$ becomes identically zero,

$$\mathcal{E}\{\epsilon_k^-\} = \mathcal{E}\{\dot{F}_k \epsilon_{k-1} + \frac{1}{2}(\alpha_k - \bar{\alpha}_k) + B_k w_k\} = 0 .$$

Averaging the nonlinear function $h_k(x_k)$ gives

$$\mathcal{E}\{h_k(x_k^-)\} = \bar{h}_k(x_k) = h_k(\hat{x}_k^-) + \frac{1}{2}\bar{\beta}_k , \tag{6.166}$$

where the expectation of β_k can be found as

$$\mathcal{E}\{\beta_k\} = \bar{\beta}_k = \sum_{j=1}^{M} e_j^M \operatorname{tr}\left\{\ddot{H}_{jk} P_k^-\right\} , \tag{6.167}$$

that allows us to obtain the error covariance as will be shown next.

Prior Error Covariance

Using (6.158) and (6.164) and taking into account that, for zero mean Gaussian noise, the vector $\mathcal{E}\{\epsilon_{k-1}(\epsilon_{k-1}^T \ddot{F}_{lk} \epsilon_{k-1})\}$ and other similar vectors are equal to zero, the *a priori* error covariance $P_k^- = \mathcal{E}\{\epsilon_k^- \epsilon_k^{-T}\}$ can be transformed as

$$\begin{aligned} P_k^- &= \mathcal{E}\{[f_k(x_{k-1}) + B_k w_k - f_k(\hat{x}_{k-1}) - \frac{1}{2}\bar{\alpha}_k][\dots]^T\} \\ &= \mathcal{E}\{[F_k \epsilon_{k-1} + B_k w_k + \frac{1}{2}(\alpha_k - \bar{\alpha}_k)][\dots]^T\} \\ &= F_k P_{k-1} F_k^T + B_k R_k B_k^T + \frac{1}{2}\bar{\mathcal{F}}_k , \end{aligned} \tag{6.168}$$

where matrix $\bar{\mathcal{F}}_k$ is specialized via its (uv)th component

$$\begin{aligned} \bar{\mathcal{F}}_{(uv)k} &= \operatorname{tr}(\ddot{F}_{uk} P_{k-1} \ddot{F}_{vk} P_{k-1}) \\ &\quad + \frac{1}{2}\operatorname{tr}(\ddot{F}_{uk} P_{k-1}) \operatorname{tr}(\ddot{F}_{vk} P_{k-1}) - \frac{1}{2}\bar{\alpha}_{uk} \bar{\alpha}_{vk}^T \\ &= \operatorname{tr}(\ddot{F}_{uk} P_{k-1} \ddot{F}_{vk} P_{k-1}). \end{aligned} \tag{6.169}$$

Posterior Error Covariance

Reasoning similarly, the *a posteriori* error covariance $P_k = \mathcal{E}\{\epsilon_k \epsilon_k^T\}$ can be transformed using extended nonlinearities if we first represent it as

$$\begin{aligned} P_k &= \mathcal{E}\{[f_k(x_{k-1}) + B_k w_k - f_k(\hat{x}_{k-1}) \\ &\quad - \frac{1}{2}\bar{\alpha}_k - K_k(y_k - \bar{h}_k(\hat{x}_k^-))][\dots]^T\} \\ &= \mathcal{E}\{[F_k \epsilon_{k-1} + B_k w_k + \frac{1}{2}(\alpha_k - \bar{\alpha}_k) - K_k(H_k \epsilon_k^- \\ &\quad + \frac{1}{2}(\beta_k - \bar{\beta}_k) + v_k)][\dots]^T\} , \end{aligned}$$

take into account that $\mathcal{E}\{\varepsilon_{k-1}\} = 0$ and $\mathcal{E}\{\varepsilon_k^-\} = 0$, provide the averaging, and obtain

$$\begin{aligned}
P_k &= F_k P_{k-1} F_k^T - F_k \mathcal{E}\{\varepsilon_{k-1}\varepsilon_k^{-T}\} H_k^T K_k^T \\
&+ \frac{1}{2}\bar{\mathcal{F}}_k - \frac{1}{4}(\mathcal{E}\{\alpha_k \beta_k^T\} - \bar{\alpha}_k \bar{\beta}_k^T) K_k^T + B_k R_k B_k^T \\
&- B_k \mathcal{E}\{w_k \varepsilon_k^{-T}\} H_k^T K_k^T - K_k H_k \mathcal{E}\{\varepsilon_k^- \varepsilon_{k-1}^T\} F_k^T \\
&- K_k H_k \mathcal{E}\{\varepsilon_k^- w_k^T\} B_k^T + K_k H_k P_k^- H_k^T K_k^T \\
&- \frac{1}{4} K_k \mathcal{E}\{\beta_k \alpha_k^T\} + \frac{1}{4} K_k \bar{\beta}_k \bar{\alpha}_k^T + \frac{1}{4} K_k \mathcal{E}\{\beta_k \beta_k^T\} K_k^T \\
&- \frac{1}{4} K_k \bar{\beta}_k \bar{\beta}_k^T K_k^T + K_k Q_k K_k^T \, .
\end{aligned}$$

Due to the symmetry of the matrices P_k, R_k, and Q_k, the following expectations can be transformed as

$$\begin{aligned}
\mathcal{E}\{\varepsilon_k^- \varepsilon_{k-1}^T\} &= \mathcal{E}\{(x_k - \hat{x}_k^-)(x_{k-1} - \hat{x}_{k-1})^T\} \\
&= \mathcal{E}\{[F_k \varepsilon_{k-1} + B_k w_k + \frac{1}{2}(\alpha_k - \bar{\alpha}_k)]\varepsilon_{k-1}^T\} = F_k P_{k-1} \, , \\
\mathcal{E}\{w_k \varepsilon_k^{-T}\} &= \mathcal{E}\{w_k [F_k \varepsilon_{k-1} + B_k w_k + \frac{1}{2}(\alpha_k - \bar{\alpha}_k)]^T\} \\
&= R_k B_k^T \, , \\
\mathcal{E}\{\varepsilon_{k-1} \varepsilon_k^{-T}\} &= P_{k-1} F_k^T \, , \\
\mathcal{E}\{\varepsilon_k^- w_k^T\} &= B_k R_k \, .
\end{aligned}$$

Then, taking into account that the expectations of other products are matrices with zero components, we finally write the covariance P_k in the form

$$\begin{aligned}
P_k &= (I - K_k H_k) P_k^- (I - K_k H_k)^T + K_k Q_k K_k^T \\
&+ \frac{1}{2}(\bar{\mathcal{F}}_k H_k^T K_k^T + K_k H_k \bar{\mathcal{F}}_k) + \frac{1}{2} K_k \bar{\mathcal{H}}_k K_k^T \\
&- \frac{1}{2}(\bar{\mathcal{M}}_k K_k^T + K_k \bar{\mathcal{M}}_k^T) \, ,
\end{aligned} \tag{6.170}$$

where, the (rg)th component of matrix $\bar{\mathcal{H}}_k$ is defined by

$$\bar{\mathcal{H}}_{(rg)k} = \text{tr}\left[\ddot{H}_{rk} P_k^- \ddot{H}_{gk} P_k^-\right] \tag{6.171}$$

and the (ur)th component of matrix $\bar{\mathcal{M}}_k$ is

$$\begin{aligned}
\bar{\mathcal{M}}_{(ur)k} &= \text{tr}[\ddot{F}_{uk} P_{k-1} F_k^T \ddot{H}_{rk} F_k P_{k-1}] + \sum_{q=1}^{K} \sum_{t=1}^{K} \ddot{H}_{rk} \\
&\times \text{tr}[\ddot{F}_{uk} P_{k-1} \ddot{F}_{qk} P_{k-1} \ddot{F}_{tk} P_{k-1}] \, .
\end{aligned} \tag{6.172}$$

It has to be remarked now that when developing the second-order extended algorithms, the authors use two forms of P_k. The complete form in (6.170) can be found in [12, 157]. On the contrary, in [72, 185] only first-order components are preserved.

The pseudocode of the EFIR-2 filtering algorithm is listed as Algorithm 16. For the given N, Q_k, and R_k, the set of auxiliary matrices is computed and updated at each k. Then all matrices and

Algorithm 16: Second-Order EFIR-2 Filtering Algorithm

Data: y_k, Q_k, R_k, N
Result: $\hat{x}_k, P_k$

begin
  for $k = N-1, N, \cdots$ **do**
    $m = k - N + 1\,, s = k - N + K\,;$
    $P_s = 0\,;$
    $\Phi_s = (H_{m,s}^T H_{m,s})^{-1}\,;$
    $\mathcal{G}_s = \mathcal{F}_s^{m+1} \Phi_s \mathcal{F}_s^{m+1^T}\,;$
    $\bar{x}_s = \mathcal{F}_s^{m+1} \Phi_s H_{m,s}^T Y_{m,s}\,;$
    for $l = s+1 : k$ **do**
      $\bar{\alpha}_l = \sum_{i=1}^{K} e_i^K \operatorname{tr}\left\{\ddot{F}_{il} P_{l-1}\right\}\,;$
      $\bar{\beta}_l = \sum_{j=1}^{M} e_j^M \operatorname{tr}\left\{\ddot{H}_{jl} P_l^-\right\}\,;$
      $\bar{\mathcal{F}}_l$ by (6.169), $\bar{\mathcal{H}}_l$ by (6.171), $\bar{\mathcal{M}}_l$ by (6.172) ;
      $P_l^- = F_l P_{l-1} F_l^T + B_l R_l B_l^T + \frac{1}{2}\bar{\mathcal{F}}_l\,;$
      $\hat{x}_l^- = f_l(\hat{x}_{l-1}) + \frac{1}{2}\bar{\alpha}_l\,;$
      $\bar{h}_l(x_l^-) = h_l(\hat{x}_l^-) + \frac{1}{2}\bar{\beta}_l\,;$
      $\mathcal{G}_l = [\dot{H}_l^T \dot{H}_l + (\dot{F}_l \mathcal{G}_{l-1} \dot{F}_l^T)^{-1}]^{-1}\,;$
      $K_l = \mathcal{G}_l H_l^T\,;$
      $\bar{x}_l = \bar{x}_l^- + K_l[y_l - \bar{h}_l(\bar{x}_l^-)]\,;$
      $P_l = (I - K_l H_l) P_l^- (I - K_l H_l)^T + K_l Q_l K_l^T\,;$
      $+\frac{1}{2}(\bar{\mathcal{F}}_l H_l^T K_l^T + K_l H_l \bar{\mathcal{F}}_l) + \frac{1}{2} K_l \bar{\mathcal{H}}_l K_l^T\,;$
      $-\frac{1}{2}(\bar{\mathcal{M}}_l K_l^T + K_l \bar{\mathcal{M}}_l^T)\,;$
    end for
    $\hat{x}_k = \bar{x}_k, P_k\,;$
  end for
end

vectors are updated iteratively, and the last updated $\hat{x}_k$ and P_k go to the output when $l = k$. Although the second-order EFIR-2 algorithm has been designed to improve accuracy, empirical studies show that its usefulness is questionable due to the following drawbacks:

- Unlike the EFIR-1 Algorithm 15, the EFIR-2 Algorithm 16 requires Q_k and R_k and is thus less robust. It thus has more in common with the EKF rather than with the EFIR-1 filter.
- As noted in [178], nothing definite can be said about the accuracy of the EFIR-2 algorithm compared to the EFIR-1 algorithm. The same conclusion was made earlier in [185] regarding the EKF-2 algorithm.

Overall, it can be said that the EFIR-2 and EKF-2 filtering algorithms do not demonstrate essential advantages over the EFIR-1 and EKF-1 algorithms. Moreover, these algorithms are more computationally complex and slower in operation.

6.8 Robustness of the UFIR Filter

Practice dictates that having a good estimator is not enough if the operating conditions are not satisfied from various perspectives, especially in unknown environments. In such a case, the required estimator must also be robust. In Chapter 1, we introduced the fundamentals of robustness, which is seen as the ability of an estimator not to respond to undesirable factors such as errors in noise statistics, model errors, and temporary uncertainties. We are now interested in investigating the trade-off in robustness between the UFIR estimator and some other available state estimators. To obtain reliable results, we will examine along the KF, which is optimal and not robust, and the robust game theory recursive H_∞ filter (3.307)–(3.310), which was developed in [185]. We will use the following state-space model

$$x_k = Fx_{k-1} + w_k \,, \tag{6.173}$$

$$y_k = Hx_k + v_k \,, \tag{6.174}$$

where the noise covariances Q and R in the noise vectors $w_k \sim \mathcal{N}(0, Q)$ and $v_k \sim \mathcal{N}(0, R)$ are not necessarily known, as required by KF. To apply the H_∞ filter, w_k and v_k will be thought of as norm-bounded with symmetric and positive definite matrices $\mathcal{Q}$ and $\mathcal{R}$.

In our scenario, there is not enough information about Q and R to optimally tune KF, and the maximum values of the matrices $\mathcal{Q}$ and $\mathcal{R}$ are also not known to make the H_∞ filter robust. Therefore, we organize a scaling test for robustness by representing any uncertain component Υ in (6.173) and (6.174) as $\Upsilon = \bar{\Upsilon} + \Delta = \gamma\bar{\Upsilon}$, where the increment $\Delta = (\gamma - 1)\bar{\Upsilon}$ is due to a positive-valued scaling factor $\gamma > 0$ such that $\gamma = 1$ means undisturbed $\Upsilon = \bar{\Upsilon}$. This test can be applied with the following substitutions: $Q \leftarrow \alpha^2 Q$, $\mathcal{Q} \leftarrow \alpha^2\mathcal{Q}$, $R \leftarrow \beta^2 R$, $\mathcal{R} \leftarrow \beta^2\mathcal{R}$, $F \leftarrow \eta F$, and $H \leftarrow \mu H$, where the set of scaling factors $\{\alpha, \beta, \eta, \mu\} > 0$ can either change matrices or not when $\{\alpha, \beta, \eta, \mu\} = 1$. For ηF, the product matrix $\mathcal{F}_r^g$ becomes

$$\mathcal{F}_r^g \leftarrow \begin{cases} (\eta F)^{r-g+1} & g \leqslant r \\ \eta I & g = r+1 \\ 0 & g > r+1 \end{cases} . \tag{6.175}$$

We can learn about the robustness of recursive estimators by examining the sensitivity of the bias correction gain to interfering factors and its immunity to these factors. Taken from the UFIR Algorithm 11, the alternate KF (3.92)–(3.96) [184], and the H_∞ filter (3.307)–(3.310) [185], the bias corrections gains can be written as, respectively,

$$K_k^{\mathrm{U}} = [H^T H + (F\mathcal{G}_{k-1}F^T)^{-1}]^{-1}H^T \,, \tag{6.176}$$

$$K_k^{\mathrm{K}} = P_k^-(I + H^T R^{-1} H P_k^-)^{-1} H^T R^{-1} \,. \tag{6.177}$$

$$K_k^{\infty} = \mathcal{P}_k^-(I - \theta\bar{S}_k\mathcal{P}_k^- + H^T\mathcal{R}^{-1}H\mathcal{P}_k^-)^{-1}H^T\mathcal{R}^{-1} \,, \tag{6.178}$$

where $P_k^- = FP_{k-1}F^T + Q$, $\mathcal{P}_k^- = F\mathcal{P}_{k-1}F^T + \mathcal{Q}$, and matrix $\bar{S}_k$ is constrained by the positive definite matrix $(\mathcal{P}_k^-)^{-1} - \theta\bar{S}_k + H^T\mathcal{R}^{-1}H > 0$. Note that the tuning factor θ is required by the game theory H_∞ filter in order to outperform KF.

The influence of $\{\alpha, \beta, \eta, \mu, \theta\}$ on (6.176)–(6.178) can be learned if we use the following rule: *if the error factor causes a decrease in K_k, then the bias errors will grow; otherwise, random errors will dominate.* The increments ΔK caused by the error factors in the bias correction gains can now be represented for each of the filters as

$$\Delta K^{\mathrm{U}}(\eta,\mu) = K^{\mathrm{U}}(\eta,\mu) - K^{\mathrm{U}}(1,1)\,,$$
$$\Delta K^{\mathrm{K}}(\alpha,\beta,\eta,\mu) = K^{\mathrm{K}}(\alpha,\beta,\eta,\mu) - K^{\mathrm{K}}(1,1,1,1)\,,$$
$$\Delta K^{\infty}(\alpha,\beta,\eta,\mu,\theta) = K^{\infty}(\alpha,\beta,\eta,\mu,\theta) - K^{\infty}(1,1,1,1,0)\,,$$

and we notice that lower ΔK means higher robustness. There is one more important preliminary remark to be made. While more tuning factors make an estimator potentially more accurate and precise, it also makes an estimator less robust due to possible tuning errors (Fig. 1.3). In this sense, the two-parameter $\{\eta,\mu\}$ UFIR filter appears to be more robust, but less accurate, while the four-parameter $\{\alpha,\beta,\eta,\mu\}$ KF and five-parameter $\{\alpha,\beta,\eta,\mu,\theta\}$ game theory H_∞ filter are less robust but potentially more accurate.

6.8.1 Errors in Noise Covariances and Weighted Matrices

The additional errors produced by estimators applied to stochastic models and caused by $\alpha \neq 1$ and $\beta \neq 1$ occur due to the practical inability to collect accurate estimates of noise covariances and norm-bounded weights, especially for LTV systems.

UFIR Filter

The UFIR filter does not require any information about zero mean noise. Thus, this filter is insensitive (robust) to variations in α and β, unlike the KF and game theory H_∞ filter.

Kalman Filter

For $\alpha \neq 1$ and $\beta \neq 1$, errors in the KF can be learned in stationary mode if we allow $P_k^- = P_{k-1}^- = P^-$. Accordingly, the error covariance can be represented using the DARE

$$FP^-F^T - P^- + \alpha^2 Q - FP^-(I + \beta^{-2}H^T R^{-1} H P^-)^{-1} \times \beta^{-2} H^T R^{-1} H P^- F^T = 0\,. \tag{6.179}$$

The solution to (6.179) does not exist in simple form. But if we accept $P_{k-1} \approx P_k^- = P^-$, which is true for low measurement noise, $R \approx 0$, then (6.179) can be replaced by the discrete Lyapunov equation [184] (A.34)

$$FP^-F^T - P^- + \alpha^2 Q = 0\,, \tag{6.180}$$

which is solvable (A.35). Although approximation (6.180) can be rough when v_k is not small, it helps to compare the KF and H_∞ filter for robustness, as will be shown next.

Lemma 6.1 Given $\alpha > 0$ and $\| F \| \leqslant a$, where $a = \rho(F) + \epsilon > 0$, $\rho(F)$ is the spectral radius of F, and $\epsilon > 0$, the solution to (6.180) can be represented for $a < 1$ and $\alpha^2/(1-a^2) \geq 1$ with negative definiteness in the form [180]

$$P^- \leqslant Q\frac{\alpha^2}{1-a^2}\,. \tag{6.181}$$

Proof: Write the solution to (6.180) as an infinite sum [83],

$$P^- = \alpha^2 \sum_{i=0}^{\infty} F^i Q F^{i^T} \tag{6.182}$$

and transform the p-norm of (6.182) as

$$\begin{aligned}\| P^- \|_p &= \alpha^2 \| \sum_{i=0}^{\infty} F^i Q F^{i^T} \|_p \,, \\ &\leqslant \alpha^2 \sum_{i=0}^{\infty} \| F^i \|_p \| Q \|_p \| F^{i^T} \|_p \,, \\ &\leqslant \alpha^2 \| Q \|_p \sum_{i=0}^{\infty} a^{2i} = \| Q \|_p \frac{\alpha^2}{1 - a^2} \,. \end{aligned} \tag{6.183}$$

Choose the gain factor so that $\alpha^2/(1-a^2) \geq 1$, go to (6.181), and complete the proof. □

When $R \approx 0$, we can assume that KF is tracking state with high precision and let $P_{k-1} \approx 0$ that gives $P_k^- \approx Q$. This gives $P_k \approx H^T R^{-1} H$ and $P_{n+1}^- \approx Q$, and we conclude that $P^- = P_k^- \approx P_{k+1}^- \approx Q$. Note that this relationship is consistent with the equality in (6.183), which was obtained for the isolated cases of $\alpha = 1$ and $a = 0$.

Referring to lemma 6.1, the bias correction gain (6.177) can now be approximated for $a < 1$ by the inequality

$$K^{\mathrm{K}} \leqslant \left[H^T R^{-1} H + \frac{\beta^2}{\alpha^2}(1 - a^2) Q^{-1} \right]^{-1} H^T R^{-1} \,, \tag{6.184}$$

which does not hold true for $a \geq 1$. It is worth noting that α and β act in opposite directions, and therefore the worst case will be when $\alpha > 1$ and $\beta < 1$ or vice versa.

Game Theory H_∞ Filter

Reasoning similarly for equal weighting errors obtained with $\bar{S}_k = I$, the bias correction gain (6.178) for the recursive game theory H_∞ filter can be approximated by the inequality [180]

$$K^{\infty} \leqslant \left[H^T \mathcal{R}^{-1} H + \frac{\beta^2}{\alpha^2}(1 - a^2) \mathcal{Q}^{-1} - \beta^2 \theta I \right]^{-1} H^T \mathcal{R}^{-1} \,. \tag{6.185}$$

It can be seen that $\theta = 0$ makes no difference with the KF gain (6.184) for Gaussian models implying $Q = \mathcal{Q}$ and $R = \mathcal{R}$. Otherwise, we can expect KF to be less accurate and the H_∞ filter to perform better, provided that both $\mathcal{Q}$ and $\mathcal{R}$ are properly maximized. Indeed, if we assume that $\alpha \neq 1$ and $\beta \neq 1$, then we can try to find some small $\theta > 0$ such that the effect of α and β is compensated. However, if θ is not set properly, the H_∞ filter can cause even more errors than the KF. Moreover, even if θ is set properly, any deviation of β from unity will additionally chance the gain through the product of $\beta^2\theta$. More information and examples on this topic can be found in [180].

What follows behind the approximations (6.184) and (6.185) is that, if properly tuned with θ, the H_∞ filter will be more accurate than the KF in the presence of errors in noise covariances. In practice, however, it can be difficult to find and set the correct θ, especially for LTV systems that imply a new θ at each point in time. If so, then the H_∞ filter may give more errors and even diverge. Thus, the relationship between the filters in terms of robustness to errors in noise statistics should be practically as follows:

$$\mathrm{KF} \lesseqgtr H_\infty < \mathrm{UFIR} \,, \tag{6.186}$$

where $\mathrm{KF} < H_\infty$ holds true for properly maximized $\mathcal{Q}$ and $\mathcal{R}$ and properly set θ. Otherwise, the KF may perform better. In turn, the UFIR filter that is $\{\alpha, \beta\}$-invariant may be a better choice, especially with large errors in noise covariances.

6.8.2 Model Errors

Permanent estimation errors arise from incorrect modeling and persistent model uncertainties, known as *mismodeling*, when the model is not fully consistent with the process over all time [69]. If no improvement is expected in the model, then the best choice may be to use robust estimators, in which case testing becomes an important subject of confirmation of the estimator accuracy and stability.

UFIR Filter

In UFIR filtering, errors caused by $\{\eta, \mu\}$ are taken into account only on the horizon $[m, k]$. Accordingly, the bias correction gain K^{U} can be expressed as

$$K^{\mathrm{U}} = \mathcal{G}H^T = (C_{m,k}^T C_{m,k})^{-1}H^T$$
$$= \frac{1}{\mu}\left[\sum_{i=0}^{N-1} \eta^{-2(N-2-i)}(F^{-N+2+i})^T H^T H F^{-N+2+i}\right]^{-1} H^T \tag{6.187}$$

and approximated using the following lemma.

Lemma 6.2 Given model (6.173) and (6.174) with $\{\eta, \mu\} > 0$ and $\| F \| \leqslant a$, where $a = \rho(F) + \epsilon > 0$, $\rho(F)$ is the spectral radius of F and $\epsilon > 0$. Then the GNPG (6.16c) of the UFIR filter is represented at $n - 1$ with

$$\mathcal{G}_{k-1}^{-1} \leqslant \begin{cases} \mu^2 \frac{1-\chi^{N-1}}{(1-\chi)\chi^{N-2}} H^T H & , \quad \chi \neq 1 \\ \mu^2 (N-1) H^T H & , \quad \chi = 1 \end{cases}, \tag{6.188}$$

where $\chi = (\eta a)^2 < 1$ and $H^T H$ can be singular.

Proof: Let $F \leftarrow \eta F$ and $H \leftarrow \mu H$. Represent $\mathcal{G}_{k-1}^{-1} = C_{m,k-1}^T C_{m,k-1}$ on $[m, k-1]$ with the finite sum as

$$\mathcal{G}_{k-1}^{-1} = \mu^2 \sum_{i=0}^{N-2} \eta^{-2(N-2-i)} (F^{-N+2+i})^T$$
$$\times H^T H F^{-N+2+i} . \tag{6.189}$$

Apply the triangle inequality to the norms, use the properties of the norms, and transform the p-norm of (6.189) as

$$\| \mathcal{G}_{k-1}^{-1} \|_p = \mu^2 \| \sum_{i=0}^{N-2} \eta^{-2(N-2-i)} (F^{-N+2+i})^T H^T H F^{-N+2+i} \|_p$$
$$\leqslant \mu^2 \sum_{i=0}^{N-2} \eta^{-2(N-2-i)} \| (F^{-N+2+i})^T \|_p \, \| H^T H \|_p \, \| F^{-N+2+i} \|_p$$
$$\leqslant \mu^2 \| H^T H \|_p \sum_{i=0}^{N-2} \chi^{-N+2+i} . \tag{6.190}$$

Now transform (6.190) using the geometrical sum $\sum_{i=0}^{N-2} r^i = \frac{1-r^{N-1}}{1-r}$ if $\chi \neq 1$, replace the result by $N-1$ if $\chi = 1$, arrive at (6.188), and complete the proof. □

The gain $K^{\mathrm{U}} = \mathcal{G}H^T$ can finally be approximated with

$$K^{\mathrm{U}} \leqslant \frac{1}{\mu}\left[H^T H + a^2 \frac{1-(\eta a)^{2(N-1)}}{(\eta a)^{2N}(1-\eta^2 a^2)} F^{-T} H^T H F^{-1}\right]^{-1} H^T , \tag{6.191}$$

$$\text{if} \quad \chi \neq 1 ,$$

$$\leqslant \frac{1}{\mu}\left[H^T H + \frac{N-1}{\eta^2} F^{-T} H^T H F^{-1}\right]^{-1} H^T , \tag{6.192}$$

$$\text{if} \quad \chi = 1 .$$

It is seen that the factor μ affects K^{U} directly and as a reciprocal. However, the same cannot be said about the factor η raised to the power N. Indeed, for $\chi = (\eta a)^2 < 1$, the gain K^{U} can significantly grow. Otherwise, when $\chi > 1$, it decreases. Thus, we conclude that the effect of errors in the system matrix on the bias correction gain of the UFIR filter is more complex than the effect of errors in the observation matrix.

Kalman Filter

When examining Kalman filter errors associated with mismodeling, $\eta \neq 1$ and $\mu \neq 1$, it should be kept in mind that such errors are usually observed along with errors in noise covariances. If we ignore errors in the description of noise, the picture will be incomplete due to multiplicative effects. Therefore, we assume that $\{\eta, \mu\} \neq 1$ and $\{\alpha, \beta\} \neq 1$, and analyze the errors similarly to the UFIR filter. To do this, we start with the discrete Lyapunov equation (6.180), the solution of which can be approximated using the following lemmas.

Lemma 6.3 Given $\{\eta, \mu\} \neq 1$, $\{\alpha, \eta\} \neq 1$, and $\| F \| \leqslant a$, where $a = \rho(F) + \epsilon > 0$, $\varepsilon > 0$, and $\rho(F)$ is the spectral radius of F. Then the prior error covariance P^- of the KF can be approximated for $\chi = (\eta a)^2 < 1$ with

$$P^- \leqslant Q \frac{\alpha^2}{1-\chi} . \tag{6.193}$$

Proof: The proof is postponed to "Problems" and can be provided similarly to lemma 6.1. □

Lemma 6.4 Given $\{\eta, \mu\} \neq 1$, $\{\alpha, \eta\} \neq 1$, and $\| F \| \leqslant a$, where $a = \rho(F) + \epsilon > 0$, $\varepsilon > 0$, and $\rho(F)$ is the spectral radius of F. Then the prior error covariance P^- of the KF can be represented for $\chi = (\eta a)^2 < 1$ with

$$P^- \leqslant \begin{cases} \infty I , & \text{if} \quad \chi \geq 1 , a \lesseqgtr 1 \\ Q\frac{\alpha^2}{1-\chi} , & \text{if} \quad \chi < 1 , a < \frac{1}{\eta} \end{cases} . \tag{6.194}$$

Proof: The proof can be found in [180] and is postponed to "Problems." □

Referring to lemma 6.3 and lemma 6.4, the bias correction gain K^{K} of the KF, which is given by (6.177), can be approximated as

$$K^K \leqslant \frac{1}{\mu}\left[H^T R^{-1} H + \frac{\beta^2(1-\eta^2 a^2)}{\alpha^2\mu^2} Q^{-1}\right]^{-1} H^T R^{-1} , \tag{6.195}$$

$$\text{if} \quad \chi < 1 ,$$

$$\leqslant \frac{1}{\mu}(H^T R^{-1} H)^{-1} H^T R^{-1} , \tag{6.196}$$

$$\text{if} \quad \chi \geq 1 ,$$

and we notice that inequality (6.196) is inapplicable to some models.

Analyzing (6.195), it can be concluded that the mismodeling errors, $\{\eta, \mu\} \neq 1$, and errors in noise covariance, $\{\alpha, \beta\} \neq 1$, cause multiplicative effects in the Kalman gain. In the best case, these effects can compensate for each another, although not completely. In the worst case, which is of practical interest, errors can grow essentially. Again, we see that η and μ act in opposite directions, as in the case of $\{\alpha, \beta\} \neq 1$.

Game Theory H_∞ Filter

The game theory H_∞ filter obtained in [185] does not take into account errors in the system and observation matrices. Therefore, it is interesting to know its robustness to such errors in comparison with the UFIR filter and KF. Referring to (6.195) and (6.196), the bias correction gain K^∞ of the H_∞ filter, which is given by (6.178), can be approximated for $\{\eta, \mu\} \neq 1$ and $\{\alpha, \beta\} \neq 1$ by

$$K^\infty \leqslant \frac{1}{\mu}\left[H^T R^{-1} H + \frac{\beta^2}{\mu^2}\left(\frac{1-\eta^2 a^2}{\alpha^2} Q^{-1} - \theta I\right)\right]^{-1} H^T R^{-1} , \tag{6.197}$$

$$\text{if} \quad \chi < 1 ,$$

$$\leqslant \frac{1}{\mu}\left(H^T R^{-1} H - \frac{\beta^2}{\mu^2}\theta I\right)^{-1} H^T R^{-1} \to 0 , \tag{6.198}$$

$$\text{if} \quad \chi \geq 1 .$$

These inequalities more clearly demonstrate the key advantage of the H_∞ filter: *a properly set θ can dramatically improve the estimation accuracy*. Indeed, if the tuning factor θ constrained by $(\mathcal{P}_k^-)^{-1} - \theta\bar{S}_k + H^T\mathcal{R}^{-1}H > 0$ is chosen such that the relationships in parentheses become zero, then it follows that the effects of $\{\alpha, \beta, \eta\}$ are reduced to zero, and errors in the observation matrix will remain the main source. Furthermore, by adjusting θ, the effect of $\{\alpha, \beta, \eta, \mu\}$ can be fully compensated.

Unfortunately, the previous remedy is effective only if the correct θ is available at any given time. Otherwise, the H_∞ filter may produce more errors than the KF, and can even diverge. For more information, the reader can read [180], where a complete analysis for robustness is presented with typical examples.

The general conclusion that can be drawn from the analysis of the impact of mismodeling errors is that, under $\{\eta, \mu\} \neq 1$ and $\{\alpha, \beta\} \neq 1$, the KF performance can degrade dramatically and that the H_∞ can improve it if we set a proper tuning factor θ at each time instant. Since this is hardly possible for many practical applications, the UFIR filter is still preferred, and the relationship between the filters in terms of robustness to model errors will generally be the same (6.186), where $H_\infty >$ KF holds true for properly maximized errors in F and H if $\mathcal{Q}$ and $\mathcal{R}$ are properly maximized and θ is set properly. Otherwise, the H_∞ filter may give large errors and go to divergence, while the KF may perform better.

6.8.3 Temporary Uncertainties

Certainly persistent model errors can seriously degrade the estimator performance, causing uncontrollable bias errors. But their effect can be mitigated by improving the model, at least theoretically. What cannot be done efficiently is the elimination of errors caused by temporary uncertainties such as jumps in phase, frequency, and velocity. Such errors are not easy to deal with due to the unpredictability and short duration of the impacts. The problem is complicated by the fact that temporary uncertainties exist in different forms, and it is hard to imagine that they have a universal model.

One way to investigate the associated estimation errors is to assume that the model has no past errors up to the time index $k-1$ and that the uncertainty caused by $\{\eta, \mu\} \neq 1$ occurs at k. The estimator that is less sensitive to such an impact caused by $\{\alpha, \beta\} \neq 1$ at k can be said to be the most robust.

UFIR Filter

Suppose the model is affected by $\{\alpha, \beta\}$ over all time and that P_{k-1} is known at $k-1$. At the next step, at k, the system experiences an unpredictable impact caused by $\eta \neq 1$ and $\mu \neq 1$. Provided that $\mathcal{G}_{k-1}$ is also known at $k-1$, the bias correction gain K_k^{U} of the UFIR filter (6.176) can be transformed to

$$K_k^{\mathrm{U}} = \frac{1}{\mu}\left[H^T H + \frac{1}{\eta^2\mu^2}(F\mathcal{G}_{k-1}F^T)^{-1}\right]^{-1} H^T . \tag{6.199}$$

As can be seen, η and μ act in the same way in (6.199) as in the case of model errors represented by (6.192). However, the effect in (6.199) turns out to be stronger because of the square of μ, in contrast to (6.192). This means that temporary data errors have a stronger effect on the UFIR filter than permanent model errors. The conclusion just made should not be unexpected, given the fact that rapid changes cause transients, the values of which usually exceed steady-state errors.

Kalman Filter

The Kalman gain can be approximated similarly to the bias correction gain of the UFIR filter. Substituting the prior estimate $P_k^- = FP_{k-1}F^T + Q$ into (6.177), where the matrices are scaled by the error factors η and μ taking into account that $\{\alpha, \beta\} \neq 1$, we obtain the approximation of the Kalman gain in the form

$$\begin{aligned} K_k^{\mathrm{K}} = \frac{1}{\mu}\Bigg[H^T R^{-1} H + \frac{\beta^2}{\mu^2\eta^2} \\ \times \left(FP_{k-1}F^T + \frac{\alpha^2}{\eta^2}Q\right)^{-1}\Bigg]^{-1} H^T R^{-1}. \end{aligned} \tag{6.200}$$

Analyzing (6.199) and (6.200), we draw attention to an important feature. Namely, the effect of the first η is partially compensated by the second η in the parentheses, which is not observed in the UFIR filter. Because of this, the KF responds to temporary uncertainty with lower values than the UFIR filter. In other words, the UFIR filter is more sensitive to temporary uncertainties. On the other hand, all transients in the UFIR filter are limited by N points, while the KF filter generates long transients due to IIR. Both these effects, which are always seen in practice, represent one of the fundamental differences between the UFIR filter and the KF [179].

Transients in the UFIR filter and KF: *Responding to the stepwise temporary uncertainty, the UFIR filter produces N points transient with a larger overshoot, and the KF produces a longer transient with a shorter overshoot.*

Examples of transients in the UFIR filter and KF are given in Fig. 4.2 and Fig. 6.3, and we notice that the difference between the overshoots is usually (30 − 50)%.

Game Theory H_∞ Filter

Providing similar transformations as for (6.200), we approximate the bias correction gain of the H_∞ filter (6.178) as

$$K_k^\infty = \frac{1}{\mu}\left\{H^T\mathcal{R}^{-1}H + \frac{\beta^2}{\mu^2\eta^2}\right.$$
$$\left.\times\left[\left(F\mathcal{P}_{n-1}F^T + \frac{\alpha^2}{\eta^2}\mathcal{Q}\right)^{-1} - \eta^2\theta I\right]\right\}^{-1} H^T\mathcal{R}^{-1}\,, \tag{6.201}$$

and notice an important difference with (6.200). In (6.201), there is an additional term containing θ, which can be chosen such that the effect of $\{\alpha, \beta, \eta, \mu\}$ is fully compensated. This is considered to be the main advantage of the game theory H_∞ filter over KF. The drawback is that the tuning factor θ is an unknown analytical function of the set $\{\alpha, \beta, \eta, \mu\}$, which contains uncertain components. Therefore, if θ cannot be set properly at every point of time, the H_∞ filter performance may significantly degrade, and the filter may even demonstrate the divergence [180]. Analyzing (6.201), it can also be noted that the problem is complicated by the higher sensitivity of K_k^∞ to θ.

All that follows from the previous analysis of estimation errors caused by temporary uncertainties is that the UFIR filter can still be considered the most robust solution. The game theory H_∞ filter can improve the performance of KF by properly setting the tuning factor θ. However, since the exact θ is not available at every point of time, this filter can become unstable and diverge.

6.9 Implementation of Polynomial UFIR Filters

Hardware implementation of digital filters is usually not required in state space. In contrast, scalar input-to-output structures have found much more applications. Researchers often prefer to use UFIR polynomial structures, whenever suboptimal filtering is required. Examples can be found in tracking, control, positioning, timekeeping, etc. If such a UFIR filter matches the model, then the group delay reaches a minimum. Otherwise, the group delay grows at a lower rate than in IIR filters. This makes UFIR structures very useful in engineering practice of suboptimal filtering. Next we will present and analyze the polynomial UFIR structures in the z domain [29] and *discrete Fourier transform* (DFT) domain [30].

6.9.1 Filter Structures in z-Domain

The transfer function $\mathcal{T}^{(i)}(z)$ of the ith degree polynomial UFIR filter is specialized by the z-transform applied to the FIR function $h_k^{(i)}$ given by (6.120) with $p = 0$; that is,

$$\mathcal{T}^{(i)}(z) = \sum_{k=0}^{N-1} h_k^{(i)} z^{-k} \tag{6.202a}$$

$$= \sum_{j=0}^{i} a_{ji} \sum_{k=0}^{N-1} k^j z^{-k}\,, \tag{6.202b}$$

where $z = e^{j\omega T}$, ω is an angular frequency, T is the sampling time, a_{ji} is given by (6.121), and conventionally we will also use j in $z = e^{j\omega T}$ as the imaginary sign $j = \sqrt{-1}$. The following properties of

$\mathcal{T}^{(i)}(z)$ can be listed in addition to the inherent 2π-periodicity, symmetry of $|\mathcal{T}^{(i)}(z)|$, and antisymmetry of arg $\mathcal{T}^{(i)}(z)$.

- **Transfer function at** $\omega = 0$. By $z = e^0 = 1$, the transfer function becomes

$$\mathcal{T}^{(i)}(e^0) = 1 \tag{6.203}$$

for all i, which means that the UFIR filter is an LP filter.

- **Impulse response at** $k = 0$. By the inverse z-transform, the value of $h_k^{(i)}$ at $k = 0$ is defined as

$$h_0^{(i)} = \frac{1}{2\pi j} \oint\limits_{C_1} \frac{\mathcal{T}^{(i)}(z)}{z} \mathrm{d}z \tag{6.204a}$$

$$= \frac{1}{2\pi} \int_0^{2\pi} \mathcal{T}^{(i)}(e^{j\omega T}) \mathrm{d}(\omega T) . \tag{6.204b}$$

Because $h_0^{(i)}$ is always positive-valued, $h_0^{(i)} > 0$, and for all i, the counterclockwise circular integration in (6.204a) always produces a positive imaginary value, which gives

$$\oint\limits_{C_1} \frac{\mathcal{T}^{(i)}(z)}{jz} \mathrm{d}z = 2\pi h_0^{(i)} > 0 . \tag{6.205}$$

- **Transfer function at** $\omega T = \pi$. By $z = e^{j\pi}$, the transfer function becomes

$$\mathcal{T}^{(i)}(e^{j\pi}) = -\frac{1}{2} \sum_{j=0}^{i} a_{ji} \left[(-1)^N E_j(N) - E_j(0) \right] , \tag{6.206}$$

where $E_j(x)$ is the Euler polynomial. For low-degree impulse responses, the Euler polynomials become $E_0(x) = 1$, $E_1(x) = x - \frac{1}{2}$, $E_2(x) = x^2 - x$, and $E_3(x) = x^3 - \frac{3}{2}x^2 + \frac{1}{4}$.

- **Energy**. By Parseval's theorem, the following relations hold

$$\frac{1}{2\pi} \int_0^{2\pi} |\mathcal{T}^{(i)}(e^{j\omega T})|^2 \mathrm{d}(\omega T) = \sum_{k=0}^{N-1} h_k^{(i)^2} \tag{6.207a}$$

$$= \sum_{j=0}^{i} a_{ji} \sum_{k=0}^{N-1} h_k^{(i)} k^j \tag{6.207b}$$

$$= a_{0i} = h_0^{(i)} , \tag{6.207c}$$

and it follows that the energy (or squared norm) of the FIR function in the z domain is equal to the value of $h_k^{(i)}$ at $k = 0$.

- **Unbiasedness in the** z **domain**. The following theorem establishes an unbiasedness condition that is fundamental to digital FIR filters in the z domain.

Theorem 6.2 Given a digital FIR filter with an ith degree transfer function $\mathcal{T}^{(i)}(z)$ and an input polynomial signal specified on $[m, k]$. Then the FIR filter will be unbiased if it satisfies the following unbiasedness condition in the z domain:

$$\int_0^{2\pi} \mathcal{T}^{(i)}(e^{j\omega T}) \mathrm{d}(\omega T) = \int_0^{2\pi} |\mathcal{T}^{(i)}(e^{j\omega T})|^2 \mathrm{d}(\omega T) . \tag{6.208}$$

Proof: The proof appears instantly when compared to (6.204b) and (6.207c). □

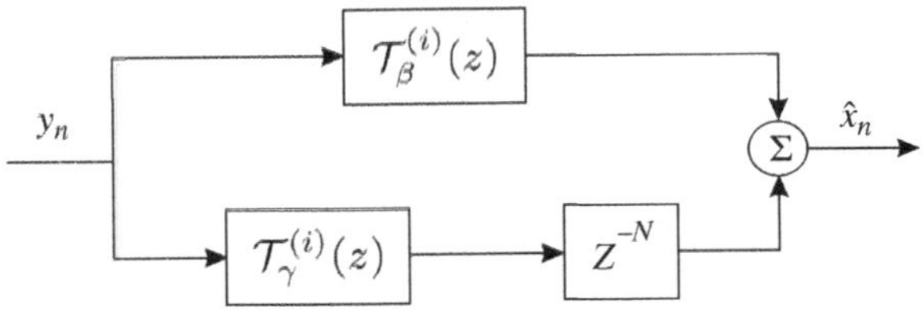

Figure 6.10 Generalized block diagram of the ith degree UFIR filter.

- **Noise power gain.** The squared norm of $h_k^{(i)}$, which represents the NPG $g^{(i)} = \| h_k^{(i)} \|_2^2$ of the UFIR filter (Table 6.2), is also the energy of $h_k^{(i)}$. By Parseval's theorem, NPG takes the following equivalent forms

$$g^{(i)} = \frac{1}{2\pi}\int_0^{2\pi} |\mathcal{T}^{(i)}(e^{j\omega T})|^2 \, \mathrm{d}(\omega T) \tag{6.209a}$$

$$= \frac{1}{2\pi}\int_0^{2\pi} \mathcal{T}^{(i)}(e^{j\omega T}) \, \mathrm{d}(\omega T) \tag{6.209b}$$

$$= h_0^{(i)} = a_{0l} \, . \tag{6.209c}$$

It is worth noting that using to the properties listed, it becomes possible to design and optimize polynomial FIR filter structures in the z domain with maximum efficiency for real-time operation.

Transfer Function in the z-Domain

Although the inner sum in (6.202b) does not have a closed form, the properties of the FIR filter in the z-domain allow one to represent $\mathcal{T}^{(i)}(z)$ as

$$\mathcal{T}^{(i)}(z) = \frac{\sum_{j=0}^{i} \beta_j z^{-j} + z^{-N} \sum_{j=0}^{i} \gamma_j z^{-j}}{1 + \sum_{j=1}^{i+1} \alpha_j z^{-j}} \, . \tag{6.210}$$

By assigning the following subtransfer functions

$$\mathcal{T}_\beta^{(i)}(z) = \left(\sum_{j=0}^{i} \beta_j z^{-j} \right) \Big/ \left(1 + \sum_{j=1}^{i+1} \alpha_j z^{-j} \right), \tag{6.211}$$

$$\mathcal{T}_\gamma^{(i)}(z) = \left(\sum_{j=0}^{i} \gamma_j z^{-j} \right) \Big/ \left(1 + \sum_{j=1}^{i+1} \alpha_j z^{-j} \right), \tag{6.212}$$

the generalized block diagram of the ith degree UFIR filter can be shown as in Fig. 6.10. The low-degree coefficients α_j, β_j, and γ_j are listed for this structure in Table 6.3 [29].

Example 6.5 ***Transfer function of the first-degree UFIR filter in z domain [29].*** If we use the coefficients given in Table 6.3, then the transfer function $\mathcal{T}^{(1)}(z)$ of the first-degree polynomial UFIR filter having a ramp impulse response function $h_k^{(1)}$ comes from (6.210) as

$$\mathcal{T}^{(1)}(z) = \frac{2}{N} \frac{\frac{a_{01}N}{2} - 2z^{-1} + z^{-N}\left(1 - \frac{N-2}{N+1} z^{-1}\right)}{(1 - z^{-1})^2} \, . \tag{6.213}$$

Simple analysis shows that the *region of convergence* (ROC) for (6.213) is all z and that the filter is both stable and causal. A block diagram corresponding to (6.213) is exhibited in Fig. 6.11, and it

Table 6.3 Coefficients α_j, β_j, and γ_j of Low-Degree UFIR Filters.

	i			
	0	**1**	**2**	**3**
β_0	$\frac{1}{N}$	a_{01}	a_{02}	a_{03}
β_1	0	$-\frac{4}{N}$	$-\frac{18(N-1)}{N(N+1)}$	$-\frac{48(N^2-2N+2)}{N(N+1)(N+2)}$
β_2	0	0	$\frac{9}{N}$	$\frac{24(2N-3)}{N(N+1)}$
β_3	0	0	0	$-\frac{16}{N}$
γ_0	$-\frac{1}{N}$	$\frac{2}{N}$	$-\frac{3}{N}$	$\frac{4}{N}$
γ_1	0	$-\frac{2(N-2)}{N(N+1)}$	$\frac{6(N-3)}{N(N+1)}$	$-\frac{12(N-4)}{N(N+1)}$
γ_2	0	0	$-\frac{3(N-2)(N-3)}{N(N+1)(N+2)}$	$\frac{12(N-3)(N-4)}{N(N+1)(N+2)}$
γ_3	0	0	0	$-\frac{4(N-2)(N-3)(N-4)}{N(N+1)(N+2)(N+3)}$
α_1	−1	−2	−3	−4
α_2	0	1	3	6
α_3	0	0	−1	−4
α_4	0	0	0	1

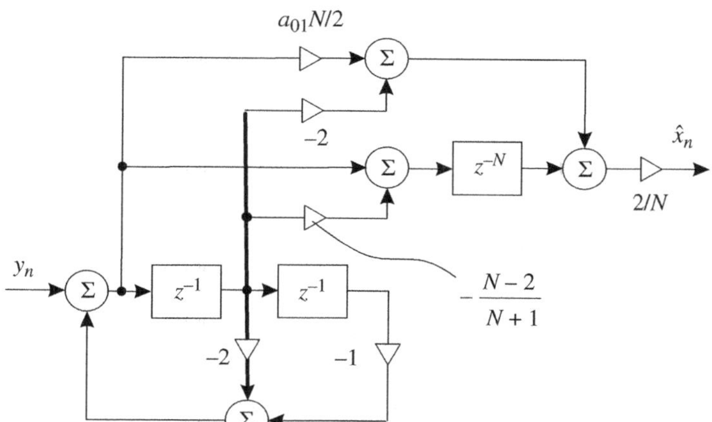

Figure 6.11 Block diagram of the first-degree (ramp) polynomial UFIR filter.

follows that the filter utilizes six multipliers, four adders, and three time-delay subfilters. It can be shown that this structure can be further optimized to have three multipliers, five adders, and three time-delay subfilters, as shown in [27]. Note that block diagrams of other low-degree polynomial UFIR filters can be found in [29]. □

The magnitude response functions $|\mathcal{T}^{(i)}(e^{j\omega T})|$ of low-degree polynomial UFIR filters are shown in Fig. 6.12. A distinctive feature of LP filters of this type is a negative slop of 20 dB per decade in the Bode plot (Fig. 6.12b) in the transient region. Another feature is that increasing the degree of the polynomial filter expands the bandwidth. It can also be seen that the transfer function has multiple intensive side lobes that are related to the shape of the impulse response (Fig. 6.5). The later property is discouraged in the design of standard LP filters. However, it is precisely this shape of $|\mathcal{T}^{(i)}(e^{j\omega T})|$ that guarantees the filter unbiasedness.

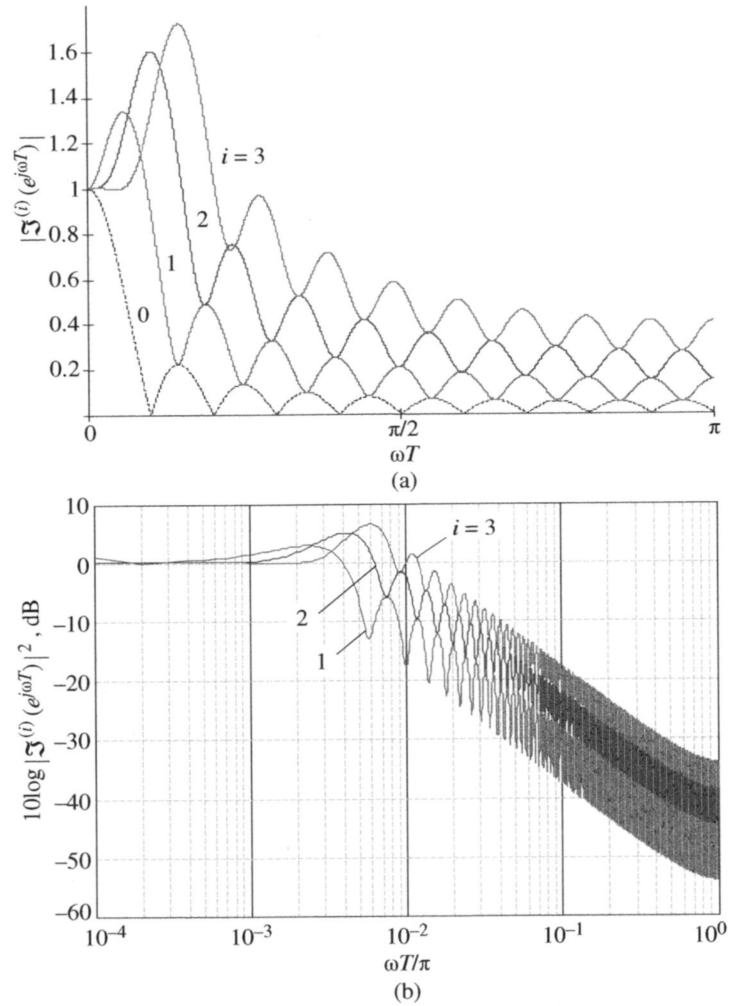

Figure 6.12 Magnitude response functions $|\mathcal{T}^{(i)}(e^{j\omega T})|$ of low-degree polynomial UFIR filters: (a) linear scale for $N = 20$ and (b) Bode plot for $N = 500$.

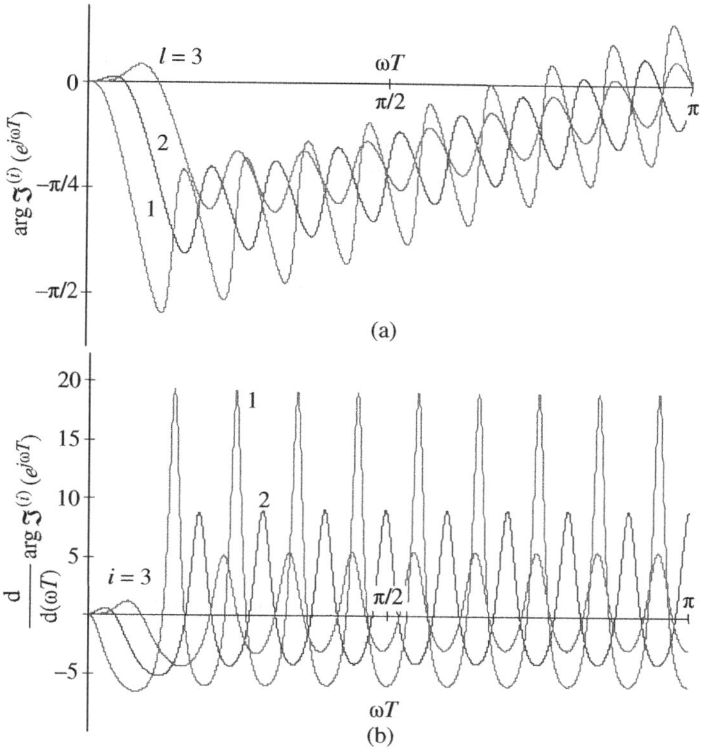

Figure 6.13 Phase response functions of the low-degree UFIR filters for $N = 20$: (a) phase response $\arg \mathcal{T}^{(i)}(e^{j\omega T})$ and (b) group delay $\frac{\mathrm{d}}{\mathrm{d}(\omega T)} \arg \mathcal{T}^{(i)}(e^{j\omega T})$.

Figure 6.12a assures that bias is eliminated at the filter output by shifting and lifting the side lobes of a zero-degree uniform FIR filter, which provides simple averaging. Accordingly, the i-degree filter passes the spectral content close to zero without change, magnifies the components falling into the first lobe, and attenuates the power of the higher-frequency components with 10 dB per decade (Fig. 6.12b).

The phase response functions $\arg \mathcal{T}^{(i)}(e^{j\omega T})$ of low-degree polynomial UFIR filters are shown in Fig. 6.13a, and the group delay functions obtained by $\frac{\mathrm{d}}{\mathrm{d}(\omega T)} \arg \mathcal{T}^{(i)}(e^{j\omega T})$ are shown in Fig. 6.13b. The phase response function of this filter is linear on average (Fig. 6.13a). However, the phase response changes following variations in the magnitude response, which, in turn, leads to changes in the group delay around a small constant value (Fig. 6.13b). From the standpoint of the design of basic LP filters, the presence of periodic variations in the phase response is definitely a disadvantage. But it is also a key condition that cannot be violated without making the filter output biased.

6.9.2 Transfer Function in the DFT Domain

The DFT of the ith degree UFIR filter impulse response $h_k^{(i)}$ is obtained as

$$\mathcal{T}_n^{(i)} = \sum_{k=0}^{N-1} h_k^{(i)} W_N^{nk} , \tag{6.214}$$

where $W_N = e^{-j\frac{2\pi}{N}}$. In addition to the inherent N-periodicity, the symmetry of $|\mathcal{T}_n^{(i)}|$, and the anti-symmetry of arg $\mathcal{T}_n^{(i)}$, the following properties are important for the implementation of UFIR filters in the DFT domain.

- **DFT value at** $n = 0$. By $n = 0$, the function W_N^{nk} becomes unity, $W_{2N}^{0k} = 1$, and, by the properties listed in Table 6.2, the DFT function (6.214) is transformed to

$$\mathcal{T}_0^{(i)} = 1 \tag{6.215}$$

  for all i, which fits with LP filtering.
- **Impulse response at** $k = 0$. For $k = 0$ and $W_{2N}^{n0} = 1$, the inverse DFT (IDFT) applied to $\mathcal{T}_n^{(i)}$ gives

$$h_0^{(i)} = \frac{1}{N}\sum_{n=0}^{N-1} \mathcal{T}_n^{(i)} . \tag{6.216}$$

  Since $h_0^{(i)} > 0$ holds for all i, the sum of the DFT coefficients is real and positive. It then follows that

$$\sum_{n=0}^{N-1} \mathcal{T}_n^{(i)} = N h_0^{(i)} > 0 . \tag{6.217}$$

- **DFT value at** $n = N/2$. At the point of symmetry $n = N/2$ for even N, the function W_{2N}^{nk} becomes $W_{2N}^{0k} = (-1)^k$ and the following relation holds

$$\mathcal{T}_{\frac{N}{2}}^{(i)} = -\frac{1}{2}\sum_{j=0}^{l} a_{ji}[E_j(N) - E_j(0)] , \tag{6.218}$$

  where $E_j(x)$ is the Euler polynomial. For low degrees, the Euler polynomials take the following form: $E_0(x) = 1$, $E_1(x) = x - \frac{1}{2}$, $E_2(x) = x^2 - x$, and $E_3(x) = x^3 - \frac{3}{2}x^2 + \frac{1}{4}$.
- **Noise power gain.** As a measure of noise reduction at the UFIR filter output, NPG $g^{(i)}$ for white Gaussian noise is determined by the energy of $h_k^{(i)}$. The NPG has the following equivalent forms in the DFT domain:

$$g^{(i)} = \frac{1}{N}\sum_{n=0}^{N-1} |\mathcal{T}_n^{(i)}|^2 \tag{6.219a}$$

$$= \frac{1}{N}\sum_{n=0}^{N-1} \mathcal{T}_n^{(i)} \tag{6.219b}$$

$$= h_0^{(i)} = a_{0i} . \tag{6.219c}$$

- **Unbiasedness condition in the DFT domain.** The following theorem establishes an unbiasedness condition that is fundamental to UFIR filters in the DFT domain.

  Theorem 6.3 Given a digital FIR filter with a ith degree transfer function $\mathcal{T}_n^{(i)}$ and an input polynomial signal specified on $[m, k]$, then the FIR filter will be unbiased if the following unbiasedness condition is satisfied in the DFT domain:

$$\sum_{n=0}^{N-1} |\mathcal{T}_n^{(i)}|^2 = \sum_{n=0}^{N-1} \mathcal{T}_n^{(i)} . \tag{6.220}$$

Proof. To prove (6.220), use the following fundamental condition for optimal filtering: *the order of the optimal and/or unbiased filter must be the same as that of the system.* Then represent $h_k^{(i)}$ by (6.124),

$$h_k^{(i)} = \sum_{j=0}^{i} a_{ji} k^j ,$$

where the coefficient a_{ji} is given by (6.125). Recall that function (6.124) has the following main properties, given in Table 6.2: the sum of its coefficients is equal to unity, and the moments zero; that is,

$$1 = \sum_{k=0}^{N-1} h_k^{(i)} , \tag{6.221}$$

$$0 = \sum_{k=0}^{N-1} h_k^{(i)} k^u , \quad 1 \leqslant u \leqslant i . \tag{6.222}$$

Use Parseval's theorem and obtain the following relationships,

$$\frac{1}{N}\sum_{n=0}^{N-1} |\mathcal{T}_n^{(i)}|^2 = \sum_{k=0}^{N-1} h_k^{(i)2} \tag{6.223a}$$

$$= \sum_{j=0}^{i} a_{ji} \sum_{k=0}^{N-1} h_k^{(i)} k^j \tag{6.223b}$$

$$= a_{0i} = h_0^{(i)} . \tag{6.223c}$$

Now observe that the IDFT applied to $\mathcal{T}_n^{(i)}$ at $k = 0$ gives $h_0^{(i)} = \frac{1}{N}\sum_{n=0}^{N-1} \mathcal{T}_n^{(i)}$ that finally leads to (6.220) and completes the proof.

- **Estimation error boundary.** The concept of NPG $g^{(i)}$ can be used to specify the bound $\bar{\varepsilon}^{(i)}$ for the estimation error in the three-sigma sense as

$$\bar{\varepsilon}^{(i)} = 3\sigma_v \sqrt{g^{(i)}} , \tag{6.224}$$

where σ_v is the standard deviation of the measurement white noise v_n.

Note that the previously discussed properties of polynomial UFIR filters in the DFT domain are preserved for any polynomial degree, but in practice only low-degree digital UFIR filters, $l \leqslant 3$, are usually found. The reason for using low-degree filters is explained earlier: increasing the degree leads to larger random errors at the filter output.

Transfer Function in the DFT Domain

Using the properties listed, the transfer function $\mathcal{T}_n^{(i)}$ of the ith degree UFIR filter can be represented in the DFT domain with the sum

$$\mathcal{T}_n^{(i)} = \sum_{j=0}^{i} a_{ji} \sum_{k=0}^{N-1} k^j W_N^{nk} , \tag{6.225}$$

in which the inner sum has no closed-form solution for an arbitrary j. However, solutions can be found for low-degree polynomials as shown in [30]. Next we will illustrate such a solution for a first-degree UFIR filter and postpone to "Problems" the search for solutions associated with higher degrees.

Example 6.6 ***First-degree UFIR filter in DFT domain [30].*** The first-degree (ramp) UFIR filter is associated with signals, which are linear on $[m, k]$. The impulse response of this filter that is specialized with the coefficients $a_{01} = \frac{2(2N-1)}{N(N+1)}$ and $a_{11} = -\frac{6}{N(N+1)}$ has the DFT

$$\mathcal{T}_n^{(1)} = -\frac{6}{N+1}\left[1 - \frac{j}{2\sin(\pi n/N)} W_{2N}^n\right] \tag{6.226a}$$

$$= \frac{3}{(N+1)\sin(\pi n/N)} e^{j\frac{\pi}{2}\left(\frac{2n}{N}-1\right)}, \tag{6.226b}$$

where $0 < n < N-1$. At $n = 0$ and $n = N$, the function $\mathcal{T}_n^{(1)}$ becomes unity. Note that for $i \in [2,3]$, $\mathcal{T}_n^{(i)}$ can be found in [30]. □

The magnitude response $|\mathcal{T}_n^{(i)}|$, Bode plot, and phase response arg $\mathcal{T}_n^{(i)}$ functions of the ith degree, $i \in [1,3]$, polynomial UFIR filter are shown in Fig. 6.14. For the 1st degree, the functions are computed by (6.226b) and for the next two low degrees the transfer functions can be found in [30]. In addition to the inherent properties of periodicity with a repetition period of N points, symmetry of $|\mathcal{T}_n^{(i)}|$, and antisymmetry of arg $\mathcal{T}_n^{(i)}$, the transfer function $\mathcal{T}_n^{(i)}$ has several other features that are of practical importance.

- The magnitude response $|\mathcal{T}_n^{(i)}|$ is a monotonically decreasing function of n, which changes from $n = 0$ to $n = N/2$ (Fig. 6.14a), and has a transition slope n^{-2} on the Bode plot (Fig. 6.14b). So we have further proof that the UFIR filter is essentially an LP filter.
- The phase response arg $\mathcal{T}_n^{(i)}$ is an antisymmetric function existing from $-\frac{\pi}{2}$ to $\frac{\pi}{2}$ on $n \in [0, N-1]$ with a positive slope, except for the transient region where an increase in the filter degree makes it more complex (Fig. 6.14c). This function is close to linear for low-degree filters, and it is strictly linear for the first-degree filter.

Finally, we come to an important practical conclusion. Since the transfer function of the polynomial UFIR filter is monotonic and does not contain the periodic variations observed in the z-domain (Fig. 6.13), it follows that the filter can be easily implemented with all of the advantages of suboptimal unbiased filtering.

6.10 Summary

By abandoning the requirement for initial values and ignoring zero mean noise, the UFIR state estimator appears to be the most robust among other linear estimators such as OFIR, OUFIR, and ML state estimators. Moreover, its iterative algorithm is more robust than the recursive KF algorithm. The only tuning factor required for the UFIR filter is the optimal horizon, which can be determined much more easily than noise statistics. Furthermore, at given averaging horizons, the UFIR state estimator becomes blind, which is very much appreciated in practice. In general, it is only within a narrow range of error factors caused by various types of uncertainties that optimal and ML estimators are superior to the UFIR estimator. For the most part, this explains the better performance of the UFIR state estimator in many real-world applications.

Like other FIR filters, the UFIR filter operates on the averaging horizon $[m, k]$ of N points, from $m = k - N + 1$ to k. Its discrete convolution-based batch resembles the LS estimator. But, the latter is not a state estimator. The main performance characteristic of a scalar input-to-output UFIR filter is NPG. If the UFIR filter is designed to work in state space, then its noise reduction properties are characterized by GNPG.

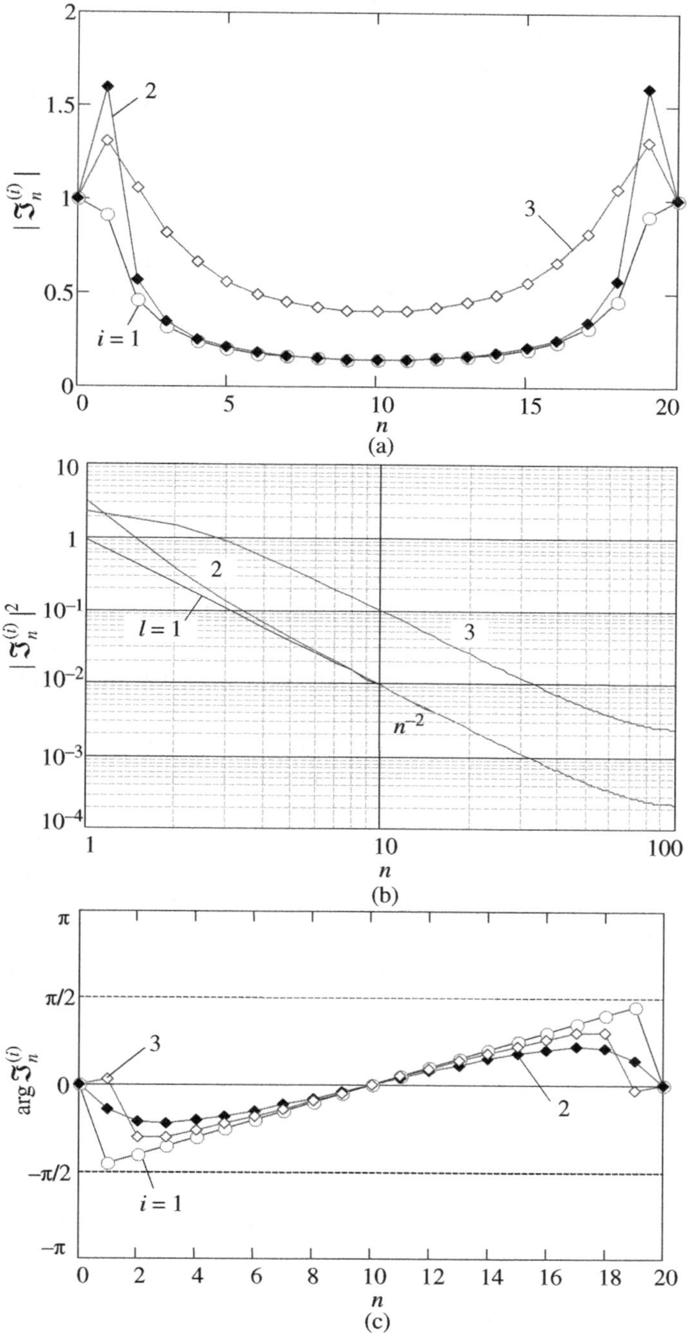

Figure 6.14 DFT of the low-degree polynomial UFIR filters: (a) magnitude response $|\mathcal{T}_n^{(i)}|$ for $N = 20$, (b) Bode plot of $|\mathcal{T}_n^{(i)}|^2$ for $N = 200$, and (c) phase response arg $\mathcal{T}_n^{(i)}$ for $N = 20$.

The recursive forms used to iteratively compute the batch UFIR estimate are not Kalman recursions. An important feature is that UFIR filter recursions serve any zero mean noise, while Kalman recursions are optimal only for white Gaussian noise. It is worth noting that the error covariance of the UFIR and Kalman filters are represented by the same Riccati equation. The difference lies in the different bias correction gains. The optimal bias correction gain of the KF is called the Kalman gain K_k, while the bias correction gain of the UFIR filter is computed in terms of GNPG as $\mathcal{G}_k H^T$.

Since the UFIR filter does not involve noise covariances to the algorithm, it minimizes the MSE on the optimal horizon of N_{opt} points (Fig. 6.1). The significance of N_{opt} is that random errors increase if the horizon length N is chosen such that $N < N_{\text{opt}}$. Otherwise, bias errors grow if $N > N_{\text{opt}}$. Using (6.49), the optimal horizon N_{opt} can be estimated through the measurement residual without using ground truth. In response to stepwise temporary uncertainties, the UFIR filter generates finite time transients at N points with larger overshoots, while the KF generates longer transients with lower overshoots. This property represents the main difference between the transients in both filters.

It turns out that, since zero mean noise is ignored by UFIR state estimators, the FFFM, FFBM, BFFM, and BFBM q-lag UFIR smoothing algorithms are equivalent. It should also be kept in mind that all unbiased, optimal, and optimal unbiased state estimators are essentially LP filters. Moreover, the zero-degree (simple average) UFIR filter is the best in the sense of the minimum produces noise. This is because an increase in the filter degree or the number of the states leads to an increase in random errors at the estimator output. It also follows that the set of degree polynomial impulse responses of the UFIR filter establishes a class of discrete orthogonal polynomials that are suitable for signal analysis and restoration due to the built-in unbiasedness.

Although the UFIR filter ignores zero mean noise, it is not suitable for colored noise that is biased on short horizons. If the colored noise is well-approximated by the Gauss-Markov process, then the general UFIR filter becomes universal for time-correlated and uncorrelated driving white noise sources. Like GKF, a general UFIR filter can be designed to work with CMN and CPN. It can also be applied to state-space models with smooth nonlinearities using the first- or second-order Taylor series expansions. However, it turns out that the second-order EFIR-2 filter has no practical advantage over the first-order EFIR-1 filter.

The implementation of polynomial UFIR filters can be most efficiently obtained in the DFT domain due to the following critical property: the DFT transfer function is smooth and does not exhibit periodic variations inherent to the z-domain.

6.11 Problems

1 Explain the difference between polynomial fitting and filtering (smoothing) of polynomial models.

2 NPG for scalar signals is given by (6.15). Give the interpretation of GNPG defined by (6.16a) for state vectors. Why does the NPG decrease with an increase in the averaging horizon?

3 **Solved problem: Recursive forms for UFIR filter**. Consider the batch UFIR filtering estimate $\hat{x}_k = (C^T_{m,k} C_{m,k})^{-1} C^T_{m,k} Y_{m,k}$ with the fundamental gain given by (6.10b). Another way to obtain recursive forms for this filter (Algorithm 11) is the following [179]. Represent the inverse of the GNPG $\mathcal{G}_k = (C_{m,k}^T C_{m,k})^{-1}$ as

$$\begin{aligned}\mathcal{G}_k^{-1} &= \sum_{i=0}^{N-1} (\mathcal{F}_k^{m+1+i})^{-T} H_{m+i}^T H_{m+i} (\mathcal{F}_k^{m+1+i})^{-1} ,\\ &= H_k^T H_k + F_k^{-T} \left[\sum_{i=0}^{N-2} (\mathcal{F}_{k-1}^{m+1+i})^{-T} H_{m+i}^T H_{m+i} (\mathcal{F}_{k-1}^{m+1+i})^{-1} \right] F_k^{-1} .\end{aligned}$$

Since $\mathcal{F}_r^g = 0$ holds for $g > r + 1$, write $\mathcal{G}_{k-1}^{-1}$ as

$$\mathcal{G}_{k-1}^{-1} = \sum_{i=0}^{N-2} (\mathcal{F}_{k-1}^{m+1+i})^{-T} H_{m+i}^T H_{m+i} (\mathcal{F}_{k-1}^{m+1+i})^{-1} ,$$

substitute to the previous relation, and arrive at the recursion (6.21) for $\mathcal{G}_k$.
Similarly, represent the product $C_{m,k}^T Y_{m,k}$ as

$$\begin{aligned}C_{m,k}^T Y_{m,k} &= H_k^T y_k + F_k^{-T} \sum_{i=0}^{N-2} (\mathcal{F}_{k-1}^{m+1+i})^{-T} H_{m+i} y_{m+i} ,\\ &= H_k^T y_k + F_k^{-T} C_{m,k-1}^T Y_{m,k-1} ,\end{aligned}$$

combine with $\mathcal{G}_k$, obtain $\hat{x}_k = \mathcal{G}_k (H_k^T y_k + F_k^{-T} C_{m,k-1}^T Y_{m,k-1})$, substitute $C_{m,k-1}^T Y_{m,k-1} = \mathcal{G}_{k-1}^{-1} \hat{x}_{k-1}$, and come up with

$$\hat{x}_k = \mathcal{G}_k (H_k^T y_k + F_k^{-T} \mathcal{G}_{k-1}^{-1} \hat{x}_{k-1}) .$$

From (6.21) find $\mathcal{G}_{k-1}^{-1} = F_k^T (\mathcal{G}_k^{-1} - H_k^T H_k) F_k$, substitute into the previous relation, and arrive at the recursion (6.29) for the UFIR estimate.

4 The error covariance of the UFIR filter is given by (6.37) as

$$\begin{aligned}P_k &= (I - \mathcal{G}_k H_k^T H_k)(F_k P_{k-1} F_k^T + B_k Q_k B_k^T)(I - \mathcal{G}_k H_k^T H_k)^T \\ &\quad + \mathcal{G}_k H_k^T R_k H_k \mathcal{G}_k .\end{aligned}$$

Compare this relation to the error covariance of the KF and explain the difference between the bias correction gain $K_k = \mathcal{G}_k H_k^T$ of the UFIR filter and the Kalman gain $K_k = P_k^- H_k^T S_k^{-1}$. Why can the Kalman gain not be less than the optimal value? Why does the bias correction gain K_k of the UFIR decrease with increasing the horizon length?

5 To minimize MSE, the averaging horizon for the UFIR filter should be optimal N_{opt} (Fig. 6.1) by solving the following minimization problem

$$N_{\text{opt}} = \arg\min_N \ \text{tr} P_k(N) .$$

Why does the OFIR filter not need an optimal horizon? Why does the error in the OFIR filter decrease with increasing the horizon length, while the UFIR filter minimizes the MSE by only N_{opt}? Give simple and intuitive explanations.

6 Measurement data transmitted over a wireless communication channel with latency are represented at the receiver with the following state-space equations

$$\begin{aligned}x_k &= F x_{k-1} + w_k ,\\ y_k &= \gamma_{0k} H x_k + \gamma_{1k} H x_{k-1} + \gamma_{2k} H x_{k-2} + v_k ,\end{aligned}$$

where γ_{ik}, $i \in [0,2]$, is a deterministic weight, which can be either unity or zero. The time-stamped data indicate a value of i for which the weight is unity and the other weights are zero. Transform the model to another without delay and develop a UFIR filter.

7 A useful property of the measurement residual S_k of the UFIR filter is that the derivative of its trace with respect to the horizon length reaches a minimum when N approaches N_{opt} (Fig. 6.2). This makes it possible to determine N_{opt} by (6.49) or by solving the minimization problem

$$N_{\text{opt}} \cong \arg\min_{N} \frac{\partial}{\partial N} \text{tr} S_k(N) .$$

Give an explanation to this fact and discuss the problems that may arise in the practical implementations of this method.

8 An open and challenging problem is to find an analytical relationship between N_{opt} and noise covariances Q_k and R_k similarly to (6.51) [153]. Find a solution to this problem for a periodical process observed as

$$y_k = a_{0k} + a_{1k} \cos\left(\frac{\pi}{12}k + \phi_1\right) + v_k ,$$

where $v_k \sim \mathcal{N}(0, \sigma_v^2)$ and a_{0k}, a_{1k}, and ϕ_1 are state variables. Write the state equation with additive white Gaussian noise.

9 The GNPG of the backward UFIR filter is defined by (6.62) as

$$\overleftarrow{\mathcal{G}}_m = \mathcal{X}_k^{m+1} (\overleftarrow{H}_{k,m}^T \overleftarrow{H}_{k,m})^{-1} \mathcal{X}_k^{m+1^T} .$$

Represent GNPG as $\overleftarrow{\mathcal{G}}_m = (\overleftarrow{C}_{k,m}^T \overleftarrow{C}_{k,m})^{-1}$ and specify $\overleftarrow{C}_{k,m}$.

10 The error covariance P_m of the backward UFIR filter is given by

$$\begin{aligned} P_m &= (I - \overleftarrow{\mathcal{G}}_m H_m^T H_m) F_{m+1}^{-1} (P_{m+1} + B_{m+1} Q_{m+1} B_{m+1}^T) \\ &\quad \times F_{m+1}^{-T} (I - \overleftarrow{\mathcal{G}}_m H_m^T H_m)^T + \overleftarrow{\mathcal{G}}_m H_m^T R_m H_m \overleftarrow{\mathcal{G}}_m . \end{aligned}$$

Reasoning similarly as for the forward UFIR filter, provide the derivation of (6.72). Analyze this relationship and find out what will happen to P_m if the noise covariance Q_{m+1} is set equal to zero.

11 The error covariance of the batch q-lag FFFM UFIR smoother is given by (6.85). Recursive computation of (6.85) is provided by (6.94), where the matrix L_q still has a batch form (6.92) given by

$$L_q = \sum_{i=1}^{q} \mathcal{F}_{k-q+i}^{k-q+2^T} H_{k-q+1}^T H_{k-q+i} (\mathcal{F}_k^{k-q+i+1})^{-1} .$$

Find a recursive form for (6.92) or, otherwise, develop a different approach for recursive computation of (6.85).

12 Using deductive reasoning, the recursive form for $\mathcal{Z}_q^{(1)} = \bar{D}_{m,k}\mathcal{Q}_{m,k}\tilde{D}_{m,k}^{(q)T}$ was found as (6.90) and for $\mathcal{Z}_q^{(3)} = \tilde{D}_{m,k}^{(q)}\mathcal{Q}_{m,k}G_{m,k}^T\hat{\mathcal{H}}_{m,k}^{\mathrm{h}^T}$ as (6.93), respectively,

$$\mathcal{Z}_q^{(1)} = \mathcal{Z}_{q-1}^{(1)} + M_q(\mathcal{F}_k^{k-q+2})^{-1} ,$$

$$\mathcal{Z}_q^{(3)} = \mathcal{Z}_{q-1}^{(3)} + M_q L_q \mathcal{G}_k ,$$

Show other possible ways to obtain recursions for (6.90) and (6.93).

13 Referring to the optimal state estimation problem and considering the transform of the UFIR filter gain $\hat{\mathcal{H}}_k = (C_{m,k}^T C_{m,k})^{-1} C_{m,k}^T$, find an explanation for the fact that the optimal, optimal unbiased, ML, and unbiased FIR filters are LP filters.

14 It follows from (6.129) that polynomial FIR functions $h_k^{(i)}$, $i \in [0, K-1]$, given by (6.124), establish a new class discrete orthogonal polynomials related by the recurrence relation (6.128). Reasoning in a similar way, examine for orthogonality the class of more general p-shift FIR functions $h_k^{(i)}(p)$, $i \in [0, K-1]$, $k \in [p, N-1+p]$, given by (6.120), and find the recurrence relation, if any.

15 Find an explanation for the fact that increasing the number of states makes the state estimator less effective in terms of noise reduction. In this sense, explain why simple averaging is best in terms of the minimum produced noise.

16 It follows from Fig. 6.5 that prediction errors grow with increasing the p-step. Which method is more efficient in predicting future states: stochastic prediction requiring future noise values? Deterministic prediction ignoring future noise? Or linear deterministic prediction?

17 General UFIR and Kalman filters serve for correlated and de-correlated Gauss-Markov process noise $w_k = \Theta_k w_{k-1} + \mu_k$ and measurement noise $v_k = \Psi_k v_{k-1} + \xi_k$, where $\mu_k \sim \mathcal{N}(0, Q_\mu)$ and $\xi_k \sim \mathcal{N}(0, R_\xi)$. Can we modify the UFIR filter and KF for non-Gaussian colored noise? If not, explain why.

18 The periodic system is represented in state space by the equations

$$x_k = \begin{bmatrix} \cos\varphi & \sin\varphi \\ -\sin\varphi & \cos\varphi \end{bmatrix} x_{k-1} + \begin{bmatrix} 1 \\ 1 \end{bmatrix} w_k ,$$

$$y_k = \begin{bmatrix} 1 & 0 \\ 0 & 1 \end{bmatrix} x_k + v_k ,$$

where φ is a constant angular measure, and w_k and v_k are zero mean noise sources. Using the derivation of the UFIR filter for polynomial models, derive the scalar FIR functions for the first state h_{1k} and the second state h_{2k} to satisfy the unbiasedness condition $\mathcal{E}\{\hat{x}_k\} = \mathcal{E}\{x_k\}$.

19 The bias correction gain of the KF is approximated by (6.184). Referring to the derivation of (6.184), give a proof of the corresponding approximation (6.185) for the H_∞ filter.

20 Lemma 6.3 states that the prior error covariance P^- of the KF is approximated by

$$P^- \leqslant Q\frac{\alpha^2}{1-\chi},$$

where α and χ are given constants. Using the proof of lemma 6.1, give a corresponding proof of the lemma 6.3.

21 Using the derivation procedure given in [180], provide a proof of approximation (6.194) of the prior error covariance P^- of the KF stated by lemma 6.4.

22 A linear digital system operates normally without uncertainties from the past to the time index $k-1$ inclusive. Then it experiences uncertainties in the coefficients η and μ at the time index k. Check that the approximate gains of the UFIR filter (6.199), KF (6.200), and H_∞ filter (6.201) are correct.

23 For the ith degree polynomial scalar FIR function $h_k^{(i)}$ existing on $[m,k]$ and the transfer function $\mathcal{T}^{(i)}(z)$, the unbiasedness constraint in the z-domain is established by theorem 6.2 as

$$\int_0^{2\pi} \mathcal{T}^{(i)}(e^{j\omega T})\mathrm{d}(\omega T) = \int_0^{2\pi} |\mathcal{T}^{(i)}(e^{j\omega T})|^2\mathrm{d}(\omega T)\,.$$

Obtain the corresponding constraint for the general UFIR filter matrix gain case $\hat{\mathcal{H}}_{m,k}$.

24 Given a digital UFIR filter with the ith degree transfer function $\mathcal{T}_n^{(i)}$ corresponding to a polynomial FIR function $h_k^{(i)}$ existing on $[m,k]$, theorem 6.3 establishes the unique unbiasedness constrain for this filter in the DFT domain as

$$\sum_{n=0}^{N-1} |\mathcal{T}_n^{(l)}|^2 = \sum_{n=0}^{N-1} \mathcal{T}_n^{(l)}\,.$$

Obtain the corresponding constraint for the UFIR filter matrix gain $\hat{\mathcal{H}}_{m,k}$.

25 The transfer function $\mathcal{T}^{(i)}$ of the ith degree polynomial UFIR filter is given in the z-domain by (6.210), which for arbitrary i has no closed-form solution. Using the properties of the UFIR filters in the z-domain and Table 6.3, find the closed forms of this function for the second and third degrees and represent the filter with block diagrams similar to that shown in Fig. 6.11.

26 The transfer function $\mathcal{T}_n^{(i)}$ of the ith degree polynomial UFIR filter is given by (6.225) and has no closed-form solution. Find solutions in closed-form for special cases of the second and third degrees.

7

FIR Prediction and Receding Horizon Filtering

When the number of factors coming into play in a phenomenological complex is too large, scientific method in most cases fails. One need only think of the weather, in which case the prediction even for a few days ahead is impossible.

Albert Einstein, Science, Philosophy and Religion (1879–1955)

7.1 Introduction

A one-step state predictive FIR approach called RH FIR filtering was developed for MPC. As an excerpt from [106] says, the term *receding horizon* was introduced "since the horizon recedes as time proceeds." Therefore, it can be applied to any FIR structure and is thus redundant. But, with due respect to this still-used technical jargon, we will keep it for one-step FIR predictive filtering. Note that the theory of bias-constrained (not optimal) RH FIR filtering was developed by W. H. Kwon and his followers [106].

The idea behind RH FIR filtering is to use an FE-based model and derive an FIR predictive filter to obtain an estimate at k over the horizon $[m-1, k-1]$ of most recent past observations. Since the predicted state can be used at the current discrete point, the properties of such filters are highly valued in digital state feedback control. Although an FIR filter that gives an estimate at $k-1$ over $[m-1, k-1]$ can also be used with this purpose by projecting the estimate to k, the RH FIR filter does it directly. Equivalently, the estimate obtained over $[m, k]$ can be projected to $k+1$ using the one-step FIR predictor.

In this chapter, we elucidate the modern theory of FIR prediction and RH FIR filtering. Some of the most interesting RH FIR solutions will also be considered, and we notice that the zero time step makes the RH FIR and FIR estimators equal in continuous time. Since the FIR predictor and the RH FIR filter can be transformed into each other by changing the time index, we will mainly focus on the FIR predictors due to their simpler forms.

7.2 Prediction Strategies

The current object state can most accurately be estimated at k as $\hat{x}_{k|k}$ using the *a posteriori* OFIR filter or KF. Since the estimate $\hat{x}_{k|k}$ may be too biased for state feedback control at the next time index $k+1$, a one-step predicted estimate $\tilde{x}_{k+1} \triangleq \tilde{x}_{k+1|k}$ is required. There are two basic strategies for solving this problem (Fig. 7.1):

Optimal and Robust State Estimation: Finite Impulse Response (FIR) and Kalman Approaches, First Edition.
Yuriy S. Shmaliy and Shunyi Zhao.

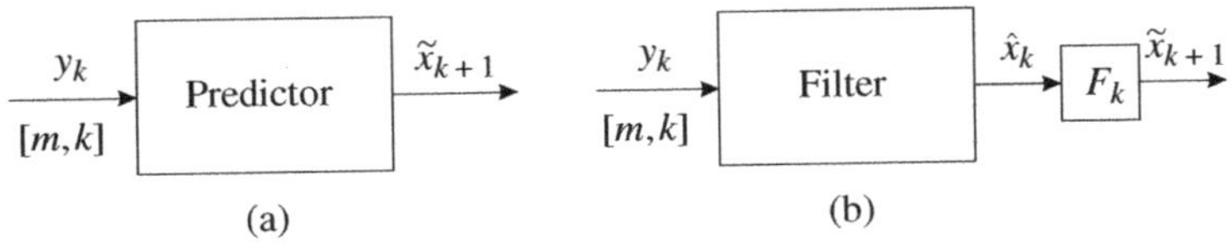

Figure 7.1 Two basic strategies to obtain the predicted estimate $\tilde{x}_{k+1}$ at $k+1$ over $[m,k]$: (a) prediction and (b) projection from k to $k+1$.

- Use the OFIR predictor and obtain the estimate $\tilde{x}_{k+1}$ over $[m,k]$ as shown in Fig. 7.1a. This also implies that, by replacing k with $k-1$, the estimate $\hat{x}_k$ can be obtained over $[m-1,k-1]$ using the RH *Kalman predictor* (KP) [103] or RH FIR filtering.
- Use the OFIR filter, obtain the *a posteriori* estimate $\hat{x}_{k|k}$, and then project it to $k+1$ as shown in Fig. 7.1b.

Both these strategies are suitable for state feedback control but suffer from an intrinsic drawback: *the predicted estimate is less accurate than the filtered one*. Note that KP can be used here as a *limited memory predictor* (LMP) operating on $[m,k]$ to produce an estimate at $k+1$.

7.2.1 Kalman Predictor

The general form of the KP for LTV systems appears if we consider the discrete-time state-space model [100]

$$x_{k+1} = F_k x_k + E_k u_k + B_k w_k \,, \tag{7.1}$$

$$y_k = H_k x_k + v_k \,, \tag{7.2}$$

where $x_k \in \mathbb{R}^K, u_k \in \mathbb{R}^L, y_k \in \mathbb{R}^P, F_k \in \mathbb{R}^{K\times K}, H_k \in \mathbb{R}^{P\times K}, E_k \in \mathbb{R}^{K\times L}, B_k \in \mathbb{R}^{K\times M}, w_k \sim \mathcal{N}(0,Q_k) \in \mathbb{R}^M$, and $v_k \sim \mathcal{N}(0,R_k) \in \mathbb{R}^P$. The prior predicted estimate can be extracted from (7.1) as

$$\tilde{x}^-_{k+1} = F_k \tilde{x}_k + E_k u_k \,, \tag{7.3}$$

where $\tilde{x}_k$ is the predicted estimate at k, and the measurement residual $s_k = y_k - H_k\tilde{x}_k = H_k\epsilon_k + v_k$ gives the innovation covariance

$$S_k = H_k P_k H_k^T + R_k \,. \tag{7.4}$$

Referring to (7.3), the prediction $\tilde{x}_{k+1}$ at $k+1$ can be written as

$$\begin{aligned}\tilde{x}_{k+1} &= \tilde{x}^-_{k+1} + K_k s_k \,,\\ &= F_k\tilde{x}_k + E_k u_k + K_k(y_k - H_k\tilde{x}_k)\end{aligned} \tag{7.5}$$

and then, for the estimation error

$$\begin{aligned}\epsilon_{k+1} &= x_{k+1} - \tilde{x}_{k+1}\\ &= (F_k - K_k H_k)\epsilon_k + B_k w_k - K_k v_k \,,\end{aligned} \tag{7.6}$$

the error covariance can be found to be

$$P_{k+1} = (F_k - K_k H_k)P_k(F_k - K_k H_k)^T + B_k Q_k B_k^T + K_k R_k K_k^T \,. \tag{7.7}$$

Further minimizing the trace of (7.7) by K_k gives the optimal bias correction gain (KP gain)

$$K_k = F_k P_k H_k^T S_k^{-1} \,, \tag{7.8}$$

where the innovation covariance S_k is given by (7.4). Using (7.8), the error covariance (7.7) can finally be written as

$$P_{k+1} = (F_k - K_k H_k) P_k F_k^T + B_k Q_k B_k^T \,. \tag{7.9}$$

Thus, the estimates are updated in the KP algorithm for the given x_0 and P_0 as follows [103]:

$$S_k = H_k P_k H_k^T + R_k \,, \tag{7.10}$$

$$K_k = F_k P_k H_k^T S_k^{-1} \,, \tag{7.11}$$

$$\tilde{x}_{k+1} = F_k \tilde{x}_k + E_k u_k + K_k (y_k - H_k \tilde{x}_k) \,, \tag{7.12}$$

$$P_{k+1} = (F_k - K_k H_k) P_k F_k^T + B_k Q_k B_k^T \,, \tag{7.13}$$

and we notice that KP does not require the prior error covariance and operates only with P_k. The KP can also work as an LMP on $[m, k]$ to obtain an estimate at $k+1$ for the given initial $\tilde{x}_{m-1}$ and P_{m-1} at $k-1$.

We can now start looking at FIR predictors, which traditionally require extended state and observation equations.

7.3 Extended Predictive State-Space Model

Reasoning along similar lines as for the state-space equations (4.1) and (4.2), introducing an extended predictive state vector

$$X_{m+1,k+1} = [x_{m+1}^T \; x_{m+2}^T \; \dots \; x_{k+1}^T]^T \,,$$

and taking other extended vectors from (4.4)–(4.6), (4.12), and (4.13), we extend the model (7.1) and (7.2) as

$$X_{m+1,k+1} = F_{m,k}^p x_m + S_{m,k} U_{m,k} + D_{m,k} W_{m,k} \,, \tag{7.14}$$

$$Y_{m,k} = H_{m,k}^p x_m + L_{m,k}^p U_{m,k} + G_{m,k}^p W_{m,k} + V_{m,k} \,, \tag{7.15}$$

where the extended matrices are given as $F_{m,k}^p = F_{m,k} F_m$,

$$H_{m,k}^p = \bar{H}_{m,k} F_{m-1,k-1} \,, \tag{7.16}$$

$$L_{m,k}^p = \bar{H}_{m,k} S_{m,k}^p \,, \tag{7.17}$$

$$G_{m,k}^p = \bar{H}_{m,k} D_{m,k}^p \,, \tag{7.18}$$

$$S_{m,k}^p = \begin{bmatrix} 0 & 0 & \dots & 0 & 0 \\ E_m & 0 & \dots & 0 & 0 \\ \vdots & \vdots & \ddots & \vdots & \vdots \\ \mathcal{F}_{k-2}^{m+1} E_m & \mathcal{F}_{k-2}^{m+2} E_{m+1} & \dots & 0 & 0 \\ \mathcal{F}_{k-1}^{m+1} E_m & \mathcal{F}_{k-1}^{m+2} E_{m+1} & \dots & E_{k-1} & 0 \end{bmatrix} , \tag{7.19}$$

matrix $D_{m,k}^p$ has the same structure and components as $S_{m,k}^p$ if we substitute E_k with B_k, $S_{m,k}$ is given by (4.9), $D_{m,k}$ becomes $S_{m,k}$ if we substitute E_k with B_k, and matrix $\bar{H}_{m,k} = \text{diag}\left(H_m \; H_{m+1} \; \dots \; H_k\right)$ is diagonal. Hereinafter, the superscript "p" is used to denote matrices in prediction models.

As with the FE-based model, extended equations (7.14) and (7.15) will be used to derive FIR predictors and RH FIR filters.

7.4 UFIR Predictor

The UFIR predictor can be derived if we define the prediction as

$$\tilde{x}_{k+1} \triangleq \tilde{x}_{k+1|k} = \hat{\mathcal{H}}^{p}_{m,k} Y_{m,k} + \hat{\mathcal{H}}^{pf}_{m,k} U_{m,k} , \tag{7.20}$$

and extract from (7.14) the model

$$x_{k+1} = \mathcal{F}^{m}_{k} x_m + \bar{S}_{m,k} U_{m,k} + \bar{D}_{m,k} W_{m,k} , \tag{7.21}$$

where $\bar{S}_{m,k}$ is the last row vector in $S_{m,k}$ and so is $\bar{D}_{m,k}$ in $D_{m,k}$.

The unbiasedness condition $\mathcal{E}\{\tilde{x}_{k+1}\} = \mathcal{E}\{x_{k+1}\}$ applied to (7.20) and (7.21) gives two unbiasedness constraints

$$\mathcal{F}^{m}_{k} = \hat{\mathcal{H}}^{p}_{m,k} H^{p}_{m,k} , \tag{7.22}$$

$$\hat{\mathcal{H}}^{pf}_{m,k} = \bar{S}_{m,k} - \hat{\mathcal{H}}^{p}_{m,k} L^{p}_{m,k} , \tag{7.23}$$

which have the same forms as (4.21) and (4.22) previously found for the OFIR filter, but with modified matrices and gains $\hat{\mathcal{H}}^{p}_{m,k}$ and $\hat{\mathcal{H}}^{pf}_{m,k}$.

7.4.1 Batch UFIR Predictor

In batch form, the UFIR predictor appears by solving (7.22) for the fundamental gain $\hat{\mathcal{H}}^{p}_{m,k}$, which gives

$$\hat{\mathcal{H}}^{p}_{m,k} = \mathcal{F}^{m}_{k} (H^{p^T}_{m,k} H^{p}_{m,k})^{-1} H^{p^T}_{m,k} \tag{7.24a}$$

$$= (C^{p^T}_{m,k} C^{p}_{m,k})^{-1} C^{p^T}_{m,k} \tag{7.24b}$$

$$= \mathcal{G}_k C^{p^T}_{m,k} , \tag{7.24c}$$

where $C^{p}_{m,k} = H^{p}_{m,k} \mathcal{F}^{m^{-1}}_{k}$ is the auxiliary block matrix and the GNPG matrix $\mathcal{G}_k = (C^{p^T}_{m,k} C^{p}_{m,k})^{-1}$ is square and symmetric. Using (7.23), the UFIR predicted estimate can be written in the batch form as

$$\tilde{x}_{k+1} = \hat{\mathcal{H}}^{p}_{m,k} Y_{m,k} + (\bar{S}_{m,k} - \hat{\mathcal{H}}^{p}_{m,k} L^{p}_{m,k}) U_{m,k} , \tag{7.25}$$

where the gain $\hat{\mathcal{H}}^{p}_{m,k}$ is defined by (7.24b). The error covariance is given by

$$P_{k+1} = \mathcal{W}^{p}_{m,k} \mathcal{Q}_{m,k} \mathcal{W}^{p^T}_{m,k} + \mathcal{V}^{p}_{m,k} \mathcal{R}_{m,k} \mathcal{V}^{p^T}_{m,k} , \tag{7.26}$$

where the system and observation error residual matrices are defined as

$$\mathcal{W}^{p}_{m,k} = \bar{D}_{m,k} - \hat{\mathcal{H}}^{p}_{m,k} G^{p}_{m,k} , \tag{7.27}$$

$$\mathcal{V}^{p}_{m,k} = \hat{\mathcal{H}}^{p}_{m,k} . \tag{7.28}$$

Again we notice that the form (7.26) is unique for all bias-constrained FIR state estimators, where the individual properties are collected in the error residual matrices, such as (7.27) and (7.28).

7.4.2 Iterative Algorithm using Recursions

Recursions for the batch UFIR prediction (7.25) can be found similarly to the batch UFIR filtering estimate if we represent $\tilde{x}_{k+1}$ as the sum of the homogeneous estimate $\tilde{x}^{\mathrm{h}}_{k+1}$ and the forced estimates $\tilde{x}^{\mathrm{f}}_{k+1}$. To do this, we will use the following matrix decompositions,

$$C^p_{m,k} = \begin{bmatrix} C^p_{m,k-1} \\ H_k \end{bmatrix} F_k^{-1}, \quad L^p_{m,k} = \begin{bmatrix} L^p_{m,k-1} & 0 \\ H_k \bar{S}_{m,k-1} & 0 \end{bmatrix},$$

$$\bar{S}_{m,k} = \begin{bmatrix} F_k \bar{S}_{m,k-1} & E_k \end{bmatrix}, \quad H^p_{m,k} = \begin{bmatrix} H^p_{m,k-1} \\ H_k \mathcal{F}^m_{k-1}, \end{bmatrix},$$

$$G^p_{m,k} = \begin{bmatrix} G^p_{m,k-1} & 0 \\ H_k \bar{D}_{m,k-1} & 0 \end{bmatrix}. \tag{7.29}$$

To find a recursion for $\tilde{x}^{\mathrm{h}}_{k+1}$ using $C^p_{m,k}$ taking from (7.29), we transform the inverse of GNPG $\mathcal{G}_k = (C^{p^T}_{m,k} C^p_{m,k})^{-1}$ as

$$\begin{aligned}
\mathcal{G}_k^{-1} &= F_k^{-T} \begin{bmatrix} C^{p^T}_{m,k-1} & H_k^T \end{bmatrix} \begin{bmatrix} C^p_{m,k-1} \\ H_k \end{bmatrix} F_k^{-1} \\
&= F_k^{-T} (H_k^T H_k + \mathcal{G}_{k-1}^{-1}) F_k^{-1} .
\end{aligned}$$

This gives the following direct and inverse recursive forms,

$$\mathcal{G}_k = F_k (H_k^T H_k + \mathcal{G}_{k-1}^{-1})^{-1} F_k^T , \tag{7.30}$$

$$\mathcal{G}_{k-1}^{-1} = F_k^T \mathcal{G}_k^{-1} F_k - H_k^T H_k . \tag{7.31}$$

Likewise, we represent the product $C^{p^T}_{m,k} Y_{m,k}$ as

$$C^{p^T}_{m,k} Y_{m,k} = F_k^{-T} (\mathcal{G}_{k-1}^{-1} \hat{x}_k + H_k^T y_k) \tag{7.32}$$

and then transform the homogeneous estimate $\tilde{x}^{\mathrm{h}}_{k+1} = \mathcal{G}_k C^{p^T}_{m,k} Y_{m,k}$ to

$$\tilde{x}^{\mathrm{h}}_{k+1} = F_k \tilde{x}^{\mathrm{h}}_k + \mathcal{G}_k F_k^{-T} H_k^T (y_k - H_k \tilde{x}^{\mathrm{h}}_k) , \tag{7.33}$$

where the GNPG $\mathcal{G}_k$ is computed recursively using (7.30).

To find a recursive form for the forced estimate in (7.25), it is necessary to find recursions for the two components in (7.25) separately. To this end, we refer to (7.29) and first obtain

$$\bar{S}_{m,k} U_{m,k} = F_k \bar{S}_{m,k-1} U_{m,k-1} + E_k u_k . \tag{7.34}$$

Next, we use $\hat{\mathcal{H}}^p_{m,k} = \mathcal{G}_k \mathcal{F}^{m^{-T}}_k H^{p^T}_{m,k}$, take some decompositions from (7.29), and transform the remaining component $\hat{\mathcal{H}}^p_{m,k} L^p_{m,k} U_{m,k}$ as

$$\begin{aligned}
\hat{\mathcal{H}}^p_{m,k} L^p_{m,k} U_{m,k} = \mathcal{G}_k (&\mathcal{F}^{m^{-T}}_k H^{p^T}_{m,k-1} L^p_{m,k-1} \\
&+ F_k^{-T} H_k^T H_k \bar{S}_{m,k-1}) U_{m,k-1} .
\end{aligned} \tag{7.35}$$

Combining (7.34) and (7.35), we finally write the forced estimate in the form

$$\tilde{x}^{\mathrm{f}}_{k+1} = F_k \tilde{x}^{\mathrm{f}}_k - \mathcal{G}_k F_k^{-T} H_k^T H_k \tilde{x}^{\mathrm{f}}_k + E_k u_k . \tag{7.36}$$

The UFIR prediction $\tilde{x}_{k+1} = \tilde{x}^{\mathrm{h}}_{k+1} + \tilde{x}^{\mathrm{f}}_{k+1}$ can now be represented with the recursion

$$\tilde{x}_{k+1} = F_k\tilde{x}_k + E_k u_k + \mathcal{G}_k F_k^{-T} H_k^T (y_k - H_k\tilde{x}_k)\,, \tag{7.37}$$

and the pseudocode of the iterative UFIR prediction algorithm can be listed as Algorithm 17. This algorithm iteratively updates the estimates on $[m, k]$, starting with estimates $\mathcal{G}_s$ and $\tilde{x}_{s+1}$ computed over $[m, s]$ in short batch forms, and the final prediction appears at $k + 1$.

Algorithm 17: Iterative UFIR Prediction Algorithm

Data: y_k, u_k
begin
  for $k = N-1, N, \cdots$ **do**
    $m = k - N + 1,\ s = k - N + K$;
    $\mathcal{G}_s = (\mathcal{C}^{pT}_{m,s}\mathcal{C}^{p}_{m,s})^{-1}$;
    $\check{x}_{s+1} = \mathcal{G}_s\mathcal{C}^{pT}_{m,s}(Y_{m,s} - L^{p}_{m,s}U_{m,s}) + \bar{S}_{m,s}U_{m,s}$;
    for $i = s + 1 : k$ **do**
      $\mathcal{G}_i = F_i(H_i^T H_i + \mathcal{G}_{i-1}^{-1})^{-1}F_i^T$;
      $K_i = \mathcal{G}_i F_i^{-T} H_i^T$;
      $\check{x}_{i+1} = F_i\check{x}_i + E_i u_i + K_i(y_i - H_i\check{x}_i)$;
    end for
    $\tilde{x}_{k+1} = \check{x}_{k+1}$;
  end for
  Result: $\tilde{x}_{k+1}$
end

Now, let us again note two forms of UFIR state estimators suitable for stochastic state feedback control. One can replace k by $k - 1$ in (7.37) and consider $\tilde{x}_k$ as an RH UFIR filtering estimate. Otherwise, the same task can be accomplished by taking the UFIR filtering estimate at $k - 1$ and projecting it onto k using the system matrix F_k. While these solutions are not completely equivalent, they provide a similar control quality because both are unbiased.

Example 7.1 ***Tracking with UFIR filter and predictor.*** To compare the performances of the UFIR filter and predictor, consider the moving vehicle tracking problem. The trajectory is measured every second, $\tau = 1$ s, using a GPS navigation unit that provides the ground truth. The two-state polynomial model (6.38) and (6.39) is used with matrices specified in Example 6.1. The scalar $w_k \sim \mathcal{N}(0, \sigma_w^2)$ and $v_k \sim \mathcal{N}(0, \sigma_v^2)$ are specified with $\sigma_w = 2/\mathrm{s}$, and $\sigma_v = 3.75$ and the optimal horizon for the UFIR filter is measured as $N_{\mathrm{opt}} = 5$.

The vehicle y-coordinate (m), estimated using both UFIR structures, is shown in Fig. 7.2a. In the time span of [55…63], both the filter and the predictor match the model. Consequently, they give almost the same estimates with hardly distinguishable discrepancy. The difference between the estimates becomes clearly visible when UFIR estimators temporary mismatch the trajectory, as in the time interval of [63…65]. Here we see typical transients and notice that the predictor reacts with a delay on one point and a larger overshoot due to its basic properties. The difference between the estimates is especially neatly seen in Fig. 7.2b, which shows tracking errors. Over the entire time scale, the predictor generates larger errors than the filter, which, as can be seen, fits with the generalized errors shown in Fig. 5.1. □

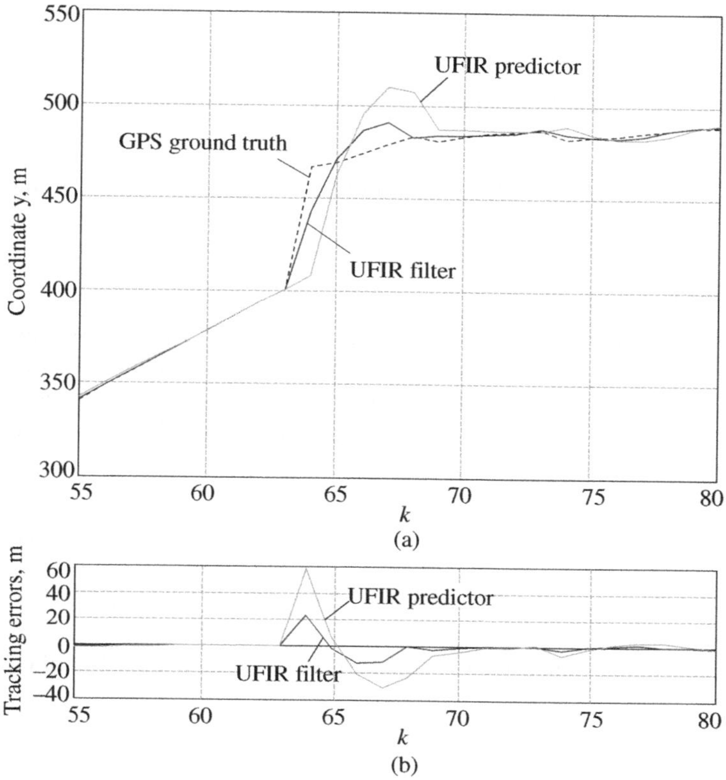

Figure 7.2 Moving vehicle tracking in y-coordinate (m) using UFIR filter and UFIR predictor: (a) estimates and (b) estimation errors.

Equivalence of Projected and Predicted Estimates

So far, we have looked at LTV systems, for which the BE- and FE-based state models cannot be converted into each other, and the projected and predicted UFIR estimates are not equivalent. However, in the special case of the LTI system without input, the equivalence of such structures can be shown.

Consider (6.10a), assume that all matrices are time-invariant, substitute the subscript m, k in matrices with N, and write

$$\hat{x}_k = \hat{\mathcal{H}}_N Y_{m,k} \,,$$

where $\hat{\mathcal{H}}_N = F^{N-1}(H_N^T H_N)^{-1} H_N^T$ is the UFIR filter gain. Then the projected estimate $\hat{x}_{k+1}^{\mathrm{pj}}$ can be written as

$$\begin{aligned}\hat{x}_{k+1}^{\mathrm{pj}} &= F\hat{x}_k = F\hat{\mathcal{H}}_N Y_{m,k} \\ &= F^N (H_N^T H_N)^{-1} H_N^T Y_{m,k}\end{aligned}$$

and the predicted estimate $\tilde{x}_{k+1}^{pr}$ can be written as

$$\tilde{x}_{k+1}^{pr} = F^N (H_N^{p^T} H_N^p)^{-1} H_N^{p^T} Y_{m,k} \,.$$

It is easy to show now that for LTI systems the matrices H_N and H_N^p are identical and thus the predicted and projected UFIR estimates are equivalent, $\tilde{x}_{k+1}^{pr} = \hat{x}_{k+1}^{\mathrm{pj}}$. It also follows that, for

LTI systems without input, the UFIR prediction can be organized using the projected estimate as $\tilde{x}^{pr}_{k+1} = F\hat{x}_k$.

> ***UFIR prediction:*** *For LTI systems, the UFIR predicted estimate $\tilde{x}_{k+1}$ and projected estimate $\hat{x}_{k+1} = F\hat{x}_k$ are equivalent.*

This statement was confirmed by Example 7.1, where it was numerically demonstrated that the predicted and projected UFIR estimates are identical.

7.4.3 Recursive Error Covariance

There are two ways to find recursive forms for the error covariance of the UFIR predictor. We can start with (7.25), find recursions for each of the batch components, and then combine them in the final form. Otherwise, we can obtain P_{k+1} using recursion (7.37). To make sure that these approaches lead to the same results, next we give the most complex derivation and postpone the simplest to "Problems."

Consider the batch error covariance (7.26) and represent it as

$$P_{k+1} = P^{(1)}_k - P^{(2)}_k - P^{(3)}_k + P^{(4)}_k + P^{(5)}_k \,, \tag{7.38}$$

where the matrices are: $P^{(1)}_k = \bar{D}_{m,k}\mathcal{Q}_{m,k}\bar{D}^T_{m,k}$, $P^{(2)}_k = \bar{D}_{m,k}\mathcal{Q}_{m,k}G^{p^T}_{m,k}\hat{\mathcal{H}}^{p^T}_{m,k}$, $P^{(3)}_k = \hat{\mathcal{H}}^p_{m,k}G^p_{m,k}\mathcal{Q}_{m,k}\bar{D}^T_{m,k}$, $P^{(4)}_k = \hat{\mathcal{H}}^p_{m,k}G^p_{m,k}\mathcal{Q}_{m,k}G^{p^T}_{m,k}\hat{\mathcal{H}}^{p^T}_{m,k}$, and $P^{(5)}_k = \hat{\mathcal{H}}^p_{m,k}\mathcal{R}_{m,k}\hat{\mathcal{H}}^{p^T}_{m,k}$.

Following the derivation procedure applied in Chapter 6 to the error covariance of the UFIR filter, we represent components of (7.38) as

$$\begin{aligned}
P^{(1)}_k &= F_k P^{(1)}_{k-1} F^T_k + B_k Q_k B^T_k \,, \\
P^{(2)}_k &= F_k P^{(2)}_{k-1} F^T_k - F_k P^{(2)}_{k-1} H^T_k H_k F^{-1}_k \mathcal{G}_k \\
&\quad + F_k P^{(1)}_{k-1} H^T_k H_k F^{-1}_k \mathcal{G}_k \,, \\
P^{(3)}_k &= F_k P^{(3)}_{k-1} F^T_k - \mathcal{G}_k F^{-T}_k H^T_k H_k P^{(3)}_{k-1} F^T_k \\
&\quad + \mathcal{G}_k F^{-T}_k H^T_k H_k P^{(1)}_{k-1} F^T_k \,, \\
P^{(4)}_k &= F_k P^{(4)}_{k-1} F^T_k - F_k P^{(4)}_{k-1} H^T_k H_k F^{-1}_k \mathcal{G}_k - \mathcal{G}_k F^{-T}_k H^T_k H_k P^{(4)}_{k-1} F^T_k \\
&\quad + \mathcal{G}_k F^{-T}_k H^T_k H_k P^{(4)}_{k-1} H^T_k H_k F^{-1}_k \mathcal{G}_k + F_k P^{(3)}_{k-1} H^T_k H_k F^{-1}_k G_k \\
&\quad - \mathcal{G}_k F^{-T}_k H^T_k H_k P^{(3)}_{k-1} H^T_k H_k F^{-1}_k \mathcal{G}_k + \mathcal{G}_k F^{-T}_k H^T_k H_k P^{(2)}_{k-1} F^T_k \\
&\quad - \mathcal{G}_k F^{-T}_k H^T_k H_k P^{(2)}_{k-1} H^T_k H_k F^{-1}_k \mathcal{G}_k \\
&\quad + \mathcal{G}_k F^{-T}_k H^T_k H_k P^{(1)}_{k-1} H^T_k H_k F^{-1}_k \mathcal{G}_k \\
P^{(5)}_k &= F_k P^{(5)}_{k-1} F^T_k - F_k P^{(5)}_{k-1} H^T_k H_k F^{-1}_k \mathcal{G}_k - \mathcal{G}_k F^{-T}_k H^T_k H_k P^{(5)}_{k-1} F^T_k \\
&\quad + \mathcal{G}_k F^{-T}_k H^T_k H_k P^{(5)}_{k-1} H^T_k H_k F^{-1}_k \mathcal{G}_k + \mathcal{G}_k F^{-T}_k H^T_k R_k H_k F^{-1}_k \mathcal{G}_k \,.
\end{aligned}$$

Combining these matrices in (7.38), we obtain the following recursive form for the error covariance,

$$\begin{aligned}
P_{k+1} &= (F_k - \mathcal{G}_k F^{-T}_k H^T_k H_k) P_k (F_k - \mathcal{G}_k F^{-T}_k H^T_k H_k)^T \\
&\quad + \mathcal{G}_k F^{-T}_k H^T_k R_k H_k F^{-1}_k \mathcal{G}_k + B_k Q_k B^T_k \,.
\end{aligned} \tag{7.39}$$

All that follows from (7.39) is that it is unified by the KP error covariance (7.7) if we introduce the bias correction gain $K_k = \mathcal{G}_k F^{-T}_k H^T_k$ instead of the Kalman gain. It should also be noted that recursion (7.39) can be obtained much easier if we start with the recursive estimate (7.37). The corresponding derivation is postponed to "Problems."

7.5 Optimal FIR Predictor

In the discrete convolution-based batch form, the OFIR predictive estimate can be defined similarly to the OFIR filtering estimate as

$$\tilde{x}_{k+1} \triangleq \tilde{x}_{k+1|k} = \mathcal{H}^{p}_{m,k} Y_{m,k} + \mathcal{H}^{pf}_{m,k} U_{m,k} , \tag{7.40}$$

where the gains $\mathcal{H}^{p}_{m,k}$ and $\mathcal{H}^{pf}_{m,k}$ are to be found by minimizing the MSE with respect to the model

$$x_{k+1} = \mathcal{F}^{m}_{k} x_m + \bar{S}_{m,k} U_{m,k} + \bar{D}_{m,k} W_{m,k} , \tag{7.41}$$

which is given by the last row vector in (7.14) and where $\bar{S}_{m,k}$ is the last row vector in $S_{m,k}$ and so is $\bar{D}_{m,k}$ in $D_{m,k}$.

7.5.1 Batch Estimate and Error Covariance

For the estimation error $\varepsilon_{k+1} = x_{k+1} - \tilde{x}_{k+1}$, determined taking into account (7.40) and (7.41), we apply the orthogonality condition as

$$\mathcal{E}\{(x_{k+1} - \mathcal{H}^{p}_{m,k} Y_{m,k} - \mathcal{H}^{pf}_{m,k} U_{m,k}) Y^T_{m,k}\} = 0 \tag{7.42}$$

and transform it to

$$\begin{aligned} &\mathcal{B}^{p}_{m,k} \mathcal{X}_m H^{p^T}_{m,k} + \mathcal{W}^{p}_{m,k} \mathcal{Q}_{m,k} G^{p^T}_{m,k} - \mathcal{V}^{p}_{m,k} \mathcal{R}_{m,k} \\ &\quad = (\mathcal{H}^{p}_{m,k} L_{m,k} - \bar{S}_{m,k} + \mathcal{H}^{pf}_{m,k}) \Psi_{m,k} L^{p^T}_{m,k} , \end{aligned} \tag{7.43}$$

where the error residual matrices given by

$$\mathcal{B}^{p}_{m,k} = \mathcal{F}^{m}_{k} - \mathcal{H}^{p}_{m,k} H^{p}_{m,k} , \tag{7.44}$$

$$\mathcal{W}^{p}_{m,k} = \bar{D}_{m,k} - \mathcal{H}^{p}_{m,k} G^{p}_{m,k} , \tag{7.45}$$

$$\mathcal{V}^{p}_{m,k} = \mathcal{H}^{p}_{m,k} \tag{7.46}$$

ensure optimal cancellation of regular (bias) and random errors at the OFIR predictor output.

For zero input, $\Psi_{m,k} = 0$, relation (7.43) gives the fundamental gain $\mathcal{H}^{p}_{m,k}$ for the OFIR predictor,

$$\mathcal{H}^{p}_{m,k} = (\mathcal{F}^{m}_{k} \mathcal{X}_m H^{p^T}_{m,k} + \mathcal{Z}^{p}_{1})(\mathcal{Z}^{p}_{\mathcal{X}} + \mathcal{Z}^{p}_{2} + \mathcal{R}_{m,k})^{-1} , \tag{7.47a}$$

$$= (\hat{\mathcal{H}}^{p}_{m,k} \mathcal{Z}^{p}_{\mathcal{X}} + \mathcal{Z}^{p}_{1})(\mathcal{Z}^{p}_{\mathcal{X}} + \mathcal{Z}^{p}_{2} + \mathcal{R}_{m,k})^{-1} , \tag{7.47b}$$

where $\mathcal{Z}^{p}_{\mathcal{X}} = H^{p}_{m,k} \mathcal{X}_m H^{p^T}_{m,k}$, $\mathcal{Z}^{p}_{1} = \bar{D}_{m,k} \mathcal{Q}_{m,k} G^{p^T}_{m,k}$, $\mathcal{Z}^{p}_{2} = G^{p}_{m,k} \mathcal{Q}_{m,k} G^{p^T}_{m,k}$, and the UFIR predictor gain $\hat{\mathcal{H}}^{p}_{m,k}$ is given by (7.24b).

Since the forced impulse response $\mathcal{H}^{pf}_{m,k}$ is determined by constraint (7.23), the batch OFIR predictor (7.40) eventually becomes

$$\tilde{x}_{k+1} = \mathcal{H}^{p}_{m,k} Y_{m,k} + (\bar{S}_{m,k} - \mathcal{H}^{p}_{m,k} L^{p}_{m,k}) U_{m,k} , \tag{7.48}$$

where $Y_{m,k}$ and $U_{m,k}$ are real vectors containing data collected on $[m, k]$. It can be seen that, for deterministic models with zero noise, we have $\mathcal{H}^{p}_{m,k} = \hat{\mathcal{H}}^{p}_{m,k}$, and thus the OFIR predictor becomes the UFIR predictor.

The batch error covariance P_{k+1} for the OFIR predictor is given by

$$P_{k+1} = \mathcal{B}^p_{m,k}\chi_m\mathcal{B}^{p^T}_{m,k} + \mathcal{W}^p_{m,k}\mathcal{Q}_{m,k}\mathcal{W}^{p^T}_{m,k} + \mathcal{V}^p_{m,k}\mathcal{R}_{m,k}\mathcal{V}^{p^T}_{m,k} , \quad (7.49)$$

where the error residual matrices are provided with (7.44)–(7.46). Next, we will find recursive forms required to develop an iterative OFIR predictive algorithm based on (7.49).

7.5.2 Recursive Forms and Iterative Algorithm

Using expansions (7.29), the recursions for the OFIR predictor can be found similarly to the OFIR filter, as stated in the following theorem.

Theorem 7.1 Given a batch OFIR predictor (7.48), its iterative computation on $[m, k]$ for given $\tilde{x}_m$ and P_m is provided by recursions

$$S_i = H_iP_iH_i^T + R_i , \quad (7.50)$$

$$K_i = F_iP_iH_i^TS_i^{-1} , \quad (7.51)$$

$$\tilde{x}_{i+1} = F_i\tilde{x}_i + E_iu_i + K_i(y_i - H_i\tilde{x}_i) , \quad (7.52)$$

$$P_{i+1} = (F_i - K_iH_i)P_iF_i^T + B_iQ_iB_i^T , \quad (7.53)$$

where the iterative variable i changes from m to k, and the estimates $\tilde{x}_{k+1}$ and P_{k+1} are taken when $i = k$.

Proof: To obtain recursive forms for (7.48), first find recursions for the gain $\mathcal{H}^p_{m,k} = \mathcal{H}^{1p}_{m,k}\mathcal{H}^{2p}_{m,k}$ defined by (7.47b).

Referring to (7.29), represent matrix $\mathcal{H}^{1p}_{m,k}$ recursively as

$$\begin{aligned}
\mathcal{H}^{1p}_{m,k} &= \mathcal{F}^m_k\chi_mH^{p^T}_{m,k} + \bar{D}_{m,k}\mathcal{Q}_{m,k}G^{p^T}_{m,k} \\
&= \mathcal{F}^m_k\chi_m\begin{bmatrix} H^{p^T}_{m,k-1} & \mathcal{F}^{m^T}_{k-1}H_k^T \end{bmatrix} \\
&\quad + \begin{bmatrix} F_k\bar{D}_{m,k-1} & B_k \end{bmatrix}\mathcal{Q}_{m,k}\begin{bmatrix} G^{p^T}_{m,k-1} & \bar{D}^T_{m,k-1}H_k^T \\ 0 & 0 \end{bmatrix} \\
&= F_k\begin{bmatrix} \mathcal{H}^{1p}_{m,k-1} & M_{k-1}H_k^T \end{bmatrix} ,
\end{aligned} \quad (7.54)$$

where the matrix M_k is given by

$$M_k = \mathcal{F}^m_k\chi_m\mathcal{F}^{m^T}_k + \bar{D}_{m,k}\mathcal{Q}_{m,k}\bar{D}^T_{m,k} . \quad (7.55)$$

Next, consider $\mathcal{H}^{2p}_{m,k} = (Z_\chi + Z_2 + \mathcal{R}_{m,k})^{-1}$, take into account that

$$\mathcal{H}^{2p^{-1}}_{m,k} - \mathcal{R}_{m,k} = H^p_{m,k}\chi_mH^{p^T}_{m,k} + G^p_{m,k}\mathcal{Q}_{m,k}G^{p^T}_{m,k} ,$$

and transform $Z_\chi + Z_2$ as

$$Z_\chi + Z_2 = \begin{bmatrix} H^p_{m,k-1} \\ H_k\mathcal{F}^m_{k-1} \end{bmatrix}\chi_m\begin{bmatrix} H^{p^T}_{m,k-1} & \mathcal{F}^{m^T}_{k-1}H_k^T \end{bmatrix}$$

$$+\begin{bmatrix} G^p_{m,k-1} & 0 \\ H_k\bar{D}_{m,k-1} & 0 \end{bmatrix} Q_{m,k} \begin{bmatrix} G^{p^T}_{m,k-1} & \bar{D}^T_{m,k-1}H^T_k \\ 0 & 0 \end{bmatrix}$$

$$= \begin{bmatrix} \mathcal{H}^{2p^{-1}}_{m,k-1} - \mathcal{R}_{m,k-1} & \tilde{\mathcal{H}}^{2p^T}_{m,k-1} \\ \tilde{\mathcal{H}}^{2p}_{m,k-1} & H_kM_{k-1}H^T_k \end{bmatrix},$$

where $\tilde{\mathcal{H}}^{2p}_{m,k-1} = H_k\mathcal{H}^{1p}_{m,k-1}$. Now, for $\mathcal{R}_{m,k} = \text{diag}\,(\mathcal{R}_{m,k-1}\ R_k)$, represent $Z_\chi + Z_2 + \mathcal{R}_{m,k}$ as

$$Z_\chi + Z_2 + \mathcal{R}_{m,k} = \begin{bmatrix} \mathcal{H}^{2p^{-1}}_{m,k-1} & \tilde{\mathcal{H}}^{2p^T}_{m,k-1} \\ \tilde{\mathcal{H}}^{2p}_{m,k-1} & H_kM_kH^T_k + R_k \end{bmatrix},$$

follow the derivation of (4.35), substitute the previous matrix with the sum of two block matrices

$$\tilde{Z}_{m,k} = \begin{bmatrix} 0 & \tilde{\mathcal{H}}^{2p^T}_{m,k-1} \\ \tilde{\mathcal{H}}^{2p}_{m,k-1} & H_kM_kH^T_k \end{bmatrix}, \quad \bar{Z}_{m,k} = \text{diag}(\mathcal{H}^{2p^{-1}}_{m,k-1}\ R_k),$$

decompose $\mathcal{H}^{2p}_{m,k}$ as

$$\mathcal{H}^{2p}_{m,k} = \bar{Z}^{-1}_{m,k}\hat{Z}^{-1}_{m,k}, \tag{7.56}$$

where $\hat{Z}_{m,k} = I + \tilde{Z}_{m,k}\bar{Z}^{-1}_{m,k}$, use the Schur complement (A.23), and represent $\hat{Z}^{-1}_{m,k}$ as

$$\hat{Z}^{-1}_{m,k} = \begin{bmatrix} \mathcal{Z}_{11} & \mathcal{Z}_{12} \\ \mathcal{Z}_{21} & \mathcal{Z}_{22} \end{bmatrix}, \tag{7.57}$$

where the components are given by

$$\mathcal{Z}_{11} = I - \mathcal{Z}_{12}\tilde{\mathcal{H}}^{2p}_{m,k-1}\mathcal{H}^{2p}_{m,k-1}, \quad \mathcal{Z}_{12} = -\tilde{\mathcal{H}}^{2p^T}_{m,k-1}R^{-1}_k\Omega^{-1}_k,$$
$$\mathcal{Z}_{21} = -\Omega^{-1}_k\tilde{\mathcal{H}}^{2p}_{m,k-1}\mathcal{H}^{2p}_{m,k-1}, \quad \mathcal{Z}_{22} = \Omega^{-1}_k,$$

in which use the following matrices

$$\Omega_k = I + H_k\Lambda^-_kH^T_kR^{-1}_k, \tag{7.58}$$

$$\Lambda_k = M_{k-1} - \mathcal{H}^p_{m,k-1}\mathcal{H}^{1p^T}_{m,k-1}. \tag{7.59}$$

Substituting (7.57) into (7.56) and combining with (7.54) give the recursion

$$\mathcal{H}^p_{m,k} = [(F_k - K_kH_k)\mathcal{H}^p_{m,k-1}\ \ K_k], \tag{7.60}$$

where the bias correction gain is given by

$$K_k = F_k\Lambda_kH^T_k(H_k\Lambda_kH^T_k + R_k)^{-1} \tag{7.61}$$

and Λ_k is defined by (7.59). Next, substitute (7.61) into (7.48) for $u_k = 0$ and obtain the recursive form

$$\tilde{x}^{\text{h}}_{k+1} = F_k\tilde{x}^{\text{h}}_k + K_k(y_k - H_k\tilde{x}^{\text{h}}_k). \tag{7.62}$$

Now, consider $\tilde{x}^{\text{f}}_k$ defined by (7.48), refer to (7.29), derive another recursion

$$\tilde{x}^{\text{f}}_{k+1} = F_k\tilde{x}^{\text{f}}_k - K_kH_k\tilde{x}^{\text{f}}_k + E_ku_k, \tag{7.63}$$

combine it with (7.62), and end up with the recursive OFIR prediction

$$\tilde{x}_{k+1} = F_k\tilde{x}_k + E_ku_k + K_k(y_k - H_k\tilde{x}_k). \tag{7.64}$$

Reasoning similarly and referring to (7.60) and (7.54), substitute Λ_k given by (7.59) with P_k, represent recursively as

$$P_{k+1} = (F_k - K_k H_k) P_k F_k^T + B_k Q_k B_k^T , \tag{7.65}$$

and complete the proof. □

A simple glance at the result reveals that the iterative OFIR prediction algorithm (theorem 7.1), operating on $[m, k]$, employs the KP recursions given by (7.10)–(7.13). We wish to note this property as fundamental, since all estimators that minimize MSE in the same linear stochastic model can be transformed into each other. It also follows, as an extension, that (7.48) represents the batch KP by setting the starting point to zero, $m = 0$.

Example 7.2 ***Tracking by OFIR filter and predictor.*** Now we wish to solve the same tracking problem as in Example 7.1, but using an OFIR filter and predictor. To make a difference using the same model, we choose a different measured trajectory, where we deliberately select an area with a rapidly changed direction.

The vehicle y-coordinate (m), estimated using the OFIR filter and predictor, is shown in Fig. 7.3 along with the GPS-based ground truth. What this experiment assures is that the OFIR predictor is

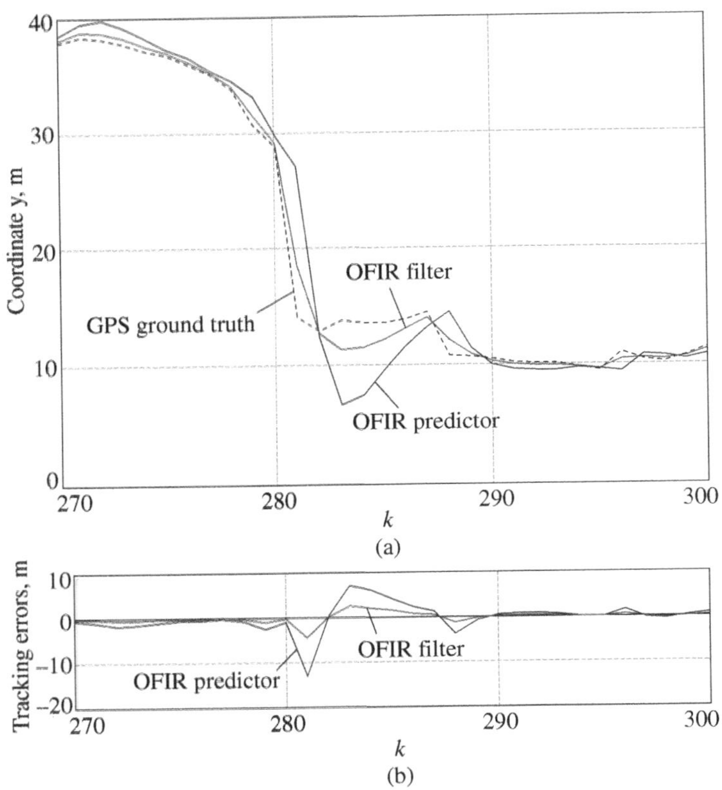

Figure 7.3 Tracking a moving vehicle along the y-coordinate (m) using an OFIR filter and predictor: (a) estimates and (b) estimation errors.

less accurate than the OFIR filter when the process temporary undergoes rapid changes. Outside of this are and in general, it follows that the OFIR predictor, like the UFIR predictor discussed in Example 7.1, is inherently less accurate than the OFIR filter. □

It is also worth mentioning that the OFIR projector and predictor give almost identical estimates (Example 7.2). Thus, it follows that a simple projection of the state from k to $k+1$ through the system matrix can effectively serve not only for unbiased prediction but also for suboptimal prediction.

7.6 Receding Horizon FIR Filtering

Suboptimal RH FIR filters subject to the unbiasedness constraint were originally obtained for stationary stochastic processes in [105] and for nonstationary stochastic processes in [106]. Both solutions were called the *minimum variance FIR* (MVF) filter. Therefore, to distinguish the difference, we will refer to them as the MVF-I filter and the MVF-II filter, respectively. Note that MVF filters turned out to be the first practical solutions in the family of FIR state estimators, although their derivation draws heavily on the early work [97]. Next, we will derive both MVF filters, keeping the original ideas and derivation procedures, but accepting the definitions given in this book.

7.6.1 MVF-I Filter for Stationary Processes

The MVF-I filter resembles the OUFIR predictive filter but ignores the process dynamics and thus has most in common with the weighted LS estimate (3.61), which is suitable for stationary processes.

To obtain the MVF-I solution, let us start with the model in (7.1) and (7.2). Since the MVF-I filter ignores the process dynamics, it only needs the observation equation, which can be extended on the horizon $[k-N, k-1]$ as

$$Y_{k-1} = C_{k-1}x_k - L_{k-1}U_{k-1} - G_{k-1}W_{k-1} + V_{k-1} \,, \tag{7.66}$$

where the following extended vectors were introduced,

$$\begin{aligned}
Y_{k-1} &\triangleq Y_{k-N,k-1} = \begin{bmatrix} y_{k-N}^T & y_{k-N+1}^T & \cdots & y_{k-1}^T \end{bmatrix}^T , \\
U_{k-1} &\triangleq U_{k-N,k-1} = \begin{bmatrix} u_{k-N}^T & u_{k-N+1}^T & \cdots & u_{k-1}^T \end{bmatrix}^T , \\
W_{k-1} &\triangleq W_{k-N,k-1} = \begin{bmatrix} w_{k-N}^T & w_{k-N+1}^T & \cdots & w_{k-1}^T \end{bmatrix}^T , \\
V_{k-1} &\triangleq V_{k-N,k-1} = \begin{bmatrix} v_{k-N}^T & v_{k-N+1}^T & \cdots & v_{k-1}^T \end{bmatrix}^T ,
\end{aligned}$$

and the extended matrices $C_{k-1} \triangleq C_{k-N,k-1}$, $L_{k-1} \triangleq L_{k-N,k-1}$, $G_{k-1} \triangleq G_{k-N,k-1}$, and $\bar{H}_{k-1} \triangleq \bar{H}_{k-N,k-1}$ are defined as

$$C_{k-1} = \bar{H}_{k-1} \begin{bmatrix} (\mathcal{F}_{k-1}^{k-N})^{-1} \\ (\mathcal{F}_{k-1}^{k-N+1})^{-1} \\ \vdots \\ F_{k-1}^{-1} \end{bmatrix} , \tag{7.67}$$

$$L_{k-1} = \bar{H}_{k-1} \begin{bmatrix} F_{k-N}^{-1} E_{k-N} & \mathcal{F}_{m}^{k-N^{-1}} E_m & \dots & \mathcal{F}_{k-2}^{k-N^{-1}} E_{k-2} & \mathcal{F}_{k-1}^{k-N^{-1}} E_{k-1} \\ 0 & F_m^{-1} E_m & \dots & \mathcal{F}_{k-2}^{m^{-1}} E_{k-2} & \mathcal{F}_{k-1}^{m^{-1}} E_{k-1} \\ \vdots & \vdots & \ddots & \vdots & \vdots \\ 0 & 0 & \dots & F_{k-2}^{-1} E_{k-2} & \mathcal{F}_{k-1}^{k-2^{-1}} E_{k-1} \\ 0 & 0 & \dots & 0 & F_{k-1}^{-1} E_{k-1} \end{bmatrix}, \tag{7.68}$$

$$\bar{H}_{k-1} = \text{diag}(\ H_{k-N} \ \ H_{k-N+1} \dots H_{k-1}\ ) \ . \tag{7.69}$$

Note that matrix G_{k-1} becomes equal to matrix L_{k-1} if we replace E_k by B_k.

We can now define the RH FIR filtering estimate as

$$\tilde{x}_k = \mathcal{H}_{k-1} Y_{k-1} + \mathcal{H}_{k-1}^{\mathrm{f}} U_{k-1} \tag{7.70}$$

$$\begin{aligned} &= \mathcal{H}_{k-1} C_{k-1} x_k + (\mathcal{H}_{k-1}^{\mathrm{f}} - \mathcal{H}_{k-1} L_{k-1}) U_{k-1} \\ &\quad - \mathcal{H}_{k-1} G_{k-1} W_{k-1} + \mathcal{H}_{k-1} V_{k-1} \ , \end{aligned} \tag{7.71}$$

provide the averaging of both sides of (7.71), and obtain two unbiasedness constraints,

$$I = \mathcal{H}_{k-1} C_{k-1} \ , \tag{7.72}$$

$$\mathcal{H}_{k-1}^{\mathrm{f}} = \mathcal{H}_{k-1} L_{k-1} \ . \tag{7.73}$$

Now, substituting (7.72) and (7.73) into (7.71) gives

$$\tilde{x}_k = x_k - \mathcal{H}_{k-1}(G_{k-1} W_{k-1} - V_{k-1}) \ ,$$

the estimation error becomes

$$\varepsilon_k = x_k - \tilde{x}_k = \mathcal{H}_{k-1}(G_{k-1} W_{k-1} - V_{k-1}) \ ,$$

and the error covariance $P_k = \mathcal{E}\{\varepsilon_k \varepsilon_k^T\}$ can be written as

$$P_k = \mathcal{H}_{k-1}(G_{k-1} \mathcal{Q}_{k-1} G_{k-1}^T + \mathcal{R}_{k-1}) \mathcal{H}_{k-1}^T \tag{7.74a}$$

$$= \mathcal{W}_{k-1} \mathcal{Q}_{k-1} \mathcal{W}_{k-1}^T + \mathcal{V}_{k-1} \mathcal{R}_{k-1} \mathcal{V}_{k-1}^T \ , \tag{7.74b}$$

where $\mathcal{W}_k = \mathcal{H}_k G_k$ and $\mathcal{V}_k = \mathcal{H}_k$ are the error residual matrices.

To find the gain $\mathcal{H}_{k-1}$ subject to constraint (7.72), the trace of P_k can be minimized with $\mathcal{H}_{k-1}$ using the Lagrange multiplier method as

$$\begin{aligned} \frac{\partial}{\partial \mathcal{H}_{k-1}} \text{tr}[&\mathcal{H}_{k-1}(G_{k-1} \mathcal{Q}_{k-1} G_{k-1}^T + \mathcal{R}_{k-1}) \mathcal{H}_{k-1}^T \\ &+ \Lambda(I - \mathcal{H}_{k-1} C_{k-1})] = 0 \end{aligned}$$

that, if we introduce $\Omega_{k-1} = G_{k-1} \mathcal{Q}_{k-1} G_{k-1}^T + \mathcal{R}_{k-1}$, gives

$$\begin{aligned} \Lambda^T C_{k-1}^T &= 2 \mathcal{H}_{k-1} \Omega_{k-1} \ , \\ C_{k-1} \Lambda &= 2 \Omega_{k-1} \mathcal{H}_{k-1}^T \ . \end{aligned} \tag{7.75}$$

Multiplying both sides of (7.75) by the nonzero $C_{k-1}^T \Omega_{k-1}^{-1}$ from the left-hand side and using (7.72), we obtain the Lagrange multiplier as

$$\begin{aligned} \Lambda &= 2(C_{k-1}^T \Omega_{k-1}^{-1} C_{k-1})^{-1} C_{k-1}^T \mathcal{H}_{k-1}^T \\ &= 2(C_{k-1}^T \Omega_{k-1}^{-1} C_{k-1})^{-1} \ , \end{aligned}$$

and then the substitution of Λ in (7.75) gives the gain [105]

$$\mathcal{H}_{k-1} = (C_{k-1}^T \Omega_{k-1}^{-1} C_{k-1})^{-1} C_{k-1}^T \Omega_{k-1}^{-1} \,. \tag{7.76}$$

We finally represent the MVF-I filter (7.70) with

$$\tilde{x}_k = \mathcal{H}_{k-1}(Y_{k-1} - L_{k-1}U_{k-1}) \tag{7.77a}$$

$$= (C_{k-1}^T \Omega_{k-1}^{-1} C_{k-1})^{-1} C_{k-1}^T \Omega_{k-1}^{-1} (Y_{k-1} - L_{k-1}U_{k-1}) \tag{7.77b}$$

and notice that the error covariance for (7.77b) is defined by (7.74b).

It can be seen that the MVF-I filter has the form of the ML-I FIR filter (4.98a) and thus belongs to the family of ML state estimators. The difference is that the error residual matrix $\mathcal{W}_k$ in (7.74b) does not include the matrix $\bar{D}$ containing information of the process dynamics. Therefore, the MVF-I filter is most suitable for stationary and quasistationary processes. For LTI systems, recursive computation of (7.77b) is provided in [105, 106]. In the general case of LTV systems, recursions can be found using the OUFIR-II filter derivation procedure, and we postpone it to "Problems."

7.6.2 MVF-II Filter for Nonstationary Processes

A more general MVF-II filter was derived in [106] using a similar procedure as for the OUFIR-II filter. However, to obtain an estimate at k, the MVF-II filter takes data from $[k-N, k-1]$, while the OUFIR-II filter from $[m, k]$.

The MVF-II filter can be obtained using the model in (7.14) and (7.15), if we keep the definitions for MVF-I, introduce a time shift, and write

$$x_k = \mathcal{F}_{k-1}^{k-N} x_{k-N} + \bar{S}_{k-1}U_{k-1} + \bar{D}_{k-1}W_{k-1} \,, \tag{7.78}$$

$$Y_{k-1} = H_{k-1}x_{k-N} + L_{k-1}U_{k-1} + G_{k-1}W_{k-1} + V_{k-1} \,, \tag{7.79}$$

where the extended matrices are given by

$$H_{k-1} = \bar{H}_{k-1}F_{m-1,k-1}F_{k-N} \,, \tag{7.80}$$

$$L_{k-1} = \bar{H}_{k-1}S_{k-1} \,, \tag{7.81}$$

$$G_{k-1} = \bar{H}_{k-1}D_{k-1} \,, \tag{7.82}$$

$$S_{k-1} = \begin{bmatrix} 0 & 0 & \dots & 0 & 0 \\ E_{k-N} & 0 & \dots & 0 & 0 \\ \vdots & \vdots & \ddots & \vdots & \vdots \\ \mathcal{F}_{k-3}^{m}E_{k-N} & \mathcal{F}_{k-3}^{m+1}E_m & \dots & 0 & 0 \\ \mathcal{F}_{k-2}^{m}E_{k-N} & \mathcal{F}_{k-2}^{m+1}E_m & \dots & E_{k-2} & 0 \end{bmatrix} , \tag{7.83}$$

matrix D_{k-1} is equal to matrix S_{k-1} by replacing E_k with B_k, $\bar{S}_{k-1}$ is the last row vector in S_{k-1} and so is $\bar{D}_{m,k}$ in $D_{m,k}$, and matrix $\bar{H}_{m,k} = \mathrm{diag}\left(H_{k-N}\; H_{k-N+1}\; \dots\; H_{k-1}\right)$ is diagonal.

We can now define the MVF-II estimate and transform using (7.79) as

$$\tilde{x}_k = \mathcal{H}_{k-1}Y_{k-1} + \mathcal{H}_{k-1}^{\mathrm{f}}U_{k-1} \tag{7.84}$$

$$= \mathcal{H}_{k-1}H_{k-1}x_{k-N} + (\mathcal{H}_{k-1}L_{k-1} + \mathcal{H}_{k-1}^{\mathrm{f}})U_{k-1}$$
$$+ \mathcal{H}_{k-1}G_{k-1}W_{k-1} + \mathcal{H}_{k-1}V_{k-1} \,. \tag{7.85}$$

Introducing $C_{k-1} = H_{k-1}(\mathcal{F}_{k-1}^{k-N})^{-1}$ and applying the unbiasedness conditions to (7.85) and (7.78), we next obtain the unbiasedness constraints

$$I = \mathcal{H}_{k-1}C_{k-1} , \tag{7.86}$$

$$\mathcal{H}_{k-1}^{\mathrm{f}} = \bar{S}_{k-1} - \mathcal{H}_{k-1}L_{k-1} , \tag{7.87}$$

define the estimation error as $\varepsilon_{k-1} = x_{k-1} - \tilde{x}_{k-1}$, use the constraints (7.86) and (7.87), transform ε_{k-1} to

$$\varepsilon_{k-1} = (\bar{D}_{k-1} - \mathcal{H}_{k-1}G_{k-1})W_{k-1} - \mathcal{H}_{k-1}V_{k-1} , \tag{7.88}$$

and find the error covariance

$$P_k = (\bar{D}_{k-1} - \mathcal{H}_{k-1}G_{k-1})\mathcal{Q}_{k-1}(\bar{D}_{k-1} - \mathcal{H}_{k-1}G_{k-1})^T$$
$$+ \mathcal{H}_{k-1}\mathcal{R}_{k-1}\mathcal{H}_{k-1}^T . \tag{7.89}$$

To embed the unbiasedness, we solve the optimization problem

$$J = \underset{\mathcal{H}_{k-1},\Lambda}{\arg\min}\ \mathrm{tr}\ [P_k + \Lambda(I - \mathcal{H}_{k-1}C_{k-1})] \tag{7.90}$$

by putting to zero the derivatives of the trace of the matrix function with respect to $\mathcal{H}_{k-1}$ and Λ, and obtain

$$C_{k-1}\Lambda = -2G_{k-1}Q_{k-1}\bar{D}_{k-1}^T + 2\Omega_{k-1}\mathcal{H}_{k-1}^T . \tag{7.91}$$

Then multiplying both sides of (7.91) from the left-hand side with $C_{k-1}^T\Omega_{k-1}^{-1}$ and using (7.86) gives the Lagrange multiplier

$$\Lambda = -2(C_{k-1}^T\Omega_{k-1}^{-1}C_{k-1})^{-1}C_{k-1}^T\Omega_{k-1}^{-1}G_{k-1}Q_{k-1}\bar{D}_{k-1}^T$$
$$+ 2(C_{k-1}^T\Omega_{k-1}^{-1}C_{k-1})^{-1} . \tag{7.92}$$

We finally substitute (7.92) into (7.91), find the gain $\mathcal{H}_{k-1}$ in the form

$$\mathcal{H}_{k-1} = (C_{k-1}^T\Omega_{k-1}^{-1}C_{k-1})^{-1}C_{k-1}^T\Omega_{k-1}^{-1} + \bar{D}_{k-1}Q_{k-1}G_{k-1}^T\Omega_{k-1}^{-1}$$
$$\times [I - C_{k-1}(C_{k-1}^T\Omega_{k-1}^{-1}C_{k-1})^{-1}C_{k-1}^T\Omega_{k-1}^{-1}] , \tag{7.93}$$

represent the MVF-II filtering estimate (7.84) as

$$\tilde{x}_k = \mathcal{H}_{k-1}Y_{k-1} + (\bar{S}_{k-1} - \mathcal{H}_{k-1}L_{k-1})U_{k-1} , \tag{7.94}$$

where $\mathcal{H}_{k-1}$ is given by (7.93), and write the error covariance (7.89) in the standard form

$$P_k = \mathcal{W}_{k-1}\mathcal{Q}_{k-1}\mathcal{W}_{k-1}^T + \mathcal{V}_{k-1}\mathcal{R}_{k-1}\mathcal{V}_{k-1}^T , \tag{7.95}$$

where the error residual matrices are defined by

$$\mathcal{W}_{k-1} = \bar{D}_{k-1} - \mathcal{H}_{k-1}G_{k-1} , \tag{7.96}$$

$$\mathcal{V}_{k-1} = \mathcal{H}_{k-1} . \tag{7.97}$$

It can now be shown that for LTI systems the gain (7.93) is equivalent to the gain (4.69a) of the OUFIR-II filter. Since the OUFIR-II gain is identical to the ML-I FIR gain (4.113b), it follows that the MVF-II filter also belongs to the class of ML estimators. Therefore, the recursions for the MVF-II filter can be taken from Algorithm 8, not forgetting that this algorithm has the following disadvantages: 1) exact initial values are required to initialize iterations, 2) it is less accurate than the OFIR algorithm, and 3) it is more complex than the OFIR algorithm. All of this means that optimal unbiased recursions persisting for an MVF-II filter will have limited advantages over Kalman recursions. Finally, the recursive forms found in [106] for the MVF-II filter are much more complex than those in Algorithm 8.

7.7 Maximum Likelihood FIR Predictor

Like the standard ML FIR filter, the ML FIR predictor has two possible algorithmic implementations. We can obtain the ML FIR prediction at $k+1$ on the horizon $[m,k]$ in what we will call the ML-I FIR predictor. We can also first obtain the ML FIR backward estimate at the start point m over $[m,k]$ and then project it unbiasedly onto $k+1$ in what we will call the ML-II FIR predictor. Because the ML approach is unbiased, the unbiased projection in the ML-II FIR predictor is justified for practical purposes.

7.7.1 ML-I FIR Predictor

To derive the ML-I FIR predictor, we consider the state-space model

$$x_{k+1} = F_k x_k + B_k w_k \,, \tag{7.98}$$

$$y_k = H_k x_k + v_k \,, \tag{7.99}$$

extend it as (7.14) and (7.15), extract x_{k+1}, and obtain

$$x_{k+1} = \mathcal{F}_k^m x_m + \bar{D}_{m,k} W_{m,k} \,, \tag{7.100}$$

$$Y_{m,k} = H_{m,k}^p x_m + G_{m,k}^p W_{m,k} + V_{m,k} \,. \tag{7.101}$$

The ML-I FIR predicted estimate can now be determined for data taken from $[m,k]$ by maximizing the likelihood $p(Y_{m,k}|x_{k+1})$ of x_{k+1} as

$$\tilde{x}_{k+1|k} = \arg\max_{x_{k+1}} p(Y_{m,k}|x_{k+1}) \,. \tag{7.102}$$

The solution to the maximization problem (7.102) can be found if we extract x_m from (7.100) as $x_m = \mathcal{F}_k^{m^{-1}} x_{k+1} - \mathcal{F}_k^{m^{-1}} \bar{D}_{m,k} W_{m,k}$, substitute into (7.101), and then represent (7.101) as

$$Y_{m,k} - C_{m,k}^p x_{k+1} = \mathcal{N}_{m,k}^p \,, \tag{7.103}$$

where $C_{m,k}^p = H_{m,k}^p (\mathcal{F}_k^m)^{-1}$ and all random components are combined in

$$\mathcal{N}_{m,k}^p = (G_{m,k}^p - C_{m,k}^p \bar{D}_{m,k}) W_{m,k} + V_{m,k} \,. \tag{7.104}$$

For a multivariate normal distribution, the likelihood of x_{k+1} is given by

$$p(Y_{m,k}|x_{k+1}) \propto \exp\left\{-\frac{1}{2}(Y_{m,k} - C_{m,k}^p x_{k+1})^T \Sigma_{m,k}^{p^{-1}}(\dots)\right\} \,, \tag{7.105}$$

where the covariance matrix is defined as

$$\Sigma_{m,k}^p = \mathcal{E}\{\mathcal{N}_{m,k}^p \mathcal{N}_{m,k}^{p^T}\} \tag{7.106a}$$

$$= (G_{m,k}^p - C_{m,k}^p \bar{D}_{m,k}) \mathcal{Q}_{m,k} (\dots)^T + \mathcal{R}_{m,k} \,. \tag{7.106b}$$

The maximization problem (7.102) can now be equivalently replaced by the minimization problem

$$\tilde{x}_{k+1|k} = \underset{x_{k+1}}{\arg\min} \left\{ -\frac{1}{2}(Y_{m,k} - C^p_{m,k}x_{k+1})^T \Sigma^{p^{-1}}_{m,k}(\dots) \right\} , \tag{7.107}$$

which assumes minimization of the quadratic form. Referring to (7.106b), we find a solution to (7.107) in the form

$$\tilde{x}_{k+1|k} = (C^{p^T}_{m,k}\Sigma^{p^{-1}}_{m,k}C^p_{m,k})^{-1}C^{p^T}_{m,k}\Sigma^{p^{-1}}_{m,k}Y_{m,k} \tag{7.108}$$

and write the error covariance using (7.104) as

$$P_{k+1} = (G^p_{m,k} - C^p_{m,k}\bar{D}_{m,k})\mathcal{Q}_{m,k}(G^p_{m,k} - C^p_{m,k}\bar{D}_{m,k})^T + \mathcal{R}_{m,k} . \tag{7.109}$$

What finally comes is that the prediction (7.108) differs from the previously obtained ML-I FIR filtering estimate (4.98a) only in the modified matrices $C^p_{m,k}$ and $\Sigma^p_{m,k}$, which reflect the features of the state model (7.98) and the predictive estimate (7.102).

7.7.2 ML-II FIR Predictor

As mentioned earlier, the ML-II FIR prediction appears if we first estimate the initial state at m over $[m, k]$ and then project it forward to $k + 1$. The first part of this procedure has already been supported by (4.114)–(4.119). Applied to the model in (7.98) and (7.99), this gives the *a posteriori* ML FIR estimate

$$\hat{x}_{m|k} = (H^{p^T}_{m,k}\Omega^{-1}_{m,k}H^p_{m,k})^{-1}H^{p^T}_{m,k}\Omega^{-1}_{m,k}Y_{m,k} . \tag{7.110}$$

In turn, the unbiased projection of (7.110) onto $k + 1$ can be obtained as

$$\begin{aligned} \tilde{x}_{k+1|k} &= \mathcal{F}^m_k\hat{x}_{m|k} \\ &= \mathcal{F}^m_k(H^{p^T}_{m,k}\Omega^{-1}_{m,k}H^p_{m,k})^{-1}H^{p^T}_{m,k}\Omega^{-1}_{m,k}Y_{m,k} \\ &= (C^{p^T}_{m,k}\Omega^{-1}_{m,k}C^p_{m,k})^{-1}C^{p^T}_{m,k}\Omega^{-1}_{m,k}Y_{m,k} \end{aligned} \tag{7.111}$$

to be an ML-II FIR predictive estimate, the error covariance of which is given by

$$P_{k+1} = G^p_{m,k}\mathcal{Q}_{m,k}G^{p^T}_{m,k} + \mathcal{R}_{m,k} . \tag{7.112}$$

It is now easy to show that the ML-II FIR predictor (7.111) has the same structure of the error covariance (7.112) as the structure (7.74b) of the MVF-I filter (7.77b), and we conclude that these estimates can be converted to each other by introducing a time shift.

7.8 Extended OFIR Prediction

When state feedback control is required for nonlinear systems, then linear estimators generally cannot serve, and extended predictive filtering or prediction is used. To develop an EOFIR predictor, we assume that the process and its observation are both nonlinear and represent them with the following state and observation equations

$$x_{k+1} = f_k(x_k) + w_k , \tag{7.113}$$

$$y_k = h_k(x_k) + v_k . \tag{7.114}$$

Now the EOFIR predictor can be obtained similarly to the EOFIR filter if we expand the nonlinear functions with the Taylor series. Assuming that the functions $f_k(x_k)$ and $h_k(x_k)$ are sufficiently smooth, we expand them around the available estimate $\hat{x}_k$ using the second-order Taylor series as

$$f_k(x_k) \cong f_k(\hat{x}_k) + \dot{F}_k \epsilon_k + \frac{1}{2}\alpha_k \,, \tag{7.115}$$

$$h_k(x_k) \cong h_k(\hat{x}_k) + \dot{H}_k \epsilon_k + \frac{1}{2}\beta_k \,, \tag{7.116}$$

where the increment $\epsilon_k = x_k - \hat{x}_k$ is equivalent to the estimation error, the Jacobian matrices are given by

$$\dot{F}_k = \left.\frac{\partial f_k}{\partial x}\right|_{x=\hat{x}_k} , \tag{7.117}$$

$$\dot{H}_k = \left.\frac{\partial h_k}{\partial x}\right|_{x=\hat{x}_k} , \tag{7.118}$$

the second-order terms are determined as [15]

$$\alpha_k = \sum_{i=1}^{K} e_i^K \epsilon_k^T \ddot{F}_{ik} \epsilon_k \,, \tag{7.119}$$

$$\beta_k = \sum_{j=1}^{M} e_j^M \epsilon_k^T \ddot{H}_{jk} \epsilon_k \,, \tag{7.120}$$

and the Hessian matrices are defined by

$$\ddot{F}_{ik} = \left.\frac{\partial^2 f_{ik}}{\partial x^2}\right|_{x=\hat{x}_k} , \tag{7.121}$$

$$\ddot{H}_{jk} = \left.\frac{\partial^2 h_{jk}}{\partial x^2}\right|_{x=\hat{x}_k} , \tag{7.122}$$

where f_{ik}, $i \in [1, K]$, and h_{jk}, $j \in [1, M]$, are the ith and jth components of $f_k(x_k)$ and $h_k(x_k)$, respectively. Also, $e_i^K \in \mathbb{R}^K$ and $e_j^M \in \mathbb{R}^M$ are Cartesian basis vectors with ones in the ith and jth components and zeros elsewhere.

The nonlinear model in (7.113) and (7.114) can thus be linearized as

$$x_{k+1} = \dot{F}_k x_k + \eta_k + w_k \,, \tag{7.123}$$

$$\tilde{y}_k = \dot{H}_k x_k + v_k \,, \tag{7.124}$$

where $\tilde{y}_k = y_k - \psi_k$ is the modified observation, in which

$$\psi_k = h_k(\hat{x}_k) - \dot{H}_k \hat{x}_k + \frac{1}{2}\beta_k \tag{7.125}$$

is a correction vector, and η_k given by

$$\eta_k = f_k(\hat{x}_k) - \dot{F}_k \hat{x}_k + \frac{1}{2}\alpha_k \tag{7.126}$$

plays the role of an input signal.

It follows from this model that the second-order additions α_k and β_k affect only η_k and ψ_k. If α_k and β_k have little effect on the prediction, they can be omitted as in the EOFIR-1 predictor. Otherwise, one should use the EOFIR-II predictor.

The pseudocode of the EOFIR prediction algorithm serving both options is listed as Algorithm 18, where the matrix ψ_k is computed by (7.125) using the Taylor series expansions. It can be seen that the nonlinear functions $f_k(x_k)$ and $h_k(x_k)$ are only used here to update the prediction, while the error covariance matrix is updated using the extended matrices. Another feature is that Algorithm 18 is universal for both EOFIR-I and EOFIR-II predictors. Indeed, in the case of the EOFIR-I predictor, the terms β_k and α_k vanish in the matrix ψ_k, and for the EOFIR-II predictor they must be preserved. It should also be noted that the more sophisticated second-order EOFIR-II predictor does not demonstrate clear advantages over the EOFIR-I predictor, although we have already noted this earlier.

Algorithm 18: Extended OFIR Prediction Algorithm

Data: $y_k, \hat{x}_m, P_m, Q_k, R_k, N$
Result: $\hat{x}_{k+1}$

begin
  for $k = 1, 2, \cdots$ **do**
    $m = k - N + 1$ if $k > N - 1$ and $m = 0$ otherwise ;
    for $i = m + 1 : k$ **do**
      $S_l = \dot{H}_l P_l \dot{H}_l^T + R_l$;
      $K_l = \dot{F}_l P_l \dot{H}_l^T S_l^{-1}$;
      $\bar{x}_{l+1} = f_l(\bar{x}_l) + K_l\{y_l - \psi_i - h_i[f_i(\bar{x}_i)]\}$;
      $P_{l+1} = (\dot{F}_l - K_l \dot{H}_l) P_l \dot{F}_l^T + Q_i$;
    end for
    $\tilde{x}_{k+1} = \bar{x}_{k+1}$;
  end for
end

7.9 Summary

Digital stochastic control requires predictive estimation to provide effective state feedback control, since filtering estimates may be too biased at the next time point. Prediction or predictive filtering can be organized using any of the available state estimation techniques. The requirement for such structure is that they must provide one-step prediction with maximum accuracy. This allows for suboptimal state feedback control, even though the predicted estimate is less accurate than the filtering estimate. Next we summarize the most important properties of the FIR predictors and RH FIR filters.

To organize one-step prediction at k, the RH FIR predictive filter can be used to obtain an estimate over the data FH $[m - 1, k - 1]$. Alternatively, we can use any type of FIR predictor to obtain an estimate at $k + 1$ over $[m, k]$. If necessary, we can change the time variable to obtain an estimate at k over $[m - 1, k - 1]$. We can also use the FIR filtering estimate available at $k - 1$ and project it to k using the system matrix.

The iterative OFIR predictor uses the KP recursions. The difference between the OFIR predictor and OFIR filter estimates is poorly discernible when these structures fully fit the model. But in the presence of uncertainties, the predictor can be much less accurate than the filter. For LTI systems without input, the UFIR prediction is equivalent to a one-step projected UFIR filtering estimate. The UFIR and OFIR predictors can be obtained in batch form and in an iterative form using recursions.

The MVF-I filter and the ML-II FIR predictor have the same batch form, and both these estimators belong to the family of ML state estimators. The MVF-I filter is suitable for stationary and quasistationary stochastic processes, while the MVF-II filter is suitable for nonstationary stochastic processes. The EOFIR predictor can be obtained similarly to the extended OFIR filter using first- or second-order Taylor series expansions.

7.10 Problems

1 Following the derivation of the KP algorithm (7.10)–(7.13), obtain the LMP algorithm at $k+1$ over $[m, k]$ and at k over $[m-1, k-1]$.

2 Given the state-space model in (7.1) and (7.2), using the Bayesian approach, derive the KP and show its equivalence to the algorithm (7.10)–(7.13).

3 The risk function of the RH FIR estimate $\tilde{x}_{k|k-1}$ is given by

$$J_k = \mathcal{E}[(x_k - \tilde{x}_{k|k-1})^T(x_k - \tilde{x}_{k|k-1})] \,.$$

Show that the inequality $J_{k+1} - J_k > 0$ holds if J_k satisfies the cost function $J_k = \underset{N}{\arg\min}\ [(x_k - \tilde{x}_{k|k-1})^T(x_k - \tilde{x}_{k|k-1})]$.

4 A system is represented in state space with the following equations: $x_{k+1} = Fx_k + E_x u_k + w_k$ and $y_k = Hx_k + D_y u_k + v_k$. Extend this model on $[m, k]$ and derive the UFIR predictor.

5 Consider the system described in item 4 and derive the OFIR predictor.

6 Given the following harmonic two-state space model

$$x_{k+1} = \begin{bmatrix} \cos\frac{\pi}{64} + \delta_k & \sin\frac{\pi}{64} \\ -\sin\frac{\pi}{64} & \cos\frac{\pi}{64} + \delta_k \end{bmatrix} x_k + \begin{bmatrix} 0.3 \\ 0.1 \end{bmatrix} u_k + \begin{bmatrix} 1 \\ 1 \end{bmatrix} w_k \,,$$
$$y_k = [\,1 \quad 0\,]x_k + v_k \,,$$

where w_k and v_k are white Gaussian with the covariances $\sigma_w^2 = 1$ and $\sigma_v^2 = 100$ and the disturbance $\delta_k = 0.04$ is induced from 350 to 400, simulate this process for the initial state $x_k^T = [\,100 \quad 0.01\,]$ and estimate x_{k+1} numerically using different FIR predictors. Select the most and less accurate predictors among the predictors OFIR, UFIR, ML-I FIR, and ML-II FIR.

7 Consider the problem described in item 6, estimate x_k using the MVF-I and MVF-II predictive filters, and compare the errors. Explain the difference between the predictive estimates.

8 The error covariances $P_k^{(1)}$ and $P_k^{(2)}$ of two FIR predictors are given by the solutions of the following DDREs

$$P_{k+1}^{(1,2)} = FP_k^{(1,2)}F^T - FP_k^{(1,2)}H^T(HP_k^{(1,2)}H^T + R)^{-1} \times HP_k^{(1,2)}F^T + Q^{(1,2)} \,.$$

Find the difference $\Delta P_k = P_k^{(2)} - P_k^{(1)}$ and analyze the dependence of ΔP_k on the system noise covariances $Q^{(1)}$ and $Q^{(2)}$.

9 Consider the error residual matrices $\mathcal{W}^p_{m,k} = \bar{D}_{m,k} - \hat{\mathcal{H}}^p_{m,k}G^p_{m,k}$ and $\mathcal{V}^p_{m,k} = \hat{\mathcal{H}}^p_{m,k}$ of the UFIR predictor (7.25) and explain why a decrease in GNPG $= \hat{\mathcal{H}}^p_{m,k}\hat{\mathcal{H}}^{p^T}_{m,k}$ improves measurement noise reduction.

10 Explain why an FIR filter and an FIR predictor that satisfy the same cost function are equivalent in continuous time.

11 Given the state-space model, $x_{k+1} = Fx_k + B_k w_k$ and $y_k = Hx_k + v_k$, and the KP and KF estimates, respectively,

$$\begin{aligned} \tilde{x}_{k+1} &= F\tilde{x}_k + K_k(y_k - H\tilde{x}_k)\,, \\ \hat{x}_k &= F\hat{x}_{k-1} + K_k(y_k - HF\hat{x}_{k-1})\,. \end{aligned} \tag{7.127}$$

under what conditions do these estimates 1) become equivalent and 2) cannot be converted into each other?

12 Given the MVF-I filter (7.77b), following the derivation of the OUFIR-II filter, find recursive forms for the MVF-I filter and design an iterative algorithm.

13 The recursive form for the batch error covariance (7.26), which corresponds to the batch UFIR predictor (7.25), is given by (7.39) as

$$\begin{aligned} P_{k+1} = {} & (F_k - \mathcal{G}_k F_k^{-T} H_k^T H_k) P_k (F_k - \mathcal{G}_k F_k^{-T} H_k^T H_k)^T \\ & + \mathcal{G}_k F_k^{-T} H_k^T R_k H_k F_k^{-1} \mathcal{G}_k + B_k Q_k B_k^T\,. \end{aligned}$$

Obtain this recursion using the recursive UFIR predictor estimate

$$\tilde{x}_{k+1} = F_k \tilde{x}_k + E_k u_k + \mathcal{G}_k F_k^{-T} H_k^T (y_k - H_k \tilde{x}_k)\,.$$

14 The MVF-II filter is represented by the fundamental gain (7.93). Referring to the similarity with the OUFIR filter, find recursive forms and design an iterative algorithm for the MVF-II filter.

15 Consider the state-space model $x_{k+1} = Fx_k + Eu_k + w_k$ and $y_k = Hx_k + v_k$. Define the MVF predictive filtering estimate as

$$\tilde{x}_k = \sum_{i=k-N}^{k-1} \mathcal{H}_{k-i} y_i + \sum_{i=k-N}^{k-1} \mathcal{H}^{\mathrm{f}}_{k-i} u_i$$

and obtain the gains $\mathcal{H}_k$ and $\mathcal{H}^{\mathrm{f}}_k$ for $k \in [1, N]$.

16 The state of the LTI system without input is estimated using OFIR filtering as $\hat{x}_k$ (4.30b). Project this estimate to $k+1$ as $\hat{x}^{\mathrm{pj}}_{k+1} = F\hat{x}_k$. Also consider the OFIR predicted estimate $\tilde{x}^{pr}_{k+1} = \mathcal{H}^p_{m,k} Y_{m,k}$. Under what conditions do the projected and predicted estimates 1) become identical and 2) cannot be converted into one another?

17 The batch UFIR prediction is given by

$$\tilde{x}^{pr}_{k+1} = F^N (H_N^{p^T} H_N^p)^{-1} H_N^{p^T} Y_{m,k}\,.$$

Compare this prediction with the LS prediction and highlight the differences.

18 A nonlinear system is represented in state space with the equations $x_{k+1} = f_k(x_k, u_k, w_k)$ and $y_k = h_k(x_k, v_k)$. Suppose that the white Gaussian noise vectors w_k and v_k have low intensity components and that the input u_k is known. Apply the first-order Taylor series expansion and obtain an EOFIR predictor.

19 Consider the previously described problem, apply the same conditions, and obtain a first-order EFIR predictor and an RH EFIR filter.

20 The gain of the RH MVF filter is specified by (7.76) as

$$\mathcal{H}_{k-1} = (C_{k-1}^T \Omega_{k-1}^{-1} C_{k-1})^{-1} C_{k-1}^T \Omega_{k-1}^{-1} \; .$$

Can this gain be applied when the system matrix F_k is singular? If not, modify this gain to be applicable for a singular matrix F_k.

21 A wireless network is represented with the following state-space model,

$$x_{k+1} = F_k x_k + E_k u_k + B_k w_k \; ,$$
$$y_k = H_k x_k + v_k \; ,$$

where $y_k = [y_k^{(1)^T} \; \dots \; y_k^{(n)^T}]^T$ is the observation provided by n sensors and $H_k = [H_k^{(1)^T} \; \dots \; H_k^{(n)^T}]^T$. Zero mean mutually uncorrelated white Gaussian noise vectors w_k and $v_k = [v_k^{(1)^T} \; \dots \; v_k^{(n)^T}]^T$ have the covariances Q_k and $R_k = \text{diag}[R_k^{(1)^T} \; \dots \; R_k^{(n)^T}]^T$, respectively. Think about how to obtain a UFIR predictor and an RH UFIR filter with some kind of consensus in measurements to simplify the algorithm.

8

Robust FIR State Estimation Under Disturbances

Robustness signifies insensitivity to small deviations from the assumptions.

Peter J. Huber, [74] p. 1

Methods of state estimation in the presence of uncertain but bounded impacts, called *robust*, have been developed in attempts to cope with processes affected by unpredictable disturbances. The most advanced estimators of this type were obtained in the transform domain using the disturbance-to-error transfer function $\mathcal{T}$; thus, for LTI systems and zero mean persistent disturbances. In H_2 filtering, the Frobenius norm of $\mathcal{T}$ is minimized, which makes the H_2 filter similar to the optimal filter. In H_∞ filtering, the induced norm of $\mathcal{T}$ is minimized in the so-called $\mathcal{L}_2$-to-$\mathcal{L}_2$ or energy-to-energy filter. In generalized H_2 filtering, the energy-to-peak $\mathcal{T}$ is minimized in a filter that has the $\mathcal{L}_2$-to-$\mathcal{L}_\infty$ structure. Yet another $\mathcal{L}_\infty$-to-$\mathcal{L}_\infty$ estimator minimizes the peak error relative to the peak disturbance in the so-called peak-to-peak filter. Setting aside $\mathcal{T}$, the game theory suggests to design H_∞ state estimators by minimizing the ratio of the bounded estimation error and disturbance norms over finite horizons. This approach is more flexible as it can be applied to LTV systems with time-variable disturbances. Note that the aforementioned methods were mainly used for developing recursive estimators. Attempts have also been made to extend the approaches to FIR state estimation, although the results achieved so far turned out to be limited.

In this chapter, we develop the theory of robust FIR state estimation (filtering and prediction) for LTI systems operating under disturbances in the presence of initial and measurement errors. As the approach suggests, the state estimator becomes robust if it is tuned for maximum disturbance. On the other hand, since the disturbance can be of any color, the most robust state estimator is in batch form. It should be noted that most of the solutions presented in this chapter are rather theoretical, and further efforts should be made to turn them into practical algorithms.

8.1 Extended Models Under Disturbances

By solving an optimization problem with some cost function applied to the disturbance-to-error transfer function $\mathcal{T}$, FIR state estimators can be developed using FE-based and BE-based LTI state-space models. Recall that FIR prediction fits better with the FE-based model, FIR filtering with the BE-based model, and that both these models require extensions on $[m, k]$.

Optimal and Robust State Estimation: Finite Impulse Response (FIR) and Kalman Approaches, First Edition.
Yuriy S. Shmaliy and Shunyi Zhao.

Backward Euler-Based State-Space Model

The BE-based state-space model of the LTI system, which fits with *a posteriori* FIR filtering, is given by

$$x_k = Fx_{k-1} + Eu_k + Bw_k \,, \tag{8.1}$$

$$y_k = Hx_k + Dw_k + v_k \,, \tag{8.2}$$

where $x_k \in \mathbb{R}^K$, $u_k \in \mathbb{R}^L$, $y_k \in \mathbb{R}^P$, $F \in \mathbb{R}^{K\times K}$, $H \in \mathbb{R}^{P\times K}$, $E \in \mathbb{R}^{K\times L}$, $B \in \mathbb{R}^{K\times M}$, $D \in \mathbb{R}^{P\times M}$. Here, $w_k \in \mathbb{R}^M$ is some zero mean external disturbance and $v_k \in \mathbb{R}^P$ is the zero mean observation error or noise. Both w_k and v_k are assumed to be not well-known, but norm-bounded. The term with D appears in (8.2) under severe disturbance. If the order of the disturbance *de facto* is less than that of the system, then $D = 0$ and (8.2) becomes the standard observation equation. On the other hand, $D \neq 0$ means that the system is unstable or boundary stable. Therefore, when the approach cannot be applied to unstable systems, the term with D is dropped.

The extension of model (8.1) and (8.2) on $[m, k]$ is given by

$$X_{m,k} = F_N x_m + S_N U_{m,k} + D_N W_{m,k} \,, \tag{8.3}$$

$$Y_{m,k} = H_N x_m + L_N U_{m,k} + T_N W_{m,k} + V_{m,k} \,, \tag{8.4}$$

where the extended block matrices are defined as

$$F_N = \begin{bmatrix} I \\ F \\ \vdots \\ F^{N-2} \\ F^{N-1} \end{bmatrix}, \quad S_N = \begin{bmatrix} E & 0 & \dots & 0 & 0 \\ FE & E & \dots & 0 & 0 \\ \vdots & \vdots & \ddots & \vdots & \vdots \\ F^{N-2}E & F^{N-3}E & \dots & E & 0 \\ F^{N-1}E & F^{N-2}E & \dots & FE & E \end{bmatrix},$$

$$D_N = \begin{bmatrix} B & 0 & \dots & 0 & 0 \\ FB & B & \dots & 0 & 0 \\ \vdots & \vdots & \ddots & \vdots & \vdots \\ F^{N-2}B & F^{N-3}B & \dots & B & 0 \\ F^{N-1}B & F^{N-2}B & \dots & FB & B \end{bmatrix},$$

$H_N = \bar{H}_N F_N$, $L_N = \bar{H}_N S_N$, $T_N = G_N + \bar{T}_N$, $G_N = \bar{H}_N D_N$, and matrices $\bar{H}_N = \text{diag}(H\, H \dots H)$ and $\bar{T}_N = \text{diag}(D\, D \dots D)$ are diagonal.

To design an FIR filter, the state x_k can be represented by the last row vector in (8.3) as

$$x_k = F^{N-1} x_m + \bar{S}_N U_{m,k} + \bar{D}_N W_{m,k} \,, \tag{8.5}$$

where matrix $\bar{S}_N$ is the last row vector in S_N and so is $\bar{D}_N$ in D_N.

Forward Euler Method–Based Model

The FE-based LTI state-space model, which fits with prediction and predictive filtering required in state feedback control, is given by

$$x_{k+1} = Fx_k + Eu_k + Bw_k \,, \tag{8.6}$$

$$y_k = Hx_k + Dw_k + v_k \tag{8.7}$$

and its extension on $[m, k]$ is obtained as

$$X_{m+1,k+1} = F_N^p x_m + S_N U_{m,k} + D_N W_{m,k} \,, \tag{8.8}$$

$$Y_{m,k} = H_N^p x_m + L_N^p U_{m,k} + T_N^p W_{m,k} + V_{m,k} \,, \tag{8.9}$$

where the extended matrices are defined by $F_N^p = FF_N$, $H_N^p = \bar{H}_N F_N$, $L_N^p = \bar{H}_N S_N^p$, $T_N^p = G_N^p + \bar{T}_N$, $G_N^p = \bar{H}_N D_N^p$, and

$$S_N^p = \begin{bmatrix} 0 & 0 & \dots & 0 & 0 & 0 \\ E & 0 & \dots & 0 & 0 & 0 \\ FE & E & \dots & 0 & 0 & 0 \\ \vdots & \vdots & \ddots & \vdots & \vdots & \vdots \\ F^{N-3}E & F^{N-2}E & \dots & E & 0 & 0 \\ F^{N-2}E & F^{N-3}E & \dots & FE & E & 0 \end{bmatrix},$$

$$D_N^p = \begin{bmatrix} 0 & 0 & \dots & 0 & 0 & 0 \\ B & 0 & \dots & 0 & 0 & 0 \\ FB & B & \dots & 0 & 0 & 0 \\ \vdots & \vdots & \ddots & \vdots & \vdots & \vdots \\ F^{N-3}B & F^{N-2}B & \dots & B & 0 & 0 \\ F^{N-2}B & F^{N-3}B & \dots & FB & B & 0 \end{bmatrix}.$$

To obtain an FIR predictor, we specify the model of the state x_{k+1} by the last row vector in (8.8) as

$$x_{k+1} = F^N x_m + \bar{S}_N U_{m,k} + \bar{D}_N W_{m,k} \,, \tag{8.10}$$

where matrix $\bar{S}_N$ is the last row vector in matrix S_N and so is $\bar{D}_N$ in D_N.

8.2 The *a posteriori* H_2 FIR Filtering

The standard H_2 filtering problem implies minimizing the Frobenius norm of the disturbances-to-error transfer function $\mathcal{T}$ using time-frequency duality [19]. Because the H_2 problem is convex, it is mathematically tractable, and solutions can be found in closed forms. A specific feature is that the H_2 solution is sought in the transform domain, while the gain for the FIR filter is required in the time domain. It is also worth noting that minimizing the norm of $\mathcal{T}$ gives the FIR filter gain that does not depend on noise and disturbances, unlike the OFIR filter gain. Therefore, as shown in [141], the gains for H_2 state estimators should be obtained by minimizing the Frobenius norm of the weighted $\mathcal{T}$, which will be considered later.

Consider the model in (8.5) and (8.4) and define the FIR estimate as

$$\begin{aligned} \hat{x}_k &= \mathcal{H}_N Y_{m,k} + \mathcal{H}_N^{\mathrm{f}} U_{m,k} \,, \\ &= \mathcal{H}_N H_N x_m + (\mathcal{H}_N L_N + \mathcal{H}_N^{\mathrm{f}}) U_{m,k} + \mathcal{H}_N T_N W_{m,k} + \mathcal{H}_N V_{m,k} \,. \end{aligned} \tag{8.11}$$

The unbiasedness condition $\mathcal{E}\{x_k\} = \mathcal{E}\{\hat{x}_k\}$ applied to (8.5) and (8.11) yields two unbiasedness constraints

$$I = \mathcal{H}_N C_N \,, \tag{8.12}$$

$$\mathcal{H}_N^{\mathrm{f}} = \bar{S}_N - \mathcal{H}_N L_N \,, \tag{8.13}$$

where $C_N = H_N F^{-(N-1)}$. Now, using (8.5) and (8.11), we transform the estimation error $\varepsilon_k = x_k - \hat{x}_k$ to

$$\varepsilon_k = \mathcal{B}_N x_m + (\bar{S}_N - \mathcal{H}_N L_N - \mathcal{H}_N^{\mathrm{f}})U_{m,k} + \mathcal{W}_N W_{m,k} - \mathcal{V}_N V_{m,k} , \tag{8.14}$$

where the error residual matrices are defined by

$$\mathcal{B}_N = F^{N-1} - \mathcal{H}_N H_N , \tag{8.15}$$

$$\mathcal{W}_N = \bar{D}_N - \mathcal{H}_N T_N , \tag{8.16}$$

$$\mathcal{V}_N = \mathcal{H}_N . \tag{8.17}$$

Our natural desire is to have an estimator that follows the control signal exactly. Therefore, we embed the constraint (8.13) into (8.14), by removing the term with $U_{m,k}$, and represent the estimation error ε_k as the sum of the three suberrors as

$$\varepsilon_k = \bar{\varepsilon}_{xk} + \bar{\varepsilon}_{wk} + \bar{\varepsilon}_{vk} ,$$

In the transform domain we thus have a structure shown in Fig. 8.1, where $\mathcal{T}_x(z)$ is the ε_x-to-ε transfer function, $\mathcal{T}_w(z)$ is the ε_w-to-ε transfer function, and $\mathcal{T}_v(z)$ is the ε_v-to-ε transfer function.

What we can immediately notice is that the initial state error ε_x goes to ε unaltered, and thus the ε_x-to-ε transfer function is the identity matrix, $\mathcal{T}_x(z) = I$. To find the ε_w-to-ε transfer function $\mathcal{T}_w(z)$, we first transform the vector $W_{m,k}$ using the following column matrix rule [106, 196].

Lemma 8.1 ***(Column matrix rule).*** Given a block column matrix $Z_{m,k} = [\, z_m^T \; z_{m+1}^T \; \dots \; z_k^T \,]^T$ specified on $[m, k]$. It is represented recursively as

$$Z_{m,k} = A_w Z_{m-1,k-1} + B_w z_k , \tag{8.18}$$

using the following strictly sparse matrices

$$A_w = \begin{bmatrix} 0 & I & 0 & \dots & 0 \\ 0 & 0 & I & \dots & 0 \\ \vdots & \vdots & \vdots & \ddots & \vdots \\ 0 & 0 & 0 & \dots & I \\ 0 & 0 & 0 & \dots & 0 \end{bmatrix} , \quad B_w = \begin{bmatrix} 0 \\ 0 \\ \vdots \\ 0 \\ I \end{bmatrix} . \tag{8.19}$$

Proof: The proof is self obvious. □

Using (8.18), we represent matrix $W_{m,k}$ as

$$W_{m,k} = A_w W_{m-1,k-1} + B_w w_k ,$$

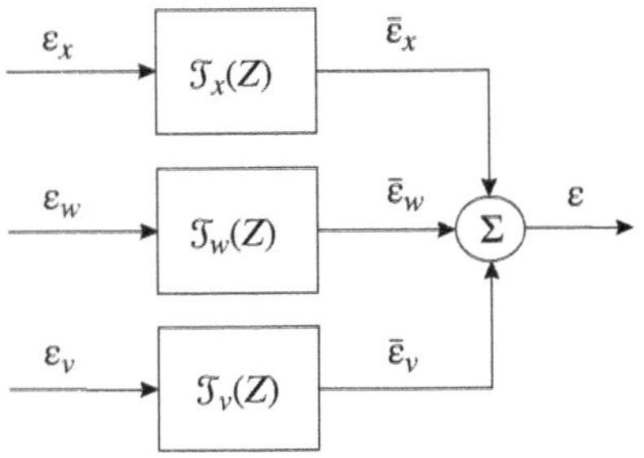

Figure 8.1 Errors in the H_2 state estimator in the z-domain.

where matrices A_w and B_w are given by (8.19). The transform applied to $W_{m,k}$ gives $W(z) = (Iz - A_w)^{-1}zB_w w(z)$. By applying the transform to $\varepsilon_{wk} = (\bar{D}_N - \mathcal{H}_N T_N)W_{m,k}$ and combining with $W(z)$, we finally obtain the ε_w-to-ε transfer function

$$\mathcal{T}_w(z) = \mathcal{W}_N(Iz - A_w)^{-1}zB_w \tag{8.20}$$

and note that (8.20) differs from $\mathcal{T}_w(z)$ found for the FE-based model [109, 232] by the operator z in front of B_w. The presence of the additional z speaks in favor of the greater stability of the BE-based model but does not affect the $\mathcal{T}_w(z)$ norm. Similarly, we represent vector $V_{m,k}$ as $V_{m,k} = A_w V_{m-1,k-1} + B_w v_k$ and write the corresponding transfer function as

$$\mathcal{T}_v(z) = -\mathcal{V}_N(Iz - A_w)^{-1}zB_w \,. \tag{8.21}$$

Now the estimation error can be written in the transform domain as $\varepsilon = \mathcal{B}_N x_m + \mathcal{T}_w(z)\varepsilon_w + \mathcal{T}_v(z)\varepsilon_v$ and we proceed to development of the *a posteriori* H_2-OFIR filter.

8.2.1 H_2-OFIR Filter

The H_2 performance is guaranteed by minimizing the squared Frobenius norm $\|\mathcal{T}(z)\|_F^2$ of the transfer function $\mathcal{T}(z)$ averaged over all frequencies [232]. This approach, supported by lemma 8.1, yields an H_2 FIR filter gain that does not depend on the disturbance or, in other words, the disturbance has a unit intensity. We come to this conclusion by considering the disturbance-invariant transfer functions (8.20) and (8.21), which involve only the model matrices and do not have the error, noise, and disturbance properties as variables. This means that proper weights must be found for these functions [141]. To better justify this, recall that the OFIR filter gain is a function of noise covariances, so we want the H_2-OFIR filter gain to have the same property.

The weight should be chosen such that the disturbance properties are completely transferred to the weighted $\mathcal{T}$. Formally, this can be done by considering the product $\mathcal{T}(z)\varpi_k$, where ϖ_k is the weighting vector. Using the property of the Frobenius norm $\|A^*A\|_F = \|AA^*\|_F$, the squared Frobenius norm of the weighted transfer function $\bar{\mathcal{T}}(z) = \mathcal{T}(z)\varpi_k$ can then be determined by averaging over both variables z and k as

$$\begin{aligned}
\|\bar{\mathcal{T}}(z)\|_F^2 &= \mathcal{E}_z\{\mathcal{E}_k\{\text{tr}[\mathcal{T}(z)\varpi_k\varpi_k^*\mathcal{T}^*(z)]\}\} \\
&= \frac{1}{2\pi}\int_0^{2\pi} \text{tr}\,[\mathcal{T}(e^{j\omega T})\mathcal{E}\{\varpi_k\varpi_k^*\}\mathcal{T}^*(e^{j\omega T})]\; d\omega T \\
&= \frac{1}{2\pi}\int_0^{2\pi} \text{tr}\,[\mathcal{T}(e^{j\omega T})\Xi\mathcal{T}^*(e^{j\omega T})]\; d\omega T \,,
\end{aligned} \tag{8.22}$$

where $\Xi = \mathcal{E}\{\varpi_k\varpi_k^*\}$ is the proper weighting matrix. By minimizing (8.22), we can obtain the H_2 filter gain as a function of Ξ. Since Ξ is required to have error, noise, and disturbance properties as variables, the weighting vector ϖ_k can be taken for each of the suberrors from the error equation (8.14) as we will show next.

For $\mathcal{T}_x(z) = I$, the weight $\Xi_x = \mathcal{E}\{\varpi_{xk}\varpi_{xk}^T\} = \mathcal{B}_N\mathcal{E}\{x_m x_m^T\}\mathcal{B}_N^T = \mathcal{B}_N \mathcal{X}_m \mathcal{B}_N^T$ corresponding to ε_x transforms (8.22) to

$$\begin{aligned}
\|\bar{\mathcal{T}}_x(z)\|_F^2 &= \frac{1}{2\pi}\int_0^{2\pi} \text{tr}(\mathcal{B}_N\mathcal{X}_m\mathcal{B}_N^T)\; d\omega T \,, \\
&= \text{tr}(\mathcal{B}_N\mathcal{X}_m\mathcal{B}_N^T) \,.
\end{aligned} \tag{8.23}$$

For ε_w, the norm (8.22) can be written as

$$\|\bar{\mathcal{T}}_w(z)\|_F^2 = \frac{1}{2\pi}\int_0^{2\pi} \text{tr}\left[\mathcal{T}_w(e^{j\omega T})\Xi_w\mathcal{T}_w^T(e^{j\omega T})\right]\; d\omega T \,, \tag{8.24}$$

where the weight Ξ_w is still to be found. Referring to [232], it can be shown that for $\mathcal{T}_w(z)$ given by (8.20) and $\Xi_w = I$ this norm becomes

$$\|\bar{\mathcal{T}}_w(z)\|_F^2 = \text{tr}(\mathcal{W}_N L \mathcal{W}_N^T) , \tag{8.25}$$

where the positive definite symmetric matrix L is a solution to the discrete Lyapunov equation

$$A_w L A_w^T + M = L . \tag{8.26}$$

Since M is allowed in (8.26) to be any positive definite matrix, we choose $M = B_w \mathcal{Q}_N B_w^T$ and write the solution to (8.26) as an infinite sum [83]

$$P = \sum_{i=0}^{\infty} A_w^i B_w \mathcal{Q}_N B_w^T A_w^{T^i} . \tag{8.27}$$

For A_w and B_w given by (8.19), the solution (8.27) gives $P = \mathcal{Q}_N$, and the norm (8.25) finally becomes

$$\|\bar{\mathcal{T}}_w(z)\|_F^2 = \text{tr}(\mathcal{W}_N \mathcal{Q}_N \mathcal{W}_N^T) . \tag{8.28}$$

Arguing similarly, we write the corresponding norm for ε_v as

$$\|\bar{\mathcal{T}}_v(z)\|_F^2 = \frac{1}{2\pi} \int_0^{2\pi} \text{tr} \left[\mathcal{T}_v(e^{j\omega T}) \Xi_v \mathcal{T}_v^T(e^{j\omega T})\right] \, d\omega T \tag{8.29}$$

and obtain

$$\|\bar{\mathcal{T}}_v(z)\|_F^2 = \text{tr}(\mathcal{V}_N \mathcal{R}_N \mathcal{V}_N^T) . \tag{8.30}$$

It is now important to note that the previously defined squared norms are equal to the squared ℓ_2 norms of the corresponding suberrors. Therefore, the trace of the error matrix $P = \mathcal{E}\{\varepsilon_k \varepsilon_k^T\}$ for mutually uncorrelated sources of errors can be represented as

$$\begin{aligned}
\text{tr } P &= \mathcal{E}\{(\varepsilon_{xk} + \varepsilon_{wk} + \varepsilon_{vk})^T (\varepsilon_{xk} + \varepsilon_{wk} + \varepsilon_{vk})\} \\
&= \mathcal{E}\{\varepsilon_{xk}^T \varepsilon_{xk}\} + \mathcal{E}\{\varepsilon_{wk}^T \varepsilon_{wk}\} + \mathcal{E}\{\varepsilon_{vk}^T \varepsilon_{vk}\} \\
&= \|\bar{\mathcal{T}}_x(z)\|_F^2 + \|\bar{\mathcal{T}}_w(z)\|_F^2 + \|\bar{\mathcal{T}}_v(z)\|_F^2
\end{aligned} \tag{8.31}$$

and transformed to

$$\text{tr}P = \text{tr}(\mathcal{B}_N \mathcal{X}_m \mathcal{B}_N^T + \mathcal{W}_N \mathcal{Q}_N \mathcal{W}_N^T + \mathcal{V}_N \mathcal{R}_N \mathcal{V}_N^T) . \tag{8.32}$$

All that follows from (8.31) is that (8.32) can be minimized by minimizing the sum of the squared Frobenius norms of the corresponding weighted transfer functions.

Since the error residual matrices $\mathcal{B}_N$, $\mathcal{W}_N$, and $\mathcal{V}_N$ defined by (8.15)–(8.17) are functions of gain $\mathcal{H}_N$ and the H_2 problem is convex, the gain for the H_2-OFIR filter can now be obtained by solving the minimization problem

$$\mathcal{H}_N = \underset{\mathcal{H}_N}{\arg\min} \; \text{tr}(\mathcal{B}_N \mathcal{X}_m \mathcal{B}_N^T + \mathcal{W}_N \mathcal{Q}_N \mathcal{W}_N^T + \mathcal{V}_N \mathcal{R}_N \mathcal{V}_N^T) , \tag{8.33}$$

which can be equivalently formulated as

$$\frac{\partial}{\partial \mathcal{H}_N} \text{tr}(\mathcal{B}_N \mathcal{X}_m \mathcal{B}_N^T + \mathcal{W}_N \mathcal{Q}_N \mathcal{W}_N^T + \mathcal{V}_N \mathcal{R}_N \mathcal{V}_N^T) = 0$$

and transformed to

$$\mathcal{H}_N (H_N \mathcal{X}_m H_N^T + T_N \mathcal{Q}_N T_N^T + \mathcal{R}_N) = \mathcal{F}^{N-1} \mathcal{X}_m H_N^T + \bar{D}_N \mathcal{Q}_N T_N^T .$$

Now note that H_2 filtering can be applied only if the LTI system is stable and, therefore, all values are bounded [106, 232]. The stability is guaranteed by $D = 0$ in (8.7,) and thus $T_N = G_N$. Accordingly, the gain of the H_2-OFIR filter becomes

$$\mathcal{H}_N = (F^{N-1}\chi_m H_N^T + \bar{D}_N \mathcal{Q}_N G_N^T) \times (H_N \chi_m H_N^T + G_N \mathcal{Q}_N G_N^T + \mathcal{R}_N)^{-1} \tag{8.34}$$

and we note that (8.34) has the same structure as (4.28b) of the homogeneous gain of the *a posteriori* OFIR filter modified for LTI systems. The essential difference is that the block error matrices $\mathcal{Q}_N$ and $\mathcal{R}_N$ in the H_2-OFIR filter can have all nonzero components, whereas in the OFIR filter obtained for white Gaussian noise, they are diagonal.

> ***H_2-OFIR filter****: The a posteriori H_2-OFIR filter generalizes the a posteriori OFIR filter in the special case of LTI processes with white Gaussian and mutually uncorrelated disturbances.*

The *a posteriori* H_2-OFIR filtering estimate can thus be defined as

$$\hat{x}_k = \mathcal{H}_N Y_{m,k} + (\bar{S}_N - \mathcal{H}_N L_N) U_{m,k} \, . \tag{8.35}$$

The corresponding error covariance is given for mutually uncorrelated χ_m, $\mathcal{Q}_N$, and $\mathcal{R}_N$ by

$$P_k = \mathcal{B}_N \chi_m \mathcal{B}_N^T + \mathcal{W}_N \mathcal{Q}_N \mathcal{W}_N^T + \mathcal{V}_N \mathcal{R}_N \mathcal{V}_N^T \, , \tag{8.36}$$

where the error residual matrices $\mathcal{B}_N$, $\mathcal{W}_N$, and $\mathcal{V}_N$ are specified by (8.15)–(8.17). It is worth noting that, in view of the supposedly unknown nature of non-Gaussian matrices $\mathcal{Q}_N$ and $\mathcal{R}_N$, iterative computation of (8.35) is generally unavailable.

8.2.2 Optimal Unbiased H_2 FIR Filter

The bias-constrained *a posteriori* H_2-OUFIR filter appears by minimizing (8.32) subject to constraint (8.12), which removes the requirement of initial state. The Lagrangian cost function for this problem is

$$J = \mathrm{tr}(\mathcal{W}_N \mathcal{Q}_N \mathcal{W}_N^T + \mathcal{V}_N \mathcal{R}_N \mathcal{V}_N^T) + \mathrm{tr}\Lambda(I - \mathcal{H}_N C_N) \tag{8.37}$$

and the minimization problem can be stated as

$$\mathcal{H}_N = \arg\min_{\mathcal{H}_N, \Lambda} J(\mathcal{H}_N, \Lambda) \, . \tag{8.38}$$

The solution to (8.38) is given by solving two equations

$$\frac{\partial}{\partial \mathcal{H}_N} J = -2\bar{D}_N \mathcal{Q}_N G_N^T + 2\mathcal{H}_N \Omega_N - \Lambda^T C_N^T = 0 \, , \tag{8.39}$$

$$\frac{\partial}{\partial \Lambda} J = I - C_N^T \mathcal{H}_N^T = 0 \, , \tag{8.40}$$

where $\Omega_N = G_N \mathcal{Q}_N G_N^T + \mathcal{R}_N$, and we notice that (8.40) is equivalent to constraint (8.13).

From (8.39) we obtain

$$C_N \Lambda = 2(\Omega_N \mathcal{H}_N^T - G_N \mathcal{Q}_N G_N^T) \, . \tag{8.41}$$

By multiplying (8.41) from the left-hand sides by $C_N^T \Omega_N^{-1}$ and using (8.40), we then find the Lagrange multiplier

$$\Lambda = 2(C_N^T \Omega_N^{-1} C_N)^{-1}(I - C_N^T \Omega_N^{-1} G_N \mathcal{Q}_N \bar{D}_N^T) \, ,$$

which should be used when extracting the gain $\mathcal{H}_N$ from (8.41). This finally gives $\mathcal{H}_N$ in the form

$$\mathcal{H}_N = (C_N^T \Omega_N^{-1} C_N)^{-1} C_N^T \Omega_N^{-1} + \bar{D}_N \mathcal{Q}_N \bar{D}_N^T \Omega_N^{-1} \times [I - C_N (C_N^T \Omega_N^{-1} C_N)^{-1} C_N^T \Omega_N^{-1}] . \quad (8.42)$$

Again we note the easily seen similarity of the gain $\mathcal{H}_N$ (8.42) of the *a posteriori* H_2-OUFIR filter to the gain (4.69a) of the *a posteriori* OUFIR-II filter, which is also the gain (4.113a) of the ML-I FIR filter. The difference is that (8.42) is valid for LTI systems with arbitrary block error matrices $\mathcal{Q}_N$ and $\mathcal{R}_N$, which may have all nonzero components, whereas OUFIR filtering is applied to white Gaussian processes with diagonal $\mathcal{Q}_N$ and $\mathcal{R}_N$.

> ***H_2-OUFIR filter**: The a posteriori H_2-OUFIR filter generalizes the a posteriori OUFIR filter in the special case of LTI processes with white Gaussian and mutually uncorrelated disturbances.*

The H_2-OUFIR filtering estimate has the same form as (8.35),

$$\hat{x}_k = \mathcal{H}_N Y_{m,k} + (\bar{S}_N - \mathcal{H}_N L_N) U_{m,k} , \quad (8.43)$$

where the gain matrix $\mathcal{H}_N$ is given by (8.42). The error covariance for the H_2-OUFIR filter is defined as

$$P_k = \mathcal{W}_N \mathcal{Q}_N \mathcal{W}_N^T + \mathcal{V}_N \mathcal{R}_N \mathcal{V}_N^T , \quad (8.44)$$

where the error residual matrices $\mathcal{W}_N$ and $\mathcal{V}_N$ are determined by (8.16) and (8.17) using the gain (8.42). It should also be noted that the H_2-OUFIR filter belongs to the class of ML estimators and, as in H_2-OFIR filtering, iterative computation of (8.43) is not available for full block matrices $\mathcal{Q}_N$ and $\mathcal{R}_N$.

Example 8.1 ***Moving vehicle tracking in nonwhite noise.*** Consider a vehicle moving in the suburban area. The vehicle coordinates, measured every second, $\tau = 1$ s, using a GPS navigator, are considered ground truth. All along, the vehicle coordinates are also measured by a radar adding uncertain colored noise v_k. The vehicle tracking problem is represented with two-state equations,

$$x_k = F x_{k-1} + B w_k ,$$
$$y_k = H x_k + v_k ,$$

where the matrices are given by

$$F = \begin{bmatrix} 1 & \tau \\ 0 & 1 \end{bmatrix}, \quad B = \begin{bmatrix} \tau \\ 1 \end{bmatrix}, \quad H = \begin{bmatrix} 1 & 0 \end{bmatrix},$$

and $w_k \sim \mathcal{N}(0, \sigma_w^2)$ is the velocity noise. The colored noise v_k is considered Gauss-Markov $v_k = \psi v_{k-1} + \xi_k$, where $\xi_k \sim \mathcal{N}(0, \sigma_\xi^2)$ and the coloredness factor $0 < \psi < 1$ is chosen for stability. For the average velocity of 36 km/hour or 10 m/s and error of 20%, the standard deviation of the velocity noise is set as $\sigma_w = 2$ m/s. The GPS positioning service guarantees errors less than 15 m with a probability of 95% in the 2σ sense. Therefore, the standard deviation of noise ξ_k is set as $\sigma_\xi = 3.75$ m. The FIR filters is tuned to $N = 5$, and the error matrix $\mathcal{R}_N(\psi)$ is measured for $\psi = 0.05$ and $\psi = 0.95$ as, respectively,

$$\mathcal{R}_N(0.05) = \begin{bmatrix} \mathbf{18.23} & 0.526 & 0.118 & 0.297 & -0.072 \\ 0.526 & \mathbf{18.24} & 0.529 & 0.122 & 0.302 \\ 0.118 & 0.529 & \mathbf{18.24} & 0.532 & 0.129 \\ 0.297 & 0.122 & 0.532 & \mathbf{18.24} & 0.54 \\ -0.072 & 0.302 & 0.129 & 0.54 & \mathbf{18.25} \end{bmatrix},$$

$$\mathcal{R}_N(0.95) = \begin{bmatrix} \mathbf{178.15} & 168.81 & 160.32 & 152.19 & 144.17 \\ 168.81 & \mathbf{178.18} & 168.85 & 160.38 & 152.26 \\ 160.32 & 168.85 & \mathbf{178.23} & 168.91 & 160.46 \\ 152.19 & 160.38 & 168.91 & \mathbf{178.30} & 169.01 \\ 144.17 & 152.26 & 160.46 & 169.01 & \mathbf{178.44} \end{bmatrix}.$$

As can be seen, $\mathcal{R}_N$ is symmetric with nonzero nondiagonal components, even when noise is near white Gaussian. For KF, the measurement noise covariance in both cases is taken as $R = (\mathcal{R}_N)_{1,1}$. Since the velocity standard deviation $\sigma_w = 2$ m/s may not apply to specific vehicles on all roads and in all areas, the question arises of robustness to errors.

In Fig. 8.2 we show RMSEs produced by the KF and H_2-OFIR, H_2-OUFIR, and UFIR filters under CMN as functions of σ_w for two limiting cases of $\psi = 0.05$ and $\psi = 0.95$. A robust UFIR filter ignores any information about zero mean noise and is therefore used as a benchmark. Being insensitive to σ_w, this filter generates a constant and low RMSE value across the entire scale. The nonrobust KF demonstrates the worst performance. The H_2-OFIR filter, which we run at each point m on $[m, k]$ with KF, behaves better than KF. Nevertheless, this filter gives large errors when σ_w is set less than the actual $\sigma_{w(\text{actual})}$. For $\psi = 0.05$ (near white noise) and $\sigma_w < \sigma_{w(\text{actual})}$, the H_2-OUFIR filter is almost as accurate as the robust UFIR filter. It gives intermediate errors when $\sigma_w = \sigma_{w(\text{actual})}$ and converges to KF when $\sigma_w > \sigma_{w(\text{actual})}$. For strictly CMN with $\psi = 0.95$, the H_2-OUFIR filter performs better than others on the whole scale, except for the H_2-OFIR filter, which is slightly more accurate when $\sigma_w = \sigma_{w(\text{actual})}$. □

Example 8.1 clearly demonstrates the advantage in robustness of the batch H_2-OFIR and H_2-OUFIR filters operating with full block measurement error matrices $\mathcal{R}_N$.

Example 8.2 ***Robust tracking under disturbances.*** Consider the vehicle tracking problem described in example 8.1, where the white Gaussian measurement noise $v_k \sim \mathcal{N}(0, \sigma_v^2)$ has the standard deviation $\sigma_v = 10$ m. The vehicle trajectory is affected by the Gauss-Markov process $w_k = \phi w_{k-1} + \zeta_k$, where the scalar disturbance factor ϕ is chosen as $0 < \phi < 1$, and the driving noise $\zeta_k \sim \mathcal{N}(0, \sigma_w^2)$ has a standard deviation $\sigma_w = 0.3$ m/s.

To compare the estimation errors produced by the batch H_2-OFIR, recursive Kalman, and iterative UFIR filter, we sequentially generate a disturbance process by changing ϕ from zero to 0.95 with a step 0.05. Next, we consider the following scenarios of filter optimal tuning for 1) ϕ and $N_{\text{opt}}(\phi)$, 2) $\phi = 0$ and $N_{\text{opt}} = 26$, and 3) $\phi = 0.95$ and $N_{\text{opt}} = 10$. Typical tracking RMSEs are shown in Fig. 8.3 as functions of ϕ, and we note the following features:

- *Case 1 (theoretical): Tuning for ϕ and $N_{\text{opt}}(\phi)$.* When the filters are optimally tuned for ϕ, their RMSEs reach the lowest possible values, as shown in Fig. 8.3a. As ϕ increases, errors in all filters also increase. The H_2-OFIR and UFIR filters produce consistent errors that grow at a low rate, while the errors in KF grow with a high rate. Since setting the filter for each current ϕ is hardly possible in practice, this case can be considered *theoretical*.
- *Case 2 (regular): Tuning for $\phi = 0$ and $N_{\text{opt}} = 26$.* When the disturbance is not specified, all filters are usually tuned to white noise. Obviously, this tuning gives the smallest errors when $\phi = 0$. However, increasing ϕ causes all errors to increase at a high rate, as shown in Fig. 8.3b.
- *Case 3 (robust): Tuning for $\phi = 0.95$ and $N_{\text{opt}} = 10$.* When the disturbance boundary is known, all filter can be tuned for $\phi = 0.95$. This drastically lowers the RMSEs in all filters compared to tuning for $\phi = 0$ (Fig. 8.3b). Herewith, all filters tuned for $\phi = 0.95$ give practically the same

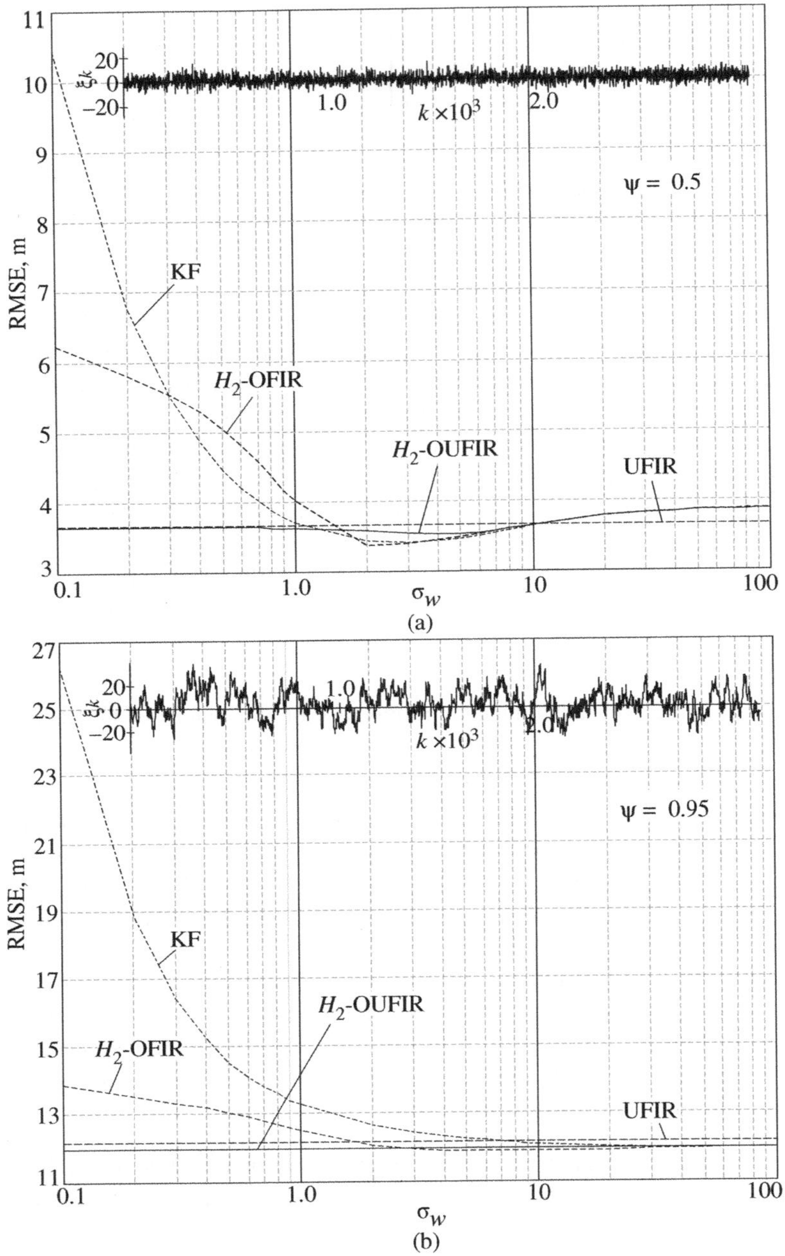

Figure 8.2 RMSEs produced in the east direction by Kalman, H_2-OFIR, H_2-OUFIR, and UFIR filters under CMN as functions of the velocity standard deviation σ_w in the algorithms: (a) $\psi = 0.05$ and $\psi = 0.95$.

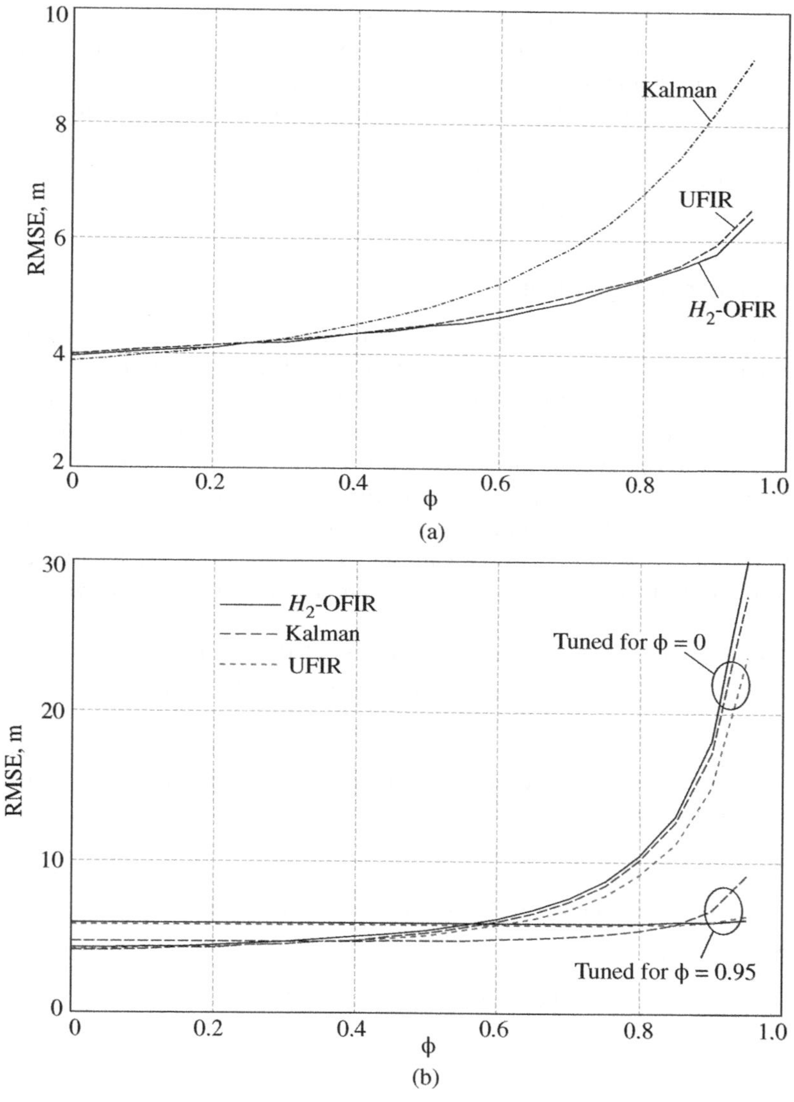

Figure 8.3 Typical RMSEs generated by the H_2-OFIR, Kalman, and UFIR filters as functions of the disturbance factor ϕ after optimally tuned to: (a) ϕ and $N_{opt}(\phi)$ and (b) $\phi = 0$, $N_{opt} = 26$ and $\phi = 0.95$, $N_{opt} = 10$. Robust mode is when all filters are tuned for maximum disturbance with $\phi = 0.95$.

errors for all values of ϕ, which means robustness to ϕ. More precisely, here the batch H_2-OFIR filter is the most robust, the iterative UFIR filter is slightly less robust, and the recursive KF is the worst.

Comparing the RMSEs shown in Fig. 8.3a and Fig. 8.3b (tuning for $\phi = 0.95$), we conclude that tuning for each ϕ gives the smallest errors, but can hardly be implemented practically. On the contrary, tuning for $\phi = 0.95$ gives slightly more errors, but this case is feasible and robust. □

Example 8.2 clearly illustrates the idea behind filter robustness. If an estimator is tuned to white noise ($\phi = 0$), then any deviation from this point due to $\phi > 0$ causes 1) an increase in errors and 2) an increase in disturbance. Since both these effects are statistically summarized, the RMSE can grow significantly, as in Fig. 8.3b. In contrast, if an estimator is tuned to the maximized disturbance ($\phi = 0.95$), then any deviation from this point caused by $\phi < 0.95$ leads to 1) an increase in errors and 2) a decrease in disturbance. Since both these effects cancel each other out, the RMSE can remain almost unchanged, as in Fig. 8.3b, which means robustness.

8.2.3 Suboptimal H_2 FIR Filtering Algorithms

Another way to solve the problem of state estimation under disturbances is to use LMI [200]. The approach has emerged in attempts to solve mathematically intractable optimization problems using a convex constraint called LMI. Over time, it was shown that a very wide range of problems arising in systems, signal processing, and control theory can be reduced to a few standard convex or quasiconvex optimization problems, involving LMI [22]. The general idea behind the approach is to reformulate and solve an optimization problem with a convex objective function and LMI constraints using efficient numerical algorithms. While the LMI-based solution does not look elegant, it does exist. With regard to FIR filtering, this means that the filter gain can be determined numerically in matrix form [106]. If we use such a gain, the state estimate will appear without actually tracing how.

Returning to H_2 FIR state estimation, we note that even though analytical solutions are available here, LMI-based forms are necessary for the development of suboptimal hybrid structures.

Suboptimal H_2 FIR Filter

Let us consider the error matrix P_k represented by the sum of three quadratic forms as (8.32). Let us also introduce an auxiliary matrix $\mathcal{Z}$ such that

$$\mathcal{Z} > \mathcal{B}_N \chi_m \mathcal{B}_N^T + \mathcal{W}_N \mathcal{Q}_N \mathcal{W}_N^T + \mathcal{V}_N \mathcal{R}_N \mathcal{V}_N^T , \tag{8.45}$$

where the error residual matrices are defined by (8.15)–(8.17). We can rewrite (8.45) as

$$\begin{aligned}&\mathcal{Z} - (\mathcal{H}_N H_N - F^{N-1})\chi_m(\mathcal{H}_N H_N - F^{N-1})^T \\ &-(\mathcal{H}_N G_N - \bar{D}_N)\mathcal{Q}_N(\mathcal{H}_N G_N - \bar{D}_N)^T - \mathcal{H}_N \mathcal{R}_N \mathcal{H}_N^T > 0\end{aligned}$$

and transform to

$$\mathcal{Z} - \mathcal{A} + \mathcal{C}^T \mathcal{H}_N^T + \mathcal{H}_N \mathcal{C} - \mathcal{H}_N (H_N \chi_m H_N^T + \Omega_N)\mathcal{H}_N^T > 0 , \tag{8.46}$$

where the following matrices are introduced: $\mathcal{A} = F^{N-1}\chi_m F^{N-1^T} + \bar{D}_N \mathcal{Q}_N \bar{D}_N^T$, $\mathcal{C}^T = F^{N-1}\chi_m H_N^T + \bar{D}_N \mathcal{Q}_N G_N^T$, and $\Omega_N = G_N \mathcal{Q}_N G_N^T + \mathcal{R}_N$.

As we can seen, the inequality (8.46) is nonlinear with respect to the gain $\mathcal{H}_N$. But if we use the Schur complement, we can equivalently replace (8.46) with the LMI as

$$\begin{bmatrix} \mathcal{Z} - \mathcal{A} + \mathcal{C}^T \mathcal{H}_N^T + \mathcal{H}_N \mathcal{C} & \mathcal{H}_N \\ \mathcal{H}_N^T & (H_N \chi_m H_N^T + \Omega_N)^{-1} \end{bmatrix} > 0 . \tag{8.47}$$

We can then numerically determine the suboptimal gain for the H_2 FIR filter by solving the following minimization problem

$$\mathcal{H}_N = \min_{\mathcal{H}_N, \mathcal{Z}} \operatorname{tr} \mathcal{Z} \tag{8.48}$$

subject to (8.47).

The best candidate for initializing the minimization procedure is of course the UFIR filter gain $\hat{\mathcal{H}}_N = (C_N^T C_N)^{-1} C_N^T$. Since the tuned UFIR filter is less accurate than the tuned optimal filter, it is assumed that its gain is always greater then $\mathcal{H}_N$. Provided that $\mathcal{H}_N$ is obtained with (8.48), the suboptimal H_2 filtering estimate can be computed using (8.35) with the error matrix (8.36).

Bias-Constrained Suboptimal H_2 FIR Filter

Without any innovation, we can also look at (8.44) and obtain a suboptimal numerical algorithm for the bias-constrained H_2 FIR filter. Accordingly, we introduce an auxiliary matrix $\mathcal{Z}$, start with

$$\mathcal{Z} > \mathcal{W}_N \mathcal{Q}_N \mathcal{W}_N^T + \mathcal{V}_N \mathcal{R}_N \mathcal{V}_N^T , \tag{8.49}$$

and represent (8.49) as

$$\mathcal{Z} - \bar{D}_N \mathcal{Q}_N \bar{D}_N^T + \mathcal{K}^T \mathcal{H}_N^T + \mathcal{H}_N \mathcal{K} - \mathcal{H}_N \Omega_N \mathcal{H}_N^T > 0 , \tag{8.50}$$

$$\begin{bmatrix} \mathcal{Z} - \bar{D}_N \mathcal{Q}_N \bar{D}_N^T + \mathcal{K}^T \mathcal{H}_N^T + \mathcal{H}_N \mathcal{K} & \mathcal{H}_N \\ \mathcal{H}_N^T & \Omega_N^{-1} \end{bmatrix} > 0 , \tag{8.51}$$

where $\mathcal{K}^T = \bar{D}_N \mathcal{Q}_N G_N^T$.

The gain for the bias-constrained suboptimal H_2 FIR filter can then be numerically computed by solving the minimization problem

$$\mathcal{H}_N = \min_{H_N, Z} \operatorname{tr} \mathcal{Z} \tag{8.52}$$
$$\text{subject to (8.51) and } \mathcal{H}_N H_N = F^{N-1} .$$

Traditionally, the minimization procedure (8.52) can be initialized using the UFIR filter gain $\hat{\mathcal{H}}_N = (C_N^T C_N)^{-1} C_N^T$. Provided that $\mathcal{H}_N$ is numerically available from (8.52), the estimate can be obtained by (8.43) and the error covariance by (8.44).

Minimum Variance FIR H_2 Filter

The RH MVF H_2 predictive filter, previously obtained in [109] by minimizing the squared Frobenius norm of the unweighed disturbance-to-error transfer function $\mathcal{T}$, is valid for $\mathcal{R}_N = 0$ and $\mathcal{Q}_N = I$. For this filter, the inequalities (8.50) and (8.51) become

$$\mathcal{Z} - \mathcal{H}_N^p (G_N^p + \bar{T}_N)(G_N^p + \bar{T}_N)^T \mathcal{H}_N^{p^T} > 0 , \tag{8.53}$$

$$\begin{bmatrix} \mathcal{Z} & \mathcal{H}_N^p (G_N^p + \bar{T}_N) \\ (G_N^p + \bar{T}_N)^T \mathcal{H}_N^{p^T} & I \end{bmatrix} > 0 , \tag{8.54}$$

and the filter gain can be determined by solving the minimization problem

$$\mathcal{H}_N^p = \min_{\mathcal{H}_N^p, Z} \operatorname{tr} \mathcal{Z} \tag{8.55}$$
$$\text{subject to (8.54) and } \mathcal{H}_N^p C_N^p = I .$$

The UFIR filter gain $\hat{\mathcal{H}}_N^p$ was also suggested in [109] to initialize the minimization procedure, and we notice that the RH MVF H_2 predictive filtering estimate is computed by $\hat{x}_k = \mathcal{H}_N^p Y_{m-1,k-1}$.

Summing up, we note that the *a posteriori* H_2 FIR filter can exist in different forms depending on the cost function and methods of solving the minimization problem. If no constraints are imposed on the minimization problem, then it follows that the H_2-OFIR filter has the same structure as the OFIR filter. With the built-in unbiasedness constraint, this filter becomes the H_2-OUFIR filter, which belongs to the class of ML estimators [70]. It is worth noting that the H_2-OFIR state estimation problem is mathematically tractable and, therefore, has closed-form solutions. However, numerical computation of the gain for this filter using LMI is required to design hybrid suboptimal state estimators, such as the H_2/H_∞ FIR filter.

8.3 H_2 FIR Prediction

Predictive estimates help in solving some signal processing problems and are needed for state feedback control. Based on the FE-based model (8.8) and (8.9), the FIR prediction can be defined as

$$\begin{aligned}\tilde{x}_{k+1} &= \mathcal{H}_N^p Y_{m,k} + \mathcal{H}_N^{pf} U_{m,k} \,, \\ &= \mathcal{H}_N H_N^p x_m + (\mathcal{H}_N L_N^p + \mathcal{H}_N^{f}) U_{m,k} + \mathcal{H}_N T_N^p W_{m,k} + \mathcal{H}_N V_{m,k} \,.\end{aligned} \tag{8.56}$$

Then the unbiasedness condition $\mathcal{E}\{x_{k+1}\} = \mathcal{E}\{\tilde{x}_{k+1}\}$ applied to (8.10) and (8.56) yields two unbiasedness constraints

$$I = \mathcal{H}_N^p C_N^p \,, \tag{8.57}$$

$$\mathcal{H}_N^{pf} = \bar{S}_N - \mathcal{H}_N^p L_N^p \,, \tag{8.58}$$

where $C_N^p = H_N F^{-N}$, and we obtain the error $\varepsilon_{k+1} = x_{k+1} - \tilde{x}_{k+1}$ as

$$\varepsilon_{k+1} = \mathcal{B}_N^p x_m + (\bar{S}_N - \mathcal{H}_N^p L_N^p - \mathcal{H}_N^{pf}) U_{m,k} + \mathcal{W}_N^p W_{m,k} - \mathcal{V}_N^p V_{m,k} \,, \tag{8.59}$$

where the residual error matrices are defined by

$$\mathcal{B}_N^p = F^N - \mathcal{H}_N^p H_N^p \,, \tag{8.60}$$

$$\mathcal{W}_N^p = \bar{D}_N - \mathcal{H}_N^p T_N^p \,, \tag{8.61}$$

$$\mathcal{V}_N^p = \mathcal{H}_N^p \,. \tag{8.62}$$

Substituting (8.58) into (8.59), we represent (8.59) as the sum of the following remaining suberrors,

$$\begin{aligned}&\varepsilon_{x(k+1)} = x_m - \hat{x}_m = \mathcal{B}_N^p x_m \,, \quad \varepsilon_{w(k+1)} = \mathcal{W}_N^p W_{m,k} \,, \\ &\varepsilon_{v(k+1)} = -\mathcal{V}_N^p V_{m,k} \,.\end{aligned} \tag{8.63}$$

Then we note that, as in the case of the H_2 FIR filter, the initial error $\varepsilon_{x(k+1)}$ translates directly into the estimation error ε_{k+1}. Therefore, the $(x_m - \hat{x}_m)$-to-ε_{k+1} transfer function is the identity matrix, $\mathcal{T}_x(z) = I$.

To find the w_k-to-ε_{k+1} transfer function $\mathcal{T}_w(z)$ and the v_k-to-ε_{k+1} transfer function $\mathcal{T}_v(z)$, we first represent vectors $W_{m,k}$ and $V_{m,k}$ using the rule (8.18) and then apply the z transform. This allows us to represent the required transfer functions $\mathcal{T}_w(z)$ and $\mathcal{T}_v(z)$ in the forms [232]

$$\mathcal{T}_w(z) = \mathcal{W}_N (Iz - A_w)^{-1} B_w \,, \tag{8.64}$$

$$\mathcal{T}_v(z) = \mathcal{V}_N (Iz - A_w)^{-1} B_w \,. \tag{8.65}$$

We now see that functions (8.64) and (8.65) differ from functions (8.20) and (8.21) found for the BE-based model. The difference is that there is no z operator in front of matrix B_w. However, this feature does not affect the Frobenius norms, despite that FE-based solutions are less stable.

8.3.1 H_2-OFIR Predictor

We can now follow the norm definitions given by (8.22)–(8.24) and provide the norms for the H_2-OFIR predictor,

$$\|\bar{\mathcal{T}}_x(z)\|_F^2 = \frac{1}{2\pi}\int_0^{2\pi} \text{tr}\,[\mathcal{B}_N^p \mathcal{X}_m \mathcal{B}_N^{p^T}]\,\mathrm{d}\,\omega T\,, \tag{8.66}$$

$$\|\bar{\mathcal{T}}_{\left\{\substack{w\\v}\right\}}(z)\|_F^2 = \frac{1}{2\pi}\int_0^{2\pi} \text{tr}\left[\mathcal{T}_{\left\{\substack{w\\v}\right\}}(e^{j\omega T})\begin{Bmatrix}\mathcal{Q}_N\\\mathcal{R}_N\end{Bmatrix}\mathcal{T}^*_{\left\{\substack{w\\v}\right\}}(e^{j\omega T})\right]\mathrm{d}\,\omega T\,. \tag{8.67}$$

Arguing similarly to (8.63), we write the trace of the error matrix in the form

$$\begin{aligned}
\text{tr}\,P &= \mathcal{E}\{(\varepsilon_{x(k+1)} + \varepsilon_{w(k+1)} + \varepsilon_{v(k+1)})^T(\dots)\}\\
&= \mathcal{E}\{\varepsilon^T_{x(k+1)}\varepsilon_{x(k+1)}\} + \mathcal{E}\{\varepsilon^T_{w(k+1)}\varepsilon_{w(k+1)}\} + \mathcal{E}\{\varepsilon^T_{v(k+1)}\varepsilon_{v(k+1)}\}\\
&= \|\bar{\mathcal{T}}_x(z)\|_F^2 + \|\bar{\mathcal{T}}_w(z)\|_F^2 + \|\bar{\mathcal{T}}_v(z)\|_F^2\\
&= \text{tr}(\mathcal{B}_N^p\mathcal{X}_m\mathcal{B}_N^{p^T} + \mathcal{W}_N^p\mathcal{Q}_N\mathcal{W}_N^{p^T} + \mathcal{V}_N^p\mathcal{R}_N\mathcal{V}_N^{p^T})\,.
\end{aligned} \tag{8.68}$$

The gain for the H_2-OFIR predictor can now be computed numerically by solving the minimization problem

$$\mathcal{H}_N^p = \underset{\mathcal{H}_N^p}{\arg\min}\ \text{tr}(\mathcal{B}_N^p\mathcal{X}_m\mathcal{B}_N^{p^T} + \mathcal{W}_N^p\mathcal{Q}_N\mathcal{W}_N^{p^T} + \mathcal{V}_N^p\mathcal{R}_N\mathcal{V}_N^{p^T})\,, \tag{8.69}$$

which the equivalent form

$$\frac{\partial}{\partial\mathcal{H}_N^p}\text{tr}(\mathcal{B}_N^p\mathcal{X}_m\mathcal{B}_N^{p^T} + \mathcal{W}_N^p\mathcal{Q}_N\mathcal{W}_N^{p^T} + \mathcal{V}_N^p\mathcal{R}_N\mathcal{V}_N^{p^T}) = 0$$

gives

$$\mathcal{H}_N^p = (F^N\mathcal{X}_m H_N^{p^T} + \bar{D}_N\mathcal{Q}_N G_N^{p^T})(H_N^p\mathcal{X}_m H_N^{p^T} + \Omega_N^p)^{-1}\,, \tag{8.70}$$

where $\Omega_N^p = G_N^p\mathcal{Q}_N G_N^{p^T} + \mathcal{R}_N$.

Analyzing (8.70), we come to the same conclusions as for the H_2-OFIR filter. Indeed, (8.70) has the same structure as the homogeneous gain (7.47b) of the OFIR predictor modified for LTI systems. The difference is in the block matrices $\mathcal{Q}_N$ and $\mathcal{R}_N$, which in the H_2-OFIR predictor can have all nonzero components, whereas in the OFIR predictor they are diagonal.

> ***H_2-OFIR predictor**: The H_2-OFIR predictor generalizes the OFIR predictor in the special case of LTI systems with white Gaussian and mutually uncorrelated disturbances.*

We finally write the H_2-OFIR prediction as

$$\tilde{x}_{k+1} = \mathcal{H}_N^p Y_{m,k} + (\bar{S}_N - \mathcal{H}_N^p L_N)U_{m,k} \tag{8.71}$$

and the error covariance as

$$P_{k+1} = \mathcal{B}_N^p\mathcal{X}_m\mathcal{B}_N^{p^T} + \mathcal{W}_N^p\mathcal{Q}_N\mathcal{W}_N^{p^T} + \mathcal{V}_N^p\mathcal{R}_N\mathcal{V}_N^{p^T}\,, \tag{8.72}$$

where the error residual matrices $\mathcal{B}_N^p$, $\mathcal{W}_N^p$, and $\mathcal{V}_N^p$ are given by (8.60)–(8.62).

8.3.2 Bias-Constrained H_2-OUFIR Predictor

Now it is just a matter of using similar transformations to obtain the bias-constrained H_2-OUFIR predictor and suboptimal numerical algorithms. To do this, we consider the trace (8.68), build in the constraint (8.58) that removes the term with χ_m, and then minimize the trace by subjecting it to (8.57). The Lagrangian cost associated with this problem is

$$J = \text{tr}(\mathcal{W}_N^p \mathcal{Q}_N \mathcal{W}_N^{p^T} + \mathcal{V}_N^p \mathcal{R}_N \mathcal{V}_N^{p^T}) + \text{tr}\Lambda(I - \mathcal{H}_N^p C_N^p) , \tag{8.73}$$

and we formulate the minimization as

$$\mathcal{H}_N^p = \arg\min_{\mathcal{H}_N^p, \Lambda} J(\mathcal{H}_N^p, \Lambda) . \tag{8.74}$$

By repeating steps (8.39)–(8.41), we obtain the gain $\mathcal{H}_N^p$ for the H_2-OUFIR predictor in the form

$$\begin{aligned}\mathcal{H}_N^p &= (C_N^{p^T} \Omega_N^{p^{-1}} C_N^p)^{-1} C_N^{p^T} \Omega_N^{p^{-1}} + \bar{D}_N \mathcal{Q}_N G_N^{p^T} \Omega_N^{p^{-1}} \\ &\quad \times [I - C_N^p (C_N^{p^T} \Omega_N^{p^{-1}} C_N^p)^{-1} C_N^{p^T} \Omega_N^{p^{-1}}] ,\end{aligned} \tag{8.75}$$

write the prediction as

$$\tilde{x}_{k+1} = \mathcal{H}_N^p Y_{m,k} + (\bar{S}_N - \mathcal{H}_N^p L_N^p) U_{m,k} , \tag{8.76}$$

and specify the error covariance for uncorrelated χ_m, $\mathcal{Q}_N$, and $\mathcal{R}_N$ by

$$P_k = \mathcal{W}_N^p \mathcal{Q}_N \mathcal{W}_N^{p^T} + \mathcal{V}_N^p \mathcal{R}_N \mathcal{V}_N^{p^T} , \tag{8.77}$$

where the residual error matrices $\mathcal{W}_N^p$ and $\mathcal{V}_N^p$ are defined by (8.61) and (8.62). We finally notice that the H_2-OUFIR predictor belongs to the class of ML estimators.

8.3.3 Suboptimal H_2 FIR Predictive Algorithms

Numerical suboptimal algorithms can also be designed to compute the gains for H_2 FIR predictors using LMI. The derivation procedure is very similar to that developed for the H_2 FIR filters. Therefore, we will mainly present the final results without going into details.

Suboptimal H_2 FIR Predictor

Consider (8.72), introduce an auxiliary matrix $\mathcal{Z}$, write

$$\mathcal{Z} > B_N^p \chi_m B_N^{p^T} + \mathcal{W}_N^p \mathcal{Q}_N \mathcal{W}_N^{p^T} + \mathcal{V}_N^p \mathcal{R}_N \mathcal{V}_N^{p^T} ,$$

and transform this inequality to

$$\mathcal{Z} - \mathcal{A}^p + \mathcal{C}^{p^T} \mathcal{H}_N^{p^T} + \mathcal{H}_N^p \mathcal{C}^p - \mathcal{H}_N^p (H_N^p \chi_m H_N^{p^T} + \Omega_N^p) \mathcal{H}_N^{p^T} > 0 ,$$

where $\mathcal{A}^p = F^N \chi_m F^{N^T} + \bar{D}_N \mathcal{Q}_N \bar{D}_N^T$ and $\mathcal{C}^{p^T} = F^N \chi_m H_N^{p^T} + \bar{D}_N \mathcal{Q}_N G_N^T$.

Use Schur's complement and represent the previous inequality with LMI as

$$\begin{bmatrix} \mathcal{Z} - \mathcal{A}^p + \mathcal{C}^{p^T} \mathcal{H}_N^{p^T} + \mathcal{H}_N^p \mathcal{C}^p & \mathcal{H}_N^p \\ \mathcal{H}_N^{p^T} & (H_N^p \chi_m H_N^{p^T} + \Omega_N^p)^{-1} \end{bmatrix} > 0 . \tag{8.78}$$

Solve the minimization problem

$$\mathcal{H}_N^p = \min_{\mathcal{H}_N^p, \mathcal{Z}^p} \text{tr}\, \mathcal{Z}^p \quad \text{subject to (8.78)} \tag{8.79}$$

starting with the UFIR predictor gain $\hat{\mathcal{H}}_N^p = (C_N^{p^T} C_N^p)^{-1} C_N^{p^T}$, and predict the state as $\tilde{x}_{k+1} = \mathcal{H}_N^p Y_{m,k}$.

Bias-Constrained Suboptimal H_2 FIR Predictor

Consider (8.77), introduce a matrix $\mathcal{Z}$ such that

$$\mathcal{Z} > \mathcal{W}_N^p \mathcal{Q}_N \mathcal{W}_N^{p^T} + \mathcal{V}_N^p \mathcal{R}_N \mathcal{V}_N^{p^T} ,$$

and represent this inequality with

$$\mathcal{Z} - \bar{D}_N \mathcal{Q}_N \bar{D}_N^T + \mathcal{C}^{p^T} \mathcal{H}_N^{p^T} + \mathcal{H}_N^p \mathcal{C}^p - \mathcal{H}_N^p \Omega_N^p \mathcal{H}_N^{p^T} > 0 ,$$

$$\begin{bmatrix} \mathcal{Z} - \bar{D}_N \mathcal{Q}_N \bar{D}_N^{p^T} + \mathcal{C}^{p^T} \mathcal{H}_N^{p^T} + \mathcal{H}_N^p \mathcal{C}^p & \mathcal{H}_N^p \\ \mathcal{H}_N^{p^T} & \Omega_N^{p^{-1}} \end{bmatrix} > 0 , \tag{8.80}$$

where $\mathcal{C}^p = \bar{D}_N \mathcal{Q}_N G_N^{p^T}$. Solve the minimization problem

$$\mathcal{H}_N^p = \min_{\mathcal{H}_N^p, \mathcal{Z}} \operatorname{tr} \mathcal{Z} \tag{8.81}$$

$$\text{subject to (8.80) and } \mathcal{H}_N^p H_N^p = F^N$$

by initializing the minimization procedure with $\tilde{\mathcal{H}}_N^p = (C_N^{p^T} C_N^p)^{-1} C_N^{p^T}$ and compute the prediction as $\tilde{x}_{k+1} = \mathcal{H}_N^p Y_{m,k}$.

8.3.4 Receding Horizon H_2-MVF Filter

The bias-constrained RH H_2 FIR predictive filter was obtained in [109] for stationary processes, ignoring the state dynamics. Traditionally we will call this filter the H_2-MVF filter and note that it matches the FE-based state model on $[m-1, k-1]$.

To develop the H_2-MVF filter, we refer to (7.66) and write the extended observation equation as

$$Y_{k-1} = C_N x_k - L_N U_{k-1} - T_N W_{k-1} + V_{k-1} , \tag{8.82}$$

where the time-invariant augmented matrices are defined as $T_N = G_N + \bar{T}_N, \bar{T}_N = \operatorname{diag}(D\ D\ \dots\ D)$,

$$C_N = \begin{bmatrix} HF^{-N} \\ HF^{-(N-1)} \\ \vdots \\ HF^{-1} \end{bmatrix} ,$$

$$L_N = \begin{bmatrix} HF^{-1}E & HF^{-2}E & \dots & HF^{-N+1}E & HF^{-N}E \\ 0 & HF^{-1}E & \dots & HF^{-N+2}E & HF^{-N+1}E \\ \vdots & \vdots & \ddots & \vdots & \vdots \\ 0 & 0 & \dots & HF^{-1}E & HF^{-2}E \\ 0 & 0 & \dots & 0 & HF^{-1}E \end{bmatrix} ,$$

and the matrix G_N can be written as L_N if we replace E with B.

Since the H_2 approach was developed for stabile systems implying $D = 0$ and $T_N = G_N$, we write the H_2-MVF estimate $\tilde{x}_k$ as

$$\tilde{x}_k = \mathcal{H}_N Y_{k-1} + \mathcal{H}_N^{\mathrm{f}} U_{k-1} \tag{8.83}$$

$$= \mathcal{H}_N C_N x_k + (\mathcal{H}_N^{\mathrm{f}} - \mathcal{H}_N L_N) U_{k-1}$$

$$- \mathcal{H}_N G_N W_{k-1} + \mathcal{H}_N V_{k-1} . \tag{8.84}$$

Then the unbiasedness condition $\mathcal{E}\{x_k\} = \mathcal{E}\{\tilde{x}_k\}$ gives two constraints

$$I = \mathcal{H}_N C_N , \tag{8.85}$$

$$\mathcal{H}_N^{\mathrm{f}} = \mathcal{H}_N L_N . \tag{8.86}$$

Substituting (8.85) and (8.86) into (8.84), we obtain $\tilde{x}_k = x_k - \mathcal{H}_N(G_N W_{k-1} - V_{k-1})$ and transform the estimation error to

$$\varepsilon_k = x_k - \tilde{x}_k = \mathcal{W}_N W_{k-1} - \mathcal{V}_N V_{k-1} ,$$

where $\mathcal{W}_N - \mathcal{H}_N G_N$ and $\mathcal{V} = \mathcal{H}_N$ are the error residual matrices.

Now, the Lagrangian cost (8.37) and the solution to the minimization problem (8.38) give

$$\frac{\partial}{\partial \mathcal{H}_N} J = 2\mathcal{H}_N \Omega_N - \Lambda^T C_N^T = 0 , \tag{8.87}$$

$$\frac{\partial}{\partial \Lambda} J = I - C_N^T \mathcal{H}_N^T = 0 \tag{8.88}$$

and, from (8.87) and referring to (8.88), we find the Lagrange multiplier $\Lambda = -2(C_N^T \Omega_N^{-1} C_N)^{-1}$. Finally, we substitute Λ into (8.87), obtain the gain $\mathcal{H}_N$ in the form

$$\mathcal{H}_N = (C_N^T \Omega_N^{-1} C_N)^{-1} C_N^T \Omega_N^{-1} , \tag{8.89}$$

and notice that this gain is equivalent to that originally derived for this filter in [109]. The H_2-MVF estimate can thus be computed by (8.76) and the error covariance by (8.77) if we use (8.89).

It can also be shown that the gain (8.89) is equal to the gain of the previously obtained ML-II FIR filter (4.120) and the RH MVF-I filter (7.76) [106]. A specific feature is that the field of applications of this gain is strictly limited to stationary and quasistationary processes. As in the case of H_2 FIR filtering, the gain (8.89) is applicable to LTI systems when the block error matrices $\mathcal{Q}_N$ and $\mathcal{R}_N$ have all nonzero components. Therefore, (8.89) generally cannot be computed using recursions. Recall that ML-II FIR and RH MVF-I filters were developed for white Gaussian and uncorrelated noise sources with diagonal matrices $\mathcal{Q}_N$ and $\mathcal{R}_N$, for which recursions are available.

Example 8.3 ***H_2-MVF filter for unweighted $\mathcal{T}(z)$ [109]*** Consider the case of $\mathcal{R}_N = 0$. Obtained in [109] by minimizing the squared Frobenius norm of the unweighted $\mathcal{T}(z)$, the gain for the H_2-MVF filter turns out to be equivalent to (8.89), but with $\Omega_N = (G_N + \bar{T}_N)(G_N + \bar{T}_N)^T$. Like the gain of the UFIR filter, this gain is also robust because it has no tuning factors. On the contrary, this gain can be very inaccurate and imprecise because its scope is limited to $\mathcal{Q}_N = I$, and therefore any disturbance that has $\mathcal{Q}_N \neq I$ can cause large errors. □

To conclude this section, it is worth noting that the gain for the H_2 FIR predictor can also be computed numerically using LMI, as for the H_2 FIR filter. This results in suboptimal algorithms like (8.48), (8.52), and (8.55) and corresponding LMIs (8.47), (8.51), and (8.54). What one needs to do is just take matrices from the FE-based model.

8.4 H_∞ FIR State Estimation

Before discussing H_∞ FIR filtering, recall that the H_2 FIR filter minimizes the squared Frobenius norm of the weighted disturbance-to-error transfer function averaged over all frequencies (8.22). Thereby, it provides optimal H_2 performance, but does not guarantee that possible peaks in $\mathcal{T}$ will also be suppressed by averaging. Moreover, if the H_2 filter is not properly tuned, the peak errors in its output may grow due to bias errors, as in the KF and OFIR filter.

The H_∞ filtering approach was developed to minimize the H_∞ norm of the disturbance-to-error (ς-to-ε) transfer function $\|\mathcal{T}\|_\infty = \sup \sigma_{\max}[\mathcal{T}(z)]$, where $\sigma_{\max}[\mathcal{T}(z)]$ is the maximum singular value of $\mathcal{T}(z)$. A feature of the H_∞ norm is that it minimizes the highest peak value of $\mathcal{T}(z)$ in the Bode plot. In H_∞ filtering, the induced H_∞ norm

$$\|\mathcal{T}\|_\infty = \sup_{\varsigma\neq 0} \frac{\|\mathcal{T}\varsigma\|_2}{\|\varsigma\|_2} = \sup_{\varsigma\neq 0} \frac{\|\varepsilon\|_2}{\|\varsigma\|_2} \tag{8.90}$$

of the ς-to-ε transfer function $\mathcal{T}$ [70] is commonly minimized, where the squared norms of the disturbance $\|\varsigma\|_2^2 = \sum_{i=m}^{k} \varsigma_i^* \varsigma_i$ and the estimation error $\|\varepsilon\|_2^2 = \sum_{i=m}^{k} \varepsilon_i^* \varepsilon_i$ are equal to their energies on $[m, k]$. Note that (8.90) holds due to Parseval's theorem. Therefore, the H_∞ approach applies in both the time domain and the transform domain. Since $\|\mathcal{T}\|_\infty^2$ represents the maximum energy gain from ς to ε, then it follows that the H_∞ norm reflects the worst estimator case and its minimization results in a robust estimator. Moreover, for stable systems the H_∞ norm coincides with the $\mathcal{L}_2$ induced norm of the disturbance-to-error operator [159]. Therefore, it is also referred to as $\|\mathcal{T}\|_\infty = \|\mathcal{T}\|_{2,2}$.

In the standard formulation of H_∞ filtering [70], the robust H_∞ FIR filtering problem can be formulated as follows. Find the fundamental gain $\mathcal{H}_N$ for the H_∞ FIR filter to minimize $\|\mathcal{T}\|_\infty$, given by (8.90) on the horizon $[m, k]$, by solving the following optimization problem,

$$\mathcal{H}_N = \inf_{\mathcal{H}_N} \sup_{\varsigma\neq 0} \frac{\sum_{i=m}^{k} \varepsilon_i^T P_\varepsilon \varepsilon_i}{\sum_{i=m}^{k} \varsigma_i^T P_\varsigma \varsigma_i}, \tag{8.91}$$

where P_ε and P_ς are some proper weights. Since closed-form solutions for (8.91) can be found only in some special cases, consider the following problem

$$\mathcal{H}_N \Leftarrow \sup_{\varsigma\neq 0} \frac{\sum_{i=m}^{k} \varepsilon_i^T P_\varepsilon \varepsilon_i}{\sum_{i=m}^{k} \varsigma_i^T P_\varsigma \varsigma_i} < \gamma^2 , \tag{8.92}$$

which allows us to define $\mathcal{H}_N$ numerically for a given small positive $\gamma > 0$ and develop suboptimal algorithms. Note that the factor γ^2, which indicates the fraction of the disturbance energy that goes into the estimator error, should preferably be small. But because γ^2 cannot be too small for stable estimators, its value should be constrained.

Example 8.4 *H_∞ **filtering problem.*** Consider a two-state polynomial LTI model. Numerically computed for $N = 6$ the squared norms $\|\mathcal{T}\|^2$ of the disturbance-to-error transfer function of the original UFIR filter and another UFIR filter tuned to have the H_∞ performance are shown in Fig. 8.4.

In an effort to study the effect without actually solving the H_∞ problem, we experimentally tune the UFIR filter to reduce the main (first) peak in the transform domain, which is easily seen in the transfer function. But we could not reach the goal without increasing in other (smaller) peaks. Therefore, the decrease in the main peak was stopped when it dropped to the level of other spectral irregularities, and we came to the following conclusion. Using the H_∞ approach, suppression of the main spectral peak should be provided until it reached the level of other spectral peaks. □

Noting that various robust H_∞ FIR state estimators can be designed by solving the H_∞ problem using FE- or BE-based state-space models, we start with the *a posteriori* H_∞ FIR filter.

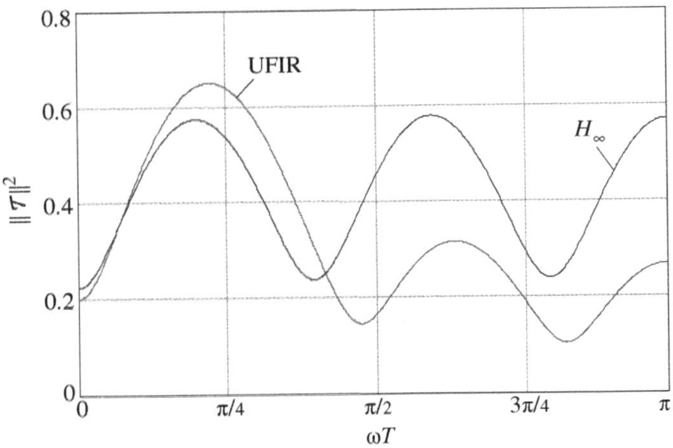

Figure 8.4 Squared norms of the disturbance-to-error transfer functions of the original UFIR filter and another UFIR filter tuned to have the H_∞ performance.

8.4.1 The *a posteriori* H_∞ FIR Filter

Consider the BE-based state-space model (8.1) and (8.2), modify it as

$$x_k = Fx_{k-1} + Bw_k \,, \tag{8.93}$$

$$y_k = Hx_k + Dw_k \tag{8.94}$$

and extend on $[m, k]$ as

$$x_k = F^{N-1}x_m + \bar{D}_N W_{m,k} \,, \tag{8.95}$$

$$Y_{m,k} = H_N x_m + T_N W_{m,k} \,. \tag{8.96}$$

To derive the H_∞ FIR filter, we need the following bounded real lemma (BRL).

Lemma 8.2 ***Bounded real lemma (filtering).*** Given model (8.93) and (8.94). Let $\gamma > 0$ and $S = HB + D$. If there exists a matrix $X > 0$ such that the following LMI is soluble,

$$\begin{bmatrix} -X^{-1} & F & B & 0 \\ F^T & -X & 0 & F^T H^T \\ B^T & 0 & -\gamma P_w & S^T \\ 0 & HF & S & -\gamma P_y^{-1} \end{bmatrix} < 0 \,, \tag{8.97}$$

then the following inequality holds on $[m, k]$,

$$\frac{\sum_{i=m}^{k} y_i^T P_y y_i}{\sum_{i=m}^{k} w_i^T P_w w_i} < \gamma^2 \,. \tag{8.98}$$

Proof: Consider the dissipativity inequality on an FH $[m, k]$,

$$V(x_k) - V(x_m) < \sum_{i=m}^{k} s(w_i, y_i) \,, \tag{8.99}$$

where $V(x_k)$ is the Lyapunov (storage) function representing the energy stored in the system at k and $s(w_i, y_i)$ is a supply function representing the energy that is supplied to the system at i. Choose the storage function $V(x_k) = x_k^T K x_k$ and, referring to (8.98), assign $s(w_i, y_i) = \gamma^2 w_i^T P_w w_i - y_i^T P_y y_i > 0$ to be the supply function. Then rewrite (8.99) as

$$\sum_{i=m}^{k} (y_i^T P_y y_i - \gamma^2 w_i^T P_w w_i) + V(x_k) - V(x_m) < 0 ,$$

substitute y_i taken from (8.94) and x_i from (8.93), assign $S = HB + D$, go to

$$\sum_{i=m}^{k} [(HFx_{i-1} + Sw_i)^T P_y (HFx_{i-1} + Sw_i) - \gamma^2 w_i^T P_w w_i]$$
$$+ x_k^T K x_k - x_m^T K x_m < 0 ,$$

note that values beyond $[m, k]$, namely at $m - 1$, are not available for FIR filtering, change the lower limit in the sum to $m + 1$, unite all components in the sum, and come up with

$$\sum_{i=m+1}^{k} [(HFx_{i-1} + Sw_i)^T P_y (HFx_{i-1} + Sw_i) - \gamma^2 w_i^T P_w w_i$$
$$+ x_i^T K x_i - x_{i-1}^T K x_{i-1}] < 0 . \tag{8.100}$$

To eliminate variables, rearrange the terms and rewrite (8.100) as

$$\sum_{i=m+1}^{k} \begin{bmatrix} x_{i-1} \\ w_i \end{bmatrix}^T \Theta \begin{bmatrix} x_{i-1} \\ w_i \end{bmatrix} < 0 ,$$

which is satisfied if the following LMI holds,

$$\Theta = \begin{bmatrix} F^T KF + F^T H^T P_y HF - K & F^T KB + F^T H^T P_y S \\ B^T KF + S^T P_y HF & B^T KB + S^T P_y S - \gamma^2 P_w \end{bmatrix} < 0 . \tag{8.101}$$

Then decompose (8.101) as

$$\begin{bmatrix} -K & 0 \\ 0 & -\gamma^2 P_w \end{bmatrix} + \begin{bmatrix} F^T & F^T H^T \\ B^T & S^T \end{bmatrix} \begin{bmatrix} K & 0 \\ 0 & P_y \end{bmatrix} \begin{bmatrix} F & B \\ HF & S \end{bmatrix} < 0 ,$$

consider it as a Schur's complement, and represent with another LMI

$$\begin{bmatrix} -K & 0 & F^T & F^T H^T \\ 0 & -\gamma^2 P_w & B^T & S^T \\ F & B & -K^{-1} & 0 \\ HF & S & 0 & -P_y^{-1} \end{bmatrix} < 0 . \tag{8.102}$$

Now multiply (8.102) from the left-hand and right-hand sides with the following matrices, respectively,

$$\begin{bmatrix} 0 & 0 & I & 0 \\ I & 0 & 0 & 0 \\ 0 & I & 0 & 0 \\ 0 & 0 & 0 & I \end{bmatrix} , \quad \begin{bmatrix} 0 & I & 0 & 0 \\ 0 & 0 & I & 0 \\ I & 0 & 0 & 0 \\ 0 & 0 & 0 & I \end{bmatrix}$$

and obtain

$$\begin{bmatrix} -K^{-1} & F & B & 0 \\ F^T & -K & 0 & F^T H^T \\ B^T & 0 & -\gamma^2 P_w & S^T \\ 0 & HF & S & -P_y^{-1} \end{bmatrix} < 0 . \tag{8.103}$$

Finally multiply (8.103) from both sides with diag($\gamma^{0.5}I$ $\gamma^{-0.5}I$ $\gamma^{-0.5}I$ $\gamma^{0.5}I$), introduce a new variable $X = K/\gamma$, arrive at (8.97), and complete the proof. □

To obtain the *a posteriori* H_∞ FIR filter using lemma 8.2, we first rewrite the estimation error (8.14) for $u_k = 0$ as

$$\varepsilon_k = \mathcal{B}_N x_m + \mathcal{W}_N W_{m,k} - \mathcal{V}_N V_{m,k} , \tag{8.104}$$

where $\mathcal{B}_N = F^{N-1} - \mathcal{H}_N H_N$, $\mathcal{W}_N = \bar{D}_N - \mathcal{H}_N T_N$, and $\mathcal{V}_N = \mathcal{H}_N$ and use the following matrix forms given by lemma 8.1,

$$W_{m,k} = A_w W_{m-1,k-1} + B_w w_k , \tag{8.105}$$

$$V_{m,k} = A_w V_{m-1,k-1} + B_w v_k , \tag{8.106}$$

where matrices A_w and B_w are defined by (8.19).

We now introduce two augmented vectors $z_k = [\, W_{m,k}^T \;\; V_{m,k}^T \;\; i_k^T \,]^T$, where $i_k = x_m$, and $\xi_k = [\, w_k^T \;\; v_k^T \,]^T$, and combine them in the following state-space model:

$$z_k = \tilde{F}_\varsigma z_{k-1} + \tilde{B}_\varsigma \xi_k , \tag{8.107}$$

$$\varepsilon_k = \tilde{C}_\varsigma z_k , \tag{8.108}$$

in which the newly introduced block matrices have the form

$$\tilde{F}_\varsigma = \begin{bmatrix} A_w & 0 & 0 \\ 0 & A_w & 0 \\ 0 & 0 & I \end{bmatrix}, \quad \tilde{B}_\varsigma = \begin{bmatrix} B_w & 0 \\ 0 & B_w \\ 0 & 0 \end{bmatrix},$$

$$\tilde{C}_\varsigma = \begin{bmatrix} \bar{D}_N - \mathcal{H}_N G_N & -\mathcal{H}_N & F^{N-1} - \mathcal{H}_N H_N \end{bmatrix} . \tag{8.109}$$

Note that the strictly sparse matrices $\tilde{F}_\varsigma$ and $\tilde{B}_\varsigma$ can significantly reduce computational complexity.

An important property of the model (8.107) and (8.108) follows immediately: all error residual matrices are combined into a new observation matrix $\tilde{C}_\varsigma$, which is thus completely responsible for the H_∞ filter performance. Using (8.107) and (8.108) and applying lemma 8.2, we can now obtain the *a posteriori* H_∞ FIR filter in various suboptimal algorithmic forms. Each of these forms gives the gain $\mathcal{H}_N$ for a stable H_∞ FIR filter, and it is assumed that the best performance is obtained with the lowest possible $\gamma > 0$. The following options can be considered:

1. Apply lemma 8.2 to (8.107) and (8.108) and notice that $D = 0$, $P_y = P$, and $P_w = P_\xi$. For some symmetric invertible matrix $X > 0$ and the smallest possible γ find the gain $\mathcal{H}_N$, satisfying the following LMI

$$\begin{bmatrix} -X^{-1} & \tilde{F}_\varsigma & \tilde{B}_\varsigma & 0 \\ \tilde{F}_\varsigma^T & -X & 0 & \tilde{F}_\varsigma^T \tilde{C}_\varsigma^T \\ \tilde{B}_\varsigma^T & 0 & -\gamma P_\xi & \tilde{B}_\varsigma^T \tilde{C}_\varsigma^T \\ 0 & \tilde{C}_\varsigma \tilde{F}_\varsigma & \tilde{C}_\varsigma \tilde{B}_\varsigma & -\gamma P^{-1} \end{bmatrix} < 0 , \quad \gamma > 0 . \tag{8.110}$$

Start solving the problem (8.110) with the UFIR filter gain $\hat{\mathcal{H}}_N = (C_N^T C_N)^{-1} C_N^T$. If necessary, remove the bias error term by setting the third component in the matrix $\tilde{C}_\varsigma$ (8.109) to zero.

2. To avoid the inversion of X, pre- and post-multiply the matrix (8.97) with diag($X\ I\ I\ I$) and obtain another form of the LMI in lemma 8.2

$$\begin{bmatrix} -X & XF & XB & 0 \\ F^TX & -X & 0 & F^TH^T \\ B^TX & 0 & -\gamma P_w & S^T \\ 0 & HF & S & -\gamma P_y^{-1} \end{bmatrix} < 0 \tag{8.111}$$

that gives another form of (8.110)

$$\begin{bmatrix} -X & X\tilde{F}_\varsigma & X\tilde{B}_\varsigma & 0 \\ \tilde{F}_\varsigma^TX & -X & 0 & \tilde{F}_\varsigma^T\tilde{C}_\varsigma^T \\ \tilde{B}_\varsigma^TX & 0 & -\gamma P_\xi & \tilde{B}_\varsigma^T\tilde{C}_\varsigma^T \\ 0 & \tilde{C}_\varsigma\tilde{F}_\varsigma & \tilde{C}_\varsigma\tilde{B}_\varsigma & -\gamma P^{-1} \end{bmatrix} < 0\,. \tag{8.112}$$

Then find the gain $\mathcal{H}_N$ as in item 1.

3. Consider (8.101), introduce $\Psi = K + \tilde{C}_\varsigma^T P \tilde{C}_\varsigma$, and find $\mathcal{H}_N$ by solving the following LMI problem

$$\begin{bmatrix} \tilde{F}_\varsigma^T\Psi\tilde{F}_\varsigma - K & \tilde{F}_\varsigma^T\Psi\tilde{B}_\varsigma \\ \tilde{B}_\varsigma^T\Psi\tilde{F}_\varsigma & \tilde{B}_\varsigma^T\Psi\tilde{B}_\varsigma - \gamma^2 P_\xi \end{bmatrix} < 0\,, \tag{8.113}$$

$$\tilde{B}_\varsigma^T\Psi\tilde{B}_\varsigma - \gamma^2 P_\xi < 0\,.$$

4. Find $\mathcal{H}_N$ by solving the following DARI

$$\tilde{F}_\varsigma^T\Psi\tilde{F}_\varsigma - K - \tilde{F}_\varsigma^T\Psi\tilde{B}_\varsigma(\tilde{B}_\varsigma^T\Psi\tilde{B}_\varsigma - \gamma^2 P_\xi)^{-1}\tilde{B}_\varsigma^T\Psi\tilde{F}_\varsigma < 0\,, \tag{8.114}$$

$$\tilde{B}_\varsigma^T\Psi\tilde{B}_\varsigma - \gamma^2 P_\xi < 0\,.$$

Using the gain $\mathcal{H}_N$, numerically determined by one of the described methods, we can obtain the *a posteriori* H_∞ filtering estimate and error covariance for uncorrelated w_k, v_k, and x_m as, respectively,

$$\hat{x}_k = \mathcal{H}_N Y_{m,k}\,, \tag{8.115}$$

$$P = \mathcal{B}_N \mathcal{X}_m \mathcal{B}_N^T + \mathcal{W}_N \mathcal{Q}_N \mathcal{W}_N^T + \mathcal{V}_N \mathcal{R}_N \mathcal{V}_N^T\,, \tag{8.116}$$

where the error residual matrices are given after (8.104). It should be noted that all of the previous options, giving $\mathcal{H}_N$, can be used in the development of practical algorithms. However, the second option based on (8.112) is considered the most elaborate.

8.4.2 H_∞ FIR Predictor

As has been repeatedly noted, the FE-based state-space model is better suited for developing predictors and RH filters. Moreover, the FIR gain obtained for LTI systems can be used equivalently in FIR prediction and RH FIR filtering. To numerically find the gain for the H_∞ FIR predictor and the H_∞ RH FIR filter, we consider the following FE-based state-space model

$$x_{k+1} = Fx_k + Bw_k\,, \tag{8.117}$$

$$y_k = Hx_k + Dw_k\,, \tag{8.118}$$

and extend it using (8.8) and (8.9) as

$$x_{k+1} = F^N x_m + \bar{D}_N W_{m,k} , \tag{8.119}$$

$$Y_{m,k} = H_N^p x_m + T_N^p W_{m,k} . \tag{8.120}$$

The following BRL (predictive) will be required to develop suboptimally the H_∞ FIR predictor.

Lemma 8.3 ***Bounded real lemma (prediction).*** Given model (8.117) and (8.118), let $\gamma > 0$, if there exists a symmetric invertible matrix $X > 0$ such that the following LMI is soluble,

$$\begin{bmatrix} -X^{-1} & F & B & 0 \\ F^T & -X & 0 & H^T \\ B^T & 0 & -\gamma P_w & D^T \\ 0 & H & D & -\gamma P_y^{-1} \end{bmatrix} < 0 , \tag{8.121}$$

then the following inequality holds on $[m, k]$,

$$\frac{\sum_{i=m}^{k} y_i^T P_y y_i}{\sum_{i=m}^{k} w_i^T P_w w_i} < \gamma^2 . \tag{8.122}$$

Proof: Consider the dissipativity inequality

$$V(x_{k+1}) - V(x_m) < \sum_{i=m}^{k} s(w_i, y_i) , \tag{8.123}$$

choose the Lyapunov function $V(x_k) = x_k^T K x_k$ and the supply function $s(w_i, y_i) = \gamma^2 w_i^T P_w w_i - y_i^T P_y y_i > 0$, and rewrite (8.123) as

$$\sum_{i=m}^{k} (y_i^T P_y y_i - \gamma^2 w_i^T P_w w_i) + V(x_{k+1}) - V(x_m) < 0 .$$

Replace x_{k+1} in $V(x_{k+1})$ with (8.117), combine the terms in the sum, and come up with

$$\sum_{i=m}^{k} [(x_i^T H^T + w_i^T D^T) P_y (H x_i + D w_i) - \gamma^2 w_i^T P_w w_i$$
$$+ (x_i^T F^T + w_i^T B^T) K (F x_i + B w_i) - x_i^T K x_i] < 0 . \tag{8.124}$$

To eliminate variables, rearrange the terms and rewrite (8.124) as

$$\sum_{i=m}^{k} \begin{bmatrix} x_i \\ w_i \end{bmatrix}^T \Theta \begin{bmatrix} x_i \\ w_i \end{bmatrix} < 0 ,$$

which is satisfied if the following LMI holds,

$$\Theta = \begin{bmatrix} H^T P_y H + F^T K F - K & H^T P_y D + F^T K B \\ D^T P_y H + B^T K F & D^T P_y D + B^T K B - \gamma^2 P_w \end{bmatrix} < 0 . \tag{8.125}$$

Decompose (8.125) as

$$\begin{bmatrix} -K & 0 \\ 0 & -\gamma^2 P_w \end{bmatrix} + \begin{bmatrix} F^T & H^T \\ B^T & D^T \end{bmatrix} \begin{bmatrix} K & 0 \\ 0 & P_y \end{bmatrix} \begin{bmatrix} F & B \\ H & D \end{bmatrix} < 0 ,$$

consider it as a Schur's complement, and represent with LMI as

$$\begin{bmatrix} -K & 0 & F^T & H^T \\ 0 & -\gamma^2 P_w & B^T & D^T \\ F & B & -K^{-1} & 0 \\ H & D & 0 & -P_y^{-1} \end{bmatrix} < 0 . \tag{8.126}$$

Multiply (8.126) from both sides as in lemma 8.1, obtain (8.121), and complete the proof. □

There is another useful form of (8.121),

$$\begin{bmatrix} -X & XF & XB & 0 \\ F^TX & -X & 0 & H^T \\ B^TX & 0 & -\gamma P_w & D^T \\ 0 & H & D & -\gamma P_y^{-1} \end{bmatrix} < 0 , \tag{8.127}$$

which avoids the inversion of X and appears by multiplying (8.121) on both sides by the diagonal matrix diag($X\ I\ I\ I$). Note that (8.121) and (8.127) stated by lemma 8.3 are standard LMIs for H_∞ prediction [22, 41, 46, 106, 232]. In various forms, (8.121) can also be found in [31].

Using lemma 8.3, the H_∞ FIR predictor can be obtained similarly to the *a posteriori* H_∞ filter. Accordingly, we first write the estimation error (8.59) for $u_k = 0$ as

$$\varepsilon_{k+1} = \mathcal{B}^p_N x_m + \mathcal{W}^p_N W_{m,k} - \mathcal{V}^p_N V_{m,k} , \tag{8.128}$$

where the error residual matrices are given by $\mathcal{B}^p_N = F^N - \mathcal{H}^p_N H^p_N$, $\mathcal{W}^p_N = \bar{D}_N - \mathcal{H}^p_N T^p_N$, and $\mathcal{V}^p_N = \mathcal{H}^p_N$. Referring to lemma 8.1, we represent vectors $W_{m,k}$ and $V_{m,k}$ as (8.105) and (8.106),

$$W_{m,k} = A_w W_{m-1,k-1} + B_w w_k ,$$

$$V_{m,k} = A_w V_{m-1,k-1} + B_w v_k ,$$

where matrices A_w and B_w are given by (8.19). Using the augmented vectors $\xi_k = [\ w_k^T\ v_k^T\ ]^T$ and $z_k = [\ W_{m,k}^T\ V_{m,k}^T\ i_k^T\ ]^T$, where $i_k = x_m$, the disturbance state model can now be written as (8.107),

$$z_k = \tilde{F}_\varsigma z_{k-1} + \tilde{B}_\varsigma \xi_k ,$$

where the matrices $\tilde{F}_\varsigma$ and $\tilde{B}_\varsigma$ are given by (8.109).

At this point, we reorganize the estimation error (8.128) to have a compact matrix form $\varepsilon_{k+1} = \hat{C}^p_\varsigma z_k$ and define the matrix $\hat{C}^p_\varsigma$ as

$$\begin{aligned} \hat{C}^p_\varsigma &= \begin{bmatrix} \mathcal{W}^p_{m,k} & -\mathcal{V}^p_{m,k} & \mathcal{B}^p_{m,k} \end{bmatrix} \\ &= \begin{bmatrix} \bar{D}_N - \mathcal{H}^p_N G^p_N & -\mathcal{H}^p_N & F^N - \mathcal{H}^p_N H^p_N \end{bmatrix} . \end{aligned} \tag{8.129}$$

By combining $\varepsilon_k = \hat{C}^p_\varsigma z_{k-1}$ with $z_{k-1} = \tilde{F}_\varsigma^{-1} z_k - \tilde{F}_\varsigma^{-1}\tilde{B}_\varsigma \xi_k$ taken from (8.107), we finally obtain the error equation in the form

$$\varepsilon_k = \hat{C}^p_\varsigma \tilde{F}_\varsigma^{-1} z_k - \hat{C}^p_\varsigma \tilde{F}_\varsigma^{-1}\tilde{B}_\varsigma \xi_k . \tag{8.130}$$

Since the FE- and BE-based models generally cannot be converted to each other, we replace the time index k in (8.107) by $k+1$, accept the approximation $\xi_k = \xi_{k-1}$, and write the disturbance-to-error state-space model using (8.130) in the desired form (8.117) and (8.118) as

$$z_{k+1} = \tilde{F}_\varsigma z_k + \tilde{B}_\varsigma \xi_k , \tag{8.131}$$

$$\varepsilon_k = \tilde{C}^p_\varsigma z_k + \tilde{D}^p_\varsigma \xi_k , \tag{8.132}$$

where the strictly sparse matrices $\tilde{F}_\varsigma$ and $\tilde{B}_\varsigma$ are given by (8.109) and the matrices $\tilde{C}^p_\varsigma$ and $\tilde{D}^p_\varsigma$ are defined as

$$\tilde{C}^p_\varsigma = \hat{C}^p_\varsigma \tilde{F}^{-1}_\varsigma , \quad \tilde{D}^p_\varsigma = -\tilde{C}^p_\varsigma \tilde{B}_\varsigma = -\hat{C}^p_\varsigma \tilde{F}^{-1}_\varsigma \tilde{B}_\varsigma . \tag{8.133}$$

Lemma 8.3 can finally be applied to the model in (8.131) and (8.132), and the gain $\mathcal{H}^p_N$ can be computed numerically using one of the following options:

1. Apply lemma 8.3 to (8.131) and (8.132), and notice that $P_y = P$ and $P_w = P_\xi$. For some symmetric invertible matrix $X > 0$ find the smallest $\gamma > 0$ such that the following inequality holds

$$\begin{bmatrix} -X^{-1} & \tilde{F}_\varsigma & \tilde{B}_\varsigma & 0 \\ \tilde{F}^T_\varsigma & -X & 0 & \tilde{C}^{p^T}_\varsigma \\ \tilde{B}^T_\varsigma & 0 & -\gamma P_\xi & \tilde{D}^{p^T}_\varsigma \\ 0 & \tilde{C}^p_\varsigma & \tilde{D}^p_\varsigma & -\gamma P^{-1} \end{bmatrix} < 0 , \quad \gamma > 0 . \tag{8.134}$$

   Solve the problem (8.134) starting with $\hat{\mathcal{H}}^p_N = (C^{p^T}_N C^p_N)^{-1} C^{p^T}_N$. To make the gain $\mathcal{H}^p_N$ unbiased, set the third component in matrix $\hat{C}^p_\varsigma$ (8.129) to zero and embed the unbiasedness constraint as shown in [39].
2. Solve the problem posed in item 1 for LMI (8.127) represented as

$$\begin{bmatrix} -X & X\tilde{F}_\varsigma & X\tilde{B}_\varsigma & 0 \\ \tilde{F}^T_\varsigma X & -X & 0 & \tilde{C}^{p^T}_\varsigma \\ \tilde{B}^T_\varsigma X & 0 & -\gamma P_\xi & \tilde{D}^{p^T}_\varsigma \\ 0 & \tilde{C}^p_\varsigma & \tilde{D}^p_\varsigma & -\gamma P^{-1} \end{bmatrix} < 0 , \quad \gamma > 0 . \tag{8.135}$$

3. Refer to (8.125) and solve the following LMI problem

$$\begin{bmatrix} \tilde{C}^{p^T}_\varsigma P \tilde{C}^p_\varsigma + \tilde{F}^T_\varsigma K \tilde{F}_\varsigma - K & \tilde{C}^{p^T}_\varsigma P \tilde{D}^p_\varsigma + \tilde{F}^T_\varsigma K \tilde{B}_\varsigma \\ \tilde{D}^{p^T}_\varsigma P \tilde{C}^p_\varsigma + \tilde{B}^T_\varsigma K \tilde{F}_\varsigma & \tilde{D}^{p^T}_\varsigma P \tilde{D}^p_\varsigma + \tilde{B}^T_\varsigma K \tilde{B}_\varsigma - \gamma^2 P_\xi \end{bmatrix} < 0 , \tag{8.136}$$

   subject to $\tilde{D}^{p^T}_\varsigma P \tilde{D}^p_\varsigma + \tilde{B}^T_\varsigma K \tilde{B}_\varsigma - \gamma^2 P_\xi < 0$.
4. Solve the following DARI

$$\begin{aligned} &\tilde{C}^{p^T}_\varsigma P \tilde{C}^p_\varsigma + \tilde{F}^T_\varsigma K \tilde{F}_\varsigma - K - (\tilde{C}^{p^T}_\varsigma P \tilde{D}^p_\varsigma + \tilde{F}^T_\varsigma K \tilde{B}_\varsigma) \\ &\times (\tilde{D}^{p^T}_\varsigma P \tilde{D}^p_\varsigma + \tilde{B}^T_\varsigma K \tilde{B}_\varsigma - \gamma^2 P_\xi)^{-1} (\tilde{D}^{p^T}_\varsigma P \tilde{C}^p_\varsigma + \tilde{B}^T_\varsigma K \tilde{F}_\varsigma) < 0 , \end{aligned} \tag{8.137}$$

   subject to $\tilde{D}^{p^T}_\varsigma P \tilde{D}^p_\varsigma + \tilde{B}^T_\varsigma K \tilde{B}_\varsigma - \gamma^2 P_\xi < 0$.

It follows from the previous relationships that the gain $\mathcal{H}^p_N$ is a variable of two matrices $\tilde{C}^p_\varsigma$ and $\tilde{D}^p_\varsigma$ specified by (8.133) and (8.129). Provided that $\mathcal{H}^p_N$ is determined numerically and suboptimally, the H_∞ FIR prediction $\hat{x}_{k+1}$ and the error matrix P are obtained for uncorrelated w_k, v_k, and x_m as

$$\tilde{x}_{k+1} = \mathcal{H}^p_N Y_{m,k} , \tag{8.138}$$

$$P = \mathcal{B}^p_N \mathcal{X}_m \mathcal{B}^{p^T}_N + \mathcal{W}^p_N \mathcal{Q}_N \mathcal{W}^{p^T}_N + \mathcal{V}^p_N \mathcal{R}_N \mathcal{V}^{p^T}_N , \tag{8.139}$$

where the residual error matrices $\mathcal{B}^p_N$, $\mathcal{W}^p_N$, and $\mathcal{V}^p_N$ are given after (8.128).

8.5 H_2/H_∞ FIR Filter and Predictor

A prediction with improved robustness can be obtained if we recall that the H_2 FIR state estimator minimizes the squared Frobenius norm of $\bar{\mathcal{T}}$ and the H_∞ FIR state estimator minimizes the peak error in the transform domain. What next can be done is to develop a hybrid H_2/H_∞ FIR estimator that can minimize errors to meet both these criteria. Such an H_2/H_∞ FIR structure is considered more robust than the H_2 and H_∞ FIR state estimators. Without going into the details that can be found in [106], we will look at the following two most elaborate H_2/H_∞ FIR algorithms.

Hybrid *a posteriori* H_2/H_∞ FIR Filter

Solve the following minimization problem

$$\mathcal{H}_N \Leftarrow \inf_{\mathcal{H}_N, \mathcal{Z}} \{\mathrm{tr}\mathcal{Z}, \gamma > 0\} \tag{8.140}$$

subject to

$$\begin{bmatrix} \mathcal{Z} - \mathcal{A} + C^T\mathcal{H}_N^T + \mathcal{H}_N C & \mathcal{H}_N \\ \mathcal{H}_N^T & (H_N \mathcal{X}_m H_N^T + \Omega_N)^{-1} \end{bmatrix} > 0\,,$$

$$\begin{bmatrix} -X & X\tilde{F}_\varsigma & X\tilde{B}_\varsigma & 0 \\ \tilde{F}_\varsigma^T X & -X & 0 & \tilde{F}_\varsigma^T \tilde{C}_\varsigma^T \\ \tilde{B}_\varsigma^T X & 0 & -\gamma P_\xi & \tilde{B}_\varsigma^T \tilde{C}_\varsigma^T \\ 0 & \tilde{C}_\varsigma \tilde{F}_\varsigma & \tilde{C}_\varsigma \tilde{B}_\varsigma & -\gamma P^{-1} \end{bmatrix} < 0\,,$$

where $\tilde{C}_\varsigma(\mathcal{H}_N)$ is given by (8.109) and $X > 0$ is some symmetric matrix. Initialize the minimization with $\mathcal{H}_N = (C_N^T C_N)^{-1} C_N^T$.

Hybrid H_2/H_∞ FIR Predictor

Solve the following minimization problem

$$\mathcal{H}_N^p \Leftarrow \inf_{\mathcal{H}_N^p, \mathcal{Z}^p} \{\mathrm{tr}\mathcal{Z}^p, \gamma > 0\} \tag{8.141}$$

subject to

$$\begin{bmatrix} \mathcal{Z}^p - \bar{D}_N \mathcal{Q}_N \bar{D}_N^{p^T} + C^{p^T} \mathcal{H}_N^{p^T} + \mathcal{H}_N^p C^p & \mathcal{H}_N^p \\ \mathcal{H}_N^{p^T} & \Omega_N^{-1} \end{bmatrix} > 0\,,$$

$$\begin{bmatrix} -X & X\tilde{F}_\varsigma & X\tilde{B}_\varsigma & 0 \\ \tilde{F}_\varsigma^T X & -X & 0 & \tilde{C}_\varsigma^{p^T} \\ \tilde{B}_\varsigma^T X & 0 & -\gamma P_\xi & \tilde{D}_\varsigma^{p^T} \\ 0 & \tilde{C}_\varsigma^p & \tilde{D}_\varsigma^p & -\gamma P^{-1} \end{bmatrix} < 0\,,$$

where $\tilde{C}_\varsigma^p(\mathcal{H}_N^p)$ and $\tilde{D}_\varsigma^p(\mathcal{H}_N^p)$ are given by (8.133) and $X > 0$ is some symmetric matrix. Initialize the minimization with $\mathcal{H}_N^p = (C_N^{p^T} C_N^p)^{-1} C_N^{p^T}$.

8.6 Generalized H_2 FIR State Estimation

Another robust state estimation approach that minimizes the peak error for the maximized disturbance energy is referred to as *generalized H_2 filtering*. The idea was originally formulated by Wilson in [213] and later developed in [188] into the *energy-to-peak* or $\mathcal{L}_2$-to-$\mathcal{L}_\infty$ estimation algorithms using LMI, in both continuous and discrete time. The $\mathcal{L}_2$-to-$\mathcal{L}_\infty$ state estimator appears from the energy-to-peak lemma proved in [188] and used by many authors [5, 31, 144, 159]. Since our concern is in FIR filtering and predictive solutions, we will prove this lemma for the FE- and BE-based state models, starting with dissipativity inequalities and ending with a more general derivation. Note that the $\mathcal{L}_2$-to-$\mathcal{L}_\infty$ filter can be derived as shown in [212].

8.6.1 Energy-to-Peak Lemma

The idea behind the energy-to-peak filtering [213] arose in an attempt to minimize the squared peak error over the bounded disturbance energy or, equivalently, to minimize the error infinity norm[1] [188] on $[m, k]$ as

$$\|\varepsilon\|_\infty \triangleq \sup_{i\in[m,k]} \|\varepsilon_i\|_2 = \sqrt{\sup_{i\in[m,k]} (\varepsilon_i^T P_\varepsilon \varepsilon_i)}$$

over the bounded disturbance norm $\|w\|_2 = \sqrt{\sum_{i=m}^{k} w_i^T P_w w_i}$. Using this approach, the gain for the FIR filter can be determined by solving the following optimization problem

$$\mathcal{H}_N = \inf_{\mathcal{H}_N} \sup_{\|w\|_2<\infty} \frac{\|\varepsilon\|_\infty}{\|w\|_2} \,. \tag{8.142}$$

To solve (8.142), we need an energy-to-peak lemma [188], which, as we will show later, can be universally applied to develop algorithms for FIR filtering and FIR prediction.

BE-Based State Model

To prove the energy-to-peak lemma for the *a posteriori* $\mathcal{L}_2$-to-$\mathcal{L}_\infty$ FIR filter, we consider the following BE-based state-space model,

$$x_k = Fx_{k-1} + Bw_k \,, \tag{8.143}$$

$$y_k = Hx_k + Dw_k \,, \tag{8.144}$$

where w_k is the norm-bounded disturbance, $\|w_k\|_2 < \infty$. The extension of this model to $[m, k]$ is as follows,

$$x_k = F^{N-1}x_m + \bar{D}_N W_{m,k} \,, \tag{8.145}$$

$$Y_{m,k} = H_N x_m + T_N W_{m,k} \,, \tag{8.146}$$

where all matrices are specified after (8.8).

It is worth noting that the model in (8.143) and (8.144) is introduced to describe the conversion of the disturbance to the estimation error and, therefore, has nothing to do with the real process and its state space representation. Later, we will replace this model with a disturbance-to-error state space model and apply the energy-to-peak lemma to develop numerical algorithms for $\mathcal{L}_2$-to-$\mathcal{L}_\infty$ FIR state estimators using LMI.

1 Note that the error infinity norm $\|\varepsilon\|_\infty$ has another meaning in state estimation than the standard vector infinity norm.

Lemma 8.4 ***(Energy-to-peak lemma).*** Given model (8.143) and (8.144), let $\gamma > 0$ be a scalar and w_k a bounded disturbance, $\|w\|_2 < \infty$. The $\mathcal{L}_2 - \mathcal{L}_\infty$ performance is satisfied by

$$\sup_{\|w\|_2<\infty} \frac{\|y\|_\infty}{\|w\|_2} < \gamma \,, \tag{8.147}$$

if there exists a positive definite matrix $X > 0$ such that

$$\begin{bmatrix} DD^T & H \\ H^T & -X^{-1} \end{bmatrix} \leqslant \gamma^2 P_y^{-1} \,, \tag{8.148}$$

$$\begin{bmatrix} BP_wB^T - X & F \\ F^T & -X^{-1} \end{bmatrix} < 0 \,. \tag{8.149}$$

Proof: Consider the dissipativity inequality (8.99). For the supply function $s(w_k, y_k) = s(w_k) = w_k^T P_w w_k$ [159] given by (8.147), the inequality (8.99) becomes

$$V(x_k) - V(x_m) < \sum_{i=m}^{k} w_i^T P_w w_i \,. \tag{8.150}$$

Since the model in (8.143) and (8.144) is LTI, consider (8.150) on $[0, N-1]$. As suggested in [188], assume that nothing is stored at zero, assign $V_{N-1} = x_{N-1}^T X^{-1} x_{N-1}$, rewrite (8.150) as $V(x_{N-1}) < \sum_{i=0}^{N-1} w_i^T P_w w_i = \|w\|_2^2$, and represent with an LMI as

$$\begin{bmatrix} \|w\|_2^2 & x_{N-1}^T \\ x_{N-1} & X \end{bmatrix} > 0 \,. \tag{8.151}$$

The LMI (8.151) has disturbance and state as variables, but we need disturbance and output. Conversion to the required form can be provided if we follow [188] and consider

$$\begin{bmatrix} 1 & 0 \\ 0 & H \end{bmatrix} \begin{bmatrix} \|w\|_2^2 & x_{N-1}^T \\ x_{N-1} & X \end{bmatrix} \begin{bmatrix} 1 & 0 \\ 0 & H^T \end{bmatrix} + \begin{bmatrix} w_{N-1}^T \\ D \end{bmatrix} \begin{bmatrix} w_{N-1} & D^T \end{bmatrix} > 0 \,,$$

which after several transformation steps becomes

$$\begin{bmatrix} \sum_{i=0}^{N-1} \|w_i\|_2^2 + w_{N-1}^T w_{N-1} & x_{N-1}^T H^T + w_{N-1}^T D^T \\ Hx_{N-1} + Dw_{N-1} & HXH^T + DD^T \end{bmatrix} > 0 \,,$$

$$\begin{bmatrix} \sum_{i=0}^{N-1} \|w_i\|_2^{2+} & y_{N-1}^T \\ y_{N-1} & HXH^T + DD^T \end{bmatrix} > 0 \,, \tag{8.152}$$

where $\|w\|_2^{2+} = \sum_{i=0}^{N-1} \|w_i\|_2^2 + w_{N-1}^T w_{N-1}$. Use the Schur complement, and rewrite (8.152) as

$$y_{N-1}^T (HXH^T + DD^T)^{-1} y_{N-1} < \|w\|_2^{2+} \,. \tag{8.153}$$

Now represent (8.147) as

$$\frac{1}{\gamma^2} y_{N-1}^T P_y y_{N-1} < \|w\|_2^{2+} \,, \tag{8.154}$$

compare to (8.153), arrive at the inequality

$$HXH^T + DD^T < \gamma^2 P_y^{-1} \,, \tag{8.155}$$

and conclude that (8.147) and (8.148) hold if (8.155) holds.

To specify the matrix P for (8.155), write the dissipativity inequality at an arbitrary time index k in the forms

$$V(x_k) - V(x_{k-1}) < \|w_k\|_2^2 ,$$

$$x_{k-1}^T X^{-1} x_{k-1} - x_k^T X^{-1} x_k + w_k^T P_w w_k > 0 ,$$

substitute x_{k-1} taken from (8.143), and go to the quadratic form inequality

$$\begin{bmatrix} x_{k-1} \\ w_k \end{bmatrix}^T \begin{bmatrix} X^{-1} - F^T X^{-1} F & -F^T X^{-1} B \\ -B^T X^{-1} F & P_w - B^T X^{-1} B \end{bmatrix} \begin{bmatrix} x_{k-1} \\ w_k \end{bmatrix} > 0$$

that holds if the following LMI holds,

$$\begin{bmatrix} X^{-1} - F^T X^{-1} F & -F^T X^{-1} B \\ -B^T X^{-1} F & P_w - B^T X^{-1} B \end{bmatrix} > 0 . \tag{8.156}$$

Using the Schur complement, rewrite (8.156) as

$$X^{-1} - F^T [X^{-1} + X^{-1} B (P_w - B^T X^{-1} B)^{-1} B^T X^{-1}] F > 0 ,$$

apply the Kailath variant (A.10), provide the transformations, and end up with the Lyapunov inequality

$$X - B P_w^{-1} B^T - F X F^T > 0 . \tag{8.157}$$

Finally, represent (8.157) as (8.149) and complete the proof. □

FE-Based State Model

To obtain the $\mathcal{L}_2$-to-$\mathcal{L}_\infty$ FIR predictor similarly to the $\mathcal{L}_2$-to-$\mathcal{L}_\infty$ FIR filter, we also need a lemma for the following FE-based state-space model

$$x_{k+1} = F x_k + B w_k , \tag{8.158}$$

$$y_k = H x_k + D w_k , \tag{8.159}$$

where w_k is the bounded disturbance, $\|w_k\|_2 < \infty$. The corresponding extended model is given by

$$x_k = F^N x_m + \bar{D}_N W_{m,k} , \tag{8.160}$$

$$Y_{m,k} = H_N^p x_m + T_N^p W_{m,k} , \tag{8.161}$$

where all matrices are specified after (8.4). As in the case of filtering, we will first use the model in (8.158) and (8.159) to prove an energy-to-peak lemma and then substitute this model with the disturbance-to-error state space model to develop a numerical algorithm for the $\mathcal{L}_2$-to-$\mathcal{L}_\infty$ FIR predictor using LMI.

By repeating the necessary steps to prove lemma 8.4, we conclude that this lemma is universally applicable to the BE-based model in (8.143) and (8.144) and to the FE-based model in (8.158) and (8.159). Since the corresponding proof is similar to the proof given by (8.150)–(8.157), we avoid details and show the main differences.

We start with the dissipativity inequality (8.150),

$$V(x_{k+1}) - V(x_m) < \sum_{i=m}^{k} w_i^T P_w w_i , \tag{8.162}$$

transform (8.162) to

$$\begin{bmatrix} \|w\|_2^2 & x_N^T \\ x_N & X \end{bmatrix} > 0\,,$$

$$\begin{bmatrix} \sum_{i=0}^{N} \|w_i\|_2^{2+} & y_N^T \\ y_N & HXH^T + DD^T \end{bmatrix} > 0\,,$$

where $\|w\|_2^{2+} = \sum_{i=0}^{N-1} \|w_i\|_2^2 + w_N^T w_N$, and represent with

$$y_N^T (HXH^T + DD^T)^{-1} y_N < \|w\|_2^{2+}\,. \tag{8.163}$$

Reasoning similarly to (8.153)–(8.155), we arrive at (8.148). To find X for (8.163), we consider the dissipativity inequality

$$V(x_{k+1}) - V(x_k) < \|w_k\|_2^2\,,$$

substitute x_{k+1} with (8.158), provide the transformations, arrive at the Lyapunov inequality (8.149), and complete the proof.

8.6.2 $\mathcal{L}_2$-to-$\mathcal{L}_\infty$ FIR Filter and Predictor

Using lemma 8.4, the *a posteriori* $\mathcal{L}_2$-to-$\mathcal{L}_\infty$ FIR filter and the $\mathcal{L}_2$-to-$\mathcal{L}_\infty$ FIR predictor can now be derived if we employ the disturbance-to-error state-space models associated with the FE and BE methods.

The *a posteriori* $\mathcal{L}_2$-to-$\mathcal{L}_\infty$ FIR Filter

To derive the *a posteriori* $\mathcal{L}_2$-to-$\mathcal{L}_\infty$ FIR filter, we consider the estimation error associated with the model in (8.93) and (8.94) as a function of the disturbance, measurement error, and initial error

$$\varepsilon_k = \mathcal{B}_N x_m + \mathcal{W}_N W_{m,k} - \mathcal{V}_N V_{m,k}\,, \tag{8.164}$$

where $\mathcal{B}_N = F^{N-1} - \mathcal{H}_N H_N$, $\mathcal{W}_N = \bar{D}_N - \mathcal{H}_N T_N$, and $\mathcal{V}_N = \mathcal{H}_N$. Using lemma 8.1, we represent a vector $W_{m,k}$ as

$$W_{m,k} = A_w W_{m-1,k-1} + B_w w_k\,, \tag{8.165}$$

where the matrices A_w and B_w are defined by (8.19). Similarly, we represent a vector $V_{m,k}$ as (8.106).

We then use the augmented vector $z_k = [\, W_{m,k}^T \; V_{m,k}^T \; i_k^T \,]^T$, where $i_k = x_m$, and a vector $\xi_k = [\, w_k^T \; v_k^T \,]^T$, and represent the disturbance-to-error model in discrete-time state space as

$$z_k = \tilde{F}_\varsigma z_{k-1} + \tilde{B}_\varsigma \xi_k\,, \tag{8.166}$$

$$\varepsilon_k = \tilde{C}_\varsigma z_k\,, \tag{8.167}$$

where the matrices are given by

$$\tilde{F}_\varsigma = \begin{bmatrix} A_w & 0 & 0 \\ 0 & A_w & 0 \\ 0 & 0 & I \end{bmatrix}, \quad \tilde{B}_\varsigma = \begin{bmatrix} B_w & 0 \\ 0 & B_w \\ 0 & 0 \end{bmatrix},$$

$$\tilde{C}_\varsigma = \begin{bmatrix} \bar{D}_N - \mathcal{H}_N G_N & -\mathcal{H}_N & F^{N-1} - \mathcal{H}_N H_N \end{bmatrix}. \tag{8.168}$$

Now we have the model in (8.166) and (8.167), which is equivalent to the model in (8.143) and (8.144) and can apply lemma 8.4. Noticing that $\tilde{D}_\varsigma = 0$ in (8.167), we finally come to an algorithm for numerically computing the gain $\mathcal{H}_N$ for the *a posteriori* $\mathcal{L}_2$-to-$\mathcal{L}_\infty$ FIR filter operating under disturbances, measurement errors, and initial errors.

Given some positive definite matrix X, solve the minimization problem

$$\mathcal{H}_N \Leftarrow \inf_{\mathcal{H}_N} \gamma > 0 \quad (8.169)$$

$$\text{subject to}$$

$$\tilde{C}_\varsigma X \tilde{C}_\varsigma^T \leqslant \gamma^2 P^{-1} ,$$

$$\begin{bmatrix} \tilde{B}_\varsigma P_\xi^{-1} \tilde{B}_\varsigma^T - X & \tilde{F}_\varsigma \\ \tilde{F}_\varsigma^T & -X^{-1} \end{bmatrix} < 0 ,$$

where the sparse matrices $\tilde{F}_\varsigma$ and $\tilde{B}_\varsigma$ and matrix $\tilde{C}_\varsigma$ are given by (8.168). Initialize the minimization with $\hat{\mathcal{H}}_N = (C_N^T C_N)^{-1} C_N^T$. Provided that $\mathcal{H}_N$ is computed numerically, we obtain the *a posteriori* $\mathcal{L}_2$-to-$\mathcal{L}_\infty$ FIR filtering estimate by $\hat{x}_k = \mathcal{H}_N Y_{m,k}$ and the error matrix as

$$P = \mathcal{B}_N \mathcal{X}_m \mathcal{B}_N^T + \mathcal{W}_N \mathcal{Q}_N \mathcal{W}_N^T + \mathcal{V}_N \mathcal{Q}_N \mathcal{V}_N^T , \quad (8.170)$$

where the error residual matrices $\mathcal{B}_N$, $\mathcal{W}_N$, and $\mathcal{V}_N$ are specified after (8.164).

$\mathcal{L}_2$-to-$\mathcal{L}_\infty$ FIR Predictor

The $\mathcal{L}_2$-to-$\mathcal{L}_\infty$ FIR predictor can be obtained similarly if we start with the FE-based state-space model (8.117) and (8.118) and consider the estimation error (8.128) at $k+1$,

$$\varepsilon_{k+1} = \mathcal{B}_N^p x_m + \mathcal{W}_N^p W_{m,k} - \mathcal{V}_N^p V_{m,k} , \quad (8.171)$$

where $\mathcal{B}_N^p = F^N - \mathcal{H}_N^p H_N^p$, $\mathcal{W}_N^p = \bar{D}_N - \mathcal{H}_N^p T_N^p$, and $\mathcal{V}_N^p = \mathcal{H}_N^p$ are the error residual matrices, and the matrices $W_{m,k}$ and $V_{m,k}$ are represented by (8.105) and (8.106) as

$$W_{m,k} = A_w W_{m-1,k-1} + B_w w_k ,$$

$$V_{m,k} = A_w V_{m-1,k-1} + B_w v_k ,$$

where the sparse matrices A_w and B_w are defined by (8.19).

Using the augmented vectors $\xi_k = [\, w_k^T \; v_k^T \,]^T$ and $z_k = [\, W_{m,k}^T \; V_{m,k}^T \; i_k^T \,]^T$, where $i_k = x_m$, we write the disturbance-to-error state model as (8.166) and represent the error (8.171) in a compact matrix form as $\varepsilon_{k+1} = \tilde{C}_\varsigma z_k$. By combining $\varepsilon_k = \tilde{C}_\varsigma z_{k-1}$ and $z_{k-1} = \tilde{F}_\varsigma^{-1} z_k - \tilde{F}_\varsigma^{-1} \tilde{B}_\varsigma \xi_k$ taken from (8.166), we write the error equation as $\varepsilon_k = \tilde{C}_\varsigma \tilde{F}_\varsigma^{-1} z_k - \tilde{C}_\varsigma \tilde{F}_\varsigma^{-1} \tilde{B}_\varsigma \xi_k$. Then we replace the time index k in (8.166) by $k+1$, accept $\xi_k = \xi_{k-1}$, and transform the disturbance-to-error state-space model to the desired form (8.158) and (8.159) to have

$$z_{k+1} = \tilde{F}_\varsigma z_k + \tilde{B}_\varsigma \xi_k , \quad (8.172)$$

$$\varepsilon_k = \tilde{C}_\varsigma^p z_k + \tilde{D}_\varsigma^p \xi_k , \quad (8.173)$$

where the sparse matrices $\tilde{F}_\varsigma$ and $\tilde{B}_\varsigma$ are given by (8.168) and the matrices $\tilde{C}_\varsigma^p$ and $\tilde{D}_\varsigma^p$ are defined as

$$\tilde{C}_\varsigma^p = \begin{bmatrix} \bar{D}_N - \mathcal{H}_N^{ph} G_N^p & -\mathcal{H}_N^{ph} & F^N - \mathcal{H}_N^{ph} H_N^p \end{bmatrix} \tilde{F}_\varsigma^{-1} , \quad (8.174)$$

$$\tilde{D}_\varsigma^p = -\tilde{C}_\varsigma^p \tilde{B}_\varsigma . \quad (8.175)$$

Finally, we apply lemma 8.4 to the model in (8.172) and (8.173) and end up with the following numerical algorithm to compute the gain $\mathcal{H}_N^p$ for an $\mathcal{L}_2$-to-$\mathcal{L}_\infty$ FIR predictor.

For some positive definite matrix X, solve the minimization problem

$$\mathcal{H}_N^p \Leftarrow \inf_{\mathcal{H}_N^p} \gamma > 0 \tag{8.176}$$

subject to

$$\begin{bmatrix} \tilde{C}_\varsigma^p \tilde{B}_\varsigma \tilde{B}_\varsigma^T \tilde{C}_\varsigma^{p^T} & \tilde{C}_\varsigma^p \\ \tilde{C}_\varsigma^{p^T} & -X^{-1} \end{bmatrix} \leqslant \gamma^2 P^{-1} ,$$

$$\begin{bmatrix} \tilde{B}_\varsigma P_\xi^{-1} \tilde{B}_\varsigma^T - X & \tilde{F}_\varsigma \\ \tilde{F}_\varsigma^T & -X^{-1} \end{bmatrix} < 0 ,$$

where the matrix $\tilde{C}_\varsigma^p$ is specified by (8.174). Initialize the minimization with $\hat{\mathcal{H}}_N^p = (C_N^{p^T} C_N^p)^{-1} C_N^{p^T}$.

Provided that $\mathcal{H}_N^p$ is obtained numerically by (8.176), we compute the $\mathcal{L}_2$-to-$\mathcal{L}_\infty$ FIR prediction as $\tilde{x}_{k+1} = \mathcal{H}_N^p Y_{m,k}$ and the error matrix by

$$P_{k+1} = \mathcal{B}_N^p \mathcal{X}_m \mathcal{B}_N^{p^T} + \mathcal{W}_N^p \mathcal{Q}_N \mathcal{W}_N^{p^T} + \mathcal{V}_N^p \mathcal{R}_N \mathcal{V}_N^{p^T} , \tag{8.177}$$

where the error residual matrices $\mathcal{B}_N^p$, $\mathcal{W}_N^p$, and $\mathcal{V}_N^p$ are given after (8.171).

It has to be remarked now that several $\mathcal{L}_2$-to-$\mathcal{L}_\infty$ RH FIR predictive filters have been developed earlier for data taken from $[k-N, k-1]$. For these solutions, (8.82) serves as the observation model, and an estimate is computed by $\tilde{x}_k = \mathcal{H}_N^p Y_{k-N,k-1}$. The predictive filter designed in [106] by solving a minimax problem was called the $\mathcal{L}_2$-*E* FIR (LEF) filter, and the corresponding gain was found analytically in the ML form (8.89). Therefore, the LEF filter does not demonstrate the strict $\mathcal{L}_2$-to-$\mathcal{L}_\infty$ properties and, as we mentioned after (8.89), applies to stationary and quasistationary processes. Using lemma 8.3, the gain for the $\mathcal{L}_2$-to-$\mathcal{L}_\infty$ RH FIR predictive filter was obtained in [5] as a special case of (8.176). In [7], this filter was later reinterpreted using a different approach by decomposing the filter gain $\mathcal{H}_N^p$ into a block row matrix containing a subgain at each horizon step. The transformation rule suggested in [7] results in an LMI condition, avoiding the use of lemma 8.3. Note that in the previous works measurement errors were not taken into account.

8.7 $\mathcal{L}_1$ FIR State Estimation

When a system operates under impulsive external attacks, minimizing the peak error for maximized disturbance energy may not be enough to achieve good performance. More efficient should be an algorithm that minimizes the maximum error for maximum disturbance in what is now known as robust *peak-to-peak* filtering, or $\mathcal{L}_\infty$-to-$\mathcal{L}_\infty$ filtering, or, most generally, $\mathcal{L}_1$ state estimation. The problem was originally formulated and solved in [208] and then reformulated in [1, 133], developed for discrete-time systems in [209], and reconsidered in [144]. In RH FIR filtering, the corresponding solution was originally found in [6]. Following the principles of the approach described in [6, 209], next we will develop $\mathcal{L}_\infty$-to-$\mathcal{L}_\infty$ FIR state estimators using the FE- and BE-based state-space models.

The *peak-to-peak* FIR state estimation problem can be viewed as the minimization of the $\mathcal{L}_1$ norm of the transfer function $\mathcal{T}$ of the peak disturbance w to the peak error ε channel on the horizon $[m, k]$; that is,

$$\|\mathcal{T}\|_{\mathcal{L}_1} = \sup_{\|w\|<\infty} \frac{\|\varepsilon\|_\infty}{\|w\|_\infty} = \sup_{\|w\|<\infty} \frac{\sqrt{\sup_{i\in[m,k]} \|\varepsilon_i\|_2}}{\sqrt{\sup_{i\in[m,k]} \|w_i\|_2}} . \tag{8.178}$$

The FIR filter gain $\mathcal{H}_N$ can thus be determined by solving the following optimization problem,

$$\mathcal{H}_N = \inf_{\mathcal{H}_N} \sup_{\|w\|<\infty} \frac{\sup_{i\in[m,k]} \|\varepsilon_i\|_2^2}{\sup_{i\in[m,k]} \|w_i\|_2^2} \,. \tag{8.179}$$

Since the optimal solution for (8.179) is mathematically untractable, some assumptions must be made. Then note that $\|\varepsilon_k\|$ at k cannot exceed its upper bound $\sup_{i\in[m,k]} \|\varepsilon_i\|_2$ and substitute $\sup_{i\in[m,k]} \|\varepsilon_i\|_2^2 = \alpha_\varepsilon \|\varepsilon_k\|_2^2$, where α_ε is a scaling scalar. Likewise, substitute $\sup_{i\in[m,k]} \|w_i\|_2^2 = \alpha_w \|w_k\|_2^2$, where α_w is a scaling scalar. To solve the optimization problem suboptimally, we thus can rewrite (8.179) as

$$\mathcal{H}_N \Leftarrow \sup_{\|w\|<\infty} \frac{\|\varepsilon_k\|_2^2}{\mu\|w_k\|_2^2} < \gamma^2 \,, \tag{8.180}$$

where $\mu = \alpha_w/\alpha_\varepsilon$ is a scaling scalar, and set a small $\gamma > 0$ such that $\mathcal{H}_N$ becomes suboptimal. To design $\mathcal{L}_\infty$-to-$\mathcal{L}_\infty$ FIR filtering and predictive algorithms, we traditionally need peak-to-peak lemmas for BE-based and FE-based state-space models.

8.7.1 Peak-to-Peak Lemma

Proofs of the peak-to-peak lemma associated with the FE-based state model can be found in various forms in many papers. Using (8.180), next we will give this lemma for two $\mathcal{L}_\infty$-to-$\mathcal{L}_\infty$ FIR state estimators: *a posteriori* filter (BE-based model) and predictor (FE-based model).

Lemma 8.5 ***Peak-to-peak lemma (filtering).*** Given the state-space model

$$x_k = Fx_{k-1} + Bw_k \,,$$
$$y_k = Hx_k + Dw_k \,.$$

Let $\gamma > 0$ be a small real scalar number and w_k a norm-bounded disturbance, $\|w\|_\infty < \infty$. The $\mathcal{L}_\infty$–$\mathcal{L}_\infty$ filtering performance is guaranteed by satisfying

$$\sup_{\|w\|_\infty<\infty} \frac{\|y\|_\infty^2}{\|w\|_\infty^2} < \mu\gamma^2 \,, \tag{8.181}$$

if there exists a positive definite matrix $K > 0$ and real scalar numbers $\mu > 0$ and $\lambda > 0$ such that

$$\begin{bmatrix} (1+\lambda)F^TKF - K & (1+\lambda)F^TKB \\ (1+\lambda)B^TKF & (1+\lambda)B^TKF - \mu P_w \end{bmatrix} < 0 \,, \tag{8.182}$$

$$\begin{bmatrix} K & 0 & H^T \\ 0 & \mu(\gamma^2 - \lambda^{-1})P_w & D^T \\ H & D & P_y^{-1} \end{bmatrix} > 0 \,. \tag{8.183}$$

Proof: Introduce the Lyapunov function $V(x_k) = x_k^TKx_k$ and take a disturbance term $\mu\|w_k\|_2^2$ from (8.180). Consider the first-order differential inequality of the Lyapunov function [159], and represent it in discrete time as

$$V(x_k) - V(x_{k-1}) + \lambda V(x_k) < \mu w_k^T P_w w_k \,. \tag{8.184}$$

In FIR filtering all transients are finished at the end of the horizon $[m, k]$, and the steady-state solution to (8.184) at the horizon end gives the stability condition

$$x_k^T K x_k < \frac{\mu}{\lambda} w_k^T P_w w_k \,. \tag{8.185}$$

Now, represent (8.184) using several equivalent forms as

$$x_k^T K x_k - x_{k-1}^T K x_{k-1} + \lambda x_k^T K x_k - \mu w_k^T P_w w_k < 0 \,,$$
$$(1+\lambda)(F x_{k-1} + B w_k)^T K (F x_{k-1} + B w_k) - x_{k-1}^T K x_{k-1} - \mu w_k^T w_k < 0 \,,$$
$$\begin{bmatrix} x_{k-1} \\ w_k \end{bmatrix}^T \begin{bmatrix} (1+\lambda)F^T K F - K & (1+\lambda)F^T K B \\ (1+\lambda)B^T K F & (1+\lambda)B^T K B - \mu P_w \end{bmatrix} \begin{bmatrix} x_{k-1} \\ w_k \end{bmatrix} < 0 \,,$$

and note that the last inequality holds if (8.182) holds.

To prove (8.183), rewrite (8.181) as $y_k^T P_y y_k < \mu\gamma^2 w_k^T P_w w_k$, add and extract a scalar $\frac{\mu}{\lambda} w_k^T P_w w_k$,

$$y_k^T P_y y_k < \mu\gamma^2 w_k^T P_w w_k + \frac{\mu}{\lambda} w_k^T P_w w_k - \frac{\mu}{\lambda} w_k^T P_w w_k \,,$$

embed the stability condition (8.185) by substituting the first $\frac{\mu}{\lambda} w_k^T P_w w_k$ with $x_k^T K x_k$, and come up with the inequality

$$x_k^T K x_k + \mu(\gamma^2 - \lambda^{-1}) w_k^T P_w w_k - y_k^T P_y y_k > 0 \,.$$

Take y_k from (8.144) and transform this inequality as

$$[x_k^T K x_k + \mu(\gamma^2 - \lambda^{-1}) w_k^T P_w w_k]$$
$$- [x_k^T H^T P_y H x_k + x_k^T H^T P_y D w_k + w_k^T D^T P_y H x_k + w_k^T D^T P_y D w_k] > 0 \,.$$
$$\begin{bmatrix} x_k \\ w_k \end{bmatrix}^T \begin{bmatrix} K & 0 \\ 0 & \mu(\gamma^2 - \lambda^{-1})P_w \end{bmatrix} \begin{bmatrix} x_k \\ w_k \end{bmatrix} - \begin{bmatrix} x_k \\ w_k \end{bmatrix}^T \begin{bmatrix} H^T \\ D^T \end{bmatrix} P_y \begin{bmatrix} H & D \end{bmatrix} \begin{bmatrix} x_k \\ w_k \end{bmatrix} > 0 \,.$$

Observe that the last inequality holds if the following inequality holds

$$\begin{bmatrix} K & 0 \\ 0 & \mu(\gamma^2 - \lambda^{-1})P_w \end{bmatrix} - \begin{bmatrix} H^T \\ D^T \end{bmatrix} P_y \begin{bmatrix} H & D \end{bmatrix} > 0 \,.$$

Finally, use the Schur complement, transform this inequality to the LMI (8.183), and complete the proof. □

Now note that if we set $\mu = 1$ in (8.180), then (8.183) can be obtained in another form shown in [159]. However, this cannot be done directly by putting $\mu = 1$ in (8.183).

Lemma 8.6 ***Peak-to-peak lemma* (prediction).** Given the state space model

$$x_{k+1} = F x_k + B w_k \,,$$
$$y_k = H x_k + D w_k \,.$$

Let $\gamma > 0$ be a small real scalar number and w_k the bounded disturbance, $\|w\|_\infty < \infty$. The $\mathcal{L}_\infty$–$\mathcal{L}_\infty$ performance is guaranteed by satisfying

$$\sup_{\|w\|_\infty < \infty} \frac{\|y\|_\infty^2}{\|w\|_\infty^2} < \mu\gamma^2 \,, \tag{8.186}$$

if there exists a positive definite matrix $K > 0$ and real scalar numbers $\mu > 0$ and $\lambda > 0$ such that

$$\begin{bmatrix} F^TKF - K + \lambda K & F^TKB \\ B^TKF & B^TKF - \mu P_w \end{bmatrix} < 0 , \tag{8.187}$$

$$\begin{bmatrix} K & 0 & H^T \\ 0 & \mu(\gamma^2 - \lambda^{-1})P_w & D^T \\ H & D & P_y^{-1} \end{bmatrix} > 0 . \tag{8.188}$$

Proof: Introduce the Lyapunov function $V(x_k) = x_k^TKx_k$ and the disturbance term $\mu\|w_k\|_2^2$, consider the first-order differential inequality of the Lyapunov function [159], and represent it in discrete time as

$$V(x_{k+1}) - V(x_k) + \lambda V(x_k) < \mu w_k^T P_w w_k . \tag{8.189}$$

In the steady state arrive at (8.185). Next, transform (8.189) as

$$x_{k+1}^TKx_{k+1} - x_k^TKx_k + \lambda x_k^TKx_k - \mu w_k^TP_w w_k < 0 ,$$

$$\begin{bmatrix} x_k \\ w_k \end{bmatrix}^T \begin{bmatrix} F^TKF - K + \lambda K & F^TKB \\ B^TKF & B^TKB - \mu P_w \end{bmatrix} \begin{bmatrix} x_k \\ w_k \end{bmatrix} < 0 ,$$

which holds if (8.187) holds. Finally, find the derivation of (8.188) in lemma 8.5 and complete the proof. □

Note that the inequality (8.187) can be found in [6]. In continuous time, the inequality that is equivalent to (8.187) can be found in many works [31, 159]. The inequality (8.188), which serves for (8.181), does not appear in other works that consider (8.181) with $\mu = 1$. Finally, the obvious advantage of (8.188) is that it combines all tuning variables in one term and is thus more computationally efficient.

8.7.2 $\mathcal{L}_\infty$-to-$\mathcal{L}_\infty$ FIR Filtering and Prediction

Employing lemma 8.5 and lemma 8.6, the robust $\mathcal{L}_\infty$-to-$\mathcal{L}_\infty$ *a posteriori* FIR filter and FIR predictor can now be formalized using the earlier described disturbance-to-error state-space models.

The *a posteriori* $\mathcal{L}_\infty$-to-$\mathcal{L}_\infty$ FIR Filter

Consider the disturbance-to-error state space model (8.166) and (8.167),

$$z_k = \tilde{F}_\varsigma z_{k-1} + \tilde{B}_\varsigma \xi_k ,$$

$$\varepsilon_k = \tilde{C}_\varsigma z_k ,$$

where matrices $\tilde{F}_\varsigma$, $\tilde{B}_\varsigma$, and $\tilde{C}_\varsigma$ associated with the original state-space model (8.143) and (8.144) are given by (8.168). Then apply lemma 8.5 and arrive at the following algorithm to compute numerically a gain $\mathcal{H}_N$ for the robust *a posteriori* $\mathcal{L}_\infty$-to-$\mathcal{L}_\infty$ FIR filter.

Solve the minimization problem,

$$\mathcal{H}_N \Leftarrow \inf_{\mathcal{H}_N} \gamma > 0 \tag{8.190}$$

subject to

$$\begin{bmatrix} (1+\lambda)\tilde{F}_\varsigma^TK\tilde{F}_\varsigma - K & (1+\lambda)\tilde{F}_\varsigma^TK\tilde{B}_\varsigma \\ (1+\lambda)\tilde{B}_\varsigma^TK\tilde{F}_\varsigma & (1+\lambda)\tilde{B}_\varsigma^TK\tilde{F}_\varsigma - \mu P_\xi \end{bmatrix} < 0 , \tag{8.191}$$

$$\begin{bmatrix} K & 0 & \tilde{C}_{\varsigma}^T \\ 0 & \mu(\gamma^2 - \lambda^{-1})P_{\xi} & 0 \\ \tilde{C}_{\varsigma} & 0 & P^{-1} \end{bmatrix} > 0 \,, \quad \lambda > 0 \,, \quad \mu > 0 \,, \tag{8.192}$$

where matrix $\tilde{C}_{\varsigma}$ and strictly sparse matrices $\tilde{F}_{\varsigma}$ and $\tilde{B}_{\varsigma}$ are given by (8.109). Initialize the minimization with $\hat{\mathcal{H}}_N = (C_N^T C_N)^{-1} C_N^T$ and note that no other matrix but $\tilde{C}_{\varsigma}$ is a function of $\mathcal{H}_N$. Provided numerically $\mathcal{H}_N$, compute the *a posteriori* $\mathcal{L}_\infty$-to-$\mathcal{L}_\infty$ FIR filtering estimate $\hat{x}_k$ and error matrix P_k by

$$\hat{x}_k = \mathcal{H}_N Y_{m,k} \,, \tag{8.193}$$

$$P = \mathcal{B}_N \mathcal{X}_m \mathcal{B}_N^T + \mathcal{W}_N \mathcal{Q}_N \mathcal{W}_N^T + \mathcal{V}_N \mathcal{R}_N \mathcal{V}_N^T \,, \tag{8.194}$$

where the error residual matrices $\mathcal{B}_N$, $\mathcal{W}_N$, and $\mathcal{V}_N$ are specified after (8.104).

$\mathcal{L}_\infty$-to-$\mathcal{L}_\infty$ FIR Predictor

Consider the disturbance-to-error state-space model in (8.172) and (8.173),

$$z_{k+1} = \tilde{F}_{\varsigma} z_k + \tilde{B}_{\varsigma} \xi_k \,,$$
$$\varepsilon_k = \tilde{C}_{\varsigma}^p z_k + \tilde{D}_{\varsigma}^p \xi_k \,,$$

where all of the matrices $\tilde{F}_{\varsigma}$ and $\tilde{B}_{\varsigma}$ can be taken from (8.168), and the matrices $\tilde{C}_{\varsigma}^p$ and $\tilde{D}_{\varsigma}^p$ are defined by (8.133) as associated with the original state-space model in (8.158) and (8.159),

$$x_{k+1} = F x_k + B w_k \,,$$
$$y_k = H x_k + D w_k \,.$$

Apply lemma 8.6 to the model (8.172) and (8.173). To determine numerically the gain $\mathcal{H}_N^p$ for the $\mathcal{L}_\infty$-to-$\mathcal{L}_\infty$ FIR predictor, solve the following minimization problem,

$$\mathcal{H}_N^p \Leftarrow \inf_{\mathcal{H}_N^p} \gamma > 0 \tag{8.195}$$

subject to

$$\begin{bmatrix} \tilde{F}_{\varsigma}^T K \tilde{F}_{\varsigma} - K + \lambda K & \tilde{F}_{\varsigma}^T K \tilde{B}_{\varsigma} \\ \tilde{B}_{\varsigma}^T K \tilde{F}_{\varsigma} & \tilde{B}_{\varsigma}^T K \tilde{F}_{\varsigma} - \mu P_{\xi} \end{bmatrix} < 0 \,, \tag{8.196}$$

$$\begin{bmatrix} K & 0 & \tilde{C}_{\varsigma}^{p^T} \\ 0 & \mu(\gamma^2 - \lambda^{-1})P_{\xi} & \tilde{D}_{\varsigma}^{p^T} \\ \tilde{C}_{\varsigma}^p & \tilde{C}_{\varsigma}^p 0 & P^{-1} \end{bmatrix} > 0 \,, \quad \lambda > 0 \,, \quad \mu > 0 \,. \tag{8.197}$$

where the sparse matrices $\tilde{F}_{\varsigma}$ and $\tilde{B}_{\varsigma}$ are given by (8.168) and the matrix $\tilde{C}_{\varsigma}^p$ is specified with (8.174). Note that $\tilde{C}_{\varsigma}^p$ is the only matrix that depends on $\mathcal{H}_N^p$, and initialize the minimization with $\mathcal{H}_N^p = (C_N^{p^T} C_N^p)^{-1} C_N^{p^T}$.

Provided that $\mathcal{H}_N^p$ is numerically available from (8.195), the $\mathcal{L}_\infty$-to-$\mathcal{L}_\infty$ FIR prediction and error covariance are obtained by

$$\tilde{x}_{k+1} = \mathcal{H}_N^p Y_{m,k} \,, \tag{8.198}$$

$$P = \mathcal{B}_N^p \mathcal{X}_m \mathcal{B}_N^{p^T} + \mathcal{W}_N^p \mathcal{Q}_N \mathcal{W}_N^{p^T} + \mathcal{V}_N^p \mathcal{R}_N \mathcal{V}_N^{p^T} \,, \tag{8.199}$$

where the error residual matrices $\mathcal{B}_N^p$ and $\mathcal{W}_N^p$ are given after (8.128).

8.8 Game Theory FIR State Estimation

The game theory robust approach, originally developed by Banavar in [14] for H_∞ filtering, can also be seen as an artificial intelligence tool because it obeys some game rule rather than dealing with the estimator transfer function and impulse response. Since the rules of the game can be different, this can lead to different state estimation solutions. The significance of Banavar's game rule [14] is that it was adopted almost unchanged, and the H_∞ filter developed in [163] and some other works has led to an elegant Kalman-like recursive algorithm (3.307)–(3.310) proposed by Simon [185]. It is worth noting that this has become possible because the game theory approach proposed in [14] has much in common with the general H_∞ filter theory developed in [70].

Game theory H_∞ filters were originally developed using Kalman recursions by minimizing the energy of the estimation error for the maximized squared norms of the initial, system, and measurement errors. Since the estimation error ε_k of the FIR filter is computed over $[m, k]$, the computation of the error energy requires twice the horizon, which significantly increases the mathematical load, especially when N is large and recursions are not available. Therefore, Banavar's H_∞ rule does not agree well with FIR filtering, and it is more reasonable to minimize the error power computed over $[m, k]$ in what can be called the energy-to-power FIR filter.

The bias-constrained energy-to-power FIR filter was originally developed in the ML FIR form in [106] by solving the minimax problem for a norm-bounded disturbance without accounting for measurement errors. In this section, we develop a more general theory of the game theory energy-to-power FIR state estimator and find the gains for the corresponding suboptimal FIR filter and FIR predictor using LMI.

8.8.1 The *a posteriori* Energy-to-Power FIR Filter

Consider the state-space model (8.1)–(8.4) with zero input $u_k = 0$, define the *a posteriori* FIR estimate as (8.11), and refer to the estimation error (8.14) with the error residual matrices defined by (8.15)–(8.17). Using the game theory approach, we can determine the gain for the *a posteriori* energy-to-power FIR filter by solving the following optimization problem

$$\mathcal{H}_N = \inf_{\mathcal{H}_N} \sup_{\varepsilon_m, W_{m,k}, V_{m,k}} \frac{\|\varepsilon_k\|^2}{\|\varepsilon_m\|^2_{P_m^{-1}} + \|w\|^2_{Q_{m,k}^{-1}} + \|v\|^2_{\mathcal{R}_{m,k}^{-1}}} , \tag{8.200}$$

where $\|\varepsilon_k\|^2 \triangleq \|\varepsilon_k\|_2^2$ is the squared ℓ_2 norm of vector $\varepsilon_k = x_k - \hat{x}_k$ computed over $[m, k]$ and the weighted squared norms of the initial error and disturbances are given as $\|\varepsilon_m\|^2_{P_m^{-1}} = \varepsilon_m^T P_m^{-1} \varepsilon_m$, $\|w\|^2_{Q_{m,k}^{-1}} = \sum_{i=m}^{k} w_i^T Q_{m,k}^{-1} w_i$, and $\|v\|^2_{\mathcal{R}_{m,k}^{-1}} = \sum_{i=m}^{k} v_i^T \mathcal{R}_{m,k}^{-1} v_i$. We wish to have the gain $\mathcal{H}_N$ such that the estimation error norm is minimized for the maximized error norms.

Since the solution to (8.200) is generally mathematically untractable, a suboptimal $\mathcal{H}_N$ can be found numerically by introducing a small scalar γ and minimizing the squared norms ratio in (8.200) as

$$\mathcal{H}_N \Leftarrow \frac{\|\varepsilon_k\|^2}{\|\varepsilon_m\|^2_{P_m^{-1}} + \|w\|^2_{Q_{m,k}^{-1}} + \|v\|^2_{\mathcal{R}_{m,k}^{-1}}} < \gamma^2 . \tag{8.201}$$

Now, we can modify (8.201) using the error residual matrices (8.15)–(8.17) and rewrite (8.201) in terms of the extended state-space model as

$$\frac{x_m^T \mathcal{B}_n^T \mathcal{B}_N x_m + W_{m,k}^T \mathcal{W}_N^T \mathcal{W}_N W_{m,k} + V_{m,k}^T \mathcal{V}_N^T \mathcal{V}_N V_{m,k}}{\varepsilon_m^T P_m^{-1} \varepsilon_m + W_{m,k}^T \mathcal{Q}_N^{-1} W_{m,k} + V_{m,k}^T \mathcal{R}_N^{-1} V_{m,k}} < \gamma^2 . \tag{8.202}$$

Substituting $x_m = \varepsilon_m + \hat{x}_m$, we next transform (8.202) to

$$\hat{x}_m^T \mathcal{B}_N^T \mathcal{B}_N \hat{x}_m + \hat{x}_m^T \mathcal{B}_N^T \mathcal{B}_N \varepsilon_m + \varepsilon_m^T \mathcal{B}_N^T \mathcal{B}_N \hat{x}_m + \varepsilon_m^T (\mathcal{B}_N^T \mathcal{B}_N - \gamma^2 P_m^{-1}) \varepsilon_m$$
$$W_{m,k}^T (\mathcal{W}_N^T \mathcal{W}_N - \gamma^2 \mathcal{Q}_N^{-1}) + V_{m,k}^T (\mathcal{V}_N^T \mathcal{V}_N - \gamma^2 \mathcal{R}_N^{-1}) < 0 \,.$$

To eliminate variables, we represent the previous inequality with

$$\begin{bmatrix} x_m \\ \varepsilon_m \\ W_{m,k} \\ V_{m,k} \end{bmatrix}^T \begin{bmatrix} \mathcal{B}_N^T \mathcal{B}_N & \mathcal{B}_N^T \mathcal{B}_N & 0 & 0 \\ \mathcal{B}_N^T \mathcal{B}_N & \mathcal{B}_N^T \mathcal{B}_N - \gamma^2 P_m^{-1} & 0 & 0 \\ 0 & 0 & \mathcal{Z}_w & 0 \\ 0 & 0 & 0 & \mathcal{Z}_v \end{bmatrix} \begin{bmatrix} x_m \\ \varepsilon_m \\ W_{m,k} \\ V_{m,k} \end{bmatrix} < 0 \,,$$

where $\mathcal{Z}_w = \mathcal{W}_N^T \mathcal{W}_N - \gamma^2 \mathcal{Q}_N^{-1}$ and $\mathcal{Z}_v = \mathcal{V}_N^T \mathcal{V}_N - \gamma^2 \mathcal{R}_N^{-1}$, and note that, in the new form, this inequality holds only if the following matrix is negative definite,

$$\begin{bmatrix} \mathcal{B}_N^T \mathcal{B}_N & \mathcal{B}_N^T \mathcal{B}_N & 0 & 0 \\ \mathcal{B}_N^T \mathcal{B}_N & \mathcal{B}_N^T \mathcal{B}_N - \gamma^2 P_m^{-1} & 0 & 0 \\ 0 & 0 & \mathcal{W}_N^T \mathcal{W}_N - \gamma^2 \mathcal{Q}_N^{-1} & 0 \\ 0 & 0 & 0 & \mathcal{V}_N^T \mathcal{V}_N - \gamma^2 \mathcal{R}_N^{-1} \end{bmatrix} < 0 \,. \tag{8.203}$$

We now recall that the leading principal minors of a negative definite matrix are all negative. The first-order leading principal minor $T = \mathcal{B}_N^T \mathcal{B}_N$ will be negative definite, if we introduce a new variable Z such that $Z - \mathcal{B}_N^T \mathcal{B}_N < 0$ and solve a corresponding LMI. The second-order leading principle minor

$$\begin{bmatrix} \mathcal{B}_N^T \mathcal{B}_N & \mathcal{B}_N^T \mathcal{B}_N \\ \mathcal{B}_N^T \mathcal{B}_N & \mathcal{B}_N^T \mathcal{B}_N - \gamma^2 P_m^{-1} \end{bmatrix}$$

will be negative definite if the following Schur complement is negative,

$$T - T(T - \gamma^2 P_m^{-1}) T < 0 \,.$$

Using the Woodbury identity, we transform this inequality to $T^{-1} - \gamma^{-2} P_m > 0$ and notice that it is also valid for the third leading principal minor and the matrix (8.203) itself. Thus, the leading principal minors give two conditions:

$$\begin{bmatrix} Z & \mathcal{B}_N^T \\ \mathcal{B}_N & I \end{bmatrix} < 0 \,, \quad \begin{bmatrix} \gamma^2 P_m^{-1} & \mathcal{B}_N^T \\ \mathcal{B}_N & I \end{bmatrix} > 0 \,. \tag{8.204}$$

Since the diagonal elements are all negative in the negative definite matrix, the remaining matrices in (8.203) are also negative definite, and we represent then with the following LMIs

$$\begin{bmatrix} \gamma \mathcal{Q}_N^{-1} & \mathcal{W}_N^T \\ \mathcal{W}_N & \gamma I \end{bmatrix} > 0 \,, \quad \begin{bmatrix} \gamma \mathcal{R}_N^{-1} & \mathcal{V}_N^T \\ \mathcal{V}_N & \gamma I \end{bmatrix} > 0 \,. \tag{8.205}$$

Using the error residual matrices (8.15)–(8.17), we finally formulate the robust *a posteriori* energy-to-power FIR filtering algorithm as follows. Solve the following minimization problem by initializing the minimization with $\mathcal{H}_N = (C_N^T C_N)^{-1} C_N^T$,

$$\mathcal{H}_N \Leftarrow \min_{\mathcal{H}_N} \gamma > 0 \tag{8.206}$$

subject to

$$\begin{bmatrix} Z & (F^{N-1} - \mathcal{H}_N H_N)^T \\ F^{N-1} - \mathcal{H}_N H_N & I \end{bmatrix} < 0 \,,$$
$$\begin{bmatrix} \gamma^2 P_m^{-1} & (F^{N-1} - \mathcal{H}_N H_N)^T \\ F^{N-1} - \mathcal{H}_N H_N & I \end{bmatrix} > 0 \,, \tag{8.207}$$

$$\begin{bmatrix} \gamma \mathcal{Q}_N^{-1} & (\bar{D}_N - \mathcal{H}_N \mathcal{T}_N)^T \\ \bar{D}_N - \mathcal{H}_N \mathcal{T}_N & \gamma I \end{bmatrix} > 0 , \tag{8.208}$$

$$\begin{bmatrix} \gamma \mathcal{R}_N^{-1} & \mathcal{H}_N^T \\ \mathcal{H}_N & \gamma I \end{bmatrix} > 0 . \tag{8.209}$$

Provided that the gain $\mathcal{H}_N$ is numerically available from (8.206), the *a posteriori* energy-to-power FIR filtering estimate and the error covariance can be computed as

$$\hat{x}_k = \mathcal{H}_N Y_{m,k} , \tag{8.210}$$

$$P = \mathcal{B}_N \mathcal{X}_m \mathcal{B}_N^T + \mathcal{W}_N \mathcal{Q}_N \mathcal{W}_N^T + \mathcal{V}_N \mathcal{R}_N \mathcal{V}_N^T , \tag{8.211}$$

where the error residual matrices are given by (8.15)–(8.17).

8.8.2 Energy-to-Power FIR Predictor

Similarly, the game theory energy-to-power FIR predictor can be obtained if we consider the state-space model (8.6)–(8.9) with zero input $u_k = 0$, define the predicted FIR estimate as (8.56), and consider the estimation error (8.59) with the error residual matrices given by (8.60)–(8.62). Then we can determine the gain for this predictor by solving numerically the following optimization problem

$$\mathcal{H}_N^p = \inf_{\mathcal{H}_N^p} \sup_{\epsilon_m, W_{m,k}, V_{m,k}} \frac{\|\epsilon_{k+1}\|^2}{\|\epsilon_m\|^2_{P_m^{-1}} + \|w\|^2_{\mathcal{Q}^{-1}_{m,k}} + \|v\|^2_{\mathcal{R}^{-1}_{m,k}}} . \tag{8.212}$$

Since problem (8.212) is generally mathematically untractable, we can find the gain suboptimally by solving the inequality

$$\mathcal{H}_N^p \Leftarrow \frac{\|\epsilon_{k+1}\|^2}{\|\epsilon_m\|^2_{P_m^{-1}} + \|w\|^2_{\mathcal{Q}^{-1}_{m,k}} + \|v\|^2_{\mathcal{R}^{-1}_{m,k}}} < \gamma^2 . \tag{8.213}$$

The next steps to deal with (8.213) are similar to the game theory energy-to-power FIR filter derived earlier, and thus the gain for the corresponding FIR predictor can be found as follows. Solve the following minimization problem starting the minimization with $\mathcal{H}_N^p = (C_N^{p^T} C_N^p)^{-1} C_N^{p^T}$,

$$\mathcal{H}_N^p \Leftarrow \min_{\mathcal{H}_N^p} \gamma > 0 \tag{8.214}$$

subject to

$$\begin{bmatrix} Z & (F^N - \mathcal{H}_N^p H_N^p)^T \\ F^N - \mathcal{H}_N^p H_N^p & I \end{bmatrix} < 0 ,$$

$$\begin{bmatrix} \gamma^2 P_m^{-1} & (F^N - \mathcal{H}_N^p H_N^p)^T \\ F^N - \mathcal{H}_N^p H_N^p & I \end{bmatrix} > 0 , \tag{8.215}$$

$$\begin{bmatrix} \gamma \mathcal{Q}_N^{-1} & (\bar{D}_N - \mathcal{H}_N^p \mathcal{T}_N^p)^T \\ \bar{D}_N - \mathcal{H}_N^p \mathcal{T}_N^p & \gamma I \end{bmatrix} > 0 , \tag{8.216}$$

$$\begin{bmatrix} \gamma \mathcal{R}_N^{-1} & \mathcal{H}_N^{p^T} \\ \mathcal{H}_N^p & \gamma I \end{bmatrix} > 0 . \tag{8.217}$$

Provided that $\mathcal{H}_N^p$ is available from (8.214), compute the game theory energy-to-power FIR prediction and the corresponding error covariance as

$$\tilde{x}_{k+1} = \mathcal{H}_N^p Y_{m,k} , \tag{8.218}$$

$$P = \mathcal{B}_N^p \chi_m \mathcal{B}_N^{p^T} + \mathcal{W}_N^p \mathcal{Q}_N \mathcal{W}_N^{p^T} + \mathcal{V}_N^p \mathcal{R}_N \mathcal{V}_N^{p^T} , \tag{8.219}$$

using the error residual matrices given by (8.60)–(8.62).

We have already shown how to make the H_2 FIR state estimator bias-constrained or optimal unbiased. It now remains to note that all the other robust FIR state estimation numerical algorithms obtained in this chapter can also be modified to be bias-constrained. This can be done if we remove the term with χ_m and subject the LMI-based algorithms to the unbiasedness constraint, as in (8.81). Modifications can also be provided for more general state-space models with control inputs, and we postpone it to "Problems."

8.9 Recursive Computation of Robust FIR Estimates

A common feature of robust FIR estimators discussed in this chapter is that the suboptimal time-invariant gain $\mathcal{H}_N \in \mathbb{R}^{K\times N}$ is preliminary obtained numerically at the test stage. The gain has K rows and N columns, and we can represent it in the block matrix form

$$\mathcal{H}_N = \begin{bmatrix} h_{N-1} & h_{N-2} & \dots & h_0 \end{bmatrix} , \tag{8.220}$$

where $h_i \in \mathbb{R}^K$, $i \in [0, N-1]$, is a known column matrix. Using (8.220), we can also obtain recursive forms for systems with and without the input (control) signal u_k. It is worth noting that such forms can be used to compute all available FIR filtering estimates, provided that the filter gain $\mathcal{H}_N$ is obtained numerically as (8.220).

8.9.1 Uncontrolled Processes

When solving signal processing problems, we often deal with disturbed uncontrolled processes represented in state space without input. In this case, the FIR filtering estimate is given by

$$\hat{x}_k = \mathcal{H}_N Y_{m,k} . \tag{8.221}$$

We now consider the general LTV case, substitute (8.220) with

$$\mathcal{H}_{m,k} = \begin{bmatrix} h_m & h_{m+1} & \dots & h_k \end{bmatrix} = \begin{bmatrix} \mathcal{H}_{m,k-1} & h_k \end{bmatrix} , \tag{8.222}$$

and similarly decompose $Y_{m,k}$ as $Y_{m,k} = \begin{bmatrix} Y_{m,k-1}^T & y_k^T \end{bmatrix}^T$. This gives the following simple recursive form

$$\begin{aligned} \hat{x}_k &= \mathcal{H}_N Y_{m,k} \\ &= \begin{bmatrix} \mathcal{H}_{m,k-1} & h_k \end{bmatrix} \begin{bmatrix} Y_{m,k-1} \\ y_k \end{bmatrix} \\ &= \hat{x}_{k-1} + h_k y_k . \end{aligned} \tag{8.223}$$

Note that we can always go back to the LTI case by extracting the time variable k from (8.222). This concludes the discussion, and we move on to the general case of systems with inputs, the recursive algorithm of which will be associated with (8.223) in a particular case.

8.9.2 Controlled Processes

For LTI systems with input u_k, the robust FIR filtering estimate $\hat{x}_k$ is given by a familiar relation

$$\begin{aligned}\hat{x}_k &= \mathcal{H}_N Y_{m,k} + (\bar{S}_N - \mathcal{H}_N L_N) U_{m,k} \\ &= \hat{x}_k^{\mathrm{h}} + \hat{x}_k^{\mathrm{f}} \,.\end{aligned} \tag{8.224}$$

To find recursive forms for (8.224), we also start with the general LTV case, replace $\bar{S}_N$ with $\bar{S}_{m,k}$ and L_N with $L_{m,k}$, assign $U_{m,k} = \begin{bmatrix} U_{m,k-1}^T & u_k^T \end{bmatrix}^T$, refer to the decompositions introduced in Chapter 4, and write

$$\bar{S}_{m,k} = \begin{bmatrix} F_k \bar{S}_{m,k-1} & E_k \end{bmatrix}, \quad L_{m,k} = \begin{bmatrix} L_{m,k-1} & 0 \\ H_k F_k \bar{S}_{m,k-1} & H_k E_k \end{bmatrix} . \tag{8.225}$$

Noticing that the recursion for $\hat{x}_k^{\mathrm{h}}$ is given by (8.223), we represent $\hat{x}_k^{\mathrm{f}}$ as

$$\begin{aligned}\hat{x}_k^{\mathrm{f}} &= \bar{S}_{m,k} U_{m,k} - \mathcal{H}_{m,k} L_{m,k} U_{m,k} \\ &= \begin{bmatrix} F_k \bar{S}_{m,k-1} & E_k \end{bmatrix} \begin{bmatrix} U_{m,k-1} \\ u_k \end{bmatrix} \\ &\quad - \begin{bmatrix} \mathcal{H}_{m,k-1} & h_k \end{bmatrix} \begin{bmatrix} L_{m,k-1} & 0 \\ H_k F_k \bar{S}_{m,k-1} & H_k E_k \end{bmatrix} \begin{bmatrix} U_{m,k-1} \\ u_k \end{bmatrix} \\ &= F_k \bar{S}_{m,k-1} U_{m,k-1} + E_k u_k - \mathcal{H}_{m,k-1} L_{m,k-1} U_{m,k-1} \\ &\quad - h_k H_k F_k \bar{S}_{m,k-1} U_{m,k-1} - h_k H_k E_k u_k \\ &= \hat{x}_{k-1}^{\mathrm{f}} - \bar{S}_{m,k-1} U_{m,k-1} + \bar{S}_{m,k} U_{m,k} \\ &\quad - h_k H_k F_k \bar{S}_{m,k-1} U_{m,k-1} - h_k H_k E_k u_k \,,\end{aligned} \tag{8.226}$$

introduce an auxiliary vector

$$\bar{u}_k = \bar{S}_{m,k} U_{m,k} = F_k \bar{u}_{k-1} + E_k u_k \,, \tag{8.227}$$

and transform (8.226) to

$$\hat{x}_k^{\mathrm{f}} = \hat{x}_{k-1}^{\mathrm{f}} - h_k H_k E_k \bar{u}_{k-1} - h_k H_k E_k u_k + \bar{u}_k - \bar{u}_{k-1} \,. \tag{8.228}$$

By combining (8.223) and (8.228), we finally arrive at the recursions

$$\bar{u}_k = F_k \bar{u}_{k-1} + E_k u_k \,, \tag{8.229}$$

$$\hat{x}_k = \hat{x}_{k-1} + h_k (y_k - H_k \bar{u}_k) + \bar{u}_k - \bar{u}_{k-1} \,, \tag{8.230}$$

which serve on $[m, k]$ for the initial values $\bar{u}_m = 0$ and $\hat{x}_m = h_m y_m$.

The pseudocode of the general iterative robust FIR filtering algorithm for LTV systems with input u_k is listed as Algorithm 19. The computation assumes that the time-varying gain $\mathcal{H}_{m,k}$ is available for each horizon $[m, k]$. Since this is not the case of the robust FIR state estimators obtained in the transform domain, we also provide the pseudocode (Algorithm 20) of the robust FIR filtering algorithm associated with the transform domain and related to LTI systems. This algorithm employs a time-invariant gain (8.220) and starts iterating for an initial $\bar{u}_m = 0$ and $\hat{x}_m = h_{N-1} y_m$. One can use Algorithm 19 or Algorithm 20 to reduce the computational complexity of robust batch estimates, especially when $N \gg 1$.

Algorithm 19: Iterative FIR Filtering Algorithm for LTV Models

Data: y_k, u_k
begin
  for $k = 1, 2, \cdots$ **do**
    $m = k - N + 1$ if $k > N - 1$ and $m = 0$ otherwise;
    $\bar{u}_m = 0, x_m = h_m y_m$;
    for $i = m + 1 : k$ **do**
      $\bar{u}_i = F_i \bar{u}_{i-1} + E_i u_i$;
      $\bar{x}_i = \bar{x}_{i-1} + h_i (y_i - H_i \bar{u}_i) + \bar{u}_i - \bar{u}_{i-1}$;
    end for
    $\hat{x}_k = \bar{x}_k$;
  end for
  Result: $\hat{x}_k$
end

Algorithm 20: Iterative FIR Filtering Algorithm for LTI Models

Data: y_k, u_k
begin
  for $k = 1, 2, \cdots$ **do**
    $m = k - N + 1$ if $k > N - 1$ and $m = 0$ otherwise;
    $\bar{u}_m = 0, x_m = h_{N-1} y_m$;
    for $i = m + 1 : k$ **do**
      $\bar{u}_i = F \bar{u}_{i-1} + E u_i$;
      $\bar{x}_i = \bar{x}_{i-1} + h_{i-k} (y_i - H \bar{u}_i) + \bar{u}_i - \bar{u}_{i-1}$;
    end for
    $\hat{x}_k = \bar{x}_k$;
  end for
  Result: $\hat{x}_k$
end

8.10 FIR Smoothing Under Disturbances

Anyone serious about signal processing is interested in smoothed estimates, which are often required in the presence of disturbances. Several robust RH FIR smoothers have been developed for disturbed processes. The energy-to-error RH FIR smoother subject to the unbiasedness constraint was obtained in [67] for deterministic discrete models under disturbances, and the continuous-time minimax RH FIR smoother was proposed in [68]. The H_∞ RH FIR smoother derived in [3] was obtained by computing the gain numerically using LMI.

In batch form, the derivation of robust FIR smoothers can be provided by combining the OFIR smoothing approach developed in chapter 5 with robust criteria discussed in this chapter. We do not pay attention to these solutions and postpone derivation of the relevant algorithms to "Problems."

8.11 Summary

The main theme of this chapter was to introduce the reader to various types of robust FIR state estimators for systems operating under disturbances in the presence of data and initial errors, all norm-bounded. We have shown that the estimator becomes robust if it is tuned to the maximized disturbance. In this case, any deviation from the optimal (minimum error) tuning point causes an increase in errors and a decrease in disturbance, which compensate for each other, and hence robustness. The corresponding algorithms can be obtained and studied quite simply by using the extended state-space models and the corresponding lemmas. To keep the results more memorable, next we list the most critical remarks and conclusions related to H_2, H_∞, $\mathcal{L}_2$-to-$\mathcal{L}_\infty$, $\mathcal{L}_\infty$-to-$\mathcal{L}_\infty$, and hybrid FIR structures operated under norm-bounded disturbances and errors.

Robust FIR state estimators of this type are developed to minimize the disturbance-to-error weighted transfer function $\bar{\mathcal{T}}(z)$ in the transform domain or the disturbance-to-error norms ratio γ in the time domain using different cost criteria in the presence of norm-bounded measurement and initial errors. These estimators are batch-based, operate on full error matrices, and generally do not have recursive forms.

The H_2 FIR state estimator minimizes the squared Frobenius norm of the weighted disturbance-to-error transfer function $\bar{\mathcal{T}}(z)$. By Parseval's theorem, this is equivalent to minimizing MSE in the time domain. Therefore, the H_2 problem is convex and has closed-form optimal solutions. The *a posteriori* H_2-OFIR filter generalizes the *a posteriori* OFIR filter in the special case of LTI processes with white Gaussian and uncorrelated disturbances and errors. The later also holds for H_2-OFIR and OFIR predictors. For diagonal Gaussian error matrices, the *a posteriori* H_2-OFIR filter and the H_2-OFIR predictor become, respectively, the *a posteriori* Kalman filter and the Kalman predictor.

The H_∞ FIR estimator minimizes the H_∞ induced norm of $\mathcal{T}(z)$, which reflects the worst estimation case. Thereby, it minimizes the highest peak value of $\mathcal{T}(z)$ in the Bode plot. Since the H_∞ norm coincides with the $\mathcal{L}_2$ induced norm, the H_2 FIR state estimator is also the energy-to-energy or $\mathcal{L}_2$-to-$\mathcal{L}_2$ FIR state estimator. To improve the performance, the H_∞ FIR algorithm can be combined with the H_2 FIR algorithm.

The $\mathcal{L}_2$-to-$\mathcal{L}_\infty$ FIR estimator also known as the generalized H_2 FIR estimator minimizes the energy-to-peak norms ratio by maximizing the disturbance $\mathcal{L}_2$ norm and minimizing the error $\mathcal{L}_\infty$ norm. This estimator can also be combined with other FIR structures in hybrid algorithms. The $\mathcal{L}_1$ FIR estimator minimizes the peak error for the maximized peak disturbance and is called the peak-to-peak or $\mathcal{L}_\infty$-to-$\mathcal{L}_2$ FIR estimator. This estimator is useful when the system is affected by impulsive attacks and energy minimization does not provide sufficient accuracy and stability.

For systems operating under disturbances, measurement errors, and initial errors, robust FIR state estimators can also be developed using a game theory approach. Examples are the energy-to-power FIR filter and predictor presented in this chapter. Since the game theory can be viewed as an artificial intelligence tool, many other efficient FIR solutions can be expected in the near future.

All of the approaches discussed in this chapter can be used to develop robust FIR smoothing algorithms. Since such smoothers are obtained very similarly to filters and predictors, we postpone them to "Problems."

8.12 Problems

1 Consider the H_2 FIR filtering problem. Obtain the gain $\mathcal{H}_N$ for the non-weighted transfer function $T(z)$ and compare it with the optimal gain (8.34). Explain the difference and outline drawbacks.

2 The H_2 FIR filter gain

$$\mathcal{H}_N = (C_N^T \Omega_N^{-1} C_N)^{-1} C_N^T \Omega_N^{-1}$$

is given in [109] with $\Omega_N = (G_N + \bar{T}_N)(G_N + \bar{T}_N)^T$. Compare this gain to (8.42) and analyze the differences. What practical limitations does this gain have?

3 Consider two discrete-time systems represented by the equations $x_k = Fx_{k-1} + Bw_k$ and $y_k = Hx_k + Dw_k$ and matrices

$$1)\quad F = \begin{bmatrix} 1 & 1 & 0.5 \\ 0 & 1 & 1 \\ 0 & 0 & 1 \end{bmatrix}, \quad B = \begin{bmatrix} 0.5 & 0 \\ 1 & 0 \\ 0 & 1 \end{bmatrix}, \quad H = \begin{bmatrix} 1 & 0 & 0 \\ 0 & 1 & 0 \end{bmatrix}, \quad D = \begin{bmatrix} 0 & 1 \\ 0 & 0 \end{bmatrix}.$$

$$2)\quad F = \begin{bmatrix} \sin\varphi & \cos\varphi \\ -\cos\varphi & \sin\varphi \end{bmatrix}, \quad B = \begin{bmatrix} 1 & 0 \\ 0 & 1 \end{bmatrix}, \quad H = \begin{bmatrix} 1 & 0 \end{bmatrix}, \quad D = \begin{bmatrix} 1 \end{bmatrix}.$$

Compute the H_2 and H_∞ norms of the transfer functions of both systems. Investigate the effect of the angle φ on the second system.

4 Given the model in (8.158) and (8.159), let $\gamma > 0$ be a real scalar and w_k a bounded peak disturbance, $\|w\|_\infty < \infty$. Find the solution, if any, to the peak-to-energy or $\mathcal{L}_\infty$–$\mathcal{L}_2$ filtering problem

$$\mathcal{H}_N \Leftarrow \sup_{\|w\|_\infty < \infty} \frac{\|y\|_2^2}{\|w\|_\infty^2} < \gamma^2 ,$$

Otherwise, argue that the gain $\mathcal{H}_N$ cannot be derived.

5 Given the matrices A, $V \in \mathbb{R}^{n\times n}$, $P > 0$, $0 < X \in \mathbb{R}^{n\times n}$, $B \in \mathbb{R}^{n\times m}$, $C \in \mathbb{R}^{p\times n}$, $D \in \mathbb{R}^{p\times m}$, $0 < R \in \mathbb{R}^{m\times m}$, and $0 < S \in \mathbb{R}^{p\times p}$, where X, R, and S are positive definite. Show that the matrix inequality [31]

$$\begin{bmatrix} -V - V^T & VA + P & VB & 0 & V \\ * & -2P + X & 0 & C^T & 0 \\ * & * & -R & D^T & 0 \\ * & * & * & -S & 0 \\ * & * & * & * & -X \end{bmatrix} < 0 ,$$

where "$*$" means the transform of the relevant symmetric term, implies the following LMI [31]

$$\begin{bmatrix} PA + A^T P & PB & C^T \\ * & -R & D^T \\ * & * & -S \end{bmatrix} < 0.$$

6 Given the discrete Lyapunov equation $F^TPF - P + Q = 0$. Represent this equation in various LMI forms.

7 Show that for $A \in \mathbb{R}^{n\times n}$, $B \in \mathbb{R}^{n\times m}$, $C \in \mathbb{R}^{p\times n}$, $D \in \mathbb{R}^{p\times m}$, positive definite $0 < P \in \mathbb{R}^{n\times n}$ and $0 < Q \in \mathbb{R}^{n\times n}$, and a scalar $\gamma > 0$ the following matrix inequalities are equivalent

$$\begin{bmatrix} PA + A^TP & PB & C^T \\ * & -\gamma I & D^T \\ * & * & -\gamma I \end{bmatrix} < 0 , \quad \begin{bmatrix} AQ + QA^T & B & QC^T \\ * & -\gamma I & D^T \\ * & * & -\gamma I \end{bmatrix} < 0 .$$

8 Consider the DARI

$$F^TPF - F^TPB(R + B^TPB)^{-1}BPF + Q - P > 0 .$$

Using Schur's complement, show that this inequality can be equivalently represented by the following matrix inequalities

$$\begin{bmatrix} F^TPF - P + Q & F^TPB \\ * & R + B^TPB \end{bmatrix} > 0 ,$$

$$\begin{bmatrix} Q & 0 & F^TP & P \\ * & R & B^TP & 0 \\ * & * & P & 0 \\ * & * & * & P \end{bmatrix} > 0 , \quad \begin{bmatrix} Q & 0 & F^T & I \\ * & R & B^T & 0 \\ * & * & X & 0 \\ * & * & * & X \end{bmatrix} > 0 ,$$

if there exist $P > 0$, $Q > 0$, and $X > 0$.

9 A continuous-time LTI system operating under the bounded disturbance $w(t)$ is given in state space with the equations $\frac{d}{dt}x(t) = Ax(t) + Bw(t)$ and $y(t) = Cx(t) + Dw(t)$. The supply function $s(w, y)$ is given by

$$s(w, y) = \begin{bmatrix} w \\ y \end{bmatrix}^T \begin{bmatrix} Q & S \\ S^T & R \end{bmatrix} \begin{bmatrix} w \\ y \end{bmatrix} = w^TQw + w^TSy + y^TS^Tw + y^TRy ,$$

where the positive definite symmetric matrix P is

$$P = \begin{bmatrix} Q & S \\ S^T & R \end{bmatrix} .$$

Transform this state-space model to discrete time and define the supply function.

10 Using the derivation of the minimax RH FIR smoother given in [68], obtain the minimax OUFIR smoother.

11 The H_∞ RH FIR smoother was obtained in [3] by solving the LMI. Following this approach, obtain a suboptimal H_∞ FIR smoother for systems affected by disturbances and measurement errors.

9

Robust FIR State Estimation for Uncertain Systems

A statistical analysis, properly conducted, is a delicate dissection of uncertainties, a surgery of suppositions.

Michael J. Moroney [132], p. 3.

In physical systems, various uncertainties occur naturally and are usually impossible to deal with. An example is the sampling time that is commonly set constant but changes due to frequency drifts in low-accuracy oscillators of timing clocks. To mitigate the effect of uncertainties, more process states can be involved that, however, can cause computational errors and latency. Therefore, robust estimators are required [79, 161]. Most of works developing estimators for uncertain systems follow the approach proposed in [51], where the system and observation uncertainties are represented via a single strictly bounded unknown matrix and known real constant matrices. For uncertainties considered as multiplicative errors, the approach was developed in [56], and for uncertainties coupled with model matrices with scalar factors, some results were obtained in [180]. In early works on robust FIR filtering for uncertain systems [98, 99], the problem was solved using recursive forms that is generally not the case. In the convolution-based batch form, several solutions were originally found in [151, 152] for some special cases.

Like in the case of disturbances, robust FIR estimators can be designed using different approaches by minimizing estimation errors for maximized uncertainties. Moreover, we will show that effects caused by uncertainties and disturbances can be accounted for as an unspecified impact. Accordingly, the idea behind the estimator robustness can be illustrated as shown in Fig. 9.1, which is supported by Fig. 8.3. Assume that an error factor η exists from $\eta_{\min}$ to $\eta_{\max}$ and causes an unspecified impact $\psi(\eta)$ (uncertainty, disturbance, etc.) to grow from point A to point C. Suppose that optimal tuning mitigates the effect by a factor α and consider two extreme cases. By tuning an estimator to $\eta_{\min}$, we go from point A to point B. Then an increase in η will cause an increase in tuning errors and in $\psi(\eta)$, and the estimation error can significantly grow. Now, we tune an estimator to $\eta_{\max}$ and go from point C to D. Then a decrease in η will cause an increase in tuning errors and a decrease in $\psi(\eta)$. Since both these effects compensate for each other, the estimate becomes robust.

In this chapter, we develop the theory of robust FIR state estimation for uncertain systems operating under disturbances with initial and measurement errors. In this regard, such estimators can be considered the most general, since they unify other robust FIR solutions in particular cases. However, further efforts need to be made to turn most of these estimators into practical algorithms.

Optimal and Robust State Estimation: Finite Impulse Response (FIR) and Kalman Approaches, First Edition.
Yuriy S. Shmaliy and Shunyi Zhao.

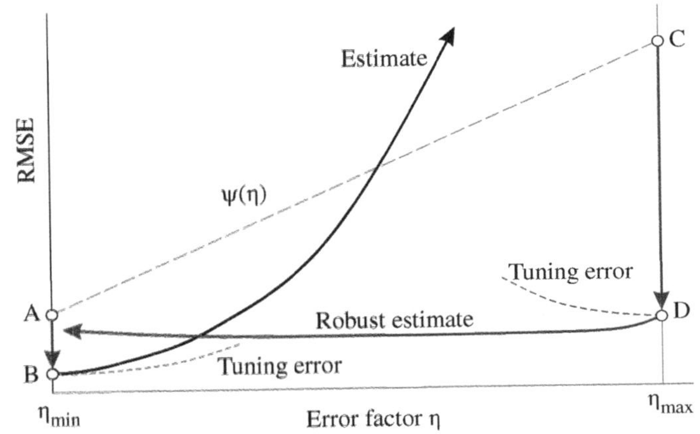

Figure 9.1 Errors caused by optimal tuning an estimator to η_{min} and η_{max}: tuning to η_{max} makes the filter robust.

9.1 Extended Models for Uncertain Systems

Traditionally, we will develop robust FIR state estimators for uncertain systems using either a BE-based model that is suitable for *a posteriori* FIR filtering or an FE-based model that is suitable for FIR prediction and FIR predictive filtering. For clarity, we note once again that these solutions are generally inconvertible. Since the most elaborated methods, which guarantee robust performance, have been developed for uncertain LTI systems in the transform domain, we start with BE- and FE-based state space models and their extensions on $[m, k]$.

Backward Euler Method–Based Model

Consider an uncertain linear system and represent it in discrete-time state-space with the following equations,

$$x_k = F_k^u x_{k-1} + E_k^u u_k + B_k^u w_k \,, \tag{9.1}$$

$$y_k = H_k^u x_k + D_k^u w_k + v_k \,, \tag{9.2}$$

where $F_k^u = F + \Delta F_k$, $E_k^u = E + \Delta E_k$, $B_k^u = B + \Delta B_k$, $H_k^u = H + \Delta H_k$, and $D_k^u = D + \Delta D_k$. The time-varying increments ΔF_k, ΔE_k, ΔB_k, ΔH_k, and ΔD_k represent bounded parameter uncertainties, w_k is the disturbance, and v_k is the measurement error. Hereinafter, we will use the superscript "u" to denote uncertain matrices.

We assume that all errors in (9.1) and (9.2) are norm-bounded, have zero mean, and can vary arbitrailry over time; so we cannot know their exact distributions and covariances. Note that the zero mean assumption matters, because otherwise a nonzero mean will cause regular bias errors and the model will not be considered correct.

To extend (9.1) and (9.2) to $[m, k]$, we separate the regular and zero mean uncertain components and represent the model in standard form

$$x_k = F x_{k-1} + E u_k + \xi_k \,, \tag{9.3}$$

$$y_k = H x_k + \zeta_k \,, \tag{9.4}$$

where the zero mean uncertain vectors are given by

$$\xi_k = \Delta F_k x_{k-1} + \Delta E_k u_k + B_k^u w_k \,, \tag{9.5}$$

$$\zeta_k = \Delta H_k x_k + D_k^u w_k + v_k \,. \tag{9.6}$$

Then, similarly to (8.1), the model in (9.3) can be extended as

$$X_{m,k} = F_N x_m + S_N U_{m,k} + \hat{F}_N \Xi_{m,k} \,, \tag{9.7}$$

where matrices F_N and S_N are defined after (8.4) and the extended error vector $\Xi_{m,k}$ and matrix $\hat{F}_N$ are given by

$$\Xi_{m,k} = \begin{bmatrix} \xi_m \\ \xi_{m+1} \\ \vdots \\ \xi_{k-1} \\ \xi_k \end{bmatrix}, \quad \hat{F}_N = \begin{bmatrix} I & 0 & \dots & 0 & 0 \\ F & I & \dots & 0 & 0 \\ \vdots & \vdots & \ddots & \vdots & \vdots \\ F^{N-2} & F^{N-3} & \dots & I & 0 \\ F^{N-1} & F^{N-2} & \dots & F & I \end{bmatrix}. \tag{9.8}$$

We next extend the uncertain vector ξ_k to $[m, k]$ as

$$\Xi_{m,k} = F^{\Delta}_{m,k} x_m + S^{\Delta}_{m,k} U_{m,k} + (\bar{B}_N + D^{\Delta}_{m,k}) W_{m,k} \,, \tag{9.9}$$

where the uncertain block matrices are defined by

$$F^{\Delta}_{m,k} = \begin{bmatrix} 0 \\ \Delta F_{m+1} \\ \vdots \\ \Delta F_{k-1} \tilde{\mathcal{F}}^{m+1}_{k-2} \\ \Delta F_k \tilde{\mathcal{F}}^{m+1}_{k-1} \end{bmatrix}, \quad \bar{B}_N = \mathrm{diag}(\underbrace{B \ B \ \dots \ B}_{N}),$$

$$S^{\Delta}_{m,k} = \begin{bmatrix} \Delta E_m & 0 & \dots & 0 & 0 \\ \Delta F_{m+1} E^u_m & \Delta E_{m+1} & \dots & 0 & 0 \\ \vdots & \vdots & \ddots & \vdots & \vdots \\ \Delta F_{k-1} \tilde{\mathcal{F}}^{m+1}_{k-2} E^u_m & \Delta F_{k-1} \tilde{\mathcal{F}}^{m+2}_{k-2} E^u_{m+1} & \dots & \Delta E_{k-1} & 0 \\ \Delta F_k \tilde{\mathcal{F}}^{m+1}_{k-1} E^u_m & \Delta F_k \tilde{\mathcal{F}}^{m+2}_{k-1} E^u_{m+1} & \dots & \Delta F_k E^u_{k-1} & \Delta E_k \end{bmatrix},$$

$$D^{\Delta}_{m,k} = \begin{bmatrix} \Delta B_m & 0 & \dots & 0 & 0 \\ \Delta F_{m+1} B^u_m & \Delta B_{m+1} & \dots & 0 & 0 \\ \vdots & \vdots & \ddots & \vdots & \vdots \\ \Delta F_{k-1} \tilde{\mathcal{F}}^{m+1}_{k-2} B^u_m & \Delta F_{k-1} \tilde{\mathcal{F}}^{m+2}_{k-2} B^u_{m+1} & \dots & \Delta B_{k-1} & 0 \\ \Delta F_k \tilde{\mathcal{F}}^{m+1}_{k-1} B^u_m & \Delta F_k \tilde{\mathcal{F}}^{m+2}_{k-1} B^u_{m+1} & \dots & \Delta F_k B^u_{k-1} & \Delta B_k \end{bmatrix}, \tag{9.10}$$

in which $E^u_i = E + \Delta E_i$ and $B^u_i = B + \Delta B_i$ hold for $i \in [m, k]$, and matrix $\tilde{\mathcal{F}}^g_r$ of the uncertain product is specified with

$$\tilde{\mathcal{F}}^g_r = \begin{cases} F^u_r F^u_{r-1} \dots F^u_g \,, & g < r+1 \,, \\ I \,, & g = r+1 \\ 0 \,, & g > r+1 \end{cases} . \tag{9.11}$$

By combining (9.7) with (9.9) and referring to the identity $\hat{F}_N\bar{B}_N = D_N$, where matrix D_N is defined after (8.4), we rewrite model (9.7) as

$$\begin{aligned} X_{m,k} &= (F_N + \tilde{F}_{m,k})x_m + (S_N + \tilde{S}_{m,k})U_{m,k} \\ &\quad + (D_N + \tilde{D}_{m,k})W_{m,k} \,, \end{aligned} \tag{9.12}$$

where $\tilde{F}_{m,k} = \hat{F}_N F^{\Delta}_{m,k}$, $\tilde{S}_{m,k} = \hat{F}_N S^{\Delta}_{m,k}$, and $\tilde{D}_{m,k} = \hat{F}_N D^{\Delta}_{m,k}$. We now notice that, for systems without uncertainties, $\hat{F}_N F^{\Delta}_{m,k} = 0$, $\hat{F}_N S^{\Delta}_{m,k} = 0$, and $\hat{F}_N D^{\Delta}_{m,k} = 0$ bring (9.12) to the standard form (8.3).

The system current state x_k can now be expressed in terms of the last row vector in (9.12) as

$$\begin{aligned} x_k &= (F^{N-1} + \bar{\tilde{F}}_{m,k})x_m + (\bar{S}_N + \bar{\tilde{S}}_{m,k})U_{m,k} \\ &\quad + (\bar{D}_N + \bar{\tilde{D}}_{m,k})W_{m,k} \,, \end{aligned} \tag{9.13}$$

where $\bar{\tilde{F}}_{m,k}$, $\bar{\tilde{S}}_{m,k}$, and $\bar{\tilde{D}}_{m,k}$ are the last row vectors in $\tilde{F}_{m,k}$, $\tilde{S}_{m,k}$, and $\tilde{D}_{m,k}$, respectively.

We also extend the observation model (9.4) as

$$Y_{m,k} = H_N x_m + L_N U_{m,k} + M_N \Xi_{m,k} + \Pi_{m,k}, \tag{9.14}$$

where matrices H_N and L_N are defined after (8.4), $M_N = \bar{H}_N \hat{F}_N$, and $\Pi_{m,k} = [\, \zeta_m^T \;\; \zeta_{m+1}^T \;\; \cdots \;\; \zeta_k^T \,]^T$ is the vector of uncertain observation errors, which has the following extension to $[m, k]$,

$$\begin{aligned} \Pi_{m,k} &= N^{\Delta}_{m,k} x_m + L^{\Delta}_{m,k} U_{m,k} + M^{\Delta}_{m,k} \Xi_{m,k} \\ &\quad + (\bar{T}_N + \bar{T}^{\Delta}_{m,k})W_{m,k} + V_{m,k} \,, \end{aligned} \tag{9.15}$$

for which the uncertain matrices are given by

$$N^{\Delta}_{m,k} = \begin{bmatrix} \Delta H_m \\ \Delta H_{m+1} F \\ \vdots \\ \Delta H_{k-1} F^{N-2} \\ \Delta H_k F^{N-1} \end{bmatrix}, \quad \bar{T}^{\Delta}_{m,k} = \mathrm{diag}(\, \Delta D_m \;\; \Delta D_{m-1} \;\; \cdots \;\; \Delta D_k \,),$$

$$M^{\Delta}_{m,k} = \begin{bmatrix} \Delta H_m & 0 & \cdots & 0 & 0 \\ \Delta H_{m+1} F & \Delta H_{m+1} & \cdots & 0 & 0 \\ \vdots & \vdots & \ddots & \vdots & \vdots \\ \Delta H_{k-1} F^{N-2} & \Delta H_{k-1} F^{N-3} & \cdots & \Delta H_{k-1} & 0 \\ \Delta H_k F^{N-1} & \Delta H_k F^{N-2} & \cdots & \Delta H_k F & \Delta H_k \end{bmatrix}, \tag{9.16}$$

$L^{\Delta}_{m,k} = M^{\Delta}_{m,k} \bar{E}_N$, $S^{\Delta}_{m,k}$ is specified after (9.10), and $\bar{E}_N = \mathrm{diag}(\underbrace{E, E \,\ldots\, E}_{N})$ and $\bar{T}_N = \mathrm{diag}(\underbrace{D, D \,\ldots\, D}_{N})$ are diagonal.

By combining (9.14) and (9.15), we finally represent the extended observation equation in the form

$$\begin{aligned} Y_{m,k} &= (H_N + \tilde{H}_{m,k})x_m + (L_N + \tilde{L}_{m,k})U_{m,k} \\ &\quad + (T_N + \tilde{T}_{m,k})W_{m,k} + V_{m,k}, \end{aligned} \tag{9.17}$$

where the uncertain matrices are defined in terms of the matrices introduced previously as

$$\tilde{H}_{m,k} = N^{\Delta}_{m,k} + (M_N + M^{\Delta}_{m,k})F^{\Delta}_{m,k} \,, \tag{9.18}$$

$$\tilde{L}_{m,k} = L^{\Delta}_{m,k} + (M_N + M^{\Delta}_{m,k})S^{\Delta}_{m,k} , \tag{9.19}$$

$$\tilde{T}_{m,k} = M^{\Delta}_{m,k}\bar{B}_N + (M_N + M^{\Delta}_{m,k})D^{\Delta}_{m,k} + \bar{T}^{\Delta}_{m,k} . \tag{9.20}$$

It can now be shown that exact modeling with $\tilde{H}_{m,k} = 0$, $\tilde{L}_{m,k} = 0$, and $\tilde{T}_{m,k} = 0$ makes (9.17) the standard model (8.4).

Thus, the BE-based state-space model in (9.1) and (9.2), extended to $[m, k]$ for organizing *a posteriori* FIR filtering of uncertain systems under bounded disturbances with initial and data errors, is given by (9.13) and (9.17).

Forward Euler Method–Based Model

Keeping the definitions of vectors and matrices introduced for (9.1) and (9.2), we now write the FE-based state-space model for uncertain systems as

$$x_{k+1} = F^u_k x_k + E^u_k u_k + B^u_k w_k , \tag{9.21}$$

$$y_k = H^u_k x_k + D^u_k w_k + v_k . \tag{9.22}$$

By reorganizing the terms, we next represent this model in the standard form

$$x_{k+1} = Fx_k + Eu_k + \xi^p_k , \tag{9.23}$$

$$y_k = Hx_k + \zeta^p_k , \tag{9.24}$$

where the uncertain vectors associated with the prediction are denoted by the superscript "p" and are given by

$$\xi^p_k = \Delta F_k x_k + \Delta E_k u_k + B^u_k w_k , \tag{9.25}$$

$$\zeta^p_k = \Delta H_k x_k + D^u_k w_k + v_k . \tag{9.26}$$

Obviously, extensions of vectors (9.23) and (9.24) to $[m, k]$ can be provided similarly to the BE-based model. Referring to (8.8), we first represent (9.23) on $[m, k]$ with respect to the prediction vector $X_{m+1,k+1}$ as

$$X_{m+1,k+1} = F^p_N x_m + S_N U_{m,k} + \hat{F}_N \Xi^p_{m,k} , \tag{9.27}$$

where $\Xi^p_{m,k} = [\ \xi^{p^T}_m \ \ \xi^{p^T}_{m+1} \ \cdots \ \xi^{p^T}_k \ ]^T$, $F^p_N = FF_N$, and F_N, S_N, and D_N are defined after (8.4). Similarly to (9.9), we also express the vector $\Xi^p_{m,k}$ on $[m, k]$ as

$$\Xi^p_{m,k} = F^{p\Delta}_{m,k} x_m + S^{\Delta}_{m,k} U_{m,k} + (\bar{B}_N + D^{\Delta}_{m,k})W_{m,k} , \tag{9.28}$$

where matrix $F^{p\Delta}_{m,k}$ is defined by

$$F^{p\Delta}_{m,k} = \begin{bmatrix} \Delta F_m \\ \Delta F_{m+1} F^u_m \\ \vdots \\ \Delta F_{k-1} \tilde{\mathcal{F}}^m_{k-2} \\ \Delta F_k \tilde{\mathcal{F}}^m_{k-1} \\ \end{bmatrix} . \tag{9.29}$$

Combining (9.27) and (9.28), we finally obtain the extended state equation

$$X_{m+1,k+1} = (F_N^p + \tilde{F}_{m,k}^p)x_m + (S_N + \tilde{S}_{m,k})U_{m,k} + (D_N + \tilde{D}_{m,k})W_{m,k} , \tag{9.30}$$

where $\tilde{F}_{m,k}^p = \hat{F}_N F_{m,k}^{p\Delta}$, $\tilde{S}_{m,k} = \hat{F}_N S_{m,k}^{\Delta}$, and $\tilde{D}_{m,k} = \hat{F}_N \tilde{D}_{m,k}^{\Delta}$, and notice that zero uncertainties make equation (9.30) equal to (8.8).

The predicted state x_{k+1} can now be extracted from (9.30) to be

$$x_{k+1} = (F^N + \bar{\tilde{F}}_{m,k}^p)x_m + (\bar{S}_N + \bar{\tilde{S}}_{m,k})U_{m,k} + (\bar{D}_N + \bar{\tilde{D}}_{m,k})W_{m,k} , \tag{9.31}$$

where $\bar{\tilde{F}}_{m,k}^p$ is the last row vectors in $\tilde{F}_{m,k}^p$ and $\bar{\tilde{S}}_{m,k}$ and $\bar{\tilde{D}}_{m,k}$ are defined after (9.13).

Without any innovation, we extend the observation equation (9.24) to $[m, k]$ as

$$Y_{m,k} = H_N^p x_m + L_k^p U_{m,k} + M_N^p \Xi_{m,k}^p + \Pi_{m,k}^p , \tag{9.32}$$

where $H_N^p = \bar{H}_N F_N$, $L_N^p = M_N^p \bar{E}_N$, and

$$M_N^p = \begin{bmatrix} 0 & 0 & \dots & 0 & 0 \\ H & 0 & \dots & 0 & 0 \\ \vdots & \vdots & \ddots & \vdots & \vdots \\ HF^{N-3} & HF^{N-4} & \dots & 0 & 0 \\ HF^{N-2} & HF^{N-3} & \dots & H & 0 \end{bmatrix}, \quad \Pi_{m,k}^p = \begin{bmatrix} \zeta_m^p \\ \zeta_{m+1}^p \\ \vdots \\ \zeta_{k-1}^p \\ \zeta_k^p \end{bmatrix} . \tag{9.33}$$

Similarly, we extend the vector $\Pi_{m,k}^p$ as

$$\Pi_{m,k}^p = N_{m,k}^{\Delta} x_m + L_{m,k}^{p\Delta} U_{m,k} + M_{m,k}^{p\Delta} \Xi_{m,k}^p (\bar{T}_N + \bar{T}_{m,k}^{\Delta})W_{m,k} + V_{m,k} , \tag{9.34}$$

where $L_{m,k}^{p\Delta} = M_{m,k}^{p\Delta} \bar{E}_N$,

$$N_{m,k}^{\Delta} = \begin{bmatrix} \Delta H_m \\ \Delta H_{m+1} F \\ \vdots \\ \Delta H_{k-1} F^{N-2} \\ \Delta H_k F^{N-1} \end{bmatrix},$$

$$M_{m,k}^{p\Delta} = \begin{bmatrix} 0 & 0 & \dots & 0 & 0 \\ \Delta H_{m+1} & 0 & \dots & 0 & 0 \\ \vdots & \vdots & \ddots & \vdots & \vdots \\ \Delta H_{k-1} F^{N-3} & \Delta H_{k-1} F^{N-4} & \dots & 0 & 0 \\ \Delta H_k F^{N-2} & \Delta H_k F^{N-3} & \dots & \Delta H_k & 0 \end{bmatrix},$$

and matrices $\bar{T}_N$ and $\bar{T}_{m,k}^{\Delta}$ are defined previously.

Finally, substituting (9.28) and (9.34) into (9.32) and reorganizing the terms, we obtain the extended observation equation

$$Y_{m,k} = (H_N^p + \tilde{H}_{m,k}^p)x_m + (L_k^p + \tilde{L}_{m,k}^p)U_{m,k} + (T_N^p + \tilde{T}_{m,k}^p)W_{m,k} + V_{m,k} , \tag{9.35}$$

where the uncertain matrices are given by

$$\tilde{H}^{p}_{m,k} = N^{\Delta}_{m,k} + (M^{p}_{N} + M^{p\Delta}_{m,k})F^{p\Delta}_{m,k} , \quad (9.36)$$

$$\tilde{L}^{p}_{m,k} = L^{p\Delta}_{m,k} + (M^{p}_{N} + M^{p\Delta}_{m,k})S^{\Delta}_{m,k} , \quad (9.37)$$

$$\tilde{T}^{p}_{m,k} = M^{p\Delta}_{m,k}\bar{B}_N + (M^{p}_{N} + M^{p\Delta}_{m,k})D^{\Delta}_{m,k} + \bar{T}^{\Delta}_{m,k} , \quad (9.38)$$

and all other definitions can be found earlier. The last thing to notice is that without uncertainties, (9.35) becomes the standard equation (8.9).

Now that we have provided extended models for uncertain systems, we can start developing FIR filters and FIR predictors.

9.2 The *a posteriori* H_2 FIR Filtering

Various types of *a posteriori* H_2 FIR filters (optimal, optimal unbiased, ML, and suboptimal) can be obtained for uncertain systems represented by the BE-based model. Traditionally, we start with the *a posteriori* FIR filtering estimate defined using (9.17) as

$$\begin{aligned}\hat{x}_k &= \mathcal{H}_N Y_{m,k} + \mathcal{H}^{\mathrm{f}}_N U_{m,k} \\ &= \mathcal{H}_N(H_N + \tilde{H}_{m,k})x_m + \mathcal{H}_N(L_N + \tilde{L}_{m,k})U_{m,k} + \mathcal{H}^{\mathrm{f}}_N U_{m,k} \\ &\quad + \mathcal{H}_N(G_N + \tilde{T}_{m,k})W_{m,k} + \mathcal{H}_N V_{m,k} ,\end{aligned} \quad (9.39)$$

where the uncertain matrices $\tilde{H}_{m,k}$, $\tilde{L}_{m,k}$, and $\tilde{T}_{m,k}$ are given by (9.18)–(9.20).

Under the assumption that all error factors, including the uncertainties, have zero mean, the unbiasedness condition $\mathcal{E}\{\hat{x}_k\} = \mathcal{E}\{x_k\}$ applied to (9.13) and (9.39) gives two unbiasedness constraints,

$$F^{N-1} = \mathcal{H}_N H_N , \quad (9.40)$$

$$\mathcal{H}^{\mathrm{f}}_N = \bar{S}_N - \mathcal{H}_N L_N . \quad (9.41)$$

We now write the estimation error $\epsilon_k = x_k - \hat{x}_k$ as

$$\begin{aligned}\epsilon_k &= (F^{N-1} - \mathcal{H}_N H_N + \bar{\tilde{F}}_{m,k} - \mathcal{H}_N\tilde{H}_{m,k})x_m \\ &\quad + (\bar{S}_N - \mathcal{H}_N L_N - \mathcal{H}^{\mathrm{f}}_N + \bar{\tilde{S}}_{m,k} - \mathcal{H}_N\tilde{L}_{m,k})U_{m,k} \\ &\quad + (\bar{D}_N - \mathcal{H}_N T_N + \bar{\tilde{D}}_{m,k} - \mathcal{H}_N\tilde{T}_{m,k})W_{m,k} - \mathcal{H}_N V_{m,k}\end{aligned} \quad (9.42)$$

and generalize with

$$\begin{aligned}\epsilon_k &= (\mathcal{B}_N + \tilde{\mathcal{B}}_{m,k})x_m + (\mathcal{U}_N + \tilde{\mathcal{U}}_{m,k})U_{m,k} \\ &\quad + (\mathcal{W}_N + \tilde{\mathcal{W}}_{m,k})W_{m,k} - \mathcal{V}_N V_{m,k} ,\end{aligned} \quad (9.43)$$

where the regular error residual matrices $\mathcal{B}_N$, $\mathcal{W}_N$, and $\mathcal{V}_N$ are given by (8.15)–(8.17), $\mathcal{U}_N = \bar{S}_N - \mathcal{H}_N L_N - \mathcal{H}^{\mathrm{f}}_N$ is the regular bias caused by the input signal and removed in optimal and optimal unbiased filters by the constraint (9.41), and the uncertain error residual matrices are defined as

$$\tilde{\mathcal{B}}_{m,k} = \bar{\tilde{F}}_{m,k} - \mathcal{H}_N\tilde{H}_{m,k} , \quad (9.44)$$

$$\tilde{\mathcal{U}}_{m,k} = \bar{\tilde{S}}_{m,k} - \mathcal{H}_N\tilde{L}_{m,k} , \quad (9.45)$$

$$\tilde{\mathcal{W}}_{m,k} = \bar{\tilde{D}}_{m,k} - \mathcal{H}_N\tilde{T}_{m,k} . \quad (9.46)$$

By introducing the disturbance-induced errors

$$\bar{\varepsilon}_{xk} = \mathcal{B}_N x_m \,, \quad \bar{\varepsilon}_{wk} = \mathcal{W}_N W_{m,k} \,,$$
$$\bar{\varepsilon}_{vk} = \mathcal{V}_N V_{m,k} \tag{9.47}$$

and the uncertainty-induced errors

$$\tilde{\varepsilon}_{xk} = \tilde{\mathcal{B}}_{m,k} x_m \,, \quad \tilde{\varepsilon}_{wk} = \tilde{\mathcal{W}}_{m,k} W_{m,k} \,,$$
$$\tilde{\varepsilon}_{uk} = \tilde{\mathcal{U}}_{m,k} U_{m,k} \,, \tag{9.48}$$

and then neglecting regular errors by embedding the constraint (9.41), we represent the estimation error as the sum of the sub errors as

$$\varepsilon_k = \bar{\varepsilon}_{xk} + \bar{\varepsilon}_{wk} + \bar{\varepsilon}_{vk} + \tilde{\varepsilon}_{xk} + \tilde{\varepsilon}_{wk} + \tilde{\varepsilon}_{uk} \,, \tag{9.49}$$

where the components are given by (9.47) and (9.48).

In the transform domain, we now have the structure shown in Fig. 9.2, where we recognize two types of errors caused by 1) disturbance and errors and 2) uncertainties, and the corresponding transfer functions:

- $\mathcal{T}_{\bar{x}}(z)$ is the ε_x-to-$\bar{\varepsilon}_x$ transfer function (initial errors).
- $\mathcal{T}_{\bar{w}}(z)$ is the ε_w-to-$\bar{\varepsilon}_w$ transfer function (disturbance).
- $\mathcal{T}_{\bar{v}}(z)$ is the ε_v-to-$\bar{\varepsilon}_v$ transfer function (measurement errors).
- $\mathcal{T}_{\tilde{x}}(z)$ is the ε_x-to-$\tilde{\varepsilon}_x$ transfer function (initial uncertainty).
- $\mathcal{T}_{\tilde{w}}(z)$ is the ε_w-to-$\tilde{\varepsilon}_w$ transfer function (system uncertainty).
- $\mathcal{T}_{\tilde{u}}(z)$ is the ε_u-to-$\tilde{\varepsilon}_u$ transfer function (input uncertainty).

Using the definitions of the specific errors and the transfer functions presented earlier and in Fig. 9.2, different types of FIR filters can be obtained for uncertain systems operating under disturbances, initial errors, and data errors. Next we start with the *a posteriori* H_2-OFIR filter.

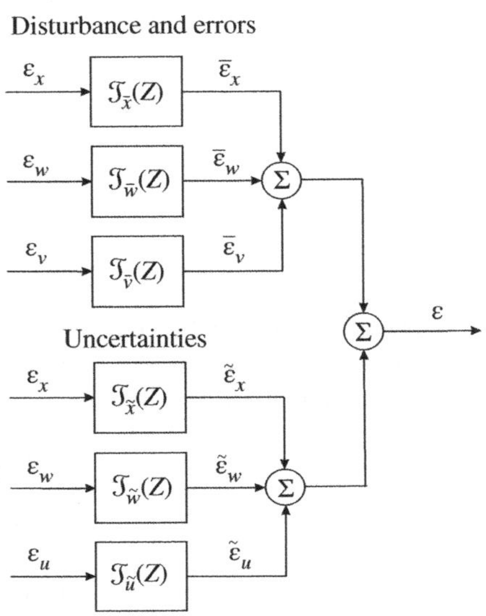

Figure 9.2 Errors in the H_2-OFIR state estimator caused by uncertainties, disturbances, and errors in the z-domain.

9.2.1 H_2-OFIR Filter

To obtain the *a posteriori* H_2-OFIR filter for uncertain systems, we will need the following lemma.

Lemma 9.1 ***(Frobenius norm).*** Given the system $\begin{bmatrix} A_w & B_w \\ C_w & 0 \end{bmatrix}$, where the sparse matrices A_w and B_w are defined by (8.19) and C_w is the real matrix, the transfer function $\mathcal{T}(z) = C_w(Iz - A_w)^{-1}zB_w$, and a symmetric positive definite weighting matrix Ξ, then the squared Frobenius norm of the weighted transfer function $\bar{\mathcal{T}}(z)$ is

$$\| \bar{\mathcal{T}}(z) \|_F^2 = \frac{1}{2\pi} \int_0^{2\pi} \text{tr}\ [\mathcal{T}(e^{j\omega T})\Xi\mathcal{T}^*(e^{j\omega T})]\ d\omega T \tag{9.50}$$

$$= \text{tr}(C_w \Xi C_w^T) . \tag{9.51}$$

Proof: Assume that $\Xi = I$. Refer to [232] and show that (9.50) can be represented equivalently for the nonweighted transfer function $\mathcal{T}(z)$ as $\| \mathcal{T}(z) \|_F^2 = \text{tr}(C_w L C_w^T)$, where the controllability Grammian L is a solution to the discrete Lyapunov equation $A_w L A_w^T - L = M$ given by (8.26), in which M is a positive definite matrix.

Write the solution to the Lyapunov equation as the infinite sum $L = \sum_{i=0}^{\infty} A_w^i M A_w^{Ti}$. Because M is allowed to be any positive definite matrix, choose $M = B_w \Xi B_w^T$, refer to (8.19), provide the transformations, show that $L = \Xi$, arrive at (9.51), and complete the proof. □

To obtain the *a posteriori* H_2-OFIR filter using lemma 9.1, we first note that the initial state error ε_x goes to $\bar{\varepsilon}_k$ unchanged and the ε_x-to-$\bar{\varepsilon}_x$ transfer function is thus an identity matrix, $\mathcal{T}_{\bar{x}}(z) = I$. Using lemma 9.1, we then write the squared norms for the disturbances and errors as

$$\| \bar{\mathcal{T}}_{\bar{x}}(z) \|_F^2 = \text{tr}(\mathcal{B}_N \chi_m \mathcal{B}_N^T) , \quad \| \bar{\mathcal{T}}_{\bar{w}}(z) \|_F^2 = \text{tr}(\mathcal{W}_N \mathcal{Q}_N \mathcal{W}_N^T) ,$$
$$\| \bar{\mathcal{T}}_{\bar{v}}(z) \|_F^2 = \text{tr}(\mathcal{V}_N \mathcal{R}_N \mathcal{V}_N^T) . \tag{9.52}$$

The squared Frobenius norms associated with uncertain errors can be specified similarly. For the ε_x-to-$\tilde{\varepsilon}_x$ transfer function, we write the squared norm $\| \bar{\mathcal{T}}_{\tilde{x}}(z) \|_F^2$ as

$$\begin{aligned} \| \bar{\mathcal{T}}_{\tilde{x}}(z) \|_F^2 &= \text{tr}\mathcal{E}\{\tilde{\mathcal{B}}_{m,k} x_m x_m^T \tilde{\mathcal{B}}_{m,k}^T\} \\ &= \text{tr}\mathcal{E}\{(\tilde{\bar{\mathcal{F}}}_{m,k} - \mathcal{H}_N \tilde{\mathcal{H}}_{m,k}) x_m x_m^T (\tilde{\bar{\mathcal{F}}}_{m,k} - \mathcal{H}_N \tilde{\mathcal{H}}_{m,k})^T\} \\ &= \text{tr}(\tilde{\chi}_m^F - \tilde{\chi}_m^{FH} \mathcal{H}_N^T - \mathcal{H}_N \tilde{\chi}_m^{HF} + \mathcal{H}_N \tilde{\chi}_m^H \mathcal{H}_N^T) , \end{aligned} \tag{9.53}$$

where the uncertain error matrices are defined by

$$\tilde{\chi}_m^F = \mathcal{E}\{\tilde{\bar{\mathcal{F}}}_{m,k} x_m x_m^T \tilde{\bar{\mathcal{F}}}_{m,k}^T\} , \tag{9.54}$$

$$\tilde{\chi}_m^{FH} = \mathcal{E}\{\tilde{\bar{\mathcal{F}}}_{m,k} x_m x_m^T \tilde{\mathcal{H}}_{m,k}^T\} , \tag{9.55}$$

$$\tilde{\chi}_m^{HF} = \mathcal{E}\{\tilde{\mathcal{H}}_{m,k} x_m x_m^T \tilde{\bar{\mathcal{F}}}_{m,k}^T\} , \tag{9.56}$$

$$\tilde{\chi}_m^H = \mathcal{E}\{\tilde{\mathcal{H}}_{m,k} x_m x_m^T \tilde{\mathcal{H}}_{m,k}^T\} . \tag{9.57}$$

Likewise, for the ε_w-to-$\tilde{\varepsilon}_w$ transfer function we write the squared norm $\| \bar{\mathcal{T}}_{\tilde{w}}(z) \|_F^2$ as

$$\begin{aligned}\| \bar{\mathcal{T}}_{\tilde{w}}(z) \|_F^2 &= \text{tr}\ \mathcal{E}\{\tilde{\mathcal{W}}_{m,k} W_{m,k} W_{m,k}^T \tilde{\mathcal{W}}_{m,k}^T\} \\ &= \text{tr}\ \mathcal{E}\{(\bar{\tilde{D}}_{m,k} - \mathcal{H}_N \tilde{T}_{m,k}) W_{m,k} W_{m,k}^T (\bar{\tilde{D}}_{m,k} - \mathcal{H}_N \tilde{T}_{m,k})^T\} \\ &= \text{tr}(\tilde{Q}_N^D - \tilde{Q}_N^{DT} \mathcal{H}_N^T - \mathcal{H}_N \tilde{Q}_N^{TD} + \mathcal{H}_N \tilde{Q}_N^T \mathcal{H}_N^T)\,, \end{aligned} \tag{9.58}$$

where the uncertain error matrices are given by

$$\tilde{Q}_N^D = \mathcal{E}\{\bar{\tilde{D}}_{m,k} W_{m,k} W_{m,k}^T \bar{\tilde{D}}_{m,k}^T\}\,, \tag{9.59}$$

$$\tilde{Q}_N^{DT} = \mathcal{E}\{\bar{\tilde{D}}_{m,k} W_{m,k} W_{m,k}^T \tilde{T}_{m,k}^T\}\,, \tag{9.60}$$

$$\tilde{Q}_N^{TD} = \mathcal{E}\{\tilde{T}_{m,k} W_{m,k} W_{m,k}^T \bar{\tilde{D}}_{m,k}^T\}\,, \tag{9.61}$$

$$\tilde{Q}_N^T = \mathcal{E}\{\tilde{T}_{m,k} W_{m,k} W_{m,k}^T \tilde{T}_{m,k}^T\}\,. \tag{9.62}$$

Finally, for the ε_u-to-$\tilde{\varepsilon}_u$ transfer function we obtain the squared norm $\| \bar{\mathcal{T}}_{\tilde{u}}(z) \|_F^2$ as

$$\begin{aligned}\| \bar{\mathcal{T}}_{\tilde{u}}(z) \|_F^2 &= \text{tr}\mathcal{E}\{\tilde{\mathcal{U}}_{m,k} U_{m,k} U_{m,k}^T \tilde{\mathcal{U}}_{m,k}^T\} \\ &= \text{tr}\mathcal{E}\{(\bar{\tilde{S}}_{m,k} - \mathcal{H}_N \tilde{L}_{m,k}) U_{m,k} U_{m,k}^T (\bar{\tilde{S}}_{m,k} - \mathcal{H}_N \tilde{L}_{m,k})^T\} \\ &= \text{tr}(\tilde{M}_N^S - \tilde{M}_N^{SL} \mathcal{H}_N^T - \mathcal{H}_N \tilde{M}_N^{LS} + \mathcal{H}_N \tilde{M}_N^L \mathcal{H}_N^T)\,, \end{aligned} \tag{9.63}$$

using the uncertain error matrices

$$\tilde{M}_N^S = \mathcal{E}\{\bar{\tilde{S}}_{m,k} U_{m,k} U_{m,k}^T \bar{\tilde{S}}_{m,k}^T\}\,, \tag{9.64}$$

$$\tilde{M}_N^{SL} = \mathcal{E}\{\bar{\tilde{S}}_{m,k} U_{m,k} U_{m,k}^T \tilde{L}_{m,k}^T\}\,, \tag{9.65}$$

$$\tilde{M}_N^{LS} = \mathcal{E}\{\tilde{L}_{m,k} U_{m,k} U_{m,k}^T \bar{\tilde{S}}_{m,k}^T\}\,, \tag{9.66}$$

$$\tilde{M}_N^L = \mathcal{E}\{\tilde{L}_{m,k} U_{m,k} U_{m,k}^T \tilde{L}_{m,k}^T\}\,. \tag{9.67}$$

Using the previous definitions, we can now represent the trace of the estimation error matrix P associated with the estimation error (9.48) as

$$\begin{aligned}\text{tr}\ P &= \mathcal{E}\{(\bar{\varepsilon}_{xk} + \bar{\varepsilon}_{wk} + \bar{\varepsilon}_{vk} + \tilde{\varepsilon}_{xk} + \tilde{\varepsilon}_{wk} + \tilde{\varepsilon}_{uk})^T(\dots)\} \\ &= \mathcal{E}\{\bar{\varepsilon}_{xk}^T \bar{\varepsilon}_{xk}\} + \mathcal{E}\{\bar{\varepsilon}_{wk}^T \bar{\varepsilon}_{wk}\} + \mathcal{E}\{\bar{\varepsilon}_{vk}^T \bar{\varepsilon}_{vk}\} \\ &\quad + \mathcal{E}\{\tilde{\varepsilon}_{xk}^T \tilde{\varepsilon}_{xk}\} + \mathcal{E}\{\tilde{\varepsilon}_{wk}^T \tilde{\varepsilon}_{wk}\} + \mathcal{E}\{\tilde{\varepsilon}_{uk}^T \tilde{\varepsilon}_{uk}\} \\ &= \| \bar{\mathcal{T}}_{\bar{x}}(z) \|_F^2 + \| \bar{\mathcal{T}}_{\bar{w}}(z) \|_F^2 + \| \bar{\mathcal{T}}_{\bar{v}}(z) \|_F^2 \\ &\quad + \| \bar{\mathcal{T}}_{\tilde{x}}(z) \|_F^2 + \| \bar{\mathcal{T}}_{\tilde{w}}(z) \|_F^2 + \| \bar{\mathcal{T}}_{\tilde{u}}(z) \|_F^2 \end{aligned}$$

and determine the gain $\mathcal{H}_N$ for the *a posteriori* H_2-OFIR filter by solving the following minimization problem

$$\begin{aligned}\mathcal{H}_N &= \arg\min_{\mathcal{H}_N}\ \text{tr}P \\ &= \arg\min_{\mathcal{H}_N}\ \text{tr}(\mathcal{B}_N \mathcal{X}_m \mathcal{B}_N^T + \mathcal{W}_N \mathcal{Q}_N \mathcal{W}_N^T + \mathcal{V}_N \mathcal{R}_N \mathcal{V}_N^T \\ &\quad + \| \bar{\mathcal{T}}_{\tilde{x}}(z) \|_F^2 + \| \bar{\mathcal{T}}_{\tilde{w}}(z) \|_F^2 + \| \bar{\mathcal{T}}_{\tilde{u}}(z) \|_F^2)\,, \end{aligned} \tag{9.68}$$

where the norms for uncertain errors are given by (9.53), (9.58), and (9.63).

Since the H_2 filtering problem is convex, we equivalently consider instead of (9.68) the equality

$$\frac{\partial}{\partial \mathcal{H}_N} \mathrm{tr} P = 0 , \tag{9.69}$$

for which the trace tr P can be written as

$$\begin{aligned} \mathrm{tr}\, P = \mathrm{tr}[&F^{N-1}\chi_m F^{N-1^T} + \bar{D}_N \mathcal{Q}_N \bar{D}_N^T + \tilde{\chi}_m^F + \tilde{M}_N^S + \tilde{Q}_N^D \\ &- 2(F^{N-1}\chi_m H_N^T + \bar{D}_N \mathcal{Q}_N G_N^T + \tilde{\chi}_m^{FH} + \tilde{M}_N^{SL} + \tilde{Q}_N^{DT})\mathcal{H}_N^T \\ &+ \mathcal{H}_N(H_N \chi_m H_N^T + \Omega_N + \tilde{\chi}_m^H + \tilde{M}_N^L + \tilde{Q}_N^T)\mathcal{H}_N^T] , \end{aligned} \tag{9.70}$$

where $\Omega_N = G_N \mathcal{Q}_N G_N^T + \mathcal{R}_N$.

By applying the derivative (9.69) to (9.70), neglecting the correlation between different error sources, and setting $D = 0$ that gives $T_N = G_N$, we finally obtain the gain for the *a posteriori* H_2-OFIR filter applied to uncertain systems operating under disturbances, initial errors, and data errors,

$$\begin{aligned} \mathcal{H}_N = &(F^{N-1}\chi_m H_N^T + \bar{D}_N \mathcal{Q}_N G_N^T + \tilde{\chi}_m^{FH} + \tilde{M}_N^{SL} + \tilde{Q}_N^{DT}) \\ &\times (H_N \chi_m H_N^T + \Omega_N + \tilde{\chi}_m^H + \tilde{M}_N^L + \tilde{Q}_N^T)^{-1} . \end{aligned} \tag{9.71}$$

As can be seen, zero uncertain terms make (9.71) the gain (8.34) obtained for systems operating under disturbances, initial errors, and data errors. This means that the gain (9.71) is most general for LTI systems.

For the gain $\mathcal{H}_N$ obtained by (9.71), we write the *a posteriori* H_2-OFIR filtering estimate as

$$\hat{x}_k = \mathcal{H}_N Y_{m,k} + (\bar{S}_N - \mathcal{H}_N L_N) U_{m,k} \tag{9.72}$$

and specify the estimation error matrix as

$$\begin{aligned} P = &\mathcal{B}_N \chi_m \mathcal{B}_N^T + \mathcal{W}_N \mathcal{Q}_N \mathcal{W}_N^T + \mathcal{V}_N \mathcal{R}_N \mathcal{V}_N^T \\ &+ \tilde{P}_x + \tilde{P}_w + \tilde{P}_u , \end{aligned} \tag{9.73}$$

where the uncertain error matrices are defined by

$$\begin{aligned} \tilde{P}_x &= \mathcal{E}\{\tilde{\mathcal{B}}_{m,k} x_m x_m^T \tilde{\mathcal{B}}_{m,k}^T\} \\ &= \tilde{\chi}_m^F - \tilde{\chi}_m^{FH} \mathcal{H}_N^T - \mathcal{H}_N \tilde{\chi}_m^{HF} + \mathcal{H}_N \tilde{\chi}_m^H \mathcal{H}_N^T , \end{aligned} \tag{9.74}$$

$$\begin{aligned} \tilde{P}_w &= \mathcal{E}\{\tilde{\mathcal{W}}_{m,k} W_{m,k} W_{m,k}^T \tilde{\mathcal{W}}_{m,k}^T\} \\ &= \tilde{Q}_N^D - \tilde{Q}_N^{DT} \mathcal{H}_N^T - \mathcal{H}_N \tilde{Q}_N^{TD} + \mathcal{H}_N \tilde{Q}_N^T \mathcal{H}_N^T , \end{aligned} \tag{9.75}$$

$$\begin{aligned} \tilde{P}_u &= \mathcal{E}\{\tilde{\mathcal{U}}_{m,k} U_{m,k} U_{m,k}^T \tilde{\mathcal{U}}_{m,k}^T\} \\ &= \tilde{M}_N^S - \tilde{M}_N^{SL} \mathcal{H}_N^T - \mathcal{H}_N \tilde{M}_N^{LS} + \mathcal{H}_N \tilde{M}_N^L \mathcal{H}_N^T . \end{aligned} \tag{9.76}$$

Any uncertainty in system modeling leads to an increase in estimation errors, which is obvious. In this regard, using the H_2-OFIR filter with a gain (9.71) gives a chance to minimize errors under the uncertainties. However, a good filter performance is not easy to reach. Efforts should be made to determine boundaries for all uncertainties and other error matrices (9.74)–(9.76). Otherwise, mistuning can cause the filter to generate large errors and lose the advantages of robust filtering.

Equivalence with OFIR Filter

Not only for theoretical reasons, but rather for practical utility, we will now show that the gain (9.71) of the *a posteriori* H_2-OFIR filter obtained for uncertain systems is equivalent to the OFIR filter gain valid in white Gaussian environments. Indeed, for white Gaussian noise the FIR filter optimality is guaranteed by the orthogonality condition $\mathcal{E}\{\varepsilon_k Y_{m,k}^T\} = 0$ that, if we use (9.17) and (9.43), can be rewritten as

$$\begin{aligned} 0 = \mathcal{E}\{[(F^{N-1} - \mathcal{H}_N H_N)x_m + (\bar{D}_N - \mathcal{H}_N G_N)W_{m,k} - \mathcal{H}_N V_{m,k} \\ + (\tilde{\bar{F}}_{m,k} - \mathcal{H}_N \tilde{H}_{m,k})x_m + (\tilde{\bar{S}}_{m,k} - \mathcal{H}_N \tilde{L}_{m,k})U_{m,k} \\ +(\tilde{\bar{D}}_{m,k} - \mathcal{H}_N \tilde{T}_{m,k})W_{m,k}][x_m^T(H_N^T + \tilde{H}_{m,k}^T) + U_{m,k}^T(L_N^T + \tilde{L}_{m,k}^T) \\ +W_{m,k}^T(G_N^T + \tilde{T}_{m,k}^T) + V_{m,k}^T]\} . \end{aligned} \tag{9.77}$$

Assuming all error sources are independent and uncorrelated zero mean white Gaussian processes and providing the averaging, we can easily transform (9.77) to (9.71). This provides further evidence that, according to Parseval's theorem, minimizing the error spectral energy in the transform domain is equivalent to minimizing the MSE in the time domain.

9.2.2 Bias-Constrained H_2-OFIR Filter

A known drawback of optimal filters is that optimal performance cannot be guaranteed without setting correct initial values. This is especially critical for the H_2-OFIR and OFIR filters, which require initial values for each horizon $[m, k]$. To remove the requirement of the initial state in the H_2-OFIR filter, its gain must be subject to unbiasedness constraints, and then the remaining errors can be analyzed as shown in Fig. 9.3.

Referring to the previous, we can now design the H_2-OUFIR filter for uncertain systems, minimizing the trace of the error matrix (9.73) subject to the constraint (9.40). As in the case of the OUFIR filter, the gain obtained in such a way is freed from the regular errors, and its error matrix depends only on uncertainties, disturbances, and data errors, as shown in Fig. 9.3. Next we give the corresponding derivation.

First, we use (9.74)–(9.76) and represent the error matrix (9.73) as

$$\begin{aligned} P = \bar{D}_N \mathcal{Q}_N \bar{D}_N^T + \mathcal{Z}_1 - 2(\bar{D}_N \mathcal{Q}_N G_N^T + \mathcal{Z}_2)\mathcal{H}_N^T \\ + \mathcal{H}_N(\Omega_N + \mathcal{Z}_3)\mathcal{H}_N^T , \end{aligned} \tag{9.78}$$

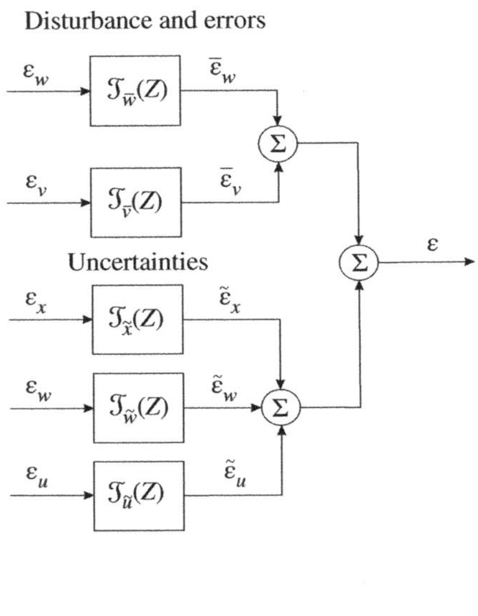

Figure 9.3 Errors in the H_2-OUFIR state estimator caused by uncertainties, disturbances, and data errors in the z-domain.

where the newly introduced uncertain matrices have the form

$$\mathcal{Z}_1 = \tilde{\chi}_m^F + \tilde{M}_N^S + \tilde{Q}_N^D \,, \tag{9.79}$$

$$\mathcal{Z}_2 = \tilde{\chi}_m^{FH} + \tilde{M}_N^{SL} + \tilde{Q}_N^{DT} \,, \tag{9.80}$$

$$\mathcal{Z}_3 = \tilde{\chi}_m^H + \tilde{M}_N^L + \tilde{Q}_N^T \,. \tag{9.81}$$

The Lagrangian cost function associated with (9.78) becomes

$$J = \text{tr}P + \text{tr}\Lambda(I - \mathcal{H}_N C_N) \tag{9.82}$$

and we determine the gain $\mathcal{H}_N$ by solving the minimization problem

$$\mathcal{H}_N = \arg\min_{\mathcal{H}_N,\Lambda} J(\mathcal{H}_N, \Lambda) \,. \tag{9.83}$$

The solution to (9.83) is available by solving two equations

$$\begin{aligned} \frac{\partial}{\partial \mathcal{H}_N} J = & -2(\bar{D}_N \mathcal{Q}_N G_N^T + \mathcal{Z}_2) + 2\mathcal{H}_N(\Omega_N + \mathcal{Z}_3) \\ & - \Lambda^T C_N^T = 0 \,, \end{aligned} \tag{9.84}$$

$$\frac{\partial}{\partial \Lambda} J = I - C_N^T \mathcal{H}_N^T = 0 \,, \tag{9.85}$$

and we notice that (9.85) is equivalent to the unbiasedness constraint (9.40).

The first equation (9.84) gives

$$C_N \Lambda = -2(G_N \mathcal{Q}_N \bar{D}_N^T + \mathcal{Z}_2^T) + 2(\Omega_N + \mathcal{Z}_3)\mathcal{H}_N^T \,. \tag{9.86}$$

Multiplying both sides of (9.86) from the left-hand side by a nonzero $C_N^T \Omega_N^{-1}$ and referring to the constraint (9.85), we obtain

$$C_N^T \Omega_N^{-1} C_N \Lambda = -2C_N^T \Omega_N^{-1}(G_N \mathcal{Q}_N \bar{D}_N^T + \mathcal{Z}_2^T) + 2(I + C_N^T \Omega_N^{-1} \mathcal{Z}_3 \mathcal{H}_N^T)$$

and retrieve the Lagrange multiplier

$$\begin{aligned} \Lambda = & \, 2(C_N^T \Omega_N^{-1} C_N)^{-1}[I - C_N^T \Omega_N^{-1}(G_N \mathcal{Q}_N \bar{D}_N^T \\ & + \mathcal{Z}_2^T - \mathcal{Z}_3 \mathcal{H}_N^T)] \,. \end{aligned} \tag{9.87}$$

Reconsidering (9.84), substituting (9.85) and (9.87), and performing some transformations, we obtain the gain for the H_2-OUFIR filter in the form

$$\begin{aligned} \mathcal{H}_N = & \{(C_N^T \Omega_N^{-1} C_N)^{-1} C_N^T + (\bar{D}_N \mathcal{Q}_N G_N^T + \mathcal{Z}_2) \\ & \times [I - \Omega_N^{-1} C_N (C_N^T \Omega_N^{-1} C_N)^{-1} C_N^T]\} \\ & \times [\Omega_N + \mathcal{Z}_3 - \mathcal{Z}_3^T \Omega_N^{-1} C_N (C_N^T \Omega_N^{-1} C_N)^{-1} C_N^T]^{-1} \,. \end{aligned} \tag{9.88}$$

Note that for zero uncertain matrices $\mathcal{Z}_2$ and $\mathcal{Z}_3$, the gain (9.88) becomes the gain (8.42) of the H_2-OUFIR filter, which is valid for systems affected by disturbances. The obvious advantage of (9.88) is that it does not require initial values and thus is more suitable to operate on $[m, k]$.

Summarizing, we note that for $\mathcal{H}_N$ determined by (9.88), the H_2-OUFIR filtering estimate $\hat{x}_k$ and error matrix P are obtained as, respectively,

$$\hat{x}_k = \mathcal{H}_N Y_{m,k} + (\bar{S}_N - \mathcal{H}_N L_N) U_{m,k} \,, \tag{9.89}$$

$$\begin{aligned} P = & \, \mathcal{W}_N \mathcal{Q}_N \mathcal{W}_N^T + \mathcal{V}_N \mathcal{R}_N \mathcal{V}_N^T \\ & + \tilde{P}_x + \tilde{P}_w + \tilde{P}_u \,, \end{aligned} \tag{9.90}$$

where the error matrices $\tilde{P}_x$, $\tilde{P}_w$, and $\tilde{P}_u$ associated with system uncertainties are given by (9.74)–(9.76). It is worth noting that the uncertain component $\tilde{P}_x$ cannot be removed by embedding unbiasedness, since it represents zero mean uncertainty in the initial state. The same can be said about the uncertain matrix $\tilde{P}_u$, which is caused by the zero mean input uncertainty.

9.3 H_2 FIR Prediction

When an uncertain system operates under disturbances, initial errors, and measurement errors, then state feedback control can be organized using a H_2-OFIR predictor, which gives robust estimates if the error matrices are properly maximized. The prediction can be organized in two ways. The one-step ahead predicted estimate can be obtained through the system matrix as $\tilde{x}_{k+1} = F\hat{x}_k$ or $\tilde{x}_k = F\hat{x}_{k-1}$. Note that there is a well-founded conclusion, drawn in [119] and corroborated in [171], that such an unbiased prediction can provide more accuracy than optimal prediction. Another way is to derive an optimal predictor that we will consider next.

Using the FE-based model, we define the one-step FIR prediction as

$$\begin{aligned}\tilde{x}_{k+1} &= \mathcal{H}_N^p Y_{m,k} + \mathcal{H}_N^{pf} U_{m,k} \\ &= \mathcal{H}_N^p (H_N^p + \tilde{H}_{m,k}^p) x_m + \mathcal{H}_N^p (L_N^p + \tilde{L}_{m,k}^p) U_{m,k} + \mathcal{H}_N^{pf} U_{m,k} \\ &\quad + \mathcal{H}_N^p (G_N^p + \tilde{T}_{m,k}^p) W_{m,k} + \mathcal{H}_N^p V_{m,k} ,\end{aligned} \tag{9.91}$$

where the uncertain matrices $\tilde{H}_{m,k}^p$, $\tilde{L}_{m,k}^p$, and $\tilde{T}_{m,k}^p$ are given by (9.36)–(9.38).

The unbiasedness condition $\mathcal{E}\{\tilde{x}_{k+1}\} = \mathcal{E}\{x_{k+1}\}$ applied to (9.31) and (9.91) yields two unbiasedness constraints,

$$F^N = \mathcal{H}_N^p H_N^p , \tag{9.92}$$

$$\mathcal{H}_N^{pf} = \bar{S}_N - \mathcal{H}_N^p L_N^p \tag{9.93}$$

and the estimation error $\varepsilon_{k+1} = x_{k+1} - \tilde{x}_{k+1}$ becomes

$$\begin{aligned}\varepsilon_{k+1} &= (F^N - \mathcal{H}_N^p H_N^p + \bar{\tilde{F}}_{m,k}^p - \mathcal{H}_N^p \tilde{H}_{m,k}^p) x_m \\ &\quad + (\bar{S}_N - \mathcal{H}_N^p L_N^p - \mathcal{H}_N^{pf} + \bar{\tilde{S}}_{m,k} - \mathcal{H}_N^p \tilde{L}_{m,k}^p) U_{m,k} \\ &\quad + (\bar{D}_N - \mathcal{H}_N^p T_N^p + \bar{\tilde{D}}_{m,k} - \mathcal{H}_N^p \tilde{T}_{m,k}^p) W_{m,k} - \mathcal{H}_N^p V_{m,k} .\end{aligned} \tag{9.94}$$

By embedding (9.93), we next generalize ε_{k+1} in the form

$$\begin{aligned}\varepsilon_{k+1} &= (\mathcal{B}_N^p + \tilde{\mathcal{B}}_{m,k}^p) x_m + \tilde{\mathcal{U}}_{m,k}^p U_{m,k} \\ &\quad + (\mathcal{W}_N^p + \tilde{\mathcal{W}}_{m,k}^p) W_{m,k} - \mathcal{V}_N^p V_{m,k} ,\end{aligned} \tag{9.95}$$

where the regular error residual matrices $\mathcal{B}_N^p$, $\mathcal{W}_N^p$, and $\mathcal{V}_N^p$ are given by (8.60)–(8.62) and the uncertain error residual matrices can be taken from (9.94) as

$$\tilde{\mathcal{B}}_{m,k}^p = \bar{\tilde{F}}_{m,k}^p - \mathcal{H}_N^p \tilde{H}_{m,k}^p , \tag{9.96}$$

$$\tilde{\mathcal{U}}_{m,k}^p = \bar{\tilde{S}}_{m,k} - \mathcal{H}_N^p \tilde{L}_{m,k}^p , \tag{9.97}$$

$$\tilde{\mathcal{W}}_{m,k}^p = \bar{\tilde{D}}_{m,k} - \mathcal{H}_N^p \tilde{T}_{m,k}^p . \tag{9.98}$$

Following Fig. 9.1, we now introduce the disturbance-induced errors

$$\begin{aligned}&\bar{\varepsilon}_{x(k+1)} = \mathcal{B}_N^p x_m , \quad \bar{\varepsilon}_{w(k+1)} = \mathcal{W}_N^p W_{m,k} , \\ &\bar{\varepsilon}_{v(k+1)} = \mathcal{V}_N^p V_{m,k}\end{aligned} \tag{9.99}$$

and the uncertainty-induced errors

$$\tilde{\varepsilon}_{x(k+1)} = \tilde{\mathcal{B}}^p_{m,k} x_m \,, \quad \tilde{\varepsilon}_{w(k+1)} = \tilde{\mathcal{W}}^p_{m,k} W_{m,k} \,,$$
$$\tilde{\varepsilon}_{u(k+1)} = \tilde{\mathcal{U}}^p_{m,k} U_{m,k} \,, \tag{9.100}$$

and represent the estimation error as

$$\varepsilon_{k+1} = \bar{\varepsilon}_{x(k+1)} + \bar{\varepsilon}_{w(k+1)} + \bar{\varepsilon}_{v(k+1)} + \tilde{\varepsilon}_{x(k+1)} + \tilde{\varepsilon}_{w(k+1)} + \tilde{\varepsilon}_{u(k+1)} \,. \tag{9.101}$$

It should now be noted that with the help of (9.101) we can develop different kinds of FIR predictors for uncertain systems operating under disturbances in the presence of initial and data errors.

9.3.1 Optimal H_2 FIR Predictor

Using lemma 9.1, it is a matter of similar transformations to show that the trace of the error matrix of the H_2-OFIR predictor is given by

$$\begin{aligned}\text{tr}\, P &= \mathcal{E}\{(\bar{\varepsilon}_{x(k+1)} + \bar{\varepsilon}_{w(k+1)} + \bar{\varepsilon}_{v(k+1)} + \tilde{\varepsilon}_{x(k+1)} + \tilde{\varepsilon}_{w(k+1)} + \tilde{\varepsilon}_{u(k+1)})^T \\ &\quad \times (\dots)\} \\ &= \mathcal{E}\{\bar{\varepsilon}^T_{x(k+1)}\bar{\varepsilon}_{x(k+1)}\} + \mathcal{E}\{\bar{\varepsilon}^T_{w(k+1)}\bar{\varepsilon}_{w(k+1)}\} + \mathcal{E}\{\bar{\varepsilon}^T_{v(k+1)k}\bar{\varepsilon}_{v(k+1)}\} \\ &\quad + \mathcal{E}\{\tilde{\varepsilon}^T_{x(k+1)}\tilde{\varepsilon}_{x(k+1)}\} + \mathcal{E}\{\tilde{\varepsilon}^T_{w(k+1)}\tilde{\varepsilon}_{w(k+1)}\} + \mathcal{E}\{\tilde{\varepsilon}^T_{u(k+1)}\tilde{\varepsilon}_{u(k+1)}\} \\ &= \| \bar{\mathcal{T}}^p_{\bar{x}}(z) \|^2_F + \| \bar{\mathcal{T}}^p_{\bar{w}}(z) \|^2_F + \| \bar{\mathcal{T}}^p_{\bar{v}}(z) \|^2_F \\ &\quad + \| \bar{\mathcal{T}}^p_{\tilde{x}}(z) \|^2_F + \| \bar{\mathcal{T}}^p_{\tilde{w}}(z) \|^2_F + \| \bar{\mathcal{T}}^p_{\tilde{u}}(z) \|^2_F \,,\end{aligned} \tag{9.102}$$

where the squared weighted sub-norms for the properly chosen weight ϖ^p_k are defined by

$$\| \bar{\mathcal{T}}^p(z) \|^2_F = \text{tr}\bar{\mathcal{T}}^p \varpi^p_k \varpi^{p^T}_k \bar{\mathcal{T}}^{p^T} \,. \tag{9.103}$$

The first three squared norms in (9.102) are given by

$$\| \bar{\mathcal{T}}^p_{\bar{x}}(z) \|^2_F = \text{tr}(\mathcal{B}^p_N \chi_m \mathcal{B}^{p^T}_N) \,, \tag{9.104}$$

$$\| \bar{\mathcal{T}}^p_{\bar{w}}(z) \|^2_F = \text{tr}(\mathcal{W}^p_N \mathcal{Q}_N \mathcal{W}^{p^T}_N) \,, \tag{9.105}$$

$$\| \bar{\mathcal{T}}^p_{\bar{v}}(z) \|^2_F = \text{tr}(\mathcal{V}^p_N \mathcal{R}_N \mathcal{V}^{p^T}_N) \,. \tag{9.106}$$

The squared norm $\| \bar{\mathcal{T}}^p_{\tilde{x}}(z) \|^2_F$ can be found using (9.100) and (9.96) to be

$$\begin{aligned}\| \bar{\mathcal{T}}^p_{\tilde{x}}(z) \|^2_F &= \text{tr}\mathcal{E}\{\tilde{\mathcal{B}}^p_{m,k} x_m x^T_m \tilde{\mathcal{B}}^{p^T}_{m,k}\} \\ &= \text{tr}\mathcal{E}\{(\tilde{\bar{\mathcal{F}}}^p_{m,k} - \mathcal{H}^p_N \tilde{\mathcal{H}}^p_{m,k}) x_m x^T_m (\tilde{\bar{\mathcal{F}}}^p_{m,k} - \mathcal{H}^p_N \tilde{\mathcal{H}}^p_{m,k})^T\} \\ &= \text{tr}(\tilde{\chi}^F_m - \tilde{\chi}^{FH}_m \mathcal{H}^{p^T}_N - \mathcal{H}^p_N \tilde{\chi}^{HF}_m + \mathcal{H}^p_N \tilde{\chi}^H_m \mathcal{H}^{p^T}_N) \,,\end{aligned} \tag{9.107}$$

where the uncertain error matrices are defined by

$$\tilde{\chi}^F_m = \mathcal{E}\{\tilde{\bar{\mathcal{F}}}^p_{m,k} x_m x^T_m \tilde{\bar{\mathcal{F}}}^{p^T}_{m,k}\} \,, \tag{9.108}$$

$$\tilde{\chi}^{FH}_m = \mathcal{E}\{\tilde{\bar{\mathcal{F}}}^p_{m,k} x_m x^T_m \tilde{\mathcal{H}}^{p^T}_{m,k}\} \,, \tag{9.109}$$

$$\tilde{\chi}^{HF}_m = \mathcal{E}\{\tilde{\mathcal{H}}^p_{m,k} x_m x^T_m \tilde{\bar{\mathcal{F}}}^{p^T}_{m,k}\} \,, \tag{9.110}$$

$$\tilde{\chi}^H_m = \mathcal{E}\{\tilde{\mathcal{H}}^p_{m,k} x_m x^T_m \tilde{\mathcal{H}}^{p^T}_{m,k}\} \,. \tag{9.111}$$

The squared norm $\| \bar{\mathcal{T}}^p_{\tilde{w}}(z) \|_F^2$ can be transformed to

$$\begin{aligned}\| \bar{\mathcal{T}}^p_{\tilde{w}}(z) \|_F^2 &= \text{tr}\mathcal{E}\{\tilde{\mathcal{W}}^p_{m,k} W_{m,k} W^T_{m,k} \tilde{\mathcal{W}}^{p^T}_{m,k}\} \\ &= \text{tr}\mathcal{E}\{(\bar{\tilde{D}}_{m,k} - \mathcal{H}^p_N \tilde{T}^p_{m,k}) W_{m,k} W^T_{m,k} (\bar{\tilde{D}}_{m,k} - \mathcal{H}^p_N \tilde{T}^p_{m,k})^T\} \\ &= \text{tr}(\tilde{Q}^D_N - \tilde{Q}^{DT}_N \mathcal{H}^{p^T}_N - \mathcal{H}^p_N \tilde{Q}^{TD}_N + \mathcal{H}^p_N \tilde{Q}^T_N \mathcal{H}^{p^T}_N) ,\end{aligned} \tag{9.112}$$

by introducing the uncertain error matrices

$$\tilde{Q}^D_N = \mathcal{E}\{\bar{\tilde{D}}_{m,k} W_{m,k} W^T_{m,k} \bar{\tilde{D}}^T_{m,k}\} , \tag{9.113}$$

$$\tilde{Q}^{DT}_N = \mathcal{E}\{\bar{\tilde{D}}_{m,k} W_{m,k} W^T_{m,k} \tilde{T}^{p^T}_{m,k}\} , \tag{9.114}$$

$$\tilde{Q}^{TD}_N = \mathcal{E}\{\tilde{T}^p_{m,k} W_{m,k} W^T_{m,k} \bar{\tilde{D}}^T_{m,k}\} , \tag{9.115}$$

$$\tilde{Q}^T_N = \mathcal{E}\{\tilde{T}^p_{m,k} W_{m,k} W^T_{m,k} \tilde{T}^{p^T}_{m,k}\} . \tag{9.116}$$

Likewise, the squared norm $\| \bar{\mathcal{T}}^p_{\tilde{u}}(z) \|_F^2$ can be represented with

$$\begin{aligned}\| \bar{\mathcal{T}}^p_{\tilde{u}}(z) \|_F^2 &= \text{tr}\mathcal{E}\{\tilde{\mathcal{U}}^p_{m,k} U_{m,k} U^T_{m,k} \tilde{\mathcal{U}}^{p^T}_{m,k}\} \\ &= \text{tr}\mathcal{E}\{(\bar{\tilde{S}}_{m,k} - \mathcal{H}^p_N \tilde{L}^p_{m,k}) U_{m,k} U^T_{m,k} (\bar{\tilde{S}}_{m,k} - \mathcal{H}^p_N \tilde{L}^p_{m,k})^T\} \\ &= \text{tr}(\tilde{M}^S_N - \tilde{M}^{SL}_N \mathcal{H}^{p^T}_N - \mathcal{H}^p_N \tilde{M}^{LS}_N + \mathcal{H}^p_N \tilde{M}^L_N \mathcal{H}^{p^T}_N) ,\end{aligned} \tag{9.117}$$

using the uncertain error matrices

$$\tilde{M}^S_N = \mathcal{E}\{\bar{\tilde{S}}_{m,k} U_{m,k} U^T_{m,k} \bar{\tilde{S}}^T_{m,k}\} , \tag{9.118}$$

$$\tilde{M}^{SL}_N = \mathcal{E}\{\bar{\tilde{S}}_{m,k} U_{m,k} U^T_{m,k} \tilde{L}^{p^T}_{m,k}\} , \tag{9.119}$$

$$\tilde{M}^{LS}_N = \mathcal{E}\{\tilde{L}^p_{m,k} U_{m,k} U^T_{m,k} \bar{\tilde{S}}^T_{m,k}\} , \tag{9.120}$$

$$\tilde{M}^L_N = \mathcal{E}\{\tilde{L}^p_{m,k} U_{m,k} U^T_{m,k} \tilde{L}^{p^T}_{m,k}\} . \tag{9.121}$$

Based upon (9.102) and using the previously determined squared sub-norms, we determine the gain for the H_2-OFIR predictor by solving the following minimization problem

$$\begin{aligned}\mathcal{H}^p_N &= \arg\min_{\mathcal{H}^p_N} \ \text{tr}P \\ &= \arg\min_{\mathcal{H}^p_N} \ \text{tr}(\mathcal{B}^p_N \chi_m \mathcal{B}^{p^T}_N + \mathcal{W}^p_N \mathcal{Q}_N \mathcal{W}^{p^T}_N + \mathcal{V}^p_N \mathcal{R}_N \mathcal{V}^{p^T}_N \\ &\quad + \| \bar{\mathcal{T}}^p_{\tilde{x}}(z) \|_F^2 + \| \bar{\mathcal{T}}^p_{\tilde{w}}(z) \|_F^2 + \| \bar{\mathcal{T}}^p_{\tilde{u}}(z) \|_F^2) ,\end{aligned} \tag{9.122}$$

where the uncertain norms are given by (9.107), (9.112), and (9.117). To find $\mathcal{H}^p_N$, we further substitute (9.122) equivalently with

$$\frac{\partial}{\partial \mathcal{H}^p_N} \text{tr}P = 0 \tag{9.123}$$

and transform the trace tr$P$ to

$$\begin{aligned}\text{tr}P = \text{tr}[&F^N \chi_m F^{N^T} + \bar{D}_N \mathcal{Q}_N \bar{D}^T_N + \tilde{\chi}^F_m + \tilde{M}^S_N + \tilde{Q}^D_N \\ &-2(F^N \chi_m H^{p^T}_N + \bar{D}_N \mathcal{Q}_N G^{p^T}_N + \tilde{\chi}^{FH}_m + \tilde{M}^{SL}_N\end{aligned}$$

$$
+\tilde{Q}_N^{DT})\mathcal{H}_N^{p^T} + \mathcal{H}_N^p(H_N^p \chi_m H_N^{p^T} + \Omega_N^p + \tilde{\chi}_m^H + \tilde{M}_N^L + \tilde{Q}_N^T)\mathcal{H}_N^{p^T}] , \tag{9.124}
$$

where $\Omega_N^p = G_N^p \mathcal{Q}_N G_N^{p^T} + \mathcal{R}_N$.

By applying the derivative (9.123) to (9.124), we obtain the gain for the H_2-OFIR predictor as

$$
\mathcal{H}_N^p = (F^N \chi_m H_N^{p^T} + \bar{D}_N \mathcal{Q}_N G_N^{p^T} + \tilde{\chi}_m^{FH} + \tilde{M}_N^{SL} + \tilde{Q}_N^{DT})
(H_N^p \chi_m H_N^{p^T} + \Omega_N^p + \tilde{\chi}_m^H + \tilde{M}_N^L + \tilde{Q}_N^T)^{-1} \tag{9.125}
$$

and notice that, by neglecting the uncertain terms, this gain becomes the gain (8.70) derived for systems operating under disturbances.

Finally, we end up with the batch H_2-OFIR prediction

$$
\tilde{x}_{k+1} = \mathcal{H}_N^p Y_{m,k} + (\bar{S}_N - \mathcal{H}_N^p L_N^p)U_{m,k} , \tag{9.126}
$$

where gain $\mathcal{H}_N^p$ is given by (9.125), and write the error matrix as

$$
P = \mathcal{B}_N^p \chi_m \mathcal{B}_N^{p^T} + \mathcal{W}_N^p \mathcal{Q}_N \mathcal{W}_N^{p^T} + \mathcal{V}_N^p \mathcal{R}_N \mathcal{V}_N^{p^T}
+\tilde{P}_x^p + \tilde{P}_w^p + \tilde{P}_u^p , \tag{9.127}
$$

where the uncertain error matrices are defined by

$$
\begin{aligned}
\tilde{P}_x^p &= \mathcal{E}\{\tilde{\mathcal{B}}_{m,k}^p x_m x_m^T \tilde{\mathcal{B}}_{m,k}^{p^T}\} \\
&= \tilde{\chi}_m^F - \tilde{\chi}_m^{FH}\mathcal{H}_N^{p^T} - \mathcal{H}_N^p \tilde{\chi}_m^{HF} + \mathcal{H}_N^p \tilde{\chi}_m^H \mathcal{H}_N^{p^T} ,
\end{aligned} \tag{9.128}
$$

$$
\begin{aligned}
\tilde{P}_w^p &= \mathcal{E}\{\tilde{\mathcal{W}}_{m,k}^p W_{m,k} W_{m,k}^T \tilde{\mathcal{W}}_{m,k}^{p^T}\} \\
&= \tilde{Q}_N^D - \tilde{Q}_N^{DT}\mathcal{H}_N^{p^T} - \mathcal{H}_N^p \tilde{Q}_N^{TD} + \mathcal{H}_N^p \tilde{Q}_N^T \mathcal{H}_N^{p^T} ,
\end{aligned} \tag{9.129}
$$

$$
\begin{aligned}
\tilde{P}_u^p &= \mathcal{E}\{\tilde{\mathcal{U}}_{m,k}^p U_{m,k} U_{m,k}^T \tilde{\mathcal{U}}_{m,k}^{p^T}\} \\
&= \tilde{M}_N^S - \tilde{M}_N^{SL}\mathcal{H}_N^{p^T} - \mathcal{H}_N^p \tilde{M}_N^{LS} + \mathcal{H}_N^p \tilde{M}_N^L \mathcal{H}_N^{p^T} .
\end{aligned} \tag{9.130}
$$

The batch form (9.126) gives an optimal prediction of the state of an uncertain system operating under disturbances, initial errors, and measurement errors. Because this algorithm operates with full block error matrices, it can provide better accuracy than the best available recursive Kalman-like scheme relying on diagonal block error matrices. Next, we will show that the gain (9.125) of the H_2-OFIR predictor (9.126) generalizes the gain of the OFIR predictor for white Gaussian processes.

Equivalence with OFIR Predictor

By Parseval's theorem, the minimization of the error spectral energy in the transform domain is equivalent to the minimization of the MSE in the time domain. When all uncertainties, disturbances, and errors are white Gaussian and uncorrelated, then the orthogonality condition $\mathcal{E}\{\varepsilon_{k+1} Y_{m,k}^T\} = 0$ applied to (9.32) and (9.101) guarantees the FIR predictor optimality. Accordingly, we have

$$
\begin{aligned}
0 = \mathcal{E}\{&[(F^N - \mathcal{H}_N^p H_N^p)x_m + (\bar{D}_N - \mathcal{H}_N^p G_N^p)W_{m,k} - \mathcal{H}_N^p V_{m,k}\} \\
&+ (\tilde{\bar{F}}_{m,k}^p - \mathcal{H}_N^p \tilde{H}_{m,k}^p)x_m + (\tilde{\bar{S}}_{m,k} - \mathcal{H}_N^p \tilde{L}_{m,k}^p)U_{m,k}
\end{aligned}
$$

$$+ (\bar{\tilde{D}}_{m,k} - \mathcal{H}_N^p \tilde{T}_{m,k}^p) W_{m,k}][x_m^T (H_N^{p^T} + \tilde{H}_{m,k}^{p^T}) + U_{m,k}^T (L_N^{p^T} + \tilde{L}_{m,k}^{p^T})$$
$$+ W_{m,k}^T (G_N^{p^T} + \tilde{T}_{m,k}^{p^T}) + V_{m,k}^T]\} . \tag{9.131}$$

Providing averaging in (9.131) for mutually independent and uncorrelated error sources, we transform (9.131) into (9.124) and note that the H_2-OFIR predictor has the same structure as the OFIR predictor. The obvious difference between these solutions resides in the fact that the H_2-OFIR predictor does not impose restrictions on the error matrices, while the OFIR predictor requires them to be white Gaussian, that is, diagonal. Then it follows that the H_2-OFIR predictor is a more general estimator for LTI systems.

9.3.2 Bias-Constrained H_2-OUFIR Predictor

Referring to the inherent disadvantage of optimal state estimation of uncertain systems, which is an initial state requirement, we note that the H_2-OFIR predictor may not be sufficiently accurate, especially for short $[m, k]$, if the initial state is not set correctly. In H_2-OUFIR prediction, this issue is circumvented by embedding the unbiasedness constraint, and now we will extend this approach to the H_2-OUFIR predictor.

Considering the error matrix (9.127) of the H_2-OFIR predictor, we first remove the term containing χ_m using the unbiasedness constraint (9.92). Then we rewrite (9.127) as

$$P = \bar{D}_N \mathcal{Q}_N \bar{D}_N^T + \mathcal{Z}_1 - 2(\bar{D}_N \mathcal{Q}_N G_N^{p^T} + \mathcal{Z}_2) \mathcal{H}_N^{p^T}$$
$$+ \mathcal{H}_N^p (\Omega_N^p + \mathcal{Z}_3) \mathcal{H}_N^{p^T} , \tag{9.132}$$

where the matrices $\mathcal{Z}_1$, $\mathcal{Z}_2$, and $\mathcal{Z}_3$ are defined as

$$\mathcal{Z}_1 = \tilde{\chi}_m^F + \tilde{M}_N^S + \tilde{Q}_N^D , \tag{9.133}$$

$$\mathcal{Z}_2 = \tilde{\chi}_m^{FH} + \tilde{M}_N^{SL} + \tilde{Q}_N^{DT} , \tag{9.134}$$

$$\mathcal{Z}_3 = \tilde{\chi}_m^H + \tilde{M}_N^L + \tilde{Q}_N^T \tag{9.135}$$

in terms of the uncertain matrices specified for the H_2-OFIR predictor.

We now write the Lagrangian cost function for (9.132),

$$J = \mathrm{tr}P + \mathrm{tr}\Lambda(I - \mathcal{H}_N^p C_N^p) , \tag{9.136}$$

and determine the gain $\mathcal{H}_N^p$ by solving the minimization problem

$$\mathcal{H}_N^p = \arg\min_{\mathcal{H}_N^p, \Lambda} J(\mathcal{H}_N^p, \Lambda) . \tag{9.137}$$

The solution to (9.137) can be found by solving two equations

$$\frac{\partial}{\partial \mathcal{H}_N^p} J = -2(\bar{D}_N \mathcal{Q}_N G_N^{p^T} + \mathcal{Z}_2) + 2\mathcal{H}_N^p (\Omega_N^p + \mathcal{Z}_3)$$
$$- \Lambda^T C_N^{p^T} = 0 , \tag{9.138}$$

$$\frac{\partial}{\partial \Lambda} J = I - C_N^{p^T} \mathcal{H}_N^{p^T} = 0 , \tag{9.139}$$

where the second equation (9.139) is equal to the constraint (9.92).

From the first equation (9.138) we find

$$\mathcal{C}_N^p \Lambda = -2(\mathcal{G}_N^p \mathcal{Q}_N \bar{\mathcal{D}}_N^T + \mathcal{Z}_2^T) + 2(\Omega_N^p + \mathcal{Z}_3)\mathcal{H}_N^{p^T} . \tag{9.140}$$

We then multiply both sides of (9.140) from the left-hand side with a nonzero $\mathcal{C}_N^{p^T} \Omega_N^{p^{-1}}$ and, using the constraint (9.92), obtain

$$\begin{aligned} \mathcal{C}_N^T \Omega_N^{p^{-1}} \mathcal{C}_N \Lambda = &-2\mathcal{C}_N^T \Omega_N^{p^{-1}} (\mathcal{G}_N \mathcal{Q}_N \bar{\mathcal{D}}_N^T + \mathcal{Z}_2^T) \\ &+ 2(I + \mathcal{C}_N^T \Omega_N^{p^{-1}} \mathcal{Z}_3 \mathcal{H}_N^{p^T}) . \end{aligned} \tag{9.141}$$

From (9.141), we extract the Lagrange multiplier

$$\begin{aligned} \Lambda = &2(\mathcal{C}_N^T \Omega_N^{p^{-1}} \mathcal{C}_N)^{-1} [I - \mathcal{C}_N^T \Omega_N^{p^{-1}} (\mathcal{G}_N \mathcal{Q}_N \bar{\mathcal{D}}_N^T \\ &+ \mathcal{Z}_2^T - \mathcal{Z}_3 \mathcal{H}_N^{p^T})] . \end{aligned} \tag{9.142}$$

Looking at (9.138) again, substituting (9.142), and providing some transformations, we finally obtain the gain for the H_2-OUFIR predictor as

$$\begin{aligned} \mathcal{H}_N^p = &\{(\mathcal{C}_N^T \Omega_N^{p^{-1}} \mathcal{C}_N)^{-1} \mathcal{C}_N^T + (\bar{\mathcal{D}}_N \mathcal{Q}_N \mathcal{G}_N^T + \mathcal{Z}_2) \\ &\times [I - \Omega_N^{p^{-1}} \mathcal{C}_N (\mathcal{C}_N^T \Omega_N^{p^{-1}} \mathcal{C}_N)^{-1} \mathcal{C}_N^T]\} \\ &\times [\Omega_N^p + \mathcal{Z}_3 - \mathcal{Z}_3^T \Omega_N^{p^{-1}} \mathcal{C}_N (\mathcal{C}_N^T \Omega_N^{p^{-1}} \mathcal{C}_N)^{-1} \mathcal{C}_N^T]^{-1} . \end{aligned} \tag{9.143}$$

As in the previous cases of state estimation of uncertain systems, we take notice that the zero uncertain matrices $\mathcal{Z}_2$ and $\mathcal{Z}_3$ make the gain (9.143) equal to the gain (8.70) of the H_2-OUFIR predictor, developed under disturbances and measurement errors. We also notice that the gain (9.143) does not require initial values and thus is more suitable for finite horizons.

Finally, the H_2-OUFIR prediction $\tilde{x}_{k+1}$ can be computed using (9.126), and the error matrix P_{k+1} can be written by neglecting χ_m as, respectively,

$$\tilde{x}_{k+1} = \mathcal{H}_N^p Y_{m,k} + (\bar{S}_N - \mathcal{H}_N^p L_N^p) U_{m,k} , \tag{9.144}$$

$$\begin{aligned} P = &\mathcal{W}_N^p \mathcal{Q}_N \mathcal{W}_N^{p^T} + \mathcal{V}_N^p \mathcal{R}_N \mathcal{V}_N^{p^T} \\ &\tilde{P}_x^p + \tilde{P}_w^p + \tilde{P}_u^p , \end{aligned} \tag{9.145}$$

where the uncertain error matrices $\tilde{P}_x^p$, $\tilde{P}_w^p$, and $\tilde{P}_u^p$ are defined by (9.128)–(9.130) and the gain $\mathcal{H}_N^p$ is given by (9.143).

To summarize, it is worth noting that, as in the H_2-OUFIR filter case, efforts should be made to specify the uncertain matrices for (9.143). If these matrices are properly maximized, then prediction over $[m, k]$ can be robust and sufficiently accurate. Otherwise, errors can grow and become unacceptably large.

9.4 Suboptimal H_2 FIR Structures Using LMI

Design of hybrid state estimators with improved robustness requires suboptimal H_2 FIR algorithms using LMI. Since hybrid FIR structures are typically designed based on different types of H_∞ estimators, the H_2 algorithm should have a similar structure using LMI. In what follows, we will consider such suboptimal FIR algorithms.

9.4.1 Suboptimal H_2 FIR Filter

To obtain the numerical gain $\mathcal{H}_N$ for a suboptimal H_2 FIR filter using LMI, we refer to (9.73) and introduce an additional positive definite matrix $\mathcal{Z}$ such that

$$\begin{aligned}\mathcal{Z} > &\mathcal{B}_N \chi_m \mathcal{B}_N + \mathcal{W}_N \mathcal{Q}_N \mathcal{W}_N^T + \mathcal{V}_N \mathcal{R}_N \mathcal{V}_N^T \\ &+ \tilde{P}_x + \tilde{P}_w + \tilde{P}_u \,.\end{aligned} \tag{9.146}$$

Substituting the error residual matrices taken from (8.15)–(8.17) and (9.74)–(9.76), we rewrite the inequality (9.146) as

$$\begin{aligned}&\mathcal{Z} - (\mathcal{H}_N H_N - F^{N-1})\chi_m(\mathcal{H}_N H_N - F^{N-1})^T \\ &- (\mathcal{H}_N G_N - \bar{D}_N)\mathcal{Q}_N(\mathcal{H}_N G_N - \bar{D}_N)^T - \mathcal{H}_N \mathcal{R}_N \mathcal{H}_N^T \\ &- \tilde{\chi}_m^F + \tilde{\chi}_m^{FH} \mathcal{H}_N^T + \mathcal{H}_N \tilde{\chi}_m^{HF} - \mathcal{H}_N \tilde{\chi}_m^H \mathcal{H}_N^T \\ &- \tilde{Q}_N^D + \tilde{Q}_N^{DT} \mathcal{H}_N^T + \mathcal{H}_N \tilde{Q}_N^{TD} - \mathcal{H}_N \tilde{Q}_N^T \mathcal{H}_N^T \\ &- \tilde{M}_N^S + \tilde{M}_N^{SL} \mathcal{H}_N^T + \mathcal{H}_N \tilde{M}_N^{LS} - \mathcal{H}_N \tilde{M}_N^L \mathcal{H}_N^T > 0\end{aligned}$$

and represent it with

$$\mathcal{Z} - \mathcal{A} + \mathcal{B}\mathcal{H}_N^T + \mathcal{H}_N \mathcal{C} - \mathcal{H}_N \mathcal{D} \mathcal{H}_N^T > 0 \,, \tag{9.147}$$

where the introduced auxiliary matrices are given by

$$\mathcal{A} = F^{N-1} \chi_m F^{N-1^T} + \bar{D}_N \mathcal{Q}_N \bar{D}_N^T + \tilde{\chi}_m^F + \tilde{Q}_N^D + \tilde{M}_N^S \,, \tag{9.148}$$

$$\mathcal{B} = H_N \chi_m F^{N-1^T} + G_N \mathcal{Q}_N \bar{D}_N^T + \tilde{\chi}_m^{FH} + \tilde{Q}_N^{DT} + \tilde{M}_N^{SL} \,, \tag{9.149}$$

$$\mathcal{C} = F^{N-1} \chi_m H_N^T + \bar{D}_N \mathcal{Q}_N G_N^T + \tilde{\chi}_m^{HF} + \tilde{Q}_N^{TD} + \tilde{M}_N^{LS} \,, \tag{9.150}$$

$$\mathcal{D} = H_N \chi_m H_N^T + \Omega_N - \tilde{\chi}_m^H - \tilde{Q}_N^T - \tilde{M}_N^L \,. \tag{9.151}$$

Using the Schur complement, we finally represent the inequality (9.147) with the following LMI

$$\begin{bmatrix} \mathcal{Z} - \mathcal{A} + \mathcal{B}\mathcal{H}_N^T + \mathcal{H}_N \mathcal{C} & \mathcal{H}_N \\ \mathcal{H}_N^T & \mathcal{D}^{-1} \end{bmatrix} > 0 \,. \tag{9.152}$$

The gain $\mathcal{H}_N$ for the suboptimal H_2 FIR filter can now be computed numerically by solving the following minimization problem

$$\mathcal{H}_N = \underset{\substack{\mathcal{H}_N, \mathcal{Z} \\ \text{subject to (9.152)}}}{\min} \operatorname{tr}\mathcal{Z} \,. \tag{9.153}$$

As in other similar cases, the best candidate for starting solving (9.153) is the UFIR filter gain $\hat{\mathcal{H}}_N = (C_N^T C_N)^{-1} C_N^T$. Provided that $\mathcal{H}_N$ is found numerically, the suboptimal H_2 FIR filtering estimate can be computed as

$$\hat{x}_k = \mathcal{H}_N Y_{m,k} \tag{9.154}$$

and the error matrix P can be computed by (9.73).

9.4.2 Bias-Constrained Suboptimal H_2 FIR Filter

In a like manner, the gain for the bias-constrained suboptimal H_2 FIR filter appears in LMI form, if we refer to (9.90) and introduce an auxiliary positive definite matrix $\mathcal{Z}$ such that

$$\mathcal{Z} > \mathcal{W}_N \mathcal{Q}_N \mathcal{W}_N^T + \mathcal{V}_N \mathcal{R}_N \mathcal{V}_N^T + \tilde{P}_x + \tilde{P}_w + \tilde{P}_u \,, \tag{9.155}$$

where the error residual matrices are given by (8.16), (8.17), and (9.74)–(9.76). Then we rewrite (9.155) as

$$\begin{aligned}
&\mathcal{Z} - (\mathcal{H}_N G_N - \bar{D}_N)\mathcal{Q}_N(\mathcal{H}_N G_N - \bar{D}_N)^T - \mathcal{H}_N \mathcal{R}_N \mathcal{H}_N^T \\
&- \tilde{\chi}_m^F + \tilde{\chi}_m^{FH} \mathcal{H}_N^T + \mathcal{H}_N \tilde{\chi}_m^{HF} - \mathcal{H}_N \tilde{\chi}_m^H \mathcal{H}_N^T \\
&- \tilde{Q}_N^D + \tilde{Q}_N^{DT} \mathcal{H}_N^T + \mathcal{H}_N \tilde{Q}_N^{TD} - \mathcal{H}_N \tilde{Q}_N^T \mathcal{H}_N^T \\
&- \tilde{M}_N^S + \tilde{M}_N^{SL} \mathcal{H}_N^T + \mathcal{H}_N \tilde{M}_N^{LS} - \mathcal{H}_N \tilde{M}_N^L \mathcal{H}_N^T > 0
\end{aligned}$$

and transform to

$$\mathcal{Z} - \mathcal{A} + \mathcal{B}\mathcal{H}_N^T + \mathcal{H}_N \mathcal{C} - \mathcal{H}_N \mathcal{D} \mathcal{H}_N^T > 0 \,, \tag{9.156}$$

using the following auxiliary matrices,

$$\mathcal{A} = \bar{D}_N \mathcal{Q}_N \bar{D}_N^T + \tilde{\chi}_m^F + \tilde{Q}_N^D + \tilde{M}_N^S \,, \tag{9.157}$$

$$\mathcal{B} = G_N \mathcal{Q}_N \bar{D}_N^T + \tilde{\chi}_m^{FH} + \tilde{Q}_N^{DT} + \tilde{M}_N^{SL} \,, \tag{9.158}$$

$$\mathcal{C} = \bar{D}_N \mathcal{Q}_N G_N^T + \tilde{\chi}_m^{HF} + \tilde{Q}_N^{TD} + \tilde{M}_N^{LS} \,, \tag{9.159}$$

$$\mathcal{D} = \Omega_N - \tilde{\chi}_m^H - \tilde{Q}_N^T - \tilde{M}_N^L \,. \tag{9.160}$$

Using the Schur complement, we represent (9.156) in the LMI form as

$$\begin{bmatrix} \mathcal{Z} - \mathcal{A} + \mathcal{B}\mathcal{H}_N^T + \mathcal{H}_N \mathcal{C} & \mathcal{H}_N \\ \mathcal{H}_N^T & \mathcal{D}^{-1} \end{bmatrix} > 0 \tag{9.161}$$

and determine the gain for the bias-constrained suboptimal H_2 FIR filter by solving numerically the following minimization problem

$$\mathcal{H}_N = \underset{\substack{\mathcal{H}_N, \mathcal{Z} \\ \text{subject to (9.152) and } I = \mathcal{H}_N C_N}}{\min} \operatorname{tr} \mathcal{Z} \,. \tag{9.162}$$

Traditionally, we start the minimization with $\hat{\mathcal{H}}_N = (C_N^T C_N)^{-1} C_N^T$. Provided that $\mathcal{H}_N$ is numerically available from (9.162), the suboptimal H_2 FIR filtering estimate is computed by

$$\hat{x}_k = \mathcal{H}_N Y_{m,k} \tag{9.163}$$

and the error matrix P is computed using (9.90). Note that the gain $\mathcal{H}_N$ obtained by solving (9.162) is more robust due to the rejection of the initial state requirement.

9.4.3 Suboptimal H_2 FIR Predictor

To find the suboptimal gain for the H_2 FIR predictor using LMI, we introduce an auxiliary positive definite matrix $\mathcal{Z}$ to satisfy the inequality

$$\mathcal{Z} > \mathcal{B}^p_N \chi_m \mathcal{B}^{p^T}_N + \mathcal{W}^p_N \mathcal{Q}_N \mathcal{W}^{p^T}_N + \mathcal{V}^p_N \mathcal{R}_N \mathcal{V}^{p^T}_N + \tilde{\mathcal{P}}^p_x + \tilde{\mathcal{P}}^p_w + \tilde{\mathcal{P}}^p_u . \tag{9.164}$$

We then use (8.60)–(8.62) and (9.128)–(9.130) and transform (9.164) to

$$\begin{aligned}
&\mathcal{Z} - (\mathcal{H}^p_N H^p_N - F^N)\chi_m(\mathcal{H}^p_N H^p_N - F^N)^T \\
&- (\mathcal{H}^p_N G^p_N - \bar{D}_N)\mathcal{Q}_N(\mathcal{H}^p_N G^p_N - \bar{D}_N)^T - \mathcal{H}^p_N \mathcal{R}_N \mathcal{H}^{p^T}_N \\
&- \tilde{\chi}^F_m + \tilde{\chi}^{FH}_m \mathcal{H}^{p^T}_N + \mathcal{H}^p_N \tilde{\chi}^{HF}_m - \mathcal{H}^p_N \tilde{\chi}^H_m \mathcal{H}^{p^T}_N \\
&- \tilde{Q}^D_N + \tilde{Q}^{DT}_N \mathcal{H}^{p^T}_N + \mathcal{H}^p_N \tilde{Q}^{TD}_N - \mathcal{H}^p_N \tilde{Q}^T_N \mathcal{H}^{p^T}_N \\
&- \tilde{M}^S_N + \tilde{M}^{SL}_N \mathcal{H}^{p^T}_N + \mathcal{H}^p_N \tilde{M}^{LS}_N - \mathcal{H}^p_N \tilde{M}^L_N \mathcal{H}^{p^T}_N > 0 ,
\end{aligned}$$

where the uncertain matrices are the same as for the H_2 FIR predictor. We next represent this inequality as

$$\mathcal{Z} - \mathcal{A} + \mathcal{B}\mathcal{H}^{p^T}_N + \mathcal{H}^p_N \mathcal{C} - \mathcal{H}^p_N \mathcal{D} \mathcal{H}^{p^T}_N > 0 , \tag{9.165}$$

where the introduced auxiliary matrices are the following

$$\mathcal{A} = F^N \chi_m F^{N^T} + \bar{D}_N \mathcal{Q}_N \bar{D}^T_N + \tilde{\chi}^F_m + \tilde{Q}^D_N + \tilde{M}^S_N , \tag{9.166}$$

$$\mathcal{B} = H^p_N \chi_m F^{N^T} + G^p_N \mathcal{Q}_N \bar{D}^T_N + \tilde{\chi}^{FH}_m + \tilde{Q}^{DT}_N + \tilde{M}^{SL}_N , \tag{9.167}$$

$$\mathcal{C} = F^N \chi_m H^{p^T}_N + \bar{D}_N \mathcal{Q}_N G^{p^T}_N + \tilde{\chi}^{HF}_m + \tilde{Q}^{TD}_N + \tilde{M}^{LS}_N , \tag{9.168}$$

$$\mathcal{D} = H^p_N \chi_m H^{p^T}_N + \Omega_N - \tilde{\chi}^H_m - \tilde{Q}^T_N - \tilde{M}^L_N . \tag{9.169}$$

Using the Schur complement, we represent (9.165) in the LMI form

$$\begin{bmatrix} \mathcal{Z} - \mathcal{A} + \mathcal{B}\mathcal{H}^{p^T}_N + \mathcal{H}^p_N \mathcal{C} & \mathcal{H}^p_N \\ \mathcal{H}^{p^T}_N & \mathcal{D}^{-1} \end{bmatrix} > 0 \tag{9.170}$$

that allows finding numerically the gain for the H_2 FIR predictor by solving the following minimization problem

$$\mathcal{H}^p_N = \min_{\mathcal{H}^p_N, \mathcal{Z}} \operatorname{tr} \mathcal{Z} \quad \text{subject to (9.170)} \tag{9.171}$$

using $\hat{\mathcal{H}}^p_N = (C^{p^T}_N C^p_N)^{-1} C^{p^T}_N$ as an initial try. The suboptimal H_2 FIR prediction can finally be computed by

$$\tilde{x}_{k+1} = \mathcal{H}^p_N Y_{m,k} \tag{9.172}$$

and the error matrix P by (9.127).

9.4.4 Bias-Constrained Suboptimal H_2 FIR Predictor

As before, to remove the requirement of the initial state, the gain for the bias-constrained suboptimal H_2 FIR predictor can be computed using LMI. To do this, we traditionally look at (9.145) and introduce a positive definite matrix $\mathcal{Z}$ such that

$$\begin{aligned}\mathcal{Z} > &\mathcal{W}_N^p \mathcal{Q}_N \mathcal{W}_N^{p^T} + \mathcal{V}_N^p \mathcal{R}_N \mathcal{V}_N^{p^T}\\ &+ \tilde{P}_x^p + \tilde{P}_w^p + \tilde{P}_u^p \,. \end{aligned} \tag{9.173}$$

We then use the error residual matrices (8.61), (8.62), and (9.128)–(9.130) and transform (9.173) to

$$\begin{aligned}&\mathcal{Z} - (\mathcal{H}_N^p G_N^p - \bar{D}_N)\mathcal{Q}_N(\mathcal{H}_N^p G_N^p - \bar{D}_N)^T - \mathcal{H}_N^p \mathcal{R}_N \mathcal{H}_N^{p^T}\\ &- \tilde{\chi}_m^F + \tilde{\chi}_m^{FH}\mathcal{H}_N^{p^T} + \mathcal{H}_N^p \tilde{\chi}_m^{HF} - \mathcal{H}_N^p \tilde{\chi}_m^H \mathcal{H}_N^{p^T}\\ &- \tilde{Q}_N^D + \tilde{Q}_N^{DT}\mathcal{H}_N^{p^T} + \mathcal{H}_N^p \tilde{Q}_N^{TD} - \mathcal{H}_N^p \tilde{Q}_N^T \mathcal{H}_N^{p^T}\\ &- \tilde{M}_N^S + \tilde{M}_N^{SL}\mathcal{H}_N^{p^T} + \mathcal{H}_N^p \tilde{M}_N^{LS} - \mathcal{H}_N^p \tilde{M}_N^L \mathcal{H}_N^{p^T} > 0 \,,\end{aligned}$$

which we further generalize as

$$\mathcal{Z} - \mathcal{A} + \mathcal{B}\mathcal{H}_N^{p^T} + \mathcal{H}_N^p \mathcal{C} - \mathcal{H}_N^p \mathcal{D}\mathcal{H}_N^{p^T} > 0 \,, \tag{9.174}$$

using the auxiliary matrices

$$\mathcal{A} = \bar{D}_N \mathcal{Q}_N \bar{D}_N^T + \tilde{\chi}_m^F + \tilde{Q}_N^D + \tilde{M}_N^S \,, \tag{9.175}$$

$$\mathcal{B} = G_N^p \mathcal{Q}_N \bar{D}_N^T + \tilde{\chi}_m^{FH} + \tilde{Q}_N^{DT} + \tilde{M}_N^{SL} \,, \tag{9.176}$$

$$\mathcal{C} = \bar{D}_N \mathcal{Q}_N G_N^{p^T} + \tilde{\chi}_m^{HF} + \tilde{Q}_N^{TD} + \tilde{M}_N^{LS} \,, \tag{9.177}$$

$$\mathcal{D} = \Omega_N - \tilde{\chi}_m^H - \tilde{Q}_N^T - \tilde{M}_N^L \,. \tag{9.178}$$

Using the Schur complement, (9.174) can now be substituted by the LMI

$$\begin{bmatrix} \mathcal{Z} - \mathcal{A} + \mathcal{B}\mathcal{H}_N^{p^T} + \mathcal{H}_N^p \mathcal{C} & \mathcal{H}_N^p \\ \mathcal{H}_N^{p^T} & \mathcal{D}^{-1} \end{bmatrix} > 0 \tag{9.179}$$

and the gain for the bias-constrained suboptimal H_2 FIR predictor can be numerically found by solving the minimization problem

$$\mathcal{H}_N^p = \underset{\substack{\mathcal{H}_N^p, \mathcal{Z} \\ \text{subject to (9.179) and } I = \mathcal{H}_N^p C_N^p}}{\min \operatorname{tr} \mathcal{Z}} \,, \tag{9.180}$$

if we start with $\hat{\mathcal{H}}_N^p = (C_N^{p^T} C_N^p)^{-1} C_N^{p^T}$. Finally, the bias-constrained suboptimal H_2 prediction can be obtained as

$$\tilde{x}_{k+1} = \mathcal{H}_N^p Y_{m,k} \tag{9.181}$$

and the error matrix P can be computed using (9.145).

9.5 H_∞ FIR State Estimation for Uncertain Systems

To obtain various robust H_∞ FIR state estimators for uncertain systems, we will start with the estimation errors of the FIR filter (9.49) and the FIR predictor (9.101) and will follow the lines previously developed to estimate the state under disturbances. Namely, using the induced norm $\| \mathcal{T} \|_\infty$ defined by (8.90), we will find the gain for the H_∞ FIR estimator corresponding to uncertain systems by numerically solving the familiar suboptimal problem

$$\mathcal{H}_N \Leftarrow \sup_{\varsigma\neq 0} \frac{\sum_{i=m}^{k} \varepsilon_i^T P_\varepsilon \varepsilon_i}{\sum_{i=m}^{k} \varsigma_i^T P_\varsigma \varsigma_i} < \gamma^2 , \tag{9.182}$$

where the scalar factor γ^2, which indicates the part of the maximized uncertainty energy that goes to the output, should preferably be small. In what follows, we will develop an H_∞ filter and an H_∞ predictor for uncertain systems operating under disturbances, initial errors, and measurement errors.

9.5.1 The *a posteriori* H_∞ FIR Filter

Consider an uncertain system operating under a bounded zero mean disturbance w_k and measurement error v_k. Put $u_k = 0$ and represent this system with the BE-based state-space model (9.1)–(9.2) as

$$x_k = F_k^u x_{k-1} + B_k^u w_k , \tag{9.183}$$

$$y_k = H_k^u x_k + D_k^u w_k + v_k , \tag{9.184}$$

where $F_k^u = F + \Delta F_k$, $B_k^u = B + \Delta B_k$, $H_k^u = H + \Delta H_k$, and $D_k^u = D + \Delta D_k$. Note that the uncertain increments ΔF_k, ΔB_k, ΔH_k, and ΔD_k represent time-varying bounded parameters specified after (9.2).

We rewrite model (9.183) and (9.184) in the form of (9.3) and (9.4) as

$$x_k = F x_{k-1} + \xi_k , \tag{9.185}$$

$$y_k = H x_k + \zeta_k , \tag{9.186}$$

where the zero mean uncertain error vectors ξ_k and ζ_k are defined by

$$\xi_k = \Delta F_k x_{k-1} + B_k^u w_k , \tag{9.187}$$

$$\zeta_k = \Delta H_k x_k + D_k^u w_k + v_k \tag{9.188}$$

to play the role of zero mean errors in the model in (9.185) and (9.186).

Following (9.13) and (9.17), we next extend the model in (9.185) and (9.186) to $[m, k]$ as

$$x_k = (F^{N-1} + \bar{\tilde{F}}_{m,k}) x_m + (\bar{D}_N + \bar{\tilde{D}}_{m,k}) W_{m,k} , \tag{9.189}$$

$$Y_{m,k} = (H_N + \tilde{H}_{m,k}) x_m + (T_N + \tilde{T}_{m,k}) W_{m,k} + V_{m,k} , \tag{9.190}$$

where $\bar{\tilde{F}}_{m,k}$ and $\bar{\tilde{D}}_{m,k}$ are the last row vectors in the uncertain matrices $\tilde{F}_{m,k} = \hat{F}_N F^u_{m,k}$ and $\tilde{D}_{m,k} = \hat{F}_N D^u_{m,k}$, respectively. Note that the matrix $\hat{F}_N$ is given by (9.8), $F^u_{m,k}$ and $D^u_{m,k}$ by (9.10), $\tilde{H}_{m,k}$ by (9.18) using (9.16), and $\tilde{T}_{m,k}$ by (9.20) using (9.16).

The H_∞ FIR filter can now be derived for uncertain systems if we write the estimation error (9.43) for $u_k = 0$ as

$$\varepsilon_k = (\mathcal{B}_N + \tilde{\mathcal{B}}_{m,k})x_m + (\mathcal{W}_N + \tilde{\mathcal{W}}_{m,k})W_{m,k} - \mathcal{V}_N V_{m,k} , \tag{9.191}$$

where the error residual matrices are given by (8.15)–(8.17), (9.44), and (9.46),

$$\mathcal{B}_N = F^{N-1} - \mathcal{H}_N H_N , \tag{9.192}$$

$$\mathcal{W}_N = \bar{D}_N - \mathcal{H}_N T_N , \tag{9.193}$$

$$\mathcal{V}_N = \mathcal{H}_N , \tag{9.194}$$

$$\tilde{\mathcal{B}}_{m,k} = \tilde{\bar{F}}_{m,k} - \mathcal{H}_N \tilde{H}_{m,k} , \tag{9.195}$$

$$\tilde{\mathcal{W}}_{m,k} = \tilde{\bar{D}}_{m,k} - \mathcal{H}_N \tilde{T}_{m,k} , \tag{9.196}$$

and the vectors $W_{m,k}$ and $V_{m,k}$ can be written using lemma 8.1 as

$$W_{m,k} = A_w W_{m-1,k-1} + B_w w_k , \tag{9.197}$$

$$V_{m,k} = A_w V_{m,k-1} + B_w v_k , \tag{9.198}$$

where the sparse matrices A_w and B_w are defined by (8.19).

Using (9.197), (9.198), and the augmented vectors $z_k = [\, W_{m,k}^T \;\; V_{m,k}^T \;\; i_k^T \,]^T$, where $i_k = x_m$, and $\xi_k = [\, w_k^T \;\; v_k^T \,]^T$, we write the uncertainty-to-error state space model in the standard form

$$z_k = \tilde{F}_\varsigma z_{k-1} + \tilde{B}_\varsigma \xi_k , \tag{9.199}$$

$$\varepsilon_k = \tilde{C}_\varsigma z_k , \tag{9.200}$$

where the sparse matrices $\tilde{F}_\varsigma$ and $\tilde{B}_\varsigma$ are given by (8.109) as

$$\tilde{F}_\varsigma = \begin{bmatrix} A_w & 0 & 0 \\ 0 & A_w & 0 \\ 0 & 0 & I \end{bmatrix}, \quad \tilde{B}_\varsigma = \begin{bmatrix} B_w & 0 \\ 0 & B_w \\ 0 & 0 \end{bmatrix} \tag{9.201}$$

and all terms containing gain $\mathcal{H}_N$ are collected in the matrix $\tilde{C}_\varsigma$,

$$\begin{aligned} \tilde{C}_\varsigma &= \begin{bmatrix} \mathcal{W}_N + \tilde{\mathcal{W}}_{m,k} & -\mathcal{V}_N & \mathcal{B}_N + \tilde{\mathcal{B}}_{m,k} \end{bmatrix} \\ &= \begin{bmatrix} (\bar{D}_N - \mathcal{H}_N G_N + \tilde{\bar{D}}_{m,k} - \mathcal{H}_N \tilde{T}_{m,k})^T \\ -\mathcal{H}_N^T \\ (F^{N-1} - \mathcal{H}_N H_N + \tilde{\bar{F}}_{m,k} - \mathcal{H}_N \tilde{H}_{m,k})^T \end{bmatrix}^T , \end{aligned} \tag{9.202}$$

which is a very important algorithmic property of this model. Indeed, there is only one matrix $\tilde{C}_\varsigma$, whose components are functions of $\mathcal{H}_N$, which makes the model in (9.199) and (9.200) computationally efficient.

Using (9.199) and (9.200), we can now apply the BRL lemma 8.2 and develop an *a posteriori* H_∞ FIR filter using LMI for uncertain systems operating under disturbances and measurement errors. Traditionally, we will look at the solutions taking notice of the necessity to have a numerical gain $\mathcal{H}_N$ such that $\gamma > 0$ satisfying (9.182) reaches a minimum for maximized uncertainties. The following options are available:

- Apply lemma 8.2 to (9.199) and (9.200) and observe that $D = 0$, $P_y = P$, and $P_w = P_\xi$. For some symmetric matrix $X > 0$, find the gain $\mathcal{H}_N$, which is a variable of the matrix $\tilde{C}_\varsigma$ (9.202), by minimizing $\gamma > 0$ to satisfy the following LMI

$$\begin{bmatrix} X & X\tilde{F}_\varsigma & X\tilde{B}_\varsigma & 0 \\ \tilde{F}_\varsigma^T X & -X & 0 & \tilde{F}_\varsigma^T \tilde{C}_\varsigma^T \\ \tilde{B}_\varsigma^T X & 0 & -\gamma P_\xi & \tilde{B}_\varsigma^T \tilde{C}_\varsigma^T \\ 0 & \tilde{C}_\varsigma \tilde{F}_\varsigma & \tilde{C}_\varsigma \tilde{B}_\varsigma & -\gamma P^{-1} \end{bmatrix} < 0 . \quad (9.203)$$

- Consider (8.101), assign $\Psi = K + \tilde{C}_\varsigma^T P \tilde{C}_\varsigma$, and solve for $\mathcal{H}_N$ the following LMI problem by minimizing γ,

$$\begin{bmatrix} \tilde{F}_\varsigma^T \Psi \tilde{F}_\varsigma - K & \tilde{F}_\varsigma^T \Psi \tilde{B}_\varsigma \\ \tilde{B}_\varsigma^T \Psi \tilde{F}_\varsigma & \tilde{B}_\varsigma^T \Psi \tilde{B}_\varsigma - \gamma^2 P_\xi \end{bmatrix} < 0 ,$$

$$\tilde{B}_\varsigma^T \Psi \tilde{B}_\varsigma - \gamma^2 P_\xi < 0 . \quad (9.204)$$

- Solve for $\mathcal{H}_N$ the following DARE by minimizing γ,

$$\tilde{F}_\varsigma^T \Psi \tilde{F}_\varsigma - K - \tilde{F}_\varsigma^T \Psi \tilde{B}_\varsigma (\tilde{B}_\varsigma^T \Psi \tilde{B}_\varsigma - \gamma^2 P_\xi)^{-1} \tilde{B}_\varsigma^T \Psi \tilde{F}_\varsigma < 0 ,$$

$$\tilde{B}_\varsigma^T \Psi \tilde{B}_\varsigma - \gamma^2 P_\xi < 0 . \quad (9.205)$$

The robust *a posteriori* H_∞ FIR filtering estimate and the error matrix can now be suboptimally computed for maximized zero mean uncertainties, disturbances, data errors, and initial errors by, respectively,

$$\hat{x}_k = \mathcal{H}_N Y_{m,k} , \quad (9.206)$$

$$\begin{aligned} P &= (\mathcal{B}_N + \tilde{\mathcal{B}}_{m,k})\mathcal{X}_m(\mathcal{B}_N + \tilde{\mathcal{B}}_{m,k})^T \\ &\quad + (\mathcal{W}_N + \tilde{\mathcal{W}}_{m,k})\mathcal{Q}_N(\mathcal{W}_N + \tilde{\mathcal{W}}_{m,k})^T \\ &\quad + \mathcal{V}_N \mathcal{R}_N \mathcal{V}_N^T \end{aligned} \quad (9.207)$$

using the gain $\mathcal{H}_N$, obtained by one of the previous algorithmic options, and the error residual matrices in (9.192)–(9.196). It is worth noting that, although all of the previous algorithmic options are feasible, the LMI form (9.203) is the most elaborate.

9.5.2 H_∞ FIR Predictor

Robust prediction based on the H_∞ FIR approach can be organized for uncertain systems in the same way as for systems operating under disturbances and measurement errors. To find a suboptimal gain for the H_∞ FIR predictor, we traditionally use the FE-based model in (9.21) and (9.22) with $u_k = 0$,

$$x_{k+1} = F_k^u x_k + B_k^u w_k , \quad (9.208)$$

$$y_k = H_k^u x_k + D_k^u w_k + v_k , \quad (9.209)$$

and represent it by reorganizing the terms as

$$x_{k+1} = F x_k + \xi_k^p , \quad (9.210)$$

$$y_k = H x_k + \zeta_k^p , \quad (9.211)$$

where the uncertain error vectors are given by

$$\xi_k^p = \Delta F_k x_k + B_k^u w_k \,, \tag{9.212}$$

$$\zeta_k^p = \Delta H_k x_k + D_k^u w_k + v_k \,. \tag{9.213}$$

We extend the model in (9.210) and (9.211) to $[m, k]$ as

$$x_{k+1} = (F^N + \bar{\tilde{F}}^p_{m,k})x_m + (\bar{D}_N + \bar{\tilde{D}}^p_{m,k})W_{m,k} \,, \tag{9.214}$$

$$Y_{m,k} = (H_N^p + \bar{\tilde{H}}^p_{m,k})x_m + (T_N^p + \bar{\tilde{T}}^p_{m,k})W_{m,k} + V_{m,k} \,, \tag{9.215}$$

where the matrix $\bar{\tilde{F}}^p_{m,k}$ is the last row vector in matrix $\tilde{F}^p_{m,k} = \hat{F}_N F^{pu}_{m,k}$ defined using (9.8) and (9.29), $\bar{\tilde{D}}^p_{m,k}$ is specified after (9.13), $\bar{\tilde{H}}^p_{m,k}$ by (9.36), and $\bar{\tilde{T}}^p_{m,k}$ by (9.38). We next take the regular error residual matrices $\mathcal{B}_N^p$, $\mathcal{W}_N^p$, and $\mathcal{V}_N^p$ from (8.60)–(8.62) and rewrite the estimation error (9.95) as

$$\varepsilon_{k+1} = (\mathcal{B}_N^p + \tilde{\mathcal{B}}^p_{m,k})x_m + (\mathcal{W}_N^p + \tilde{\mathcal{W}}^p_{m,k})W_{m,k} - \mathcal{V}_N^p V_{m,k} \,, \tag{9.216}$$

where the remaining uncertain error residual matrices are given by

$$\mathcal{B}_N^p = F^N - \mathcal{H}_N^p H_N^p \,, \tag{9.217}$$

$$\mathcal{W}_N^p = \bar{D}_N - \mathcal{H}_N^p T_N^p \,, \tag{9.218}$$

$$\mathcal{V}_N^p = \mathcal{H}_N^p \,, \tag{9.219}$$

$$\tilde{\mathcal{B}}^p_{m,k} = \bar{\tilde{F}}^p_{m,k} - \mathcal{H}_N^p \tilde{H}^p_{m,k} \,, \tag{9.220}$$

$$\tilde{\mathcal{W}}^p_{m,k} = \bar{\tilde{D}}_{m,k} - \mathcal{H}_N^p \tilde{T}^p_{m,k} \,. \tag{9.221}$$

Using lemma 8.1, we also represent $W_{m,k}$ and $V_{m,k}$ as (8.105) and (8.106),

$$W_{m,k} = A_w W_{m-1,k-1} + B_w w_k \,,$$
$$V_{m,k} = A_w V_{m-1,k-1} + B_w v_k \,,$$

where matrices A_w and B_w are defined by (8.19).

Now, we assign vectors $z_k = [\, W_{m,k}^T \;\; V_{m,k}^T \;\; i_k^T \,]^T$, where $i_k = x_m$, and $\xi_k = [\, w_k^T \;\; v_k^T \,]^T$, follow the lines developed after (9.184), and obtain the uncertainty-to-error state-space model

$$z_k = \tilde{F}_\varsigma z_{k-1} + \tilde{B}_\varsigma \xi_k \,, \tag{9.222}$$

$$\varepsilon_k = \tilde{C}_\varsigma^p z_k + \tilde{D}_\varsigma^p \xi_k \,, \tag{9.223}$$

in which the sparse matrices $\tilde{F}_\varsigma$ and $\tilde{B}_\varsigma$ are given after (8.108) and matrices $\tilde{C}_\varsigma^p$ and $\tilde{D}_\varsigma^p$ are defined by (8.133) as

$$\tilde{C}_\varsigma^p = \hat{C}_\varsigma^p \tilde{F}_\varsigma^{-1} \,, \tag{9.224}$$

$$\tilde{D}_\varsigma^p = -\tilde{C}_\varsigma^p \tilde{B}_\varsigma = -\hat{C}_\varsigma^p \tilde{F}_\varsigma^{-1} \tilde{B}_\varsigma \,, \tag{9.225}$$

where the matrix $\hat{C}_\varsigma^p$ is given by

$$\hat{C}_\varsigma^p = \begin{bmatrix} \mathcal{W}_N^p + \tilde{\mathcal{W}}^p_{m,k} & -\tilde{\mathcal{V}}^p_{m,k} & \mathcal{B}_N^p + \tilde{\mathcal{B}}^p_{m,k} \end{bmatrix}$$

$$= \begin{bmatrix} (\bar{D}_N - \mathcal{H}^p_N T^p_N - \bar{\tilde{D}}_{m,k} + \mathcal{H}^p_N \tilde{T}^p_{m,k})^T \\ -\mathcal{H}^{p^T}_N \\ (F^N - \mathcal{H}^p_N H^p_N - \bar{\tilde{F}}^p_{m,k} + \mathcal{H}^p_N \tilde{H}^p_{m,k})^T \end{bmatrix}^T . \tag{9.226}$$

This model does not reveal any new features, and we simply note that, as in H_∞ FIR filtering, the error residual matrices are collected here in the modified observation matrix $\tilde{C}^p_\varsigma$ (9.226), which is thus completely responsible for the performance of the H_∞ FIR predictor.

We can now apply the BRL lemma 8.3 to (9.222) and (9.223) and develop a suboptimal H_∞ FIR predictor by satisfying the performance criterion (9.182). By avoiding repeating the details of the steps above, we simply list the three main options here:

- Apply lemma 8.3 to (9.224) and (9.225) and note that $P_y = P$ and $P_w = P_\xi$. For some symmetric matrix X, solve for $\mathcal{H}^p_N$ the LMI problem

$$\begin{bmatrix} -X & X\tilde{F}_\varsigma & X\tilde{B}_\varsigma & 0 \\ \tilde{F}^T_\varsigma X & -X & 0 & \tilde{C}^{p^T}_\varsigma \\ \tilde{B}^T_\varsigma X & 0 & -\gamma P_\xi & \tilde{D}^{p^T}_\varsigma \\ 0 & \tilde{C}^p_\varsigma & \tilde{D}^p_\varsigma & -\gamma P^{-1} \end{bmatrix} < 0 , \quad \gamma > 0 , \tag{9.227}$$

  starting with $\hat{\mathcal{H}}^p_N = (C^{p^T}_N C^p_N)^{-1} C^{p^T}_N$.
- Refer to (8.136) and solve the following LMI problem

$$\begin{bmatrix} \tilde{C}^{p^T}_\varsigma P \tilde{C}^p_\varsigma + \tilde{F}^T_\varsigma K \tilde{F}_\varsigma - K & \tilde{C}^{p^T}_\varsigma P \tilde{D}^p_\varsigma + \tilde{F}^T_\varsigma K \tilde{B}_\varsigma \\ \tilde{D}^{p^T}_\varsigma P \tilde{C}^p_\varsigma + \tilde{B}^T_\varsigma K \tilde{F}_\varsigma & \tilde{D}^{p^T}_\varsigma P \tilde{D}^p_\varsigma + \tilde{B}^T_\varsigma K \tilde{B}_\varsigma - \gamma^2 P_\xi \end{bmatrix} < 0 ,$$
$$\tilde{D}^{p^T}_\varsigma P \tilde{D}^p_\varsigma + \tilde{B}^T_\varsigma K \tilde{B}_\varsigma - \gamma^2 P_\xi < 0 . \tag{9.228}$$

  starting with the same initial $\hat{\mathcal{H}}^p_N$ as in (9.227).
- Solve the DARE

$$\begin{aligned} &\tilde{C}^{p^T}_\varsigma P \tilde{C}^p_\varsigma + \tilde{F}^T_\varsigma K \tilde{F}_\varsigma - K - (\tilde{C}^{p^T}_\varsigma P \tilde{D}^p_\varsigma + \tilde{F}^T_\varsigma K \tilde{B}_\varsigma) \\ &\times (\tilde{D}^{p^T}_\varsigma P \tilde{D}^p_\varsigma + \tilde{B}^T_\varsigma K \tilde{B}_\varsigma - \gamma^2 P_\xi)^{-1} (\tilde{D}^{p^T}_\varsigma P \tilde{C}^p_\varsigma + \tilde{B}^T_\varsigma K \tilde{F}_\varsigma) < 0 , \end{aligned} \tag{9.229}$$

  subject to $\tilde{D}^{p^T}_\varsigma P \tilde{C}^p_\varsigma + \tilde{B}^T_\varsigma K \tilde{B}_\varsigma - \gamma^2 P_\xi < 0$.

The H_∞ FIR predictor can finally be summarized with the following estimate and error matrix, respectively,

$$\tilde{x}_{k+1} = \mathcal{H}^p_N Y_{m,k} , \tag{9.230}$$

$$\begin{aligned} P = &(\mathcal{B}^p_N + \tilde{\mathcal{B}}^p_{m,k}) \mathcal{X}_m (\mathcal{B}^p_N + \tilde{\mathcal{B}}^p_{m,k})^T \\ &+ (\mathcal{W}^p_N + \tilde{\mathcal{W}}^p_{m,k}) \mathcal{Q}_N (\mathcal{W}^p_N + \tilde{\mathcal{W}}^p_{m,k})^T \\ &+ \mathcal{V}^p_N \mathcal{R}_N \mathcal{V}^{p^T}_N , \end{aligned} \tag{9.231}$$

where all the error residual matrices are functions of the gain $\mathcal{H}^p_N$, which must be computed numerically by solving the LMI problem using one of three options (9.227)–(9.229).

9.6 Hybrid H_2/H_∞ FIR Structures

In an effort to achieve the highest robustness in estimating uncertain systems, the design of hybrid structures that combine the properties of different estimators is considered a top priority. As we will show, such structures can be created taking into account disturbances, initial errors, and data errors.

The *a posteriori* H_2/H_∞ FIR Filter

A hybrid LMI-based algorithm to numerically compute the suboptimal gain $\mathcal{H}_N$ for the *a posteriori* H_2/H_∞ FIR filter can be developed by solving the following minimization problem subject to constraints (9.152) and (9.203),

$$\begin{aligned}
\mathcal{H}_N \Leftarrow & \inf_{\mathcal{H}_N,\mathcal{Z}} \{\mathrm{tr}\mathcal{Z}, \gamma > 0\} \\
& \text{subject to} \\
0 < & \begin{bmatrix} \mathcal{Z} - \mathcal{A} + \mathcal{B}\mathcal{H}_N^T + \mathcal{H}_N\mathcal{C} & \mathcal{H}_N \\ \mathcal{H}_N^T & \mathcal{D}^{-1} \end{bmatrix}, \\
0 > & \begin{bmatrix} X & X\tilde{F}_\varsigma & X\tilde{B}_\varsigma & 0 \\ \tilde{F}_\varsigma^T X & -X & 0 & \tilde{F}_\varsigma^T \tilde{C}_\varsigma^T \\ \tilde{B}_\varsigma^T X & 0 & -\gamma P_\xi & \tilde{B}_\varsigma^T \tilde{C}_\varsigma^T \\ 0 & \tilde{C}_\varsigma \tilde{F}_\varsigma & \tilde{C}_\varsigma \tilde{B}_\varsigma & -\gamma P^{-1} \end{bmatrix},
\end{aligned} \tag{9.232}$$

for which all matrices can be taken from the definitions given for (9.152) and (9.203). Initialization must be started with some symmetric matrix $X > 0$ and $\hat{\mathcal{H}}_N = (C_N^T C_N)^{-1} C_N^T$. Since both constraints serve to minimize γ, such a hybrid FIR structure is considered more robust than either of the H_2 and H_∞ FIR filters.

Suboptimal H_2/H_∞ FIR Predictor

Similarly to the H_2/H_∞ filter, a hybrid LMI-based algorithm for numerically computing the suboptimal gain $\mathcal{H}_N^p$ for the H_2/H_∞ FIR predictor can be developed by solving the following minimization problem subject to constraints (9.170) and (9.227),

$$\begin{aligned}
\mathcal{H}_N^p \Leftarrow & \inf_{\mathcal{H}_N^p,\mathcal{Z}^p} \{\mathrm{tr}\mathcal{Z}^p, \gamma > 0\} \\
& \text{subject to} \\
0 < & \begin{bmatrix} \mathcal{Z} - \mathcal{A} + \mathcal{B}\mathcal{H}_N^{p^T} + \mathcal{H}_N^p\mathcal{C} & \mathcal{H}_N^p \\ \mathcal{H}_N^{p^T} & \mathcal{D}^{-1} \end{bmatrix}, \\
0 > & \begin{bmatrix} -X & X\tilde{F}_\varsigma & X\tilde{B}_\varsigma & 0 \\ \tilde{F}_\varsigma^T X & -X & 0 & \tilde{C}_\varsigma^{p^T} \\ \tilde{B}_\varsigma^T X & 0 & -\gamma I & \tilde{D}_\varsigma^{p^T} \\ 0 & \tilde{C}_\varsigma^p & \tilde{D}_\varsigma^p & -\gamma I \end{bmatrix},
\end{aligned} \tag{9.233}$$

for which all matrices can be taken from the definitions given for (9.170) and (9.227). Initialization of the minimization procedure must be started using some symmetric matrix $X > 0$ and $\hat{\mathcal{H}}_N^p = (C_N^{p^T} C_N^p)^{-1} C_N^{p^T}$. Like the hybrid H_2/H_∞ FIR filter, the hybrid H_2/H_∞ FIR predictor is also considered more robust than the H_2 FIR predictor and the H_∞ FIR predictor.

9.7 Generalized H_2 FIR Structures for Uncertain Systems

In Chapter 8, we developed the robust generalized H_2 approach for FIR state estimators operating under disturbances. Originally formulated in [213] and discussed in detail in [188], the approach suggests minimizing the peak error for the maximized disturbance energy in the *energy-to-peak* or $\mathcal{L}_2$-to-$\mathcal{L}_\infty$ algorithms using LMI. We also showed that using the energy-to-peak lemma the gain for the corresponding FIR state estimator can be obtained by solving the following optimization problem

$$\mathcal{H}_N = \inf_{\mathcal{H}_N} \sup_{\|w\|_2<\infty} \frac{\| \varepsilon\|_\infty}{\| w\|_2} . \tag{9.234}$$

Now it is worth noting that if we consider w in a broader sense, then the problem (9.234) can be extended to uncertain systems operating under disturbances, initial errors, and measurement errors. Following the same reasoning as for systems affected by disturbances, next we will use the FE- and BE-based state space models modified for uncertain systems and develop robust algorithms using LMI to numerically compute the gains for $\mathcal{L}_2$-to-$\mathcal{L}_\infty$ FIR filter and predictor.

9.7.1 The *a posteriori* $\mathcal{L}_2$-to-$\mathcal{L}_\infty$ FIR Filter

Unlike the standard H_2 approach, the robust generalized H_2 approach does not require the matrix D to be necessarily zero. Therefore, we modify the BE-based state-space model (8.143) and (8.144) for the system $\begin{bmatrix} F & B \\ H & D \end{bmatrix}$ and write

$$x_k = F_k^u x_{k-1} + B_k^u w_k , \tag{9.235}$$

$$y_k = H_k^u x_k + D_k^u w_k + v_k , \tag{9.236}$$

where the uncertain matrices F_k^u, B_k^u, H_k^u, and D_k^u are defined after (9.1) and (9.2). We assume that the disturbance w_k and the data error v_k are norm-bounded, $\| w_k \|< \infty$ and $\| v_k \|< \infty$, reorganize the terms, and represent the model in (9.235) and (9.236) in the standard LTI form

$$x_k = Fx_{k-1} + \xi_k , \tag{9.237}$$

$$y_k = Hx_k + \zeta_k , \tag{9.238}$$

where the newly introduced zero mean uncertain errors are given by

$$\xi_k = \Delta F_k x_{k-1} + B_k^u w_k , \tag{9.239}$$

$$\zeta_k = \Delta H_k x_k + D_k^u w_k + v_k . \tag{9.240}$$

We next extend the model in (9.237) and (9.238) to $[m,k]$ as

$$x_k = (F^{N-1} + \bar{\tilde{F}}_{m,k})x_m + (\bar{D}_N + \bar{\tilde{D}}_{m,k})W_{m,k} , \tag{9.241}$$

$$Y_{m,k} = (H_N + \tilde{H}_{m,k})x_m + (T_N + \tilde{T}_{m,k})W_{m,k} + V_{m,k} , \tag{9.242}$$

where the matrices $\bar{\tilde{F}}_{m,k}$ and $\bar{\tilde{D}}_{m,k}$ are the last row vectors in the uncertain matrices $\tilde{F}_{m,k} = \hat{F}_N F^u_{m,k}$ and $\tilde{D}_{m,k} = \hat{F}_N D^u_{m,k}$, respectively. Matrix $\hat{F}_N$ is given by (9.8), $F^u_{m,k}$ and $D^u_{m,k}$ by (9.10), $\tilde{H}_{m,k}$ by (9.18) using (9.16), and $\tilde{T}_{m,k}$ by (9.20) using (9.16).

The estimation error (9.43) can now be rewritten for $u_k = 0$ as

$$\varepsilon_k = (\mathcal{B}_N + \tilde{\mathcal{B}}_{m,k})x_m + (\mathcal{W}_N + \tilde{\mathcal{W}}_{m,k})W_{m,k} - \mathcal{V}_N V_{m,k} , \tag{9.243}$$

where the error residual matrices are given by (8.15)–(8.17), (9.44), and (9.46),

$$\begin{aligned} &\mathcal{B}_N = F^{N-1} - \mathcal{H}_N H_N , \quad \mathcal{W}_N = \bar{D}_N - \mathcal{H}_N T_N , \\ &\mathcal{V}_N = \mathcal{H}_N , \\ &\tilde{\mathcal{B}}_{m,k} = \tilde{\bar{F}}_{m,k} - \mathcal{H}_N \tilde{H}_{m,k} , \quad \tilde{\mathcal{W}}_{m,k} = \tilde{\bar{D}}_{m,k} - \mathcal{H}_N \tilde{T}_{m,k} , \end{aligned}$$

and the vectors $W_{m,k}$ and $V_{m,k}$ can be represented using lemma 8.1 by (9.197) and (9.198), respectively.

Using (9.243), (9.197), (9.198), and the augmented vectors $z_k = [\, W_{m,k}^T \;\; V_{m,k}^T \;\; i_k^T \,]^T$, where $i_k = x_m$, and $\xi_k = [\, w_k^T \;\; v_k^T \,]^T$, we obtain the uncertainty-to-error state-space model in the standard form

$$z_k = \tilde{F}_\varsigma z_{k-1} + \tilde{B}_\varsigma w_k , \tag{9.244}$$

$$\varepsilon_k = \tilde{C}_\varsigma z_k , \tag{9.245}$$

where the constant sparse matrices $\tilde{F}_\varsigma$ and $\tilde{B}_\varsigma$ are given by (9.201) and the $\mathcal{H}_N$-varying matrix $\tilde{C}_\varsigma$ is defined by (9.202) as

$$\tilde{C}_\varsigma = \begin{bmatrix} (\bar{D}_N - \mathcal{H}_N G_N + \tilde{\bar{D}}_{m,k} - \mathcal{H}_N \tilde{T}_{m,k})^T \\ -\mathcal{H}_N^T \\ (F^{N-1} - \mathcal{H}_N H_N + \tilde{\bar{F}}_{m,k} - \mathcal{H}_N \tilde{H}_{m,k})^T \end{bmatrix}^T . \tag{9.246}$$

Using the model (9.244) and (9.245) and the energy-to-peak lemma 8.4, we can finally design of a numerical algorithm using LMI for computing the suboptimal gain for the *a posteriori* $\mathcal{L}_2$-to-$\mathcal{L}_\infty$ FIR filter for uncertain systems operating under disturbances, initial errors, and measurement errors. Taking into account $\tilde{D}_\zeta = 0$ in (9.245), the algorithm becomes as follows.

Solve for $\mathcal{H}_N$ the following minimization problem,

$$\begin{aligned} &\mathcal{H}_N \Leftarrow \inf_{\mathcal{H}_N} \gamma > 0 \\ &\quad \text{subject to} \\ &\quad \begin{bmatrix} -\gamma^2 P^{-1} & \tilde{C}_\varsigma \\ \tilde{C}_\varsigma^T & -X^{-1} \end{bmatrix} \leqslant 0 , \\ &\quad \begin{bmatrix} \tilde{B}_\varsigma P_\xi^{-1} \tilde{B}_\varsigma^T - X & \tilde{F}_\varsigma \\ \tilde{F}_\varsigma^T & -X^{-1} \end{bmatrix} < 0 , \end{aligned} \tag{9.247}$$

by initializing the minimization with $\hat{\mathcal{H}}_N = (C_N^T C_N)^{-1} C_N^T$. Provided that $\mathcal{H}_N$ is numerically available, the *a posteriori* $\mathcal{L}_2$-to-$\mathcal{L}_\infty$ FIR filtering estimate and the error matrix can be computed by, respectively,

$$\hat{x}_k = \mathcal{H}_N Y_{m,k} , \tag{9.248}$$

$$\begin{aligned} P = {} & (\mathcal{B}_N + \tilde{\mathcal{B}}_{m,k})\chi_m(\mathcal{B}_N + \tilde{\mathcal{B}}_{m,k})^T \\ & + (\mathcal{W}_N + \tilde{\mathcal{W}}_{m,k})\mathcal{Q}_N(\mathcal{W}_N + \tilde{\mathcal{W}}_{m,k})^T \\ & + \mathcal{V}_N \mathcal{R}_N \mathcal{V}_N^T , \end{aligned} \tag{9.249}$$

where the error residual matrices are introduced after (9.243). What should not be left behind is that the algorithm (9.247) can also be included in hybrid structures to improve robustness.

9.7.2 $\mathcal{L}_2$-to-$\mathcal{L}_\infty$ FIR Predictor

Similar to the filtering counterpart, we can develop the $\mathcal{L}_2$-to-$\mathcal{L}_\infty$ FIR predictor for uncertain systems. To make it possible to avoid the details, we start with the FE-based state-space model (9.21) and (9.22) and rewrite it as

$$x_{k+1} = F^u x_k + B^u w_k \,, \tag{9.250}$$

$$y_k = H^u x_k + D^u w_k + v_k \,, \tag{9.251}$$

where w_k is a bounded disturbance, $\| w_k \| < \infty$. By reorganizing the terms, we represent (9.250) and (9.251) as

$$x_{k+1} = F x_k + \xi_k^p \,, \tag{9.252}$$

$$y_k = H x_k + \zeta_k^p \,, \tag{9.253}$$

where the uncertain zero mean vectors are defined by

$$\xi_k^p = \Delta F_k x_k + B_k^u w_k \,, \tag{9.254}$$

$$\zeta_k^p = \Delta H_k x_k + D_k^u w_k + v_k \,. \tag{9.255}$$

Using the same from as before, we extend the model (9.252) and (9.253) to $[m, k]$ as

$$x_{k+1} = (F^N + \bar{\tilde{F}}^p_{m,k}) x_m + (\bar{D}_N + \bar{\tilde{D}}_{m,k}) W_{m,k} \,, \tag{9.256}$$

$$Y_{m,k} = (H_N^p + \tilde{H}^p_{m,k}) x_m + (T_N^p + \tilde{T}^p_{m,k}) W_{m,k} + V_{m,k} \,, \tag{9.257}$$

where $\bar{\tilde{F}}^p_{m,k}$ is the last row vectors in the uncertain matrix $\tilde{F}^p_{m,k}$ and $\bar{\tilde{D}}_{m,k}$ is defined after (9.13). The uncertain matrices $\tilde{H}^p_{m,k}$ and $\tilde{T}^p_{m,k}$ are given by (9.36) and (9.38), respectively.

We now write the prediction error as

$$\begin{aligned} \varepsilon_{k+1} = {} & (\mathcal{B}_N^p + \tilde{\mathcal{B}}^p_{m,k}) x_m + (\mathcal{W}_N^p + \tilde{\mathcal{W}}^p_{m,k}) W_{m,k} \\ & - \mathcal{V}_N^p V_{m,k} \,, \end{aligned} \tag{9.258}$$

for which the regular and uncertain error residual matrices $\mathcal{B}_N^p$, $\mathcal{W}_N^p$, $\mathcal{V}_N^p$, $\tilde{\mathcal{B}}^p_{m,k}$, and $\tilde{\mathcal{W}}^p_{m,k}$ are listed here

$$\mathcal{B}_N^p = F^N - \mathcal{H}_N^p H_N^p \,, \tag{9.259}$$

$$\mathcal{W}_N^p = \bar{D}_N - \mathcal{H}_N^p T_N^p \,, \tag{9.260}$$

$$\mathcal{V}_N^p = \mathcal{H}_N^p \,, \tag{9.261}$$

$$\tilde{\mathcal{B}}^p_{m,k} = \bar{\tilde{F}}^p_{m,k} - \mathcal{H}_N^p \tilde{H}^p_{m,k} \,, \tag{9.262}$$

$$\tilde{\mathcal{W}}^p_{m,k} = \bar{\tilde{D}}_{m,k} - \mathcal{H}_N^p \tilde{T}^p_{m,k} \tag{9.263}$$

and the matrices $W_{m,k}$ and $V_{m,k}$ are given by (9.197) and (9.198),

$$W_{m,k} = A_w W_{m-1,k-1} + B_w w_k \,,$$
$$V_{m,k} = A_w V_{m-1,k-1} + B_w w_k \,,$$

where the sparse matrices A_w and B_w are defined by (8.19).

Combining the augmented vectors $z_k = [\, W_{m,k}^T \;\; V_{m,k}^T \;\; i_k^T \,]^T$, where $i_k = x_m$, and $\xi_k = [\, w_k^T \;\; v_k^T \,]^T$, we come up with the uncertainty-to-error state model

$$z_k = \tilde{F}_\varsigma z_{k-1} + \tilde{B}_\varsigma \xi_k \,, \tag{9.264}$$

where the matrices $\tilde{F}_\varsigma$ and $\tilde{B}_\varsigma$ are given by (8.109).

Referring to (9.259)–(9.263), we next represent the prediction error (8.158) in the compact form

$$\varepsilon_{k+1} = \hat{C}_\varsigma^p z_k \,, \tag{9.265}$$

where the matrix $\hat{C}_\varsigma^p$ is defined by

$$\hat{C}_\varsigma^p = \begin{bmatrix} \mathcal{W}_N^p + \tilde{\mathcal{W}}_{m,k}^p & -\mathcal{V}_{m,k}^p & \mathcal{B}_N^p + \tilde{\mathcal{B}}_{m,k}^p \end{bmatrix} = \begin{bmatrix} (\bar{D}_N - \mathcal{H}_N^p T_N^p + \tilde{\bar{D}}_{m,k} - \mathcal{H}_N^p \tilde{T}_{m,k}^p)^T \\ -\mathcal{H}_N^{p^T} \\ (F^N - \mathcal{H}_N^p H_N^p + \tilde{\bar{F}}_{m,k}^p - \mathcal{H}_N^p \tilde{H}_{m,k}^p)^T \end{bmatrix}^T . \tag{9.266}$$

By combining $\varepsilon_k = \hat{C}_\varsigma^p z_{k-1}$ taken from (9.265) and $z_{k-1} = \tilde{F}_\varsigma^{-1} z_k - \tilde{F}_\varsigma^{-1} \tilde{B}_\varsigma \xi_k$ taken from (9.264), we finally obtain the estimation error as

$$\varepsilon_k = \hat{C}_\varsigma^p \tilde{F}_\varsigma^{-1} z_k - \hat{C}_\varsigma^p \tilde{F}_\varsigma^{-1} \tilde{B}_\varsigma \xi_k \,. \tag{9.267}$$

At this point, we replace in (9.264) the time index k with $k+1$, accept $\xi_k = \xi_{k-1}$ that is not very critical, and arrive at the disturbance-to-error state-space model in the desired form of (8.117) and (8.118),

$$z_{k+1} = \tilde{F}_\varsigma z_k + \tilde{B}_\varsigma \xi_k \,, \tag{9.268}$$

$$\varepsilon_k = \tilde{C}_\varsigma^p z_k + \tilde{D}_\varsigma^p \xi_k \,, \tag{9.269}$$

where the sparse matrices $\tilde{F}_\varsigma$ and $\tilde{B}_\varsigma$ are given by (8.109) and the matrices $\tilde{C}_\varsigma^p$ and $\tilde{D}_\varsigma^p$ are defined by

$$\tilde{C}_\varsigma^p = \hat{C}_\varsigma^p \tilde{F}_\varsigma^{-1} \,, \quad \tilde{D}_\varsigma^p = -\tilde{C}_\varsigma^p \tilde{B}_\varsigma = -\hat{C}_\varsigma^p \tilde{F}_\varsigma^{-1} \tilde{B}_\varsigma \,. \tag{9.270}$$

Now, the $\mathcal{L}_2$-to-$\mathcal{L}_\infty$ FIR predictor can be developed for uncertain systems operating under disturbances, initial errors, and measurement errors if we apply lemma 8.4 to the model in (9.268) and (9.269). This results in the following numerical procedure to numerically compute the suboptimal gain $\mathcal{H}_N^p$.

For some positive define matrix P, solve the minimization problem

$$\mathcal{H}_N^p \Leftarrow \inf_{\mathcal{H}_N^p} \gamma > 0$$
$$\text{subject to}$$
$$\begin{bmatrix} \tilde{D}_\varsigma^p \tilde{D}_\varsigma^{p^T} & \tilde{C}_\varsigma^p \\ \tilde{C}_\varsigma^{p^T} & -X^{-1} \end{bmatrix} \leqslant \gamma^2 P^{-1} \,,$$

$$\begin{bmatrix} \tilde{B}_\varsigma P_\xi^{-1} \tilde{B}_\varsigma^T - X & \tilde{F}_\varsigma \\ \tilde{F}_\varsigma^T & -X^{-1} \end{bmatrix} < 0 , \tag{9.271}$$

where matrices $\tilde{C}_\varsigma^p$ and $\tilde{D}_\varsigma^p$ are specified by (9.270). Initialize the minimization with $\hat{\mathcal{H}}_N^p = (C_N^{p^T} C_N^p)^{-1} C_N^{p^T}$. Provided that $\mathcal{H}_N^p$ is available from (9.271), the $\mathcal{L}_2$-to-$\mathcal{L}_\infty$ FIR prediction appears as $\tilde{x}_{k+1} = \mathcal{H}_N^p Y_{m,k}$ with the error matrix

$$\begin{aligned} P &= (\mathcal{B}_N^p + \tilde{\mathcal{B}}_{m,k}^p)\chi_m(\mathcal{B}_N^p + \tilde{\mathcal{B}}_{m,k}^p)^T \\ &\quad + (\mathcal{W}_N^p + \tilde{\mathcal{W}}_{m,k}^p)\mathcal{Q}_N(\mathcal{W}_N^p + \tilde{\mathcal{W}}_{m,k}^p)^T + \mathcal{V}_N^p \mathcal{R}_N \mathcal{V}_N^{p^T} , \end{aligned} \tag{9.272}$$

where the error residual matrices defined by (9.259)–(9.263) are functions of $\mathcal{H}_N^p$ computed numerically by solving the minimization problem (9.271). Finally note that this algorithm can also be included in robust hybrid FIR predictive algorithms.

9.8 Robust $\mathcal{L}_1$ FIR Structures for Uncertain Systems

When system uncertainty is caused by sudden, unpredictable and abrupt changes [37, 112], then the H_2 and H_∞ approaches may not be as efficient as the robust $\mathcal{L}_1$ state estimation that provides *peak-to-peak* or $\mathcal{L}_\infty$-to-$\mathcal{L}_\infty$ filtering and prediction. Using the approach developed in Chapter 8 for systems operating under disturbances, next we will develop more general $\mathcal{L}_\infty$-to-$\mathcal{L}_\infty$ FIR state estimators for uncertain systems operating under disturbances, initial errors, and measurement errors.

We will view the robust *peak-to-peak* FIR state estimation problem as minimization of the $\mathcal{L}_1$ norm (8.178) of the induced $\mathcal{T}$ represented by the ratio of the squared norms of the peak uncertainty w_k and the peak error ε_k. Accordingly, we will determine the gain $\mathcal{H}_N$ for this estimator by satisfying the cost function (8.180) represented as

$$\mathcal{H}_N \Leftarrow \sup_{\|w\|<\infty} \frac{\| \varepsilon_k \|_2^2}{\mu \| w_k \|_2^2} < \gamma^2 , \tag{9.273}$$

where μ is a constant scalar and the minimum value of $\gamma > 0$ guarantees the gain suboptimality.

9.8.1 The *a posteriori* $\mathcal{L}_\infty$-to-$\mathcal{L}_\infty$ FIR Filter

To develop an $\mathcal{L}_1$ FIR filter for uncertain systems, we start with the familiar uncertainty-to-error state space model (9.244) and (9.245),

$$\begin{aligned} z_k &= \tilde{F}_\varsigma z_{k-1} + \tilde{B}_\varsigma w_k , \\ \varepsilon_k &= \tilde{C}_\varsigma z_k , \end{aligned}$$

where the sparse matrices $\tilde{F}_\varsigma$ and $\tilde{B}_\varsigma$ are defined by (9.201), the matrix $\tilde{C}_\varsigma$ is given by (9.246) as

$$\tilde{C}_\varsigma = \begin{bmatrix} (\bar{D}_N - \mathcal{H}_N G_N + \tilde{\bar{D}}_{m,k} - \mathcal{H}_N \tilde{T}_{m,k})^T \\ -\mathcal{H}_N^T \\ (F^{N-1} - \mathcal{H}_N H_N + \tilde{\bar{F}}_{m,k} - \mathcal{H}_N \tilde{H}_{m,k})^T \\ \end{bmatrix}^T , \tag{9.274}$$

and all other definitions can be adopted from (9.238).

We next apply lemma 8.5 to (9.244) and (9.245), note that $\tilde{D}_\varsigma = 0$, and arrive at the following algorithm to compute a suboptimal gain $\mathcal{H}_N$ for the robust $\mathcal{L}_\infty$-to-$\mathcal{L}_\infty$ *a posteriori* FIR filter.

Solve the minimization problem,

$$\begin{aligned}
&\mathcal{H}_N \Leftarrow \inf_{\mathcal{H}_N} \gamma > 0 \\
&\quad \text{subject to} \\
&\quad \begin{bmatrix} (1+\lambda)\tilde{F}_\varsigma^T K \tilde{F}_\varsigma - K & (1+\lambda)\tilde{F}_\varsigma^T K \tilde{B}_\varsigma \\ (1+\lambda)\tilde{B}_\varsigma^T K \tilde{F}_\varsigma & (1+\lambda)\tilde{B}_\varsigma^T K \tilde{F}_\varsigma - \mu P_\xi \end{bmatrix} < 0 , \\
&\quad \begin{bmatrix} K & 0 & \tilde{C}_\varsigma^T \\ 0 & \mu(\gamma^2 - \lambda^{-1})P_\xi & 0 \\ \tilde{C}_\varsigma & 0 & P^{-1} \end{bmatrix} > 0 , \quad \lambda > 0 , \quad \mu > 0 ,
\end{aligned} \tag{9.275}$$

where the matrix $\tilde{C}_\varsigma$ is given by (9.274) and the sparse matrices $\tilde{F}_\varsigma$ and $\tilde{B}_\varsigma$ are defined by (9.201). The initialization should be started with $\hat{\mathcal{H}}_N = (C_N^T C_N)^{-1} C_N^T$. For the gain $\mathcal{H}_N$ obtained by solving (9.275), the *a posteriori* $\mathcal{L}_\infty$-to-$\mathcal{L}_\infty$ FIR filtering estimate is computed by (9.248) and the error matrix by (9.249).

9.8.2 $\mathcal{L}_\infty$-to-$\mathcal{L}_\infty$ FIR Predictor

Similarly, we develop the $\mathcal{L}_\infty$-to-$\mathcal{L}_\infty$ FIR predictor using the uncertainty-to-error state-space model (9.268) and (9.269),

$$\begin{aligned}
z_{k+1} &= \tilde{F}_\varsigma z_k + \tilde{B}_\varsigma \xi_k , \\
\varepsilon_k &= \tilde{C}_\varsigma^p z_k + \tilde{D}_\varsigma^p \xi_k ,
\end{aligned}$$

where the sparse matrices $\tilde{F}_\varsigma$ and $\tilde{B}_\varsigma$ are given by (9.201) and the matrices $\tilde{C}_\varsigma^p$ and $\tilde{D}_\varsigma^p$ are defined by (9.270) using the matrix $\hat{C}_\varsigma^p$ defined by (9.266). The lemma 8.6 applied to the previous state-space model finally gives the algorithm to numerically compute the suboptimal gain $\mathcal{H}_N^p$ for the robust $\mathcal{L}_\infty$-to-$\mathcal{L}_\infty$ FIR predictor.

Solve the minimization problem,

$$\begin{aligned}
&\mathcal{H}_N^p \Leftarrow \inf_{\mathcal{H}_N^p} \gamma > 0 \\
&\quad \text{subject to} \\
&\quad \begin{bmatrix} \tilde{F}_\varsigma^T K \tilde{F}_\varsigma - K & \tilde{F}_\varsigma^T K \tilde{B}_\varsigma \\ \tilde{B}_\varsigma^T K \tilde{F}_\varsigma & \tilde{B}_\varsigma^T K \tilde{F}_\varsigma - \mu P_\xi \end{bmatrix} < 0 , \\
&\quad \begin{bmatrix} K & 0 & \tilde{C}_\varsigma^{p^T} \\ 0 & \mu(\gamma^2 - \lambda^{-1})P_\xi & \tilde{D}_\varsigma^{p^T} \\ \tilde{C}_\varsigma^p & \tilde{D}_\varsigma^p & P_y^{-1} \end{bmatrix} > 0 , \quad \lambda > 0 , \quad \mu > 0 ,
\end{aligned} \tag{9.276}$$

starting the minimization with $\hat{\mathcal{H}}_N^p = (C_N^{p^T} C_N^p)^{-1} C_N^{p^T}$. Provided that $\mathcal{H}_N^p$ is available from (9.276), the $\mathcal{L}_\infty$-to-$\mathcal{L}_\infty$ FIR prediction is computed by $\tilde{x}_{k+1} = \mathcal{H}_N^p Y_{m,k}$ and the error matrix by (9.272).

We finally notice that all robust FIR algorithms developed in this chapter for uncertain systems can be modified to be bias-constrained (suboptimally unbiased) if we remove the terms with χ_m and subject the LMI-based algorithms to the unbiasedness constraint. That can be done similarly to the bias-constrained suboptimal H_2 FIR state estimator. Moreover, all algorithms can be extended to general state-space models with control inputs. It is also worth noting that all of the FIR predictors discussed in this chapter become the RH FIR filters needed for state feedback control by changing the time variable from k to $k-1$.

9.9 Summary

In this chapter, we have presented various types of robust FIR state estimators, which minimize estimation errors for maximized system uncertainties and other errors. Uncertainties in systems can occur naturally and artificially due to external and internal reasons, which sometimes lead to unpredictable changes in matrices. Since uncertainties cannot be described in terms of distributions and covariances, robust state estimators are required. To cope with such effects, robust methods assume that the undefined matrix increments have zero mean and are norm-bounded. Because a robust FIR state estimation of uncertain systems must be performed in practice in the presence of possible disturbances, initial errors, and measurement errors, this approach is considered the most general. Its obvious advantage is that algorithms can be easily simplified for specific errors.

An efficient way to obtain robust FIR estimates is to reorganize the state-space model by moving all components with undefined matrices into errors. This makes it possible to use the state-space models previously created for disturbances, and the results obtained in Chapter 8 can be largely extended to uncertain systems.

The errors in such estimators are multivariate, since their variables are not only undefined matrix components, but also disturbances, initial errors, data errors, and uncertain increments in the control signal matrix. In view of that, each error residual matrix acquires an additional increment, which depends on specific uncertain components. Accordingly, the error matrix of the FIR estimator is generally combined by six submatrices associated with disturbances, errors, and uncertainties.

As other FIR structures, FIR state estimators for uncertain systems can be developed to be bias-constrained. This property is achieved by neglecting the terms with initial errors and embedding the unbiasedness constraint using the Lagrange method. Since the derivation procedure is the same for all FIR structures, we postponed the development of bias-constrained FIR estimators for uncertain systems to "Problems". Another useful observation can be made if we recall that the robust approach for uncertain systems has been developed in the transform domain. This means that by replacing k with $k-1$, all of the FIR predictors obtained in this chapter can easily be converted into the RH FIR predictive filters needed for state feedback control.

We finally notice that the algorithms presented in this chapter cover most of the robust FIR solutions available. However, higher robustness is achieved by introducing additional tuning factors, and efforts should be made to properly maximize uncertainties and other errors. Otherwise, the estimator performance can degrade dramatically.

9.10 Problems

1 An uncertain system is represented by the discrete-time state-space model in (9.1) and (9.2),

$$x_k = F_k^u x_{k-1} + E_k^u u_k + B_k^u w_k \,,$$
$$y_k = H_k^u x_k + D_k^u w_k + v_k \,,$$

where the uncertain matrices are modeled as $F_k^u = (1 + a_k)F$, $E_k^u = (1 + b_k)E$, $B_k^u = (1 + c_k)B$, $H_k^u = (1 + d_k)H$, and $D_k^u = (1 + e_k)D$. The known matrices F, E, B, H, and D are constant, and the uncertain parameters a_k, b_k, c_k, d_k, and e_k have zero mean and are norm-bounded. Extend this model to $[m, k]$ and modify the H_2 FIR filtering algorithms.

2 Consider the following discrete-time state-space model with multiplicative noise components [56],

$$x_{k+1} = (F_k + \check{F}_k\rho_k)x_k + (E_k + \check{E}_k\eta_k)u_k + B_k w_k ,$$
$$y_k = (H_k + \check{H}_k\zeta_k)x_k + D_k v_k ,$$

where w_k is a bounded disturbance and ρ_k, ζ_k, and η_k are standard scalar white noise sequences with zero mean and the properties:

$$\mathcal{E}\{\rho_k\rho_j\} = \delta_{kj} , \quad \mathcal{E}\{\zeta_k\zeta_j\} = \delta_{kj} , \quad \mathcal{E}\{\eta_k\eta_j\} = \delta_{kj} ,$$
$$\mathcal{E}\{\eta_k\rho_j\} = \beta_k\delta_{kj} , \quad \mathcal{E}\{\zeta_k\eta_j\} = \sigma_k\delta_{kj} , \quad \mathcal{E}\{\zeta_k\rho_j\} = \alpha_k\delta_{kj} ,$$

where $|\beta_k| < 1$, $|\alpha_k| < 1$, and $|\sigma_k| < 1$ and δ_{kj} is the Kronecker delta. Convert this model to a more general model in (9.1) and (9.2) and modify the suboptimal FIR predictor.

3 An uncertain LTV system is represented by the following discrete-time state-space model [98]

$$x_{k+1} = (F_k + \Delta F_k)x_k + B_k w_k ,$$
$$y_k = (H_k + \Delta H_k)x_k + D_k w_k ,$$

where w_k is a bounded disturbance and F_k, B_k, H_k, and D_k are known time-varying matrices such that

$$D_k B_k^T = 0 , \quad D_k D_k^T > 0 , \tag{9.277}$$

and ΔF_k and ΔH_k are time-varying parameter uncertainties obeying the following structure

$$\begin{bmatrix} \Delta F_k \\ \Delta H_k \end{bmatrix} = \begin{bmatrix} H_1 \\ H_2 \end{bmatrix} A_k E , \tag{9.278}$$

where A_k is an unknown real time-varying matrix satisfying $A_k^T A_k \leqslant I$ and H_1 and H_2 are known real constant matrices with appropriate dimensions. Note that this model is widely used in the design of robust recursive estimators for uncertain systems. Consider this model as a special case of the general model in (9.1) and (9.2) and show that the conditions (9.277) and (9.278) can be too strict in applications. Find a way to avoid these conditions.

4 The estimation error of an FIR filter is given by (9.43) as

$$\varepsilon_k = (\mathcal{B}_N + \tilde{\mathcal{B}}_{m,k})x_m + \tilde{\mathcal{U}}_{m,k}U_{m,k}$$
$$+ (\mathcal{W}_N + \tilde{\mathcal{W}}_{m,k})W_{m,k} - \mathcal{V}_N V_{m,k} ,$$

where the residual error matrices $\mathcal{B}_N$, $\mathcal{W}_N$, and $\mathcal{V}_N$ are given by (8.15)–(8.17) and the uncertain residual matrices specified by (9.44)–(9.46). Taking notice that the *a posteriori* H_∞ FIR filter is obtained in this chapter for $u_k = 0$, rederive this filter for $u_k \neq 0$.

5 Consider the H_∞ FIR filter (9.206), the gain $\mathcal{H}_N$ for which is computed numerically using the algorithm in (9.203)–(9.205). Modify this algorithm for the estimate to be bias-constrained and make corrections in the error matrix (9.207).

6 Solve the problem described in item 5 for the H_∞ FIR predictor and modify accordingly the inequalities (9.227)–(9.229) and the error matrix (9.231).

7 A Markov jump LTV uncertain system is represented with the following discrete-rime state-space model [228]

$$x_{k+1} = F_k(r_k)x_{k-1} + B_k(r_k)w_k \,, \tag{9.279}$$

$$y_k = H_k(r_k)x_k + v_k \,, \tag{9.280}$$

where r_k is a discrete Markov chain taking values from a finite space $\mathbb{M} = \{1,2,\ldots,M\}$ with the transition probability $\pi_{ij} \triangleq p(r_k = j|r_{k-1} = i)$ for any $i,j \in \mathbb{M}$. Matrices $F_k(r_k)$, $B_k(r_k)$, and $H_k(r_k)$ are r_k-varying and the random sequences are white Gaussian, $w_k \sim \mathcal{N}(0, Q_k)$ and $v_k \sim \mathcal{N}(0, R_k)$. Transform this model to the general form (9.21) and (9.22), extend to $[m,k]$, and develop an H_2 FIR predictor.

8 Consider the problem described in item 7 and derive the H_∞, $\mathcal{L}_2$-to-$\mathcal{L}_\infty$, and $\mathcal{L}_\infty$-to-$\mathcal{L}_\infty$ FIR predictors.

9 Given an uncertain system represented with the state-space model $x_k = F_k^u x_{k-1} + Bw_k$ and $y_k = Hx_k + v_k$, where the matrices are specified as

$$F_k^u = \begin{bmatrix} 1 & \tau_k \\ 0 & 1 \end{bmatrix}, \quad B = \begin{bmatrix} \tau \\ 1 \end{bmatrix}, \quad H = \begin{bmatrix} 1 & 0 \end{bmatrix},$$

and τ_k is an uncertain nonconstant time step. Suppose that τ_k has zero mean and is bounded. Derive the H_2-OFIR and H_2-OUFIR filters.

10 A harmonic model is given in discrete-time state-space with the following equations $x_{k+1} = F_k^u x_k + Bw_k$ and $y_k = Hx_k + v_k$, where matrices are specified as

$$F_k^u = \begin{bmatrix} \cos\varphi + \xi_k & \sin\varphi \\ -\sin\varphi & \cos\varphi + \xi_n \end{bmatrix}, \quad B = \begin{bmatrix} 1 \\ 1 \end{bmatrix}, \quad H = \begin{bmatrix} 1 & 0 \end{bmatrix},$$

φ is a constant angle, ξ_k is an undefined bounded scalar, $|\xi_k| < 0.2$, and $w_k \sim \mathcal{N}(0, \sigma_w^2)$ and $v_k \sim \mathcal{N}(0, \sigma_v^2)$ are scalar white Gaussian sequences. Derive the $\mathcal{L}_2$-to-$\mathcal{L}_\infty$ and $\mathcal{L}_\infty$-to-$\mathcal{L}_\infty$ FIR predictors for this model.

11 An uncertain system is represented with the following discrete-rime state-space model

$$x_{k+1} = Fx_{k-1} + Bw_k \,,$$
$$z_k = y_{1k} + y_{2k} \,,$$

where $y_{1k} = (H + \Delta H_{1k})x_k + v_{1k}$, $y_{2k} = (H + \Delta H_{2k})x_k + v_{2k}$, and ΔH_{1k} and ΔH_{2k} are undefined and uncorrelated norm-bounded increments. Noise sequences $w_k \sim \mathcal{N}(0, Q_k)$, $v_{1k} \sim \mathcal{N}(0, R_{1k})$, and $v_{2k} \sim \mathcal{N}(0, R_{2k})$ are white Gaussian. Convert this model to the general form (9.21) and (9.22), extend to $[m,k]$, and develop an H_∞ FIR predictor.

10

Advanced Topics in FIR State Estimation

If we would serve science, we must extend her limits, not only as far as our own knowledge is concerned but in the estimation of others.

Rudolf Virchow, in the 1880s

What we considered so far concerns rather standard approaches when optimal or robust estimation is required over measurements of some process in the presence of noise or bounded errors [118]. There are many other cases, when systems and processes require advanced state-space modeling and correspondingly modified estimators [111, 112, 216]. Examples can be found in wireless networks [117], where a process is commonly measured with a redundant number of sensors and estimation can be provided with consensus in measurements, estimates, or information. Data fusion in networks is often organized in the presence of correlated noise that also must be addressed in the design [46]. In some cases, hybrid filter structures are exploited to reach the highest estimation accuracy. Wireless communication channels often deliver information at a receiver with time-stamped or randomly delayed and missing data that requires advanced modeling and modified estimators [211]. The list of additional topics can be widened, but it is common that each advanced state-space model usually results in a significantly modified or even new estimator to provide the highest available accuracy.

In this chapter we will look at several such advanced topics and discuss the corresponding FIR solutions. Our additional goal will be to demonstrate the better robustness of FIR state estimators.

10.1 Distributed Filtering over Networks

Wireless sensor networks (WSNs) are used to provide environmental sensing, condition monitoring, and process automation of a desired quantity Q. Because Q is typically measured over a big number of nodes, distributed filtering has been introduced based on consensus in measurements [137], estimates [138], and information [17], using the Kalman approach. In what follows, we will present distributed UFIR filters for WSNs with consensus in measurements and estimates and show their better robustness.

An example of a WSN organized to cover a territory of $-40\text{ m} \leqslant x \leqslant 40\text{ m}$ and $-30\text{ m} \leqslant y \leqslant 30\text{ m}$ is shown in Fig. 10.1 [205]. The nodes are placed randomly with coordinates uniformly distributed

Optimal and Robust State Estimation: Finite Impulse Response (FIR) and Kalman Approaches, First Edition.
Yuriy S. Shmaliy and Shunyi Zhao.

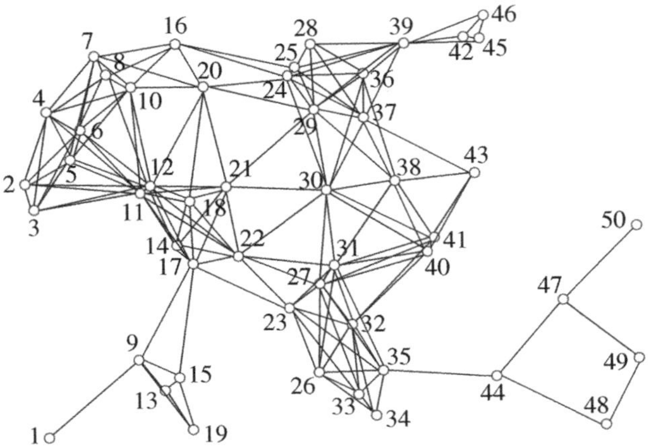

Figure 10.1 An example of a WSN with 50 nodes randomly placed in coordinates of $-40\ \text{m} \leqslant x \leqslant 40\ \text{m}$ and $-30\ \text{m} \leqslant y \leqslant 30\ \text{m}$. The number 145 of links is due to the node range of 14 m [205].

along axes x and y. The Laplace matrix of the WSN graph has 50×50 dimensions, and each node communicates with neighbors within a range of 14 m.

10.1.1 Consensus in Measurements

Suppose that a WSN such as that shown in Fig. 10.1 is organized to measure quantity Q, whose dynamics is represented with a K-state vector $x_k \in \mathbb{R}^K$. The measurement environment is created with n nodes. Each ith, $i \in [1, n]$, node provides linear measurements of Q as $y_k^{(i)} = H_k^{(i)} x_k + v_k^{(i)}$, where $H_k^{(i)} \in \mathbb{R}^{p \times K}$, integer $p \leqslant K$ is the number of the measured states, and $v_k^{(i)}$ is the measurement noise. The state-space model is

$$x_k = F_k x_{k-1} + B_k w_k \,, \tag{10.1}$$

$$y_k = H_k x_k + v_k \,, \tag{10.2}$$

where $y_k = [y_k^{(1)^T} \ \dots \ y_k^{(n)^T}]^T$ is the observation, $H_k = [H_k^{(1)^T} \ \dots \ H_k^{(n)^T}]^T$ is the augmented observation matrix, matrix B_k has proper dimensions, and $w_k \sim \mathcal{N}(0, Q_k)$ and $v_k = [v_k^{(1)^T} \ \dots \ v_k^{(n)^T}]^T \sim \mathcal{N}(0, R_k)$ are zero mean mutually uncorrelated white Gaussian noise vectors with the covariances Q_k and $R_k = \text{diag}[R_k^{(1)^T} \ \dots \ R_k^{(n)^T}]^T$.

Specified (10.1) and (10.2), one can apply any linear state estimator to determine the quantity state. The problem we are facing here is that with a large number of nodes, $n \gg 1$, a digital estimator demonstrates a computational problem. Furthermore, the estimation accuracy can be insufficient if we do not use advanced state estimation algorithms. Such algorithms are called distributed with consensus in measurements, and their design can be obtained as we will show next.

Distributed KF with Consensus in Measurements

The distributed KF approach with consensus in measurements was originally proposed and developed in [137]. The algorithm can be derived if we first consider a centralized WSN that passes data to a central station, where the state of Q is estimated by a standard KF called the centralized KF (cKF),

$$P_k^- = F_k P_{k-1} F_k^T + B_k Q_k B_k^T \,, \tag{10.3}$$

$$P_k = [(P_k^-)^{-1} + H_k^T R_k^{-1} H_k]^{-1} , \tag{10.4}$$

$$K_k = P_k H_k^T R_k^{-1} , \tag{10.5}$$

$$\hat{x}_k^- = F_k \hat{x}_{k-1} , \tag{10.6}$$

$$\hat{x}_k = \hat{x}_k^- + K_k(y_k - H_k \hat{x}_k^-) . \tag{10.7}$$

This approach has at least two serious drawbacks: 1) it requires communication between a central station and each of the nodes, and 2) estimation may be delayed due to computational complexity when H_k acquires large dimensions.

To overcome these issues, the centralized estimate (10.7) can be represented using (10.5) as

$$\begin{aligned}\hat{x}_k &= \hat{x}_k^- + K_k(y_k - H_k \hat{x}_k^-) \\ &= \hat{x}_k^- + P_k(H_k^T R_k^{-1} y_k - H_k^T R_k^{-1} H_k \hat{x}_k^-)\end{aligned} \tag{10.8}$$

and two aggregate quantities introduced: a vector z_k of fused average-consensus sensor data and fused average-consensus inverse covariance matrix S_k,

$$z_k = \frac{1}{n}\sum_{i=1}^{n} z_k^{(i)} = \frac{1}{n}\sum_{i=1}^{n} H_k^{(i)^T} R_k^{(i)^{-1}} y_k^{(i)} , \tag{10.9}$$

$$S_k = \frac{1}{n} H_k^T R_k^{-1} H_k = \frac{1}{n}\sum_{i=1}^{n} H_k^{(i)^T} R_k^{(i)^{-1}} H_k^{(i)} . \tag{10.10}$$

Next, z_k can be obtained by an LP consensus filter and S_k by a band-pass (BP) consensus filter [137]. Here n is to cover all nodes in the centralized network or several neighbouring nodes in the distributed network. The micro-KF (μKF) can then be designed for distributed filtering algorithms as

$$P_{\mu k}^- = A_k P_{\mu(k-1)} A_k^T + B_k Q_{\mu k} B_k^T , \tag{10.11}$$

$$P_{\mu k} = [(P_{\mu k}^-)^{-1} + S_k]^{-1} , \tag{10.12}$$

$$\hat{x}_k^- = A_k \hat{x}_{k-1} , \tag{10.13}$$

$$\hat{x}_k = \hat{x}_k^- + P_{\mu k}(z_k - S_k \hat{x}_k^-) , \tag{10.14}$$

where $P_{\mu k} = nP_k$ is the μKF gain that is equal to the μKF error covariance and $Q_{\mu k} = nQ_k$. It is obvious that due to the introduced z_k and S_k, the μKF operates n-times faster than the cKF while producing equal estimates by "ideal" consensus. Even so, the μKF still has the same drawback as the KF: it suffers from errors in the noise covariances and initial values, which are typically not well-known in WSNs. A more robust micro filter can be designed using the UFIR approach as we will show next.

Distributed UFIR Filter with Consensus in Measurements

Following the approach that was developed in [137] and has resulted in the μKF algorithm (10.11)–(10.14), a distributed UFIR filter for WSNs with consensus in measurements was designed in [205] to have higher robustness than the distributed KF.

Given the model in (10.1) and (10.2), the centralized UFIR (cUFIR) filtering estimate is computed by the standard iterative *a posteriori* UFIR filtering algorithm using the following recursions,

$$\mathcal{G}_l = [H_l^T H_l + (F_l \mathcal{G}_{l-1} F_l^T)^{-1}]^{-1} , \tag{10.15}$$

$$\hat{x}_l = F_l \hat{x}_{l-1} + \mathcal{G}_l H_l^T (y_l - H_l F_l \hat{x}_{l-1}) , \tag{10.16}$$

where iterations start with $l = m + K$ and end when $l = k$. The error covariance P_k for the cUFIR filter can also be computed iteratively as

$$P_l^- = A_l P_{l-1} A_l^T + B_l Q_l B_l^T \,, \tag{10.17}$$

$$P_l = (I - K_l H_l) P_l^- (I - K_l H_l)^T + K_l R_l K_l^T \tag{10.18}$$

to produce the true value when $l = k$. Although the UFIR filter is more robust than KF, the computational problem associated with WSN having a large number of nodes may be even more serious when the number of iterations is required to be large.

To design the μUFIR filter for distributed filtering, the cUFIR estimate (10.16) can be transformed as

$$\begin{aligned} \hat{x}_l &= \hat{x}_l^- + \mathcal{G}_l H_l^T (y_l - H_l \hat{x}_l^-) \\ &= \hat{x}_l^- + \mathcal{G}_l (H_l^T y_l - H_l^T H_l \hat{x}_l^-) \\ &= \hat{x}_l^- + \mathcal{G}_{\mu l} (s_l - L_l \hat{x}_l^-) \,, \end{aligned} \tag{10.19}$$

where $\hat{x}_l^- = F_l \hat{x}_{l-1}$ is the prior estimate, $\mathcal{G}_{\mu l} = n\mathcal{G}_l$ is the GNPG of the μUFIR filter, and the introduced aggregate vector s_l of fused average-consensus sensor data and a fused average-consensus matrix L_l are given by

$$s_l = \frac{1}{n} \sum_{l=1}^{n} s_l^{(i)} = \frac{1}{n} \sum_{l=1}^{n} H_l^{(i)^T} y_l^{(i)} \,, \tag{10.20}$$

$$L_l = \frac{1}{n} H_l^T H_l = \frac{1}{n} \sum_{i=1}^{n} H_l^{(i)^T} H_l^{(i)} \,. \tag{10.21}$$

We now see an important advantage. Unlike (10.10), matrix (10.21) does not require the measurement noise covariance. Therefore, the μUFIR filter needs only one consensus filter for s_l.

The μUFIR filtering algorithm can be designed as

$$\mathcal{G}_{\mu l} = [L_l + (F_l \mathcal{G}_{\mu(l-1)} F_l^T)^{-1}]^{-1} \,, \tag{10.22}$$

$$\hat{x}_l^- = A_l \hat{x}_{l-1} \,, \tag{10.23}$$

$$\hat{x}_l = \hat{x}_l^- + \mathcal{G}_{\mu l} (s_l - L_l \hat{x}_l^-) \,, \tag{10.24}$$

where the bias correction gain $\mathcal{G}_{\mu l}$ is also the GNPG of the μUFIR filter.

Fast computation of the initial $\mathcal{G}_{\mu(l-1)}$ for (10.22) and $\hat{x}_{l-1}$ for (10.23) can be provided if we use the filtered consensus values of L_k and s_k. To design the algorithm, we transform the inverse of $\mathcal{G}_s$ (6.16c) to

$$\begin{aligned} \mathcal{G}_s^{-1} &= C_{m,s}^T C_{m,s} \\ &= n \sum_{j=0}^{K-1} (\mathcal{F}_s^{m+1+j})^{-T} L_{m+j} (\mathcal{F}_s^{m+1+j})^{-1} \end{aligned} \tag{10.25}$$

and obtain

$$\mathcal{G}_{\mu s} = n\mathcal{G}_s = \left[\sum_{j=0}^{K-1} (\mathcal{F}_s^{m+1+j})^{-T} L_{m+j} (\mathcal{F}_s^{m+1+j})^{-1} \right]^{-1} \,. \tag{10.26}$$

We next transform the initial state $\hat{x}_s$ as

$$\tilde{x}_s = \mathcal{G}_s C_{m,s}^T Y_{m,s} = \frac{1}{n}\mathcal{G}_{\mu s} C_{m,s}^T Y_{m,s}$$
$$= \mathcal{G}_{\mu s}\sum_{j=0}^{K-1}(\mathcal{F}_s^{m+1+j})^{-T} s_{m+j} \,. \tag{10.27}$$

The number K of the quantity states is typically small in WSNs that makes the sums short in (10.26) and (10.27). For example, if $K = 2$, then the initial values can be computed as

$$\mathcal{G}_{\mu s} = (F_{m+1}^{-T} L_m F_{m+1}^{-1} + L_{m+1})^{-1} \,,$$
$$\tilde{x}_s = \mathcal{G}_{\mu s}(F_{m+1}^{-T} s_m + s_{m+1}) \,.$$

It is worth noticing that on given N_{opt} the μUFIR filter is a blind robust alternative to μKF and that this property is highly required for WSNs. Like in μKF, real-time operation of μUFIR filter can be supported by an LP consensus filter applied to $H_l^T y_l$ to produce s_l. The BP filter can be avoided if the same time-invariant $H_l^{(i)}$ is set to all nodes. Finally, the pseudocode of the μUFIR filtering algorithm is listed as Algorithm 21.

Algorithm 21: μUFIR Filtering Algorithm for Distributed WSNs

Data: s_k, L_k, N
Result: $\hat{x}_k$

begin
  for $k = N-1 : \infty$ **do**
    $m = k - N + 1, \quad s = m + K - 1;$
    $G_{\mu s} = \left[\sum_{j=0}^{K-1}(\mathcal{F}_s^{m+1+j})^{-T} L_{m+j} (\mathcal{F}_s^{m+1+j})^{-1}\right]^{-1};$
    $\tilde{x}_s = G_{\mu s}\sum_{j=0}^{K-1} \mathcal{F}_s^{m+1+j^{-T}} s_{m+j};$
    for $l = s+1 : k$ **do**
      $G_{\mu l} = [L_l + (A_l G_{\mu(l-1)} A_l^T)^{-1}]^{-1};$
      $\tilde{x}_l = A_l\tilde{x}_{l-1} + G_{\mu l}(s_l - L_l A_l \tilde{x}_{l-1});$
    end for
    $\hat{x}_k = \tilde{x}_k;$
  end for
end

To compute the error covariance $P_{\mu k}$ of the μUFIR filter, we transform the estimation error $\varepsilon_l = x_l - \hat{x}_l$ using (10.1) and (10.24) as

$$\varepsilon_l = A_l\varepsilon_{l-1} + B_l w_l - \mathcal{G}_{\mu l} s_l + \mathcal{G}_{\mu l} L_l A_l \hat{x}_{l-1}$$
$$= (I - \mathcal{G}_{\mu l} L_l)(A_l \varepsilon_{l-1} + B_l w_l) + \mathcal{G}_{\mu l}\frac{1}{n}\sum_{i=1}^{n} H_l^{(i)} v_l^{(i)}$$

and then represent $P_l = E\{\varepsilon_l \varepsilon_l^T\}$ with

$$P_l = (I - \mathcal{G}_{\mu l} L_l)(A_l P_{l-1} A_l^T + B_l Q_l B_l^T)(I - \mathcal{G}_{\mu l} L_l)^T + \frac{1}{n} \mathcal{G}_{\mu l} S_l \mathcal{G}_{\mu l}^T , \quad (10.28)$$

where S_l is given by (10.10). By taking $P_{\mu l} = nP_l$ and $Q_{\mu l} = nQ_l$ from μKF, we represent (10.28) in the final form of

$$P_{\mu l} = (I - \mathcal{G}_{\mu l} L_l)(A_l P_{\mu(l-1)} A_l^T + B_l Q_{\mu l} B_l^T) \times (I - \mathcal{G}_{\mu l} L_l)^T + \mathcal{G}_{\mu l} S_l \mathcal{G}_{\mu l}^T . \quad (10.29)$$

By changing l from $k - N + K + 1$ to k, matrix $P_{\mu l}$ can now be updated iteratively to produce a true value at k. We finally refer to [205] and notice that the μUFIR filter outperforms the μKF when WSN operates under disturbances in not well-specified environments.

10.1.2 Consensus in Estimates

The consensus in estimates over WSN was originally proposed in [139] and the KF modified accordingly. Soon after, the approach was used and developed by many authors and higher accuracy reported when estimates associated with each of the nodes are not equal. More recently, a distributed UFIR filter with consensus in estimates was derived in [206] and justified its higher robustness relative to distributed KF.

A general idea of the consensus in estimates can be outlined as follows. Let us look at a WSN (Fig. 10.1) as an undirected graph $\Gamma(\mathcal{V}, \mathcal{E})$ where each vertex $v^{(i)} \in \mathcal{V}$ is a node and each link is an edge of set $\mathcal{E}$, for $i \in \mathcal{I} = \{1, \ldots, n\}$ and $n = |\mathcal{V}|$. Nodes $v^{(i)}$ and $v^{(j)}$ reach an agreement if and only if the states are related as $x^{(i)} = x^{(j)}$, $\{i,j\} \in \mathcal{I}, i \neq j$ [139]. If so, then the WSN reaches a consensus with a common value called the group decision value. Because a perfect consensus is unavailable due to process noise, a consensus protocol is required to minimize a total disagreement in the WSN. It is provided by minimizing the Laplacian potential of graph $\Psi = \frac{1}{2} x^T L x$, where L is the Laplacian matrix. A linear distributed protocol for minimizing the total disagreement can thus be formulated as

$$u^{(i)} = \sum_j (x^{(j)} - x^{(i)}) . \quad (10.30)$$

Let us now suppose that a dynamic quantity Q is represented with K states and controlled at each discrete time index k. We may also assume that n nodes measure Q at each k and data are available during finite time due to limited resources. Accordingly, Q can be represented with linear state and observation equations as

$$x_k = F_k x_{k-1} + E_k u_k + B_k w_k , \quad (10.31)$$

$$y_k = H_k x_k + v_k , \quad (10.32)$$

$$y_k^{(i)} = H_k^{(i)} x_k + v_k^{(i)} , \quad (10.33)$$

where we save the same definitions as for the model in (10.1) and (10.2). The ith, $i \in [1, n]$, node measures x_k by $y_k^{(i)} \in \mathbb{R}^p$, $p \leqslant K$, with $H_k^{(i)} \in \mathbb{R}^{p \times K}$ and each node has several inclusive neighbors. Local data $y_k^{(i)}$ are united in the observation vector $y_k = [y_k^{(1)T} \ \ldots \ y_k^{(n)T}]^T$ with $H_k = [H_k^{(1)T} \ \ldots \ H_k^{(n)T}]^T$. Noise vectors w_k and $v_k = [\, v_k^{(1)T} \ \ldots \ v_k^{(n)T}]^T$ are zero mean, not obligatorily white

Gaussian, uncorrelated, and with the covariances $Q_k = E\{w_k w_k^T\}$, $R_k = \text{diag}[R_k^{(1)^T} \ \dots \ R_k^{(n)^T}]^T$, and $R_k^{(i)} = E\{v_k^{(i)} v_k^{(i)^T}\}$.

To apply FIR filtering, we expand model (10.31) and (10.32) on $[m, k]$ as (4.7) and (4.14),

$$X_{m,k} = F_{m,k} x_m + S_{m,k} U_{m,k} + D_{m,k} W_{m,k} ,$$

$$Y_{m,k} = H_{m,k} x_m + L_{m,k} U_{m,k} + G_{m,k} W_{m,k} + V_{m,k} ,$$

refer to the definitions given after (4.2), and notice that the ith local extended vector $Y_{m,k}^{(i)} = \left[y_m^{(i)^T} \ y_{m+1}^{(i)^T} \ \dots \ y_k^{(i)^T} \right]^T$ appears by depicting all variables in (4.14) with a superscript $^{(i)}$.

Batch dUFIR Filter with Consensus in Estimates

The distributed FIR filtering estimate can be assigned as (4.30b)

$$\hat{x}_k = \mathcal{H}_{m,k} Y_{m,k} + (\bar{S}_{m,k} - \mathcal{H}_{m,k} L_{m,k}) U_{m,k} ,$$

and the ith, $i \in [1, n]$, local FIR filtering estimate related to the model in (10.31) and (10.33) as

$$\hat{x}_k^{(i)} = \mathcal{H}_{m,k}^{(i)} Y_{m,k}^{(i)} + (\bar{S}_{m,k} - \mathcal{H}_{m,k}^{(i)} L_{m,k}^{(i)}) U_{m,k} , \tag{10.34}$$

where we suppose that the ith node provides a local estimate $\hat{x}_k^{(i)}$ over $Y_{m,k}^{(i)}$. Referring to [139] and (10.30), a consensus between the local estimates related to the ith node can be found by introducing a vector

$$\psi_k^{(i)} = \sum_j (\hat{x}_k^{(j)} - \hat{x}_k^{(i)}) \tag{10.35}$$

and combining $\psi_k^{(i)}$ with (10.34). This gives a consensus estimate

$$\hat{x}_k^{\text{c}} = \mathcal{H}_{m,k} Y_{m,k} + (\bar{S}_{m,k} - \mathcal{H}_{m,k} L_{m,k}) U_{m,k} + \lambda_k \psi_k^{(i)} , \tag{10.36}$$

where λ_k is a scaling factor to be optimized in the MSE sense.

To design the distributed UFIR (dUFIR) filter with this kind of consensus, the following unbiasedness condition must be satisfied, [206]

$$\mathcal{E}\{\hat{x}_k^{\text{c}}\} = \mathcal{E}\{\hat{x}_k^{(i)}\} = \mathcal{E}\{x_k\} , \tag{10.37}$$

to guarantee that the average of the estimate is equal to the average of the model, which is given by the last Nth row vector in (4.7) as (4.19),

$$x_k = \mathcal{F}_k^{m+1} x_m + \bar{S}_{m,k} U_{m,k} + \bar{D}_{m,k} W_{m,k} . \tag{10.38}$$

By defining the estimation error $\varepsilon_k^{\text{c}} = x_k - \hat{x}_k^{\text{c}}$ and considering the error covariance $P_k = \mathcal{E}\{\varepsilon_k^{\text{c}} \varepsilon_k^{\text{c}^T}\}$, the optimal factor λ_k^{opt} can be determined if we solve the minimization of problem

$$\lambda_k^{\text{opt}} = \underset{\lambda_k}{\arg\min} \ \{\text{tr}\, P_k\} . \tag{10.39}$$

The dUFIR filter for consensus in estimates can now be designed as follows. Rewrite the estimate (10.36) as

$$\begin{aligned} \hat{x}_k^{\text{c}} &= (I + n\lambda_k)[\mathcal{H}_{m,k}(Y_{m,k} - L_{m,k} U_{m,k}) + \bar{S}_{m,k} U_{m,k}] \\ &\quad - n\lambda_k [\mathcal{H}_{m,k}^{(i)}(Y_{m,k}^{(i)} - L_{m,k}^{(i)} U_{m,k}) + \bar{S}_{m,k} U_{m,k}] , \end{aligned} \tag{10.40}$$

where gains $\mathcal{H}_{m,k}$ and $\mathcal{H}^{(i)}_{m,k}$ satisfy the unbiasedness condition (10.37) and two unbiasedness constraints

$$\mathcal{H}_{m,k} = (I + n\lambda_k)\mathcal{G}_k C^T_{m,k} , \tag{10.41}$$

$$\mathcal{H}^{(i)}_{m,k} = \mathcal{G}^{(i)}_k C^{(i)^T}_{m,k} , \tag{10.42}$$

where $C_{m,k} = H_{m,k}(\mathcal{F}^{m+1}_k)^{-1}$, $\mathcal{G}_k = (C^T_{m,k}C_{m,k})^{-1}$ is the GNPG, $C^{(i)}_{m,k} = H^{(i)}_{m,k}(\mathcal{F}^{m+1}_k)^{-1}$, and $\mathcal{G}^{(i)}_k = (C^{(i)^T}_{m,k}C^{(i)}_{m,k})^{-1}$ is the GNPG of the ith filter. It now follows that information required to compute $\mathcal{H}_{m,k}$ by (10.41) and $\mathcal{H}^{(i)}_{m,k}$ by (10.42) is entirely provided by the K-state space model, which thus can be preloaded on the nodes.

Batch Optimum Consensus Factor

The optimum consensus factor λ^{opt}_k can be determined by solving the optimization problem (10.39). In doing so, we refer to (10.36) and (10.38) and transform the error covariance $P_k = \mathcal{E}\{\varepsilon_k\varepsilon^T_k\}$ for uncorrelated noise as

$$\begin{aligned} P_k &= (\bar{D}_{m,k} - \tilde{\Theta}_{m,k}G_{m,k})\mathcal{Q}_{m,k}(\bar{D}_{m,k} - \tilde{\Theta}_{m,k}G_{m,k})^T \\ &\quad + n\lambda_k(\tilde{\Theta}_{m,k}\mathcal{R}_{m,k}\tilde{\Theta}^T_{m,k} - \Theta^{(i)}_{m,k}\mathcal{R}^{(i)^T}_{m,k}\tilde{\Theta}^T_{m,k}) \\ &\quad + n[\lambda_k(\tilde{\Theta}_{m,k}\mathcal{R}_{m,k}\tilde{\Theta}^T_{m,k} - \Theta^{(i)}_{m,k}\mathcal{R}^{(i)^T}_{m,k}\tilde{\Theta}^T_{m,k})]^T \\ &\quad + n^2\lambda_k(\tilde{\Theta}_{m,k}\mathcal{R}_{m,k}\tilde{\Theta}^T_{m,k} - \tilde{\Theta}_{m,k}\mathcal{R}^{(i)}_{m,k}\Theta^{(i)^T}_{m,k})\lambda^T_k \\ &\quad + n^2\lambda_k(\Theta^{(i)}_{m,k}\mathcal{R}^{(i)^T}_{m,k}\tilde{\Theta}^T_{m,k} + \Theta^{(i)}_{m,k}\mathcal{R}^{(i)}_{m,k}\Theta^{(i)^T}_{m,k})\lambda^T_k , \end{aligned} \tag{10.43}$$

where $\tilde{\Theta}_{m,k} = \mathcal{G}_k C^T_{m,k}$, $\mathcal{Q}_{m,k} = \mathcal{E}\{W_{m,k}W^T_{m,k}\}$, $\mathcal{R}_{m,k} = \mathcal{E}\{V_{m,k}V^T_{m,k}\}$, $\mathcal{R}^{(i)}_{m,k} = \mathcal{E}\{V^{(i)}_{m,k}V^{(i)^T}_{m,k}\}$, and $\mathcal{R}^{(i)}_{m,k} = \mathcal{E}\{V_{m,k}V^{(i)^T}_{m,k}\}$.

By putting to zero the derivative applied with respect to λ_k to the trace of (10.43) using (A.3) and (A.5) and the identities $\tilde{\Theta}_{m,k}\mathcal{R}^{(i)}_{m,k}\Theta^{(i)^T}_{m,k} = (\tilde{\Theta}_{m,k}\mathcal{R}^{(i)}_{m,k}\Theta^{(i)^T}_{m,k})^T$ and $\tilde{\Theta}_{m,k}\mathcal{R}^{(i)}_{m,k}\Theta^{(i)^T}_{m,k} = \mathcal{G}_k\mathcal{G}^{(i)^{-1}}_k\Theta^{(i)}_{m,k}\mathcal{R}^{(i)}_{m,k}\Theta^{(i)^T}_{m,k}$, we obtain the optimal factor λ^{opt}_k as

$$\begin{aligned} \lambda^{\text{opt}}_k &= -\frac{1}{n}(\tilde{\Theta}_{m,k}\mathcal{R}_{m,k}\tilde{\Theta}^T_{m,k} - \mathcal{G}_k\mathcal{G}^{(i)^{-1}}_k\Theta^{(i)}_{m,k} \\ &\quad \times \mathcal{R}^{(i)}_{m,k}\Theta^{(i)^T}_{m,k})(\tilde{\Theta}_{m,k}\mathcal{R}_{m,k}\tilde{\Theta}^T_{m,k} - 2\mathcal{G}_k\mathcal{G}^{(i)^{-1}}_k \\ &\quad \times \Theta^{(i)}_{m,k}\mathcal{R}^{(i)}_{m,k}\Theta^{(i)^T}_{m,k} + \Theta^{(i)}_{m,k}\mathcal{R}^{(i)}_{m,k}\Theta^{(i)^T}_{m,k})^{-1} . \end{aligned} \tag{10.44}$$

It is worth noting that implementation of the batch estimate (10.41) with (10.44) may cause a computational problem when the dimensions of the matrices are large. The recursive forms will be considered next.

Recursive dUFIR Filtering Estimate

To compute the batch estimate (10.40) iteratively as shown in [206], $\hat{x}^c_k$ can be represented with a sum of the centralized estimate $\hat{x}_k$ and local estimate $\hat{x}^{(i)}_k$ given by (10.34) as

$$\hat{x}^c_k = (I + n\lambda_k)\hat{x}_k - n\lambda_k\hat{x}^{(i)}_k , \tag{10.45}$$

where both $\hat{x}_k$ and $\hat{x}^{(i)}_k$ can be computed using Algorithm 21 and then combined as (10.45). To make it possible, we need recursions for the factor λ^{opt}_k.

Recursive Consensus Factor

To compute the dUFIR filtering estimate iteratively, factor λ_k^{opt} (10.44) can be represented as [206]

$$\begin{aligned}\lambda_k^{\text{opt}} &= -\frac{1}{n}(\alpha_k - \mathcal{G}_k\mathcal{G}_k^{(i)^{-1}}\beta_k)\\ &\quad\times(\alpha_k - 2\mathcal{G}_k\mathcal{G}_k^{(i)^{-1}}\beta_k + \beta_k)^{-1}\,,\end{aligned} \tag{10.46}$$

where $\alpha_k = \tilde{\Theta}_{m,k}\mathcal{R}_{m,k}\tilde{\Theta}_{m,k}^T$ and $\beta_k = \Theta_{m,k}^{(i)}\mathcal{R}_{m,k}^{(i)}\Theta_{m,k}^{(i)^T}$ require recursive forms.

A recursion for $\alpha_k = \tilde{\Theta}_{m,k}\mathcal{R}_{m,k}\tilde{\Theta}_{m,k}^T$ can be found if we first write

$$\alpha_k = \mathcal{G}_k C_{m,k}\mathcal{R}_{m,k}C_{m,k}^T\mathcal{G}_k\,, \tag{10.47}$$

and then transform the product $C_{m,k}\mathcal{R}_{m,k}C_{m,k}^T$ as

$$\begin{aligned}C_{m,k}\mathcal{R}_{m,k}C_{m,k}^T &= H_k^T R_k H_k + F_k^{-T}\left[\sum_{l=0}^{N-2}(\mathcal{F}_{k-1}^{m+1+l})^{-T}H_{m+l}^T R_{m+l}H_{m+l}(\mathcal{F}_{k-1}^{m+1+l})^{-1}\right]F_k^{-1}\\ &= H_k^T R_k H_k + F_k^{-T}C_{m,k-1}\mathcal{R}_{m,k-1}C_{m,k-1}^T F_k^{-1}\,.\end{aligned} \tag{10.48}$$

An equality $\alpha_{k-1} = \mathcal{G}_{k-1}C_{m,k-1}\mathcal{R}_{m,k-1}C_{m,k-1}^T\mathcal{G}_{k-1}$ gives

$$C_{m,k-1}\mathcal{R}_{m,k-1}C_{m,k-1}^T = \mathcal{G}_{k-1}^{-1}\alpha_{k-1}\mathcal{G}_{k-1}^{-1}\,,$$

and from (10.48) we obtain a recursion for (10.47). A recursion for $\beta_k^{(i)}$ can be obtained similarly, and we finally have

$$\alpha_k = \mathcal{G}_k(H_k^T R_k H_k + F_k^{-T}\mathcal{G}_{k-1}^{-1}\alpha_{k-1}\mathcal{G}_{k-1}^{-1}F_k^{-1})\mathcal{G}_k\,, \tag{10.49}$$

$$\beta_k^{(i)} = \mathcal{G}_k^{(i)}(H_k^{(i)^T}R_k^{(i)}H_k^{(i)} + F_k^{-T}\mathcal{G}_{k-1}^{(i)^{-1}}\beta_{k-1}^{(i)}\mathcal{G}_{k-1}^{(i)^{-1}}F_k^{-1})\mathcal{G}_k^{(i)}\,, \tag{10.50}$$

where the initial values α_{k-1} and $\beta_{k-1}^{(i)}$ can be computed in short batch forms as $\alpha_s = \mathcal{G}_s C_{m,s}^T\mathcal{R}_{m,s}C_{m,s}\mathcal{G}_s^T$ and $\beta_s = \mathcal{G}_s^{(i)}C_{m,s}^{(i)^T}\mathcal{R}_{m,s}^{(i)}C_{m,s}^{(i)}\mathcal{G}_s^{(i)^T}$.

Iterative dUFIR Filtering Algorithm

The pseudocode of the iterative dUFIR filtering algorithm with consensus in estimates is listed as Algorithm 22. Given N, Algorithm 22 starts computing initial values at $s = m + K - 1$ and then updates estimates beginning at $s + 1$ until the iterative variable l reaches k. It then computes λ_k^{opt} and finishes with the consensus estimate $\hat{x}_k^{\text{c}}$. Of practical importance is that factor λ_k^{opt} is constant for LTI systems and can thus be preloaded in the node to reduce the number of operations.

10.2 Optimal Fusion Filtering Under Correlated Noise

Information fusion in WSN is one of the key problems the effective solution of which justifies the network design. To avoid the requirement of the initial state, it was proposed in [194] to provide fusion filtering of estimates received from smart sensors in the *linear unbiased minimum variance* (LUMV) sense, which means unbiased optimality inherent to OUFIR and ML filters. The approach that was developed in [227] following [194] suggests organizing fusing robust UFIR filtering of estimates received from smart sensors [204] under the time-correlated noise using an LUMV algorithm as discussed next.

Algorithm 22: Iterative dUFIR Filtering Algorithm

Data: y_k, $R_k^{(i)}$, R_k, N
Result: $\hat{x}_k$
begin
  for $k = N-1 : \infty$ **do**
    $m = k - N + 1, \quad s = m + K - 1;$
    $\mathcal{G}_s = (C_{m,s}^T C_{m,s})^{-1};$
    $\mathcal{G}_s^{(i)} = (C_{m,s}^{(i)^T} C_{m,s}^{(i)})^{-1};$
    $\tilde{x}_s = \mathcal{G}_s C_{m,s}^T (Y_{m,s} - L_{m,s} U_{m,s}) + \bar{S}_{m,s} U_{m,s};$
    $\tilde{x}_s^{(i)} = \mathcal{G}_s^{(i)} C_{m,s}^{(i)^T} (Y_{m,s}^{(i)} - L_{m,s}^{(i)} U_{m,s}) + \bar{S}_{m,s} U_{m,s};$
    $\alpha_s = \mathcal{G}_s C_{m,s}^T \mathcal{R}_{m,s} C_{m,s} \mathcal{G}_s^T;$
    $\beta_s = \mathcal{G}_s^{(i)} C_{m,s}^{(i)^T} \mathcal{R}_{m,s}^{(i)} C_{m,s}^{(i)} \mathcal{G}_s^{(i)^T};$
    for $l = s+1 : k$ **do**
      $\hat{x}_l^- = F_l \hat{x}_{l-1} + E_l u_l;$
      $\hat{x}_l^{(i)^-} = F_l \hat{x}_{l-1}^{(i)} + E_l u_l;$
      $\mathcal{G}_l = [H_l^T H_l + (F_l \mathcal{G}_{l-1} F_l^T)^{-1}]^{-1};$
      $\mathcal{G}_l^{(i)} = [H_l^{(i)^T} H_l^{(i)} + (F_l \mathcal{G}_{l-1}^{(i)} F_l^T)^{-1}]^{-1};$
      $\hat{x}_l = \hat{x}_l^- + \mathcal{G}_l H_l^T (y_l - H_l \hat{x}_l^-);$
      $\hat{x}_l^{(i)} = \hat{x}_l^{(i)^-} + \mathcal{G}_l^{(i)} H_l^{(i)^T} (y_l^{(i)} - H_l^{(i)} \hat{x}_l^{(i)^-});$
      $\alpha_l = \mathcal{G}_l (H_l^T R_l H_l + F_l^{-T} \mathcal{G}_{l-1}^{-1} \alpha_{l-1} \mathcal{G}_{l-1}^{-1} F_l^{-1}) \mathcal{G}_l;$
      $\beta_l = \mathcal{G}_l^{(i)} (H_l^{(i)^T} R_l^{(i)} H_l^{(i)} + F_l^{-T} \mathcal{G}_{l-1}^{(i)^{-1}} \beta_{l-1} \mathcal{G}_{l-1}^{(i)^{-1}} F_l^{-1}) \mathcal{G}_l^{(i)};$
    end for
    $\lambda_k^{\text{opt}} = -\frac{1}{n}(\alpha_k - \mathcal{G}_k \mathcal{G}_k^{(i)^{-1}} \beta_k)(\alpha_k - 2\mathcal{G}_k \mathcal{G}_k^{(i)^{-1}} \beta_k + \beta_k)^{-1};$
    $\hat{x}_k^c = (I + n\lambda_k)\tilde{x}_k - n\lambda_k \tilde{x}_k^{(i)};$
  end for
end

Consider a quantity Q, whose dynamics and observation provided over WSN by an ith smart sensor are represented with the state-space model

$$x_k = F x_{k-1} + E u_k + B w_k \,, \tag{10.51}$$

$$y_k^{(i)} = H^{(i)} x_k + v_k^{(i)} \,, \tag{10.52}$$

where $i \in [1, n]$ denotes the sensor index and n is the number of sensors. It is assumed that w_k and $v_k^{(i)}$ are white Gaussian with zero mean, $\mathcal{E}\{w_k\} = 0$ and $\mathcal{E}\{v_k^{(i)}\} = 0$, and the correlation property

$$\mathcal{E}\left\{ \begin{bmatrix} w_k \\ v_k^{(i)} \end{bmatrix} \begin{bmatrix} w_j^T & v_j^{(i)^T} \end{bmatrix} \right\} = \begin{bmatrix} Q & L^{(i)} \\ L^{(i)^T} & R^{(i)} \end{bmatrix} \delta_{kj} \,. \tag{10.53}$$

To de-correlate w_k and $v_k^{(i)}$, one can follow the approach described in Chapter 3 and represent the state equation (10.51) using (3.103) as

$$\begin{aligned} x_k &= F x_{k-1} + E u_k + B w_k + \Lambda (y_k^{(i)} - H^{(i)} x_k - v_k^{(i)}) \\ &= \bar{F}^{(i)} x_{k-1} + \bar{u}_k^{(i)} + \bar{w}_k^{(i)} \,, \end{aligned} \tag{10.54}$$

where $\bar{F}^{(i)} = \bar{\Lambda}^{(i)}F$, $\bar{u}_k^{(i)} = \bar{\Lambda}^{(i)}(Eu_k + \Lambda^{(i)}y_k^{(i)})$, $\bar{w}_k^{(i)} = \bar{\Lambda}^{(i)}(Bw_k - \Lambda^{(i)}v_k^{(i)})$, and $\bar{\Lambda}^{(i)} = (I + \Lambda^{(i)}H^{(i)})^{-1}$. To avoid the correlation, one can set the covariance $\mathcal{E}\{\bar{w}_k^{(i)}v_k^{(i)^T}\}$ to zero,

$$0 = \mathcal{E}\{\bar{w}_k^{(i)}v_k^{(i)^T}\} = BL^{(i)} - \Lambda^{(i)}R^{(i)} ,$$

and find $\Lambda^{(i)} = BL^{(i)}R^{(i)^{-1}}$, where matrices $L^{(i)}$ and $R^{(i)}$ are specified by (10.53) and assumed to be known. Next, $\bar{w}_k^{(i)}$ can be rewritten as

$$\bar{w}_k^{(i)} = \bar{\Lambda}^{(i)}B(w_k - L^{(i)}R^{(i)^{-1}}v_k^{(i)}) \tag{10.55}$$

and (10.54) and (10.52) can be transformed to have mutually independent and uncorrelated zero mean white Gaussian noise vectors $\xi_k^{(i)}$ and $v_k^{(i)}$ as

$$x_k = \bar{F}^{(i)}x_{k-1} + \bar{u}_k^{(i)} + B^{(i)}\xi_k^{(i)} , \tag{10.56}$$

$$y_k^{(i)} = H^{(i)}x_k + v_k^{(i)} , \tag{10.57}$$

where $B^{(i)} = \bar{\Lambda}^{(i)}B$, $\xi_k^{(i)} = w_k - L^{(i)}R^{(i)^{-1}}v_k^{(i)}$, and noise $\xi_k^{(i)}$ has the covariance $Q^{(ij)} = \mathcal{E}\{\xi_k^{(i)}\xi_r^{(j)^T}\}$ given by

$$\begin{aligned} Q^{(ij)} = [&Q - L^{(j)}(L^{(j)}R^{(j)^{-1}})^T - L^{(i)}R^{(i)^{-1}}L^{(j)^T} \\ &+ L^{(i)}R^{(i)^{-1}}R^{(ij)}(L^{(j)}R^{(j)^{-1}})^T]\delta_{kr} , \end{aligned} \tag{10.58}$$

which becomes $Q^{(ij)} = (Q - L^{(i)}R^{(i)^{-1}}L^{(i)^T})\delta_{kr}$ when $i = j$.

Provided that x_k (10.56) and $y_k^{(i)}$ (10.57) have uncorrelated white Gaussian noise vectors and that the estimates $\hat{x}_k^{(i)}$, $i \in [1, n]$, produced by smart sensors with embedded UFIR filters are completely received without loss, fusion filtering can be organized in the LUMV sense [194] as

$$\hat{x}_k = \alpha_1\hat{x}_k^{(1)} + \alpha_2\hat{x}_k^{(2)} + \cdots + \alpha_n\hat{x}_k^{(n)}, \tag{10.59}$$

where weights $\alpha_i \in \mathbb{R}^{K\times K}$, $i \in [1, n]$, are taken from the weighting matrix $\Pi = [\ \alpha_1 \ \ \alpha_2 \ \ \cdots \ \ \alpha_n\ ] \in \mathbb{R}^{K\times Kn}$ computed as

$$\Pi = \left(\Xi^T\Sigma^{-1}\Xi\right)^{-1}\Xi^T\Sigma^{-1}, \tag{10.60}$$

where $\Sigma = P^{(ij)} \in \mathbb{R}^{Kn\times Kn}$, $\{i,j\} \in [1, n]$, is a symmetric positive definite matrix, $P^{(ij)} = \mathcal{E}\{(x_k - \hat{x}_k^{(i)})(x_k - \hat{x}_k^{(j)})^T\} \in \mathbb{R}^{K\times K}$ is the error cross covariance matrix, $\Xi = [\ I_K \ \ I_K \ \ \cdots \ \ I_K\ ]^T \in \mathbb{R}^{Kn\times K}$, and $I_K \in \mathbb{R}^{K\times K}$ is the identity matrix.

To justify (10.60), the unbiasedness condition $\mathcal{E}\{\hat{x}_k\} = \mathcal{E}\{\hat{x}_k^{(i)}\} = \mathcal{E}\{x_k\}$ can be applied to (10.56) and (10.59) that gives

$$I_K = \alpha_1 + \alpha_2 + \cdots + \alpha_n \tag{10.61}$$

and then (10.60) can be rewritten as

$$\Pi\Xi - I_K = 0 . \tag{10.62}$$

The fusion estimation error $\varepsilon_k = x_k - \hat{x}_k$ can further be transformed to

$$\begin{aligned} \varepsilon_k &= x_k - (\alpha_1\hat{x}_k^{(1)} + \alpha_2\hat{x}_k^{(2)} + \cdots + \alpha_n\hat{x}_k^{(n)}) \\ &= \alpha_1(x_k - \hat{x}_k^{(1)}) + \cdots + \alpha_n(x_k - \hat{x}_k^{(n)}) \\ &= \Pi\left[(x_k - \hat{x}_k^{(1)})^T \ \ \cdots \ \ (x_k - \hat{x}_k^{(n)})^T\right]^T \end{aligned} \tag{10.63}$$

and, since the UFIR estimates are unbiased, the fusion error covariance $P = \mathcal{E}\{\varepsilon_k \varepsilon_k^T\}$ becomes

$$P = \Pi \Sigma \Pi^T . \tag{10.64}$$

The cost function for (10.64) can now be introduced as $J = \text{tr}(\Pi \Sigma \Pi^T)$ and weight Π subject to (10.62) determined by minimizing J. To this end, the Lagrangian cost can be written as

$$J^* = \text{tr}\left(\Pi \Sigma \Pi^T\right) + 2\Lambda\left(\Pi \Xi - I_K\right) , \tag{10.65}$$

where $\Lambda \in \mathbb{R}^{K\times K}$ is the Lagrange multiplier. A solution to $\partial J^*/\partial \Pi = 0$ referring to $\Sigma^T = \Sigma$ results in an equation $2\Pi\Sigma + 2\Lambda\Xi^T = 0$ that together with (10.62) gives

$$\left[\ \Pi \quad \Lambda\ \right] = \left[\ 0 \quad I_K\ \right]\begin{bmatrix} \Sigma & \Xi \\ \Xi^T & 0 \end{bmatrix}^{-1} . \tag{10.66}$$

Note that the inverse in (10.66) always exists, because matrix Σ is symmetric and positive definite. Finally, the inversion rule (A.23) transforms (10.66) to (10.61) and completes the justification.

10.2.1 Error Covariances Under Cross Correlation

It follows that optimal weights α_i can be found via matrix $\Sigma = P^{(ij)}$, $\{i,j\} \in [1,n]$, which can be determined as follows. Define the filtering error produced by the ith smart sensor as $\varepsilon_k^{(i)} = x_k - \hat{x}_k^{(i)}$, substitute the state x_k with model (8.5) and the ith estimate $\hat{x}_k^{(i)}$ using (8.43) as

$$x_k = F^{N-1}x_m + \bar{S}_N U_{m,k} + \bar{D}_N W_{m,k} , \tag{10.67}$$

$$\hat{x}_k^{(i)} = \mathcal{H}_N^{(i)} Y_{m,k}^{(i)} + (\bar{S}_N - \mathcal{H}_N^{(i)} L_N^{(i)})U_{m,k} , \tag{10.68}$$

write the extended observation vector related to the ith sensor using (8.4) as

$$Y_{m,k}^{(i)} = H_N^{(i)} x_m + L_N^{(i)} U_{m,k} + G_N^{(i)} W_{m,k} + V_{m,k}^{(i)} , \tag{10.69}$$

and transform $\varepsilon_k^{(i)}$ to

$$\begin{aligned} \varepsilon_k^{(i)} &= (F^{N-1} - \mathcal{H}_{m,k}^{(i)} H_{m,k}^{(i)})x_m \\ &\quad + (\bar{D}_N - \mathcal{H}_{m,k}^{(i)} G_N^{(i)})W_{m,k} - \mathcal{H}_{m,k}^{(i)} V_{m,k}^{(i)} . \end{aligned} \tag{10.70}$$

By subjecting to the unbiasedness constraint $F^{N-1} - \mathcal{H}_{m,k}^{(i)} H_{m,k}^{(i)}$, transform (10.70) to

$$\varepsilon_k^{(i)} = (\bar{D}_N - \mathcal{H}_{m,k}^{(i)} G_N^{(i)})W_{m,k} - \mathcal{H}_{m,k}^{(i)} V_{m,k}^{(i)} . \tag{10.71}$$

Given (10.71), the error covariance $P^{(ij)} = \mathcal{E}\{\varepsilon_k^{(i)} \varepsilon_k^{(j)^T}\}$ can now be transformed for uncorrelated $W_{m,k}^{(i)}$ and $V_{m,k}^{(i)}$ to

$$\begin{aligned} P^{(ij)} &= (\bar{B}_{m,k}^{(i)} - H_{m,k}^{(i)} F_{m,k}^{(i)}) \bar{Q}^{(ij)} (\bar{B}_{m,k}^{(j)} - H_{m,k}^{(j)} F_{m,k}^{(j)})^T \\ &\quad + H_{m,k}^{(i)} \bar{R}^{(ij)} H_{m,k}^{(j)^T} , \end{aligned} \tag{10.72}$$

where the covariance matrices are defined by

$$\begin{aligned} \bar{Q}^{(ij)} &= \mathcal{E}\{W_{m,k}^{(i)} W_{m,k}^{(j)^T}\} \\ &= \text{diag}(\,Q^{(ij)}\ Q^{(ij)}\ \dots\ Q^{(ij)}\,) , \\ \bar{R}^{(ij)} &= \mathcal{E}\{V_{m,k}^{(i)} V_{m,k}^{(j)^T}\} \\ &= \text{diag}(\,R^{(ij)}\ R^{(ij)}\ \dots\ R^{(ij)}\,) . \end{aligned}$$

The pseudocode of the biased-constrained optimal fusing filtering algorithm designed for local nodes is listed as Algorithm 23. It requires estimates $\hat{x}_k^{(i)}$, $i \in [1, n]$, produced by UFIR filters embedded to smart sensors and the cross-covariance matrix $P^{(ij)}$ specified by (10.72). It is supposed that estimates $\hat{x}_k^{(i)}$ are delivered with no latency and missing data, although time synchronization may be required. Provided weights α_i, $i \in [1, n]$, a fusion estimate $\hat{x}_k$ is computed by (10.59). It is finally worth noting that Algorithm 23 provides an ML estimate and that large components of $P^{(ij)}$ require fewer weights.

Algorithm 23: Optimal Fusion Filtering Algorithm for Local Nodes

---

Data: $\hat{x}_k^{(i)}, Q, L^{(i)}, R^{(ij)}$
Result: $\hat{x}_k$

for $i = 1 : n$ **do**
  $X^{(i)} = \bar{B}_{m,k}^{(i)} - H_{m,k}^{(i)} F_{m,k}^{(i)}$;
  for $j = 1 : n$ **do**
    $\tilde{Q}^{(ij)} = Q - L^{(j)}(L^{(j)}R^{(j)^{-1}})^T - L^{(i)}R^{(i)^{-1}}L^{(i)^T}$;
    $Q^{(ij)} = \tilde{Q}^{(ij)} + L^{(i)}R^{(i)^{-1}}R^{(ij)}(L^{(j)}R^{(j)^{-1}})^T$;
    $\bar{Q}^{(ij)} = \text{diag}(Q^{(ij)}\ Q^{(ij)}\ \cdots\ Q^{(ij)})$;
    $\bar{R}^{(ij)} = \text{diag}(R^{(ij)}\ R^{(ij)}\ \cdots\ R^{(ij)})$;
    $P^{(ij)} = X^{(i)}\bar{Q}^{(ij)}X^{(j)^T} + H_{m,k}^{(i)}\bar{R}^{(ij)}H_{m,k}^{(j)^T}$;
  end for
  $\Sigma = [P^{(ij)}]_{i,j\in[1,n]}$;
end for
$\Pi = (\Xi^T\Sigma^{-1}\Xi)^{-1}\Xi^T\Sigma^{-1} = [\,\alpha_1\ \alpha_2\ \cdots\ \alpha_n\,]$;
$\hat{x}_k = \alpha_1\hat{x}_k^{(1)} + \alpha_2\hat{x}_k^{(2)} + \cdots + \alpha_r\hat{x}_k^{(n)}$

10.3 Hybrid Kalman/UFIR Filter Structures

The design of hybrid structures refers to extreme properties of linear state estimators: the KF is *optimal but not robust*, and the UFIR filter is *robust but not optimal*. Fusing both estimates gives a new quality that we will consider next.

10.3.1 Fusing Estimates with Probabilistic Weights

Let us consider the FE-based state-space model

$$x_k = F_{k-1}x_{k-1} + E_{k-1}u_{k-1} + B_{k-1}w_{k-1}\ , \tag{10.73}$$

$$y_n = H_k x_k + v_k\ , \tag{10.74}$$

and suppose that the KF estimate and error covariance are available at $k-1$ as $\hat{x}_{k-1}^{\mathrm{K}}$ and P_{k-1}^{K}, respectively, and the UFIR estimate and error covariance as $\hat{x}_{k-1}^{\mathrm{U}}$ and P_{k-1}^{U}, respectively. To fuse these estimates optimally, a fusion filter (FF) can be designed as shown [225], and the design requires the following several steps.

Initialization

On a short initial horizon $[0, k] < N_{\text{opt}}$, the UFIR filter produces extra random errors. Therefore, a more accurate KF estimate goes to the FF output. Otherwise, when $k \geq N_{\text{opt}}$, the optimal but not robust KF estimate is fused with the robust but suboptimal UFIR estimate by introducing the Markov transition probability matrix Π and weighting vector Θ_{k-1} such that

$$\Pi = \begin{bmatrix} \pi_{11} & \pi_{12} \\ \pi_{21} & \pi_{22} \end{bmatrix}, \quad \Theta_{k-1} = \begin{bmatrix} \theta^{\mathrm{K}}_{k-1} \\ \theta^{\mathrm{U}}_{k-1} \end{bmatrix}, \tag{10.75}$$

where $\theta^{\mathrm{K}}_{k-1}$ is a probabilistic weight for the KF and $\theta^{\mathrm{U}}_{k-1}$ for the UFIR filter. The notation π_{ij}, $\{i,j\} \in [1,2]$, means the transition probability between the Kalman and UFIR subfilters.

In the FF design, matrix Π plays a critical rule to improve the performance. If the KF performs better and is thus a dominant subfilter, then the transition probabilities π_{11} and π_{21} acquire larger values. Otherwise, if the UFIR estimate is more accurate, the probabilities π_{12} and π_{22} become larger. Provided no prior information about the filter accuracy, the weights are chosen as $\pi_{11} = \pi_{22} = 0.5$ to satisfy $\sum_{j=1}^{2} \pi_{ij} = 1$ for $i = 1,2$.

For the KF, the initial state $\tilde{x}_{k-1}$ and error covariance $\tilde{P}_{k-1}$ are obtained by merging the subestimates as

$$\tilde{x}_{k-1} = \theta^{\mathrm{K}}_{k-1}\hat{x}^{\mathrm{K}}_{k-1} + \theta^{\mathrm{U}}_{k-1}\hat{x}^{\mathrm{U}}_{k-1}, \tag{10.76}$$

$$\tilde{P}_{k-1} = \theta^{\mathrm{K}}_{k-1}P^{(1)}_{k-1} + \theta^{\mathrm{U}}_{k-1}P^{(2)}_{k-1}, \tag{10.77}$$

where the suberror covariances are defined by

$$P^{(1)}_{k-1} = P^{\mathrm{K}}_{k-1} + (\tilde{x}_{k-1} - \hat{x}^{\mathrm{K}}_{k-1})(\tilde{x}_{k-1} - \hat{x}^{\mathrm{K}}_{k-1})^T, \tag{10.78}$$

$$P^{(2)}_{k-1} = P^{\mathrm{U}}_{k-1} + (\tilde{x}_{k-1} - \hat{x}^{\mathrm{U}}_{k-1})(\tilde{x}_{k-1} - \hat{x}^{\mathrm{U}}_{k-1})^T. \tag{10.79}$$

For the UFIR filter, no initialization is required, and the iterations start with the initial value $\bar{x}_s$ specified by (7.25). Iterative computation of P^{U}_k, which is not involved to the standard UFIR algorithm, begins with the initial value $\bar{P}_s$ computed by (7.26).

Prediction

Provided $\tilde{x}_{k-1}$ by (10.76) and $\tilde{P}_{k-1}$ by (10.77), the *a priori* state estimate $\hat{x}^{\mathrm{K}-}_k$ and error covariance $P^{\mathrm{K}-}_k$ are computed for the KF as

$$\hat{x}^{\mathrm{K}-}_k = F_{k-1}\tilde{x}_{k-1} + E_{k-1}u_{k-1}, \tag{10.80}$$

$$P^{\mathrm{K}-}_k = F_{k-1}\tilde{P}_{k-1}F^T_{k-1} + B_{k-1}Q_{k-1}B^T_{k-1}. \tag{10.81}$$

The weighted UFIR filter starts iterating with $\bar{x}_s$ and the true output is taken as $\hat{x}^{\mathrm{U}}_k = \bar{x}_i$ when $i = k$. The *a priori* state estimate $\hat{x}^{\mathrm{U}-}_k = \bar{x}^-_k$ is computed for the weighted UFIR filter as

$$\hat{x}^{\mathrm{U}-}_k = F_{k-1}\bar{x}_{k-1} + E_{k-1}u_{k-1}, \tag{10.82}$$

where $\bar{x}_{k-1}$ is updated iteratively using the standard UFIR procedure. The *a priori* error covariance $P^{\mathrm{U}-}_k$ is computed by

$$P^{\mathrm{U}-}_k = F_{k-1}\bar{P}_{k-1}F^T_{k-1} + B_{k-1}Q_{k-1}B^T_{k-1}, \tag{10.83}$$

where the relationship between $\bar{P}_{i-1}$ and $\bar{P}_i$ is established by the following transformation

$$\begin{aligned} \bar{P}_i &= E\left\{(x_i - \bar{x}_i)(x_i - \bar{x}_i)^T\right\} \\ &= E\left\{[F_{i-1}x_{i-1} + E_{i-1}u_{i-1} + B_{i-1}w_{i-1} - \bar{x}^-_i \right. \\ &\quad - \bar{K}_i(y_i - H_i\bar{x}^-_i)][F_{i-1}x_{i-1} + E_{i-1}u_{i-1} + B_{i-1}w_{i-1} \end{aligned}$$

$$
\begin{aligned}
&- \bar{x}_i^- - \bar{K}_i(y_i - H_i\bar{x}_i^-)]^T\} \\
&= \left(I - \bar{K}_iH_i\right)\left(F_{i-1}\bar{P}_{i-1}F_{i-1}^T + B_{i-1}Q_{i-1}B_{i-1}^T\right) \\
&\times \left(I - \bar{K}_iH_i\right)^T + \bar{K}_iR_i\bar{K}_i^T , \qquad (10.84)
\end{aligned}
$$

where $x_{i-1} - \bar{x}_{i-1}$, w_{i-1}, and v_i are pairwise independent. Since $\bar{P}_i$ is computed iteratively and $\bar{x}_i$ is not optimal for $i < k$, then $\bar{P}_k^{\mathrm{U}-}$ given by (10.83) is said to be the *a priori* upper bound of the error covariance P_k^{U}.

The *a priori* weights are defined using the Bayes' rule as

$$
\theta_{k-1}^{\mathrm{K}-} = \pi_{11}\theta_{k-1}^{\mathrm{K}} + \pi_{21}\theta_{k-1}^{\mathrm{U}} , \qquad (10.85)
$$

$$
\theta_{k-1}^{\mathrm{U}-} = \pi_{12}\theta_{k-1}^{\mathrm{K}} + \pi_{22}\theta_{k-1}^{\mathrm{U}} \qquad (10.86)
$$

via the transition probabilities specified by matrix Π.

Updating

The KF updates the estimate and error covariance recursively as

$$
\hat{x}_k^{\mathrm{K}} = \hat{x}_k^{\mathrm{K}-} + K_k^{\mathrm{K}}(y_k - H_k\hat{x}_k^{\mathrm{K}-}) , \qquad (10.87)
$$

$$
P_k^{\mathrm{K}} = (I - K_k^{\mathrm{K}}H_k)P_k^{\mathrm{K}-} , \qquad (10.88)
$$

where $K_k^{\mathrm{K}} = P_k^{\mathrm{K}-}H_k^TS_k^{-1}$ is the Kalman gain and $S_k = H_kP_k^{\mathrm{K}-}H_k^T + R_k$ is the innovation covariance. The likelihood of the KF estimate is

$$
\mathcal{L}_k^{\mathrm{K}} = \frac{1}{\sqrt{2\pi|S_k|}} \exp\left[-\frac{1}{2}(y_k - H_k\hat{x}_k^{\mathrm{K}-})S_k^{-1}(y_k - H_k\hat{x}_k^{\mathrm{K}-})^T\right] , \qquad (10.89)
$$

where $|S_k|$ is the determinant of matrix S_k.

The UFIR filter updates the estimate with

$$
\hat{x}_k^{\mathrm{U}} = \hat{x}_k^{\mathrm{U}-} + K_k^{\mathrm{U}}(y_k - H_k\hat{x}_k^{\mathrm{U}-}) , \qquad (10.90)
$$

where $K_k^{\mathrm{U}} = \mathcal{G}_kF_k^{-T}H_k^T$ is the UFIR filter bias correction gain and the GNPG $\mathcal{G}_k$ is updated by (7.30). By combining (10.83) and (10.84), the error covariance of the UFIR filter can be written as

$$
\bar{P}_k^{\mathrm{U}} = (I - K_k^{\mathrm{U}}H_k)\bar{P}_k^{\mathrm{U}-}(I - K_k^{\mathrm{U}}H_k)^T + \bar{K}_k^{\mathrm{U}}R_kK_k^{\mathrm{U}^T} \qquad (10.91)
$$

and the likelihood can be defined by

$$
\mathcal{L}_k^{\mathrm{U}} = \frac{1}{\sqrt{2\pi|\bar{S}_k|}} \exp\left[-\frac{1}{2}(y_k - H_k\hat{x}_k^{\mathrm{U}-})\bar{S}_k^{-1}(y_k - H_k\hat{x}_k^{\mathrm{U}-})^T\right] , \qquad (10.92)
$$

where $\bar{S}_k = H_k\bar{P}_k^{\mathrm{U}-}H_k^T + R_k$.

Provided $\mathcal{L}_k^{\mathrm{K}}$ and $\mathcal{L}_k^{\mathrm{U}}$, the weights $\theta_{k-1}^{\mathrm{K}-}$ and $\theta_{k-1}^{\mathrm{U}-}$ are updated as

$$
\theta_k^{\mathrm{K}} = \frac{1}{a_k}\Lambda_k^{\mathrm{K}}\theta_k^{\mathrm{K}-} , \qquad (10.93)
$$

$$
\theta_k^{\mathrm{U}} = \frac{1}{a_k}\Lambda_k^{\mathrm{U}}\theta_k^{\mathrm{U}-} , \qquad (10.94)
$$

where $a_k = \mathcal{L}_k^{\mathrm{K}}\theta_k^{\mathrm{K}-} + \mathcal{L}_k^{\mathrm{U}}\theta_k^{\mathrm{U}-}$ is a normalization constant.

FF Filtering Algorithm

Updated θ_k^{K} by (10.93) and θ_k^{U} by (10.94), the FF output estimate can finally be represented involving $\hat{x}_k^{\mathrm{K}}$ and $\hat{x}_k^{\mathrm{U}}$ as

$$\tilde{x}_k = \theta_k^{\mathrm{K}} \hat{x}_k^{\mathrm{K}} + \theta_k^{\mathrm{U}} \hat{x}_k^{\mathrm{U}} , \tag{10.95}$$

the pseudocode of the FF listed as Algorithm 24, and the following critical comments can be made:

- When $k < N_{\mathrm{opt}}$, the KF estimate goes to the FF output. Otherwise, when $k \geq N_{\mathrm{opt}}$, the KF and UFIR estimates are fused using the probabilistic weighting matrix Θ_k given by (10.75).
- The Kalman subfilter requires the initialization, while the UFIR filter does not require initial values.
- When $\theta_k^{\mathrm{K}} = 0$, the FF operates as the UFIR filter and, when $\theta_k^{\mathrm{U}} = 0$, as the KF.

Algorithm 24: Fusion Kalman/UFIR Filtering Algorithm

Data: $y_k, Q_k, R_k, \tilde{x}_0, P_0, \Pi, N$
Result: $\tilde{x}_k$

begin
  for $k = 1, 2, \cdots, N$ **do**
    Run KF to produce $\hat{x}_k$ and P_k;
  end for
  Set $\hat{x}_k^{\mathrm{U}} = \hat{x}_k$, $\bar{P}_k^{\mathrm{U}} = P_k$, and $\theta_k^{\mathrm{K}} = \theta_k^{\mathrm{U}} = 0.5$;
  for $k = N, N+1, \cdots$ **do**
    Compute $\tilde{P}_{k-1}$ by (10.77);
    $\hat{x}_k^{\mathrm{K}-} = F_{k-1}\tilde{x}_{k-1} + E_{k-1}u_{k-1}$;
    $P_k^{\mathrm{K}-} = F_k \tilde{P}_{k-1} F_k^T + B_{k-1} Q_{k-1} B_{k-1}^T$;
    $K_k^{\mathrm{K}} = P_k^{\mathrm{K}-} H_k^T (H_k P_k^{\mathrm{K}-} H_k^T + R_k)^{-1}$;
    $\hat{x}_k^{\mathrm{K}} = \hat{x}_k^{\mathrm{K}-} + K_k^{\mathrm{K}} (y_k - H_k \hat{x}_k^{\mathrm{K}-})$;
    $P_k^{\mathrm{K}} = (I - K_k^{\mathrm{K}} H_k) P_k^{\mathrm{K}-}$;
    Compute $\mathcal{L}_k^{\mathrm{K}}$ by (10.89);
    Compute $\mathcal{G}_k$ by (7.30), $\bar{x}_s^{\mathrm{U}}$ by (7.25), and $\bar{P}_s^{\mathrm{U}}$ by (7.26);
    for $i = s+1, s+2, \cdots, k$ **do**
      $\bar{x}_i^{\mathrm{U}-} = F_i \bar{x}_{i-1}^{\mathrm{U}} + E_{i-1} u_{i-1}$;
      $P_i^{\mathrm{U}} = F_i P_{i-1}^{\mathrm{U}} F_i^T + B_{i-1} Q_{i-1} B_{i-1}^T$;
      $\mathcal{G}_i = \left[H_i^T H_i + (F_i \mathcal{G}_{i-1} F_i^T)^{-1}\right]^{-1}$;
      $K_i^{\mathrm{U}} = \mathcal{G}_i H_i^T$;
      $\bar{x}_i^{\mathrm{U}} = \bar{x}_i^{\mathrm{U}-} + K_i^{\mathrm{U}} (y_i - H_i \bar{x}_i^{\mathrm{U}-})$;
      $P_i^{\mathrm{U}} = (I - K_i^{\mathrm{U}} H_i) P_i^{\mathrm{U}-} (I - K_i^{\mathrm{U}} H_i)^T + K_i^{\mathrm{U}} R_i K_i^{\mathrm{U}^T}$;
    end for
    $\hat{x}_k^{\mathrm{U}} = \bar{x}_k^{\mathrm{U}}$, $\bar{P}_k^{\mathrm{U}} = P_k^{\mathrm{U}}$;
    Compute $\mathcal{L}_k^{\mathrm{U}}$ by (10.92);
    $\tilde{x}_k = \theta_k^{\mathrm{K}} \hat{x}_k^{\mathrm{K}} + \theta_k^{\mathrm{U}} \hat{x}_k^{\mathrm{U}}$;
  end for
end

It follows from the previous that the FF ensures an optimal compromise between the accuracy of the KF and the robustness of the UFIR filter. It can also be shown that the FF can readily be modified to suit nonlinear models by fusing in the same manner the outputs of the extended filter.

10.3.2 Fusing Kalman and Weighted UFIR Estimates

Another FF has been designed in [218] by combining KF and weighted UFIR estimates. The solution refers to the model in (10.73) and (10.74), and it is supposed that KF estimates $\hat{x}^{\mathrm{K}}_{k-1}$ and P^{K}_{k-1} and weighted UFIR estimates $\hat{x}^{\mathrm{wU}}_{k-1}$ and P^{wU}_{k-1} are available at $k-1$. The upper bound $\bar{P}^{\mathrm{wU}}_{k-1}$ of P^{wU}_{k-1} is considered along with the probabilistic weights $\theta^{\mathrm{K}}_{k-1}$ and $\theta^{\mathrm{wU}}_{k-1}$ assigned for the KF and weighted UFIR filter, respectively. This FF obtains estimates in the following phases.

Initialization

Fusion is organized using the Markov transition probability matrix Π given by (10.75), but with another weighting vector Θ_{k-1}

$$\Theta_{k-1} = \begin{bmatrix} \theta^{\mathrm{K}}_{k-1} \\ \theta^{\mathrm{wU}}_{k-1} \end{bmatrix} , \tag{10.96}$$

where $\theta^{\mathrm{K}}_{k-1}$ and $\theta^{\mathrm{wU}}_{k-1}$ are probabilistic weights for the KF and weighted UFIR filter, respectively. The initial state $\tilde{x}_{k-1}$ and error covariance $\tilde{P}_{k-1}$ are obtained at $k-1$ via nearest past data as

$$\tilde{x}_{k-1} = \hat{x}^{\mathrm{K}}_{k-1} , \tag{10.97}$$

$$\tilde{P}_{k-1} = P^{\mathrm{K}}_{k-1} . \tag{10.98}$$

For the weighted UFIR filter, initial $\bar{x}_s$ and $\bar{P}_s$ are specified by

$$\bar{x}_s = \mathcal{F}^{m}_{s-1}(H^T_{m,s}\Omega_s H_{m,s})^{-1}\mathcal{F}^{m^T}_{s-1} , \tag{10.99}$$

$$\bar{P}_s = (\bar{D}_{m,s} - \mathcal{H}^{\mathrm{wU}}_s G_{m,s})\mathcal{Q}_{s,m}(\cdots)^T + \mathcal{H}^{\mathrm{wU}}_s \mathcal{R}_{s,m}\mathcal{H}^{\mathrm{wU}^T}_s , \tag{10.100}$$

where $\Omega_s = \mathrm{diag}(\omega^{-2N}, \omega^{-2N+2}, \dots, \omega^{-2N+2(s-m+1)})$, ω is the predetermined weight, and the weighted UFIR filter gain

$$\mathcal{H}^{\mathrm{wU}}_s = \mathcal{F}^{m-1}_s(H^T_{m,s}\Omega^{-2}_s H_{m,s})^{-1}H^T_{m,s}\Omega^{-2}_s \tag{10.101}$$

brings the UFIR estimate close to the ML estimate (4.119).

Prediction

Provided $\tilde{x}_{k-1}$ by (10.97) and $\tilde{P}_{k-1}$ by (10.98), the *a priori* KF state estimate $\hat{x}^{\mathrm{K}-}_k$ is computed by (10.80) and error covariance $P^{\mathrm{K}-}_k$ by (10.81). The weighted UFIR filter starts iterating with $\bar{x}_s$, and the true output is taken as $\hat{x}^{\mathrm{wU}}_k = \bar{x}_i$ when $i = k$. The *a priori* state estimate $\hat{x}^{\mathrm{wU}-}_k = \bar{x}^-_k$ is computed by the UFIR filter as

$$\hat{x}^{\mathrm{wU}-}_k = F_{k-1}\bar{x}_{k-1} + E_{k-1}u_{k-1} \tag{10.102}$$

and the error covariance $P^{\mathrm{wU}-}_k$ as

$$P^{\mathrm{wU}-}_k = F_{k-1}\bar{P}_{k-1}F^T_{k-1} + B_{k-1}Q_{k-1}B^T_{k-1} , \tag{10.103}$$

where $\bar{P}_{k-1}$ is set as $\bar{P}_s$ and the relationship between $\bar{P}_{i-1}$ and $\bar{P}_i$ is established by (10.84). The *a priori* weights $\theta^{\mathrm{K}-}_{k-1}$ and $\theta^{\mathrm{U}-}_{k-1}$ are defined by (10.85) and (10.86), respectively.

Updating

The KF updates the estimate and error covariance as (10.87) and (10.88) with the likelihood specified by (10.89). The weighted UFIR filter updates estimates with

$$\hat{x}_k^{\mathrm{wU}} = \hat{x}_k^{\mathrm{wU}-} + K_k^{\mathrm{wU}}(y_k - H_k\hat{x}_k^{\mathrm{wU}-}) , \tag{10.104}$$

where the filter gain is given by $K_k^{\mathrm{wU}} = M_k H_k^T$ and

$$M_k = H_k^T \Omega_k H_k + (\bar{F}_k M_{k-1} \bar{F}_k^T)^{-1} , \tag{10.105}$$

$$\Omega_k = \mathrm{diag}(\omega^{-2N} \ \omega^{-2N+2} \ \dots \ \omega^{-2N+2(i-s+M)}) , \tag{10.106}$$

$$\bar{F}_k = \mathcal{F}_{k-1}^m \tilde{P}_k \mathcal{F}_{k-1}^{m^T} , \tag{10.107}$$

$$\tilde{P}_k = H_{k,m+1}^T \Omega_k H_{k,m+1} . \tag{10.108}$$

Referring to (10.91) and (10.92), the error covariance of the weighted UFIR filter becomes

$$\bar{P}_k^{\mathrm{wU}} = (I - K_k^{\mathrm{wU}} H_k)\bar{P}_k^{\mathrm{wU}-}(I - K_k^{\mathrm{wU}} H_k)^T + \bar{K}_k^{\mathrm{wU}} R_k K_k^{\mathrm{wU}^T} \tag{10.109}$$

and the likelihood as given by

$$\mathcal{L}_k^{\mathrm{wU}} = \frac{1}{\sqrt{2\pi|\bar{S}_k|}} \exp\left[-\frac{1}{2}(y_k - H_k\hat{x}_k^{\mathrm{wU}-})\bar{S}_k^{-1}(y_k - H_k\hat{x}_k^{\mathrm{wU}-})^T\right] , \tag{10.110}$$

where $\bar{S}_k = H_k \bar{P}_k^{\mathrm{wU}-} H_k^T + R_k$.

Provided $\mathcal{L}_k^{\mathrm{K}}$ and $\mathcal{L}_k^{\mathrm{wU}}$, weights $\theta_{k-1}^{\mathrm{K}-}$ and $\theta_{k-1}^{\mathrm{wU}-}$ are updated as

$$\theta_k^{\mathrm{K}} = \frac{1}{a_k} \Lambda_k^{\mathrm{K}} \theta_k^{\mathrm{K}-} , \tag{10.111}$$

$$\theta_k^{\mathrm{wU}} = \frac{1}{a_k} \Lambda_k^{\mathrm{wU}} \theta_k^{\mathrm{wU}-} , \tag{10.112}$$

where $a_k = \mathcal{L}_k^{\mathrm{K}} \theta_k^{\mathrm{K}-} + \mathcal{L}_k^{\mathrm{wU}} \theta_k^{\mathrm{wU}-}$ is a normalization constant.

Using the previous modifications in the weighted UFIR estimate, the iterative FF filtering Algorithm 24 can be used straightforwardly. This algorithm provides the highest accuracy when the process is not affected by uncertainties. Otherwise, the weights should be applied to the KF and UFIR filter as in the original Algorithm 24.

10.4 Estimation Under Delayed and Missing Data

Data transmitted over communication channels often arrive at the estimator with delays due to finite propagation time and some other reasons. Moreover, nonconstant delays are accompanied by data loss called dropout or intermittence. In target tracking, latency and missing data are caused by high maneuverability of the target and failures in measurements. In smart sensors, the delay between data acquisition and their availability to the filter often occurs when data require time-consuming pre-processing. Overall, delayed and lost data are recognized as one of the main causes of instability and poor performance of control systems, and most importantly, both these phenomena cannot be addressed independently.

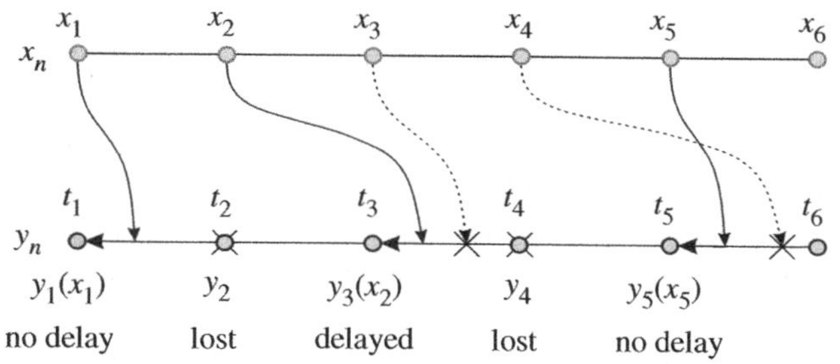

Figure 10.2 Basic scenarios with one-step-lag delayed and missing data: 1) regular, when x_1 observed as $y_1(x_1)$; 2) delayed and missing data, when x_2 is observed as $y_3(x_2)$, y_2 contains no data, and x_3 is lost; and 3) missing data with no delay, when x_5 is observed as $y_5(x_5)$, y_4 contains no data, and x_4 is lost.

Typical scenarios with delayed and missing data are sketched in Fig. 10.2 and can be outlined as follows:

- *Regular case:* No delay in the observation of state x_1 between t_1 and t_2 and x_5 between t_5 and t_6, because x_1 is related to $y_1(x_1)$ and x_5 to $y_5(x_5)$.
- *Delayed and missing data:* x_2 and x_3 are observed in $[t_3 ... t_4]$, and the earlier arrived x_2 is related to $y_3(x_2)$; thus, y_2 contains no data, and x_3 is lost.
- *Missing data with no delay:* x_4 and x_5 are observed in $[t_5 ... t_6]$, and the earlier arrived x_5 is related to $y_5(x_5)$ with no delay; thus, y_4 contains no data, and x_4 is lost.

To address the previous issues and provide imputation, two basic models of delayed and missing data are recognized:

1) *Deterministic:* Sensors are able to detect the delays and lost data, or this information is available via time-stamping.
2) *Random:* Data are transmitted with considerable, irregular, and *a priori* unknown delays; in such cases, the delay is regarded as a random process, and the best estimate is obtained by combining delayed and nondelayed data with different probabilities.

Accordingly, two types of state estimators can be developed, based either on deterministic (time-stamped) delay information or on the probability of a random delay.

10.4.1 Deterministic Delays and Missing Data

Let us consider some quantity Q measured by a smart sensor. The time-stamped measurement is transmitted over WSN channels to a central station with a known (deterministic) time-lag delay of $n_k \geq 0$ points. At the receiver, a data sensor indicates lost data with $\kappa_k = 0$ and, if data arrive safely, generates $\kappa_k = 1$. Accordingly, the state space model becomes [202]

$$x_k = Fx_{k-1} + w_k \ , \tag{10.113}$$

$$\tilde{y}_k = HFx_{k-1} \ , \tag{10.114}$$

$$y_k = \kappa_k H x_{k-n_k} + (1 - \kappa_k)\tilde{y}_k + v_k \ , \tag{10.115}$$

where, regardless of the delay, the initial state x_{k-1} is supposed to be known and can be substituted with estimate $\hat{x}_{k-1}$. When the data sensor generates $\kappa_k = 0$, the predicted data (10.114) are used and, if $\kappa_k = 1$, (10.115) reduces to the standard observation equation. The noise vectors are supposed to be white Gaussian, $w_k \sim \mathcal{N}(0, Q)$ and $v_k \sim \mathcal{N}(0, R)$, with known covariances and the property $E\{w_k v_r^T\} = 0$ for all k and r. Only one state (delayed or not) is observed at k; that is, a simultaneous observation of delayed and non-delayed states is not allowed. To design a state estimator for such a model, the observation equation (10.115) can be transformed to another one without the delay [202] as will be shown next.

Using the backward-in-time solutions, the delayed state x_{k-n_k} can be represented via x_k as

$$x_{k-n_k} = F^{-n_k}\left(x_k - \sum_{i=0}^{n_k-1} F^i w_{k-i}\right) \tag{10.116}$$

that allows transforming the observation equation for $\kappa_k = 1$ to

$$y_k = \bar{H}_k x_k + \bar{v}_k\,, \tag{10.117}$$

where $\bar{H}_k = HF^{-n_k}$ is a new n_k-varying observation matrix, vector

$$\bar{v}_k = v_k - H\sum_{i=0}^{n_k-1} F^{-n_k+i} w_{k-i} \tag{10.118}$$

is white Gaussian $\bar{v}_k \sim \mathcal{N}(0, \bar{R}_k)$ with the n_k-varying covariance

$$\bar{R}_k = E\{\bar{v}_k \bar{v}_k^T\} = R + \tilde{R}_k\,, \tag{10.119}$$

where $\tilde{R}_k = \bar{H}_k \sum_{i=0}^{n_k-1} F^i Q F^{i^T} \bar{H}_n^T$ is a n_k-varying complement.

In compact forms, (10.118) and (10.119) can now be rewritten as

$$\bar{v}_k = v_k - H\bar{B}_k W_{p_k,k}\,, \tag{10.120}$$

$$\bar{R}_k = R + H\bar{B}_k \bar{Q}_k \bar{B}_k^T H^T\,, \tag{10.121}$$

where $p_k = k - n_k + 1$, matrix $\bar{B}_k$ and noise vector $W_{p_k,k}$,

$$\bar{B}_k = \left[\, F^{-1}\; F^{-2}\; \dots\; F^{-n_k} \,\right],$$

$$W_{p_k,k} = \left[\, w_{p_k}^T\; w_{p_k+1}^T\; \dots\; w_k^T \,\right]^T,$$

are chosen such that $\bar{B}_k = 0$ and $W_{p_k,k} = 0$ when $n_k = 0$, and the covariance $\bar{Q}_k = \mathrm{diag}(\,Q\;\; Q\;\dots\; Q\,)$ has n_k diagonal components. Provided a new n_k-invariant state space model (10.113) and (10.117), the standard linear state estimators can readily be modified to deal with delayed and missing data.

UFIR Filtering Under Delayed and Missing Data

The UFIR filter can be developed for observations with time-stamped delayed data and a time-lag $n_k \geq 0$ if we start with the extended state-space model (8.3) and (8.4), write it for LTI systems with $u_k = 0$ as

$$X_{m,k} = F_N x_m + D_N W_{m,k}\,, \tag{10.122}$$

$$Y_{m,k}(\eta) = H_{m,k}(\eta) x_m + G_{m,k}(\eta) W_{m,k} + V_{m,k}(\eta)\,, \tag{10.123}$$

introduce a set $\eta \in \{n_m, n_{m+1}, \dots, n_k\}$ of delays and $p_k = k - n_k + 1$, and represent extended vectors and matrices as

$$F_N = \left[I\ F^T\ \dots\ F^{N-1^T}\right]^T , \tag{10.124}$$

$$D_N = \begin{bmatrix} I & 0 & \dots & 0 & 0 \\ F & I & \dots & 0 & 0 \\ \vdots & \vdots & \ddots & \vdots & \vdots \\ F^{N-2} & F^{N-3} & \dots & I & 0 \\ F^{N-1} & F^{N-2} & \dots & F & I \end{bmatrix} , \tag{10.125}$$

$$V_{m,k}(\eta) = \begin{bmatrix} v_m - H\bar{B}_k W_{p_m,m} \\ v_{m+1} - H\bar{B}_k W_{p_{m+1},m+1} \\ \vdots \\ v_{k-1} - H\bar{B}_k W_{p_{k-1},k-1} \\ v_k - H\bar{B}_k W_{p_k,k} \end{bmatrix} , \tag{10.126}$$

$$H_{m,k}(\eta) = \bar{H}_{m,k}(\eta) F_N , \tag{10.127}$$

$$G_{m,k}(\eta) = \bar{H}_{m,k}(\eta) D_N , \tag{10.128}$$

$$\bar{H}_{m,k}(\eta) = \mathrm{diag}(\underbrace{HF^{-n_m}\ \ HF^{-n_{m+1}}\ \dots\ HF^{-n_k}}_{N}) . \tag{10.129}$$

The unbiasedness condition $\mathcal{E}\{\hat{x}_k(\eta)\} = \mathcal{E}\{x_k\}$ applied to the batch FIR estimate $\hat{x}_k(\eta) = \mathcal{H}_N(\eta) Y_{m,k}$ and model

$$x_k = F^{N-1} x_m + \bar{D}_N W_{m,k} ,$$

where $\bar{D}_N$ is the last row vector in (10.125), results in the unbiasedness constraint

$$I = \mathcal{H}_{m,k}(\eta) C_{m,k}(\eta) , \tag{10.130}$$

where the n_k-varying block matrix $C_{m,k}(\eta)$ is defined by

$$C_{m,k}(\eta) = \begin{bmatrix} HF^{-N+1-n_m} \\ \vdots \\ HF^{-1-n_{k-1}} \\ HF^{-n_k} \end{bmatrix} . \tag{10.131}$$

A solution to (10.130) for the UFIR filter gain gives

$$\mathcal{H}_{m,k}(\eta) = [C^T_{m,k}(\eta) C_{m,k}(\eta)]^{-1} C^T_{m,k}(\eta) \tag{10.132}$$

and the batch UFIR estimate for time-stamped delayed and missing data becomes

$$\hat{x}_k = \mathcal{G}_k(\eta) C^T_{m,k}(\eta) Y_{m,k} , \tag{10.133}$$

where $Y_{m,k}$ is a vector of real data and $\mathcal{G}_k(\eta) = [C^T_{m,k}(\eta) C_{m,k}(\eta)]^{-1}$ is the n_k-varying GNPG.

Batch Error Covariance

For the estimation error defined as $\varepsilon_k(\eta) = x_k - \hat{x}_k(\eta)$, the batch error covariance $P_k(\eta) = \mathcal{E}\{\varepsilon_k(\eta)\varepsilon_k^T(\eta)\}$ appears in the following form

$$\begin{aligned}P_k(\eta) &= \mathcal{E}\{[F^{N-1}x_m + \bar{D}_N W_{m,k} - \mathcal{H}_{m,k}(\eta)Y_{m,k}(\eta)][\dots]^T\}\\ &= \mathcal{E}\{[(F^{N-1} - \mathcal{H}_{m,k}(\eta)H_{m,k}(\eta))x_m + (\bar{D}_N\\ &\quad - \mathcal{H}_{m,k}(\eta)G_{m,k}(\eta))W_{m,k} - \mathcal{H}_{m,k}(\eta)V_{m,k}(\eta)][\dots]^T\}\,. \end{aligned} \tag{10.134}$$

Referring to the constraint (10.130) and providing the averaging for uncorrelated $W_{m,k}$ and $V_{m,k}(\eta)$, (10.134) can further be represented with

$$\begin{aligned}P_k(\eta) &= [\bar{D}_N - \mathcal{H}_{m,k}(\eta)G_{m,k}(\eta)]\mathcal{Q}_N[\dots]^T\\ &\quad + \mathcal{H}_{m,k}(\eta)\mathcal{R}_N\mathcal{H}_{m,k}^T(\eta)\,,\end{aligned} \tag{10.135}$$

where $\mathcal{Q}_N = \text{diag}(Q\ Q\ \dots\ Q)$ and $\mathcal{R}_N = \text{diag}(R\ R\ \dots\ R)$ are square matrices with N diagonal elements.

Iterative UFIR Filtering Algorithm for $n_k \geq 0$

Like the standard batch UFIR filtering estimate, the n_k-varying batch estimate (10.133) can also be represented with a fast iterative algorithm using the following recursions,

$$\mathcal{G}_l(\eta) = [\bar{H}_l^T(\eta)\bar{H}_l(\eta) + (F\mathcal{G}_{l-1}(\eta)F^T)^{-1}]^{-1}\,, \tag{10.136}$$

$$\hat{x}_l = F\hat{x}_{l-1} + \mathcal{G}_l(\eta)\bar{H}_l^T(\eta)[y_l - \bar{H}_l(\eta)F\hat{x}_{l-1}]\,, \tag{10.137}$$

where $\bar{H}_l(\eta) = HF^{-n_l}$, l ranges from $s+1$ to k, $s = m + K - 1$, $m = k - N + 1$, $\mathcal{G}_s(\eta) = [C_{m,s}^T(\eta)C_{m,s}(\eta)]^{-1}$, $\hat{x}_s = \mathcal{G}_s(\eta)C_{m,s}^T(\eta)Y_{m,s}$, and the output is taken when $l = k$.

The pseudocode the UFIR filtering algorithm for $n_k \geq 0$ is listed as Algorithm 25, where for simplicity we remove the dependence of matrices on η. Although the UFIR filter is BIBO stable and not prone to divergence, latency in information delivery requires an ability to predict delayed and lost states that may cause extra errors.

Iterative Computation of Error Covariance

Fast computation of the batch error covariance (10.135) can be organized similarly to the standard UFIR error covariance by substitute x_k with (10.113) and $\hat{x}_k(\eta)$ with (10.137). By rearranging the terms and providing the averaging, a recursive form for (10.135) appears as

$$\begin{aligned}P_l(\eta) &= [I - \mathcal{G}_l(\eta)\bar{H}_l^T(\eta)\bar{H}_l(\eta)]P_l^-(\eta)[\dots]^T + \mathcal{G}_l(\eta)\bar{H}_l^T(\eta)\bar{H}_l(\eta)\\ &\quad \times \Sigma_l(\eta)\bar{H}_l^T(\eta)\bar{H}_l(\eta)\mathcal{G}_l(\eta) + \mathcal{G}_l(\eta)\bar{H}_l^T(\eta)R\bar{H}_l(\eta)\mathcal{G}_l(\eta)\,,\end{aligned} \tag{10.138}$$

where $\Sigma_l(\eta) = \sum_{i=0}^{n_l-1} F^iQF^{i^T}$, $P_l^-(\eta) = FP_{l-1}(\eta)F^T + Q$ is the prior error covariance, and l is an iterative variable. It follows from (10.138) that an increase in n_k results in a larger $\Sigma_l(\eta)$ and P_l grows accordingly. Note that there is no such an accumulative effect for R.

10.4.2 Randomly Delayed and Missing Data

The term *random delay* suggests that there may exist different models for given delay distributions and time-lags n_k. To be specific, we will consider a case when data are randomly delayed on no more than one point and that the delays are subject to the Bernoulli distribution. We will suppose

Algorithm 25: UFIR Algorithm for Time-Stamped Delayed and Missing Data

Data: y_k, n_k, N, κ_k
Result: $\hat{x}_k$
begin
  for $k = N-1 : \infty$ **do**
    $m = k - N + 1, \quad s = m + K - 1;$
    if $\kappa_n = 0$ **then**
      $y_k = HF\hat{x}_{k-1};$
    end if
    $\bar{H}_k = HF^{-n_k};$
    $\mathcal{G}_s = (C_{m,s}^T C_{m,s})^{-1};$
    $\tilde{x}_s = \mathcal{G}_s C_{m,s}^T Y_{m,s};$
    for $l = s+1 : n$ **do**
      $\mathcal{G}_l = [\bar{H}_l^T \bar{H}_l + (F\mathcal{G}_{l-1}F^T)^{-1}]^{-1};$
      $K_l = \mathcal{G}_l \bar{H}_l^T;$
      $\tilde{x}_l = F\tilde{x}_{l-1} + K_l(y_l - \bar{H}_l F\tilde{x}_{l-1});$
    end for
    $\hat{x}_k = \tilde{x}_k;$
  end for
end
† Data $y_0, y_1, \ldots, y_{N-1}$ must be available.

that dynamics of a quantity Q and its observation in a smart sensor are represented with the linear state-space model

$$x_k = Fx_{k-1} + w_k \,, \tag{10.139}$$

$$y_k = Hx_k + v_k \,, \tag{10.140}$$

where $w_k \sim \mathcal{N}(0, Q)$, $v_k \sim \mathcal{N}(0, R)$, and $\mathcal{E}\{w_k v_r^T\} = 0$ for all k and r. Assume that when transmitting over a distance, there occur binary random delays and packet dropouts. The delays are detected with some probability and lost data are predicted by (10.114) that results in two data vectors [234][203],

$$\tilde{z}_k = HF\hat{x}_{k-1} \,, \tag{10.141}$$

$$\begin{aligned} z_k = {} & \gamma_{0,k} y_k + (1-\gamma_{0,k})\,[(1-\gamma_{0,k-1})\,\gamma_{1,k} y_{k-1} \\ & + (1-(1-\gamma_{0,k-1})\gamma_{1,k})\,\tilde{z}_k] \,, \end{aligned} \tag{10.142}$$

where z_k is the received measurement, $\tilde{z}_k$ is the predicted measurement, $\hat{x}_{k-1}$ is available, $\gamma_{0,k}$ and $\gamma_{1,k}$ are binary Bernoulli distributed random scalar variables with known probabilities,

$$P\{\gamma_{0,k} = 1\} = \bar{\gamma}_{0,k} \,, \quad P\{\gamma_{0,k} = 0\} = 1 - \bar{\gamma}_{0,k} \,,$$
$$P\{\gamma_{1,k} = 1\} = \bar{\gamma}_{1,k} \,, \quad P\{\gamma_{1,k} = 0\} = 1 - \bar{\gamma}_{1,k} \,,$$

and the delay probabilities range as $0 \leqslant \{\bar{\gamma}_{0,n}, \bar{\gamma}_{1,n}\} \leqslant 1$.

It follows that data received when $\gamma_{0,k} = 1$ are equal to y_k with the probability $\bar{\gamma}_{0,k}$. When $\gamma_{0,k} = 0$, there are two possible options: latency and no data. If $\gamma_{0,k-1} = 0$ and $\gamma_{1,k} = 1$, then y_{k-1} is received with the probability $(1-\bar{\gamma}_{0,k})(1-\bar{\gamma}_{0,k-1})\bar{\gamma}_{1,k}$ and, when $\gamma_{0,k-1}=0$ and $\gamma_{1,k}=0$, with the probability $1-\bar{\gamma}_{0,k}-(1-\bar{\gamma}_{0,k})(1-\bar{\gamma}_{0,k-1})\bar{\gamma}_{1,k}$.

State-Space Model Transformation

The design of state estimators for randomly delayed and missing data can be achieved by converting the observation model (10.142) to another model without delay. To do this, we introduce the following auxiliary coefficients,

$$\alpha_{0,k} = \gamma_{0,k}\,, \tag{10.143}$$

$$\alpha_{1,k} = (1-\gamma_{0,k})(1-\gamma_{0,k-1})\gamma_{1,k}\,, \tag{10.144}$$

$$\alpha_{2,k} = (1-\gamma_{0,k})[1-(1-\gamma_{0,k-1})\gamma_{1,k}]\,. \tag{10.145}$$

Taking notice that $\alpha_{0,k}+\alpha_{1,k}+\alpha_{2,k}=1$,

$$\bar{\alpha}_{0,k} = \mathcal{E}\{\alpha_{0,k}\} = \mathcal{E}\{\alpha_{0,k}^2\}\,, \tag{10.146}$$

$$\bar{\alpha}_{1,k} = \mathcal{E}\{\alpha_{1,k}\} = \mathcal{E}\{\alpha_{1,k}^2\}\,, \tag{10.147}$$

$$\bar{\alpha}_{2,k} = \mathcal{E}\{\alpha_{2,k}\} = \mathcal{E}\{\alpha_{2,k}^2\}\,, \tag{10.148}$$

and that $\mathcal{E}\{\alpha_{i,k}\alpha_{j,k}\}=0$ holds for all $\{i,j\}\in[0,1,2]$, the observation equation (10.142) can be transformed to

$$z_k = \alpha_{0,k}y_k + \alpha_{1,k}y_{k-1} + \alpha_{2,k}\tilde{z}_k\,. \tag{10.149}$$

For one-step delay, the backward-in-time converted state (10.116) gives $x_{k-1}=F^{-1}(x_k-w_k)$. Next, by substituting into (10.149) y_k given by (10.140) and the previously defined x_{k-1}, the observation equation can be modified as

$$\begin{aligned}
z_k &= \alpha_{0,k}(Hx_k+v_k)+\alpha_{1,k}(Hx_{k-1}+v_{k-1})\\
&\quad +\alpha_{2,k}(HFx_{k-1})\\
&= \alpha_{0,k}(Hx_k+v_k)+\alpha_{1,k}[HF^{-1}(x_k-w_k)\\
&\quad +v_{k-1}]+\alpha_{2,k}H(x_k-w_k)\\
&= (\alpha_{0,k}H+\alpha_{1,k}HF^{-1}+\alpha_{2,k}H)x_k+\alpha_{0,k}v_k\\
&\quad +\alpha_{1,k}v_{k-1}-(\alpha_{1,k}HF^{-1}+\alpha_{2,k}H)w_k\\
&= \bar{H}_kx_k+\bar{v}_k\,,
\end{aligned} \tag{10.150}$$

where the modified observation matrix $\bar{H}_k$ and white Gaussian noise vector $\bar{v}_k$ are given by

$$\bar{H}_k = (\alpha_{0,k}+\alpha_{2,k})H+\alpha_{1,k}HF^{-1}\,, \tag{10.151}$$

$$\bar{v}_k = \alpha_{0,k}v_k+\alpha_{1,k}v_{k-1}-(\alpha_{1,k}HF^{-1}+\alpha_{2,k}H)w_k \tag{10.152}$$

and we notice that $\bar{v}_k$ is time-correlated with w_k.

The covariance $\bar{R}_k=\mathcal{E}\{\bar{v}_k\bar{v}_k^T\}$ can now be written as

$$\begin{aligned}
\bar{R}_k &= \bar{\alpha}_{0,k}R_k+\bar{\alpha}_{1,k}R_{k-1}+\bar{\alpha}_{1,k}HF^{-1}QF^{-T}H^T\\
&\quad +\bar{\alpha}_{2,k}HQH^T
\end{aligned} \tag{10.153}$$

and the cross covariance between $\bar{v}_k$ and w_k as

$$\mathcal{E}\{\bar{v}_kw_k^T\} = -(\bar{\alpha}_{1,k}HF^{-1}+\bar{\alpha}_{2,k}H)Q.$$

It follows from the UFIR filter design [183] that time-correlation of zero mean white noise sources should be ignored. However, this does not apply to optimal and robust state estimators, where time-correlation must be taken into account and addressed in the design.

Batch UFIR Filter

The batch UFIR filter for systems with Bernoulli-distributed randomly delayed and missing data can be designed starting with the extended model,

$$X_{m,k} = F_N x_m + D_N W_{m,k} , \tag{10.154}$$

$$Y_{m,k} = H_{m,k} x_m + G_{m,k} W_{m,k} + V_{m,k} , \tag{10.155}$$

where matrix F_N is given by (10.124), D_N by (10.125),

$$H_{m,k} = \begin{bmatrix} \bar{H}_m \\ \bar{H}_{m+1}F \\ \vdots \\ \bar{H}_k F^{k-1} \end{bmatrix} , \tag{10.156}$$

$G_{m,k} = \bar{H}_{m,k} D_N$, $\bar{H}_{m,k} = \text{diag}(\bar{H}_m \; \bar{H}_{m+1} \dots \bar{H}_k)$, and matrix $\bar{H}_k$ with random components is defined by (10.151).

We now use the standard definition of the batch UFIR filtering estimate

$$\hat{x}_k = \mathcal{G}_k C_{m,k}^T Y_{m,k} = \mathcal{H}_{m,k} Y_{m,k} , \tag{10.157}$$

where $\mathcal{G}_k = (C_{m,k}^T C_{m,k})^{-1}$ is the GNPG, $Y_{m,k}$ is a vector of real data, and matrix $C_{m,k}$ is defined as

$$C_{m,k} = \begin{bmatrix} \bar{H}_m F^{-N+1} \\ \vdots \\ \bar{H}_{k-1} F^{-1} \\ \bar{H}_k \end{bmatrix} , \tag{10.158}$$

and write the error covariance as

$$\begin{aligned} P_k = {} & (\bar{D}_k - \mathcal{H}_{m,k} Y_{m,k}) \mathcal{Q} (\bar{D}_k - \mathcal{H}_{m,k} Y_{m,k})^T \\ & + \mathcal{H}_{m,k} \mathcal{R}_k \mathcal{H}_{m,k}^T , \end{aligned} \tag{10.159}$$

where $\mathcal{Q}_k = \text{diag}(Q \; Q \cdots Q)$ and $\mathcal{R}_k = \text{diag}(\bar{R}_k \; \bar{R}_k \cdots \bar{R}_k)$ are block diagonal matrices with N diagonal matrix elements and the covariance $\bar{R}_k$ has random components specified by (10.153).

Iterative UFIR Filtering Algorithm

Iterative computation of (10.157) is not very specific compared to the standard UFIR filtering algorithm, except for the predictions option. When the data sensor generates $\kappa_k = 0$, data loss is predicted using (10.141). For given coefficients $\alpha_{0,k}$, $\alpha_{1,k}$, and $\alpha_{2,k}$, matrix $\bar{H}_n$ is computed by (10.141). Then the GNPG is computed in the batch form of (10.157) and the initial estimate $\tilde{x}_s$ by (10.157). Provided the initial values, the UFIR estimate is updated in two phases: the *a priori* state estimate is computed by $\hat{x}_l^- = F\hat{x}_{l-1}$ and the *a posteriori* estimate using

$$\mathcal{G}_l = [\bar{H}_l^T \bar{H}_l + (F\mathcal{G}_{l-1}F^T)^{-1}]^{-1} , \tag{10.160}$$

$$\hat{x}_l = \hat{x}_l^- + \mathcal{G}_l \bar{H}_l^T (y_l - \bar{H}_l \hat{x}_l^-) , \tag{10.161}$$

where an iterative variable l ranges from $s+1$ to k to produce the output when $l = k$. The pseudocode of the UFIR filtering algorithm for binary randomly delayed and missing data is listed as Algorithm 26.

Algorithm 26: Iterative UFIR Filtering Algorithm for Randomly Delayed and Missing Data

Data: y_k, $\alpha_{0,k}$, $\alpha_{1,k}$, $\alpha_{2,k}$, N, κ_k
Result: $\hat{x}_k$

begin
    for $k = N-1 : \infty$ **do**
        $m = k - N + 1, \quad s = n - N + K$;
        if $\kappa_k = 0$ **then**
            $y_k = HF\hat{x}_{k-1}$
        end if
        $\bar{H}_k = (\alpha_{0,k} + \alpha_{2,k})H + \alpha_{1,k}HF^{-1}$;
        Compute $C_{m,s}$ by (10.158);
        $\mathcal{G}_s = (C_{m,s}^T C_{m,s})^{-1}$;
        $\tilde{x}_s = \mathcal{G}_s C_{m,s}^T Y_{m,s}$;
        for $l = s+1 : k$ **do**
            $\mathcal{G}_l = [\bar{H}_l^T \bar{H}_l + (F\mathcal{G}_{l-1}F^T)^{-1}]^{-1}$;
            $K_l = \mathcal{G}_l \bar{H}_l^T$;
            $\tilde{x}_l = F\tilde{x}_{l-1} + K_l(y_l - \bar{H}_l F\tilde{x}_{l-1})$;
        end for
        $\hat{x}_k = \tilde{x}_k$;
    end for
end
† Data $y_0, y_1, \ldots$ and $Y_{m,s}$

Recursive Error Covariance

Recursive computation of the error covariance (10.159) can be provided if we substitute x_k and $\hat{x}_k$ with their recursive forms. This gives

$$P_k = (I - \mathcal{G}_k \bar{H}_k^T \bar{H}_k)(FP_{k-1}F^T - B_k Q B_k^T) \times (I - \mathcal{G}_k \bar{H}_k^T \bar{H}_k)^T + \mathcal{G}_k \bar{H}_k^T R_k \bar{H}_k \mathcal{G}_k^T \tag{10.162}$$

and we notice again that, even though the UFIR approach ignores error covariance, P_k may be required to evaluate performance. It is also worth mentioning that in a like manner the approach can be extended to arbitrary delay-lags with arbitrary distributions.

10.5 Summary

What has been presented in this chapter elucidates capabilities of FIR estimators beyond the standard state-space model. Further developments may go in any direction, since the choice of future tasks and problems is rather a complex and largely undefined goal depending on technological progress in general. Nevertheless, these and other solutions make it possible to assess the potential of the FIR approach and extend it to other tasks in signal processing and control. As a matter of fact, this was the main purpose of this chapter. Some specific generalizations follow.

The state estimation problem is best solved by transforming whenever possible an advanced state-space model to the FE- or BE-based standard state space model and applying standard FIR or IIR (Kalman-like) algorithms. The use of UFIR filters allows solving the state estimation problem over networks with consensus in measurements and estimates in the best way, since these filters do not require initial conditions and information about zero mean noise or norm-bounded errors. However, the UFIR approach is not suitable for estimation with consensus in information, since it ignores noise statistics.

Combining optimal and robust estimators in hybrid structures gives the best compromise between accuracy and robustness, and optimal fusion can be organized in different ways. Namely, all estimates or only some of them can be weighted in solving the optimal fusion task. Design of FIR state estimators for systems with randomly or deterministically delayed and missing data implies transforming the observation equation with latency to an observation equation, which does not contain the delayed state, and including a prediction option to substitute lost or incorrectly delivered data.

We finish with formulating some other problems, which are yet unsolved or supposedly can be solved using methods of FIR state estimation. This covers several traditional formulations dealing with well-known or poorly defined models and goes to some neighboring fields. Extensive development of methods of artificial intelligence dealing with limited knowledge about undergoing processes and models also gives food for designers of embedded estimators.

10.6 Problems

1 **Recursive forms**. Given diagonal block covariance matrices $\mathcal{Q}_{m,k}$ and $\mathcal{R}_{m,k}$ of white Gaussian noise, the batch OFIR filtering estimate is computed iteratively using KF recursions. Represent recursively the full (nondiagonal) block covariance matrices $\mathcal{Q}_{m,k}$ and $\mathcal{R}_{m,k}$ under colored noise, find recursive form for the OFIR and H_2 FIR filters, and design iterative filtering algorithms.

2 **Consensus in WSN**. A controlled WSN is represented with the following state-space model,

$$x_{k+1} = F_k x_k + E_k u_k + B_k w_k \, , \tag{10.163}$$

$$y_k = H_k x_k + v_k \, , \tag{10.164}$$

where $y_k = [y_k^{(1)^T} \; \dots \; y_k^{(n)^T}]^T$ is the observation due to n sensors and $H_k = [H_k^{(1)^T} \; \dots \; H_k^{(n)^T}]^T$. The zero mean noise vectors w_k and $v_k = [\, v_k^{(1)^T} \; \dots \; v_k^{(n)^T}]^T$ have nondiagonal covariances Q_k and R_k, respectively. Derive iterative UFIR algorithms with possible consensus.

3 Consider the WSN described in item 3 and derive the OFIR and ML FIR filters with consensus in measurements, estimates, and information. Derive these filters with simultaneous consensus in 1) measurements and estimates and 2) measurements and information.

4 In the model described in item 3, delayed and missing data arrive as

$$y_k = \gamma_{0k} H x_k + \gamma_{1k} H x_{k-1} + \dots + \gamma_{nk} H x_{k-n} + v_k \, , \tag{10.165}$$

where the probabilistic weights obey $\sum_{i=1}^{n} \gamma_{ik} = 1$. Design FIR state estimators with different kinds of consensus.

5 Consider the previous problem represented with the model in (10.163) and (10.165), in which γ_{ik}, $i \in [0,2]$, has the property $\gamma_{0k} + \gamma_{1k} + \gamma_{2k} = 1$. Derive the KF and UFIR filter supposing that γ_{ik} on $i \in [0,2]$ has one of the following distributions: 1) uniform, 2) Bernoulli, 3) Gaussian, 4) arbitrary.

6 **Disturbed systems.** Consider a time-invariant state-space model representing data delivered over the WSN at a central station,

$$x_{k+1} = Fx_k + Eu_k + Bw_k \ ,$$
$$y_k = Hx_k + Dw_k + v_k \ ,$$

where all vector and matrices are specified after (10.163) and (10.164). The disturbance w_k and measurement error v_k are norm-bounded with $\|w\| < \infty$ and $\|v\| < \infty$. Using the approaches developed in Chapter 8, derive the following FIR estimators: 1) H_2, 2) H_∞, 3) energy-to-peak, 4) peak-to-peak, and 5) game theory H_∞.

7 **Uncertain systems.** Consider a linear system represented with uncertain state and observation equations as

$$x_{k+1} = F_k^u x_k + E_k^u u_k + B_k^u w_k \ ,$$
$$y_k = H_k^u x_k + D_k^u w_k + v_k \ ,$$

where the uncertain matrices are defined as $F_k^u = F + \Delta F_k$, $E_k^u = E + \Delta E_k$, $B_k^u = B + \Delta B_k$, $H_k^u = H + \Delta H_k$, and $D_k^u = D + \Delta D_k$ and details can be found after (9.2). Solve various FIR state estimation problems for 1) time-correlated increments ΔF_k, ΔE_k, ΔB_k, ΔH_k, and ΔD_k and 2) for time-correlated and colored increments.

8 Given a mean square stable uncertain system with multiplicative noise components [56]

$$x_{k+1} = (F + F_1\alpha_k)x_k + (B + B_1\beta_k)w_k \qquad x_0 = 0 \ ,$$
$$y_k = (H + H_1\gamma_k)x_k + Dw_k + v_k \ ,$$
$$z_k = Lx_{k-1} \ ,$$

where w_k is the bounded disturbance, $v_k \sim \mathcal{N}(0, R)$ is measurement noise, z_k is the combination of one-lag delayed state, F, B, H, D, F_1, B_1, and H_1 are constant matrices, and α_k, β_k, and γ_k are random scalar variables with zero mean and the properties: $\mathcal{E}\{\alpha_k\alpha_l\} = \delta_{kl}$, $\mathcal{E}\{\beta_k\beta_l\} = \delta_{kl}$, $\mathcal{E}\{\gamma_k\gamma_l\} = \delta_{kl}$, $\mathcal{E}\{\alpha_k\beta_l\} = 0$, $\mathcal{E}\{\alpha_k\gamma_l\} = 0$, and $\mathcal{E}\{\beta_k\gamma_l\} = 0$ for all k and l, solve different kinds of norm-bounded FIR state estimation problems.

9 **Fusion of estimates.** Given the model in (10.163) and (10.164), where $x_k \in \mathbb{R}^K$, $u_k \in \mathbb{R}^L$, $y_k \in \mathbb{R}^P$, $w_k \sim \mathcal{N}(0, Q_k) \in \mathbb{R}^M$, $v_k \sim \mathcal{N}(0, R_k) \in \mathbb{R}^P$, $F_k \in \mathbb{R}^{K\times K}$, $H_k \in \mathbb{R}^{P\times K}$, $E_k \in \mathbb{R}^{K\times L}$, and $B_k \in \mathbb{R}^{K\times M}$, consider the KF and UFIR estimates and fuse them as (10.95),

$$\bar{x}_k = \theta_k^{\mathrm{K}} x_k^{\mathrm{K}} + \theta_k^{\mathrm{U}} x_k^{\mathrm{U}} \ .$$

Modify the pseudocode of the fusion Algorithm 23 for 1) $\theta_k^{\mathrm{K}} = I$ and $\theta_{k\mathrm{opt}}^{\mathrm{U}}$ and 2) $\theta_k^{\mathrm{U}} = I$ and $\theta_{k\mathrm{opt}}^{\mathrm{K}}$. Compare errors in the modified estimates with respect to the basic estimate (10.95).

10 The fused KF and UFIR filter estimate obtained in two different ways is given by (10.95), where the UFIR estimate is either weighted with θ_k^{wU} by (10.96) or not with θ_k^{U} by (10.75). Show advantages and drawbacks of both approaches. Why is the fused estimate (10.95) with unweighted UFIR estimate more robust?

11 **Fractional-order systems.** A *fractional-order system* is referred to as a dynamical system that is modeled by a fractional differential equation. Fractional-order state models are used to study anomalous behaviors of dynamical regular and chaotic systems. In continuous-time state-space, a fractional-order system can be represented with equations

$$
\begin{aligned}
\mathcal{D}^{\alpha}x(t) &= Ax(t) + Eu(t)\,,\\
y(t) &= Cx(t) + Du(t)\,,
\end{aligned}
$$

where $\mathcal{D}^{\alpha} = \mathrm{diag}(\,D^{\alpha_1}\ \dots\ D^{\alpha_n})$, $D^{\alpha_i} \triangleq \frac{\mathrm{d}^{\alpha_i}}{\mathrm{d}t^{\alpha_i}}$, $i \in [1, K]$, and $\alpha_1, \dots, \alpha_K$ are generally noninteger. The FIR state estimation theory is still not developed for fractional-order systems.

12 A linear system is disturbed by a scalar flicker noise $w(t)$ that has the PSD slope 1/f in the Fourier frequency f domain. The flicker noise is represented with a fractional-order differential equation $\frac{\mathrm{d}^{0.5}}{\mathrm{d}t^{0.5}}w(t) = aw(t) + \xi(t)$, where $\xi(t) \sim \mathcal{N}(0, \sigma_\xi^2)$, that does not fit with the system polynomial state-space model. Approximate this noise with a series of Gauss-Markov noise components and solve the optimal FIR state estimation problem.

13 **Pairwise Markov chains.** Given the pairwise Markov chains (PMC) model [148] represented in state space with

$$
\begin{bmatrix} x_k \\ y_k \end{bmatrix} = \begin{bmatrix} F_k & A_k \\ H_k & C_k \end{bmatrix} \begin{bmatrix} x_{k-1} \\ y_{k-1} \end{bmatrix} + \xi_k\,, \tag{10.166}
$$

where $\xi_k = \begin{bmatrix} w_k \\ v_k \end{bmatrix} \sim \mathcal{N}(0, \Sigma_k)$ has the covariance $\Sigma_k = \begin{bmatrix} Q_k & L_k \\ L_k^T & R_k \end{bmatrix}$, is this model better suited to *image processing* or *signal processing*? Find an explanation by converting (10.166) to the system difference equation.

14 **Artificial intelligence.** Under a poor knowledge of undergoing processes, artificial intelligence solves problems of reasoning, planning, learning, and object manipulating. Approaches include statistical methods and computational intelligence using tools such as machine learning, mathematical optimization, and artificial neural networks. Problems of artificial intelligence and state estimation merge under uncertainties, and room for investigation remains widely open.

15 At the initialization phase of a machine learning technique called *federated learning*, a model such as linear regression is often chosen to be trained on local nodes and initialized. To improve the model robustness, use the UFIR state estimation algorithm.

16 In *decision tree learning*, a machine learning algorithm called *bootstrap aggregated decision trees* is used to improve the stability and accuracy employing statistical classification and regression. To avoid overfitting and reduce the variance, use methods of robust FIR state estimation.

17 In *supervised learning* that solves the *machine learning* optimization task a function g is searched such that the following function is minimized,

$$J(g) = R_{\text{emp}}(g) + \lambda C(g) ,$$

where $R_{\text{emp}}(g)$ is the empirical risk of g, $C(g)$ is the penalty, and λ is a parameter to control the bias-variance trade-off. Viewing this problem as an estimation problem, find the gain for an FIR filter.

11

Applications of FIR State Estimators

Design is not how it looks like and feels like. Design is how it works.

Steve Jobs (Apple Inc. cofounder and former CEO)

This chapter is the last, and it is now a proper place to give examples of practical applications of FIR state estimators. As the approach suggests, many useful FIR engineering algorithms can be developed to solve filtering, smoothing, and prediction problems on data finite horizons under diverse operation conditions in different environments. For Gaussian processes, the batch OFIR state estimator has proven to be the most accurate in the MSE sense, and the supporting recursive KF algorithm and its various modifications have found a huge number of applications. The OUFIR state estimator and RH MVF are commonly used in the canonical ML FIR batch form, since the recursive forms for these batches are more complex than Kalman recursions. The UFIR state estimator is blind on optimal horizons (has no other tuning factors) and is thus robust, unlike the OFIR and OUFIR estimators. Therefore, the iterative UFIR filtering algorithm using recursions has found practical applications as an alternative to KF in uncertain environments. It is worth noting that there has been no further development of the original LMF idea, because LMF is nothing more than KF operating on data finite horizons. In recent decades, there has appeared a big class of norm-bounded H_2, H_∞, $\mathcal{L}_2$-to-$\mathcal{L}_\infty$, $\mathcal{L}_1$, and hybrid FIR state estimators. Such estimators are called robust, but their development for practical use has been carried out to a much lesser extent and awaits further development.

In this chapter, we provide examples of practical applications of FIR state estimators in various fields and different environments using batch and iterative algorithms.

11.1 UFIR Filtering and Prediction of Clock States

The IEEE Standard 1139-2008 [75] and International Telecommunication Union (ITU-T) recommendation G.810 [77] suggest that a clock has three states, namely, the time interval error (TIE), fractional frequency offset, and linear frequency drift rate. The problem with accurate estimation of clock state has a lot to do with the clock oscillator noise, which has a slow flicker (technical) Gaussian component having the PSD slope 1/f in the Fourier frequency f domain. Even modified for CPN [182], which has the PSD slope $1/f^2$, the KF is unable to track accurately the 1/f behaviors. In GPS-based timekeeping using one pulse per second (1PPS) signals, the problem is complicated with the uniformly distributed (non-Gaussian) measurement noise and GPS time uncertainties caused by different satellites in a view. Therefore, the UFIR filter, which ignores zero mean noise,

Optimal and Robust State Estimation: Finite Impulse Response (FIR) and Kalman Approaches, First Edition.
Yuriy S. Shmaliy and Shunyi Zhao.

better fulfils the requirement of the IEEE Standard 1139-2008 [75] that "an efficient and unbiased estimator is preferred" for clock applications [166].

11.1.1 Clock Model

In accordance with [75] and [77], the clock TIE $\alpha(t)$ caused by oscillator instabilities is modeled in continuous time t with the finite Taylor series as

$$\alpha(t) = \alpha_0 + \beta_0 t + \frac{\gamma_0}{2} t^2 + w_\alpha(t) , \tag{11.1}$$

where $\alpha_0 = \alpha(0)$ is the initial TIE (first state) at $t = 0$, $\beta_0 = \beta(0)$ is the initial fractional frequency offset (second state), and $\gamma_0 = \gamma(0)$ is the initial linear frequency drift rate (third state). Noise $w_\alpha(t) = \varphi(t)/2\pi\nu_{\text{nom}}$ is completely defined by the clock oscillator random phase deviation component $\varphi(t)$ and nominal frequency ν_{nom} in Hz. The IEEE standard [75] states that $w_\alpha(t)$ is affected by different kinds of independent phase and frequency fluctuations (white, flicker, and random walks), which PSD slopes can be measured in the Fourier frequency domain.

In state space, (11.1) is represented with the SDE

$$\frac{\mathrm{d}}{\mathrm{d}t} x(t) = Ax(t) + w(t) , \tag{11.2}$$

in which $x(t) = [\alpha(t)\ \beta(t)\ \gamma(t)]^T$ is the clock state vector and $w(t) = [w_\beta(t)\ w_\gamma(t)\ w_{\dot\gamma}(t)]^T$ is the zero mean Gaussian noise vector having the covariance $Q_w(t, \theta) = \mathcal{E}\{w(t)w^T(\theta)\}$. In this vector, $w_\beta(t)$ is the frequency noise, $w_\gamma(t)$ is the linear frequency drift noise, and $w_{\dot\gamma}(t)$ is noise in the first time derivative of the linear frequency drift. Note that $w(t)$ is nonstationary on a long time scale, although it is typically assumed to be stationary on a short time. The clock state transition matrix A is specified by

$$A = \begin{bmatrix} 0 & 1 & 0 \\ 0 & 0 & 1 \\ 0 & 0 & 0 \end{bmatrix} . \tag{11.3}$$

In discrete time index k, model (11.2) is represented with

$$x_k = Fx_{k-1} + \bar{w}_k , \tag{11.4}$$

where the clock matrix $F \triangleq F(\tau)$ is defined by the matrix exponential $F = e^{A\tau}$ to be

$$F = \begin{bmatrix} 1 & \tau & \frac{\tau^2}{2} \\ 0 & 1 & \tau \\ 0 & 0 & 1 \end{bmatrix} . \tag{11.5}$$

The zero mean Gaussian noise vector $\bar{w}_k$ and its covariance $Q_{\bar{w}}(i,j) = \mathcal{E}\{\bar{w}_i \bar{w}_j^T\}$ are defined by, respectively,

$$\bar{w}_k = \int_{t_{k-1}}^{t_k} F(\theta)w(\theta)\, \mathrm{d}\theta , \tag{11.6}$$

$$Q_{\bar{w}}(i,j) = \int_{t_{i-1}}^{t_i} \int_{t_{j-1}}^{t_j} F(\tau)Q_w(\tau, \theta)F^T(\theta)\, \mathrm{d}\theta\, d\tau \tag{11.7}$$

and we notice that (11.6) and (11.7) can be applied to all kinds of phase and frequency noise sources.

In timekeeping, only the first clock state (TIE) is commonly measured. Therefore, the observation equation becomes

$$y_k = Hx_k + v_k \ , \tag{11.8}$$

where $H = [\ 1 \quad 0 \quad 0\ ]$, and v_k is zero mean noise, which is not Gaussian in the 1PPS output of a GPS timing receiver.

11.1.2 Clock State Estimation Over GPS-Based TIE Data

To compare errors produced by the KF and UFIR filter, the TIE measurements of a clock embedded in the Frequency Counter SR620 were provided using another SR620 as shown in [179]. The 1PPS output of the GPS SynPaQ III Timing Sensor was used as a reference signal. The ground truth was obtained for the Cesium Frequency Standard CsIII time signals.

UFIR Filtering Estimate

The iterative *a posteriori* UFIR filtering Algorithm 11 can now be applied to estimate the clock state. A specific feature is that due to the extremely slowly changing TIE of a precise clock and small noise, the optimal averaging horizon appears to be very large, $N_{\text{opt}} \cong 3500$ [168].

Kalman Filtering Estimate

To apply the *a posteriori* KF algorithm, it needs specifying correctly noise covariances and initial clock state. Noise in the clock oscillator embedded into SR620 is characterized with three values of the Allan deviation: $\sigma_\beta(1s) = 2.3 \times 10^{-11}$, $\sigma_\beta(10s) = 1.0 \times 10^{-11}$, and $\sigma_\beta(100s) = 4.2 \times 10^{-11}$. Following [32], these values can be converted to the diffusion parameters q_1, q_2, and q_3 via

$$\sigma_\beta^2(\tau) = \frac{q_1}{\tau} + \frac{q_2\tau}{3} + \frac{q_3\tau^3}{20} \tag{11.9}$$

and then the noise covariance $Q_{\bar{w}}(\tau)$ specified in white Gaussian approximation as [192]

$$\frac{Q_{\bar{w}}(\tau)}{\tau} = \begin{bmatrix} q_1 + \frac{q_2\tau^2}{3} + \frac{q_3\tau^4}{20} & \frac{q_2\tau}{2} + \frac{q_3\tau^3}{8} & \frac{q_3\tau^2}{6} \\ \frac{q_2\tau}{2} + \frac{q_3\tau^3}{8} & q_2 + \frac{q_3\tau^2}{3} & \frac{q_3\tau}{2} \\ \frac{q_3\tau^2}{6} & \frac{q_3\tau}{2} & q_3 \end{bmatrix} . \tag{11.10}$$

Note that since $\sigma_y^2(\tau)$ is upper-bounded in the clock oscillator specification, it can be reduced by the factor of 2 for the KF to perform better. Finally, the variance of the sawtooth noise induced by GPS SynPaQ III Timing Sensor is measured as $Q_v = 50^2/3\ \text{ns}^2$, and the unknown initial states can be set as $\alpha(0) = \alpha_0$, $\beta(0) = 0$, and $\gamma(0) = 0$.

Typical estimates of the clock TIE α_k and fractional frequency offset β_k are sketched in Fig. 11.1 along with the GPS-based measurements of the TIE and ground truth provided by the Cesium Frequency Standard CsIII. The results reveal that efforts made to describe the clock noise covariance $Q_{\bar{w}}(\tau)$ via the Allan deviation $\sigma_\beta(\tau)$ for the KF were less successful than to measure experimentally N_{opt} for the UFIR filter. Consequently, the KF produces the worst estimates even when the UFIR filter is tuned in a wide range of N, from 1500 to 3500. It also reveals that errors in the KF can be reduced by decreasing the Allan variance $\sigma_\beta^2(\tau)$. But this takes $\sigma_\beta^2(\tau)$ values outside the scope of physical imagination and has no theoretical justification. It can also be seen that a near optimal horizon $N_{\text{opt}} = 3500$ makes the UFIR filter much more accurate than the KF. All that follows from this experiment is that the UFIR filter is more suitable for estimation of clock state than the KF.

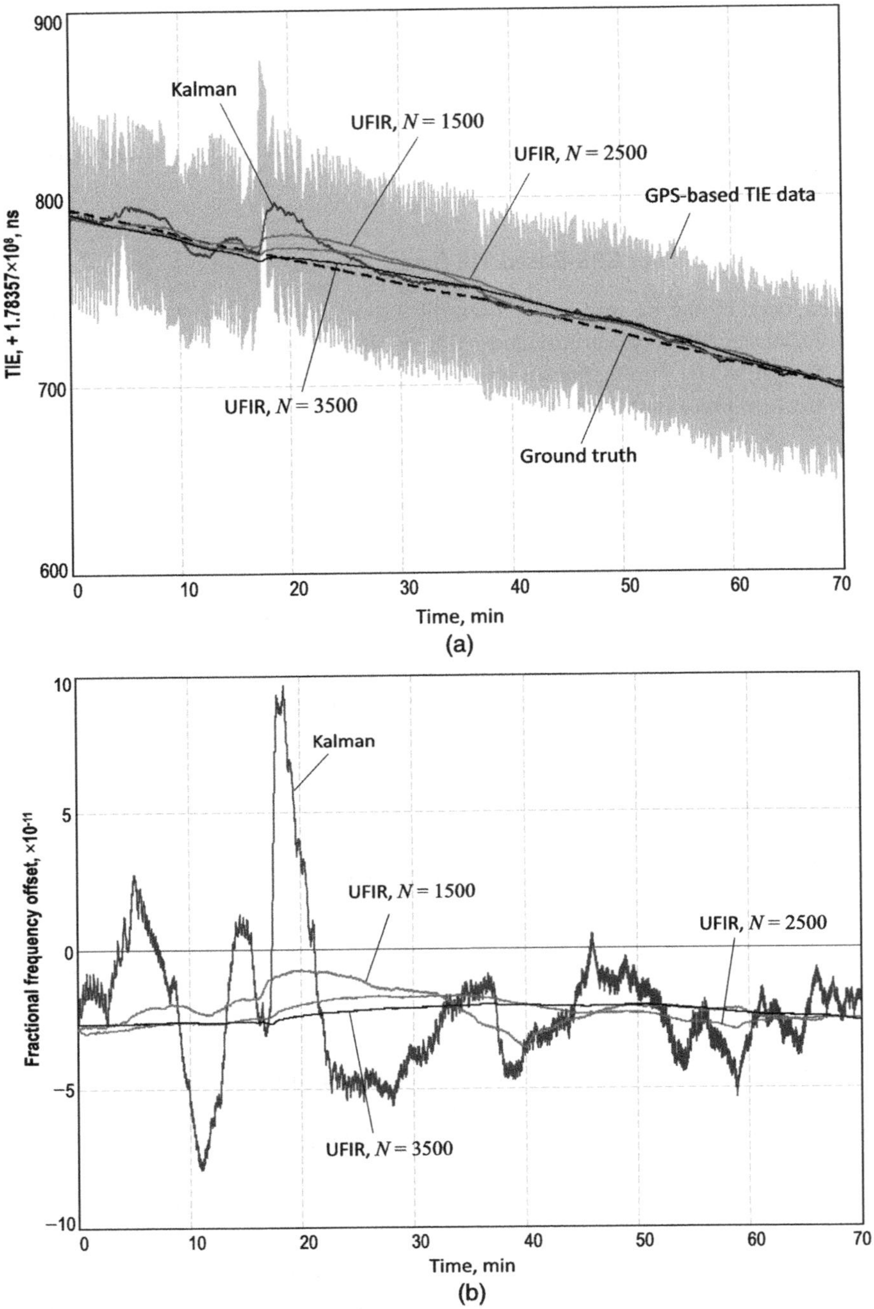

Figure 11.1 Typical estimates of the clock TIE produced by the UFIR filter and KF for the GPS-based TIE measurement and ground truth obtained by the Cesium Frequency Standard CsIII: (a) TIE and (b) fractional frequency offset.

It should be noted that the most accurate estimator for precise clocks is the batch OFIR filter, for which the experimentally measured full block covariance matrix Q_N should contain the necessary information about the clock colored noise. This matrix can be large, 3500×3500 in the previous case, and it may take time to provide the estimation. But in most cases this can be tolerated due to the slow error processes in precise clocks.

11.1.3 Master Clock Error Prediction

Error prediction in national master clocks (MCs) is required, because the International Bureau of Weights and Measures (BIPM) determines time deviations with a five-day interval as average values per day and issues the results monthly. As an example, we consider the UTC–UTC(NIST MC) time differences (285 points) measured each 10 days in 2002–2009 for the National Institute of Standards and Technology (NIST). The time scale is formed starting at $k = 0$ [52279 Modified Julian Date (MJD)] and finishing at $n = 284$ (55129 MJD) as shown in Fig. 11.2a. The MCs are commonly modeled with two states, $K = 2$. Because the NIST MC is error-corrected, we suppose that the process is stationary and employ all data available. Accordingly, we let $N = k + 1$ and use the full horizon p-step UFIR predictor to bridge a month data gap. All along we also use the KF tuned in the best way.

A highly useful property of the full horizon p-step UFIR predictor is that it needs only a step p for interpolation or extrapolation. Because the NIST MC exhibits excellent etalon properties, the initial conditions can be set to zero, $\alpha_0 = 0$ and $\beta_0 = 0$. For the resolution of 0.1 ns in the published data, the uniformly distributed digitization noise has the variance $\sigma_v^2 = 0.05^2/3\ \text{ns}^2 = 8.33 \times 10^{-4}\ \text{ns}^2$.

Figure 11.2 sketches estimates of α_k and β_k produced by the UFIR filter and KF. Current estimates of the TIE α_k are shown in Fig. 11.2a along with p-step predicted values. Predictions are depicted with arrows coming out from the points indicated with digits. Since the UFIR estimator is unbiased, the first arrow coincides in the direction with several initial measurement points. With increased k, the prediction vectors show possible behaviors extrapolated at $k + p$, $p > 0$, over measurements from 0 to k. An estimator also reveals a small positive angle resulting in the frequency offset shown in Fig. 11.2b.

Analyzing Fig. 11.2, one may conclude that the KF is less suitable for MCs than the UFIR filter. Indeed, due to transients and a small number of data points, the KF does not sketch a real error picture at the initial stage. In the intermediate region (about the point of 1×10^3 days), both filters produce consistent estimates. On a long time scale, the full horizon UFIR filter improves estimates, while the KF performs equally.

11.2 Suboptimal Clock Synchronization

The need to synchronize local time scales [76] arises with different allowed uncertainties in digital communications, bistatic radars, telephone networks, networked measurement and control systems, space systems, and computer nets. To discipline clocks, commercially available GPS timing receivers are often used to convey the reference time to the locked clock loop via the 1PPS output. The loop is organized for the clock TIE to range over time below an allowed threshold that, for digital communication networks, is specified in [78]. The GPS-based clock steering is typically organized in two ways:

- The frequency offset is adjusted in the clock oscillator, and the TIE is adjusted in the clock digital block.
- In the clock digital block, only the TIE is adjusted if the oscillator is uncontrolled.

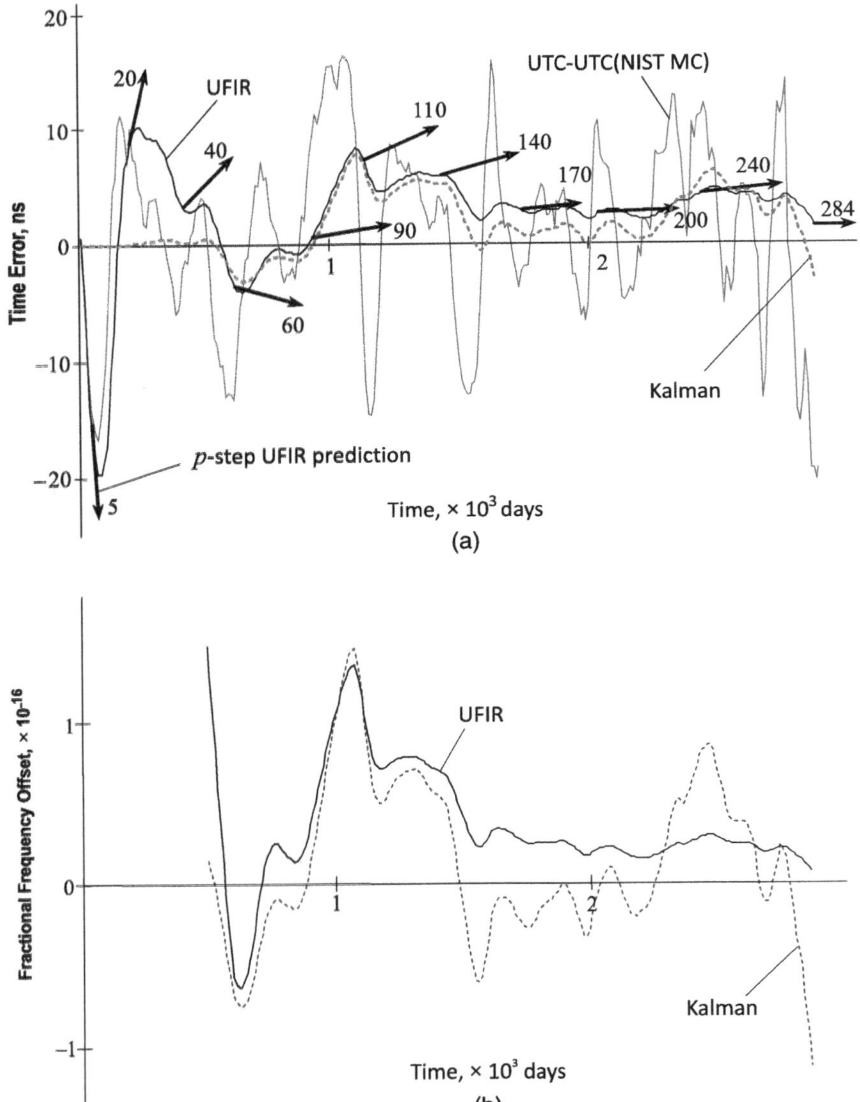

Figure 11.2 Estimates of the NIST MC current state via the UTC−UTC(NIST MC) time differences (285 points) measured in 2002–2009 each 10 days: (a) TIE prediction and (b) fractional frequency offset. Digits indicate the time points from which the arrows come out.

Many designs use the 1PPS output of a GPS timing receiver as a time reference. The time difference between the 1PPS signal, which is accurate but not precise, and the 1 s output of the local clock, which is precise but not accurate, is periodically measured using a high-resolution TIE counter. The clock TIE is then estimated by a filter to obtain a synchronizing signal intended to discipline the local clock time scale as shown in [11, 170].

In locked clocks, the TIE noise is mainly dependent on the precision of a local oscillator, and the TIE departures are due to the limited accuracy of the reference time signals. The required

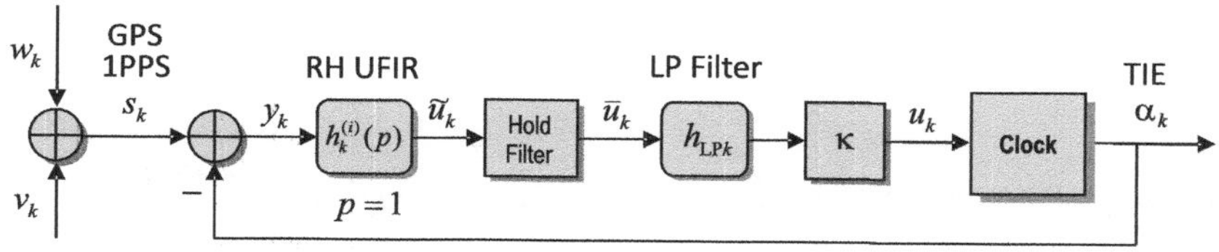

Figure 11.3 Loop model of local clock synchronization based on GPS 1PPS timing signals.

synchronization accuracy can be achieved using various filters. Averaging and LP filters are typical for commercially manufactured locked clocks. In some designs, an integrating filter, linear LS estimator, KF, or even neural network is included in a disciplining phase-locked loop (PLL) [11].

11.2.1 Clock Digital Synchronization Loop

The clock synchronization loop proposed in [11] is shown in Fig. 11.3. The GPS 1PPS timing signal s_k is represented with the GPS time uncertainty (or disturbance) w_k caused by different satellites in a view and some other random factors. It is also complicated by the uniformly distributed sawtooth noise v_k, whose values range from −50 ns to 50 ns owing to the principle of the 1PPS signal formation. Typical examples of the GPS 1PPS signal errors are given in Fig. 11.4 [11]. Random w_k (Fig. 11.4a) ranges closer to zero than the sawtooth noise v_k, which is uniformly distributed within [−50 … 50] in ns (Fig. 11.4b). Errors in the GPS 1PPS timing signal shown in Fig. 11.4c are caused by a mixture of w_k and v_k. In some GPS timing receivers, a negative sawtooth correction code is available in the protocol. If this code is applied, the 1PPS signal error becomes approximately $s_n = v_n$.

A typical function of a nonstationary TIE α_n of an unlocked crystal clock is shown in Fig. 11.5 [11]. The synchronization loop requires N neighboring past data points, from $k-N$ to $k-1$, and operates as follows. The TIE α_k is measured relative to s_k by the TIE counter, and the difference signal $y_k = s_k - \alpha_k$ is computed. To predict the disciplining signal u_k at the current point k, the RH UFIR filter is used, which has an impulse response of the ith degree $h_k^{(i)}(1)$. The predicted value $\tilde{u}_k$ is held as $\bar{u}_k$ by the hold filter, smoothed by the LP filter with the impulse response $h_{\mathrm{LP}k}$, and scaled with κ to obtain the necessary control signal $\hat{u}_k$, which disciplines the clock.

The ramp RH UFIR filter of the first degree, $i = 1$ and $p = 1$, is used with the impulse response function originally derived in [71],

$$h_k^{(1)}(1) = \begin{cases} \frac{2(2N+1)-6k}{N(N-1)} & , \quad 1 \leqslant k \leqslant N \\ 0 & , \quad \text{otherwise} \end{cases} . \tag{11.11}$$

The 1-step predictive filtering estimate is obtained as

$$\tilde{u}_k = \sum_{n=1}^{N} h_n^{(1)}(1) y_{k-n} \tag{11.12}$$

by the discrete convolution using data y_k taken from $[k-N, k-1]$.

To hold $\tilde{u}_k$ over time to continuously steer the clock between suboptimally defined values, a hold filter is used such that

$$\bar{u}_k = \tilde{u}_{\lfloor \frac{k}{M} \rfloor M} , \tag{11.13}$$

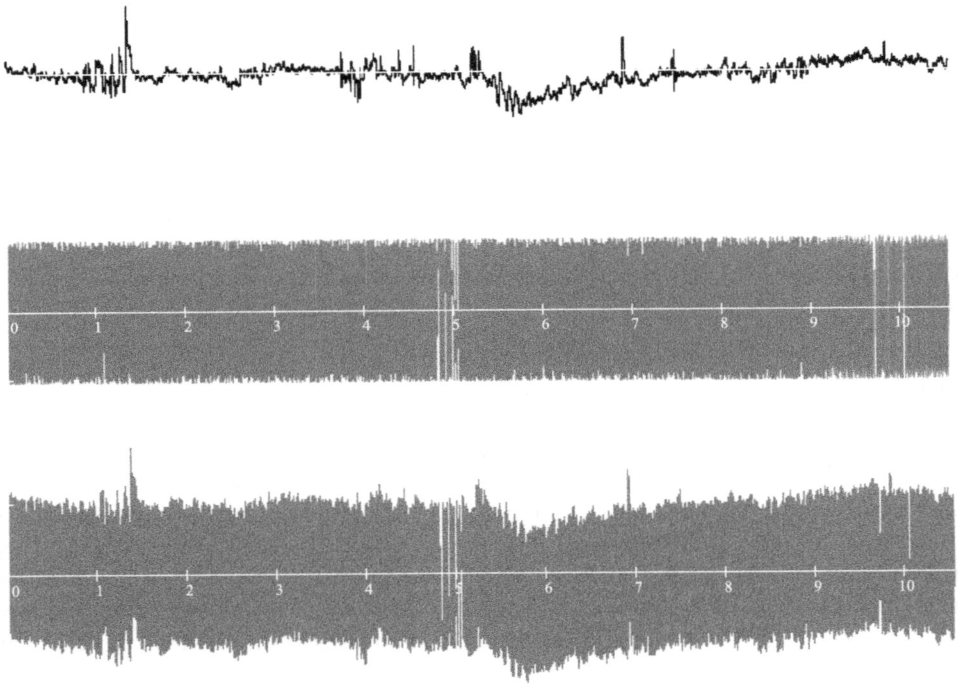

Figure 11.4 Typical errors in GPS timing receivers: (a) GPS time uncertainty w_k caused by different satellites in a view and other random factors, (b) sawtooth noise v_k induced by the receiver, and (c) time error s_k in the GPS 1PPS reference signal.

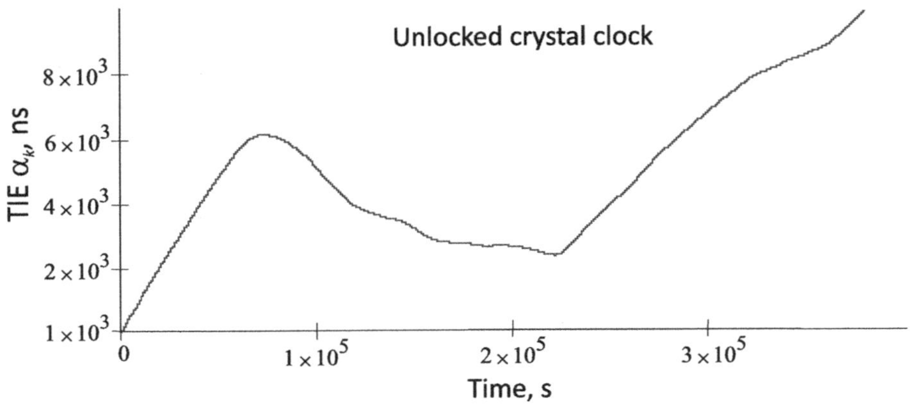

Figure 11.5 A typical function of a nonstationary TIE α_k of an unlocked crystal clock.

where $\left\lfloor \frac{k}{M} \right\rfloor$ is an integer part of k/M. By (11.13), the input and output values of the hold filter become equal when k is multiple to M. Between two such adjacent points, the output value of the hold filter is constant.

For multiple clock steering with period τM, where τ is the sampling time and M is the number of sampling intervals, the hold filter produces a step signal $\bar{u}_k$. Applied directly to the clock, $\bar{u}_k$ guarantees suboptimal steering and assures that α_k ranges within narrow bounds around zero. On the other hand, the stepwise $\bar{u}_k$ may not guarantee the required clock performance.

To smooth $\bar{u}_k$, an LP filter with the impulse response $h_{\text{LP}k}$ is used. It has been shown in [11] experimentally that the 1-order LP filter with the impulse response

$$h_{\text{LP}k} = \begin{cases} ae^{-\frac{\tau}{T}k} & , \quad k \geq 0 \\ 0 & , \quad k < 0 \end{cases}, \tag{11.14}$$

where T is the filter time constant and $a = 1 - e^{-\tau/T}$, is able to fit the demands for precise clock synchronization. Finally, the gain κ is applied to produce the control signal u_k and provide steering of clock errors.

Experimental Verification

Experimental investigations of the synchronization loop shown in Fig. 11.3 were provided in [11]. To outline the capabilities of FIR state estimators in practical applications, next we present only the most typical characteristics of a steered clock with an oven-controlled crystal oscillator (OCXO). The clock errors are presented in terms of the Allan deviation and precision time protocol (PTP) variance for $\tau = 1$ s, $M = N$, and $\kappa = 1$. The optimal horizon N_{opt} and optimal time constant T_{opt} are measured experimentally by minimizing the MSE relative to the TIE of an OCXO-based clock with and without the sawtooth. For GPS 1PPS signal with sawtooth, the optimal horizon was measured as $N_{\text{opt}} = 250$ and without the sawtooth as $N_{\text{opt}} = 150$. More details about this loop can be found in [11].

Allan Deviation

The Allan deviation of a locked OCXO-based clock was investigated in [11] as a function of the LP filter time constant T. It follows from Fig. 11.6 that the Allan deviation of the 1PPS signal with sawtooth ranges higher than that of the OCXO-based clock when the averaging time $\bar{\tau}$ is less than about 5000 sec. The suboptimal UFIR filter removes the sawtooth, improves the performance, and makes it such that the Allan deviation remains large only when $\bar{\tau} < 2300$ s. Further improvement is achieved using an LP filter, which dramatically reduces the Allan deviation to the level of an OCXO-based clock. Figure 11.6 suggests that $T \cong 2000$ s can be accepted as near optimal. Indeed, with small $\bar{\tau}$ the Allan deviation value is due to the unlocked clock and with large $\bar{\tau}$ due to the UFIR filter (or sawtooth-less measurements). The LP filter causes an excursion, who peak value reaches a minimum when $T_{\text{opt}} \cong 2000$ s, it grows and shifts to the right when $T > T_{\text{opt}}$, and it grows and shifts to the left if $T < T_{\text{opt}}$.

Precision Time Protocol Variance

Another measure of clock errors is the PTV variance. It is worth noting as a matter of notation that by $M = N$ the square root of PTP variance $\sigma^2_{\text{PTP}}(\bar{\tau})$ is equal to the time deviation (TDEV) specified in [78]. The PTP variance relates to the Allan variance $\sigma^2_{\beta}(\bar{\tau})$ as $\sigma^2_{\text{PTP}}(\bar{\tau}) = \bar{\tau}^2\sigma^2_y/3$, and [76] states that this is the main measure of locked clock errors.

The improvement in PTP variance for a GPS 1PPS signal measured without a sawtooth is illustrated in Fig. 11.7, where the recommended boundary (dashed) is taken from [78]. As can be seen,

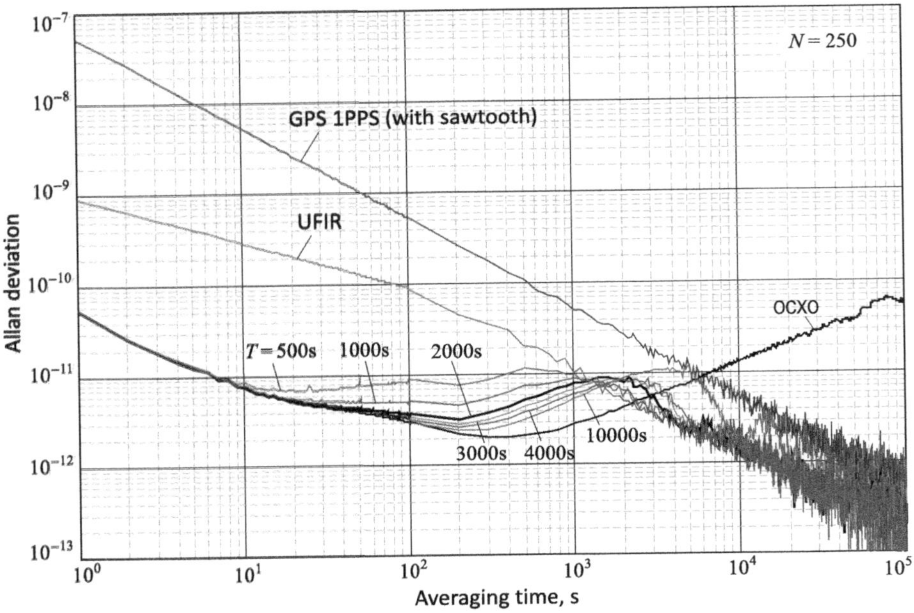

Figure 11.6 Allan deviation of the GPS-locked OCXO-based clock for different T , s of the 1-order LP filter. GPS 1PPS signal has sawtooth and $N_{opt} = 250$ [11].

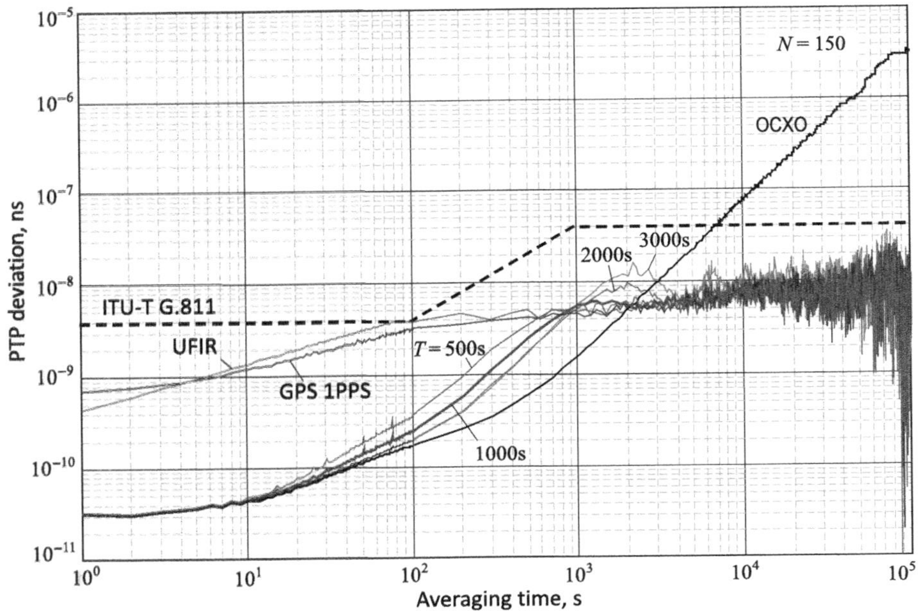

Figure 11.7 PTP deviation of a GPS locked crystal clock for different T, s of the 1-order LP filter and $N_{opt} = 150$ over GPS 1PPS signal without the sawtooth.

the PTP deviation of an OCXO-based clock exceeds the recommended boundary when $\bar{\tau} > 7000$ s. The PTP deviation of the GPS 1PPS signal without the sawtooth ranges very close to the UFIR filter output, and it follows that the UFIR filter removes the sawtooth very efficiently. Both these measures range below the recommended boundary and approach it when $\bar{\tau} \cong 100$ s. The LP filter significantly improves the performance and reduces the clock PTP deviation by a factor of 10 or more over the recommended boundary. It also follows that a near optimal time constant can be accepted as $T_{\mathrm{opt}} \cong 1000$ s.

In general, the fact remains: the iterative UFIR filtering Algorithm 11 is an efficient tool for GPS-based clock steering, since the KF is not suitable for filtering flicker noise components.

11.3 Localization Over WSNs Using Particle/UFIR Filter

A well-known disadvantage of the PF-based localization of maneuvering objects is the need to generate a large number of particles in order to avoid divergence. The process requires resampling, which assumes that particles with higher weights (i.e., high likelihoods) will be statistically selected many times. This leads to a loss of diversity among the particles so that the resultant set will contain many repeated particles. The effect is known as sample impoverishment and usually occurs when the noise is not intensive or the number of particles is small.

To make the PF-based localization more reliable by avoiding the divergence, the PF was combined in [143] with the UFIR filter. In the proposed hybrid PF/UFIR structure, the PF plays the role of the main filter, and the UFIR filter is used as the supporting filter. The PF estimates the state in normal situations, when there is sample impoverishment and no failures. Otherwise, when the PF fails, the UFIR filter restarts it.

This hybrid structure was used in [143] to provide indoor mobile robot localization using a wireless tag with a transmitter, four receivers, and a server computer as shown in Fig. 11.8. A wireless

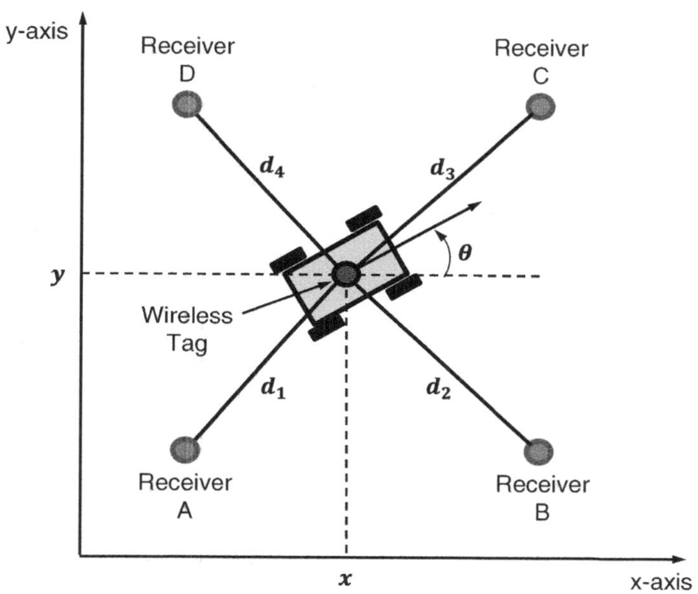

Figure 11.8 2-D schematic geometry of the mobile robot localization.

tag attached to the mobile robot transmits a signal to four receivers deployed at exactly known coordinates. The clocks of the receivers are synchronized using the synchronization line. Distances between the tag and the receivers are measured via time-of-arrival (TOA). The TOA data are transferred to the server computer to generate time-difference-of-arrival (TDOA) measurements, which are represented by the following equation,

$$\begin{bmatrix} z_{1,k} \\ z_{2,k} \\ z_{3,k} \end{bmatrix} = \begin{bmatrix} h_{1,k} \\ h_{2,k} \\ h_{3,k} \end{bmatrix} = \frac{1}{c} \begin{bmatrix} d_1 - d_2 \\ d_1 - d_3 \\ d_1 - d_4 \end{bmatrix} , \tag{11.15}$$

where $z_{1,k}$, $z_{2,k}$, and $z_{3,k}$ are the TDOA data (in units of nanoseconds) and c is the speed of light. Here, d_i, $i \in [1,4]$, is the ith distances between the mobile robot and the receivers. The distances are coupled with the robot local coordinates x_k and y_k and the ith receiver constant coordinates x_i and y_i by four nonlinear equations for $i \in [1,4]$,

$$d_i = \sqrt{(\mathrm{x}_k - \mathrm{x}_i)^2 + (\mathrm{y}_k - \mathrm{y}_i)^2} \,. \tag{11.16}$$

At the current point k, the mobile robot pose is given by the state vector $x_k = [\mathrm{x}_k \;\; \mathrm{y}_k \;\; \theta_k]^T$, where θ_k is a heading angle in the 2D plane local coordinates. Motion of the mobile robot is adjusted by the control signal $u_k = [\Delta d_k \; \Delta\theta_k]^T$, where Δd_k is the incremental distance (in meters) and $\Delta\theta_k$ is the incremental change in the heading angle (in degrees). The following difference equations represent the dynamics of the robot's movement,

$$\mathrm{x}_k = f_{1,k} = \mathrm{x}_{k-1} + \Delta d \cos\left(\theta_{k-1} + \frac{1}{2}\Delta\theta_k\right) , \tag{11.17}$$

$$\mathrm{y}_k = f_{2,k} = \mathrm{y}_{k-1} + \Delta d \sin\left(\theta_{k-1} + \frac{1}{2}\Delta\theta_k\right) , \tag{11.18}$$

$$\theta_k = f_{3,k} = \theta_{k-1} + \Delta\theta_k \,. \tag{11.19}$$

The robot is equipped with a fiber optic gyroscope (FOG), which directly measures θ_k, and the fourth required measurement becomes

$$z_{4,k} = h_{4,k} = \theta_k \,. \tag{11.20}$$

Combined the measurements, the observation vector is constructed as $y_k = [z_{1,k} \;\; z_{2,k} \;\; z_{3,k} \;\; z_{4,k}]^T$. Finally, the state equation $f_k = [\, f_{1,k} \;\; f_{2,k} \;\; f_{3,k} \,]^T$ and observation equation $h_k = [\, h_{1,k} \; h_{2,k} \; h_{3,k} \; h_{4,k} \,]^T$ formalize the localization problem in state space with

$$x_k = f_k(x_{k-1}, u_k) + w_k \,, \tag{11.21}$$

$$y_k = h_k(x_k) + v_k \,, \tag{11.22}$$

where the process noise $w_k \sim \mathcal{N}(0, Q_k)$ and measurement noise $v_k \sim \mathcal{N}(0, R_k)$ have the covariances Q_k and R_k. Given (11.21) and (11.22), the mobile robot coordinates and heading are estimated by the PF.

11.3.1 Sample Impoverishment Issue

The PF-based approach assumes that the robot coordinates are the first-order Markov processes evolving from one point to another with known initial and transition distributions and that the conditionally independent observations depend only on the robot position. To estimate x_k, y_k, and θ_k, the PF generates a set of samples at each k that approximates the distributions of the coordinates

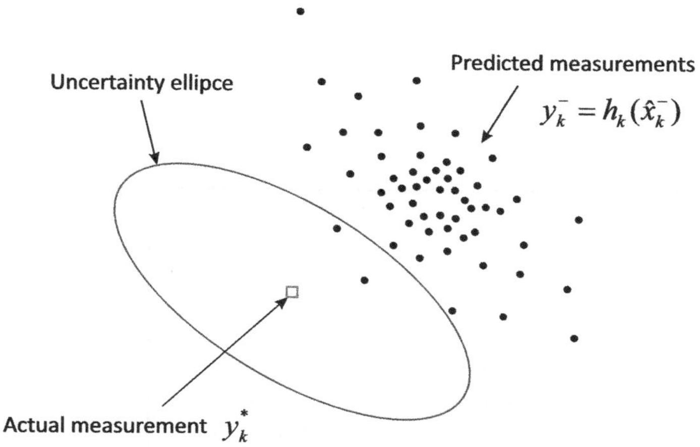

Figure 11.9 A typical scenario with the sample impoverishment.

conditioned on all past observations. This process is called *resampling* and causes an issue known as sample impoverishment.

Figure 11.9 illustrates an idea of sample impoverishment in a typical scenario. The ith dot, $i \in [1, L]$, represents a sample of the ith predicted measurement $y_{k,i}^-$ defined by

$$y_{k,i}^- = h_k(\hat{x}_{k,i}^-) , \tag{11.23}$$

where $\hat{x}_{k,i}^-$ is the ith *a priori* particle (i.e., sample of the *a priori* estimated state) and L is the number of the particles. In the Gaussian approximation, the likelihood (i.e., weight) of each particle is a reciprocal of the difference between the actual measurement y_k^* and the predicted measurement. Therefore, the closer the samples of the predicted measurement $y_{k,i}^-$ are to y_k^*, the higher the weights of the corresponding *a priori* particles.

In the example shown in Fig. 11.9, there are only two samples of predicted measurements within the measurement: uncertainty or error ellipse. Thus, only these particles will receive significant weights, and resampling will repeat them many times as the *a posteriori* particles $\hat{x}_{k,i}$. As a consequence, sample impoverishment and failures will occur. The problem is mainly associated with low-intensity process noise and/or measurement noise and/or the small number of particles required to ensure fast localization. Other factors can also cause sample impoverishment. Therefore, a great deal of efforts may be required to satisfactorily address the problem. But despite all efforts, the problem remains fundamental, and it is impossible to completely avoid sample impoverishment.

11.3.2 Hybrid Particle/UFIR Filter

To improve localization accuracy, PF has been combined in [143] with an EFIR filter in a hybrid algorithm that encompasses two key procedures: 1) detection of PF failures and 2) particles regeneration and PF resetting. PF fault detection is organized in a diagnosis algorithm using Mahalanobis distance. To regenerate new particles and reset PF, an EFIR filter is used to obtain the state estimates when the PF fails. The choice of an EFIR filter that is robust and BIBO stable was made in attempts to fulfil a basic requirement: an auxiliary filter does not have to be obligatorily accurate, but it has to be robust and stable. Note that resetting the PF with an EFIR filter instead of generating a lot particles solves another critical issue: reducing the computational complexity and time required for real-time localization.

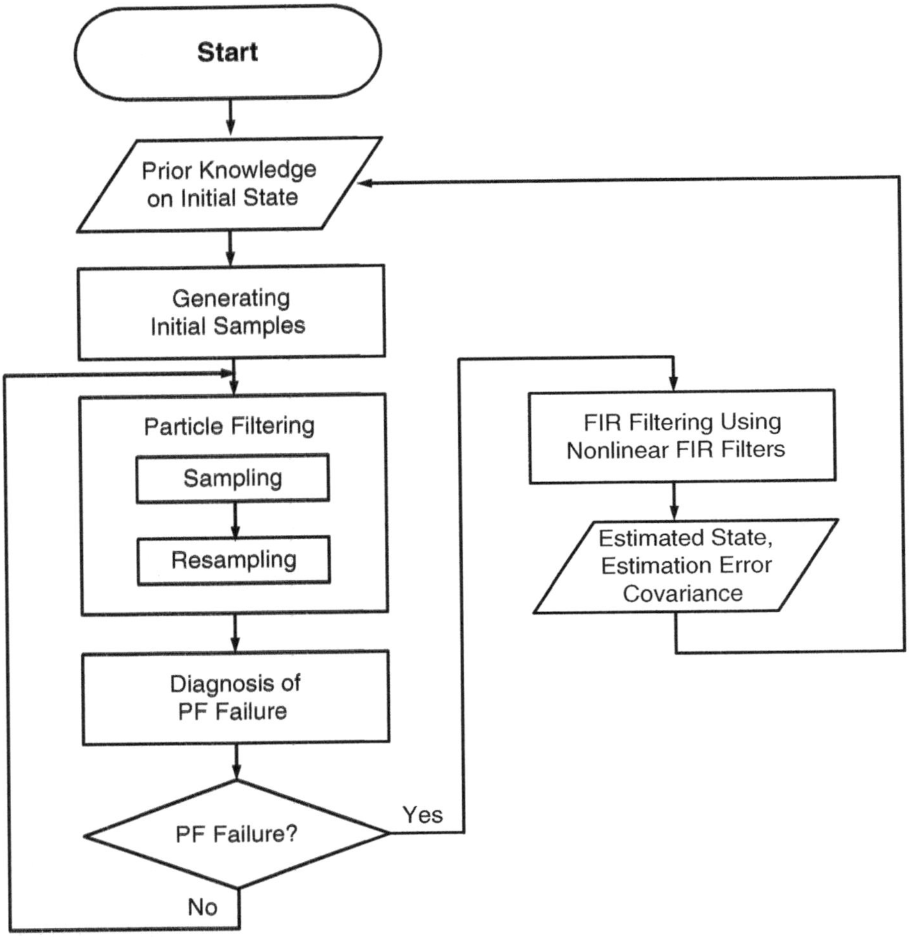

Figure 11.10 A flowchart of the hybrid PF/EFIR algorithm.

A flowchart of the hybrid PF/EFIR algorithm is given in Fig. 11.10. PF plays the role of the main filter, which provides estimates of the mobile robot coordinates and heading under normal conditions. Diagnostics of PF failures is performed continuously, and when the PF fails, the auxiliary EFIR filter provides information to reset the failed PF. Erroneous estimates at the PF output are detected taking into account the following features of sample impoverishment: 1) only a few predicted samples fall into the uncertainty ellipse, and 2) the predicted samples are usually far from actual measurement. PF rejections are detected by checking the number of samples within the uncertainty ellipse using the Mahalanobis distance.

Example 11.1 ***Localization using the RPF/EFIR algorithm [143]*** A robot starts moving from the start point (10 m, 5 m) and travels counterclockwise along a circular trajectory. The noise covariances are set as $Q_k = \mathrm{diag}(0.12 \quad 0.12 \quad 12)$ and $R_k = \mathrm{diag}(0.52 \quad 0.52 \quad 0.52 \quad 12)$. Estimation is provided using a regularized PF (RPF)/EFIR algorithm. The positioning error is computed by

$$\varepsilon_{\text{pos}} = \sqrt{(x_k - \hat{x}_k)^2 + (y_k - \hat{y}_k)^2} \,. \tag{11.24}$$

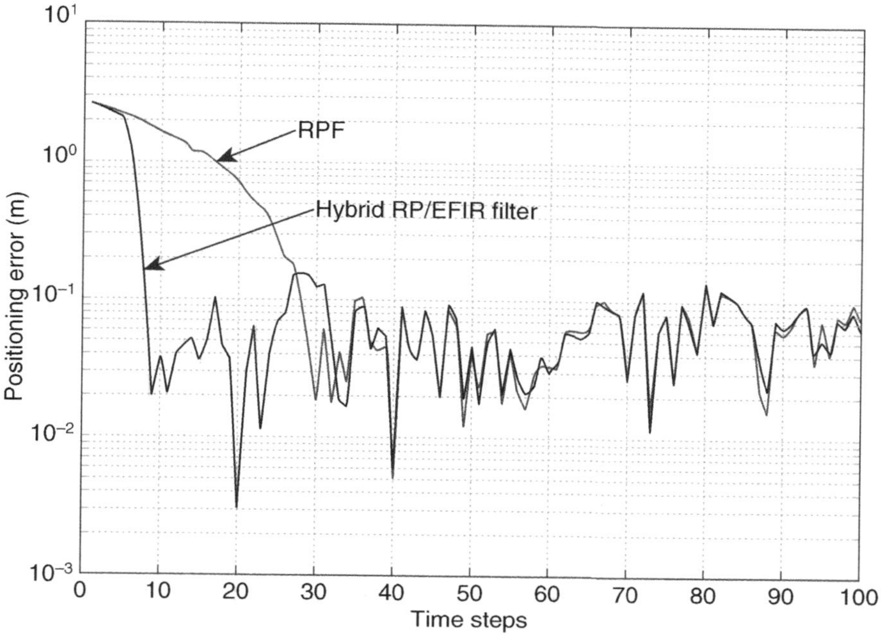

Figure 11.11 Errors of a mobile robot localization with a small number of particles, $L = 1000$ [143].

Figure 11.11 sketches typical localization errors caused by a small number of particles, $L = 1000$. As can be seen, the transient process in RPF lasts up to $k = 30$, while in the hybrid RPF/EFIR algorithm it is shorter and ends after 9 points. □

This example provides further evidence of the effectiveness of hybrid filters in solving the localization problem.

11.4 Self-Localization Over RFID Tag Grids

Radio frequency identification (RFID) tag-based networks and grids are designed to organize self-localization of various moving objects in GPS-denied indoor environments. Each tag has an ID number corresponding to a unique location and can be either active or passive. To increase awareness, information describing a local 2D or 3D surrounding can be programmed in each tag and delivered to users by request. The method is low cost and available for any purpose, provided the communication between a target and the tags. However, since low-cost measurements are usually very noisy, optimal estimators are required.

An example of the 2D schematic geometry of a vehicle traveling on an indoor floorspace nested with two RFID tags T_1 and T_2 (the number can be arbitrary) having exactly known coordinates is shown in Fig. 11.12 [149, 150]. The vehicle reader measures the distances to the tags, and the heading angle Φ is measured using a fiber-optic gyroscope (FOG); subfigures (a) and (b) are used when the vehicle and the tags are not in the same plane. The vehicle travels in direction d, and its trajectory is controlled by the left and right wheels. The incremental distances that the vehicle travels on these wheels are d_L and d_R, respectively. The distance between the left and right wheels is b, and the stabilized wheel is not shown. The vehicle moves in its own planar Cartesian coordinates

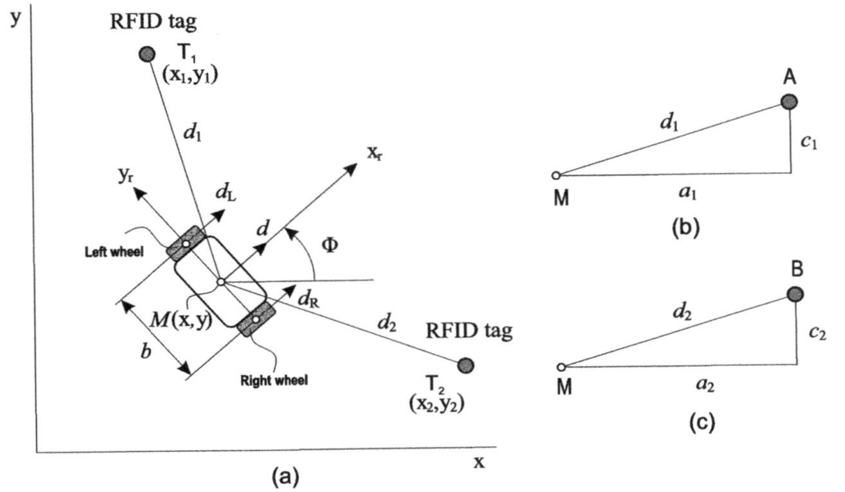

Figure 11.12 2D schematic geometry of a vehicle traveling on an indoor floorspace nested with two RFID tags, T1 and T2, having exactly known coordinates.

$(\mathrm{x_r}, \mathrm{y_r})$ with a center at $M(\mathrm{x}, \mathrm{y})$; that is, the vehicle direction always coincides with axis $\mathrm{x_r}$. The FOG measures $\boldsymbol{\Phi}$ directly.

The indoor space is commonly nested with L RFID tags $\mathrm{T}_l(\chi_l, \mu_l)$, $l \in [1, L]$, where coordinates (χ_l, μ_l) of the lth tag are exactly known. It is supposed that at each k the vehicle reader can simultaneously detect $\kappa_k \geq 2$ tags, where the number κ_k varies over time, and measures the distance d_i to the ith tag, where $i \in [1, \kappa_k]$, which can be any of the nested tags.

Referring to Fig. 11.12 and vehicle odometry, the incremental distance d_k and heading angle ϕ_k can be found as

$$d_k = \frac{1}{2}(d_{\mathrm{R}k} + d_{\mathrm{L}k}) \,, \tag{11.25}$$

$$\phi_k \cong \frac{1}{b}(d_{\mathrm{R}k} - d_{\mathrm{L}k}) \tag{11.26}$$

and the vehicle coordinates x_k and y_k and heading $\boldsymbol{\Phi}_k$ obtained by the vehicle kinematics using equations

$$f_{1k} = \mathrm{x}_k = \mathrm{x}_{k-1} + d_k \cos\left(\boldsymbol{\Phi}_{k-1} + \frac{\phi_k}{2}\right) \,, \tag{11.27}$$

$$f_{2k} = \mathrm{y}_n = \mathrm{y}_{k-1} + d_k \sin\left(\boldsymbol{\Phi}_{k-1} + \frac{\phi_k}{2}\right) \,, \tag{11.28}$$

$$f_{3k} = \boldsymbol{\Phi}_k = \boldsymbol{\Phi}_{k-1} + \phi_k \,, \tag{11.29}$$

where x_{k-1}, y_{k-1}, and $\boldsymbol{\Phi}_{k-1}$ are projected to k via the incremental distances $d_{\mathrm{L}k}$ and $d_{\mathrm{R}k}$. Note that all these values are practically not exact and undergo random variations.

11.4.1 State-Space Localization Problem

To solve the localization problem in state space, the state vector can be assigned as $x_k = [\, \mathrm{x}_k \;\; \mathrm{y}_k \;\; \boldsymbol{\Phi}_k \,]^T$ and the input vector as $u_k = [\, d_{\mathrm{L}k} \;\; d_{\mathrm{R}k} \,]^T$. Random additive components in these vectors can be supposed to be $w_k = [\, w_{\mathrm{x}k} \;\; w_{\mathrm{y}k} \;\; w_{\Phi k} \,]^T \sim \mathcal{N}(0, Q)$ and $e_k = [\, e_{\mathrm{L}k} \;\; e_{\mathrm{R}k} \,]^T \sim \mathcal{N}(0, L)$. The nonlinear state equation thus becomes

$$x_k = f_k(x_{k-1}, u_k, w_k, e_k) \,, \tag{11.30}$$

where the nonlinear function $f_k = [\, f_{1k}\ f_{2k}\ f_{3k}\,]^T$ is combined with components given by (11.27)–(11.29).

The measurement equations can be written as

$$
\begin{aligned}
d_{1k} &= \sqrt{(\mu_1 - \mathrm{y}_k)^2 + (\chi_1 - \mathrm{x}_k)^2 + c_1^2}\,, \\
&\ \ \vdots \\
d_{\kappa_k k} &= \sqrt{(\mu_{\kappa_k} - \mathrm{y}_k)^2 + (\chi_{\kappa_k} - \mathrm{x}_n)^2 + c_{\kappa_k}^2}\,, \\
\Phi_k &= \Phi_k\,.
\end{aligned}
$$

By introducing the observation vector $y_k = [\, z_{1k}\ \dots\ y_{\kappa_k k}\ y_{\phi k}\,]^T \in \mathbb{R}^{\kappa_k+1}$, nonlinear function $h_k(x_k) = [\, d_{1k}\ \dots\ d_{\kappa_k k}\ \Phi_k\,]^T \in \mathbb{R}^{\kappa_k+1}$, and measurement noise $v_k = [\, v_{1k}\ \dots\ v_{\kappa_k k}\ v_{\phi k}\,] \in \mathbb{R}^{\kappa_k+1}$, the observation equation can be rewritten in a compact form as

$$y_k = h_k(x_k) + v_k\,, \tag{11.31}$$

where noise $v_k \sim \mathcal{N}(0, R_k)$ has the covariance $R_k = \mathcal{E}\{v_k v_k^T\}$.

Extended State-Space Model

The standard procedure (3.228) applied to function $f_k \triangleq f_k(x_{k-1}, u_k, w_k, e_k)$ gives the first-order approximation

$$
\begin{aligned}
f_k &\cong f_k(\hat{x}_{k-1}, u_k, 0{,}0) + F_k(x_{k-1} - \hat{x}_{k-1}) + E_k e_k + B_k w_k \\
&= F_k x_{k-1} + \bar{u}_k + E_k e_k + B_k w_k\,,
\end{aligned} \tag{11.32}
$$

where $\hat{x}_{k-1}$ is an estimate at $k-1$, $\bar{u}_k = f_k(\hat{x}_{k-1}, u_k, 0{,}0) - F_k\hat{x}_{k-1}$ is known, and F_k, B_k, and E_k are Jacobian matrices defined by

$$
F_k = \left.\frac{\partial f_k}{\partial x}\right|_{\hat{x}_{k-1}} = \begin{bmatrix} 1 & 0 & -d_k \sin(\hat{\Phi}_{k-1} + \frac{1}{2}\phi_k) \\ 0 & 1 & d_k \cos(\hat{\Phi}_{k-1} + \frac{1}{2}\phi_k) \\ 0 & 0 & 1 \end{bmatrix}, \tag{11.33}
$$

$$
B_k = \left.\frac{\partial f_k}{\partial w}\right|_{\hat{x}_{k-1}} = F_k\,, \tag{11.34}
$$

$$
E_k = \frac{1}{2b}\begin{bmatrix} be_{ck} + d_k e_{sk} & be_{ck} - d_k e_{sk} \\ be_{sk} - d_k e_{ck} & be_{sk} + d_n e_{ck} \\ -2 & 2 \end{bmatrix}, \tag{11.35}
$$

where $e_{ck} = \cos\left(\hat{\Phi}_k^- + \frac{\phi_k}{2}\right)$ and $e_{sk} = \sin\left(\hat{\Phi}_k^- + \frac{\phi_k}{2}\right)$.

By applying (3.229) to $h_k(x_k)$, the first-order approximation becomes

$$
\begin{aligned}
h_k(x_k) &\cong h_k(\hat{x}_k^-) + \left.\frac{\partial h_k}{\partial x}\right|_{\hat{x}_k^-}(x_k - \hat{x}_k^-) \\
&= H_k x_k + \bar{y}_k\,,
\end{aligned} \tag{11.36}
$$

where $\bar{y}_k = h_k(\hat{x}_k^-) - H_k\hat{x}_k^-$ is known and H_k is a Jacobian matrix defined by

$$
H_k = \left.\frac{\partial h_k}{\partial x}\right|_{\hat{x}_k^-} = \begin{bmatrix} \frac{\hat{x}_k^- - \bar{\chi}_1}{v_{1k}} & \frac{\hat{y}_k^- - \bar{\mu}_1}{v_{1k}} & 0 \\ \vdots & \vdots & \vdots \\ \frac{\hat{x}_k^- - \bar{\chi}_{\kappa_k}}{v_{(\kappa_k k)}} & \frac{\hat{y}_k^- - \bar{\mu}_{\kappa_k}}{v_{(\kappa_k k)}} & 0 \\ 0 & 0 & 1 \end{bmatrix}, \tag{11.37}
$$

where $v_{ik} = \sqrt{(\bar{\mu}_i - \hat{y}_k^-)^2 + (\bar{\chi}_i - \hat{x}_k^-)^2 + c_i^2}$, $i \in [1, \kappa_k]$.

The first-order extended state-space model can therefore be written as

$$x_k = F_k x_{k-1} + \bar{u}_k + \tilde{e}_k + \tilde{w}_k \,, \tag{11.38}$$

$$y_k = H_k x_k + \bar{y}_k + v_k \,, \tag{11.39}$$

where $\tilde{w}_k \sim \mathcal{N}(0, \tilde{Q}_k)$ and $\tilde{e}_k \sim \mathcal{N}(0, \tilde{L}_k)$ have the covariances $\tilde{Q}_k = F_k Q F_k^T$ and $\tilde{L}_k = E_k L E_k^T$.

11.4.2 Localization Performance

Provided the approximation (11.38) and (11.39), the EKF algorithm (3.240)–(3.244) and EFIR Algorithm 15 can be directly applied to organize self-localization as shown next.

Consider a robot platform moving down an aisle in an RFID grid environment where tags are nested in the floor and ceiling (Fig. 11.13) [150].

The environment is simple because the coordinates of a tag can be easily calculated based on its geometrical position. It is supposed that the reader is able to detect eight tags at the same time at each floorspace point, but some tags may not be available due to faults, furniture isolation, low-power consumption, etc. However, at least two tags are always available.

To assess the quality of localization, noise was generated in [150] with the following standard deviations: $\sigma_x = \sigma_y = 1$ mm, $\sigma_L = \sigma_R = 0.1$ mm, and $\sigma_\Phi = 0.5°$. Since the tags on the ceiling are farthest from the platform, the standard deviations of the measurement noise are set as $\sigma_{v1} = \cdots = \sigma_{v8} = 5$ mm, $\sigma_{v9} = \cdots = \sigma_{v16} = 10$ mm, and $\sigma_\phi = 2°$. The optimal horizon is measured as $N_{\text{opt}} = 12$. The tags detected in 6 intervals (in m) along the passway are listed in Table 11.1, where tags 12 and 14–16 are not available.

Since noise statistics are typically not known exactly, an error factor $p > 0$ was introduced such that $p = 1$ makes the case ideal and the noise covariances were replaced in the algorithms by $p^2 R_k$, Q/p^2, and L/p^2. It has been shown that under ideal conditions, $p = 1$, the EKF and EFIR filter give very consistent estimates and are not prone to divergence. The situation changes dramatically for $p < 0.5$, which leads to the errors shown in Fig. 11.14. Indeed, for tag sets taken from Table 11.1 with three different realisations of the same noise, EKF demonstrates: (a) local instability, (b) single divergence, and (c) multiple divergence. In contrast, the EFIR filter turned out to be much more stable with no signs of divergence.

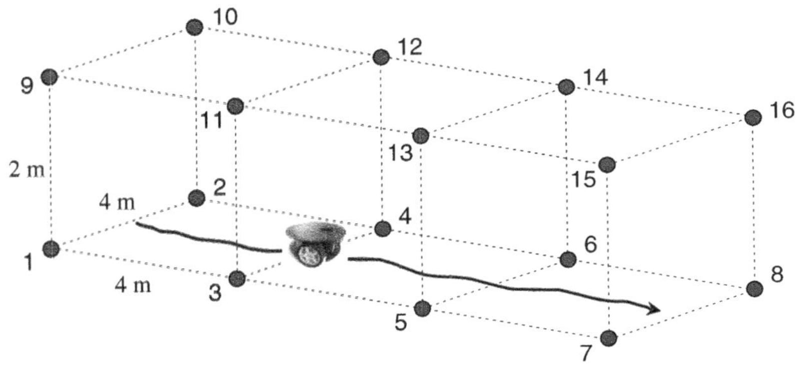

Figure 11.13 Schematic diagram of a vehicle platform traveling on an indoor passway in the RFID tag environment with eight tags mounted on a ceiling and eight tags mounted on a floor.

Table 11.1 Tags detected in six intervals (in m) along a passway shown in Fig. 11.13; tags 12 and 14–16 were not available.

m	Tag											
	1	2	3	4	5	6	7	8	9	10	11	13
0–2	x	x	–	–	–	–	–	–	x	x	–	–
2–4	x	x	x	–	–	–	–	–	x	x	–	–
4–6	–	–	x	x	–	–	–	–	–	–	x	–
6–8	–	–	x	–	x	x	–	–	–	–	–	x
8–10	–	–	–	–	x	x	–	–	–	–	–	x
10–12	–	–	–	–	–	–	x	x	–	–	–	–

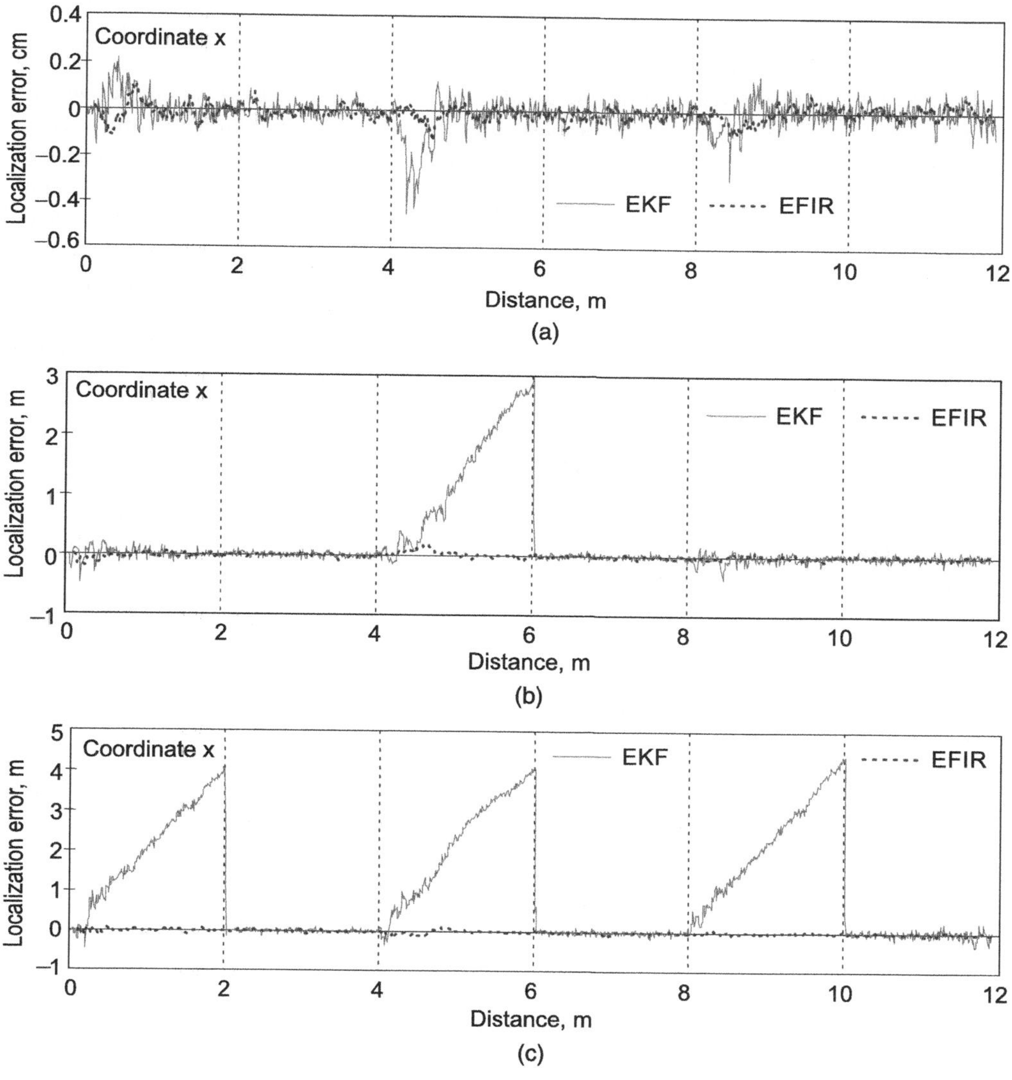

Figure 11.14 Localization errors caused by imprecisely known noise covariances for $p < 0.5$: (a) local instabilities, (b) single divergence of EKF, and (c) multiple divergence of EKF.

This investigation clearly points to another possible source of divergence in the EKF, namely, errors in the noise covariances. It also shows that increasing the number of tags interacting with the platform protects EKF from divergency so that massively nested tags can guarantee localization stability.

11.5 INS/UWB-Based Quadrotor Localization

One of the most useful modern innovations is an unmanned aerial vehicle called a *quadcopter* or *quadrotor*. Due to the ability to fulfil tasks traditionally assigned to humans, quadrotors have became irreplaceable in many branches of modern life. A prerequisite for a quadrotor to complete various tasks is accurate position information, which must be retrieved quickly and preferably optimally with sufficient robustness in complex, harsh, and volatile navigation environments.

A typical quadrotor localization scheme that combines the capabilities of an inertial navigation system (INS) and ultra wide band (UWB)–based system is shown in Fig. 11.15 [217]. A navigation environment has been created here with multiple reference nodes (RNs), and the UWB and INS units are installed on the quadrotor. The UWB-based subsystem derives the quadrotor position $L_k^{\mathrm{U}} = (\mathrm{x}_k^{\mathrm{U}}, \mathrm{y}_k^{\mathrm{U}}, \mathrm{z}_k^{\mathrm{U}})$ in the east direction $\mathrm{x}_k^{\mathrm{U}}$, north direction $\mathrm{y}_k^{\mathrm{U}}$, and vertical direction $\mathrm{z}_k^{\mathrm{U}}$ via distances d_{ik}^{U} measured between the ith RN and a quadrotor. The INS-based subsystem obtains the quadrotor position $L_k^{\mathrm{I}} = (\mathrm{x}_k^{\mathrm{I}}, \mathrm{y}_k^{\mathrm{I}}, \mathrm{z}_k^{\mathrm{I}})$. Provided L_k^{U} and L_k^{I}, the difference $\delta L_k = L_k^{\mathrm{I}} - L_k^{\mathrm{U}}$ is estimated to correct the INS position L_k^{I} and produce finally the position vector $L_k = (\mathrm{x}_k, \mathrm{y}_k, \mathrm{z}_k)$.

A specific feature of this system is that UWB data are contaminated with CMN in all of the directions as shown in Fig. 11.16. Thus, the white Gaussian approximation may not be sufficient for accurate quadrotor localization, and the GKF algorithm (3.146)–(3.151) or UFIR filtering Algorithm 13 should be applied.

11.5.1 Quadrotor State Space Model Under CMN

Based on the scheme shown in Fig. 11.15, the quadrotor dynamics can be represented with increments in the coordinates $\delta\mathrm{x}_k$, $\delta\mathrm{y}_k$, and $\delta\mathrm{z}_k$ and velocities $\delta V_{\mathrm{x}k}$, $\delta V_{\mathrm{y}k}$, and $\delta V_{\mathrm{z}k}$, which can then be united in the state equation

$$x_k = Fx_{k-1} + w_k \,, \tag{11.40}$$

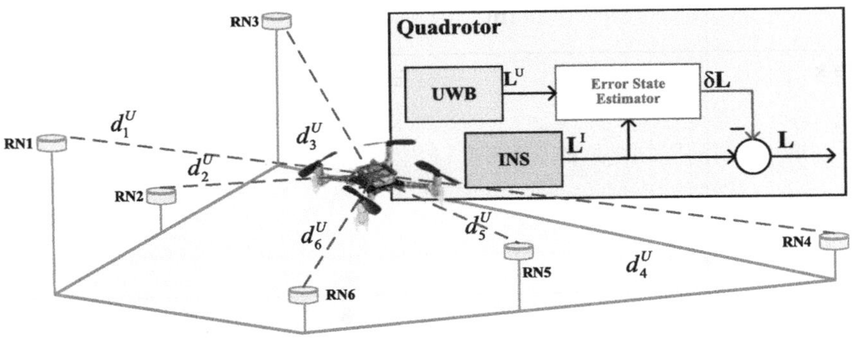

Figure 11.15 INS/UWB-integrated quadrotor localization scheme [217].

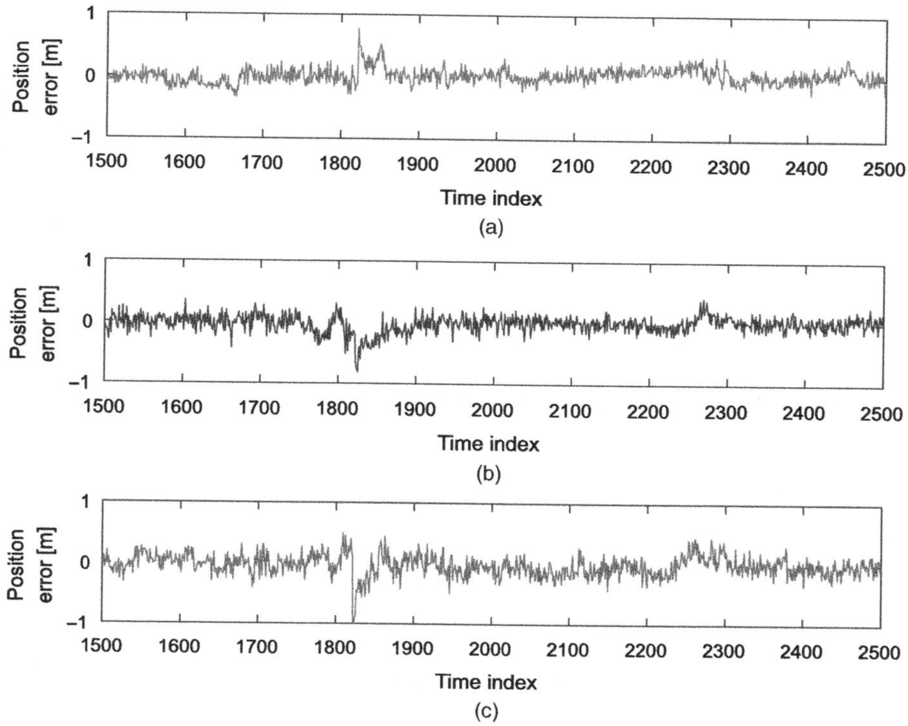

Figure 11.16 CMN in UWB-derived data: (a) east direction, (b) north direction, and (c) vertical direction.

where

$$x_k = \begin{bmatrix} \delta x_k \\ \delta V_{xk} \\ \delta y_k \\ \delta V_{yk} \\ \delta z_k \\ \delta V_{zk} \end{bmatrix}, \quad F = \begin{bmatrix} 1 & \tau & 0 & 0 & 0 & 0 \\ 0 & 1 & 0 & 0 & 0 & 0 \\ 0 & 0 & 1 & \tau & 0 & 0 \\ 0 & 0 & 0 & 1 & 0 & 0 \\ 0 & 0 & 0 & 0 & 1 & \tau \\ 0 & 0 & 0 & 0 & 0 & 1 \end{bmatrix},$$

τ is the time step, and $w_k \sim \mathcal{N}(0, Q)$.

For $\delta L_k = L_k^{\mathrm{I}} - L_k^{\mathrm{U}}$ representing measurements, the observation equation becomes

$$z_k = Hx_k + v_k \,, \tag{11.41}$$

where

$$z_k = \begin{bmatrix} \delta x_k \\ \delta y_k \\ \delta z_k \end{bmatrix}, \quad H = \begin{bmatrix} 1 & 0 & 0 & 0 & 0 & 0 \\ 0 & 0 & 1 & 0 & 0 & 0 \\ 0 & 0 & 0 & 0 & 1 & 0 \end{bmatrix},$$

where $v_k = [\, v_{xk} \;\; v_{yk} \;\; v_{zk} \,]^T$ is the CMN (Fig. 11.16) represented with the Gauss-Markov model

$$v_k = \Psi v_{k-1} + \xi_k \,, \tag{11.42}$$

in which Ψ is the coloredness factor matrix such that v_k remains stationary and noise $\xi_k \sim \mathcal{N}(0, R_\xi)$ has the covariance R_ξ.

To apply the GKF and general UFIR filter, the measurement difference is introduced and transformed as

$$\begin{aligned} y_k &= z_k - \Psi z_{k-1} \\ &= H_k + v_k - \Psi H x_{k-1} - \Psi v_{k-1} \\ &= (H - \Xi)x_k + \Xi w_k + \xi_k \\ &= C x_k + \bar{v}_k \,, \end{aligned} \tag{11.43}$$

where $\Xi = \Psi H F^{-1}$, $C = H - \Xi$, and white Gaussian $\bar{v}_k = \Xi w_k + \xi_k \sim \mathcal{N}(0, \bar{R})$ has the the covariance

$$\bar{R} = \mathcal{E}\{\bar{v}_k \bar{v}_k^T\} = \Xi\Omega + R \,, \tag{11.44}$$

where $\Omega = Q\Xi^T$. For this model, the measurement residual s_k is given by

$$\begin{aligned} s_k &= y_k - C\hat{x}_k \\ &= C x_k + \bar{v}_k - CA\hat{x}_{k-1} \\ &= CA\Delta_{k-1} + C w_k + \bar{v}_k \,, \end{aligned} \tag{11.45}$$

where $\Delta_k = x_k - \hat{x}_k$. Since the cross covariance transforms to

$$\mathcal{E}\{\bar{v}_k w_k^T\} = \mathcal{E}\{(\Xi w_k + \xi_k) w_k^T\} = \Xi Q$$

and $P_k^- = AP_{k-1}A^T + Q$ is the prior error covariance, the innovation error covariance $S_k = \mathcal{E}\{s_K s_k^T\}$ becomes

$$\begin{aligned} S_k &= CAP_{k-1}A^T C^T + CQC^T + \Xi\Omega + R + C\Omega + \Omega^T C^T \\ &= CP_n^- C^T + R + H\Omega + \Omega^T C^T \,. \end{aligned} \tag{11.46}$$

With the previous modifications, the general UFIR filtering Algorithm 13 can be applied straightforwardly. To apply the GKF algorithm (3.160)–(3.165), noise vectors w_k and $\bar{v}_k$ must be de-correlated. Alternatively, to apply the GKF algorithm (3.146)–(3.151), a new optimal Kalman gain must be derived for time-correlated noise.

11.5.2 Localization Performance

Tuning localization algorithms is an important process, which does not always end successfully due to the practical inability of measuring noise statistics in real time. To obtain reliable estimates over data arriving with a time step of $\tau = 0.02$ s, the best and worst cases were considered in [217].

Best Case: For the average quadrotor velocity of 0.6 m/s estimated with a tolerance of 20%, the noise standard deviation is set as $\sigma_w = 0.12$ m/s. Noise in the first state is ignored, and the system noise covariance is specified with

$$Q = \begin{bmatrix} \frac{\tau^2}{3} & \frac{\tau}{2} & 0 & 0 & 0 & 0 \\ \frac{\tau}{2} & 1 & 0 & 0 & 0 & 0 \\ 0 & 0 & \frac{\tau^2}{3} & \frac{\tau}{2} & 0 & 0 \\ 0 & 0 & \frac{\tau}{2} & 1 & 0 & 0 \\ 0 & 0 & 0 & 0 & \frac{\tau^2}{3} & \frac{\tau}{2} \\ 0 & 0 & 0 & 0 & \frac{\tau}{2} & 1 \end{bmatrix} \sigma_w^2 \,.$$

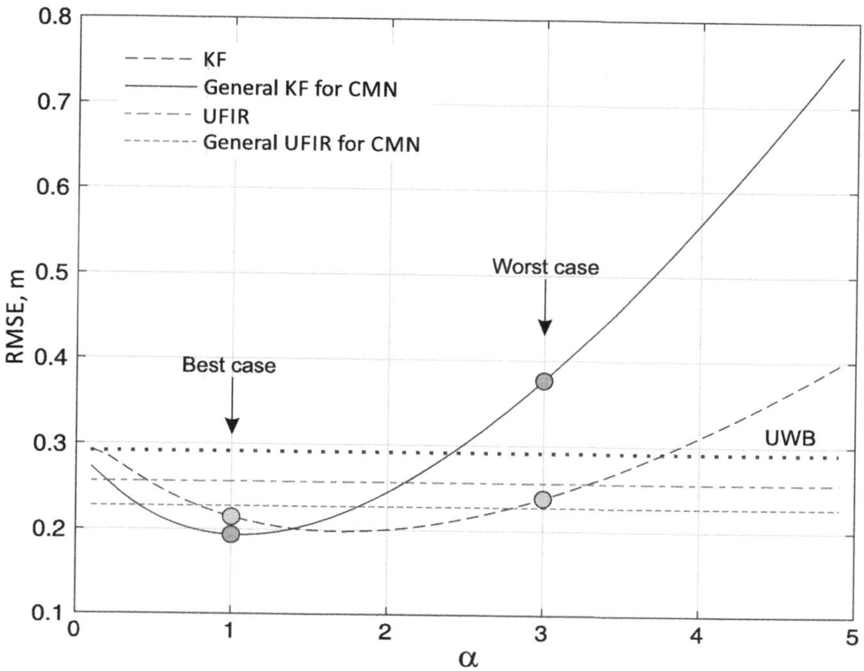

Figure 11.17 RMSEs produced by the KF, cKF, UFIR filter, and cUFIR filter. For the UWB boundary, filtering errors should be as low as possible.

The average noise variances in UWB data are measured as $\sigma_{vx}^2 = 0.16\ \text{m}^2$, $\sigma_{vy}^2 = 0.16\ \text{m}^2$, and $\sigma_{vz}^2 = 0.86\ \text{m}^2$. Accordingly, the measurement noise covariance is set as $R = \text{diag}(\sigma_{vx}^2\ \ \sigma_{vy}^2\ \ \sigma_{vz}^2)$. Experimentally, minimum localization errors were obtained by the optimal coloredness factor $\Psi_{\text{opt}} \cong 0.5$.

Worst Case: Since the actual quadrotor velocity can vary by several times and the variance of UWB noise can be several times larger than in individual measurements, the error factor α was introduced as $\frac{1}{\alpha^2}Q$ and $\alpha^2 R$, and it was supposed that in the worst case $\alpha = 3$ covers most of practical situations.

In this experiment, a quadrotor travels along a planned path considered as a reference trajectory. Typical RMSEs produced by the KF, GKF developed for CMN (cKF), UFIR filter, and general UFIR filter developed for CMN (cUFIR) are sketched in Fig. 11.17 along with the UWB error boundary. For this boundary, filtering errors should be as low as possible. What this experiment *de facto* reveals is that the properly tuned cKF is most accurate in the best case, but less accurate in the worst case. The cKF is also most sensitive to errors in the noise covariances, and its RMSE crosses the UWB boundary when $\alpha > 2.4$. The latter means that the more accurate cKF is less robust than the original KF. Thereby, this experiment confirms the earlier made conclusion that *an increase in the estimation accuracy is achieved using additional tuning factors at the expanse of robustness.* It is also seen that both UFIR filters are robust against changes in α and are reasonably accurate since their RMSEs do not exceed the UWB boundary.

11.6 Processing of Biosignals

Advances in sensor technology in recent decades have extended digital technologies to electrical and mechanical signals generated by the human body and referred to as *biosignals*. Such signals can be measured with or without contact with the body and are useful in a wide range of medical applications, from diagnostics and remote monitoring to effective prosthesis control. The best-known bioelectrical signals are: electrocardiogram (ECG), electromyogram (EMG), electroencephalogram, mechanomyogram, electrooculography, galvanic skin response, and magnetoencephalogram. Depending on the nature of the origin, biosignals can be quasiperiodic, like an ECG, or impulsive, like an EMG.

Since biosignals are generated by the body at low electrical levels, their measurements are accompanied by intense noise of little-studied origin. Another important specificity is that biosignals, as a rule, do not have simple models. Therefore, methods for optimal processing of biosignals are usually based on assumptions that cannot be theoretically substantiated, as in artificial intelligence. To illustrate possible extensions of the state estimation methods to biosignals, next we will consider examples of ECG and EMG biosignals and solve specific problems using Kalman and UFIR estimators.

11.6.1 ECG Signal Denoising Using UFIR Smoothing

The ECG signals allow detecting a wide variety of heart diseases. Since heartbeats may vary slightly from each other, accurate measurements are required. This is particularly important when data are used to extract ECG features and make decisions about the heart state using special software. Various types of noises contaminate the ECG signal during its acquisition and transmission, such as high frequency noise (electromyogram noise, additive Gaussian noise, and power line interference) and low frequency noise (baseline wandering). Because noise can cause misinterpretation of the heart state, efficient denoising is required.

A typical measured heartbeat pulse (Data) is shown in Fig. 11.18, where one can recognize the central rapid excursions, commonly referred to as the QRS-complex, slow left excursion (T-wave), slow first right excursion (T-wave), and slow second right excursion (U-wave). This suggests that the ECG signal mostly changes slow over time and is highly oversampled, with the exception of the QRS region, where it is critically sampled. Thus, in the frequency domain, the ECG signal energy is concentrated in two bands: low frequencies associated with LP filtering and high frequencies with BP filtering. By using the ith degree q-lag UFIR smoother, suboptimal denoising of the slow part can be provided on the large horizon N_{opt}, while the critically sampled fast excursions in the QRS-complex require the shortest possible horizon $N_{\mathrm{min}} = i + 1$ to avoid output bias. To enable universal noise cancellation, an adaptive iterative q-lag UFIR smoothing algorithm has been designed in [108].

Adaptive Smoothing of ECG Data

To adaptively remove noise from ECG data using an UFIR smoother, a window $[\mathrm{Q}_{\mathrm{int}}, \mathrm{S}_{\mathrm{int}}]$ is assigned to cover the QRS-complex as shown in Fig. 11.18. Since points Q and S are well detectable in the ECG pulse, the window boundaries are assigned as $\mathrm{Q}_{\mathrm{int}} = \mathrm{Q} - N_{\mathrm{opt}}$ and $\mathrm{S}_{\mathrm{int}} = \mathrm{S} + N_{\mathrm{opt}}$. The horizon N_{opt} is measured by removing the QRS-complex. By determining $\mathrm{Q}_{\mathrm{int}}$, $\mathrm{S}_{\mathrm{int}}$, N_{min}, and N_{opt}, the q-lag UFIR smoothing filter provides the following estimates in the selected ECG regions:

- $\tilde{x}_{n|n+q}(N_{\mathrm{opt}})$ in $[0 : \mathrm{Q}_{\mathrm{int}} - 1]$
- $\tilde{x}_{n|n+q}(N_{\mathrm{adt}} = N_{\mathrm{opt}} \cdots N_{\mathrm{min}})$ in $[\mathrm{Q}_{\mathrm{int}}, \mathrm{Q} - 1]$
- $\tilde{x}_{n|n+q}(N_{\mathrm{min}})$ in $[\mathrm{Q}, \mathrm{S} - 1]$
- $\tilde{x}_{n|n+q}(N_{\mathrm{adt}} = N_{\mathrm{min}} \cdots N_{\mathrm{opt}})$ in $[\mathrm{S}, \mathrm{S}_{\mathrm{int}} - 1]$
- $\tilde{x}_{n|n+q}(N_{\mathrm{opt}})$ in $[\mathrm{Q}_{\mathrm{int}} : \mathrm{End}]$

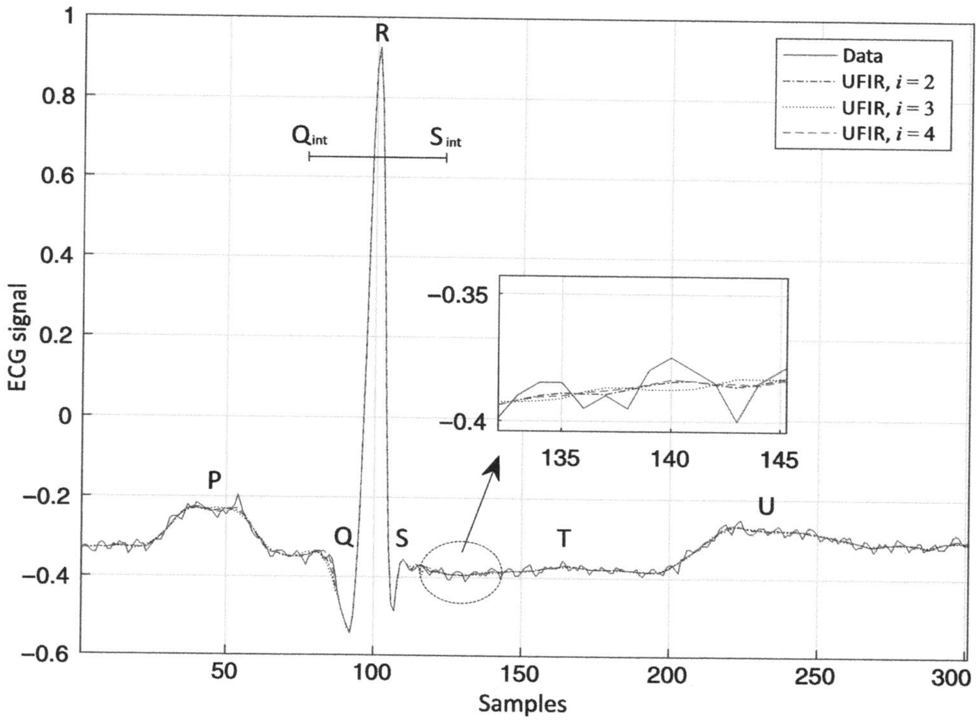

Figure 11.18 Single ECG pulse measured in the presence of noise (Data). Noise reduction is provided by an UFIR smoother of degree $i \in [2,4]$.

Here, the adaptive horizon $N_{adt} = N_{opt} \cdots N_{min}$ gradually decreases from N_{opt} to N_{min} with each new step, and $N_{adt} = N_{min} \cdots N_{opt}$ gradually increases from N_{min} to N_{opt} with each new step.

The efficiency of the low degree, $i \in [2,4]$, UFIR smoothing algorithm can be estimated from Fig. 11.18, and more details about this solution can be found in [108]. It is worth noting that Kalman smoothing cannot be effectively applied to suppress noise in ECG data due to unknown noise and rapid jumps in the QRS-complex. Note also that in view of the ECG signal quasiperiodicity the harmonic state-space model can also be used to provide denoising with UFIR smoothing.

11.6.2 EMG Envelope Extraction Using a UFIR Filter

EMG signals are records of motor unit action potentials (MUAPs) that reflect responses to electrical activity occurring in motor units (MUs) in a muscle. These signals are electrical, chemical, and mechanical in nature and are complex. Because MUAPs are acquired from different regions of the MUs, skeletal muscle contains hundreds of different types of muscle fibers, and the resulting signal is composed of several MUAPs. Measurements of EMG signals are usually performed in the presence of sensor noise and artifacts originating from various sources, such as electrocardiography interference, spurious background spikes, motion artifact, and power line interference. The MUAP characteristics depend, among other factors, on the geometric position of the electrode needle inserted into the muscle. By generating a MUAP, morphologic changes in muscle fibers can be detected and analyzed. More information on EMG signals can be found in [120] and reference therein.

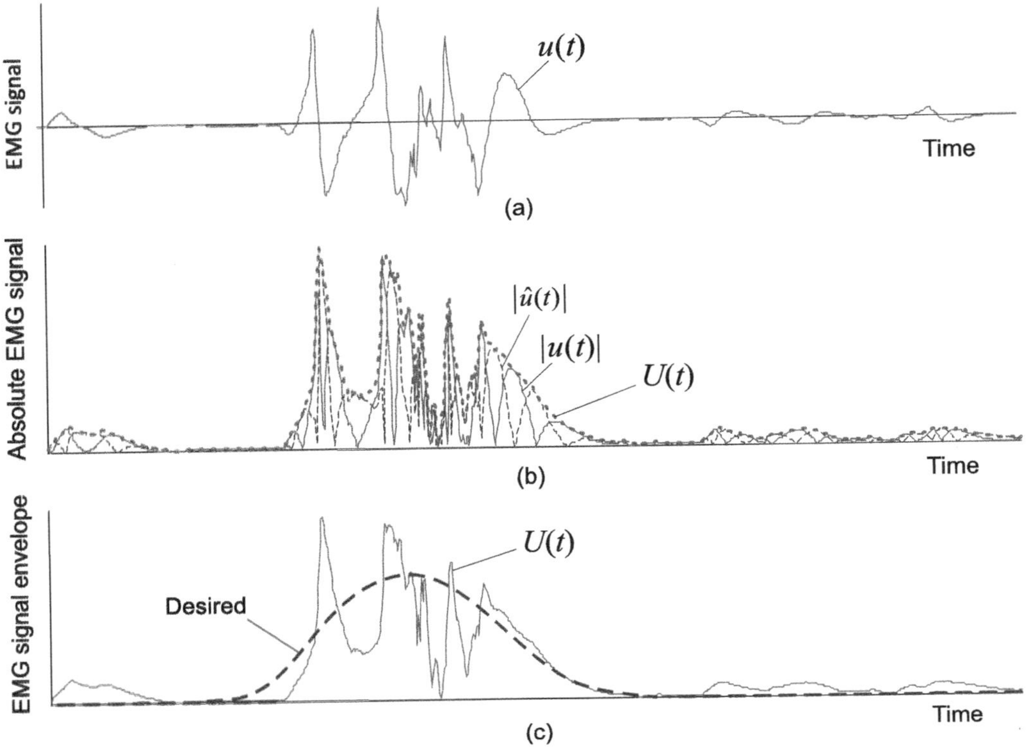

Figure 11.19 EMG signal: (a) measured EMG signal $u(t)$ composed by MUAPs, (b) Hilbert transform $\hat{u}(t)$ of $u(t)$ and envelope $U(t) = \sqrt{u^2(t) + \hat{u}^2(t)}$, and (c) extracted $U(t)$ and desired EMG signal envelope.

A typical measured EMG signal $u(t)$, composed by low-density MUAPs, is shown in Fig. 11.19a. An important function of the EMG signal is the envelope, which is used in robotic systems and prothesis control to achieve proper human-robot interaction. The envelope $U(t) = \sqrt{u^2(t) + \hat{u}^2(t)}$ can be shaped by combining $u(t)$ with the Hilbert transform $\hat{u}(t)$ as shown in Fig. 11.19b. However, this may not always lead to a smooth shape due to multiple ripples. It should be noted that for the sake of stability of the proportional control of the artificial prothesis, it is desirable that the envelope has a Gaussian shape (Fig. 11.19c).

To keep the EMG envelope as close to Gaussian as possible, it was proposed in [120] to treat the ripples on the envelope (Fig. 11.19c) as CMN and use a general UFIR filter. Accordingly, the oversampled envelope U_k was represented by the K-state polynomial model [172] under the Gauss-Markov CMN v_k assumption as

$$x_k = Fx_{k-1} + Bw_k \,, \tag{11.47}$$

$$y_k = Hx_k + v_k \,, \tag{11.48}$$

$$v_k = \psi v_{k-1} + \xi_k \,, \tag{11.49}$$

where $x_k \in \mathbb{R}^K$ is the U_k state vector and y_k is the scalar observation of U_k. For the polynomial approximation, entries of the system matrix F are obtained using the Taylor series expansion as

$$F = \begin{bmatrix} 1 & \tau & \frac{\tau^2}{2} & \dots & \frac{\tau^{K-1}}{(K-1)!} \\ 0 & 1 & \tau & \dots & \frac{\tau^{K-2}}{(K-2)!} \\ 0 & 0 & 1 & \dots & \frac{\tau^{K-3}}{(K-3)!} \\ \vdots & \vdots & \vdots & \ddots & \vdots \\ 0 & 0 & 0 & \dots & 1 \end{bmatrix} \tag{11.50}$$

and the number of states K is chosen so that the best shaping is obtained.

The observation y_k of U_k corresponds to the first state. Therefore, the observation matrix is assigned as $H = [\,1\ \ 0 \dots 0\,] \in \mathbb{R}^{1\times K}$. Matrix $B \in \mathbb{R}^{K\times P}$ projects the U_k noise $w_k \in \mathbb{R}^P$ to x_k. The scalar coloredness factor $0 < \psi < 1$ is determined during the testing phase to ensure the best shaping, and we notice that, by $\psi = 0$, noise v_k becomes white Gaussian. Noise ξ_k is zero mean and white Gaussian, $\xi_k \sim \mathcal{N}(0, \sigma^2_{\xi k})$, with the variance $E\{\xi_k^2\} = R_k = \sigma^2_{\xi k}$.

Since the noise in U_k is nonstandard, it is assumed that w_k has zero mean with uncertain statistics and distribution. To run KF, w_k is treated as zero mean and white Gaussian, $w_k \sim \mathcal{N}(0, Q)$, with the covariance $E\{w_k w_n^T\} = Q\delta_{k-n}$, where δ_k is the Kronecker symbol, and the property $E\{w_k \xi_n\} = 0$ for all k and n. It is assumed that the estimate $\hat{x}_k$ of x_k under the ripples in U_k (Fig. 11.19c) will approach the Gaussian form if we assume that ripples are due to the CMN.

EMG Signals with Low-Density MUAP

As an example, we consider an EMG signal composed of low-density MUAPs that require the Hilbert transform to smooth the envelope [120]. The model in (11.47)–(11.49) is defined by two states, $K = 2$, and matrices

$$F = \begin{bmatrix} 1 & \tau \\ 0 & 1 \end{bmatrix}, B = \begin{bmatrix} \frac{\tau^2}{2} \\ \tau \end{bmatrix}, H = \begin{bmatrix} 1 & 0 \end{bmatrix}.$$

The EMG signal is taken from the "Elbowing" database [120], which contains samples collected from 11 subjects with knee abnormality previously diagnosed by a professional and 11 normal knees. Measurements are carried out using goniometry equipment MWX8 Datalog Biometrics with the sampling frequency $f = 1$ kHz, which corresponds to $\tau = 10^{-3}$ s. Part of the database is shown in Fig. 11.20a. At the first step, the envelope is shaped using the Hilbert transform to represent "Data" as shown in Fig. 11.20b and Fig. 11.20c. Since information about noise is unavailable, the KF and UFIR filter are arbitrary tuned to produce consistent estimates with minimal variations regarding the desired smooth envelope and negligible time-delays.

The optimal horizon for a UFIR filter is measured as $N_{\text{opr}} = 74$, which means a high oversampling rate. To tune KF, several similar datasets are analyzed, and it is concluded that the assumed noise has a standard deviation of about $\sigma_\xi = 50$ μV. Since the average third state (acceleration) of the envelope is about 0.5 V/s^2 with a tolerance of about 20 % or even more, the standard deviation of the process noise is set as $\sigma_w = 0.1$ V/s^2, which makes the KF estimate consistent with the UFIR estimate. With this filter setup, the extracted envelopes are sketched in Fig. 11.20b, and it turns out that even if the filters improve the envelope without significant time delays, there are still a lots of ripples, which require further smoothing.

To get rid of the ripples in the envelope, it is next assumed that the measurement noise is colored and of Gauss-Markov origin (11.49). Then the coloredness factor is set as $\psi = 0.75$ to provide the

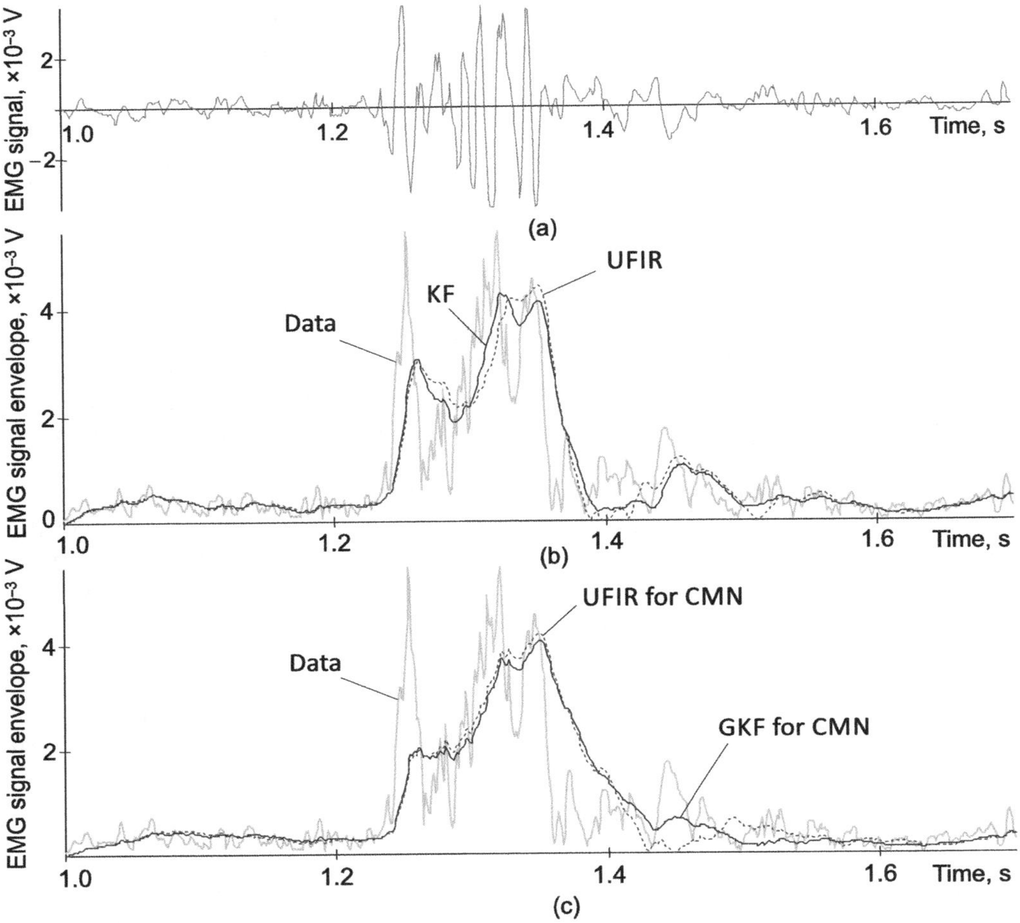

Figure 11.20 EMG signal composed with low-density MUAP and envelope extracted using the Hilbert transform (Data) [120]: (a) EMG signal, (b) KF and UFIR filtering estimate, and (c) KF and UFIR filtering estimates modified for CMN.

best smoothing without significant time delay, the optimal horizon is measured as $N_{\text{opt}} = 140$, and the GKF and UFIR filter modified for CMN are applied. What follows from the result shown in Fig. 11.20c is that the filters essentially suppress ripples and improve the envelope and that the latter finally better matches the desired Gaussian shape.

11.7 Summary

In this chapter, we looked at several practical uses of FIR state estimators, and one can deduce that the FIR approach can be extended to many other traditional and new challenging tasks. The aim was to demonstrate the difference between the FIR and IIR (KF) estimates obtained from real data. The following important points should be kept in mind.

Optimal estimators such as the OFIR filter and KF serve well when the model is process-fit and the noise is Gaussian or near Gaussian. Otherwise, the UFIR state estimator may be more accurate.

There are some boundary conditions that separate the scope of optimal and unbiased estimates in terms of accuracy and robustness. But this is a subtle matter that requires investigations in individual cases. Indeed, it is always possible to set some noise covariance, even heuristically and with no justification, to make the KF accurate, and there is always an optimal horizon N_{opt} that makes the UFIR estimate suboptimal. Therefore, the question arises, which is practically easier: to determine the noise covariances or N_{opt}? The answer depends on the specific problem, but it is obvious that it takes less efforts to determine the scalar N_{opt} than the covariance matrix. Moreover, given the horizon length of N, even not optimal, the UFIR state estimator becomes blind and therefore robust, which is highly required in many applications, especially industrial ones.

We conclude this chapter by listing several modern and nontraditional signal processing problems that can be solved using FIR state estimation technologies, and note that this list can be greatly expanded.

11.8 Problems

1 ML is pursued to estimate parameters such as mean and covariance of medical images for normality and abnormality employing a support vector machine. Since noise in images is non-Gaussian, it seems that robust UFIR/median smoothing may be more efficient.

2 In uncertain random environments, the FIR approach can help to solve the problem of estimating instantaneous speed and disturbance load torque in motors with higher robustness than standard KF-based algorithms.

3 To adapt the KF-based state estimator to uncertain environments under disturbances, artificial neural networks (ANNs) are used in what is referred to as a neuro-observer, which is an EKF-based structure augmented by an ANN to capture the unmodeled dynamics. It looks like a robust UFIR-based observer with adaptive horizon might be a more efficient solution.

4 To estimate state-of-charge of lithium-ion batteries, which is critical for electrical vehicles, adaptive H_∞ filtering is used to eliminate bias errors. Instead, the H_∞ FIR and especially UFIR filtering algorithms should be tested as more robust and stable.

5 To avoid the construction of new power plants, transmission lines, etc., the distributed generation (DG) approach has received a big attention. For optimal localization of several DGs under power losses, KF-based algorithms are used. Since noise in a DG environment is typically non-Gaussian, a robust UFIR state estimator may be the best choice.

6 To estimate the size and location of a brain tumor, a massively parallel finite difference method based on the graphics processing unit is used together with a genetic algorithm to solve the inverse problem. To provide effective noise reduction on finite intervals, FIR state estimators can be built in to achieve greater accuracy.

Appendix A

Matrix Forms and Relationships

The following matrix forms, properties, and relationships are useful in the derivation of state estimators [18, 52]. Vectors are depicted with small letters as a, x, … and matrices with capital letters as X, Y, … The matrix forms and relationships are united in groups depending on properties and applications.

A.1 Derivatives

The following derivatives of matrix and vectors products and traces of the products allow deriving state estimators in a shorter way.

$$\frac{\partial x^T a}{\partial x} = \frac{\partial a^T x}{\partial x} = a \,, \tag{A.1}$$

$$\frac{\partial x^T B x}{\partial x} = (B + B^T)x \,, \tag{A.2}$$

$$\frac{\partial}{\partial X} \operatorname{tr}(XA) = A^T \,, \tag{A.3}$$

$$\frac{\partial}{\partial X} \operatorname{tr}(X^T A) = A \,, \tag{A.4}$$

$$\frac{\partial}{\partial X} \operatorname{tr}(X^T BX) = BX + B^T X \,, \tag{A.5}$$

$$\frac{\partial}{\partial X} \operatorname{tr}(XBX^T) = XB^T + XB \,. \tag{A.6}$$

A.2 Matrix Identities

There are several matrix identities that are useful in the representation of the inverse of the sum of matrices.

The Woodbury identities [59]:

$$(A + CBC^T)^{-1} = A^{-1} - A^{-1}C(B^{-1} + C^T A^{-1} C)^{-1} C^T A^{-1} \,, \tag{A.7}$$

$$(A + UCV)^{-1} = A^{-1} - A^{-1}U(C^{-1} + VA^{-1}U)^{-1} VA^{-1} \,. \tag{A.8}$$

Optimal and Robust State Estimation: Finite Impulse Response (FIR) and Kalman Approaches, First Edition.
Yuriy S. Shmaliy and Shunyi Zhao.

For positive definite matrices P and R, there is

$$(P^{-1} + B^T R^{-1} B)^{-1} B^T R^{-1} = P B^T (B P B^T + R)^{-1} . \tag{A.9}$$

The Kailath variant:

$$(A + BC)^{-1} = A^{-1} - A^{-1} B (I + C A^{-1} B)^{-1} C A^{-1} . \tag{A.10}$$

Special cases:

$$(A + B)^{-1} = A^{-1} - A^{-1}(I + BA^{-1})^{-1} BA^{-1} , \tag{A.11}$$

$$(A^{-1} + B^{-1})^{-1} = A(A + B)^{-1} B = B(A + B)^{-1} A , \tag{A.12}$$

$$(I + A^{-1})^{-1} = A(A + I)^{-1} , \tag{A.13}$$

$$(A + BB^T)^{-1} B = A^{-1} B (I + B^T A^{-1} B)^{-1} , \tag{A.14}$$

$$A - A(A + B)^{-1} A = B - B(A + B)^{-1} B , \tag{A.15}$$

$$A^{-1} + B^{-1} = A^{-1}(A + B) B^{-1} , \tag{A.16}$$

$$(I + AB)^{-1} = I - A(I + BA)^{-1} B , \tag{A.17}$$

$$(I + AB)^{-1} A = A(I + BA)^{-1} . \tag{A.18}$$

A.3 Special Matrices

The Vandermonde matrix: This is an $m \times n$ matrix with the terms of a geometrical progression in each row,

$$V = \begin{bmatrix} 1 & \alpha_1 & \alpha_1^2 & \cdots & \alpha_1^{n-1} \\ 1 & \alpha_2 & \alpha_2^2 & \cdots & \alpha_2^{n-1} \\ 1 & \alpha_3 & \alpha_3^2 & \cdots & \alpha_3^{n-1} \\ \vdots & \vdots & \vdots & \ddots & \vdots \\ 1 & \alpha_m & \alpha_m^2 & \cdots & \alpha_m^{n-1} \end{bmatrix} . \tag{A.19}$$

The Jacobian matrix: For $f(x) \in \mathbb{R}^m$ and $x \in \mathbb{R}^n$,

$$J(x) = \frac{\partial f(x)}{\partial x} = \begin{bmatrix} \frac{\partial f_1}{\partial x} \\ \frac{\partial f_2}{\partial x} \\ \vdots \\ \frac{\partial f_m}{\partial x} \end{bmatrix} = \begin{bmatrix} \frac{\partial f_1}{\partial x_1} & \frac{\partial f_1}{\partial x_2} & \cdots & \frac{\partial f_1}{\partial x_n} \\ \frac{\partial f_2}{\partial x_1} & \frac{\partial f_2}{\partial x_2} & \cdots & \frac{\partial f_2}{\partial x_n} \\ \vdots & \vdots & \ddots & \vdots \\ \frac{\partial f_m}{\partial x_1} & \frac{\partial f_m}{\partial x_2} & \cdots & \frac{\partial f_m}{\partial x_n} \end{bmatrix} . \tag{A.20}$$

The Hessian matrix: For $f(x) \in \mathbb{R}^m$ and $x \in \mathbb{R}^n$,

$$\mathcal{H}(x) = \frac{\partial}{\partial x} \left[\frac{\partial f(x)}{\partial x} \right]^T = \begin{bmatrix} \frac{\partial^2 f}{\partial x_1^2} & \frac{\partial^2 f}{\partial x_2 \partial x_1} & \cdots & \frac{\partial^2 f}{\partial x_n \partial x_1} \\ \frac{\partial^2 f}{\partial x_1 \partial x_2} & \frac{\partial^2 f}{\partial x_2^2} & \cdots & \frac{\partial^2 f}{\partial x_n \partial x_2} \\ \vdots & \vdots & \ddots & \vdots \\ \frac{\partial^2 f}{\partial x_1 \partial x_n} & \frac{\partial^2 f}{\partial x_2 \partial x_n} & \cdots & \frac{\partial^2 f}{\partial x_n^2} \end{bmatrix} = \mathcal{H}^T(x). \tag{A.21}$$

Schur complement:

Given $A \in \mathbb{R}^{p\times p}$, $B \in \mathbb{R}^{p\times q}$, $C \in \mathbb{R}^{q\times p}$, $D \in \mathbb{R}^{q\times q}$, and a nonsingular block matrix

$$M = \begin{bmatrix} A & B \\ C & D \end{bmatrix} \in \mathbb{R}^{(p+q)\times(p+q)}. \tag{A.22}$$

If A is nonsingular, then the Schur complement of A is $M/A = D - CA^{-1}B$, and the inverse of M is computed as

$$M^{-1} = \begin{bmatrix} A^{-1} + A^{-1}B(M/A)^{-1}CA^{-1} & -A^{-1}B(M/A)^{-1} \\ -(M/A)^{-1}CA^{-1} & (M/A)^{-1} \end{bmatrix}. \tag{A.23}$$

If D is nonsingular, then the Schur complement of D is $M/D = A - BD^{-1}C$, and the inverse of M can be computed by

$$M^{-1} = \begin{bmatrix} (M/D)^{-1} & -(M/D)^{-1}BD^{-1} \\ -D^{-1}C(M/D)^{-1} & D^{-1} + D^{-1}C(M/D)^{-1}BD^{-1} \end{bmatrix}. \tag{A.24}$$

A.4 Equations and Inequalities

Several equations and inequalities are used to guarantee the stability of state estimators.

The Riccati differential equation (RDE):

$$\frac{\mathrm{d}}{\mathrm{d}t}P \triangleq P' = PA^T + AP - PCP + S\,. \tag{A.25}$$

A solution to (A.12) can be found if we assign

$$P = VU^{-1}$$

and choose matrices V and U such that

$$\begin{bmatrix} U' \\ V' \end{bmatrix} = \begin{bmatrix} F_{11} & F_{12} \\ F_{21} & F_{22} \end{bmatrix} \begin{bmatrix} U \\ V \end{bmatrix}.$$

Next, the derivative P' can be transformed as

$$\begin{aligned} P' &= (VU^{-1})' \\ &= V'U^{-1} + V(U^{-1})' \\ &= V'U^{-1} + V\left(-U^{-1}U'U^{-1}\right) \\ &= (F_{21}U + F_{22}V)U^{-1} - VU^{-1}(F_{11}U + F_{12}V)U^{-1} \\ &= F_{21} + F_{22}P - PF_{11} - PF_{12}P \end{aligned}$$

that gives an equation

$$\begin{bmatrix} U' \\ V' \end{bmatrix} = \begin{bmatrix} -A^T & C \\ S & A \end{bmatrix} \begin{bmatrix} U \\ V \end{bmatrix} = F \begin{bmatrix} U \\ V \end{bmatrix},$$

for which the solution

$$\begin{bmatrix} U(t) \\ V(t) \end{bmatrix} = e^{A(t-t_0)} \begin{bmatrix} U(t_0) \\ V(t_0) \end{bmatrix} = \begin{bmatrix} \Phi_{11}(t-t_0) & \Phi_{12}(t-t_0) \\ \Phi_{21}(t-t_0) & \Phi_{22}(t-t_0) \end{bmatrix} \begin{bmatrix} U(t_0) \\ V(t_0) \end{bmatrix}$$

yields a solution to (A.25),

$$\begin{aligned} P(t) &= [\Phi_{21}(t-t_0)U(t_0) + \Phi_{22}(t-t_0)V(t_0)] \\ &\quad \times [\Phi_{11}(t-t_0)U(t_0) + \Phi_{12}(t-t_0)V(t_0)]^{-1}. \end{aligned} \tag{A.26}$$

The continuous-time algebraic Riccati equation (CARE):

$$A^TP + PA - PBR^{-1}B^TP + Q = 0 , \tag{A.27}$$

where P is the unknown symmetric matrix and A, B, Q, and R are known real matrices.

The discrete-time algebraic Riccati equation (DARE):

$$P = A^TPA - A^TPB(R + B^TPB)^{-1}B^TPA + Q , \tag{A.28}$$

where P is the unknown symmetric matrix and A, B, Q, and R are known real matrices.

The discrete dynamic (difference) Riccati equation (DDRE):

$$P_{k-1} = Q + A^TP_kA - A^TP_kB(B^TP_kB + R)^{-1}B^TP_kA , \tag{A.29}$$

where Q is a symmetric positive semi-definite matrix and R is a symmetric positive definite matrix.

The discrete-time algebraic Riccati inequality (DARI):

$$A^TPA - A^TPB(R + B^TPB)^{-1}B^TPA + Q - P \geq 0 , \tag{A.30}$$

where P is the unknown symmetric matrix and A, B, Q, and R are known real matrices; $P > 0$, $Q \geq 0$, and $R > 0$.

The nonsymmetric algebraic Riccati equation (NARE):

$$XCX - AX - XD + B = 0 , \tag{A.31}$$

where X is the unknown nonsymmetric matrix and A, B, and C are known matrices; $P > 0$, $Q \geq 0$, and $R > 0$. The NARE is a quadratic matrix equation.

The continuous Lyapunov equation:

$$AX + XA^H + Q = 0 , \tag{A.32}$$

where Q is a Hermitian matrix and A^H is the conjugate transpose of A. The solution to (A.31) is given by

$$X = \int_0^\infty e^{A\tau}Qe^{A^H\tau}\, d\tau . \tag{A.33}$$

The discrete Lyapunov equation:

$$AXA^H - X + Q = 0 , \tag{A.34}$$

where Q is a Hermitian matrix and A^H is the conjugate transpose of A. The solution to (A.33) is given by an infinite sum as

$$X = \sum_{i=0}^{\infty} A^iQ(A^H)^i . \tag{A.35}$$

A.5 Linear Matrix Inequalities

The LMI has the form of [22]

$$F(x) = F_0 + \sum_{i=1}^{m} x_iF_i > 0 , \tag{A.36}$$

where matrix $F(x)$ is positive definite, $x \in \mathbb{R}^m$ is the variable, and the symmetric matrices $F_i = F_i^T \in \mathbb{R}^{n \times n}$, $i \in [0, m]$, are known. The following fundamental properties of the LMI (A.36) are recognized:

- It is equivalent to a set of n polynomial inequalities in x; i.e., the leading principal minors of $F(x)$ must be positive.
- It is a convex constraint on x; i.e., the set $\{x | F(x) > 0\}$ is convex.

The LMI (A.36) can represent a wide variety of convex constraints on x. In particular, this includes linear inequalities, (convex) quadratic inequalities, and matrix norm inequalities. The Riccati and Lyapunov matrix inequalities can also be cast in the form of an LMI.

When the matrices F_i are diagonal, the LMI $F(x) > 0$ is a set of linear inequalities. Nonlinear (convex) inequalities are converted to LMI form using Schur complements (A.22)–(A.24). The basic idea associated with the matrix (A.22) is as follows: the LMI

$$\begin{bmatrix} Q(x) & S(x) \\ S^T(x) & R(x) \end{bmatrix} > 0, \tag{A.37}$$

where $Q(x) = Q^T(x)$, $R(x) = R^T(x)$, and $S(x)$ depend affinely on x, is equivalent to

$$Q(x) - S(x)R^{-1}(x)S^T(x) > 0, \quad R(x) > 0\,. \tag{A.38}$$

It then follows that the set of nonlinear inequalities (A.38) can be represented as the LMI (A.36).

If the Riccati, Lyapunov, and similar equations are written as inequalities, then they can readily be represented as the LMI.

Example A.1 ***Quadratic matrix inequality.*** Given a quadratic matrix inequality

$$A^T P + PA + PBR^{-1}B^T P + Q < 0\,,$$

where A, B, $Q = Q^T$, and $R = R^T > 0$ are matrices of appropriate sizes and $P = P^T$ is the variable. Using the Schur complement, this inequality can be expressed as the LMI

$$\begin{bmatrix} A^T P + PA + Q & PB \\ B^T P & -R \end{bmatrix} < 0\,.$$

It is seen that the LMI is convex in P that does not clearly follows from the original quadratic matrix inequality.

□

A numerical solution to the LMI considered as the convex optimization problem can be found using the interior-point methods.

Appendix B

Norms

The following norms of vectors, matrices, signals, and systems are commonly used in system characterization, signal analysis, and derivation of state estimators.

B.1 Vector Norms

The *norm* is a measure of the size of a *vector*. A norm of an n-dimensional vector $x = [\, x_1 \ x_2 \ \dots \ x_n \,]^T \in \mathbb{R}^n$ is a real-valued function that is commonly denoted as $\|x\|$ and has the following basic properties [18, 52, 59]:

- $\|x\| \geqslant 0$ holds for every vector x.
- $\|\alpha x\| = |\alpha|\|x\|$ holds for every scalar α.
- $\|x + y\| \leqslant \|x\| + \|y\|$ is due to the *triangle inequality* that holds for every vector $x \in \mathbb{R}^n$ and $y \in \mathbb{R}^n$.
- $\|x\| = 0$ if $x = 0$.

The most common vector norms belong to the family of p-norms and often are denoted as the ℓ_p-norm, which is defined by

$$\|x\|_p = \left(\sum_{i=1}^{n} |x_i|^p \right)^{1/p} \tag{B.1}$$

and distinguishes several special cases most often used to measure vectors:

- The ℓ_1-norm is the sum of the absolute values of the components of x:

$$\|x\|_1 = \sum_{i=1}^{n} |x_i| = |x_1| + |x_2| + \cdots + |x_n| \,. \tag{B.2}$$

- The ℓ_2-norm or Euclidian norm:

$$\|x\|_2 = \sqrt{\sum_{i=1}^{n} |x_i|^2} = \sqrt{x^* x} \,, \tag{B.3}$$

where x^* is the conjugate transpose of x.
- The ℓ_∞-norm is the maximum of the absolute values of components of x:

$$\|x\|_\infty = \max_i |x_i| \,. \tag{B.4}$$

Other particular vector norms are less often used.

Optimal and Robust State Estimation: Finite Impulse Response (FIR) and Kalman Approaches, First Edition.
Yuriy S. Shmaliy and Shunyi Zhao.

B.2 Matrix Norms

A *matrix norm* is viewed as a vector norm in a vector space whose elements are matrices. Given matrices $A \in \mathbb{R}^{m\times n}$ and $B \in \mathbb{R}^{m\times n}$ and a scalar α, the following basic properties hold [18, 52, 59]:

- $\|A\| \geq 0$, since norm is *positive-values*.
- $\|\alpha A\| = |\alpha| \, \|A\|$ due to the property of *homogeneity*.
- $\|A + B\| \leqslant \|A\| + \|B\|$ due to the *triangle inequality*.
- $\|AB\| \leqslant \|A\|\|B\|$ for most matrices A and B in $\mathbb{R}^{n\times n}$.
- $\|A\| = 0$ if $A = 0$.

Given a vector $x \in \mathbb{R}^n$ and a matrix $A \in \mathbb{R}^{m\times n}$, the *induced matrix norm* is defined as follows:

$$\begin{aligned} \|A\| &= \sup \frac{\|Ax\|}{\|x\|} , \quad x \neq 0 , \\ &= \sup \|Ax\| , \quad \|x\| = 1 . \end{aligned} \tag{B.5}$$

Accordingly, the induced matrix p-norm is given by

$$\|A\|_p = \sup \frac{\|Ax\|_p}{\|x\|_p} , \tag{B.6}$$

and the following more general induced matrix norm is also used,

$$\|A\|_{\alpha,\beta} = \sup \frac{\|Ax\|_\alpha}{\|x\|_\beta} . \tag{B.7}$$

Given a matrix A with entries a_{ij}, the following special cases of the induced matrix norm are recognized for $\|x\| = 1$:

- The $\|A\|_1$ norm is the maximum absolute column sum of the matrix,

$$\|A\|_1 = \max_{1\leqslant j\leqslant n} \sum_{i=1}^{m} |a_{ij}| . \tag{B.8}$$

- The $\|A\|_\infty$ norm is the maximum absolute row sum of the matrix,

$$\|A\|_\infty = \max_{1\leqslant i\leqslant m} \sum_{j=1}^{n} |a_{ij}| . \tag{B.9}$$

- The $\|A\|_2$ norm is the largest singular value $\sigma_{\max}$ of the matrix,

$$\|A\|_2 = \sigma_{\max}(A) . \tag{B.10}$$

- The $\|A\|_F$ norm is the *Frobenius norm* or *Hilbert-Schmidt norm* that can be defined as

$$\|A\|_F = \sqrt{\mathrm{tr}(A^*A)} , \tag{B.11}$$

  where A^* is the conjugate transpose of A, to be equal to the square root of the sum of the absolute squares of elements of a matrix A.
- The $\|A\|_{\max}$ norm is the largest absolute value of the matrix entries,

$$\|A\|_{\max} = \max_{ij} |a_{ij}| . \tag{B.12}$$

- The $||A||_p$ Schatten p-norm is the p-norm applied to the vector of singular values σ_i of a matrix A,

$$||A||_p = \left(\sum_{i=1}^{\min\{m,n\}} \sigma_i^p \right)^{1/p} . \tag{B.13}$$

The Schatten 2-norm is equal to the Frobenius norm, and the Schatten ∞-norm is the matrix 2-norm.

- The $||A||_*$ nuclear or trace norm is the sum of the singular values of the matrix,

$$||A||_* = \mathrm{tr}\sqrt{A^*A} = \sum_{i=1}^{\min\{m,n\}} \sigma_i . \tag{B.14}$$

Other particular matrix norms are less often used.

B.3 Signal Norms

Signals are viewed as time-varying vectors. Consider a continuous-time vector signal or state vector $x(t) = [x_1(t)\ x_2(t)\ \dots\ x_n(t)]^T \in \mathbb{R}^n$ that has the spectral density $X(j\omega) = [X_1(j\omega)\ X_2(j\omega)\ \dots\ X_n(j\omega)]^T$ by the Fourier transform, $x(t) \overset{\mathcal{F}}{\Leftrightarrow} X(j\omega)$.

The $\mathcal{L}_p$-norm of a continuous-time signal is defined by

$$||x||_p = \left(\int_{-\infty}^{\infty} \sum_{i=1}^{n} |x_i(t)|^p \mathrm{d}t \right)^{1/p} , \tag{B.15}$$

and the following special cases are recognized:

- The $\mathcal{L}_1$-norm:

$$||x||_1 = \int_{-\infty}^{\infty} \sum_{i=1}^{n} |x_i(t)| \mathrm{d}t = \int_{-\infty}^{\infty} ||x(t)||_1 \mathrm{d}t = \sum_{i=1}^{n} ||x_i||_1 . \tag{B.16}$$

- The $\mathcal{L}_2$-norm or Euclidian norm:

$$||x||_2 = \sqrt{\int_{-\infty}^{\infty} x^*(t)x(t)\mathrm{d}t} \tag{B.17}$$

$$= \sqrt{\frac{1}{2\pi} \int_{-\infty}^{\infty} X^*(j\omega)X(j\omega)\mathrm{d}\omega} , \tag{B.18}$$

and (B.18) is due to the Parseval theorem.

- The RMS-norm:

$$||x||_{\mathrm{rms}} = \sqrt{\lim_{T\to\infty} \frac{1}{T} \int_{-T/2}^{T/2} x^*(t)x(t)\mathrm{d}t} . \tag{B.19}$$

- The $\mathcal{L}_\infty$-norm:

$$||x||_\infty = \sup_t \max_i |x_i(t)| = \sup_t ||x(t)||_\infty . \tag{B.20}$$

Given a discrete-time vector signal $x_k = [x_{1k}\ x_{2k}\ \dots\ x_{nk}]^T$ and its spectral density $X(e^{j\phi}) = [\,X_1(e^{j\phi})\ X_2(e^{j\phi})\ \dots\ X_n(e^{j\phi})\,]^T$ by the discrete-time Fourier transform, the norms become

$$\|x\|_p = \left(\sum_{j=0}^{\infty}\sum_{i=1}^{n}|x_{ij}|^p\right)^{1/p}, \tag{B.21}$$

$$\|x\|_1 = \sum_{j=0}^{\infty}\sum_{i=1}^{n}|x_{ij}|, \tag{B.22}$$

$$\|x\|_2 = \sqrt{\sum_{j=0}^{\infty}x_j^*x_j} = \sqrt{\frac{1}{2\pi}\int_{-\pi}^{\pi}X^*(e^{j\phi})X(e^{j\phi})\,d\phi}, \tag{B.23}$$

$$\|x\|_{\text{rms}} = \sqrt{\lim_{N\to\infty}\frac{1}{N+1}\sum_{j=0}^{N}x_j^*x_j}, \tag{B.24}$$

$$\|x\|_\infty = \sup_k \max_i |x_{ik}| = \sup_k \|x_k\|_\infty . \tag{B.25}$$

Note that the square of the $\mathcal{L}_2$-norm represents the total energy of a signal and the RMS-norm the average power of a signal.

B.4 System Norms

Two system norms called the H_2 and H_∞ norms are commonly used to characterize the time-invariant linear systems in continuous-time and discrete-time state space.

Continuous-Time State Space

Consider an LTI system represented in continuous-time state space with equations $\dot{x}(t) = Ax(t) + Bu(t)$ and $y(t) = Cx(t) + Du(t)$. Apply the Laplace transform and arrive at the $u(t)$-to-$y(t)$ transfer function

$$\mathcal{T}(s) = C(sI - A)^{-1}B + D \leftrightarrow \left[\begin{array}{c|c} A & B \\ \hline C & D \end{array}\right] \tag{B.26}$$

that is coupled with the impulse response matrix $\mathcal{H}(t)$ as $\mathcal{T}(s) \overset{\mathcal{L}}{\Leftrightarrow} \mathcal{H}(t)$. The following basic norms are used to characterize such systems:

- The H_2 norm of a stable system $\left[\begin{array}{c|c} A & B \\ \hline C & 0 \end{array}\right]$ is defined as [232]

$$\|\mathcal{T}\|_2 = \|\mathcal{H}\|_2 = \sqrt{\frac{1}{2\pi}\int_{-\infty}^{\infty}\text{tr}\,[\mathcal{T}^*(j\omega)\mathcal{T}(j\omega)]\,d\omega} \tag{B.27}$$

$$= \sqrt{\int_{-\infty}^{\infty}\text{tr}\,[\mathcal{H}^*(t)\mathcal{H}(t)]\,dt}\,. \tag{B.28}$$

Referring to (B.26), the H_2 norm (B.27) can be computed as

$$\|\mathcal{T}\|_2^2 = \text{tr}(B^*L_oB) = \text{tr}(CL_cC^*)\,, \tag{B.29}$$

where L_o and L_c are observability and controllability Gramians,

$$L_o = \int_0^\infty e^{A^*t} C^* C e^{At} \, dt \, , \quad L_c = \int_0^\infty e^{At} B B^* e^{A^*t} \, dt \, ,$$

which can also be obtained from the following Lyapunov equations:

$$A^* L_o + L_o A + C^* C = 0 \, , \quad A L_c + L_c A^* + B B^* = 0 \, .$$

- The H_∞ norm of a linear system $\left[\begin{array}{c|c} A & B \\ \hline C & D \end{array}\right]$ is defined as [70]

$$\|\mathcal{T}\|_\infty = \sup_\omega \sigma_{\max} [\mathcal{T}(j\omega)] \, , \tag{B.30}$$

where $\sigma_{\max}$ is the maximum singular value of $\mathcal{T}(s)$ and it follows that the H_∞ system norm is the highest peak value in the Bode plot. By considering the 2-induced norm of a matrix and referring to the system input $u(t)$ and output $y(t)$, it can be shown that the H_∞ system norm is obtained by

$$\|\mathcal{T}\|_\infty = \sup_{\|u\|_2 < \infty} \frac{\|y\|_2}{\|u\|_2} = \sup_{\|u\|_2 < \infty} \frac{\|\mathcal{T} u\|_2}{\|u\|_2} \tag{B.31}$$

that represents the worst case of the system gain. The H_∞ norm of a transfer function $\mathcal{T}(s)$ thus equals the 2-induced norm of the input-to-output or *energy-to-energy* mapping.

Discrete-Time FE-Based State Space

For an LTI system represented in discrete-time state space with equations $x_{k+1} = Fx_k + Bu_k$ and $y_k = Hx_k + Du_k$ using the FE method, the following input-to-output transfer function $\mathcal{T}(z)$ is produced by the z-transform,

$$\mathcal{T}(z) = H(zI - F)^{-1} B + D \, , \tag{B.32}$$

and the H_2 and H_∞ system norms are defined as:

- The H_2 norm of a stable system with $D = 0$ is given by

$$\|\mathcal{T}(z)\|_2 = \sqrt{\frac{1}{2\pi} \int_0^{2\pi} \text{tr} \, [\mathcal{T}^*(e^{j\phi}) \mathcal{T}(e^{j\phi})] \, d\phi} \tag{B.33}$$

$$= \text{tr}(B^* L_o B) = \text{tr}(H L_c H^*) \, , \tag{B.34}$$

where L_o and L_c are the unique solutions of discrete-time Lyapunov equations:

$$F^* L_o F - L_o + H^* H = 0 \, , \quad F L_c F^* - L_c + B B^* = 0 \, .$$

- The H_∞ norm is defined by

$$\|\mathcal{T}(z)\|_\infty = \sup_{-\pi \leqslant \phi \leqslant \pi} \sigma_{\max} [\mathcal{T}(e^{j\phi})] \tag{B.35}$$

and can also be computed by (B.31) that holds universally for continuous-time and discrete-time systems.

Discrete-Time BE-Based State Space

A specific feature of an LTI system represented in discrete-time state space using the BE method with equations $x_k = Fx_{k-1} + Bu_k$ and $y_k = Hx_k + Du_k$ is in another transfer function

$$\mathcal{T}(z) = H(zI - F)^{-1}zB + D \tag{B.36}$$

that, however, does not affect the H_2 and H_∞ norms, which are still defined by (B.32)–(B.34) like for the FE-based model.

Appendix C

Matlab Codes

The following Matlab codes can be used in the design of batch and iterative FIR state estimators.

C.1 Batch UFIR Filter

```
% A: state transaction matrix
% B: control input matrix
% U: Control Input
% Y: Measurement
% Q: Process noise covariance
% R: Measurement noise covariance
% H: Measurement matrix
% The state-space model is x = A x + B u + G w
% and y = H x + v
% Please note Q and R are only involved in
% the computation of error covariance.

% Important:
% U: Control Input collected within [m,k]
% Y: Measurements collected within [m,k]
% N: Estimation Horizon Length
% The Extended State-Space Model is:
% X = ExA  * x_{m-1} + ExB * U  + ExG * W
% Y = ExC  * x_{m-1} + ExD * U  + ExH * W  + V

Call [m,P]= Batch_TIVM_UFIR(Q,R,A,B,H,G,Y,U,N);

% This is the main part of the UFIR filter
function [m,P] = Batch_TIVM_UFIR(Q,R,A,B,H,G,Y,U,N)
    dim_x = size(A,1);
    dim_u = size(B,2);
    ExC = H ; ExB = B; ExQ = Q; ExR = R; ExG = G; DiagC = H;
for i = 2: N
        ExC = [C*A^i; ExC];
        ExB = blkdiag(B,ExB);
        ExG = blkdiag(G,ExG);
        ExQ = blkdiag(Q,ExQ);
        ExR = blkdiag(R,ExR);
        DiagC = blkdiag(C,DiagC);
end

for j=1:N-1
```

Optimal and Robust State Estimation: Finite Impulse Response (FIR) and Kalman Approaches, First Edition.
Yuriy S. Shmaliy and Shunyi Zhao.

```
        for i = 1 : j
            W=eye(dim_x);
            for h=i:j
                W= W * A;
            end
           ExB(:,j*dim_u+1:(j+1)*dim_u)=ExB(:,j*dim_u+1:(j+1)*dim_u)
           +[zeros((i-1)*dim_x,dim_u); W*B; zeros((N-i)*dim_x,dim_u)];
        end
    end

for j=1:N-1                                 % Extended G
    for i=1:j
        W=eye(dim_x);
        for h=i:j
            W= W * A;
        end
       ExG(:,j*dim_x+1:(j+1)*dim_x)=ExG(:,j*dim_x+1:(j+1)*dim_x)
       +[zeros((i-1)*dim_x,dim_x); W*G; zeros((N-i)*dim_x,dim_x)];
    end
end
%   Calculate the state estimate
    ExD  =  DiagC  *  ExB;
    ExH  =  DiagC  *  ExG;
    Gain =  A^(N-1) * ExC' * inv(ExC' * ExC);
    m    =  Gain * Y + (ExB(1 : dim_x,:) - Gain * ExD) * U;
    P    =  (ExG(1 : dim_x, :) - Gain * ExH) * ExQ
         *  (ExG(1 : dim_x, :) - Gain * ExH)'
         +  Gain * ExR * Gain';
end
```

C.2 Iterative UFIR Filtering Algorithm

```
Call [m,G,P]= UFIR_Initionlization(Q,R,A,B,H,G,Y,U,N);

for n = N+dim_x : N
[m, PowG] = UFIR_filter(m,A,H,B,u,G,Y(:,n));
end

function[m,G,P] = UFIR_Initionlization(A,H,B,U,Y,Q,R)
    dim_x = size(A,1);
    dim_u = size(B,2);
    ExH = H * A;
    ExB = B;
    ExQ = Q;
    ExR = R;
    ExE = eye(dim_x);
    DiagH = H;

  for i = 2: dim_x
        ExH = [H*A^i; ExH];
        ExB = blkdiag(B,ExB);
        ExE = blkdiag(eye(dim_x),ExE);
        ExQ = blkdiag(Q,ExQ);
        ExR = blkdiag(R,ExR);
        DiagH = blkdiag(H,DiagH);
   end
```

```
  N= dim_x;
  for j=1:N-1
      for i=1:j
          W=eye(dim_x);
          for h=i:j
              W= W * A;
          end
         ExB(:,j*dim_u+1:(j+1)*dim_u)=ExB(:,j*dim_u+1:(j+1)*dim_u)
         +[zeros((i-1)*dim_x,dim_u); W*B; zeros((N-i)*dim_x,dim_u)];
      end
  end

 N= dim_x;
 % Compute the extended matrices
 for j=1:N-1
      for i=1:j
          W=eye(dim_x);
          for h=i:j
              W= W * A;
          end
         ExE(:,j*dim_x+1:(j+1)*dim_x)=ExE(:,j*dim_x+1:(j+1)*dim_x)
         +[zeros((i-1)*dim_x,dim_x); W; zeros((N-i)*dim_x,dim_x)];
      end
 end

      ExEE = DiagH * ExE;      % H matric in paper
      L    = DiagH * ExB;
      G    = A ^dim_x * inv(ExH' * ExH) * (A^(dim_x))';
      % Noise Power Gain
      Gain = A^(dim_x) * (ExH' * ExH) ^-1 * ExH';
      % Bias Correction
      m    = Gain * Y + (ExB(1:dim_x,:) - Gain * L) * U;
      % State Estimate
      P    = (Gain * ExEE  -ExE(1:dim_x,:) ) * ExQ
           * (Gain * ExEE  -ExE(1:dim_x,:))'
           + Gain * ExR * Gain';
      % Noise Covariance
end
% This part is the main part.
function [m, PowG] = UFIR_filter(m,A,H,B,u,G,Y)
Pm   = A * m + B * u;
PowG = A * PowG *A'-A * PowG *A'*inv(eye(size(H'*H *A * PowG *A'))
     + H'*H *A * PowG *A')* H'* H *A * PowG *A';
Gain = PowG * H';
m    = Pm + Gain * (Y - H * Pm);
end
```

C.3 Batch OFIR Filter

```
function [m,P] = Batch_TIVM_OFIR(Q,R,A,B,H,G,Y,U,N,Init_x,Init_P)
     dim_x = size(A,1);
     dim_u = size(B,2);
     ExC = H ; ExB = B; ExQ = Q; ExR = R; ExG = G; DiagC = H;
for i = 2: N
         ExC = [C*A^i; ExC];
         ExB = blkdiag(B,ExB);
         ExG = blkdiag(G,ExG);
         ExQ = blkdiag(Q,ExQ);
```

```
        ExR = blkdiag(R,ExR);
        DiagC = blkdiag(C,DiagC);
end

for j= 1 : N-1
    for i = 1 : j
        W=eye(dim_x);
        for h=i:j
            W= W * A;
        end
       ExB(:,j*dim_u+1:(j+1)*dim_u)=ExB(:,j*dim_u+1:(j+1)*dim_u)
       +[zeros((i-1)*dim_x,dim_u); W*B; zeros((N-i)*dim_x,dim_u)];
    end
 end

for j=1 : N-1                              %   Extended G
    for i=1:j
        W=eye(dim_x);
        for h=i:j
            W= W * A;
        end
       ExG(:,j*dim_x+1:(j+1)*dim_x)=ExG(:,j*dim_x+1:(j+1)*dim_x)
       +[zeros((i-1)*dim_x,dim_x); W*G; zeros((N-i)*dim_x,dim_x)];
    end
end
%   Calculate the state estimate
    ExD    =  DiagC  *  ExB;
    ExH    =  DiagC  *  ExG;
    Lamda  =  Init_x * Init_x' + Init_P;
    Z      =  ExC * Lamda * ExC'+ ExH * ExQ * ExH' + ExR;
    Gain   =  (A^(N-1) * Lamda * ExC' + ExG(1 : dim_x,:)* ExQ * ExH')   * inv(Z);
    m      =  Gain * Y + (ExB(1 : dim_x,:) - Gain * ExD) * U;
    P      =  (A^(N-1) - Gain * ExC) * Lamda * (A^(N-1) - Gain * ExC)'
           +  (ExG(1 : dim_x, :) - Gain * ExH) * ExQ
           *  (ExG(1 : dim_x, :) - Gain * ExH)'
           +  Gain * ExR * Gain';
end
```

C.4 Iterative OFIR Filtering Algorithm

```
l = n - N + 1;
m = Init_x;
P = Init_P;
for i = l + 1 : n
  [m,P] = Iterative_TIVM_OFIR(Q,R,A,B,H,G,Y(i),U(i),m,P)
end

% Iterative_TIVM_OFIR is given as follows

function [m,P] = Iterative_TIVM_OFIR(Q,R,A,B,H,G,Y,U,N,m,P)
Pred_x = A * m + B * u;
Pred_P = A * P * A + G * Q * G';
S      = H * Pred_P * H' + R;
Gain   = Pred_P * H' * inv(S);
m      = Pred_x + Gain * (Y - H * Pred_x);
P      = Pred_P  -  Gain * H * Pred_P;
end
```

C.5 Batch OUFIR Filter

```
function [m,P] = Batch_TIVM_OUFIR(Q,R,A,B,H,G,Y,U)
    dim_x = size(A,1);
    dim_u = size(B,2);
    ExC = H ; ExB = B; ExQ = Q; ExR = R; ExG = G; DiagC = H; N = dim_x;
for i = 2: N
        ExC = [C*A^i; ExC];
        ExB = blkdiag(B,ExB);
        ExG = blkdiag(G,ExG);
        ExQ = blkdiag(Q,ExQ);
        ExR = blkdiag(R,ExR);
        DiagC = blkdiag(C,DiagC);
end
for j= 1 : N-1
    for i = 1 : j
        W=eye(dim_x);
        for h=i:j
            W= W * A;
        end
       ExB(:,j*dim_u+1:(j+1)*dim_u)=ExB(:,j*dim_u+1:(j+1)*dim_u)
       +[zeros((i-1)*dim_x,dim_u); W*B; zeros((N-i)*dim_x,dim_u)];
    end
 end
for j=1:N-1                          %   Extended G
    for i=1:j
        W=eye(dim_x);
        for h=i:j
            W= W * A;
        end
       ExG(:,j*dim_x+1:(j+1)*dim_x)=ExG(:,j*dim_x+1:(j+1)*dim_x)
       +[zeros((i-1)*dim_x,dim_x); W*G; zeros((N-i)*dim_x,dim_x)];
    end
end
%   Calculate the state estimate
    ExD    =  DiagC  *  ExB;
    ExH    =  DiagC  *  ExG;
    Z      =  ExH * ExQ * ExH' + ExR;
    GPart1 =  A^(N-1) * (ExC' * inv(Z) * ExC)^-1 * ExC' * inv(Z);
    GPart2 =  ExG(1:dim_x,:) * ExQ * ExH' * inv(Z);
    Gpart3 =  ExG(1:dim_x,:) * ExQ * ExH'* inv(Z)
           *  ExC * inv(ExC' * inv(Z) * ExC) * ExC' * inv(Z);
    Gain   =  GPart1 + GPart2 + GPart3;
    m      =  Gain * Y + (ExB(1 : dim_x,:) - Gain
           * ExD) * U;
    P      =  (ExG(1 : dim_x, :) - Gain * ExH) * ExQ
           *  (ExG(1 : dim_x, :) - Gain * ExH)'
           +  Gain * ExR * Gain';
end
```

C.6 Iterative OUFIR Filtering Algorithm

This code has two parts.

```
% Part-I
function [m,Delta,BarP,TilC] = OUFIR_Initialization(Q,R,A,B,H,G,Y,U)
    dim_x = size(A,1);
    dim_u = size(B,2);
```

```
    ExC = H ; ExB = B; ExQ = Q; ExR = R; ExG = G;
    DiagC = H; N = dim_x;
for i = 2: N
        ExC = [C*A^i; ExC];
        ExB = blkdiag(B,ExB);
        ExG = blkdiag(G,ExG);
        ExQ = blkdiag(Q,ExQ);
        ExR = blkdiag(R,ExR);
        DiagC = blkdiag(C,DiagC);
end
for j= 1 : N-1
    for i = 1 : j
        W=eye(dim_x);
        for h=i:j
            W= W * A;
        end
       ExB(:,j*dim_u+1:(j+1)*dim_u)=ExB(:,j*dim_u+1:(j+1)*dim_u)
       +[zeros((i-1)*dim_x,dim_u); W*B; zeros((N-i)*dim_x,dim_u)];
    end
end
for j=1:N-1                              %   Extended G
    for i=1:j
        W=eye(dim_x);
        for h=i:j
            W= W * A;
        end
       ExG(:,j*dim_x+1:(j+1)*dim_x)=ExG(:,j*dim_x+1:(j+1)*dim_x)
       +[zeros((i-1)*dim_x,dim_x); W*G; zeros((N-i)*dim_x,dim_x)];
    end
end
%   Calculate the state estimate
    ExD     =  DiagC   *   ExB;
    ExH     =  DiagC   *   ExG;
    Z       =  ExH * ExQ * ExH' + ExR;
    TilC    =  ExC' * inv(Z) * ExC;
    GPart1  =  A^(N-1) * (ExC' * inv(Z) * ExC)^-1 * ExC' * inv(Z);
    GPart2  =  ExG(1:dim_x,:) * ExQ * ExH' * inv(Z);
    Gpart3  =  ExG(1:dim_x,:) * ExQ * ExH'* inv(Z)
            *  ExC * inv(ExC' * inv(Z) * ExC) * ExC' * inv(Z);
    Gain    =  GPart1 + GPart2 + GPart3;
    BarP    =  A^(dim_x-1) - A * ExG(1 : dim_x, :)
            *  ExQ * ExG' * inv(TilC) * ExH;
    m       =  Gain * Y + (ExB(1 : dim_x,:) - Gain
            * ExD) * U;
    Delta   =  (ExG(1 : dim_x, :) - Gain * ExH) * ExQ
            *  (ExG(1 : dim_x, :) - Gain * ExH)'
            +  Gain * ExR * Gain';
end

% Part-II
function [m, Delta] = Iterative_OUFIR(Q,R,A,B,H,G,Y,U,m,Delta,BarP,TilC)

Predx = A * m  + B * u;
BarL = P  * H' + inv(H * P * H; + R);
PredDelta = A * Delta * A' +  G * Q * G';

P   =  A * P * A' +  G * Q * G' - A * BarL * H
    * P * A';
```

```
BarL = P * H' * inv(H * P * H' + R);
BarP = (A - A * BarL * H)* BarP;
TilC = TilC + BarP * H' * inv(H*P*H'+R)*H*BarP;
TilL = (eye(size(BarL*H))- BarL*H)*BarP
     *inv(TilC)* BarP' *H'
     * inv(H*P*H+ R);

Gain  =  BarL + TilL;
m     =  Predx  + Gain * (Y - H * Predx);

Delta = (eye(size(Gain * H)) - Gain * H) * PredDelta
      * (eye(size(Gain * H)) - Gain * H)'
      +  Gain * R * Gain';

end
```

C.7 Batch q-Lag UFIR Smoother

```
function [m,P,NPG] = UFIR_TIV_Smoother(Q,R,A,B,H,G,Y,U,N,q)
    dim_x = size(A,1);
    dim_u = size(B,2);
    ExC = H;
    ExB = B;
    ExQ = Q;
    ExR = R;
    ExG = G;
    DiagC = H;
 for i = 1: N-1
        ExC = [H*A^i; ExC];
        ExB = blkdiag(B,ExB);
        ExG = blkdiag(G,ExG);
        ExQ = blkdiag(Q,ExQ);
        ExR = blkdiag(R,ExR);
        DiagC = blkdiag(H,DiagC);
 end
 % Extend B
 for j=1:N-1
    for i=1:j
        W=eye(dim_x);
        for h=i:j
            W= W * A;
        end
       ExB(:,j*dim_u+1:(j+1)*dim_u)=ExB(:,j*dim_u+1:(j+1)*dim_u)
       +[zeros((i-1)*dim_x,dim_u);W*B;zeros((N-i)*dim_x,dim_u)];
    end
 end
% Extend G
for j=1:N-1
    for i=1:j
        W=eye(dim_x);
        for h=i:j
            W= W * A;
        end
       ExG(:,j*dim_x+1:(j+1)*dim_x)=ExG(:,j*dim_x+1:(j+1)*dim_x)
       +[zeros((i-1)*dim_x,dim_x);W*G;zeros((N-i)*dim_x,dim_x)];
    end
end
```

```
    ExD    =  DiagC    *  ExB;
    ExH    =  DiagC    *  ExG;
    GainH =  A^(N-q-1)  *  (ExC' * ExC)^-1 * ExC';
    NPG    =  GainH* GainH';
    GainF =  ExB(dim_x*(q)+1 : dim_x*(q+1),:) - GainH  * ExD;
    m      =  GainH * Y + GainF * U;         % State Estimate
    P      =  (ExG(dim_x*(q)+1:dim_x*(q+1),:) - GainH * ExH)
           *  ExQ *(ExG(dim_x*(q)+1:dim_x*(q+1),:)-GainH * ExH)'
           +  GainH * ExR * GainH';
end
```

C.8 Batch *q*-Shift FFFM OFIR Smoother

```
function [m,P] = Batch_TIVM_FFFM(Q,R,A,B,H,G,Y,U,N,q,x,P)
    dim_x = size(A,1);
    dim_u = size(B,2);
    ExC = H;
    ExB = B;
    ExQ = Q;
    ExR = R;
    ExG = G;
    DiagC = H;
  for i = 2: N
        ExC = [H*A^i; ExC];
        ExB = blkdiag(B,ExB);
        ExG = blkdiag(G,ExG);
        ExQ = blkdiag(Q,ExQ);
        ExR = blkdiag(R,ExR);
        DiagC = blkdiag(H,DiagC);
  end
  %  Extend B  when j =1, i =1,   W = A, ExB(:,2:2)  = ExB(:,2:2)
  %  +[A*B; zeros((N-1)* dim_x, dim_u)] when j =2, i ={1,2},
  %  W={A, A^2}, ExB(:,3:3)  = ExB(:,3:3)  +[A^2*B; zeros]
 for j=1:N-1
   for i = 1 : j
        W=eye(dim_x);
        for h=i:j
            W= W * A;
        end
       ExB(:,j*dim_u+1:(j+1)*dim_u)=ExB(:,j*dim_u+1:(j+1)*dim_u)
       +[zeros((i-1)*dim_x,dim_u);W*B;zeros((N-i)*dim_x,dim_u)];
    end
 end
% Extend  G
for j=1:N-1
    for i=1:j
        W=eye(dim_x);
        for h=i:j
            W= W * A;
        end
       ExG(:,j*dim_x+1:(j+1)*dim_x)=ExG(:,j*dim_x+1:(j+1)*dim_x)
       +[zeros((i-1)*dim_x,dim_x);W*G;zeros((N-i)*dim_x,dim_x)];
    end
end

    ExD    =  DiagC    *  ExB;
```

```
    ExH    =  DiagC    *  ExG;
    Lamda  =  x * x'   + P;
    Z      =   ExC * Lamda * ExC'+ ExH * ExQ * ExH' + ExR;
    Gain   =   (A^(N-q) * Lamda * ExC'
           + ExG(dim_x*(q+1)-1:dim_x*(q+1),:)* ExQ * ExH')
* inv(Z);
    m      =   Gain * Y + (ExB(dim_x*(q+1)-1:dim_x*(q+1),:)
           - Gain * ExD) * U;      % State Estimate
    P      =   (A^(N-q) - Gain * ExC) * Lamda * (A^(N-q) - Gain * ExC)'
           +  (ExG(dim_x*(q+1)-1:dim_x*(q+1),:) - Gain * ExH)
           * ExQ *(ExG(dim_x*(q+1)-1:dim_x*(q+1),:)
           - Gain * ExH)' +  Gain * ExR * Gain'; % Estimation Error Covariance
end
```

C.9 Batch *q*-Lag FFBM OFIR Smoother

The code of this algorithm has two parts.

```
% Part I
function[BarB, BarG]=Batch_TIVM_FFBM(Q,R,A,B,H,G,N,q)
    dim_x = size(A,1);
    dim_u = size(B,2);
    ExC = H * eye(dim_x);
    ExB = inv(A)*B;
    ExQ = Q;
    ExR = R;
    ExG = inv(A)*G;
    DiagC = H;
 for i = 1 : N-1
     ExC    =  [ExC; H*A^-i];
     ExQ    =  blkdiag(Q,ExQ);
     ExR    =  blkdiag(R,ExR);
     DiagC =  blkdiag(H,DiagC);
 end
 % One part of the extended matrices
     NewN = N-1;
    for i = 1 : NewN-1                                     ExB = blkdiag(inv(A)*B,ExB);
        ExG = blkdiag(inv(A)*G,ExG);
    end

for j= NewN - 1 : -1 : 1
    for i = 1 : (NewN - j)
        W=eye(dim_x);
        for h= (N - j) : -1 : i
            W= W * inv(A);
        end
 ExB(:,(j-1)*dim_u+1 : j*dim_u) = ExB(:,(j-1)*dim_u+1 : j*dim_u)
 +[zeros((NewN - i)*dim_x,dim_u); W*B; zeros((i-1)*dim_x,dim_u)];
    end
 end
 ExB = [[zeros(dim_x, (N-1)*dim_u); ExB] zeros(dim_x * N, dim_u)];
for j = NewN - 1 : -1 : 1
%  Extend  G
for i= 1 : (NewN - j)
     W=eye(dim_x);
     for h= (N - j) : -1 : i
         W= W * inv(A);
     end
```

```
    ExG(:,(j-1)*dim_x+1 : j*dim_x)= ExG(:,(j-1)*dim_x+1 : j*dim_x)
        +[zeros((NewN-i)*dim_x,dim_x);W*G;zeros((i-1)*dim_x,dim_x)];
     end
end
ExG   = [[zeros(dim_x,  (N-1)*dim_x); ExG] zeros(dim_x  * N, dim_x)];
BarB =   ExB(dim_x*(q)  + 1  : dim_x*(q+1), :);
BarG =   ExG(dim_x*(q)  + 1  : dim_x*(q+1), :);
end

% Part II
function [m,P] = Batch_FFFM_Model(Q,R,A,B,H,G,Y,U,N,q,x,P,BarB, BarG)
    dim_x = size(A,1);
    dim_u = size(B,2);
    ExC = H * A;
    ExB = B;
    ExQ = Q;
    ExR = R;
    ExG = G;
    DiagC = H;
for i = 2: N
        ExC = [H*A^i; ExC];
        ExB = blkdiag(B,ExB);
        ExG = blkdiag(G,ExG);
        ExQ = blkdiag(Q,ExQ);
        ExR = blkdiag(R,ExR);
        DiagC = blkdiag(H,DiagC);
end

for j=1:N-1
   for i = 1 : j
        W=eye(dim_x);
        for h=i:j
            W= W * A;
        end
       ExB(:,j*dim_u+1:(j+1)*dim_u)=ExB(:,j*dim_u+1:(j+1)*dim_u)
       +[zeros((i-1)*dim_x,dim_u);W*B;zeros((N-i)*dim_x,dim_u)];
    end
 end

for j=1:N-1
    for i=1:j
        W=eye(dim_x);
        for h=i:j
            W= W * A;
        end
       ExG(:,j*dim_x+1:(j+1)*dim_x)=ExG(:,j*dim_x+1:(j+1)*dim_x)
       +[zeros((i-1)*dim_x,dim_x);W*G;zeros((N-i)*dim_x,dim_x)];
    end
end
    ExD    =  DiagC    *  ExB;
    ExH    =  DiagC    *  ExG;
    Lamda =  x * x'   + P;
    Z      =   ExC * Lamda * ExC'+ ExH * ExQ * ExH' + ExR;
    BackS = A^-(q)  * ExB(1  : dim_x,:)   - BarB;
    BackD = A^-(q)  * ExG(1  : dim_x,:)   - BarG;
    Gain   =   (A^(N-q-1)  * Lamda * ExC'  + BackD * ExQ * ExH')
 * inv(Z);
    m      =   Gain * Y + (BackS - Gain * ExD) * U;
```

```
     P       =   (A^(N-q-1) - Gain * ExC) * Lamda * (A^(N-q-1) - Gain * ExC)'
             +   (BackD - Gain * ExH) * ExQ *(BackD- Gain * ExH)'
             +   Gain * ExR * Gain';
end
```

C.10 Batch *q*-Lag BFFM OFIR Smoother

The code of this algorithm has two parts.

```
% Part-I
function [ForExG_g, ForExB_g] = Batch_Forward(Q,R,A,B,H,G,N,g)
     dim_x = size(A,1);
     dim_u = size(B,2);
     ExC = H * A;
     ExB = B;
     ExQ = Q;
     ExR = R;
     ExG = G;
     DiagC = H;
for i = 2: N
          ExC   = [H*A^i; ExC];
          ExB   = blkdiag(B,ExB);
          ExG   = blkdiag(G,ExG);
          ExQ   = blkdiag(Q,ExQ);
          ExR   = blkdiag(R,ExR);
        DiagC = blkdiag(H,DiagC);
end

for j=1:N-1
     for i = 1 : j
          W=eye(dim_x);
          for h=i:j
               W= W * A;
          end
      ExB(:,j*dim_u+1:(j+1)*dim_u)=ExB(:,j*dim_u+1:(j+1)*dim_u)
      +[zeros((i-1)*dim_x,dim_u);W*B;zeros((N-i)*dim_x,dim_u)];
     end
end

for j=1:N-1
     for i=1:j
          W=eye(dim_x);
          for h=i:j
               W= W * A;
          end
        ExG(:,j*dim_x+1:(j+1)*dim_x)=ExG(:,j*dim_x+1:(j+1)*dim_x)
        +[zeros((i-1)*dim_x,dim_x);W*G;zeros((N-i)*dim_x,dim_x)];
     end
end
     ForExG_g = ExG(dim_x*(N-g-1)+1  :  dim_x *  (N-g),  :);
     ForExB_g = ExB(dim_u*(N-g-1)+1  :  dim_u *  (N-g),  :);
end

% Part-II
function [m,P] = Batch_TIVM_BFFM(Q,R,A,B,H,G,Y,U,N,g,x,P,ForExG_g,ForExB_g)
     dim_x = size(A,1);
     dim_u = size(B,2);
```

```
    ExC    = H * eye(dim_x);
    ExB    = inv(A)*B;
    % This is very important x_m = x_m + + w;
    ExQ    = zeros(2,2);
    ExR    = R;
    ExG    = inv(A)*G;
    DiagC = H;
for i = 1 : N-1
     ExC    =   [ExC; H*inv(A^i)];
     ExQ    =   blkdiag(Q,ExQ);
     ExR    =   blkdiag(R,ExR);
     DiagC =   blkdiag(C,DiagC);
end

    NewN = N-1;
    for i = 1 : NewN-1
        ExB = blkdiag(inv(A)*B,ExB);
        ExG = blkdiag(inv(A)*G,ExG);
    end

for j= NewN - 1 : -1 : 1
    for i = 1 : (NewN - j)
        W=eye(dim_x);
        for h= (N - j) : -1 : i
            W= W * inv(A);
        end
       ExB(:,(j-1)*dim_u+1 : j*dim_u) = ExB(:,(j-1)*dim_u+1 : j*dim_u)
       +[zeros((NewN - i)*dim_x,dim_u); W*B; zeros((i-1)*dim_x,dim_u)];
    end
end

ExB = [[zeros(dim_x, (N-1)*dim_u); ExB] zeros(dim_x * N, dim_u)];
for j = NewN - 1 : -1 : 1
    for i= 1 : (NewN - j)
        W=eye(dim_x);
        for h= (N - j) : -1 : i
            W= W * inv(A);
        end
       ExG(:,(j-1)*dim_x+1 : j*dim_x)=ExG(:,(j-1)*dim_x+1 : j*dim_x)
       +[zeros((NewN-i)*dim_x,dim_x); W*G; zeros((i-1)*dim_x,dim_x)];
    end
end
ExG   = [[zeros(dim_x, (N-1)*dim_x); ExG] zeros(dim_x * N, dim_x)];
ExD   =  DiagC   *  ExB;
ExH   =  DiagC   *  ExG;
Lamda =   x * x'  + P;
Z     =    ExC * Lamda * ExC'+ ExH * ExQ * ExH' + ExR;
Gain  =   (A^-(N-1-g) * Lamda * ExC' - (ForExG_g - A^(g)
       * ExG(dim_x * (N-1)+1 : dim_x * N, :) )
       * ExQ * ExH') * inv(Z);  % g is the (N-g) the raw vector
m     =   Gain * Y  + (ForExB_g - A^(g) * ExB(dim_x
       * (N-1)+1 : dim_x * N,:)
       + Gain * ExD) * U;
MatB  =  A^-(N-1-g) - Gain * ExC ;
MatW  = ForExG_g - A^g * ExG(dim_x * (N-1)+1 : dim_x * N,:) + Gain * ExH;
P     =    MatB * Lamda *  MatB' + MatW * ExQ *  MatW' +  Gain * ExR * Gain';
end
```

C.11 Batch q-Lag BFBM OFIR Smoother

```
function [m,P] = Batch_TIVM_BFBM(Q,R,A,B,H,G,Y,U,N,g,x,P)
    dim_x = size(A,1);
    dim_u = size(B,2);
    ExC   = H * eye(dim_x);
    ExB   = inv(A)*B;
    ExQ   = Q;
    ExR   = R;
    ExG   = inv(A)*G;
    DiagC = H;
for i = 1 : N-1
     ExC  =  [ExC; H*inv(A^i)];
     ExQ  =  blkdiag(Q,ExQ);
     ExR  =  blkdiag(R,ExR);
    DiagC =  blkdiag(H,DiagC);
end
    NewN = N-1;
    for i = 1 : NewN-1
      ExB = blkdiag(inv(A)*B,ExB);
      ExG = blkdiag(inv(A)*G,ExG);
    end
 for j= NewN - 1 : -1 : 1
    for i = 1 : (NewN - j)
    W=eye(dim_x);
        for h= (N - j) : -1 : i
            W= W * inv(A);
        end
      ExB(:,(j-1)*dim_u+1 : j*dim_u) = ExB(:,(j-1)*dim_u+1 : j*dim_u)
        +[zeros((NewN - i)*dim_x,dim_u); W*B; zeros((i-1)*dim_x,dim_u)];
    end
 end
ExB = [[zeros(dim_x, (N-1)*dim_u); ExB] zeros(dim_x * N, dim_u)];
for j = NewN - 1 : -1 : 1
    for i= 1 : (NewN - j)
        W=eye(dim_x);
        for h= (N - j) : -1 : i
            W= W * inv(A);
        end
      ExG(:,(j-1)*dim_x+1 : j*dim_x)=ExG(:,(j-1)*dim_x+1 : j*dim_x)
      +[zeros((NewN-i)*dim_x,dim_x); W*G; zeros((i-1)*dim_x,dim_x)];
    end
end
ExG   = [[zeros(dim_x, (N-1)*dim_x); ExG] zeros(dim_x * N, dim_x)];
ExD   =  DiagC   *  ExB;
ExH   =  DiagC   *  ExG;
Lamda =  x * x'  + P;
Z     =   ExC * Lamda * ExC'+ ExH * ExQ * ExH' + ExR;
Gain  =  (A^-(N-g-1) * Lamda * ExC' + ExG(dim_x*(N-g-1) + 1 : dim_x*(N-g), :)
      * ExQ * ExH')  * inv(Z);
 m    =   Gain * Y
      + (ExB(dim_x*(N-g-1)+1:dim_x*(N-g),:)
      + Gain*ExD) * U;
 P    =  (A^-(N-g-1) - Gain * ExC) * Lamda * (A^-(N-g-1) - Gain * ExC)'
      + (ExG(dim_x*(N-g-1)+1:dim_x*(N-g),:)-Gain * ExH) * ExQ
      * (ExG(dim_x*(N-g-1)+1:dim_x*(N-g),:)
      - Gain * ExH)' +  Gain * ExR * Gain';
end
```

References

1 Abedor J, Nagpal K, Poola K (1996) A linear matrix inequality approach to peak-to-peak gain minimization. Int J Robust Nonl Contr 6:899–927

2 Ahmed NA, Gokhale DV (1989) Entropy expressions and their estimators for multivariate distributions. IEEE Trans Inf Theory 35(3):688–692

3 Ahn CK, Han SH (2008) New H_∞ FIR smoother for linear discrete-time state-space models. IEICE Trans Commun E91.B(3):896–899

4 Ahn CK, Kim PS (2008) Fixed-lag maximum likelihood FIR smoother for state-space models. IEICE Electron Express 5(1):11–16

5 Ahn CK, Kim PS (2013) New energy-to-peak FIR filter design for systems with disturbances: a convex optimization approach. Int J Inn Comp Inform Contr 9(5):1988–1993

6 Ahn CK (2014) A new solution to induced l_∞ FIR filtering problem based on two matrix inequalities. Int J Contr 87(2):404–409

7 Ahn CK, Shmaliy YS, Shi P, Zhao S (2017) Receding-horizon $\ell_2 - \ell_\infty$ FIR filter with embedded deadbeat property. IEEE Trans Circ Syst–II Express Briefs 64(2):211–215

8 Alspach D, Sorenson H (1972) Nonlinear Bayesian estimation using Gaussian sum approximation. IEEE Trans Autom Contr 17(4):439–448

9 Anderson BDO, Moore JB (1979) Optimal Filtering. Prentice-Hall, Englewood Cliffs, NJ

10 Arasaratnam I, Haykin S, Elliott RJ (2007) Discrete-time nonlinear filtering algorithms using Gauss-Hermite quadrature. Proc IEEE 95(5):953–977

11 Arceo-Miquel L, Shmaliy YS, Ibarra-Manzano O (2009) Optimal synchronization of local clocks by GPS 1PPS signals using predictive FIR filters. IEEE Trans Inst Meas 58(6):1833–1840

12 Athans M, Wishner RP, Bertolini A (1968) Suboptimal state estimation for continuous-time nonlinear systems from discrete noisy measurements. IEEE Trans Autom Contr AC-13:504–514

13 Aysal TC, Barner KE (2007) Meridian filtering for robust signal processing. IEEE Trans Signal Process 55(8):3949–3962

14 Banavar RN, Speyer JL (1991) A linear-quadratic game approach to estimation and smoothing. IEEE Am Contr Conf, Boston, MA, USA, 26–28 June 1991

15 Bar-Shalom Y, Li XR, Kirubarajan T (2001) Estimation with applications to tracking and navigation: theory algorithms and software, Wiley-Interscience, New York

16 Basu AK (2005) Introduction to Stochastic Process. Alpha Science, Harrow, UK

17 Battistelli G, Chisci L (2014) Kullback–Leibler average, consensus on probability densities, and distributed state estimation with guaranteed stability. Automatica 50(3):707–718

18 Bellman R (1960) Introduction to Matrix Analysis. McGraw-Hill, New York

Optimal and Robust State Estimation: Finite Impulse Response (FIR) and Kalman Approaches, First Edition.
Yuriy S. Shmaliy and Shunyi Zhao.

19 Bello P (1964) Time-frequency duality. IEEE Trans Inform Theory IT-10:18–33

20 Bierman GJ (1977) Factorization Methods for Discrete Sequential Estimation. Academic, New York

21 Box GEP (1953) Non-normality and tests on variances. Biometrika 40(3-4):318–335

22 Boyd S, Ghaoui LE, Feron E (1994) Linear Matrix Inequalities in System and Control Theory. SIAM, Philadelphia

23 Bruckstein A., Kailath T (1985) Recursive limited memory filtering and scattering theory. IEEE Trans Inform Theory 31(3):440–443

24 Bryson AE, Johansen D (1965) Linear filtering for time-varying systems using measurements containing colored noise. IEEE Trans Autom Contr 10(1):4–10

25 Bryson AE, Henrikson LJ (1968) Estimation using sampled data containing sequentially correlated noise. J. Spacecraft Rockets 5(6):662–665

26 Butcher JC (2003) Numerical Methods for Ordinary Differential Equations. John Wiley & Sons, New York

27 Campbell TG, Neuvo Y (1991) Predictive FIR filters with low computational complexity. IEEE Trans Circ Syst 38(9):1067–1071

28 Carmen C (1999) Ordinary Differential Equations with Applications. Springer-Verlag, New York

29 Castro-Tinttori P, Ibarra-Manzano O, Shmaliy YS (2012) Implementation of digital FIR filters with polynomial impulse responses. Circ Syst Signal Process 31(2):611–626

30 Castro-Tinttori P, Ibarra-Manzano O, Shmaliy YS (2012) Computationally efficient FIR filtering of polynomial signals in DFT domain. Circ Syst Signal Process 31(6):2153–2166

31 Caverly RJ, Forbes JR, (2019) LMI properties and applications in systems, stability, and control theory. arXiv:1903.08599x2, 12 Jun 2019

32 Chaffee JW (1987) Relating the Allan variance to the diffusion coefficients of a linear stochastic differential equation model for precision oscillators. IEEE Trans Ultrason Ferroel Freq Contr 34(6):655–658

33 Chang G (2014) On Kalman filter for linear system with colored measurement noise. J Geodesy 88(12):1163–1170

34 Chang XH (2014) Robust Output Feedback H-infinity Control and Filtering for Uncertain Linear Systems. Springer-Verlag, Berlin

35 Chen YL, Chen BS (1994) Minimax robust deconvolution filters under stochastic parametric and noise uncertainties. IEEE Trans Signal Process 42(1):32–45

36 Chui CK, Chen G (1987) Kalman Filtering with Real-Time Applications. Springer-Verlag, Berlin

37 Costa OLV, Fragoso MD, Marques RP (2005) Discrete-Time Markov Jump Linear Systems. Springer, London

38 Cox H (1964) On the estimation of state variables and parameters for noisy dynamic systems. IEEE Trans Autom Contr AC-9:5–12

39 Darouach M (2009) H_∞ unbiased filtering for linear descriptor systems via LMI. IEEE Trans Autom Contr 54(8):1966–1972

40 Daum F (2005) Nonlinear filters: Beyond the Kalman filter. IEEE A&E Syst Mag 20(8):57–69

41 Ding SX (2013) Model-Based Fault Diagnosis Techniques: Design Schemes, Algorithms and Tools. Springer, Berlin

42 Dong Z, You Z (2006) Finite-horizon robust Kalman filtering for uncertain discrete time-varying systems with uncertain-covariance white noises. IEEE Signal Process Lett 13(8):493–496

43 Doucet A, Godsill S, Andrieu C (2000) On sequential Monte Carlo sampling methods for Bayesian filtering. Statist Comput 10(3):197–208
44 Del Moral P 1996() Non linear filtering: interacting particle solution. Markov Processes Related Fields 2(4):555–580
45 Doob JL (1953) Stochastic Processes, Wiley, New York
46 Du C, Xie L (2002) H_∞ Control and Filtering of Two-Dimensional Systems. Springer, Berlin
47 Feller W (1968) An Introduction to Probability Theory and its Applications. Vol 1, 3rd Ed, Wiley, New York
48 Feller W (1970) An Introduction to Probability Theory and its Applications. Vol 2, Wiley, New York
49 Fitzgerald R (1971) Divergence of the Kalman filter. IEEE Trans Autom. Cont. 16(6):736–747
50 Fu JB, Sun J, Gao F, Lu S (2014) Maneuvering target tracking with improved unbiased FIR filter. Int Radar Conf (Radar), Lille, 13–17 October 2014
51 Fu M, de Souza CE, Xie L (1992) H_∞ estimation for uncertain systems. Int J Robust Nonl Contr 2:87–105
52 Gantmacher FR (1959) Applications of the Theory of Matrices. Tr. by Brenner JI, Wiley-Interscience, New York
53 Gauss CF (1809) Theory of Motion of the Heavenly Bodies Moving About the Sun in Conic Sections: A Translation (2004) of Theoria Motus. Mineola, Dover, New York
54 Gelb A (ed) (1974) Applied Optimal Estimation. MIT Press, Cambridge, USA
55 Geromel JC (1999) Optimal linear filtering under Parameter uncertainty. IEEE Trans Signal Process 47(1):168–175
56 Gershon E, Shaked U, Yaesh I (2005) H_∞ Control and Estimaiton of State-multiplicative Linear Systems. Springer-Verlag, London
57 Ghaoui LE, Calafiore G (2001) Robust filtering for discrete-time systems with bounded noise and parametric uncertainty. IEEE Trans Autom Contr 46(7):1084–1089
58 Gibbs BP (2011) Advanced Kalman Filtering, Least'Squares and Modeling: A Practical Handbook. John Wiley & Sons, Hoboken
59 Golub GH, van Loan CF (1996) Matrix Computations. 3rd Ed, John Hopkins Univ Press
60 Gonzalez JG, Arce GR (2001) Optimality of the myriad filter in practical impulsive-noise environments. IEEE Trans Signal Process 49(2):438–441
61 Gonzalez G, Nava R, Escalante-Ramirez B (2018) A comparative study on discrete Shmaliy moments and their texture-based applications. Math. Probl Eng Vol 2018, art. 1673283:1–17
62 Gordon NJ, Salmond DJ, Smith AFM (1993) Novel approach to nonlinear/non-Gaussian Bayessian state estimation. IEE Proc 140(2):107–113
63 Grewal MS, Andrews AP (2008) Kalman Filtering: Theory and Practice using MATLAB. 3rd Ed, John Wiley & Sons, Hoboken
64 Gubner JA (2002) Probability and Random Processes for Electrical and Computer Engineering. Cambridge Univ. Press, Cambridge
65 Haddad WM, Bernstein DS (1988) Robust, reduced-order, nonstrictly proper state estimation via the optimal projection equations with guaranteed cost bounds. IEEE Trans Autom Contr 33(6):591–595
66 Han SH, Kwon WH, Kim PS (2001) Receding-horizon FIR filters for continuous-time state-space models without *a priori* initial state information. IEEE Trans Autom Contr 46(5):766–770
67 Han S, Kwon WH (2007) L_2-E FIR smoothers for deterministic discrete-time state-space signal models. IEEE Trans Autom Contr 52(5):927–931

68 Han S, Kwon BK, Kwon WH (2009) Minimax FIR smoothers for deterministic continuous-time state space models. Automatica 45(6):1561–1566

69 Hanlon P, Maybeck PS (2000) Characterization of Kalman filter residuals in the presence of mismodeling. IEEE Trans Aero Elect Syst 36(1):114–131

70 Hassibi B, Sayed AH, Kailath T (1999) Indefinite-Quadratic Estimation and Control: A Unified Approcah to H^2 and H^∞ Theories. SIAM, Philadelphia

71 Heinonen P, Neuvo Y (1988) FIR-median hybrid filters with predictive FIR substructures. IEEE Trans Acoust Speech Signal Process 36(6):892–899

72 Hu JS, Yang CH (2011) Second-order extended H_∞ filter for nonlinear discrete-time systems using quadratic error matrix approximation. IEEE Trans Signal Process 59:3110–3119

73 Huber PJ (1964) Robust estimation of a local parameter. Ann Math Statist 35(1):73–101

74 Huber PJ (1981) Robust Statistics. Jonh Wiley & Sons, New York

75 IEEE Standard 1139-2008 (2009) Definitions of Physical Quantities for Fundamental Frequency and Time Metrology – Random Instabilities, IEEE, Piscataway, NJ

76 IEEE Standard $1588^T M$-2002 for a precision clock synchronization protocol for networked measurement and control systems, IEEE, Piscataway, NJ

77 ITU-T Recommendation G.810 (1996) Definitions and terminology for synchronization networks, Geneva, Switzerland

78 ITU-T Recommendation G.811 (1997) Timing characteristics of primary reference clocks, Geneva, Switzerland

79 Jazwinski AH (1970) Stochastic Processes and Filtering Theory. Academic Press, Cambridge

80 Jordan DW, Smith P (1999) Nonlinear Ordinary Differential Equations: An Introduction to Dynamical Systems. 3rd ed. Oxford Univ. Press, New York

81 Julier S, Uhlmann J, Durrant-Whyte HF (1995) A new approach for filtering nonlinear systems, Amer Contr Conf, Seattle, Washington, 21–23 June 1995

82 Julier S, Uhlmann J, Durrant-Whyte, HF (2000) A new method for the nonlinear transformation of means and covariances in nonlinear filters. IEEE Trans Autom Contr 45(3):477–482

83 Kailath T, Sayed AH, Hassibi B (2000) Linear Estimation. Prentice Hall, Upper Saddle River

84 Kalman RE (1960) A new approach to linear filtering and prediction problems. Trans ASME–J Basic Eng 82(D):35–45

85 Kalman RE, Bucy RS (1961) New results in linear filtering and prediction theory. Trans ASME–J Basic Eng 83(1):95–108

86 Kassam SA (1985) Signal Detection in Non-Gaussian Noise. Springer-Verlag, New York

87 Kassam SA, Poor HV (1985) Robust techniques for signal processing: a survey, Proc. IEEE 73(3):433–481

88 Kay SM (1993) Fundamentals of Statistical Signal Processing: Estimation Theory. Prentice-Hall, Upper Saddle River

89 Khargonekar PP, Petersen IR, Zhou K (1990) Robust stabilization of uncertain linear systems: quadratic stabilizability and H_∞ control theory. IEEE Trans Autom Contr 35(3):356–361

90 Kitagawa G (1987) Non-Gaussian state-space modeling of nonstationary time series. J Amer Statist Ass 82(400):1032–1041

91 Knill O (2009) Probability and Stochastic Processes with Applications. Overseas Press India, New Delhi

92 Kushner H (1967) Nonlinear filtering: The exact dynamical equations satisfied by the conditional mode. IEEE Trans Autom Contr 12(3):262–267

93 Kwon BK, Han S, Kwon WH (2007) Minimum variance FIR smoothers for continuous-time state space signal models. IEEE Signal Process Lett 14(12):1024–1027

94 Kwon BK, Han S, Kwon OK, Kwon WH (2007) Minimum variance FIR smoothers for discrete-time state space models. IEEE Signal Process Lett 14(8):557–560

95 Kwon BK, Choi JW, Park JH, Han S, Kwon WH (2009) A Best lag size of minimum variance FIR smoothers. IEEE Signal Process Lett 16(4):307–310

96 Kwon BK, Han S (2014) An optimal fixed-lag FIR smoother for discrete time-varying state space models. J Inst Contr Robot Syst 20(1):1–5

97 Kwon OK, Kwon WH, Lee KS (1989) FIR filters and recursive forms for discrete-time state-space models. Automatica 25(5):715–728

98 Kwon OK, Ryu HS, Yoo KS (1996) Robust H_∞ FIR filter for uncertain systems. 13th Trienial World Congress, San Francisco, USA, 30 June–5 July 1996

99 Kwon OK, de Souza CE, Ryu HS (1996) Robust H_∞ FIR filter for discrete-time uncertain systems. 35th Conf Dec Contr, Kobe, Japan, 11-13 Dec 1996

100 Kwon WH, Lee KS, Kwon OK (1990) Optimal FIR filters for time-varying state-space models. IEEE Trans Aero Elect Syst 26(6):1011–1021

101 Kwon WH, Suh YS, Lee YI, Kwon OK (1994) Equivalence of finite memory filters, IEEE Trans Aero Elect Syst 30(3):968–972

102 Kwon WH, Lee KS, Lee JH (1994) Fast algorithms for optimal FIR filter and smoother of discrete-time state-space models. Automatica 30(3):489–492

103 Kwon WH, Kim PS, Park P (1999) A receding horizon Kalman FIR filter for discrete time-invariant systems. IEEE Trans Autom Contr 44(9):1787–1791

104 Kwon WH, Kim PS, Han SH (2001) Best linear unbiased FIR filters for continuous-time state-space models, Asian J Control 3(1):1–9

105 Kwon WH, Kim PS, Han S (2002) A receding horizon unbiased FIR filter for discrete-time state space models. Automatica 38(3):545–551

106 Kwon WH, Han S (2005) Receding Horizon Control: Model Predictive Control for State Models. Springer, London

107 Lancaster P, Rodman L (1995) Algebraic Riccati Equations, Clarendon Press, Oxford

108 Lastre-Dominguez CM, Shmaliy YS, Ibarra-Manzano O, Vazquez-Olguin MA (2019) Denoising and features extraction of ECG signals in state space using unbiased FIR smoothing. IEEE Access 7:152166–152178

109 Lee YS, Han SH, Kwon WH (2006) H_2/H_∞ FIR filters for discrete-time state space models. Int J Contr Autom Syst 4(5):645–652

110 Lei B, Xu G, Feng M, Zou Y et al (2017) Classification, Parameter Estimation and State Estimation. John Wiley & Sons, Hoboken

111 Lewis FL, Xie L, Popa D (2008) Optimal and Robust Estimation with an Introduction to Stochastic Control Theory. CRC Press, Boca Raton

112 Li F, Shi P, Wu L (2017) Control and Filtering for Semi-Markovian Jump Systems, Springer Int Publ, Cham, Switzerland

113 Li H, Fu M (1997) A linear matrix inequality approach to robust H_∞ filtering. IEEE Trans Signal Process 45(9):2338–2350

114 Li T, Sun S, Sattar TP, Corchado JM (2000) Fight sample degeneracy and impoverishment in particle filters: A review of inteligent approaches Art Intell 128(1–2):99–141

115 Lim J (2014) Tutorial–game theory-based extended H_∞ filtering approach to nonlinear problems in signal processing. Dig Sign Process 34:1–15

116 Liu JS, Che, R (1998) Sequential Monte Carlo methods for dynamic systems. J Amer Stat Ass 93(443):1032–1044

117 Liu Q, Wang Z, He X (2019) Stochastic Control and Filtering over Constrained Communication Networks. Springer Nature, Switzerland

118 Ma L, Wang Z, Bo Y (2019) Nonlinear Control and Filtering for Stochastic Networked Systems. CRC Press, Boca Raton

119 Makhoul J (1975) Linear prediction: A tutorial review. Proc IEEE 63(4):561–580

120 Marquez-Figueroa S, Shmaliy YS, Ibarra-Manzano O (2020) Optimal extraction of EMG signal envelope and artifacts removal assuming colored measurement noise Biomed Sign Process Contr 57(101679:1–8

121 Martin RD, Masreliez CJ (1975) Robust estimation via stochastic approximation. IEEE Trans Inf Theory 21(3):263–271

122 Martin RD, Mintz M (1983) Robust filtering and prediction for linear systems with uncertain dynamics: a game-theoretic approach. IEEE Trans Autom Contr AC-28(9):888–896

123 Martin RD, Raftery AE (1987) Robustness, computation and non Euclidean models, J Am Stat Assoc 82(400):1044–1050

124 Masreliez CJ, Martin RD (1975) Robust Bayesian estimation for the linear model and robustifying the Kalman filter. IEEE Trans Autom Contr AC-22(3):361–371

125 Maybeck PS (1979) Stochastic Models, Estimation, and Control. Academic Press, New York

126 Maybeck P (1982) Stochastic Models, Estimation and Control. Vol. 2, Academic Press, New York

127 Meditch JS (1973) A survey of data smoothing for linear and nonlinear dynamic systems. Automatica 9(2):151–162

128 Miller JH, Thomas JB (1972) Detectors for discrete-time signals in non-Gaussian noise. IEEE Trans Inf Theory 18(2):241–250

129 Mohamed SMK, Nahavandi S (2012) Robust finite-horizon Kalman filtering for uncertain discrete-time systems. IEEE Trans Autom Contr 57(6):1548–1552

130 Moore JB (1973) Discrete-time fixed-lag smoothing algorithms. Automatica 9(2):163–173

131 Morales-Mendoza LJ, Gamboa-Rosales H, Shmaliy YS (2013) A new class of discrete orthogonal polynomials for blind fitting of finite data, Signal Processing. 93(7):1785–1793

132 Moroney MJ (1951) Facts from Figures. Penguin, London

133 Nagpal K, Abedor J, Poola K (1994) An LMI approach to peak-to-peak gain minimization: filtering and control. Am Contr Conf, Baltimore, Maryland, 29 June–1 July 1994

134 Nikias CL, Shao M (1995) Signal Processing With Alpha-Stable Distributions and Applications. Wiley, New York

135 Nisar MD (2011) Minimax Robustness in Signal Processing for Communications, Shaker Verlag, Aachen

136 Norgaard M, Poulsen NK, Ravn O (2000) New developments in state estimation for nonlinear systems. Automatica 36(11):1627–1638

137 Olfati-Saber R (2005) Distributed Kalman filter using embedded consensus filters. 44th IEEE Conf Dec Contr 2005, Seville, Spain, 12-15 Dec 2005

138 Olfati-Saber R, Shamma JS (2005) Consensus filters for sensor networks and distributed sensor fusion. 44th Conf Dec Contr, Seville, Spain, 12-15 Dec 2005

139 Olfati-Saber R (2007) Distributed Kalman filtering for sensor networks. 46th IEEE Conf Dec Contr, New Orleans, LA, 12-14 Dec 2007

140 Øksendal B (2014) Stochastic Differential Equations: An Introduction with Applications. 6th Ed, Springer, New York

141 Ortega-Contreras J, Pale-Ramon E, Shmaliy YS, Xu Y (2021) A novel approach to H2 FIR prediction under disturbances and measurement errors. IEEE Sign Process Lett 28:150–154

142 Packard A, Doyle J (1990) Quadratic stability with real and complex perturbations. IEEE Trans Autom Contr 35(2):198–201

143 Pak JM, Ahn CK, Shmaliy YS, Lim MT (2015) Improving reliability of particle filter-based localization in wireless sensor networks via hybrid particle/FIR filtering. IEEE Trans Ind Inf 11(5):1089–1098

144 Palhares RM, Peres PLD (2000) Robust filtering with guaranteed energy-to-peak performance–an LMI approach. Automatica 36:851–858

145 Papoulis A (2002) Probability, Random Variables, and Stochastic Processes 5th Ed, McGraw-Hill, New York

146 Pearson K (1902) On the systematic fitting of curves to observations and measurements. Biometrika 1(3):265–303

147 Petovello MG, O'Keefe K, Lachapelle G, Cannon ME (2009) Consideration of time-correlated errors in a Kalman filter applicable to GNSS. J. of Geodesy 83(1):51–53

148 Pieczynski W (2003) Pairwise Markov chains. IEEE Trans Pattern Anal Machine Intel 25(5):634–639

149 Pomarico-Franquiz J, Shmaliy YS (2014) Accurate self-localization in RFID tag information grids using FIR filtering. IEEE Trans Ind Inf 10(2):1317–1326

150 Pomarico-Franquiz J, Granados-Cruz M, Shmaliy YS (2015) Self-localization over RFID tag grids excess channels using extended filtering techniques. IEEE J Sel Topics Sign Process 9(2):229–238

151 Quab Z, Han S, Kwon WH (2007) Robust FIR filters for linear continuous-time state-space models with uncertainies. 26th Chinese Contr Conf, Zhangjiajie, Hunan, 26-31 July 2007

152 Quab Z, Han S, Kwon WH (2007) A robust FIR filter for linear discrete-time state-space signal models with uncertainies IEEE Sign Process Lett 14(8):553–556

153 Ramirez-Echeverria F, Sarr A, Shmaliy YS (2014) Optimal memory for discrete-time FIR filters in state-space. IEEE Trans Sign Process 62(3):557–561

154 Rauch HE, Tung F, Striebel CT (1965) Maximum likelihood estimates of linear dynamic systems. AIAA J 3(8):1445–1450

155 Ristic B, Arulampalam S, Gordon N (2004) Beyond the Kalman Filter: Particle Filters for Tracking Applications. Arctech House, Norwood

156 Ross SM (1996) Stochastic Processes John Wiley & Sons, New York

157 Sage AP, Melsa JL (1971) Estimation Theory with Applications to Communications and Control. McGraw-Hill, New York

158 Savitzky A, Golay MJE (1964) Smoothing and differentiation of data by simplified least squares procedures. Analyt Chem 36(8):1627–1639

159 Scherer C, Weiland S (2015) Linear Matrix Inequalities in Control. University of Stuttgart, Stuttgart

160 Schick IC, Mitter SK (1994) Robust recursive estimation in the presence of heavy-tailed observation noise. Ann. Stat 22(2):1045–1080

161 Schweppe FC (1973) Uncertain Dynamic Systems. Prentice-Hall, Englewood Cliffs

162 Shakibaei Asli BH, Flusser J (2017) New discrete orthogonal moments for signal analysis. Sign Process 141:57–73

163 Shen X, Deng L (1997) Game theory approach to discrete H_∞ filter design. IEEE Trans Sign Process 45(4):1092–1095

164 Shen X, Deng L (2014) A new solution to induced l_∞ FIR filtering problem based on two matrix inequalities. Int J Contr 87(2):404–409

165 Shmaliy YS (2006) Continuous-Time Signals. Springer, Dordrecht

166 Shmaliy YS (2006) An unbiased FIR filter for TIE model of a local clock in applications to GPS-based timekeeping. IEEE Trans Ultrason Ferroel Freq Contr 53(5):862–870

167 Shmaliy YS (2007) Continuous-Time Systems. Springer, Dordrecht

168 Shmaliy YS, Muñoz-Diaz J, Arceo-Miquel L (2008) Optimal horizons for a one-parameter family of unbiased FIR filters. Dig Sign Process 18(5):739–750

169 Shmaliy YS, Muñoz-Diaz J, Arceo-Miquel L (2008) Optimal horizons for a one-parameter family of unbiased FIR filters/ Dig Sign Process 18(5):739–750

170 Shmaliy YS (2009) GPS-based Optimal FIR Filtering of Clock Models. Nova Science Publ, New York

171 Shmaliy YS (2009) An unbiased *p*-step predictive FIR filter for a class of noise-free discrete-time models with independently observed states. Sign Image Video Process 3(2):127–135

172 Shmaliy YS (2009) Unbiased FIR filtering of discrete time polynomial state space models. IEEE Trans Sign Process 57(4):1241–1249

173 Shmaliy YS (2010) Linear optimal FIR estimation of discrete time-invariant state-space models. IEEE Trans Sign Process 58(6):3086–3096

174 Shmaliy YS, Morales-Mendoz, L (2010) FIR smoothing of discrete-time polynomial models in state space. IEEE Trans Sign Process 58(5):2544–2555

175 Shmaliy YS (2011) An iterative Kalman-like algorithm ignoring noise and initial conditions IEEE Trans Sign Process 59(6):2465–2473

176 Shmaliy YS, Ibarra-Manzano O (2012) Time-variant linear optimal finite impulse response estimator for discrete state-space models. Int J Adapt Contr Sign Process 26(2):95–104

177 Shmaliy YS, Ibarra-Manzano O (2012) Optimal finite impulse response estimation of linear models in receiver channels with imbedded digital signal processing units. IET Sign Process 6(4):281–287

178 Shmaliy YS (2012) Suboptimal FIR filtering of nonlinear models in additive white Gaussian noise. IEEE Trans Sign Process 60(10):5519–5527

179 Shmaliy YS, Zhao S, Ahn CK (2017) Unbiased FIR filtering: an iterative alternative to Kalman filtering ignoring noise and initial conditions. IEEE Contr Syst Mag 37(5):70–89

180 Shmaliy YS, Lehmann F, Zhao S, Ahn CK (2018) Comparing robustness of the Kalman, H_∞, and UFIR filters. IEEE Trans Sign Process 66(13):3447–3458

181 Shmaliy YS, Neuvo Y, Khan S (2018) Review of unbiased FIR filters, smoothers, and predictors for polynomial signals. Front Sign Process 2(1):1–29

182 Shmaliy YS, Zhao S, Ahn CK (2019) Optimal and unbiased filtering with colored process noise using state differencing. IEEE Sign Process Lett 26(4):548–551

183 Shmaliy YS, Zhao S, Ahn CK (2020) Kalman and UFIR state estimation with coloured measurement noise using backward Euler method. IET Sign Process 14(2):64–71

184 Sima V (1996) Algorithms for Linear Quadratic Optimization. Marcel Dekker, New York

185 Simon D (2006) Optimal State Estimaiton: Kalman, H_∞, and Nonlinear Approaches. John Wiley & Sons, Hoboken

186 Simon D, Shmaliy YS (2013) Unified forms for Kalman and finite impulse response filtering and smoothing. Automatica 49(6):1892–1899

187 Simon MK (2006) Probability Distributions Involving Gaussian Random Variables. Springer, New York

188 Skelton RE, Iwasaki T, Grigoriadis KM (1998) A Unified Algebraic Approach to Linear Control Design. Taylor & Francis, London

189 Smith SW (1999) The Scientist and Engineer's Guide to Digital Signal Processing. 2nd Ed, California Tech Publ, CA

190 de Souza CE, Shaked U, Fu M (1995) Robust H_∞ filtering for continuous time varying uncertain systems with deterministic input signals. IEEE Trans Sign Process 43(3):709–719

191 Stark H, Woods JW (1994) Probability, Random Processes, and Estimation Theory for Engineers. Prentice Hall, Upper Saddle River

192 Stein SR, Filler RL (1988) Kalman filter analysis for real time applications of clocks and oscillators. 42th Ann Freq Contr Symp, Baltimore, Maryland, 1-3 June 1988

193 Stratonovich RL (1963) Topics in the Theory of Random Noise. Vol. 1, Gordon & Breach, New York

194 Sun S, Deng Z (2004) Multi-sensor optimal information fusion Kalman filter. Automatica 40(6):1017–1017

195 Sun J, Fu JB, Wang J (2014) Improved maneuvering target tracking method based on unbiased finite impulse response (UFIR) filter. US Patent 103 500 455 A, Jan. 8 2014

196 Tan Z, Soh YC, Xie L (1999) Envelope-constrained H_2 FIR filter design. Circ Syst Sign Process 18(6):539–551

197 Toda M, Patel RV (1980) Bounds on estimation errors of discrete-time filters under modeling uncertainty. IEEE Trans Autom Contr AC-25(6):1115–1121

198 Trench WF (1961) A general class of discrete time-invariant filters. J Soc Ind Appl Math 9(3):405–421

199 Uhlmann JK (1994) Simultaneous map building and localization for real time applications II. Eng Sc Report, Oxford University, UK

200 Van Antwerp JG, Braatz RD (2000) A tutorial on linear and bilinear matrix inequalities. J Process Contr 10:363–385

201 Van der Merwe R, Doucet A, de Freitas N, Wan E (2000) The unscented particle filter, Tech Rep CUED/F-INFENG/TR 380, Cambridge University Eng Dept

202 Uribe-Murcia K, Shmaliy YS, Ahn CK, Zhao S (2020) Unbiased FIR filtering for time-stamped discretely delayed and missing data. IEEE Trans Autom Contr 65(5):2155–2162

203 Uribe-Murcia K, Shmaliy YS, Andrade-Lucio JA (2021) Unbiased FIR, Kalman, and game theory $H + \infty$ filtering under Bernoulli distributed random delays and packet dropouts Neurocomputing 442:89–97

204 Vazquez-Olguin M, Shmaliy YS, Ahn CK, Ibarra-Manzano O (2017) Blind robust estimation with missing data for smart sensors using UFIR filtering. IEEE Sensors J 17(6):1819–1819

205 Vazquez-Olguin M, Shmaliy YS, Ibarra-Manzano O (2017) Distributed unbiased FIR filtering with average consensus on measurements for WSNs. IEEE Trans Ind Inf 13(3):1440–1447

206 Vazquez-Olguin M, Shmaliy YS, Ibarra-Manzano O (2020) Distributed UFIR filtering over WSNs with consensus on estimates. IEEE Trans Ind Inf 16(3):1645–1654

207 Verdú S, Poor HV (1984) On minimax robustness: a general approach and applications. IEEE Trans Inf Theory 30(2):328–340

208 Vidyasagar M (1986) Optimal rejection of persistent bounded disturbances. IEEE Trans Autom Contr AC-31(6):527–534

209 Vincent T, Abedor J, Nagpal K, Khardonekar PP (1996) Discrete-time estimators with guaranteed peak-to-peak performance. 13th Triennial World Congress, San Franciaco, 30 June–5 July 1996

210 Wang C, Walker RA, Moody MP (2007) Single antenna attitude algorithm for nonuniform antenna gain patterns. J Spacecraft Rockets 44(1):221–229

211 Wang D, Shi P, Wang W (2013) Robust Filtering and Fault Detection of Switched Delay Systems, Springer, London

212 Watson JT, Grigoriadis KM (1997) Optimal unbiased filtering via linear matrix inequalities. Am Contr Conf, Albuquerque, New Mexico, 4–6 June 1997

213 Wilson D (1989) Convolution and Hankel operator norms for linear systems. IEEE Trans Autom Contr 34(1):94–97

214 Xie L, de Souza CE, Fu M (1991) H_∞ estimation for discrete-time linear uncertain systems. Int J Robust Nonlinear Contr 1(2):111–123

215 Xie L, Soh YC, de Souza CE (1994) Robust Kalman filtering for uncertain discrete-time systems. IEEE Trans Autom Contr 39(6):1310–1314

216 Xu J, Xie L (2014) Control and estimation of piecewise affine systems. Elsevier, Cambridge

217 Xu Y, Shmaliy YS, Zhang YD, Shen T, Sun M (2021) INS/UWB-based quadrotor localization under colored measurement noise. IEEE Sensors J (to be published)

218 You SH, Ahn CK, Shmaliy YS, Zhao S (2020) Fusion Kalman and weighted UFIR state estimator with improved accuracy. IEEE Trans Ind Electron 67(12):10713–10722

219 Yuan JT, Stuller JA (1994) Order-recursive FIR smoothers. IEEE Trans Sign Process 42(5):1242–1246

220 Zaknich A (2005) Pronciples of Adaptive Filters and Self-learning Systems. Springer-Verlag, London

221 Zhao S, Shmaliy YS, Huang B, Liu F (2015) Minimum variance unbiased FIR filter for discrete time-variant systems. Automatica 53(3):355–361

222 Zhao S, Shmaliy YS, Liu F (2016) Fast Kalman-like optimal unbiased FIR filtering with applications. IEEE Trans Sign Process 64(9):2284–2297

223 Zhao S, Shmaliy YS (2016) Unified maximum likelihood form for bias constrained FIR filters. IEEE Sign Process Lett 23(12):1848–1852

224 Zhao S, Shmaliy YS (2016) Unified maximum likelihood form for bias constrained FIR filters. IEEE Sign Process Lett 23(12):1848–1852

225 Zhao S, Shmaliy YS, Shi P, Ahn CK (2017) Fusion Kalman/UFIR filter for state estimation with uncertain parameters and noise statistics. IEEE Trans Ind Electron 64(4):3075–3083

226 Zhao S, Shmaliy YS, Liu F (2017) On the iterative computation of error matrix in unbiased FIR filtering. IEEE Sign Process Lett 24(5):555–558

227 Zhao S, Shmaliy YS, Ahn CK (2018) Bias-constrained optimal fusion filtering for decentralized WSN with correlated noise sources. IEEE Trans Sign Inf Process over Networks 4(4):727–735

228 Zhao S, Ahn CK, Shi P, Shmaliy YS, Liu F (2019) Bayesian state estimation for Markovian jump systems. IEEE Syst Man Cyber Mag 5(2):27–36

229 Zhao S, Shmaliy YS, Liu F (2021) Optimal FIR filter for discrete-time LTV systems and fast iterative algorithm. IEEE Trans Circ Syst II, Express Briefs (to be published)

230 Zhao S, Shmaliy YS, Andrade-Lucio JA (2021) Backward optimal FIR filtering and recursive forms for discrete LTV processes. Signal Process (to be published)

231 Zhao S, Shmaliy YS, Andrade Lucio JA, Liu F (2021) Multi-pass optimal FIR filtering for processes with unknown initial states and temporary mismatches. IEEE Trans Ind Informat (to be published)

232 Zhou K, Doyle JC, Glover K (1996) Robust and Optimal Control. Prentice-Hall, Upper Saddle River

233 Zhu X, Soh YC, Xie L (2002) Design and analysis of discrete-time robust Kalman filters Automatica 38(6):1069–1077

234 Zhu C, Xia Y, Xie L, Yan L (2013) Optimal linear estimation for systems with transmission delays and packet dropouts. IET Sign Process 7(9):814–823

Index

Optimal and Robust State Estimation: Finite Impulse Response (FIR) and Kalman Approaches, First Edition.
Yuriy S. Shmaliy and Shunyi Zhao.

Printed and bound by CPI Group (UK) Ltd, Croydon, CR0 4YY
17/09/2026
14964259-0005